Advance of Polymers Applied to Biomedical Applications: Cell Scaffolds

Advance of Polymers Applied to Biomedical Applications: Cell Scaffolds

Special Issue Editors

Insung S. Choi
João F. Mano

MDPI • Basel • Beijing • Wuhan • Barcelona • Belgrade

MDPI

Special Issue Editors

Insung S. Choi
KAIST
Korea

João F. Mano
University of Aveiro
Portugal

Editorial Office
MDPI
St. Alban-Anlage 66
Basel, Switzerland

This is a reprint of articles from the Special Issue published online in the open access journal *Polymers* (ISSN 2073-4360) from 2016 to 2017 (available at: http://www.mdpi.com/journal/polymers/special_issues/polymer_cell_scaffolds)

For citation purposes, cite each article independently as indicated on the article page online and as indicated below:

LastName, A.A.; LastName, B.B.; LastName, C.C. Article Title. *Journal Name* **Year**, *Article Number*, Page Range.

ISBN 978-3-03897-033-0 (Pbk)
ISBN 978-3-03897-034-7 (PDF)

Contents

About the Special Issue Editors

Insung S. Choi is a Professor of Chemistry and of Bio and Brain Engineering at KAIST, Korea, and the Director of the Center for Cell-Encapsulation Research (Creative Research Initiative; 2012-date). He obtained his BS and MS degrees in Chemistry at Seoul National University in 1991 and 1993 respectively, and completed his PhD degree in Chemistry at Harvard University in 2000, under the supervision of Professor George M. Whitesides. After carrying out postdoctoral work with Robert Langer at the Department of Chemical Engineering of MIT, he joined the faculty at KAIST in 2002. He was awarded the KCS-Wily Young Chemist Award (2003), the Thieme Journal Award (2003), the Presidential Young Scientist Award (2004; KAST), and the JANG SEHEE Research Achievement Award (2013; KCS). His research interests include biomimetic science and neurochemistry. He has published over 250 peer-reviewed papers. He is an editorial board member of Chemistry—An Asian Journal (Wiley-VCH), ChemNanoMat (Wiley-VCH), and Polymers (MDPI), and an editorial advisory board member of Advanced Healthcare Materials (Wiley-VCH).

João F. Mano is a Full Professor at the Chemistry Department of the University of Aveiro, Portugal, and Director of the Compass Research Group, in the Associated Laboratory CICECO-Aveiro Institute of Materials. His research interests include the use of biomaterials and cells for the progress of transdisciplinary concepts to be employed in regenerative and personalized medicine. In particular, he has been applying biomimetic and nano/micro-technology approaches to polymer-based biomaterials and surfaces in order to develop biomedical devices with improved structural and (multi-)functional properties, and in the engineering of microenvironments to control cell behaviour and organization, to be exploited clinically in advanced therapies or in drug screening. João F. Mano has authored over 570 papers in international journals (19000 citations, h = 69). He has been a member of various scientific societies and editorial boards of international journals. He has coordinated or been involved in many national and European research projects, including an Advanced Grant from the European Research Council.

Preface to "Advance of Polymers Applied to Biomedical Applications: Cell Scaffolds"

Cells in vivo sense and respond to signals from their environment (e.g., other cells and extracellular matrices) for their orchestrated behaviour and function. These signals exist in various forms, including biochemical, electrochemical, structural, and mechanical signals. In vitro mimicry of juxtacrine interactions has been achieved mainly through the use of polymers in regenerative medicine and cell therapy. The notion of scaffolds, in particular, has been closely linked with three-dimensional (3D) or semi-3D structures onto which or into which the cells grow and proliferate. The primary function of scaffolds was initially structural support, providing physical sites for cell adhesion and growth in three dimensions. However, recent scientific and technological developments have advanced to enable the manipulation of cell behaviour and function, ultimately controlling cell fate by the use of specific polymers, including supramolecular hydrogels.

The purpose of this Special Issue is to highlight the recent achievements in the use of polymers as cell scaffolds on a broad scale, examining not only the use of polymers as cell culture scaffolds, but also including polymer-based approaches for controlling interfacial interactions of cells in vitro. Herein are proposed new biomaterials as cell supports, such as injectable copolymers or gelatin-based hydrogels for 3D bioprinting. In other reports, the incorporation of inorganic or organic fillers into polymer matrices has been attempted to improve the structure and function of biomaterials, exemplified by the reinforcement of poly("-caprolactone) (PCL) with calcium-containing mesoporous particles or tricalcium phosphate; poly(lactic-co-glycolic) (PLGA) with biphasic calcium phosphate or nanohydroxyapatite; poly(lactic acid) (PLA) with Fe_3O_4 nanoparticles; and poly(3-hydroxybutyrate-co-3-hydroxyvalerate) with calcium silicate. Natural polymers also have been combined with distinct fillers: Human mesenchymal stem cells (hMSCs) differentiation could be regulated by the addition of hydroxyapatite to chitosan-based scaffolds; and alginate hydrogels may be reinforced with cellulose nanofibrils. More complex systems were envisaged for some specific cases, such as the infiltration of photo-crosslinkable hyaluronic acid hydrogels in fused deposition-manufactured composite scaffolds of hydroxyapatite and poly(ethylene glycol)-b-PCL.

Scaffold geometries beyond the conventional porous 3D supports are also explored in this Special Issue. Membranes made of biodegradable or natural polymers are proposed, including multilayered free-standing films made of polysaccharides as a potential platform for the formation of human adipose-derived stem cell aggregates. Quasi-3D PCL supports were prepared by electrospinning for breast cancer cell culture. In another study, tubular PLGA-based structures were developed to be used as scaffolds for small-diameter vascular tissue engineering, and micro-needle cuffs were investigated for perivascular drug delivery. Surface modifications, another fundamental approach to enhance cell interactions to scaffolds, are also presented: catechol-conjugated dextran is reported as a surface coating material, showing antiplatelet capabilities; plasma treatment was studied in the modification of polyethylene terephthalate. In another study, atomic force microscopy is suggested as a powerful method for characterizing the scaffold surfaces. The manipulation of individual cells is suggested as an emerging technique in cell scaffold research: the 3D single-cell assembly allowed for structure formation without the need for artificial scaffolds; cells could be included as individual entities or in the form of aggregates into hydrogels, exhibiting distinct behaviour in cartilaginous tissue formation. In another paper, the coating of individual red blood cells was shown to radically change their behaviour, including immunogenicity.

In short, this Special Issue deals with important recent advances in cell-polymer interactions, ranging from the design, fabrication, and modification of cell scaffolds to single-cell manipulation, by contribution from experts in their repective fields. We believe that this issue conveys useful information to scientists and engineers in various fields including polymer scientists and biomedical engineers

Insung S. Choi, João F. Mano
Special Issue Editors

polymers

MDPI

Review
Characterization of Cell Scaffolds by Atomic Force Microscopy

Jagoba Iturri * and José L. Toca-Herrera

Institute for Biophysics, Department of NanoBiotechnology, University of Natural Resources and Life Sciences, Muthgasse 11, 1190 Wien, Austria; jose.toca-herrera@boku.ac.at
* Correspondence: jagoba.iturri@boku.ac.at; Tel.: +43-1-47654-80339

Received: 30 July 2017; Accepted: 16 August 2017; Published: 21 August 2017

Abstract: This review reports on the use of the atomic force microscopy (AFM) in the investigation of cell scaffolds in recent years. It is shown how the technique is able to deliver information about the scaffold surface properties (e.g., topography), as well as about its mechanical behavior (Young's modulus, viscosity, and adhesion). In addition, this short review also points out the utilization of the atomic force microscope technique beyond its usual employment in order to investigate another type of basic questions related to materials physics, chemistry, and biology. The final section discusses in detail the novel uses that those alternative measuring modes can bring to this field in the future.

Keywords: cell scaffolds; atomic force microscopy

1. Introduction

Although the first articles date from the mid-late 1990s and early 2000s [1–3], a quick search by Scopus database for the terms "Atomic Force Microscopy (AFM) + scaffolds + cells" brings, interestingly, only around 300 articles referring to them. From these results, it can be deduced that the field has still a long way to go. In the same line, the trend followed by the number of publications all along these years speaks about a timid periodical increase in 5 years periods that, even though it does not seem to definitely take off, stays as a very promising topic (see Figure 1).

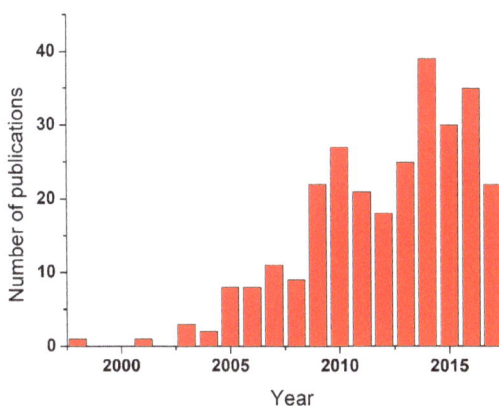

Figure 1. Time evolution of the number of published works within the field of cell scaffolds combined with Atomic Force Microscopy.

The design and development of cell scaffolds for tissue engineering purposes has undeniably gained, on its own, more and more relevance in the last 30 years. The required biocompatibility of the materials employed for constructing the scaffolds, accompanied by their ease for being either degraded or metabolized, has become a challenge for researchers in both the biomedicine and materials science fields. Scaffolds have to provide an architecture on which seeded cells can organize and develop into the desired organ or tissue prior to implantation [4]. By means of this tissue engineering process the risks of immunological responses (rejections or viral infections) are certainly decreased and almost supressed. Among the requirements that scaffolds should fulfil for having an optimal function, they must pay attention to their topography and mechanical properties, factors which undoubtedly influence the way cells proliferate, organize, and give rise to a new tissue [5–8].

In this regard, atomic force microscopy is a well-suited and popular technique to analyze topographical features and mechanics at the nanoscale. Up to date AFM has been widely used, due to its unmatched resolution, in material science, physical chemistry, chemical physics, surface science, and almost all the fields related to the molecular scale [9–11]. However, over the last 20 years AFM has also become a standard tool in life sciences for studying biological phenomena, varying from surface topography [12–14], to structural studies [15], to the quantification of interactions between biomolecules [16,17]. Then, AFM results in a highly valuable complement to other characterization techniques currently employed by the cell scaffolds community.

In this document, we perform a brief overview of the different published results that have covered so far this exciting topic of AFM together with scaffolds of a different nature, and of the way these were experimentally obtained. Special attention is paid to the material/system forming the matrix of the scaffold, as well as to the AFM measuring mode employed. For the reader's guidance, the covered scaffold examples have been divided, merely attending to the authors' personal criteria, into five major topics. The first four represent very general systems, where the usage of quite standard polymers as building blocks (i.e., polyurethane or polylactide) is in a majority. The fifth group, in turn, presents the use of diverse biopolymers and biomolecules, either with the Arg-Gly-Asp tripeptide (commonly known as RGD) moieties or without them. The decision to make a group apart with them derives from their capability, in some cases, to undergo self-assembly, together with their unique conditions for specific cell recognition. By the mentioned spontaneous arrangement they are capable of forming diverse structures that would hardly fit into one of the other categories exclusively. In turn, the focus of AFM measuring modes has principally been set on the different mechanical studies covered by the literature rather than on topography imaging, since the former are less extended and well-known. These mechanical analyses of cell scaffolds include elastic moduli, adhesive properties, and more sophisticated analyses like single chain stretching and/or the so-called single-cell probe force spectroscopy, where a living cell acts as a measuring probe, allowing quantitative characterization of what the cell-to-scaffold affinity can be. Bibliographic references regarding all these systems and measuring methodologies will be covered in this report. Furthermore, for the sake of a better understanding of these concepts, we will very briefly introduce the atomic force microscope as both an imaging and a mechanical machine, and discuss different methods that might be future interest for the community.

2. Atomic Force Microscopy, a Versatile Tool

Since its appearance in the mid-80s, in consequence of the first probe microscopes developed [18], atomic force microscopy has become a very popular and useful tool to characterize matter at both the nano- and micro-scale levels. Through the years, together with its continuous evolution and technical implementation, AFM has shown high versatility and an almost on demand adaptive capability. The systems covered in this 30-year range, i.e., from Deoxyribonucleic acid (DNA) strands up to eukaryotic cells, traversing many soft/hard materials, have contributed to its extended reputation and its presence in many (nano) biotechnology laboratories. Nevertheless, its potential use is still sometimes underestimated and certainly limited to only a few of its operating features. We then consider it crucially important to emphasize the role that this technique could play, in co-operation with other complementary techniques, in a field of growing interest like that of cell scaffolds.

2.1. AFM as an Imaging Machine

Atomic Force Microscopy has proven to be a solid alternative to other microscopy techniques (i.e., transmission and scanning electron microscopes) when investigating nano-scale structural features of (bio)materials, under an aqueous environment and at different temperatures [19,20]. In the current case, the sensing element is a flexible cantilever with a sharp tip at the end, as shown schematically in Figure 2A. The deflection of this cantilever is brought by tip-sample interaction and yields a continuous shift in the reflected laser beam, as measured with a position detector (photodiode). In addition, a piezo scanner moves the cantilever along the three dimensions. The topography of the sample is thus obtained from translation of the tip/sample interaction-derived voltage variation, and the final image is delivered by computer processing.

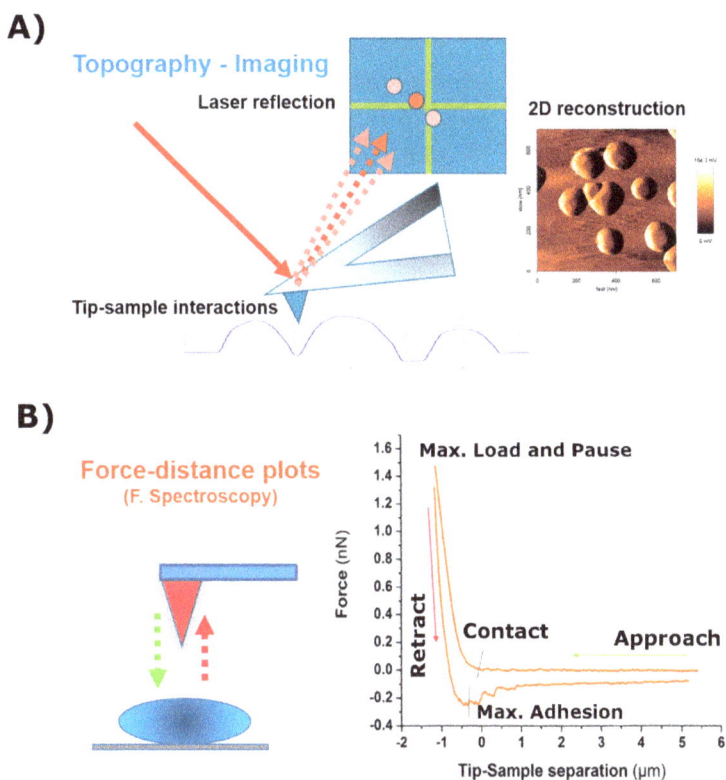

Figure 2. Scheme showing the main measuring modes in Atomic Force Microscopy. (**A**) Imaging mode and (**B**) Force Spectroscopy mode.

The most popular ways of obtaining topography imaging are contact and tapping modes. In contact mode, the value of the repulsive force between tip and sample is kept fixed during the scanning of the sample. In tapping mode, in turn, the cantilever oscillates at its resonant frequency (or close to) and when approaching the sample, the tip comes into intermittent contact with the surface. This induces a reduction in the amplitude of the oscillations resulting from the tip-sample interactions. The amplitude is then used as a feedback signal for topographic imaging. In this way, by recording the difference between the phase of the (set) drive signal and the phase of the cantilever response, a "phase" image is obtained. For instance, this type of imaging has been used to deliver information about the viscoelastic and adhesive properties of the sample [21,22] or, more recently, to resolve 3D structures formed by electrolyte solutions near a solid surface [23]. The tapping mode, in comparison with the contact mode, presents the advantage of reducing friction forces when scanning (soft) samples. In any case, the scanning force should be always kept as low as possible to avoid sample damaging, unless required, which implies the use of cantilevers of low (<0.1 N/m) spring constant values. Furthermore, the current commercial AFM design and the available thermally-controlled fluid cells allow measuring at 37 °C in different buffer solutions, in order to mimic body conditions and provide a stable system for biological samples.

2.2. AFM as a Mechanical Machine

Imaging is not the only feature from atomic force microscopes, despite its extended use. AFM devices can also be used as a "mechanical" machine allowing the investigation of adhesion and surface forces [24,25], polymer elasticity [26,27], ligand-receptor forces [28,29], or cell mechanics [30–33]. In these experiments, an AFM-tip or a colloidal probe [34] is extended and retracted towards/from the sample of interest by following exclusively the Z-axis. Such motion takes place under controlled displacement speeds, as governed by the movement of the piezo. During this process, the deflection of the cantilever is determined as a function of the displacement of the piezo-scanner [20], and the force sensed by the cantilever is calculated using Hooke's law (which is equal to the cantilever deflection times of its spring constant). Therefore, the spring constant of the cantilever should be calculated in every experiment. A review of the different experimental calibrating methods can be found in [35].

The force-distance curves recorded in this way (Figure 2B) can be divided into three clearly distinguishable segments (approach, contact with the sample, and retraction). They can be described as follows:

- The approach curve delivers information about the existing repulsive or attractive forces between the tip/colloidal probe and the sample (e.g., electrostatic, van der Waals, hydration, or entropic forces). These type of measurements have been crucial for the understanding of molecular and colloidal interactions [11,36–38].

- The second part of the curve, during contact between the cantilever and the sample, provides information about rheology-related properties (e.g., Young´s Modulus, stiffness, relaxation time, and viscosity). The estimation of both the sample stiffness and elastic modulus (E) has been described in [39,40]. In this regard, the Hertz model—in which the contact between two linear, elastic spheres is described—is one of the most commonly used models to calculate the Young's modulus from an AFM force-distance curve [41]. However, this model presents some limitations, mostly related to the omission of the adhesive forces, which limits its applicability on sticky materials. Alternatively, the Derjaguin-Mfiller-Toporov (DMT) and Johnson-Kendall-Roberts (JKR) theories were developed to overcome such limitations [42]. Also, the employment of indenter geometries different than spheres has given rise to additional adjustments, as with the Sneddon model, which is applied for conical shapes [43] and the derivative equations developed for quadratic pyramids or flat indenters [44–46]. In addition, more information can be obtained by keeping close contact between the tip and the material for a certain observation time ($t_{observation}$), which is normally denoted as the Dwell time. Depending on the measurement performed, either the Z position of the head (Relaxation) or the load applied (Creep Compliance) are fixed during the contact. This induces the material to undergo structural rearrangement in response to the load-induced deformation which, by extension, allows the obtaining of the compressive moduli and viscosities of the material tested [47].

- Finally, the segment depicting the retraction motion relates to adhesive forces, the existence of tethers, and possible molecular unfolding events. The maximum adhesion (F_{adh}) parameter, or pull-off force, is indicative of the stickiness of the sample. It is brought by the minimum of the peak in the retraction segment. Additional pulling shows the recovery path followed until achievement of the non-contact state. This tip-sample retraction can take place either via tether formation, in the shape of uniform rupture events distanced by plateaus of zero force variation [48], or when capturing individual molecules/chains, by means of saw-like adhesion peaks to be fitted by a worm-like chain (WLC) model [49,50].

It is worth mentioning that any interaction force depends on the distance between the interacting bodies. The same applies for the estimation of the indentation caused by external pressure. Therefore, an optimal detection of the contact point between tip and sample favors quantification of the type of interaction or physical quantity considered. In order to solve such an issue, it is worth highlighting the work of Benitez et al. with R programming language [51], in which an algorithm covering the statistical analysis of the slope changes of the curve is presented.

3. AFM and Cell Scaffolds: Literature Review

Based on the aforementioned measuring methods, some authors in the field of cell scaffolds have already explored the usefulness of Atomic Force Microscopy for testing different featuring aspects of their engineered materials. A broad majority of these studies obtained an advantage of the topography imaging aspects of the technique. In them, the reported research showed the analysis of a large range of materials such as natural and synthetic polymers [52–54], collagen [55], silk fibroin [56–58], cellulose [59], peptides [60,61], graphene and carbon nanotubes (CNT) [62,63], particles [64,65], or patterned structures [66,67]. This list, kept short for compact design reasons, could include more field-related articles. However, the ones here cited can undoubtedly be considered an optimal representation of each of the systems shown. A more detailed description of the cases commented has been included as a Supporting Material (Table S1).

In the following paragraphs, the description will be focused on those research works offering a more alternative use of the AFM technique, in which characterization is tackled from the point of view of mechanical properties. As already explained in the Introduction section, the studies under consideration have been divided into five main sections, attending to the matrix forming the scaffold, and are presented as follows:

3.1. Fibres

The group of fibre-like cell scaffolds is the one that presents, by far, the largest number of contributions. In the recent years, through the development of new materials and the re-definition of some existing techniques, i.e., electrospinning, researchers have found an easy way of fabricating nanometric fibres of diverse composition and multiple potential applications. In this process, AFM has become very useful in order to perform different mechanical characterization tests, ranging from classical nanoindentation up to the trapping and stretching of individual fibres, which leads to discrimination between segments of the force-distance plot required for the respective analysis.

Thus, Dunne et al. (2014) performed indentation studies (approach segment) to determine the apparent elastic modulus of dielectrophoresis-aligned nanofibrous silk fibroin-chitosan scaffolds under alternating current frequencies and the presence of ions [68]. The same protocol was subsequently followed to assess the apparent elastic modulus of human umbilical vein endothelial cells (HUVECs) seeded onto these type of scaffolds. Horimizu et al. (2013) went for thick periosteal sheets with enhanced cell layering [69]. Nano-indentation by AFM allowed distinguishing between the central and peripheral regions by means of the peak stiffness values of cells. In a similar manner, Liu et al. (2014) compared extra- to intra-fibre-mineralized collagen scaffolds. They showed how these two systems exhibit a different nanostructure. Furthermore, the intra-fibrillar-mineralized scaffold, presenting a bone-like hierarchy, was featured by a significantly increased Young's modulus in both dry and wet conditions if compared to the other one [70].

The retract segment was also used by Guarino et al. (2016) to characterize the influence of hydroxyapatite crystals on PCL composite scaffold properties with relevant effects on their biological response [71]. Hence, AFM yielded a significant increase of the pull-off (adhesion) forces from 33.7% to 48.7%, and almost a two-fold increase in roughness, as determined by imaging. These results could be also compared to complementary differential scanning calorimetry (DSC) analyses indicating a reduction of the crystallization heat, from 66.75 to 43.05 $J \cdot g^{-1}$. An analysis of the adhesive properties was also used by Firkowska et al. in other alternative, fibre-like scaffolds like that of the multi-walled carbon nanotube (CNT)-based scaffolds [72]. This system represents a novel type of matrix composed of layer-by-layer assembled polyelectrolytes together with crosslinked carbon nanotubes with regular nano-topography. AFM investigation of the adhesion behaviour of the attached cells on the films assessed their cytocompatibility.

In a higher level of refinement and complexity, the experiments done by Baker et al. included a combined atomic force microscope (AFM)/optical microscope setup to study the mechanical properties of individual electrospun fibres. These were made of either poly-ε-caprolactone (PCL) (440–1040 nm of diameter) [73] or fibrinogen (diameter = 30–200 nm) [74]. The AFM was used to stretch individual fibres and to monitor factors such as elasticity and extension capability under both dry and wet conditions which, as the authors claim, might provide enough mechanical data to guide the construction of scaffolds and other biomedical devices based on these components. Such fine control over the tip location was also employed by Spurlin et al. (2009) for absolutely different purposes [75]. In their case, the AFM tip was applied over a defined region of enzyme tissue transglutaminase 2 (TG2)-treated fibrils in order to perform loading and shearing force studies. Results indicated a 3-fold higher resistance of treated vs. untreated fibrils, which apparently alters the contractile and proliferative response of the material. The perturbation produced can be seen in Figure 3A.

3.2. Patterned Structures

The last publication mentioned above, where the AFM tip is locally utilized to produce structural changes through shear forces under high loads, opens the path to introduce the use of this technique into another group of systems, that of patterned surfaces. Hence, by means of controlling both the force applied and the exact location of the interacting tip, researchers have already envisaged the use of force microscopy to induce the formation of well-defined patterns.

One could, for instance, employ the tip as a lithographic crayon. This was presented by Podestà et al. (2005) in order to design chemical and morphological micro- and nanoscale modification of poly(2-hydroxyethyl methacrylate) (PHEMA) hydrogels produced by conventional radical polymerization [76]. In their work, the lateral motion of the tip, in combination with a high load, is applied to create different topographies by scratching the underlying polymer surface both in the presence and the absence of carbon nano-particles. Another approach was that applied by D'Acunto et al. (2007) to create patterned polymer films with tailored length scales [77]. Thus, the morphology of poly(ε-caprolactone) diol (PCL) and poly(ethylene terephthalate) (PET) is changed in the shape of ordered ripple structures by following a novel methodology, just by a single AFM scan and for relatively low applied loads. As explained by the authors, such ripple structures can be modulated and modified by changing the applied load, scanning velocity, and angle. In this way, it is possible to obtain sinusoidal structures with suitable amplitude, periodicity, and orientation. Additionally, the method is claimed to be used for the nano-patterning of large areas making it of high usefulness in many applications.

Of course, besides the more specific uses as those listed above, other researches utilize AFM in a more standard mechanical way by performing nanoindentation on the sample. This is the case of the work of McCracken al. (2016) who determined the mechanics of different PHEMA-based 3D hydrogel cellular microcultures, under different curable monomer compositions, and which are prepared via direct ink writing [78]. Their observations revealed that depending on the polymer to monomer ratios (M_r), the prepared inks yielded mesh gels of a different opening due to a lower degree of entanglement which, by extension, could impact the mechanical performance of the films. In turn, Park et al. (2015) analysed the mechanical responses of cancer cells on top of the nano-scaffolds employed to control their adhesion size [79]. On the one hand, they found that prostate cancer cells showed a linearly decreasing proliferation rate and mechanical stiffness as the size of the imprinted nanoislands decreased. This mechanical signature was exacerbated for less metastatic prostate cancer cells. On the other hand, breast cancer cells showed no dependence of mechanical responses on the geometric properties of the nano-scaffolds, despite the acute inhibition of adhesion and the abrupt mechanical changes.

3.3. Particles

Among the materials employed for cell scaffold fabrication, (nano) particles are not as extended as other building blocks but still have found their own space. It is pretty common to find them as part of composites, as shown by Yang et al. (2016) for the nanofibers of silk fibroin (SF) with up to a 40 wt % of uniformly dispersed hydroxyapatite particles (HAp) in their composition [80]. The dependency of the mechanical moduli on the content of the HAp nanoparticles was analyzed at the AFM using a three-point bending method by means of a tipless cantilever. The SF/HAp composite fibers at varying mixing ratios were electrospun over micrometer scale channels, and so the suspended SF/HAp nanofibers could be gently pressed at a controlled speed to record the corresponding forces and deflections at each location. Based on the force and deflection data, authors calculated the bending modulus of each of the SF/HAp composite nanofibers. From their results, single nanofibers became stiffer with a higher content of the HAp nanoparticles up to 20 wt % of Hap, while the further addition of the HAp nanoparticles reduced the mechanical strengths.

But particles can also be employed as an individual entity to build a scaffold. This is the case of the work published by Hong et al. (2010) [81], where mono-dispersed bioactive glass nanospheres (200–350 nm in diameter) were shown to induce rapid deposition of an apatite layer in simulated body fluid. This speaks about their excellent bio-mineralization capability. Then, authors used AFM to investigate the effect of such bioactive glass on the biomechanical properties of various mammalian cells, such as bone marrow stem cells and/or bovine aortic endothelial cells seeded on top, with diverse results depending on the cell line.

3.4. Hydrogels

In a similar level to fibrous scaffolds, hydrogels represent a largely exploited system for the design of cell scaffolds. These soft polymeric networks, of either natural or synthetic origin, feature by having a huge water content which turns them into a perfect model for biological tissue mimicking [82]. However, those factors contributing to their good properties as supporting scaffolds are not beneficial for obtaining an optimal high-resolution characterization of their structure by AFM. For instance, sample softness and high stickiness represent an extreme limitation when performing topographical imaging. It is then when the Force Spectroscopy mode becomes of primary importance, so the mechanical behaviour of the scaffold can be determined and then combined with other visualization techniques.

Abuelfilat et al. (2015) employed Scanning Electron Microscopy (SEM)-based X-ray ultramicroscopy (XuM) to visualise, reconstruct and analyse 3D porous structures of hydroxypropyl cellulose methacrylate (HPC-MA) hydrogels [83]. Then, by incorporating AFM measurements, the elastic modulus of the hydrogel was determined, and mechanical modelling of individual pores and the bulk scaffold also proved to be feasible. In similar terms, Credi et al. (2014) measured mechanical properties of divinyl sulfone (DVS)-crosslinked hyaluronic acid hydrogels [84]. By varying both DVS content and curing time, the control over the crosslinking process can be achieved in order to prepare biocompatible hydrogels with mechanical properties closely approximating those of the extracellular matrix (ECM) of natural stem cells niches.

One of the rare cases in which AFM could also deliver images of the surfaces topography was presented by Ohya et al. (2005) for Poly(*N*-isopropylacrylamide) (PNIPAM)-grafted gelatin hydrogel surfaces, a temperature-induced scaffold at a physiological temperature [85]. In this study, the authors aimed at determining the effect of the graft architecture of thermoresponsive PNIPAM-gelatin on the surface topography and elastic modulus of the hydrogels. Also, mechanical properties were determined by AFM.

3.5. Peptides and RGD Sequence

Last but not least, the fifth group in which scaffolds have been divided is devoted to peptides and assemblies containing RGD derivatives. As already mentioned above, the particularity of these biopolymers turns them into an own entity, not really matching into any of the already listed ones but, at the same time, being part of all of them. The best example of this observation is that of self-assembling peptides, which have been reported to form either well-defined fibrous structures [61,86] or more hydrogel-like films [87,88]. An additional interesting feature of peptides, and more specifically those containing RGD-moieties, is the capability of inducing specific recognition of cells based on the selective capture of integrins forming part of their Extra-Cellular Matrix (ECM) [89]. This factor has promoted the development of a large number of research studies in which the binding properties of either fibronectin, fibrinogen, or alternative synthetic peptide chains with the RGD sequence in their structure have been exploited.

In many cases, such peptides are easily combined with another matrix, like polymers, in order to activate the cell capture of systems that otherwise would remain almost inert [90]. Sabater i Serra et al. (2016), for instance, assembled fibronectin (FN) on top of 2D Poly(ethyl acrylate) (PEA) films [91]. These could be tuned by varying amounts of a crosslinker (1–10% of ethylene glycol dimethacrylate). Even though PEA has a good ability to organize FN, under an increasing crosslinker density the segmental mobility was shown to be significantly reduced. In turn, Gentsch et al. (2011) presented a single step procedure to fabricate bio-functional fibres composed of the polymer-(Cys-Gly-Gly-Arg-Gly-Asp-Ser) peptide conjugate (poly(lactic acid)-block-CGGRGDS) combined with poly(lactic-*co*-glycolic acid) (PLGA) [92]. Authors then could test the surface accessibility of the peptide on the nanoscale by AFM measurements probing the electrostatic interaction between CGGRGDS surface functionalities and a colloidal silica probe via atomic force microscopy, and the corresponding comparison with pure PLGA fibres. A short scheme of the experiment is shown in Figure 3B. The repulsive forces of pure PLGA contrast with the adhesion measured for CGGRGDS-containing samples (9 and 18 wt %).

Regarding the use of integrin for specific recognition purposes, a very different approach can also be attempted: an integrin coating would activate AFM cantilevers (either with a tip or tip-less) so, under optimal and controlled contact maintenance, they can be used to bind and test living cells. Thus, this living cell would act now as a measuring probe. This new cell-cantilever joint system can be applied to measuring interactive forces with a surface/scaffold of choice, in a process known as Single-Cell Probe Force Spectroscopy (SCFPS). Of course, the success of the process also depends on the employment of a substrate of known low affinity towards cell binding, so cells stay as round and intact as possible before being captured. Such a type of experiment was presented by Moreno-Cencerrado et al (2016) [93]. They could design a new setup consisting of two different half-surfaces coated either with a recrystallized SbpA bacterial cell surface layer proteins (S-layers) or integrin binding Fibronectin, on which Michigan Cancer Foundation (MCF)-7 breast cancer cells were incubated (see Figure 3C). As shown in their work, quantitative results about SbpA-cell and Fibronectin-cell adhesion forces as a function of the contact time could be described by this technique. At the same line, Taubenberger et al. (2014) reported about a full bunch of alternative components and factors that might be considered when designing SCFPS experiments [94].

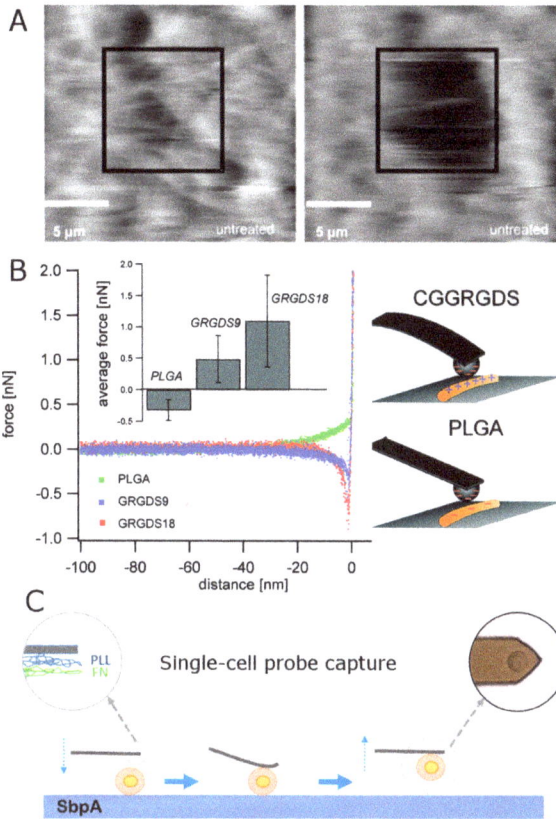

Figure 3. (**A**) Topographic scanning of collagen fibrils before (left) and after (right) scanning a 10 µm × 10 µm area at 5 nN. Scale bar represents 5 µm. Black boxes indicate the region where the increased normal force was applied. Imaging was performed under phosphate buffer at 1 nN loading force (Adapted reprint with permission from [75]. Copyright (2011) American Chemical Society). (**B**) Force−displacement curves resulting from colloidal interaction probing with a silica bead on (Gly-Arg-Gly-Asp-Ser) (GRGDS)-functional fiber compared to poly(lactic-*co*-glycolic acid) (PLGA) fiber, as schematically shown on the right sketch. The inlet shows adhesion force values as a function of PLLA-*b*-CGGRGDS content (0, 9, and 18 wt %) (Adapted reprint from [92], Copyright (2011), with permission from Elsevier). (**C**) The cell attachment process performed on an S-layer film. The fibronectin-functionalized tipless cantilever approaches towards the substrate and keeps contact for 10 min. Afterwards, the cantilever retracts slowly from the surface with the selected cell attached to it (an optical photograph shows the cell attached to the cantilever) (Reprinted from [93] with permission from the authors).

4. Conclusions and Outlook

The analysis of cell scaffolds by atomic force microscopy has gradually gained more and more relevance over the last years. The presence of an AFM at nearly all academic and research institutions has contributed to it favorably. However, it has to be said that the potential applications of AFM are still waiting to be used in research topics such as those discussed in this short review. As the reader can see in the previous sections, AFM has had a minor role so far, acting as a complement to other measuring techniques. Perhaps, a slight redefinition and description of its real potential uses may trigger its use in future studies.

For instance, most of the literature found merely regards topographical imaging of diverse systems. It is true that such a type of topographical characterization of scaffolds becomes important when studying samples in which a defined pattern or a crystal-like structure on the sub-micron scale is involved, but its interest drops significantly in the other cases. Indeed, scanning electron microscopy (SEM) could cover this part of the characterization of the sample of interest with similar results. One idea is to extend the classical concept of imaging (height, error deflection, and phase) by combining the raster scanning with force-distance curves. This is possible because the AFM functioning relies on the tip-sample interacting forces and, therefore, a topography image is already a result of those forces. Then, by considering this overall force-dependence concept, one can obtain and develop new imaging modes which depend on diverse but concrete interactions. Examples of it are Chemical AFM and the Fluorescence sensing approaches, in which the tip is employed for obtaining another type of material-derived information (e.g., surface chemistry, electrical and hydrophobic properties, etc.).

At the so-called chemical force microscopy, the chemical modification of the tip (which can have different geometries), or even of the whole cantilever body if employing tip-less sensors, allows detecting specific and/or non-specific interactions when the tip is close enough to the sample (note that the forces involved are of the molecular or colloidal range). This is also denoted as molecular recognition [95,96]. If such chemical force microscopy is usually thought of for specific interactions, an alternative methodology that also explores the chemical environment of the sample of interest would consist of the use of fluorescent tips (or colloidal probes). The idea behind this would be to utilize a probe which is sensitive to pH changes. Hence, knowing the variation that changes in the pH induces in the emission the intensity of the fluorophore, as determined by a master curve, and the fluorescent tip could act as accurate pH sensor of the substrate. A schematic sketch is shown in Figure 4A. In this way it would be possible to have a full chemical map of the sample, attending to its surface pH, which in turn relates to its electrical properties (i.e., electrostatic surface potential).

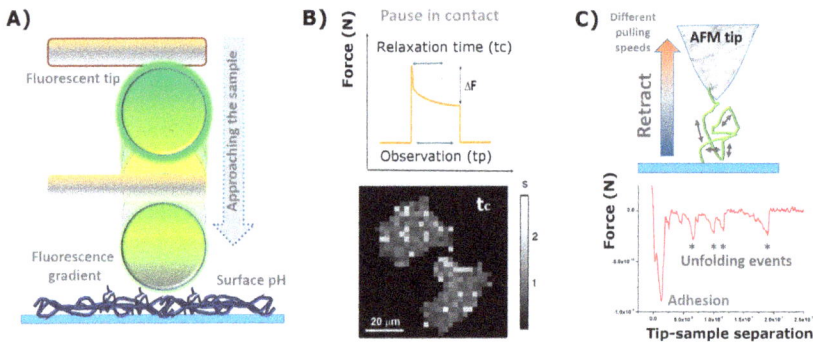

Figure 4. Potential uses of atomic force microscopy (AFM) according to the focus on each of the individual three main segments of a force-distance curve (note that time is an implicit variable in a force-distance curve and therefore can be also represented as a force-time curve). (**A**) Scheme of Fluorescence-based surface pH sensor mode, in which a stained colloidal probe is approached to the surface of interest. At sufficiently close distances the change in the pH induces quenching of the fluorescence emission (changes in surface pH as a function of time can be detected in "pause in contact mode"). (**B**) An example of Stress Relaxation imaging mode obtained on MCF-7 cells from mapping their relaxation time (t_c) values, in seconds, according to the coloured scale on the right (*adapted reprint from* [97] *with permission of the authors*). The plot on top represents the usual shape obtained for a stress relaxation experiment, keeping the contact in constant height, where ΔF denotes the force decay and t_p and t_c refer to the observation and relaxation times, respectively. (**C**) Scheme of a single chain pulling experiment. The pulling rate (speed) can be adapted to demand. Below, a representative retraction plot showing the most relevant unfolding events measured and the adhesion force peak.

However, besides the applications mentioned above, future predictions about the employment of AFM on cell scaffolds mostly stay on the side of the Force Spectroscopy measuring mode, either as a "single" or as a more "global" analysis. While the former does not deliver an image, the latter still allows the obtaining of a full mapping over the area of interest. In any of the cases, all the parts the force-distance plot is composed of (Figure 2B) should be taken into consideration for an optimal characterization. The usual nanoindentation experiments for determining the elastic modulus of the sample are sometimes not sufficient to complete the mechanical description of the sample. As explained in a previous section, extraction of the instantaneous Young's modulus only corresponds to a partial analysis of the force-distance curve. The omission of the "pause in contact" and retraction directly discards a full bunch of parameters that could become a perfect complement for completing the characterization.

- The first part of the plot (approach motion) can be used for determining the presence of repulsive/attractive forces between the tip and the simple (electrostatic, steric, and/or entropic, etc.). Examples of it were already reported in the literature, among others, by Borkovek [24], and Melzak [98]. A similar process is also described by Gentsch et al. in reference [92], and depicted in Figure 3B.

- Attending to the contact segment of the plot (or pause in contact) or studying the deformation of the scaffold under constant force or stress relaxation experiments can describe the rheology of the material for sufficiently large observation (Dwell) times. This was successfully applied to characterize the response of breast carcinoma MCF-7 cells seeded on borosilicate glass substrates [97]. In their work, the authors also presented a new type of AFM imaging called Stress Relaxation imaging, based on the local relaxation times measured after subdividing the cell into multiple domains (mapping in Figure 4B).

- The retraction motion (third segment) yields, in turn, very useful information about adhesive forces and the work required to recover the non-contact state. A good example of it would be the aforementioned "single" force microscopy mode for the stretching of individual coiled chains adhered to the tip (as in the case of proteins) [99–101]. Hence, quantification of the occurring intermediate rupture events, as the cantilever moves perpendicularly away from the surface, might deliver a precise fingerprint of the forces which govern the internal arrangement of the coils (Figure 4C).

Latest developments have also considered the use of dynamic AFM methods (based on the tapping mode) to promote the nanoscale characterization (topographical and mechanical properties) of the cell scaffolds to another level. For instance, dynamic multi-frequency AFM methods (multi-harmonic, Bimodal) and Peak Force Tapping (PFT) are pushing the AFM field to a faster, quantitative, nanomechanical characterization of complex biological systems in a higher resolution [102,103].

These methods described so far imply the discriminate characterization of either the scaffold or the cells seeded on top. Nevertheless, it might also be very interesting to directly measure the interaction between the cells and the scaffold itself (which gives an idea about the compatibility cell-scaffold). This possibility is brought by the so-called Single-Cell Probe Force Spectroscopy (SCPFS). As previously described, it combines measurements in the Force Spectroscopy mode with chemical modification of the cantilever in order to capture living cells so they can be used as a measuring probe. SCPFS provides, as the main advantage, a sensitive approach to characterize not only the interaction of a whole cell with the underlying substrate but also that of single molecules, attending to host-guest interactions [88]. The versatility of this technique leans on the number of cell lines to be employed, either prokaryotic [61,90] or eukaryotic [84,89], and the infinite possibilities for scaffold design. A glimpse over the number of works published shows the increasing interest of the community into this methodology and, therefore, a promising future can be envisaged.

Almost at the end of this review, we could ask why AFM is not extensively used when investigating cell scaffolds. Or, what are the challenges that prevent AFM from becoming more popular in characterizing cell scaffolds? The reading of the manuscripts commented on and included throughout this review already reveal some hints to address such issue. Most of the researchers investigating cell scaffolds are not true AFM specialists. Our experience shows that the mastering of AFM requires at least: (i) very good experimental and technical skills; and (ii) a correct interpretation of the AFM results. This point is not trivial since it requires image processing to complicate calculations based on the physical models. Therefore, non-specialised AFM laboratories producing cell scaffolds might take a pragmatic approach by focusing mostly on the application and the technological effects of the scaffold (e.g., the influence on the cell fate). Our opinion is that in the years to come, a refinement in scaffold bioengineering will lead to a better understanding of the scaffold/cell interface and its corresponding biotechnological implications. Thus, a complete physico-chemical characterization either of the scaffold or the total system would be crucial, and AFM may be the right experimental tool to achieve that goal.

Supplementary Materials: The following are available online at www.mdpi.com/2073-4360/9/8/383/s1, Table S1: Literature compilation regarding the exclusive use of Atomic Force Microscopy for topography imaging purposes.

Acknowledgments: Authors thank Alberto Moreno-Cencerrado for his help with the proof-read of this review.

Author Contributions: J.I. and J.L.T.-H. conceived and designed the content of the paper. J.I. and J.L.T.-H. wrote the paper.

Conflicts of Interest: The authors declare no conflict of interest.

References

1. Han, D.K.; Park, K.D.; Hubbell, J.A.; Kim, Y.H. Surface characteristics and biocompatibility of lactide-based poly(ethylene glycol) scaffolds for tissue engineering. *J. Biomater. Sci. Polym. Ed.* **1998**, *9*, 667–680. [PubMed]
2. Irvine, D.J.; Ruzette, A.V.G.; Mayes, A.M.; Griffith, L.G. Nanoscale clustering of rgd peptides at surfaces using comb polymers. 2. Surface segregation of comb polymers in polylactide. *Biomacromolecules* **2001**, *2*, 545–556. [CrossRef] [PubMed]
3. Marshall, G.W., Jr.; Balooch, M.; Kinney, J.H.; Marshall, S.J. Atomic force microscopy of conditioning agents on dentin. *J. Biomed. Mater. Res.* **1995**, *29*, 1381–1387. [CrossRef] [PubMed]
4. Stock, U.A.; Vacanti, J.P. Tissue engineering: Current state and prospects. *Annu. Rev. Med.* **2001**, *52*, 443–451. [CrossRef] [PubMed]
5. Discher, D.E.; Janmey, P.; Wang, Y.-L. Tissue cells feel and respond to the stiffness of their substrate. *Science* **2005**, *310*, 1139–1143. [CrossRef] [PubMed]
6. Prauzner-Bechcicki, S.; Raczkowska, J.; Madej, E.; Pabijan, J.; Lukes, J.; Sepitka, J.; Rysz, J.; Awsiuk, K.; Bernasik, A.; Budkowski, A.; et al. PDMS substrate stiffness affects the morphology and growth profiles of cancerous prostate and melanoma cells. *J. Mech. Behav. Biomed. Mater.* **2015**, *41*, 13–22. [CrossRef] [PubMed]
7. Vogel, V.; Sheetz, M. Local force and geometry sensing regulate cell functions. *Nat. Rev. Mol. Cell Biol.* **2006**, *7*, 265–275. [CrossRef] [PubMed]
8. Yim, E.K.; Darling, E.M.; Kulangara, K.; Guilak, F.; Leong, K.W. Nanotopography-induced changes in focal adhesions, cytoskeletal organization, and mechanical properties of human mesenchymal stem cells. *Biomaterials* **2010**, *31*, 1299–1306. [CrossRef] [PubMed]
9. Hinterdorfer, P.; Gruber, H.J.; Kienberger, F.; Kada, G.; Riener, C.; Borken, C.; Schindler, H.J. Surface attachment of ligands and receptors for molecular recognition force microscopy. *Colloids Surf. B* **2002**, *23*, 115–123. [CrossRef]
10. Sugimoto, Y.; Pou, P.; Abe, M.; Jelinek, P.; Perez, R.; Morita, S.; Custance, O. Chemical identification of individual surface atoms by atomic force microscopy. *Nature* **2007**, *446*, 64–67. [CrossRef] [PubMed]
11. Weisenhorn, A.L.; Maivald, P.; Butt, H.J.; Hansma, P.K. Measuring adhesion, attraction, and repulsion between surfaces in liquids with an atomic-force microscope. *Phys. Rev. B* **1992**, *45*, 11226–11232. [CrossRef]
12. Arnoldi, M.; Fritz, M.; Bauerlein, E.; Radmacher, M.; Sackmann, E.; Boulbitch, A. Bacterial turgor pressure can be measured by atomic force microscopy. *Phys. Rev. E* **2000**, *62*, 1034–1044. [CrossRef]
13. Charras, G.T.; Lehenkari, P.P.; Horton, M.A. Atomic force microscopy can be used to mechanically stimulate osteoblasts and evaluate cellular strain distributions. *Ultramicroscopy* **2001**, *86*, 85–95. [CrossRef]

14. Fotiadis, D.; Scheuring, S.; Müller, S.A.; Engel, A.; Müller, D.J. Imaging and manipulation of biological structures with afm. *Micron* **2002**, *33*, 385–397. [CrossRef]
15. Leporatti, S.; Gerth, A.; Kohler, G.; Kohlstrunk, B.; Hauschildt, S.; Donath, E. Elasticity and adhesion of resting and lipopolysaccharide-stimulated macrophages. *FEBS Lett.* **2006**, *580*, 450–454. [CrossRef] [PubMed]
16. Butt, H.-J.; Cappella, B.; Kappl, M. Force measurements with the atomic force microscope: Technique, interpretation and applications. *Surf. Sci. Rep.* **2005**, *59*, 1–152. [CrossRef]
17. Leckband, D. Novel recognition mechanisms in biological adhesion. *Curr. Opin. Colloid Interface Sci.* **2001**, *6*, 498–505. [CrossRef]
18. Binnig, G.; Quate, C.F.; Gerber, C. Atomic force microscope. *Phys. Rev. Lett.* **1986**, *56*, 930–933. [CrossRef] [PubMed]
19. Alessandrini, A.; Facci, P. AFM: A versatile tool in biophysics. *Meas. Sci. Technol.* **2005**, *16*, R65–R92. [CrossRef]
20. Moreno-Flores, S.; Toca-Herrera, J.L. *Hybridizing Surface Probe Microscopes: Toward a Full Description of the Meso- and Nanoworlds*; CRC Press: Boca Raton, FL, USA, 2013.
21. García, R.; Tamayo, J.; San Paulo, A. Phase contrast and surface energy hysteresis in tapping mode scanning force microscopy. *Surf. Interface Anal.* **1999**, *27*, 312–316. [CrossRef]
22. Radmacher, M.; Tillmann, R.W.; Fritz, M.; Gaub, H.E. From molecules to cells: Imaging soft samples with the atomic force microscope. *Science* **1992**, *257*, 1900–1905. [CrossRef] [PubMed]
23. Martin-Jimenez, D.; Chacon, E.; Tarazona, P.; Garcia, R. Atomically resolved three-dimensional structures of electrolyte aqueous solutions near a solid surface. *Nat. Commun.* **2016**, *7*, 12164. [CrossRef] [PubMed]
24. Borkovec, M.; Szilagyi, I.; Popa, I.; Finessi, M.; Sinha, P.; Maroni, P.; Papastavrou, G. Investigating forces between charged particles in the presence of oppositely charged polyelectrolytes with the multi-particle colloidal probe technique. *Adv. Colloid Interface Sci.* **2012**, *179–182*, 85–98. [CrossRef] [PubMed]
25. Puech, P.H.; Poole, K.; Knebel, D.; Muller, D.J. A new technical approach to quantify cell-cell adhesion forces by AFM. *Ultramicroscopy* **2006**, *106*, 637–644. [CrossRef] [PubMed]
26. Bornschlögl, T.; Rief, M. Single-molecule protein unfolding and refolding using atomic force microscopy. In *Methods in Molecular Biology*; Springer: Berlin, Germany, 2011; Volume 783, pp. 233–250.
27. Fisher, T.E.; Oberhauser, A.F.; Carrion-Vazquez, M.; Marszalek, P.E.; Fernandez, J.M. The study of protein mechanics with the atomic force microscope. *Trends Biochem. Sci.* **1999**, *24*, 379–384. [CrossRef]
28. Florin, E.L.; Moy, V.T.; Gaub, H.E. Adhesion forces between individual ligand-receptor pairs. *Science* **1994**, *264*, 415–417. [CrossRef] [PubMed]
29. Hinterdorfer, P.; Dufrêne, Y.F. Detection and localization of single molecular recognition events using atomic force microscopy. *Nat. Methods* **2006**, *3*, 347–355. [CrossRef] [PubMed]
30. Melzak, K.A.; Lázaro, G.R.; Hernández-Machado, A.; Pagonabarraga, I.; Cárdenas Díaz De Espada, J.M.; Toca-Herrera, J.L. AFM measurements and lipid rearrangements: Evidence from red blood cell shape changes. *Soft Matter* **2012**, *8*, 7716–7726. [CrossRef]
31. Radmacher, M. Measuring the elastic properties of living cells by the atomic force microscope. In *Methods in Cell Biology*; Elsevier: Amsterdam, The Netherlands, 2002; Volume 2002, pp. 67–90.
32. Scheuring, S.; Dufrêne, Y.F. Atomic force microscopy: Probing the spatial organization, interactions and elasticity of microbial cell envelopes at molecular resolution. *Mol. Microbiol.* **2010**, *75*, 1327–1336. [CrossRef] [PubMed]
33. Vargas-Pinto, R.; Gong, H.; Vahabikashi, A.; Johnson, M. The effect of the endothelial cell cortex on atomic force microscopy measurements. *Biophys. J.* **2013**, *105*, 300–309. [CrossRef] [PubMed]
34. Ducker, W.A.; Senden, T.J.; Pashley, R.M. Direct measurement of colloidal forces using an atomic force microscope. *Nature* **1991**, *353*, 239–241. [CrossRef]
35. Cumpson, P.J.; Clifford, C.A.; Portoles, J.F.; Johnstone, J.E.; Munz, M. Cantilever spring-constant calibration in atomic force microscopy. *Nano Sci. Technol.* **2008**, *8*, 289–314.
36. JPK. A Practical Guide to AFM Force Spectroscopy and Data Analysis. In *JPK Instruments Technical Notes*; JPK Instruments AG: Berlin, Germany. Available online: http://www.jpk.com/afm.230.en.html (accessed on 15 May 2017).
37. Butt, H.J. Measuring electrostatic, van der waals, and hydration forces in electrolyte solutions with an atomic force microscope. *Biophys. J.* **1991**, *60*, 1438–1444. [CrossRef]

38. García, R.; San Paulo, A. Attractive and repulsive tip-sample interaction regimes in tapping-mode atomic force microscopy. *Phys. Rev. B* **1999**, *60*, 4961–4967. [CrossRef]

39. Dimitriadis, E.K.; Horkay, F.; Maresca, J.; Kachar, B.; Chadwick, R.S. Determination of elastic moduli of thin layers of soft material using the atomic force microscope. *Biophys. J.* **2002**, *82*, 2798–2810. [CrossRef]

40. Domke, J.; Radmacher, M. Measuring the elastic properties of thin polymer films with the atomic force microscope. *Langmuir* **1998**, *14*, 3320–3325. [CrossRef]

41. Hertz, H. Uber die berührung fester, elastischer körper. *J. Reine Angew. Math.* **1882**, *92*, 156–171.

42. Cappella, B.; Dietler, G. Force-distance curves by atomic force microscopy. *Surf. Sci. Rep.* **1999**, *34*, 1–104. [CrossRef]

43. Sneddon, I.A. The relation between load and penetration in the axisymmetric boussinesq problem for a punch of arbitrary profile. *Int. J. Eng. Sci.* **1965**, *3*, 47–57. [CrossRef]

44. JPK. Determining the Elastic Modulus of Biological Samples Using Atomic Force Microscopy. In *JPK Instruments Technical Notes*; JPK Instruments AG: Berlin, Germany. Available online: http://www.jpk.com/afm.230.en.html (accessed on 15 May 2017).

45. Alcaraz, J.; Buscemi, L.; Grabulosa, M.; Trepat, X.; Fabry, B.; Farre, R.; Navajas, D. Microrheology of human lung epithelial cells measured by atomic force microscopy. *Biophys. J.* **2003**, *84*, 2071–2079. [CrossRef]

46. Gimenez, A.; Uriarte, J.J.; Vieyra, J.; Navajas, D.; Alcaraz, J. Elastic properties of hydrogels and decellularized tissue sections used in mechanobiology studies probed by atomic force microscopy. *Microsc. Res. Tech.* **2017**, *80*, 85–96. [CrossRef] [PubMed]

47. Moreno-Flores, S.; Benitez, R.; Vivanco, M.D.M.; Toca-Herrera, J.L. Stress relaxation and creep on living cells with the atomic force microscope: A means to calculate elastic moduli and viscosities of cell components. *Nanotechnology* **2010**, *21*, 44. [CrossRef] [PubMed]

48. Sun, M.; Graham, J.S.; Hegedus, B.; Marga, F.; Zhang, Y.; Forgacs, G.; Grandbois, M. Multiple membrane tethers probed by atomic force microscopy. *Biophys. J.* **2005**, *89*, 4320–4329. [CrossRef] [PubMed]

49. Hugel, T.; Seitz, M. The study of molecular interactions by afm force spectroscopy. *Macromol. Rapid Commun.* **2001**, *22*, 989–1016. [CrossRef]

50. Yamamoto, S.; Tsujii, Y.; Fukuda, T. Atomic force microscopic study of stretching a single polymer chain in a polymer brush. *Macromolecules* **2000**, *33*, 5995–5998. [CrossRef]

51. Benítez, R.; Moreno-flores, S.; Bolós, V.J.; Toca-Herrera, J.L. A new automatic contact point detection algorithm for afm force curves. *Microsc. Res. Tech.* **2013**, *76*, 870–876. [CrossRef] [PubMed]

52. Al Rez, M.F.; Binobaid, A.; Alghosen, A.; Mirza, E.H.; Alam, J.; Fouad, H.; Hashem, M.; Alsalman, H.; Almalak, H.M.; Mahmood, A.; et al. Tubular poly(ε-caprolactone)/chitosan nanofibrous scaffold prepared by electrospinning for vascular tissue engineering applications. *J. Biomater. Tissue Eng.* **2017**, *7*, 427–436. [CrossRef]

53. Chen, J.; Dong, R.; Ge, J.; Guo, B.; Ma, P.X. Biocompatible, biodegradable, and electroactive polyurethane-urea elastomers with tunable hydrophilicity for skeletal muscle tissue engineering. *ACS Appl. Mater. Interfaces* **2015**, *7*, 28273–28285. [CrossRef] [PubMed]

54. Chen, R.; Huang, C.; Ke, Q.; He, C.; Wang, H.; Mo, X. Preparation and characterization of coaxial electrospun thermoplastic polyurethane/collagen compound nanofibers for tissue engineering applications. *Colloids Surf. B Biointerfaces* **2010**, *79*, 315–325. [CrossRef] [PubMed]

55. Stylianou, A.; Yova, D. Surface nanoscale imaging of collagen thin films by atomic force microscopy. *Mater. Sci. Eng. C* **2013**, *33*, 2947–2957. [CrossRef] [PubMed]

56. Farokhi, M.; Mottaghitalab, F.; Hadjati, J.; Omidvar, R.; Majidi, M.; Amanzadeh, A.; Azami, M.; Tavangar, S.M.; Shokrgozar, M.A.; Ai, J. Structural and functional changes of silk fibroin scaffold due to hydrolytic degradation. *J. Appl. Polym. Sci.* **2014**, *131*. [CrossRef]

57. Pillai, M.M.; Gopinathan, J.; Indumathi, B.; Manjoosha, Y.R.; Santosh Sahanand, K.; Dinakar Rai, B.K.; Selvakumar, R.; Bhattacharyya, A. Silk–PVA hybrid nanofibrous scaffolds for enhanced primary human meniscal cell proliferation. *J. Membr. Biol.* **2016**, *249*, 813–822. [CrossRef] [PubMed]

58. Zhang, K.; Mo, X.; Huang, C.; He, C.; Wang, H. Electrospun scaffolds from silk fibroin and their cellular compatibility. *J. Biomed. Mater. Res. A* **2010**, *93*, 976–983. [CrossRef] [PubMed]

59. Jia, B.; Li, Y.; Yang, B.; Xiao, D.; Zhang, S.; Rajulu, A.V.; Kondo, T.; Zhang, L.; Zhou, J. Effect of microcrystal cellulose and cellulose whisker on biocompatibility of cellulose-based electrospun scaffolds. *Cellulose* **2013**, *20*, 1911–1923. [CrossRef]

60. Babolmorad, G.; Emtiazi, G.; Ghaedi, K.; Jodeiri, M. Enhanced PC12 cells proliferation with self-assembled s-layer proteins scaffolds. *Appl. Biochem. Biotechnol.* **2015**, *175*, 223–231. [CrossRef] [PubMed]
61. Silva, D.; Natalello, A.; Sanii, B.; Vasita, R.; Saracino, G.; Zuckermann, R.N.; Doglia, S.M.; Gelain, F. Synthesis and characterization of designed bmhp1-derived self-assembling peptides for tissue engineering applications. *Nanoscale* **2013**, *5*, 704–718. [CrossRef] [PubMed]
62. Zhang, K.; Zheng, H.; Liang, S.; Gao, C. Aligned plla nanofibrous scaffolds coated with graphene oxide for promoting neural cell growth. *Acta Biomater.* **2016**, *37*, 131–142. [CrossRef] [PubMed]
63. Zhao, X.; Luo, J.; Fang, C.; Xiong, J. Investigation of polylactide/poly(ε-caprolactone)/multi-walled carbon nanotubes electrospun nanofibers with surface texture. *RSC Adv.* **2015**, *5*, 99179–99187. [CrossRef]
64. Chen, C.; Lv, G.; Pan, C.; Song, M.; Wu, C.; Guo, D.; Wang, X.; Chen, B.; Gu, Z. Poly(lactic acid) (PLA) based nanocomposites—A novel way of drug-releasing. *Biomed. Mater.* **2007**, *2*, L1–L4. [CrossRef] [PubMed]
65. Hung, H.S.; Chang, C.H.; Chang, C.J.; Tang, C.M.; Kao, W.C.; Lin, S.Z.; Hsieh, H.H.; Chu, M.Y.; Sun, W.S.; Hsu, S.H. In vitro study of a novel nanogold-collagen composite to enhance the mesenchymal stem cell behavior for vascular regeneration. *PLoS ONE* **2014**, *9*, 104019. [CrossRef] [PubMed]
66. Li, M.; Cui, T.; Mills, D.K.; Lvov, Y.M.; McShane, M.J. Comparison of selective attachment and growth of smooth muscle cells on gelatin- and fibronectin-coated micropatterns. *J. Nanosci. Nanotechnol.* **2005**, *5*, 1809–1815. [CrossRef] [PubMed]
67. Marszalek, J.E.; Simon, C.G., Jr.; Thodeti, C.; Adapala, R.K.; Murthy, A.; Karim, A. 2.5d constructs for characterizing phase separated polymer blend surface morphology in tissue engineering scaffolds. *J. Biomed. Mater. Res. A* **2013**, *101 A*, 1502–1510. [CrossRef] [PubMed]
68. Dunne, L.W.; Iyyanki, T.; Hubenak, J.; Mathur, A.B. Characterization of dielectrophoresis-aligned nanofibrous silk fibroin-chitosan scaffold and its interactions with endothelial cells for tissue engineering applications. *Acta Biomater.* **2014**, *10*, 3630–3640. [CrossRef] [PubMed]
69. Horimizu, M.; Kawase, T.; Tanaka, T.; Okuda, K.; Nagata, M.; Burns, D.M.; Yoshie, H. Biomechanical evaluation by AFM of cultured human cell-multilayered periosteal sheets. *Micron* **2013**, *48*, 1–10. [CrossRef] [PubMed]
70. Liu, Y.; Luo, D.; Liu, S.; Fu, Y.; Kou, X.; Wang, X.; Sha, Y.; Gan, Y.; Zhou, Y. Effect of nanostructure of mineralized collagen scaffolds on their physical properties and osteogenic potential. *J. Biomed. Nanotechnol.* **2014**, *10*, 1049–1060. [CrossRef] [PubMed]
71. Guarino, V.; Veronesi, F.; Marrese, M.; Giavaresi, G.; Ronca, A.; Sandri, M.; Tampieri, A.; Fini, M.; Ambrosio, L. Needle-like ion-doped hydroxyapatite crystals influence osteogenic properties of PCL composite scaffolds. *Biomed. Mater.* **2016**, *11*, 015018. [CrossRef] [PubMed]
72. Firkowska, I.; Godehardt, E.; Giersig, M. Interaction between human osteoblast cells and inorganic two-dimensional scaffolds based on multiwalled carbon nanotubes: A quantitative AFM study. *Adv. Funct. Mater.* **2008**, *18*, 3765–3771. [CrossRef]
73. Baker, S.R.; Banerjee, S.; Bonin, K.; Guthold, M. Determining the mechanical properties of electrospun poly-ε-caprolactone (PCL) nanofibers using AFM and a novel fiber anchoring technique. *Mater. Sci. Eng. C* **2016**, *59*, 203–212. [CrossRef] [PubMed]
74. Baker, S.; Sigley, J.; Helms, C.C.; Stitzel, J.; Berry, J.; Bonin, K.; Guthold, M. The mechanical properties of dry, electrospun fibrinogen fibers. *Mater. Sci. Eng. C* **2012**, *32*, 215–221. [CrossRef] [PubMed]
75. Spurlin, T.A.; Bhadriraju, K.; Chung, K.H.; Tona, A.; Plant, A.L. The treatment of collagen fibrils by tissue transglutaminase to promote vascular smooth muscle cell contractile signaling. *Biomaterials* **2009**, *30*, 5486–5496. [CrossRef] [PubMed]
76. Podestà, A.; Ranucci, E.; Macchi, L.; Bongiorno, G.; Ferruti, P.; Milani, P. Micro- and nanoscale modification of poly(2-hydroxyethyl methacrylate) hydrogels by afm lithography and nanoparticle incorporation. *J. Nanosci. Nanotechnol.* **2005**, *5*, 425–430. [CrossRef] [PubMed]
77. D'Acunto, M.; Napolitano, S.; Pingue, P.; Giusti, P.; Rolla, P. Fast formation of ripples induced by AFM. A new method for patterning polymers on nanoscale. *Mater. Lett.* **2007**, *61*, 3305–3309. [CrossRef]
78. McCracken, J.M.; Badea, A.; Kandel, M.E.; Gladman, A.S.; Wetzel, D.J.; Popescu, G.; Lewis, J.A.; Nuzzo, R.G. Programming mechanical and physicochemical properties of 3d hydrogel cellular microcultures via direct ink writing. *Adv. Healthc. Mater.* **2016**, *5*, 1025–1039. [CrossRef] [PubMed]
79. Park, S.; Bastatas, L.; Matthews, J.; Lee, Y.J. Mechanical responses of cancer cells on nanoscaffolds for adhesion size control. *Macromol. Biosci.* **2015**, *15*, 851–860. [CrossRef] [PubMed]

80. Yang, D.; Kim, H.; Lee, J.; Jeon, H.; Ryu, W. Direct modulus measurement of single composite nanofibers of silk fibroin/hydroxyapatite nanoparticles. *Compos. Sci. Technol.* **2016**, *122*, 113–121. [CrossRef]

81. Hong, Z.; Luz, G.M.; Hampel, P.J.; Jin, M.; Liu, A.; Chen, X.; Mano, J.F. Mono-dispersed bioactive glass nanospheres: Preparation and effects on biomechanics of mammalian cells. *J. Biomed. Mater. Res. A* **2010**, *95*, 747–754. [CrossRef] [PubMed]

82. Deligkaris, K.; Tadele, T.S.; Olthuis, W.; van den Berg, A. Hydrogel-based devices for biomedical applications. *Sens. Actuators B Chem.* **2010**, *147*, 765–774. [CrossRef]

83. Abuelfilat, A.Y.; Kim, Y.; Miller, P.; Hoo, S.P.; Li, J.; Chan, P.; Fu, J. Bridging structure and mechanics of three-dimensional porous hydrogel with X-ray ultramicroscopy and atomic force microscopy. *RSC Adv.* **2015**, *5*, 63909–63916. [CrossRef]

84. Credi, C.; Biella, S.; De Marco, C.; Levi, M.; Suriano, R.; Turri, S. Fine tuning and measurement of mechanical properties of crosslinked hyaluronic acid hydrogels as biomimetic scaffold coating in regenerative medicine. *J. Mech. Behav. Biomed. Mater.* **2014**, *29*, 309–316. [CrossRef] [PubMed]

85. Ohya, S.; Kidoaki, S.; Matsuda, T. Poly(*n*-isopropylacrylamide) (PNIPAM)-grafted gelatin hydrogel surfaces: Interrelationship between microscopic structure and mechanical property of surface regions and cell adhesiveness. *Biomaterials* **2005**, *26*, 3105–3111. [CrossRef] [PubMed]

86. Sun, J.; Zheng, Q.; Wu, Y.; Liu, Y.; Guo, X.; Wu, W. Culture of nucleus pulposus cells from intervertebral disc on self-assembling kld-12 peptide hydrogel scaffold. *Mater. Sci. Eng. C* **2010**, *30*, 975–980. [CrossRef]

87. Cinar, G.; Ceylan, H.; Urel, M.; Erkal, T.S.; Deniz Tekin, E.; Tekinay, A.B.; Dâna, A.; Guler, M.O. Amyloid inspired self-assembled peptide nanofibers. *Biomacromolecules* **2012**, *13*, 3377–3387. [CrossRef] [PubMed]

88. Secchi, V.; Franchi, S.; Fioramonti, M.; Polzonetti, G.; Iucci, G.; Bochicchio, B.; Battocchio, C. Nanofibers of human tropoelastin-inspired peptides: Structural characterization and biological properties. *Mater. Sci. Eng. C* **2017**, *77*, 927–934. [CrossRef] [PubMed]

89. Bellis, S.L. Advantages of rgd peptides for directing cell association with biomaterials. *Biomaterials* **2011**, *32*, 4205–4210. [CrossRef] [PubMed]

90. Hersel, U.; Dahmen, C.; Kessler, H. RGD modified polymers: Biomaterials for stimulated cell adhesion and beyond. *Biomaterials* **2003**, *24*, 4385–4415. [CrossRef]

91. Sabater i Serra, R.; León-Boigues, L.; Sánchez-Laosa, A.; Gómez-Estrada, L.; Gómez Ribelles, J.L.; Salmeron-Sanchez, M.; Gallego Ferrer, G. Role of chemical crosslinking in material-driven assembly of fibronectin (nano)networks: 2d surfaces and 3d scaffolds. *Colloids Surf. B Biointerfaces* **2016**, *148*, 324–332. [CrossRef] [PubMed]

92. Gentsch, R.; Pippig, F.; Schmidt, S.; Cernoch, P.; Polleux, J.; Börner, H.G. Single-step electrospinning to bioactive polymer nanofibers. *Macromolecules* **2011**, *44*, 453–461. [CrossRef]

93. Moreno-Cencerrado, A.; Iturri, J.; Pecorari, I.; Vivanco, M.D.M.; Sbaizero, O.; Toca-Herrera, J.L. Investigating cell-substrate and cell-cell interactions by means of single-cell-probe force spectroscopy. *Microsc. Res. Tech.* **2017**, *80*, 124–130. [CrossRef] [PubMed]

94. Taubenberger, A.V.; Hutmacher, D.W.; Muller, D.J. Single-cell force spectroscopy, an emerging tool to quantify cell adhesion to biomaterials. *Tissue Eng. B Rev.* **2014**, *20*, 40–55. [CrossRef] [PubMed]

95. McKendry, R.; Theoclitou, M.-E.; Rayment, T.; Abell, C. Chiral discrimination by chemical forcemicroscopy. *Nature* **1997**, *391*, 566–568.

96. Senapati, S.; Lindsay, S. Recent progress in molecular recognition imaging using atomic force microscopy. *Acc. Chem. Res.* **2016**, *49*, 503–510. [CrossRef] [PubMed]

97. Moreno-Flores, S.; Benitez, R.; Vivanco, M.D.; Toca-Herrera, J.L. Stress relaxation microscopy: Imaging local stress in cells. *J. Biomech.* **2010**, *43*, 349–354. [CrossRef] [PubMed]

98. Melzak, K.A.; Mateescu, A.; Toca-Herrera, J.L.; Jonas, U. Simultaneous measurement of mechanical and surface properties in thermoresponsive, anchored hydrogel films. *Langmuir* **2012**, *28*, 12871–12878. [CrossRef] [PubMed]

99. Best, R.B.; Brockwell, D.J.; Toca-Herrera, J.L.; Blake, A.W.; Smith, D.A.; Radford, S.E.; Clarke, J. Force mode atomic force microscopy as a tool for protein folding studies. *Anal. Chim. Acta* **2003**, *479*, 87–105. [CrossRef]

100. Oberhauser, A.F.; Marszalek, P.E.; Carrion-Vazquez, M.; Fernandez, J.M. Single protein misfolding events captured by atomic force microscopy. *Nat. Struct. Biol.* **1999**, *6*, 1025–1028. [PubMed]

101. Schlierf, M.; Li, H.; Fernandez, J.M. The unfolding kinetics of ubiquitin captured with single-molecule force-clamp techniques. *Proc. Natl. Acad. Sci. USA* **2004**, *101*, 7299–7304. [CrossRef] [PubMed]

102. Cartagena-Rivera, A.X.; Wang, W.H.; Geahlen, R.L.; Raman, A. Fast, multi-frequency, and quantitative nanomechanical mapping of live cells using the atomic force microscope. *Sci. Rep.* **2015**, *5*, 11692. [CrossRef] [PubMed]

103. Heu, C.; Berquand, A.; Elie-Caille, C.; Nicod, L. Glyphosate-induced stiffening of hacat keratinocytes, a peak force tapping study on living cells. *J. Struct. Biol.* **2012**, *178*, 1–7. [CrossRef] [PubMed]

![polymers logo]

Review

Gelatin-Based Hydrogels for Organ 3D Bioprinting

Xiaohong Wang [1,2,*], Qiang Ao [1], Xiaohong Tian [1], Jun Fan [1], Hao Tong [1], Weijian Hou [1] and Shuling Bai [1]

[1] Department of Tissue Engineering, Center of 3D Printing & Organ Manufacturing, School of Fundamental Sciences, China Medical University (CMU), No. 77 Puhe Road, Shenyang North New Area, Shenyang 110122, China; aoqiang00@163.com (Q.A.); xhtian@cmu.edu.cn (X.T.); jfan@cmu.edu.cn (J.F.); tongh007@cmu.edu.cn (H.T.); wjhou@cmu.edu.cn (W.H.); baishuling@cmu.edu.cn (S.B.)

[2] Center of Organ Manufacturing, Department of Mechanical Engineering, Tsinghua University, Beijing 100084, China

* Correspondence: wangxiaohong709@163.com or wangxiaohong@tsinghua.edu.cn; Tel./Fax: +86-24-3190-0983

Received: 30 June 2017; Accepted: 8 August 2017; Published: 30 August 2017

Abstract: Three-dimensional (3D) bioprinting is a family of enabling technologies that can be used to manufacture human organs with predefined hierarchical structures, material constituents and physiological functions. The main objective of these technologies is to produce high-throughput and/or customized organ substitutes (or bioartificial organs) with heterogeneous cell types or stem cells along with other biomaterials that are able to repair, replace or restore the defect/failure counterparts. Gelatin-based hydrogels, such as gelatin/fibrinogen, gelatin/hyaluronan and gelatin/alginate/fibrinogen, have unique features in organ 3D bioprinting technologies. This article is an overview of the intrinsic/extrinsic properties of the gelatin-based hydrogels in organ 3D bioprinting areas with advanced technologies, theories and principles. The state of the art of the physical/chemical crosslinking methods of the gelatin-based hydrogels being used to overcome the weak mechanical properties is highlighted. A multicellular model made from adipose-derived stem cell proliferation and differentiation in the predefined 3D constructs is emphasized. Multi-nozzle extrusion-based organ 3D bioprinting technologies have the distinguished potential to eventually manufacture implantable bioartificial organs for purposes such as customized organ restoration, high-throughput drug screening and metabolic syndrome model establishment.

Keywords: 3D bioprinting; gelatin-based hydrogels; rapid prototyping (RP); organ manufacturing; implantable bioartificial organs

---

1. Introduction

In the human body, an organ is a collection of tissues that unite as a structural entity to serve one or several common functions [1]. Ordinarily, an organ is composed of more than three tissues, i.e., the main (or parenchymal) tissue as well as stromal, and sporadic tissues, in which the main tissue is specific for unique functions. For example, in the heart, the main tissue is myocardium with blood-pumping function, while the sporadic tissues include the blood vessels, nerves, and connective tissues which help the myocardium to perform the blood-pumping function [2]. At present, the only effective therapy for an organ defect/failure is through allograft organ transplantation.

Allograft organ transplantation is a medical procedure in which an organ is removed from one individual to another (i.e., the recipient or patient), to replace a damaged/missing (i.e., defect/failure) organ [3]. This is one of the major breakthroughs in modern surgery. Through transplant operations, numerous patients have regained their lives with the transplanted organs. Nevertheless, the success rate in some vital organ transplants, such as the heart, lung and liver, turn out to be far from satisfactory [4]. One of the greatest problems in these organ transplants is the acute or chronic rejection, which means that the body tends to fight off, or reject, the organs from another body [5]. Over the last several decades, numerous scientists throughout the world have explored various technologies to manufacture bioartificial organs to repair/restore the defect/failure organs [6–10].

Three-dimensional (3D) bioprinting is the process of creating cell patterns in a confined space using 3D printing technologies, where cell function and viability are preserved within the printed construct [11,12]. The major advantage of 3D bioprinting technologies in "organ manufacturing" is to automatically produce bioartificial organs using heterogeneous cell types or stem cells along with other biomaterials to repair, replace or regenerate the defect/failure organs [13–16].

Gelatin is a water-soluble protein derived from natural polymer collagen. Gelatin-based hydrogels, such as gelatin, gelatin/alginate, gelatin/chitosan, gelatin/hyaluronan, gelatin/fibrinogen, gelatin/alginate/fibrinogen, and gelatin/alginate/fibrinogen/hyaluronan, have unique features, such as excellent biocompatibilities, rapid biodegradabilities and nonimmunogenicities, in clinical applications. These hydrogels have acted as the extracellular matrices (ECMs) in organ 3D bioprinting technologies and played a critical role in various pre-defined physical (i.e., structural and morphological), chemical, and biological functionality realization [17–24].

In this review, the intrinsic/extrinsic properties of the gelatin-based hydrogels in organ 3D bioprinting technologies are outlined [25–35]. The state of the art of the physical and chemical crosslinking techniques to improve the mechanical properties of the gelatin-based hydrogels are highlighted. Among all the 3D bioprinting technologies, multi-nozzle extrusion-based 3D bioprinting has become a powerful tool for manufacturing large scale-up vascularized organs with hierarchical internal/external structures, gradient material constituents (or components) and multiple physiological functions, which have the potential to be widely used in many biomedical areas, such as, the scale-up and customized organ regenerative medicine, high throughput drug screening and energy metabolic syndrome analysis [36–40].

2. Properties of the Gelatin-Based Hydrogels

2.1. Origin of Gelatin

Gelatin is a biodegradable polypeptide derived from the partial hydrolysis of collagen, which is a fibrous protein located mainly in the connective tissues within the body [41]. A schematic illustration of the collagen hydrolysis to gelatin is shown in Figure 1. An important function of collagen is to maintain the integrity of the connective tissues, such as the bones, cartilage, corneas, tendons, ligaments, blood vessels and dentin. There are at least 16 different collagen types in the human body. The most prominent types are the Type I, II and III (or collagen I, II and III), which make up approximately 80 to 90% of all the collagens in the body. Typically, type I collagen consists of three spiral polypeptide chains, about 30 nm in length, and 1.5 nm in diameter [42]. The typical triple helix structure of the type I collagen is composed of two α-chains and a β-chain [43,44]. Additionally, collagen has low antigenicity (i.e., immunogenicity), which arises from the determinant polypeptide structures in the three spiral chains as well as the central areas of the molecules. This has greatly limited its application in biomedical fields.

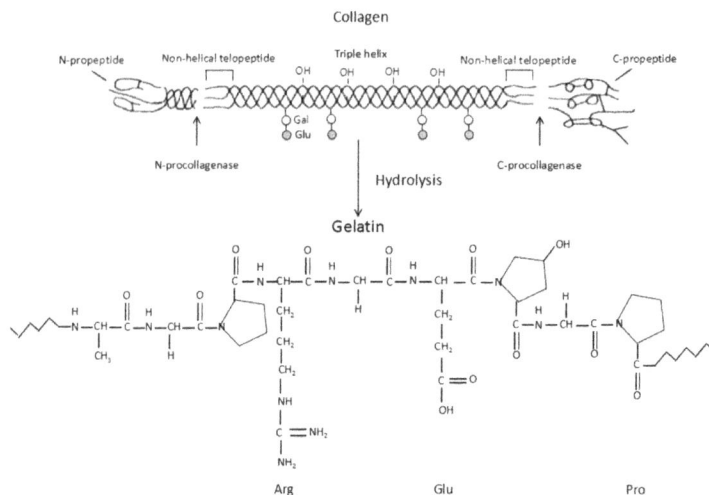

Figure 1. Schematic description of the collagen hydrolysis to gelatin.

2.2. Properties of Gelatin

Several types of gelatin exist with different compositions depending on the source of the collagen and the hydrolytic treatment methods. For example, mammalian derived gelatins from pig and bovine have been widely exploited in regenerative medicine over the last several decades [45]. These gelatins have similar polypeptide structures with human beings. Meanwhile, fish derived gelatins have a significantly lower melting point, lower gelling temperature, lower thermal stability and higher viscosity. These gelatins, therefore, have a relatively lower content of peptide repetitions (or amino acid residues), such as prolinamide and hydroxyprolinamide, in their polypeptide chains.

The hydrolytic processes of collagen I can be classified into three groups: physical, chemical and enzymatic. The entire procedure from collagen I to gelatin can be defined into three stages (or phases): pretreatment of the raw material, extraction of the gelatin and purification and/or drying of the product [46]. Depending on the protocols used for collagen I pretreatment prior to the extraction process, two main types of gelatin can be produced: Type A and Type B. Type A gelatin, with an isoelectric point of 9.0, is derived from the acidic hydrolysis of collagen I, using sulfuric or hydrochloric acid. Type B gelatin, with an isoelectric point of 5.0, is derived from the alkaline hydrolysis of collagen I, using alkaline liquid (NaOH). In these cases, asparagine and glutamine amide groups in the collagen I molecules are hydrolyzed into carboxyl groups and resulted in aspartate and glutamate residues in the gelatin molecules [47].

During the hydrolytic processes, the typical triple helix structure of the collagen I is partly broken into single-stranded polymeric molecules (single chains). The average molecular weight of gelatin is between 15,000 and 400,000 Daltons [45]. Theoretically, all gelatins, no matter where they come from, are composed of Glycine-X-Y peptide triplets repetitions, where, X and Y can be any amino acid, but proline for X- and hydroxyproline for Y- positions are the most common formulations. The amino acid composition and sequence in the single chains can vary depending largely on the origin of the gelatin, and this in turn influences its final properties [46].

Thus, gelatin is a mixture of polypeptides, in which about 20 amino acids are connected by peptide bonds. The average length and molecular weight of the polypeptide gelatins depend on several factors, such as the origins of the raw materials, the pretreatment methods and the hydrolytic processing parameters (e.g., the pH, temperature and time) [45]. There are strong non-covalent interactions, such as van der Waal forces, hydrogen bonds, and electrostatic and hydrophobic interactions among the individual gelatin chains. Compared to its progenitor collagen I, the immunogenicity of gelatin is much lower due to the damage of the triple helix structures as well as the degradation of the propeptides [48].

2.3. Properties of Gelatin-Based Hydrogels

Unlike its progenitor collagen I, which is weakly acid soluble, gelatin is a water soluble natural polymer that can absorb 5–10 times the equivalent weight of water. When the temperature is increased, the dissolution speed can be accelerated to certain degree. Gelatin solution has an amphoteric behavior because of the presence of alkaline amino acids and acid functional groups. The electrostatic charge of gelatin solution varies depending on the treatment protocols used for gelatin extraction, which gives the gelatin solution different isoelectric points [49].

The behavior of a gelatin solution depends on several factors, such as temperature, pH, concentration and preparation method. A typical property of the gelatin solution is the capability to be gelled at low temperature (about 20–30 °C) by cooling to form hydrogels. This is a sol-gel transition process, which refers to gelation. During the gelation process, locally ordered regions among the gelatin molecules take place, that are subsequently joined by non-specific bonds, such as hydrogen, electrostatic and hydrophobic bonds. The resultant hydrogel has a unique thermo-reversible (i.e., thermally responsive or thermo-sensitive) character, since the non-specific bonds can be easily broken by heating. The unique property of the gelatin solution provides the gelatin-based hydrogels with a unique property, i.e., to be printed and stacked (overlapped or piled up) based on the computer aided design (CAD) model in a controlled manner [17–22].

3. 3D Bioprinting Technologies

3.1. Introduction of Rapid Prototyping (RP)

First developed in the 1980s, RP technology advanced using several methods that apply CAD files to create 3D objects [50]. These methods are unique in their layer-by-layer adding and bonding material fashions to form solid 3D objects [51]. Through the history, there are many different names for RP technologies, such as the additive manufacturing (AM), solid freeform fabrication (SFF), increasing material manufacturing (IMM), stereolithography (SLA) and layered manufacturing (LM). Recently, these have been substituted by 3D printing or 3D bioprinting in some research areas, such as, the medical engineering and tissue engineering [52–54]. Particularly, 3D bioprinting technology prints living cells together with other biomaterials, such as polymers and growth factors, under the instructions of CAD models. However, the term RP is still popular in the industry for digital manufacturing procedures. No matter in what format it appears, the basic principle of the RP technologies is to assemble materials in layers under the instruction of CAD files [55–60].

3.2. Classification of 3D Bioprinting Technologies

According to the classification of RP technologies, 3D bioprinting technologies can been divided into three main classes based on the working principles: extrusion-based 3D bioprinting, inkjet-based 3D bioprinting and laser-assisted based 3D bioprinting. The main objective of these 3D bioprinting technologies is to print living cells in layers using CAD models to produce bioactive constructs. Recent advances in the development of the 3D bioprinters have significantly enhanced their applications in producing scale-up tissues and organs, such as the skin, myocardium, nerve, liver, cartilage, bone, and blood vessel [61–63]. Each of the 3D bioprinting technologies has its own intrinsic advantages and disadvantages in organ 3D bioprinting with respect to the automaticity, printability, scalability, accuracy, and availability of ideal "bioinks" [64–68].

3.3. Advanced Organ 3D Bioprinting Technologies

Among all the available 3D bioprinting technologies, extrusion-based 3D bioprinting technology has been advanced in various organ manufacturing areas due to the wide integrated, inestimable sophisticated and extensive automated properties [13–15]. Some of the multi-nozzle extrusion-based 3D bioprinting technologies have unique and tremendous capabilities in many biomedical applications, such as, defect/failure organ restoration, pathological organ modeling, disease organ diagnoses, and drug efficacy tests (Figures 2 and 3) [13–15,69–72].

Year	2004	2007	2009	2009	2013	2014	2016
Key event	Yan Y. *et al.* create the first extrusion organ 3D printer and assemble cell-laden gelatin-based hydrogels into large scale-up tissues [17,18].	Li S. *et al.* develop the first two-nozzle organ 3D printer and construct two cell types into complex organs with a branched vascular template [25,26].	Xu M. *et al.* build a complex (vasular) organ through an extrusion 3D printer and step-by-step adipose-derived stem cell differentiation [27,28].	Sui S. *et al.* preserve the large scale-up 3D printed tissues /organs using cryopreser-vation agents, suh as glycerol, dimethyl sulphoxide and dextran-40 [29-31].	Wang X. *et al.* print an anti-suture bioartifi-cial organ using a double-nozzle low-termpera-ture 3D printer [32,33].	Wang J. *et al.* make the first combined four-nozzle 3D printer and produce an implanta-ble liver system containing three tissues and an anti-suture overcoat [34].	Kang HW. *et al.* adapt the gelatin-based hydrogels for bone, cartilage and smooth muscle tissue 3D printing [35].

Figure 2. Histoical events of organ 3D bioprinting technologies.

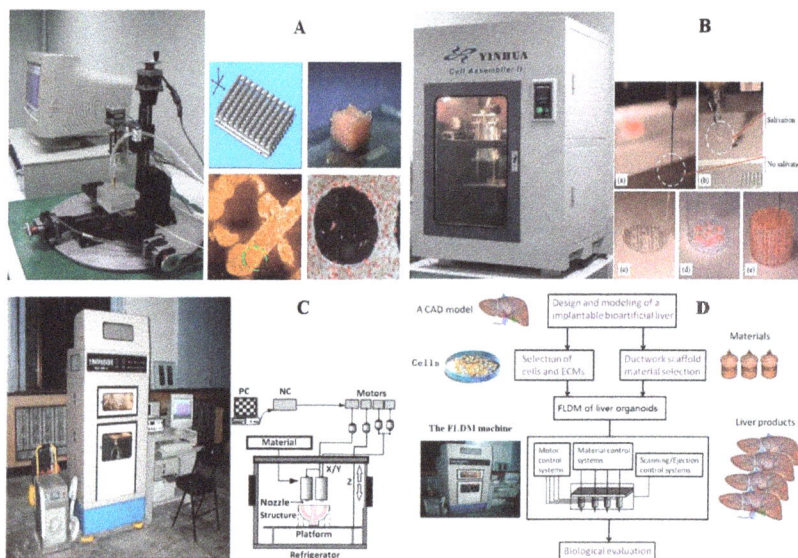

Figure 3. A schematic description of several pioneered 3D bioprinters made in Tsinghua Unversity, Prof. Wang's laboratory: (**A**) hepatocytes and/or adipose-derived stem cells (ADSCs) in the gelatin-based hydrogels were first printed into large scale-up tissues in 2004 using the single-nozzle 3D bioprinter [17–19]; (**B**) two cell types in the gelatin-based hydrogels were printed simultaneously into large scale-up organs in 2007 [24]; (**C**) both cell containing natural gelatin-based hydrogel and synthetic polymer systems were printed into large scale-up vascularized organs with a branched vascular template, that can be sutured to the host vasculatures, using the home-made double-nozzle low-temperature deposition manufacturing (DLDM) system (i.e., DLDM 3D bioprinter). An elliptical hybrid hierarchical polyurethane and cell/hydrogel construct was first produced using the DLDM 3D bioprinter [30,31]; (**D**) a schematic description of the modeling and manufacturing processes of four liver constructs with a four-nozzle low-temperature 3D bioprinter [15].

In 2004, for the first time, cells encapsulated in the biodegradable gelatin-based hydrogels were printed to large scale-up grid 3D structures with nutrient and metabolite exchanging channels (Figure 3A) [17,18]. Subsequently, a protocol for organ 3D bioprinting with multiple steps of polymer crosslinking and cocktail stem cell engagement in a predefined construct was established [27,28]. Adipose-derived stem cells (ADSCs) in the grid 3D printed construct were first induced into both endothelial and adipose tissues with the spatial effects. It is therefore the first 3D bioprinted artificial organ with a confluent endothelial cell layer on the surface of the go-through channels which could not be realized through other technologies at that period. This is no doubt a long historical dream of human beings to manufacture large vascularized tissues and/or organs.

In 2007, it was the first time that a bioartificial organ (i.e., prevascularized hepatic tissue) was printed using a two-nozzle extrusion-based 3D bioprinter [25,26]. It was also the first report that multiple cell types, such as ADSCs and hepatocytes, were assembled into large vascularized liver tissues with a uniaxial branched vascular system. In 2009, it was the first time that 3D bioprinting technologies were combined with cell cryopreservation techniques successfully to store and preserve the large scale-up bioartificial tissues and/or organs [29–31]. This can greatly save the manual labour, material resources and financial expenses in human organ manufacturing areas. In 2013, it was the first research that natural cell-laden gelatin-based hydrogel and synthetic polymer systems were printed into hierarchical constructs with a predesigned vascular template for in vivo implantation [32,33]. With the incorporation of synthetic polymers, large scale-up implantable organ manufacture comes true with anti-suture and biostable capabilities. Later in 2014, it was the first time that a more complicated organ with more than three cell types and/or material systems in a predefined construct was developed using a combined multi-nozzle 3D bioprinter [34]. This is a long-awaited breakthrough in biomedical fields.

Besides organ manufacturing, these extrusion-based 3D bioprinting technologies provide additional high-tech platforms for other biomedical purposes. For example, drug screening has experienced a long history. One of the major obstacles in drug discovery for metabolic syndrome (MS) is the lack of in vitro models that capture the complex features of the disease. This was the first time, an energy metabolism model was established using an extrusion-based 3D bioprinting technology (Figures 4 and 5) [27,28]. ADSCs encapsulated in the gelatin-based (i.e., gelatin/alginate/fibrinogen) hydrogels were printed and induced to differentiate into adipocytes and endothelial cells in a predefined 3D construct. Pancreatic islets were then deposited at the designated places in the 3D construct to mimic the adipoinsular axis with the induced adipocytes. The results indicated that drugs known to have effects on MS were shown to have an accordant effect in this model. Compared with traditional two-dimensional cell cultures, cell states in the 3D construct were much more similar to those in the native human organs. Many more physiological and pathophysiological features of energy metabolism as well as MS in vivo could be captured by the multicellular 3D construct. It is also the first large scale-up 3D printed vascularized bioartificial organ in history with a fully confluent endoderm on the surface of the predefined channels.

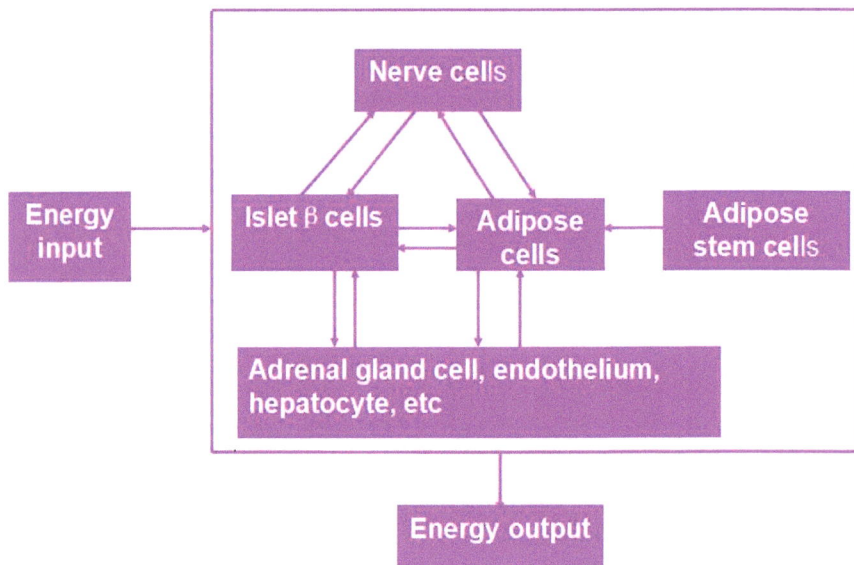

Figure 4. An energy metabolism model established through the adipose-derived stem cell (ADSC) laden gelatin-based hydorgel double-nozzle 3D bioprinting technology.

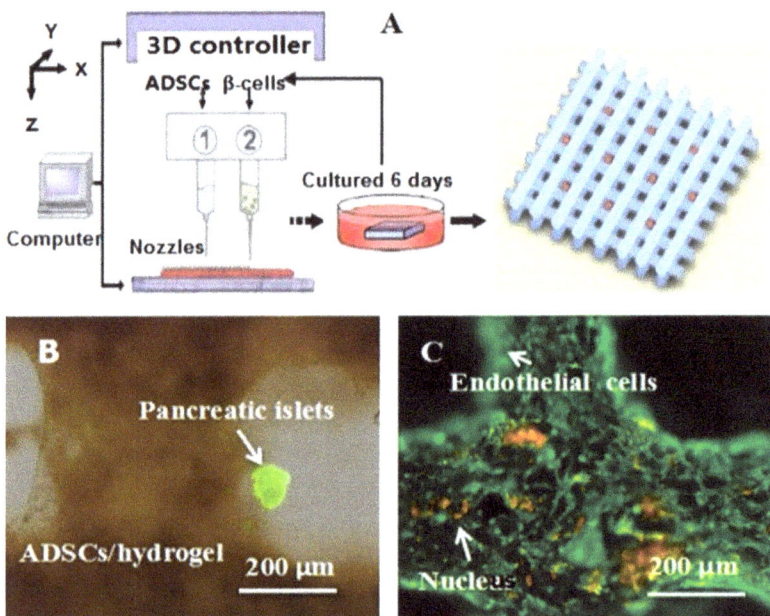

Figure 5. *Cont.*

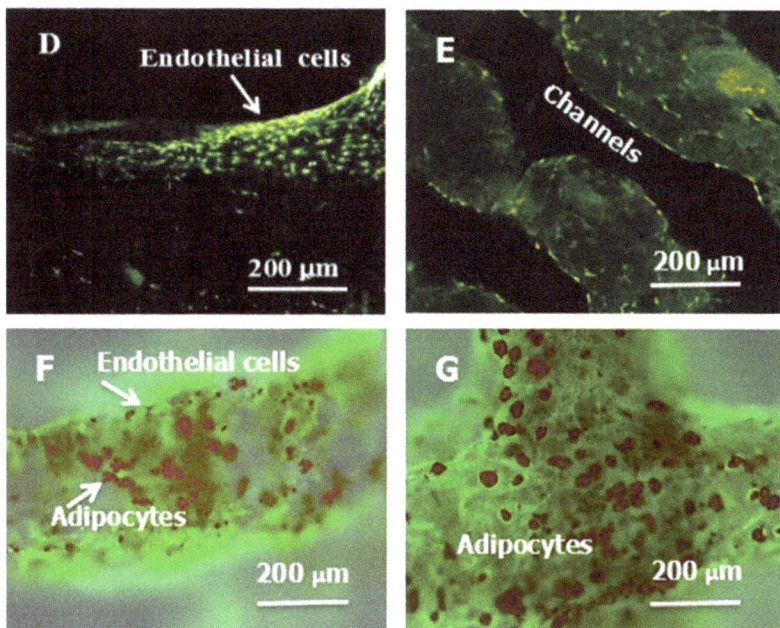

Figure 5. A large scale-up vascularized organ (i.e., vascularized adipose tissue incorporation with pancreatic islets) constructed through the double-nozzle gelatin-based hydrogel organ 3D bioprinting technology [27,28]: (**A**) the construction processes of the large scale-up vascularized organ (i.e., multicellular organ), based on the adipose derived stem cell (ADSC) laden gelatin/alginate/fibrin hydrogel 3D bioprinting and the subsequent β-cells (in the pancreatic islets) seeding procedures; (**B**) a multicellular construct, containing both ADSCs encapsulated in the gelatin/alginate/fibrin hydrogel before epidermal growth factor (EGF) engagement and pancreatic islet seeding in the predefined channels (immunostaining with anti-insulin in green); (**C**) immunostaining of the 3D construct with mAbs for CD31+ cells (i.e., mature endothelial cells from the ADSC differentiation after 3 days culture with EGF added in the culture medium) in green and pyrindine (PI) for cell nuclei (nucleus) in red; (**D**) immunostaining of the 3D construct with mAbs for CD31+ cells (i.e., mature endothelial cells) in green, having a fully confluent layer of endothelial cells (i.e., endothelium) on the surface of the predefined channels; (**E**) a vertical image of the 3D construct showing the fully confluent endothelium (i.e., endothelial cells) and the predefined go-through channels; (**F**) immunostaining of the 3D construct with mAbs for CD31+ cells in green and Oil red O staining for adipocytes in red, showing both the heterogeneous tissues coming from the ADSC differentiation after a cocktail growth factor engagement (i.e., on the surface of the channels the endothelium coming from the ADSCs differentiation after being treated with EGF for 3 days, deep inside the gelatin/alginate/fibrin hydrogel the adipose tissue coming from the ADSCs differentiation after being subsequently treated with insulin, dexamethasone and isobutylmethylxanthine (IBMX) for another 3 days. Spatial effect is prominent for the 3D printed constructs); (**G**) a control of (**F**), showing all the ADSCs in the 3D construct differentiated into target adipose tissue after 3 days treatment with insulin, dexamethasone and IBMX, but no EGF.

Back to 2003, the first center of organ manufacturing was set up in Tsinghua University by Professor Wang, the founder of "organ manufacturing", with the purpose to produce bioartificial organs using advanced building technologies. The term "organ manufacturing" can be defined in a broad sense as the generation of organ substitutes using available biomaterials either with or without cells. "Organ manufacturing" can also be defined in a narrow sense, that is to produce bioartificial organs with living cells, along with other biomaterials. A prominent character of "organ manufacturing" is that the physical and/or chemical properties of the starting materials change during and/or after the construction processes. While those processes without any changes in the physical and/or chemical properties of the starting materials can only be called fabrication, processing or machining [13–15,64–68,73–77].

Over the last decade, two distinguished "organ manufacturing" strategies have been extensively explored by Prof. Wang in Tsinghua Univeristy [13–15,64–68,73–77]. These strategies include: (1) the combined multi-nozzle 3D bioprinting technologies [64–68]; and (2) the orderly addition of combined mold systems [73–77]. Either of the strategies can potentially bring about or has brought about a great revolution in biomedical fields with regard to improving the quality of life and extending the average life span (or expectancy) of human beings.

Consequently, a great deal of pioneering work has been conducted by Prof. Wang and her students. More than 120 articles have been published, with respect to the two series of new technologies, theories and principles in complex organ manufacturing, such as (1) "organ manufacturing" like building a nuclear power station, advanced construction tools, such as the combined multi-nozzle 3D bioprinting technologies and the orderly addition of combined mold systems are the key factors [13–15,64–68,73–77]; (2) a functional vasculature, including arteries, capillaries, and veins, with a confluent endoderm can be constructed using the stem cell engagement techniques [13–15]; (3) multiple physiological functions of a complex organ, such as, the liver, heart and kidney, can be achieved using both natural and synthetic polymers [13–15,64–77].

For instance, some arbitrary bioartificial organs with highly organized geometries, tailored "bioinks", and main physiological functions have been constructed automatically and interactively. These technologies are especially useful in large scale-up and customized organ 3D bioprinting, in which CAD models established from expert experiences, or patient-specific data, including micro computed tomography (μCT), magnetic resonance imaging (MRI) and X-ray information, can be employed to guide the 3D bioprinting procedures [39,78–82].

4. Principles for Organ 3D Bioprinting

4.1. Basic Requirements for Organ 3D Bioprinting

As mentioned above, different 3D bioprinting technologies have different requirements for organ manufacturing. The fundamental issues for a typical organ 3D bioprinting is to recapitulate the essential morphological, physical (mainly mechanical and environmental), chemical, and biological properties of a native organ (i.e., its counterpart). To produce an organ substitute with a whole spectrum of physiological functions, several basic issues must be addressed first: (1) a powerful construction tool, such as the multi-nozzle 3D printer and orderly addition of combined mold system, is necessary to assemble multiple biomaterials, including diverse cell types, growth factors, and other bioactive agents, into a predefined 3D structure; (2) a functional soft-/hard-ware is preferential to automatically accomplish the intricate geometrical points (or patterns) of the target organ, including the vascular and nerve networks; (3) a large enough source (amount) of cells, especially multipotential stem cells as well as pertinent growth factors; (4) implantable natural and synthetic polymers, such as the gelatin, polyurethane (PU) and poly(lactic acid-*co*-glycolic acid) (PLGA), with excellent biocompatibilities to support multiple cellular activities, diverse tissue formation, multiple tissue coordination, and anti-suture anastomoses; (5) the ability to grasp the main characters of the target organs, with respect to the cellular environments, architectural structures and biological functions. Among all the basic issues, a powerful 3D printer, printable polymeric "bioinks" and enough cell sources are the key elements for a fully automated organ manufacturing process [13–15,64–68,73–77].

4.2. Organ 3D Bioprinting Procedures

A successful organ 3D bioprinting procedure requires several steps to implement: (1) structural blueprint predesigning; (2) starting material and construction tool preparation, including a capable multi-nozzle 3D printer and polymerous "bioinks" (including cells, polymers and growth factors); (3) 3D bioprinting; (4) post-printing organ maturation [13–15,64–68,73–77].

4.3. Blueprint Design

In the human body, each organ has its own anatomical structures, tissue types and physiological functions. Each organ needs to be particularly designed with special cell types, polymeric components, growth factors, and hierarchical architectures. To our knowledge, the main task of blueprint designing is to grasp the main characters of the target organ, but not all the minor details. For example, when we consider the liver construction, the most indispensable parts are the large vascular and bile duct networks that cannot be omitted in a liver lobe. Meanwhile, the detailed structures of capillary network can be neglected. In another words, every particular lobular architecture, does not need to be mimicked in detail. The capillary network can form spontaneously by self-assembling suitable cells/hydrogels under specific conditions if there is enough space. Thus, the incorporation of the bile duct network in the construct is much more important than the elaborate arrangement of every cell or each cell type in a particular lobular. This means that the macroscale properties of an organ are more important than the microscale features [13–15,64–68,73–77].

In general, the most indispensable parts (that cannot be omitted) for a vascularized organ, such as the liver, kidney and stomach, are the parenchymal tissues and the vascular networks. Tuned growth and remodeling frameworks of different tissues should be treated seriously. The coordination design for heterogeneous tissues in a 3D construct is extremely important for the full functionality realization [13–15,64–68,73–77]. Sometimes, a nervous network is also needed to be incorporated to support the vascular network as well as other tissues [32]. The CAD models containing a structured tree of information have played important roles for each organ blueprint designing. During the blueprint designing stage, physical (including morphological and architectural), chemical, and biological information of an organ may be collected from the clinical μCT, MRI or X-ray data. This may also come from expert experience. The digital data of a 3D image need to be simplified and translated into an actual physical entity. This CAD model can be used to guide the 3D bioprinting process. For a more complex organ, such as the brain, heart and skin, the vascular and nervous networks all need to be particularly designed with tuned growth and remodeling frameworks [13–15,64–68,73–77].

4.4. Mutinozzle 3D Printer and Polymerous "Bioink" Preparation

Each individual organ in the human body has its own requirements for 3D bioprinting. Every organ 3D bioprinter has its own prerequisites with respect to the viscosity, gelation kinetics and rheological properties of the employed "bioinks". A powerful muti-nozzle 3D printer is the kernel for a successful organ 3D bioprinting with respect to the predesigned architectures (or shapes), cellular (or histological) components, and biological (or physiological) functions. The powerful muti-nozzle 3D printer can control the following processes: (1) homogeneous and/or heterogeneous distribution of multiple cell types and/or cell-laden biomaterials (including stem cells, growth factors, and other bioactive agents) in a predefined 3D construct; (2) a perfusable vascular network, nervous network and/or other special structures in the particular 3D construct; (3) multiple functionality realization with both natural and synthetic polymers for various biomedical or clinical purposes, such as multicellular accommodation or delivery, structural integrity maintenance and anti-suture anastomoses.

For each organ 3D bioprinting, the selection of appropriate polymerous "bioinks" (i.e., starting biomaterials), such as 3D printable polymers and cell types is essential to obtain desired biological functions. As stated above, the printability of the cell-laden gelatin-based hydrogels depends largely on the physical and chemical properties of the gelatin-based hydrogels (this was demonstrated in our previous studies). The solidification property of the gelatin solution at certain temperature (below 28 °C) plays a critical role in providing the 3D structural integrity as well as facilitating the cell viability [13–15,64–68,73–77].

4.5. 3D Bioprinting Process

During the organ 3D bioprinting process, a multi-nozzle 3D printer can successively layer the two-dimensional (2D) sections to produce 3D constructs with sufficient precision and accuracy. It is imperative that the polymerous "bioinks" are injected (or extruded), deposited, solidified, and layered with certain spatial resolution. Chemical crosslinking of the pertinent polymers before, during or after the 3D bioprinting processes is necessary to keep the integrity of the 3D constructs. Associated printing parameters, such as the shear stress, extrusion speed and nozzle size may have negative effects on cellular behaviors. A comprehensive understanding each of the bioprinting parameters impacting on the cellular behaviors and optimization the printing parameters are fundamental for a special organ 3D bioprinting process.

Cell density is another critical factor for the effective cell-cell, cell-environment communications and biological functionality realization. Sometimes it is necessary to let the cells undergo proliferation and/or differentiation phases before forming homogeneous or heterogeneous tissues. Especially, stem cells encapsulated in the 3D construct can be cultured for a relatively longer period to achieve an ideal density or population [13–15,64–68,73–77].

4.6. Post-Printing Organ Maturation

The term "maturation" was borrowed from botany by Prof. Wang and others in 2003. The maturation processes of different organs vary from each other depending largely on the main parenchymal tissues and the complexity of the internal/external hierarchical architectures. After 3D bioprinting, cells in the 3D construct need time to form tissues. Meanwhile, multiple tissues need time to coordinate and to be functional. Post-culture of the 3D construct may be necessary for the vascular network formation and in situ perfusion. Pulsatile bioreactor is a promising candidate for the orientational arrangement of living cells. The role of physical force for different tissue remodeling may totally differ in different organs. It may take a long period of time to coordinate the maturation frameworks of different tissues in a particular organ [13–15,64–68,73–77].

By taking advantage of stem cells/growth factors and multi-nozzle 3D bioprinters, heterogeneous vascularized tissues and/or organs have been first generated in Prof. Wang's laboratory according to the specific locations in natural organs under the instructions of CAD models (Figure 5). This pioneering work is a historical mark which shows that "organ manufacturing" has entered a brand new era, when all the bottleneck problems, such as the in vitro large tissue generation, stem cell full differentiation, circular vasculature construction, branched nervous network incorporation, anti-suture anastomosis implantation, in situ nutrient supply and metabolite elimination, which have perplexed (daunted or confused) "tissue engineers" for more than two decades, have been overcome one by one by this unique group [13–15,64–68,73–77]. It is believed that these remarkable breakthroughs will bring huge benefit to all aspects of human health.

5. Gelatin-Based Hydrogels for Organ 3D Bioprinting

5.1. Gelatin-Based Hydrogels for 3D Bioprinting

Over the last decade, extrusion-based 3D bioprinting technologies have been emerged in various organ 3D bioprinting areas, while the gelatin-based hydorgels have received a lot of attention due to their unique physical, chemical, biological and clinical properties (Table 1) [83–92]. The gelatin-based hydrogels can be printed either as sacrificed (or fugitive) "bioinks" for channel or pore creations or solid constructs for cell survival accommodations. The extrusion-based 3D bioprinting technologies have been advanced in using the gelatin-based hydrogels to produce solid 3D structures with numerous physical, chemical, material, biological, physiological and clinical functions. Based on the data in Table 1, it can be concluded that a slight change in the ingredients of the gelatin-based hydrogels can significantly alter the final properties of the 3D printed objects.

Table 1. Resume of gelatin-based hydrogels for organ 3D printing.

3D Bioprinting Technology	"Bioink" Formulation	Crosslinking Method	Application	Morphology	Ref.
One nozzle extrusion-based 3D bioprinting developed in Tsinghua University Prof. Wang's laboratory	Gelatin/hepatocyte	2.5% glutaraldehyde solution	Large scale-up hepatic tissues		[17]
	Gelatin/chitosan/hepatocyte	3% sodium tripolyphosphate (TPP) solution	Large scale-up liver tissues		[18]
	Gelatin/alginate/hepatocyte; Gelatin/alginate/chondrocyte	10% calcium chloride ($CaCl_2$ or Ca^{2+} ion) solution	Large scale-up hepatic and cartilage tissues		[19]
	Gelatin/fibrinogen/hepatocyte; gelatin/fibrinogen/human neonatal dermal fibroblast and mesenchymal stem cell	Thrombin induced polymerization	Large scale-up hepatic tissues; vascular channels		[20]
	Gelatin/hyluronan	2% glutaraldehyde solution	Brain defect repair; cell attachment		[21]
	Gelatin/alginate/adipose-derived stem cell (ADSC)	5% $CaCl_2$ solution	Vascular networks		[22]

Table 1. *Cont.*

3D Bioprinting Technology	"Bioink" Formulation	Crosslinking Method	Application	Morphology	Ref.
	Gelatin/alginate/ADSC-laden microcapsule	Double crosslinking (100 mM/L CaCl$_2$ for ADSC-laden microcapsule; 5% CaCl$_2$ for microcapsule containing grid structure)	Vascularized tissues and organs		[23]
Two-nozzle extrusion-based 3D printing developed in Tsinghua University Prof. Wang's laboratory	Gelatin/alginate/fibrinogen/ ADSC–gelatin/alginate/fibrinogen/hepatocyte; Gelatin/alginate/fibrinogen/endothelial cell–gelatin/alginate/fibrinogen/muscle smooth cell	Double crosslinking with CaCl$_2$ and thrombin solutions	Vascularized liver and adipose tissues		[25,26]
	Gelatin/alginate/fibrinogen/ADSC–gelatin/ alginate/fibrinogen/pancreatic islet	Double crosslinking with CaCl$_2$ and thrombin solutions	Vascularized adipose, hepatic and cardiac tissues		[27,28]

Polymers **2017**, *9*, 401

Table 1. *Cont.*

3D Bioprinting Technology	"Bioink" Formulation	Crosslinking Method	Application	Morphology	Ref.
Two-nozzle low-temperature extrusion-based 3D printing developed in Tsinghua University Prof. Wang's laboratory	Gelatin/alginate/fibrinogen/HepG2; gelatin/alginate/fibrinogen/hepatocyte or gelatin/alginate/fibrinogen/hepatocyte/ADSC	Double crosslinking with CaCl$_2$ and thrombin solutions	In vitro liver tumor model establishment and anti-cancer drug screening		[38,39]
	Gelatin/lysine and polyurethane (PU) either being printed overlapped or alternated	Freeze drying (or lyophilization) for solvent sublimation (or structural stabilization) and 0.25% glutaraldehyde for gelatin/lysine crosslinking	Bioartificial organ manufacturing with expected (or controlled) mechanical properties and interconnected channels		[68]
	PU-ADSC-PU; PU- ADSC/gelatin/ alginate/fibrinogen hydrogel	Double crosslinking with CaCl$_2$ and thrombin solutions	Tubular and sandwich-like PU-ADSC/hydrogel-PU; implantable branched vascular templates		[32,33]
Dual-syringe Fab@Home printing device	Gelatin ethanolamide methacrylate (GE-MA)-methacrylated hyaluronic acid (HA-MA) (GE-MA-HA-MA)/HepG2 C3A, NIH 3T3, or Int-407 cell	Ultraviolet (UV) light (365 nm, 180 mW/cm^2) photocrosslinking	Tubular hydrogel structures for cell attachment		[83]

Table 1. *Cont.*

3D Bioprinting Technology	"Bioink" Formulation	Crosslinking Method	Application	Morphology	Ref.
One-nozzle extrusion-based 3D bioprinting	Gelatin/alginate/myoblast	CaCl₂ solution	Muscles		[84]
Fab@HomeTM (one-syringe extrusion-based 3D printing)	Gelatin/alginate/smooth muscle cell (SMC)/aortic valve leaflet interstitial cell (VIC)	10% CaCl₂ solution	Aortic valve conduits		[85]
NovoGen MMXTM, Organovo (one-nozzle extrusion-based 3D printing)	Gelatin-methacrylate or methacrylated gelatin (GelMA)	Photopolymerization by exposing GelMA precursors to UV light (360–480 nm) at 850 mW (Lumen Dynamics) using 0.5% (w/v) 2-hydroxy-1-(4-(hydroxyethox) phenyl)-2-methyl-1-propanone photo intiator	Branched vascular templates; vascularized osteogenic tissue		[86,87]
An inkjet-based 3D bioprinter	Gelatin and human umbilical vein endothelial cell (HUVEC) mixture act as a fugitive template	None	A hollow for HUVEC attachment		[88]
One-syringe extrusion-based 3D printing	Nanosilicate/GelMA	UV light (320–500 nm) for 60 s at an intensity of 6.9 mW/cm²	Electrical conductive		[89]
EnvisionTEC 3D-Bioplotter®	Polyethylene glycol (PEG)/gelatin-PEG/fibrinogen	Gelatin scaffolds were cross-linked with 15 mM EDC and 6 mM NHS, fibrinogen-containing samples were treated post-printing with 10 U/mL thrombin in 40 mM CaCl₂ for ~30 min	Grid structures for cell seeding		[90]

Table 1. *Cont.*

3D Bioprinting Technology	"Bioink" Formulation	Crosslinking Method	Application	Morphology	Ref.
Combined four-nozzle 3D bioprinting developed in Tsinghua University Prof. Wang's laboratory	Poly(lactic acid-*co*-glycolic acic) (PLGA)-gelatin/alginate/fibrinogen/ ADSC-gelatin/chitosan/hepatocyte-gelatin/ hyaluronate/Schwann cell	Double crosslinking with CaCl$_2$ and thrombin solutions	Implantable vascularized and innervated hepatic tissues		[34]
Two-syringe Fab@Home printing device	A sacrificed multi-layer (six layers) lattice gelatin/glucose construct, each layer covered with a layer of hepatocyte containing alginate hydrogel	Crosslinking with CaCl$_2$ solution	Large scale-up tissues		[91]
Multiple cartridge extrusion-based 3D printer	Polycaprolactone (PCL)-gelatin/fibrinogen/hyaluronic acid/glycerol	Thrombin induced fibrinogen polymerization	Bone, cartilage and skeletal muscle tissues		[35]
A multilayered coaxial extrusion system	A specially designed cell-responsive bioink consisting of GelMA, alginate, and 4-arm poly(ethylene glycol)-tetra-acrylate (PEGTA)	First ionically crosslinked by calcium ions (Ca^{2+} ion) followed by covalent photocrosslinking of GelMA and PEGTA	Perfusable vasculature		[92]

As stated above, the melting point of the gelatin-based hydrogels is about 28 °C. This special thermally responsive characteristic of gelatin-based hydrogels is vital important for the extrusion-based organ 3D bioprinting technology in which the cell-laden gelatin-based hydrogels take different shapes at a certain temperature (such as below 10 °C).

During the extrusion-based organ 3D bioprinting processes, cells and other biomaterials (including bioactive agents) are mixed before being printed into 3D constructs with predefined internal/external topological features (Figures 3 and 5). There are several factors that determine the gelatin-based hydrogels to be printed and layered during the 3D printing processes. Among all the factors, mechanical properties of the gelatin-based hydrogels determine whether the printed 3D constructs can be handled properly in vitro and/or implanted stably in vivo.

Generally, mechanical properties of the physically crosslinked gelatin-based hydrogels depend largely on the concentration of the primary gelatin solution. As a thermo-sensitive polymer, the tensile strength of the gelatin solution reduces sharply when the temperature is enhanced above 28 °C. The rigidity of the gelatin solution is not determined only by the temperature. The concentration, pH, and the presence of any additive all have effects on the value. It is the molecular structure, water content (or viscosity), together with the average molecular weight (mass) that define the value [41, 42]. Nevertheless, the gelation of the gelatin-based solutions is a physical crosslinking process in which a sol-gel transformation occurs due to the conformational transition of the gelatin molecules. The mechanical properties of the physically crosslinked gelatin-based hydrogels are extremely weak. When the 3D printed constructs are put into the culture medium at room temperature, they collapse immediately due to their special thermo-responsive property.

To date, various approaches have been explored to improve the mechanical strengths and structural stability before, during or after the 3D bioprinting processes. One of the effective approaches is physical blending. For example, gelatin, blended with methacrylate (i.e., methacrylamide/gelatin, GelMA), effectively improves the viscosity of the composite (or hybrid) hydrogel. Van Den Bulcke, et al., reported that the increase of gelatin or GelMA proportion can elevate the viscosity of the hybrid alginate/gelatin or alginate/GelMA polymers and the printability of the 3D bioprinting processes [93]. Photopolymerization of GelMA with water-soluble photoinitiator and UV-light can significantly increase the mechanical strengths of the composite hydrogels and the shape fidelity of the 3D printed constructs [86]. Schuurman et al., printed chondrocyte-laden GelMA hydrogel into multiple-layer structures with a cell viability of 83% [94]. Wüst et al., created a hollow structure to facilitate cells to migrate, proliferate and differentiate [95]. The GelMA hybrid hydrogels have been employed in skin, bone, cartilage and vasculature 3D bioprinting. However, the photopolymerized GelMA and its similars or derives, such as 4-arm poly(-ethylene glycol)-tetra-acrylate (PEGTA), and gelatin ethanolamine methacrylate (GE-MA)-methacrylated hyaluronic acid (HA-MA) (GE-MA-HA-MA), are non-biodegradable superpolymers (or supermolecules) which normally have very strong hardness and limited usage in bioartificial organ 3D bioprinting areas.

Another effective approach is to chemically or biochemically (i.e., enzymatically) crosslink the polymer chains (i.e., molecular strands). The polypeptide chains of the gelatin molecules can be chemically or biochemically crosslinked to form larger macromolecules. Most of the enzymatic crosslinks (such as tyrosinase, transferase, and tyrosinase) are catalyzed at neutral pH, and benign temperature. These crosslinks maybe helpful in the future organ 3D bioprinting areas. Some chemical agents, such as aldehydes (e.g., glutaraldehyde or glyceraldehyde), genipin, polyepoxides and isocyanates, have been utilized to crosslink the gelatin-based hydrogels [96,97]. When the polymeric chains are chemically crosslinked, covalent bonds formed between the free amino groups of lysine and hydroxylysine or carboxyl groups of glutamate and aspartate in the gelatin molecules, ensuring a relative stable 3D construct. Nevertheless, some of the chemical crosslinking agents (or reagents), such as the aldehydes, are toxic to cells. Caution should be taken before cells are surrounded by the toxic crosslinkers even embedded in the hydrogels. Furthermore, one should be aware that large quantities of toxic reagents, such as the glutaraldehyde, glyceraldehyde and isocyanate molecules, may be released into the body and cause vice reactions during the in vivo degradation of the chemically crosslinked gelatin-based hydrogels.

5.2. Successful Gelatin-Based Organ 3D Bioprinting Technologies

Utilizing the advantages of multi-nozzle extrusion-based 3D printers and thermo-reversible properties of the gelatin-based hydrogels, Prof. Wang and others have developed a series of organ 3D bioprinting technologies that enables us to directly deposit multiple living cells, supportive materials, growth factors and/or other bioactive agents in the right position, at the right time, and with the right amount, to form 3D living organs for in vitro culture and/or in vivo implantation (Figure 3). In these approaches, the digital data of a 3D image defined in the predesigned CAD models can be translated into actual entities. Consequently, bioartificial organs possessing essential biological, biochemical and physiological functions, such as nutrient supply, protein secretion, waste discharge, anti-suture implantation, have been completed. Various perfusable vascular networks, that integrate the merits of both natural and synthetic polymers for effective cellular accommodation, tissue formation and organ maturation have been produced successfully [13–15,64–68,73–77].

As an alternative approach to using the toxic crosslinking agents to crosslink the gelatin molecules, several chemical crosslinkable polymers, such as alginate, chitosan, fibrinogen and hyluronan, have been added to the gelatin solutions and made the crosslinking procedures more cell friendly. The gelatin-based hydrogels, such as gelatin/alginate, gelatin/fibrin, gelatin/hyaluronan and gelatin/alginate/fibrinogen, developed by Prof. Wang in Tsinghua University have profound influence in the extrusion-based organ 3D bioprinting technologies. For example, alginate is a family of polysaccharides obtained from brown alginate which can be chemically crosslinked by divalent cations, such as calcium (Ca^{2+}), strontium (Sr^{2+}), and barium (Ba^{2+}) ions [19]. Chitosan, derived from the deacetylation of chitin, is a positively charged amino polysaccharide (poly-1, 4 D-glucoamine), which can form a polyionic complex with the negatively charged gelatin-based hydrogels with ionically interacts and can be crosslinked by sodium tripolyphosphate (TPP) [18]. Hyaluronic acid or hyaluronan (HA) is another major constituent of the ECMs in the human tissues, such as the skin, lens, and cartilages. It can bind other large glycosaminoglycans (GAGs) and proteoglycans through specific HA-protein interactions and can promote the migration and growth of cells. HA can be enzymatically crosslinked [21]. Fibrinogen is a monomeric protein with a molecular weight of about 340,000 Daltons, generated when the fibrinogen peptide A and B are resected by thrombin in the coagulation process, which have been widely used as haemostatic wound dressing materials in the forms of sponge, film, powder or sheet [20,27,28]. These natural polymers have been successfully applied in the extrusion-based 3D bioprinting technologies to produce biocompatible, biodegradable, and biostable 3D constructs with predefined architectures [13–15,64–68,73–77]. The predefined architectures act as temporary organ generation templates for guiding the cellular activities, such as migration, proliferation, self-organization and differentiation (Figure 5).

5.3. Challenges for Complex Organ 3D Bioprinting with a Whole Spectrum of Physiological Functions

Organ 3D bioprinting is a comprehensive process comprised of many physical and chemical changes of the starting materials. It is also a developing and crossing multidiscipline that involves many sciences and technologies, such as materials, biology, physics, chemistry, mechanics, computer and medicine. Like the gelatin-based hydrogels which have been used in various 3D bioprinting technologies, each organ can be printed in some special ways with regard to the 3D printer types, "bioink" compositions, and clinical applications. To print a bioartificial organ with a whole spectrum of physiological functions, some challenges still exist. These challenges can be analyzed from the following three aspects.

Firstly, from the 3D printer aspect, two of the major hurdles are the out of date soft- and hardware. At present, most of the existing 3D bioprinters can only print one or two "bioinks" (i.e., starting materials) at one time with limited structural, material, and biological functions. Especially the most powerful extrusion-based 3D printers equipped with only one or two nozzles, which have greatly limited their usage in multiple material selection, hierarchical structure construction and critical function implementation. It is realized that different nozzles may have different usages in the extrusion-based 3D bioprinting. Meanwhile, different 3D bioprinting technologies may have different merits in complex organ manufacturing areas. To overcome these hurdles, combined multi-nozzle 3D printers with updated soft- and hardware are expected to provide new feasible ways for highly complicated organ 3D bioprinting. These combined multi-nozzle 3D printers should be endowed with additional capabilities to fulfill extra tasks in a target organ 3D bioprinting [79,98].

Secondly, from the "bioink" selection aspect, two of the major drawbacks until now are the insufficient tissue/organ growth abilities (with enough cell density and population) and the unmatched mechanical properties of the vascular networks. This includes the polymer systems and cell resources. Only a few existing polymers, like the PU and gelatin-based hydrogels, can meet all the basic requirements for a successful organ 3D bioprinting with excellent biocompatibilities, exclusive processibilities, desired mechanical properties and tunable biodegradabilities. Many more new polymers or polymer combinations need to be exploited. Using sacrificial polymers is an intelligent strategy to provide extra structures in a target organ. Polymers, whether biodegradable or non-biodegradable, can be used for artificial organ manufacturing. Non-biodegradable polymers have limited usage in bioartificial organ manufacturing areas. It is also expected that one type of stem cells derived from the patient can solve all the cell source problems. When stem cells are used in the "bioinks", the carcinogenic and teratogenic potentials of homologous and heterologous stem cells should be properly controlled. Those growth factors, whether they are mixed with the stem cells before 3D bioprinting or applied to the post-printing culture medium, should be optimized for a typical organ 3D bioprinting technology [81,99].

Thirdly, from the physiological function aspect, the major challenge to date is still the establishment of the perfusable hierarchical vascular networks. A perfusable vascular network is vitally important to obtain bioartificial organs that can be implanted in vivo. A fully confluent endothelial layer on the inner surface of the vascular lumen is a prerequisite for the antithrombotic properties. A further issue is the adaption of the vascular network to the host to allow in situ perfusion of the 3D construct and assure the survival of the cellular constituents within. Caution should be taken that the successful in vitro models cannot guarantee the in vivo clinical applications. Difficult as it is, the hybrid PU/gelatin-based cell-laden hydrogels or the synthetic polymer, such as PU and PLGA, encapsulated gelatin-based cell-laden hydrogels have been demonstrated to have a promising future [100,101].

6. Conclusions and Perspectives

Generally, there are four steps in a typical organ 3D bioprinting process: blueprint predesigning, 3D printer and "bioink" preparation, processing and post-printing maturation. Multi-nozzle extrusion-based organ 3D bioprinting technologies have demonstrated many distinguished advantages (such as automated, sophisticated and integrated) in "organ manufacturing" areas. Gelatin-based hydrogels, such as the gelatin/fibrinogen, gelatin/alginate/fibrinogen, and gelatin/alginate/ fibrinogen/hyaluronan, with remarkable physical, chemical, biological and medical features, have played a vitally important role in each of the successful extrusion-based organ 3D bioprinting technologies. A combined multi-nozzle 3D printer can be used to manufacture highly complicated organs with a variety of desired morphological structures, cellular components, and physiological functions, such as vascular and nervous networks. A biologically mimicking CAD model derived from patient or experience can be used to guide the customized or high throughput organ 3D bioprinting processes.

In the future, bioartificial human organ manufacturing, by using the combined multi-nozzle extrusion-based 3D bioprinting technology will be very popular in many laboratories throughout the world. More sophisticated techniques, including μCT, MRI and X-ray techniques can be integrated for the clinical datum collection, analysis, usage and storage. Modern imaging techniques, such as caged fluorophores and quantum dots will allow scientists to visualize the morphological changes of cells, tissues and organs occurring in a predefined 3D construct in real time. Much more structural, material, pathological and clinical features in each individual organ, such as the nephric tubules in the kidney, alveolus pulmonis in the lung and lactiferous ducts in the breast, will be incorporated to ensure the mass cells are functional. Computers will make the datum storage, manufacture process and architecture predesign much easier. Bioartificial organs automatically produced through the 3D bioprinting technologies with much more hierarchical structures, particular cell types, bioactive ingredients and physiological functions will play a pivotal role in regenerative medicine and other pertinent biomedical fields, such as defect/failure organ restoration, high-throughput drug screening and metabolic syndrome model establishment. Numerous patients will benefit significantly from these prominent organ 3D bioprinting technologies, theories and principles. The average longevity of human beings will definitely be prolonged miraculously through the defect/failure organ restoration applications, effective drug screening protocols and disease metabolism controlling templates.

Acknowledgments: The work was supported by grants from the National Natural Science Foundation of China (NSFC) (No. 81571832 & 81271665 & 31600793 & 81571919) and the International Cooperation and Exchanges NSFC and Japanese Society for the Promotion of Science (JSPS) (No. 81411140040).

Author Contributions: Xiaohong Wang conceived, designed and wrote the main content; Qiang Ao, Xiaohong Tian, Jun Fan, Weijian Hou, Hao Tong and Shuling Bai contributed some detailed techniques.

Conflicts of Interest: The authors declare no conflict of interest. The founding sponsors had no role in the design of the study; in the collection, analyses, or interpretation of data; in the writing of the manuscript, and in the decision to publish the results.

References

1. Widmaier, E.P.; Raff, H.; Strang, K.T. *Vander's Human Physiology the Mechanisms of Body Function*; McGraw-Hill Education: New York, NY, USA, 2016; pp. 14–15.
2. Xu, Y.; Li, D.; Wang, X. Current trends and challenges for producing artificial hearts. In *Organ Manufacturing*; Wang, X., Ed.; Nova Science Publishers Inc.: Hauppauge, NY, USA, 2015; pp. 101–125.
3. Yacoub, M.H.; Banner, N.R.; Khaghani, A.; Fitzgerald, M., Madden, B.; Tsang, V.; Radley-Smith, R.; Hodson, M. Heart-lung transplantation for cystic fibrosis and subsequent domino heart transplantation. *J. Heart Transplant.* **1990**, *9*, 459–466. [PubMed]
4. Moniaux, N.; Faivre, J.A. Reengineered liver for transplantation. *J. Hepatol.* **2011**, *54*, 386–387. [CrossRef] [PubMed]
5. Frohn, C.; Fricke, L.; Puchta, J.C.; Kirchner, H. The effect of HLA-C matching on acute renal transplant rejection. *Nephrol. Dial. Transplant.* **2001**, *16*, 355–360. [CrossRef] [PubMed]
6. Langer, R.; Vacanti, J.P. Tissue engineering. *Science* **1993**, *260*, 920–926. [CrossRef] [PubMed]
7. Wang, X.; Yan, Y.; Liu, F.; Xiong, Z.; Wu, R.; Zhang, R.; Lu, Q.P. Preparation and characterization of a collagen/chitosan/heparin matrix for an implantable bioartificial liver. *J. Biomater. Sci. Polym. Ed.* **2005**, *16*, 1063–1080. [CrossRef] [PubMed]
8. Mao, A.S.; Mooney, D.J. Regenerative medicine: Current therapies and future directions. *Proc. Natl. Acad. Sci. USA* **2015**, *112*, 14452–14459. [CrossRef] [PubMed]
9. Liu, L.; Wang, X. Organ manufacturing. In *Organ Manufacturing*; Wang, X., Ed.; Nova Science Publishers Inc.: Hauppauge, NY, USA, 2015; pp. 1–28.
10. Henzler, T.; Chai, L.; Wang, X. Integrated model for organ manufacturing: A systematic approach from medical imaging to rapid prototyping. In *Organ Manufacturing*; Wang, X., Ed.; Nova Science Publishers Inc.: Hauppauge, NY, USA, 2015; pp. 171–200.
11. Chua, C.K.; Yeong, W.Y. *Bioprinting: Principles and Applications*; World Scientific Publishing Co.: Singapore, 2015; p. 296.
12. Wang, X.; Tuomi, J.; Mäkitie, A.A.; Poloheimo, K.-S.; Partanen, J.; Yliperttula, M. The integrations of biomaterials and rapid prototyping techniques for intelligent manufacturing of complex organs. In *Advances in Biomaterials Science and Applications in Biomedicine*; Lazinica, R., Ed.; In Tech: Rijeka, Croatia, 2013; pp. 437–463.

13. Wang, X.; Yan, Y.; Zhang, R. Rapid prototyping as tool for manufacturing bioartificial livers. *Trends Biotechnol.* **2007**, *25*, 505–513. [CrossRef] [PubMed]

14. Wang, X.; Yan, Y.; Zhang, R. Recent trends and challenges in complex organ manufacturing. *Tissue Eng. B* **2010**, *16*, 189–197. [CrossRef] [PubMed]

15. Wang, X. Intelligent freeform manufacturing of complex organs. *Artif. Org.* **2012**, *36*, 951–961. [CrossRef] [PubMed]

16. Wang, X.; Zhang, Q. Overview on "Chinese-Finnish workshop on biomanufacturing and evaluation techniques". *Artif. Org.* **2011**, *35*, E191–E193. [CrossRef] [PubMed]

17. Wang, X.; Yan, Y.; Pan, Y.; Xiong, Z.; Liu, H.; Cheng, J.; Liu, F.; Lin, F.; Wu, R.; Zhang, R.; et al. Generation of three-dimensional hepatocyte/gelatin structures with rapid prototyping system. *Tissue Eng.* **2006**, *12*, 83–90. [CrossRef] [PubMed]

18. Yan, Y.; Wang, X.; Pan, Y.; Liu, H.; Cheng, J.; Xiong, Z.; Lin, F.; Wu, R.; Zhang, R.; Lu, Q. Fabrication of viable tissue-engineered constructs with 3D cell-assembly technique. *Biomaterials* **2005**, *26*, 5864–5871. [CrossRef] [PubMed]

19. Yan, Y.; Wang, X.; Xiong, Z.; Liu, H.; Liu, F.; Lin, F.; Wu, R.; Zhang, R.; Lu, Q. Direct construction of a three-dimensional structure with cells and hydrogel. *J. Bioact. Compat. Polym.* **2005**, *20*, 259–269. [CrossRef]

20. Xu, W.; Wang, X.; Yan, Y.; Zheng, W.; Xiong, Z.; Lin, F.; Wu, R.; Zhang, R. Rapid prototyping three-dimensional cell/gelatin/fibrinogen constructs for medical regeneration. *J. Bioact. Compat. Polym.* **2007**, *22*, 363–377. [CrossRef]

21. Zhang, T.; Yan, Y.; Wang, X.; Xiong, Z.; Lin, F.; Wu, R.; Zhang, R. Three-dimensional gelatin and gelatin/hyaluronan hydrogel structures for traumatic brain injury. *J. Bioact. Compat. Polym.* **2007**, *22*, 19–29. [CrossRef]

22. Yao, R.; Zhang, R.; Yan, Y.; Wang, X. In vitro angiogenesis of 3D tissue engineered adipose tissue. *J. Bioact. Compat. Polym.* **2009**, *24*, 5–24.

23. Yao, R.; Zhang, R.; Wang, X. Design and evaluation of a cell microencapsulating device for cell assembly technology. *J. Bioact. Compat. Polym.* **2009**, *24*, 48–62. [CrossRef]

24. Wang, X.; Yan, Y.; Zhang, R. Gelatin-based hydrogels for controlled cell assembly. In *Biomedical Applications of Hydrogels Handbook*; Ottenbrite, R.M., Ed.; Springer: New York, NY, USA, 2010; pp. 269–284.

25. Li, S.; Yan, Y.; Xiong, Z.; Weng, C.; Zhang, R.; Wang, X. Gradient hydrogel construct based on an improved cell assembling system. *J. Bioact. Compat. Polym.* **2009**, *24*, 84–99. [CrossRef]

26. Li, S.; Xiong, Z.; Wang, X.; Yan, Y.; Liu, H.; Zhang, R. Direct fabrication of a hybrid cell/hydrogel construct by a double-nozzle assembling technology. *J. Bioact. Compat. Polym.* **2009**, *24*, 249–265.

27. Xu, M.; Yan, Y.; Liu, H.; Yao, Y.; Wang, X. Control adipose-derived stromal cells differentiation into adipose and endothelial cells in a 3-D structure established by cell-assembly technique. *J. Bioact. Compat. Polym.* **2009**, *24*, 31–47. [CrossRef]

28. Xu, M.; Wang, X.; Yan, Y.; Yao, R.; Ge, Y. A cell-assembly derived physiological 3D model of the metabolic syndrome, based on adipose-derived stromal cells and a gelatin/alginate/fibrinogen matrix. *Biomaterials* **2010**, *31*, 3868–3877. [CrossRef] [PubMed]

29. Sui, S.; Wang, X.; Liu, P.; Yan, Y.; Zhang, R. Cryopreservation of cells in 3D constructs based on controlled cell assembly processes. *J. Bioact. Compat. Polym.* **2009**, *24*, 473–487. [CrossRef]

30. Wang, X.; Paloheimo, K.-S.; Xu, H.; Liu, C. Cryopreservation of cell/hydrogel constructs based on a new cell-assembling technique. *J. Bioact. Compat. Polym.* **2010**, *25*, 634–653. [CrossRef]

31. Wang, X.; Xu, H. Incorporation of DMSO and dextran-40 into a gelatin/alginate hydrogel for controlled assembled cell cryopreservation. *Cryobiology* **2010**, *61*, 345–351. [CrossRef] [PubMed]

32. Wang, X.; He, K.; Zhang, W. Optimizing the fabrication processes for manufacturing a hybrid hierarchical polyurethane-cell/hydrogel construct. *J. Bioact. Compat. Polym.* **2013**, *28*, 303–319. [CrossRef]

33. Huang, Y.; He, K.; Wang, X. Rapid Prototyping of a hybrid hierarchical polyurethane-cell/hydrogel construct for regenerative medicine. *Mater. Sci. Eng. C* **2013**, *33*, 3220–3229. [CrossRef] [PubMed]

34. Wang, J.; Wang, X. Development of a Combined 3D Printer and Its Application in Complex Organ Construction. Master's Thesis, Tsinghua University, Beijing, China, 2014.

35. Kang, H.-W.; Lee, S.J.; Ko, I.K.; Kengla, C.; Yoo, J.J.; Atala, A. A 3D bioprinting system to produce human-scale tissue constructs with structural integrity. *Nat. Biotechnol.* **2016**, *34*, 312–319. [CrossRef] [PubMed]

36. Wang, X. Editorial: Drug delivery design for regenerative medicine. *Curr. Pharm. Des.* **2015**, *21*, 1503–1505. [CrossRef] [PubMed]

37. Xu, Y.; Wang, X. Fluid and cell behaviors along a 3D printed alginate/gelatin/fibrin channel. *Biotechnol. Bioeng.* **2015**, *112*, 1683–1695, (Highlighted by'Global Medical Discovery'in 2015). [CrossRef] [PubMed]

38. Zhao, X.; Du, S.; Chai, L.; Xu, Y.; Liu, L.; Zhou, X.; Wang, J.; Zhang, W.; Liu, C.-H.; Wang, X. Anti-cancer drug screening based on an adipose-derived stem cell/hepatocyte 3D printing technique. *J. Stem Cell Res. Ther.* **2015**, *5*, 273.

39. Zhou, X.; Liu, C.; Zhao, X.; Wang, X. A 3D bioprinting liver tumor model for drug screening. *World J. Pharm. Pharm. Sci.* **2016**, *5*, 196–213.

40. Wang, X. 3D printing of tissue/organ analogues for regenerative medicine. In *Handbook of Intelligent Scaffolds for Regenerative Medicine*, 2nd ed.; Pan Stanford Publishing: Palo Alto, CA, USA, 2016; pp. 557–570.

41. Hulmes, D.J. Building collagen molecules, fibril, and suprafibrillar structures. *J. Struct. Biol.* **2002**, *37*, 2–10. [CrossRef] [PubMed]

42. Schrieber, R.; Gareis, H. From Collagen to Gelatine. In *Gelatine Handbook: Theory and Industrial Practice*; Wiley-VCH Verlag GmbH & Co. KGaA: Weinheim, Germany, 2007; pp. 45–117.

43. Sui, S.; Wang, X. Researches on the Fabrication and Training of Bioartificial Blood Vessels. Master's Thesis, Tsinghua University, Beijing, China, 2009.

44. Huang, Y.; Wang, X. Rotational Combined Molds for the Construction of an Organ Precursor with a Multi-Branched Vascular System. Master's Thesis, Tsinghua University, Beijing, China, 2013.

45. Foox, M.; Zilberman, M. Drug delivery from gelatin-based systems. *Expert Opin. Drug Deliv.* **2015**, *12*, 1547–1563. [CrossRef] [PubMed]

46. De Wolf, F.A. Chapter V Collagen and gelatin. *Prog. Biotechnol.* **2002**, *23*, 133–218.

47. Karim, A.A.; Bhat, R. Fish gelatin: Properties, challenges, and prospects as an alternative to mammalian gelatins. *Food Hydrocoll.* **2009**, *23*, 563–576. [CrossRef]

48. Su, K.; Wang, C. Recent advances in the use of gelatin in biomedical research. *Biotechnol. Lett.* **2015**, *37*, 2139–2145. [CrossRef] [PubMed]

49. Babin, H.; Dickinson, E. Influence of transglutaminase treatment on the thermoreversible gelation of gelatin. *Food Hydrocoll.* **2001**, *15*, 271–276. [CrossRef]

50. Hull, C.W. Apparatus for Production of Three-Dimensional Objects by Stereolithography. U.S. Patent 4,575,330, 11 March 1986.

51. Webb, P.A.; Webb, P.A. A review of rapid prototyping (RP) techniques in the medical and biomedical sector. *J. Med. Eng. Technol.* **2000**, *24*, 149–153. [CrossRef] [PubMed]

52. Murphy, S.V.; Atala, A. 3D bioprinting of tissues and organs. *Nat. Biotechnol.* **2014**, *32*, 773–785. [CrossRef] [PubMed]

53. Park, J.H.; Hong, J.M.; Ju, Y.M.; Jung, J.W.; Kang, H.-W.; Lee, S.J.; Yoo, J.J.; Kim, S.W.; Kim, S.H.; Cho, D.-W. A novel tissue-engineered trachea with a mechanical behavior similar to native trachea. *Biomaterials* **2015**, *62*, 106–115. [CrossRef] [PubMed]

54. Wang, X.; Rijff, B.L.; Khang, G. A building block approach into 3D printing a multi-channel organ regenerative scaffold. *J. Stem Cell Res. Ther.* **2015**, *11*, 1403–1411.

55. Derby, B. Printing and prototyping of tissues and scaffolds. *Science* **2012**, *338*, 921–926. [CrossRef] [PubMed]

56. Mak, H.C. Trends in computational biology-2010. *Nat. Biotechnol.* **2011**, *29*, 45–49. [CrossRef] [PubMed]

57. Xu, Y.; Li, D.; Wang, X. Liver manufacturing approaches: The thresholds of cell manipulation with bio-friendly materials for multifunctional organ regeneration. In *Organ Manufacturing*; Wang, X., Ed.; Nova Science Publishers Inc.: Hauppauge, NY, USA, 2015; pp. 201–225.

58. Cui, T.; Yan, Y.; Zhang, R.; Liu, L.; Xu, W.; Wang, X. Rapid prototyping of a double layer polyurethane-collagen conduit for peripheral nerve regeneration. *Tissue Eng. C* **2009**, *15*, 1–9. [CrossRef] [PubMed]

59. Wang, X.; Cui, T.; Yan, Y.; Zhang, R. Peroneal nerve regeneration along a new polyurethane-collagen guide conduit. *J. Bioact. Compat. Polym.* **2009**, *24*, 109–127. [CrossRef]

60. He, K.; Wang, X. Rapid prototyping of tubular polyurethane and cell/hydrogel construct. *J. Bioact. Compat. Polym.* **2011**, *26*, 363–374.

61. Nakamura, M.; Kobayashi, A.; Takagi, F.; Watanabe, A.; Hiruma, Y.; Ohuchi, K.; Iwasaki, Y.; Horie, M.; Morita, I.; Takatani, S. Biocompatible inkjet printing technique for designed seeding of individual living cells. *Tissue Eng.* **2005**, *11*, 1658–1666. [CrossRef] [PubMed]

62. Cooper, G.M.; Miller, E.D.; DeCesare, G.E.; Usas, A.; Lensie, E.L.; Bykowski, M.R.; Huard, J.; Weiss, L.E.; Losee, J.E.; Campbell, P.G. Inkjet-based biopatterning of bone morphogenetic protein-2 to spatially control calvarial bone formation. *Tissue Eng. A* **2010**, *16*, 1749–1759. [CrossRef] [PubMed]

63. Sundaramurthi, D.; Rauf, S.; Hauser, C. 3D bioprinting technology for regenerative medicine applications. *Int. J. Bioprint.* **2016**, *2*, 9. [CrossRef]

64. Zhao, X.; Wang, X. Preparation of an adipose-derived stem cell (ADSC)/fibrin-PLGA construct based on a rapid prototyping technique. *J. Bioact. Compat. Polym.* **2013**, *28*, 191–203. [CrossRef]

65. Zhao, X.; Liu, L.; Wang, J.; Xu, Y.F.; Zhang, W.M.; Khang, G.; Wang, X. In vitro vascularization of a combined system based on a 3D bioprinting technique. *J. Tissue Eng. Regen. Med.* **2014**, *10*, 833–842. [CrossRef] [PubMed]

66. Schrepfer, I.; Wang, X. Progress in 3D printing technology in health care. In *Organ Manufacturing*; Wang, X., Ed.; Nova Science Publishers Inc.: Hauppauge, NY, USA, 2015; pp. 29–74.

67. Xu, W.; Wang, X.; Yan, Y.; Zhang, R. Rapid prototyping of polyurethane for the creation of vascular systems. *J. Bioact. Compat. Polym.* **2008**, *23*, 103–114. [CrossRef]

68. Xu, W.; Wang, X.; Yan, Y.; Zhang, R. A polyurethane-gelatin hybrid construct for the manufacturing of implantable bioartificial livers. *J. Bioact. Compat. Polym.* **2008**, *23*, 409–422. [CrossRef]

69. Xu, Y.; Wang, X. Application of 3D biomimetic models for drug delivery and regenerative medicine. *Curr. Pharm. Des.* **2015**, *21*, 1618–1626. [CrossRef] [PubMed]

70. Liu, L.; Zhou, X.; Xu, Y.; Zhang, W.M.; Liu, C.-H.; Wang, X.H. Controlled release of growth factors for regenerative medicine. *Curr. Pharm. Des.* **2015**, *21*, 1627–1632. [CrossRef] [PubMed]

71. Wang, X.; Ao, Q.; Tian, X.; Fan, J.; Wei, Y.; Hou, W.; Tong, H.; Bai, S. 3D bioprinting technologies for hard tissue and organ engineering. *Materials* **2016**, *9*, 802. [CrossRef] [PubMed]

72. Lei, M.; Wang, X. Biodegradable polymers and stem cells for bioprinting. *Molecules* **2016**, *21*, 539. [CrossRef] [PubMed]

73. Wang, X.; Sui, S.; Yan, Y.; Zhang, R. Design and fabrication of PLGA sandwiched cell/fibrin constructs for complex organ regeneration. *J. Bioact. Compat. Polym.* **2010**, *25*, 229–240. [CrossRef]

74. Wang, X.; Sui, S. Pulsatile culture of a poly(DL-lactic-*co*-glycolic acid) sandwiched cell/hydrogel construct fabricated using a step-by-step mold/extraction method. *Artif. Organs* **2011**, *35*, 645–655. [CrossRef] [PubMed]

75. Wang, X.; Mäkitie, A.A.; Poloheimo, K.-S.; Tuomi, J.; Paloheimo, M.; Sui, S.C.; Zhang, Q.Q. Characterization of a PLGA sandwiched cell/fibrin tubular construct and induction of the adipose derived stem cells into smooth muscle cells. *Mater. Sci. Eng. C* **2011**, *31*, 801–808. [CrossRef]

76. Wang, X.; Wang, J. Vascularization and adipogenesis of a spindle hierarchical adipose-derived stem cell/collagen/alginate-PLGA construct for breast manufacturing. *IJITEE* **2015**, *4*, 1–8.

77. Wang, X.; Huang, Y.W.; Liu, C. A combined rotational mold for manufacturing a functional liver system. *J. Bioact. Compat. Polym.* **2015**, *39*, 436–451. [CrossRef]

78. Liu, L.; Wang, X. Creation of a vascular system for complex organ manufacturing. *Int. J. Bioprint.* **2015**, *1*, 77–86.

79. Lei, M.; Wang, X. Uterus Bioprinting. In *Organ Manufacturing*; Wang, X., Ed.; Nova Science Publishers Inc.: Hauppauge, NY, USA, 2015; pp. 335–355.

80. Zhou, X.; Wang, X. Breast Engineering. In *Organ Manufacturing*; Wang, X., Ed.; Nova Science Publishers Inc.: Hauppauge, NY, USA, 2015; pp. 357–384.

81. Wang, X. Overview on biocompatibilies of implantable biomaterials. In *Advances in Biomaterials Science and Biomedical Applications in Biomedicine*; Lazinica, R., Ed.; In Tech: Rijeka, Croatia, 2013; pp. 111–155.

82. Wang, X. Spatial effects of stem cell engagement in 3D printing constructs. *J. Stem Cells Res. Rev. Rep.* **2014**, *1*, 5–9.

83. Skardal, A.; Zhang, J.; McCoard, L.; Xu, X.; Oottamasathinen, S.; Prestwich, G.D. Photocrosslinkable hyaluronan-gelatin hydrogels for two-step bioprinting. *Tissue Eng. A* **2010**, *16*, 2675–2685. [CrossRef] [PubMed]

84. Chung, J.H.; Naficy, S.; Yue, Z.; Kapsa, R.; Quigley, A.; Moulton, S.E.; Wallace, G.G. Bio-ink properties and printability for extrusion printing living cells. *Biomater. Sci.* **2013**, *1*, 763–773. [CrossRef]

85. Duan, B.; Hockaday, L.A.; Kang, K.H.; Butcher, J.T. 3D bioprinting of heterogeneous aortic valve conduits with alginate/gelatin hydrogels. *J. Biomed. Mater. Res. A* **2013**, *101*, 1255–1264. [CrossRef] [PubMed]

86. Bertassoni, L.E.; Cecconi, M.; Manoharan, V.; Nikkhah, M.; Hjortnaes, J.; Cristino, A.L.; Barabaschi, G.; Demarchi, D.; Dokmeci, M.R.; Yang, Y.; et al. Hydrogel bioprinted microchannel networks for vascularization of tissue engineering constructs. *Lab Chip* **2014**, *14*, 2202–2211. [CrossRef] [PubMed]

87. Kolesky, D.B.; Truby, R.L.; Gladman, A.S.; Busbee, T.A.; Homan, K.A.; Lewis, J.A. 3D bioprinting of vascularized, heterogeneous cell-laden tissue constructs. *Adv. Mater.* **2014**, *26*, 3124–3130. [CrossRef] [PubMed]

88. Lee, V.K.; Kim, D.Y.; Ngo, H.; Lee, Y.; Seo, L.; Yoo, S.-S.; Vincent, P.A.; Dai, G. Creating perfused functional vascular channels using 3D bio-printing technology. *Biomaterials* **2014**, *33*, 8092–8102. [CrossRef] [PubMed]

89. Xavier, J.R.; Thakur, T.; Desai, P.; Jaiswal, M.K.; Sears, N.; Cosgriff-Hernandez, E.; Kaunas, R.; Gaharwar, A.K. Bioactive nanoengineered hydrogels for bone tissue engineering: A growth-factor-free approach. *ACS Nano* **2015**, *9*, 3109–3118. [CrossRef] [PubMed]

90. Rutz, A.L.; Hyland, K.E.; Jakus, A.E.; Burghardt, W.R.; Shah, R.N. A multi-material bioink method for 3D printing tunable, cell-compatible hydrogels. *Adv. Mater.* **2015**, *27*, 1607–1614. [CrossRef] [PubMed]

91. Leonard, J.F.; Chai, L.; Wang, X. Application of gelatin and glucose as sacrificial structures in organ manufacturing. In *Organ Manufacturing*; Wang, X., Ed.; Nova Science Publishers Inc.: Hauppauge, NY, USA, 2015; pp. 269–300.

92. Jia, W.; Gungor-Ozkerim, P.S.; Zhang, Y.S.; Yue, K.; Zhu, K.; Liu, W.; Pi, Q.; Byambaa, B.; Dokmeci, M.R.; Shin, S.R.; et al. Direct 3D bioprinting of perfusable vascular constructs using a blend bioink. *Biomaterials* **2016**, *106*, 58–68. [CrossRef] [PubMed]

93. Van Den Bulcke, A.I.; Bogdanov, B.; De Rooze, N.; Schacht, E.H.; Cornelissen, M.; Berghmans, H. Structural and rheological properties of methacrylamide modified gelatin hydrogels. *Biomacromolecules* **2000**, *1*, 31–38. [CrossRef] [PubMed]

94. Schuurman, W.; Levett, P.A.; Pot, M.W.; van Weeren, P.R.; Dhert, W.J.; Hutmacher, D.W.; Melchels, F.P.; Klein, T.J.; Malda, J. Gelatin-methacrylamide hydrogels as potential biomaterials for fabrication of tissue-engineered cartilage constructs. *Macromol. Biosci.* **2013**, *13*, 551–561. [CrossRef] [PubMed]

95. Wüst, S.; Godla, M.E.; Müller, R.; Hofmann, S. Tunable hydrogel composite with two-step processing in combination with innovative hardware upgrade for cell-based three-dimensional bioprinting. *Acta Biomater.* **2014**, *10*, 630–640. [CrossRef] [PubMed]

96. De Clercq, K.; Schelfhout, C.; Bracke, M.; De Wever, O.; Van Bockstal, M.; Ceelen, W.; Remon, J.P.; Vervaet, C. Genipin-crosslinked gelatin microspheres as a strategy to prevent postsurgical peritoneal adhesions: In vitro and in vivo characterization. *Biomaterials* **2016**, *96*, 33–46. [CrossRef] [PubMed]

97. Sisson, K.; Zhang, C.; Farach-Carson, M.C.; Chase, D.B.; Rabolt, J.F. Evaluation of cross-linking methods for electrospun gelatin on cell growth and viability. *Biomacromolecules* **2009**, *10*, 1675–1680. [CrossRef] [PubMed]

98. Liu, L.; Wang, X. Artificial blood vessels and vascular systems. In *Organ Manufacturing*; Wang, X., Ed.; Nova Science Publishers Inc.: Hauppauge, NY, USA, 2015; pp. 75–99.

99. Lei, M.; Wang, X. Rapid prototyping of artificial stomachs. In *Organ Manufacturing*; Wang, X., Ed.; Nova Science Publishers Inc.: Hauppauge, NY, USA, 2015; pp. 153–169.

100. Zhou, X.; Wang, X. Artificial Kidney Manufacturing. In *Organ Manufacturing*; Wang, X., Ed.; Nova Science Publishers Inc.: Hauppauge, NY, USA, 2015; pp. 227–244.

101. Xu, Y.; Li, D.; Wang, X. The construction of vascularized pancreas based on 3D printing techniques. In *Organ Manufacturing*; Wang, X., Ed.; Nova Science Publishers Inc.: Hauppauge, NY, USA, 2015; pp. 245–268.

![polymers logo]

Review

Hydrogels for Cartilage Regeneration, from Polysaccharides to Hybrids

Daniela Anahí Sánchez-Téllez [1,2,*], Lucía Téllez-Jurado [1] and Luís María Rodríguez-Lorenzo [2,3]

1   Instituto Politécnico Nacional-ESIQIE, Depto. Ing. en Metalurgia y Materiales, UPALM-Zacatenco,
    Mexico City 07738, Mexico; ltellezj@ipn.mx
2   Networking Biomedical Research Centre in Bioengineering, Biomaterials and Nanomedicine,
    Centro de Investigación Biomédica en Red—Bioingeniería, Biomateriales y Nanomedicina (CIBER-BBN),
    Av. Monforte de Lemos 3-5, Pabellón 11, Planta 0, 28029 Madrid, Spain; luis.rodriguez-lorenzo@ictp.csic.es
3   Department Polymeric Nanomaterials and Biomaterials, ICTP-CSIC, Juan de la Cierva 3,
    28006 Madrid, Spain
*   Correspondence: danielatellez06@gmail.com; Tel.: +34-695-198-353 or +521-55-29-54-32-91

Received: 14 October 2017; Accepted: 29 November 2017; Published: 4 December 2017

Abstract: The aims of this paper are: (1) to review the current state of the art in the field of cartilage substitution and regeneration; (2) to examine the patented biomaterials being used in preclinical and clinical stages; (3) to explore the potential of polymeric hydrogels for these applications and the reasons that hinder their clinical success. The studies about hydrogels used as potential biomaterials selected for this review are divided into the two major trends in tissue engineering: (1) the use of cell-free biomaterials; and (2) the use of cell seeded biomaterials. Preparation techniques and resulting hydrogel properties are also reviewed. More recent proposals, based on the combination of different polymers and the hybridization process to improve the properties of these materials, are also reviewed. The combination of elements such as scaffolds (cellular solids), matrices (hydrogel-based), growth factors and mechanical stimuli is needed to optimize properties of the required materials in order to facilitate tissue formation, cartilage regeneration and final clinical application. Polymer combinations and hybrids are the most promising materials for this application. Hybrid scaffolds may maximize cell growth and local tissue integration by forming cartilage-like tissue with biomimetic features.

Keywords: cartilage regeneration; polymeric hydrogels; polysaccharides; hybrid hydrogels; hybrid scaffolds

---

1. Introduction: Current Clinical Approaches and the Need for New Developments

The aim of this paper is to review the current state of the art of materials for cartilage substitution and regeneration. Section 1 describes the current state of the art in clinical treatments. Polymeric hydrogels used in cartilage regeneration and the reasons hindering their clinical success are reviewed in Section 2. The preparation techniques using polysaccharides and the resulting hydrogel properties are described in Section 3. Finally, future trends are explored in Section 4.

Several reviews about hydrogels for cartilage regeneration have been published for the last 10 years, focusing on preparation and characterization, natural and synthetic polymer precursors, gelation kinetics, cell and drug delivery, growth factors, mechanical properties and biocompatibility. Nevertheless, most of those reviews do not propose new alternatives to improve hydrogels properties which can fulfill the real clinical needs in terms of tissue regeneration, mechanical properties and degradation kinetics. Therefore, this paper reviews relevant literature published during 2013–2017, related to the application, fabrication, characterization, in vitro and in vivo assays of biomaterials based on hydrogels for cartilage regeneration. The studies selected are articles and reviews written in English.

Currently, there are no clinical satisfactory solutions for cartilage tissue regeneration [1–3]. The most widely used clinical procedure to heal cartilage injury involves penetrating the wound to the subchondral bone, allowing the access of blood flow and new biological material [4–7]. However, such clinical treatments often result in the formation of fibrocartilaginous tissue (Figure 1), which is weaker than the original one, failing to integrate properly with surrounding tissue, and degrading over a period between 6 to 12 months [8–12].

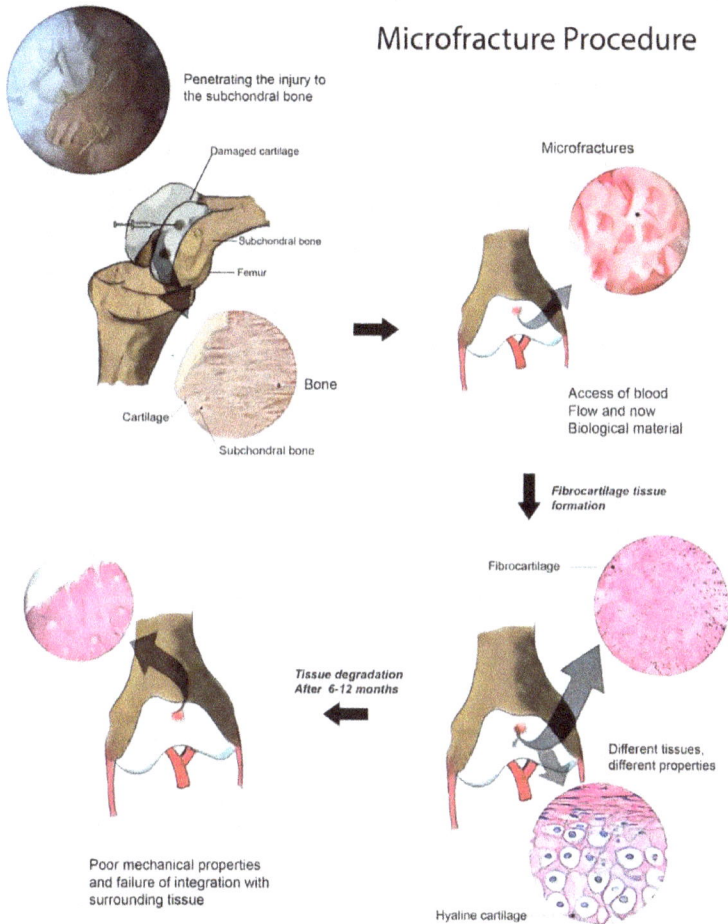

Figure 1. Common clinical procedure used to heal cartilage injury. Original illustration designed and provided by the authors.

During the last few years, material scientists and tissue engineers have tried to help clinicians by confronting the challenge of manufacturing porous 3D scaffolds which resemble the chemical composition and architecture (Figure 2) of the extracellular matrix (ECM) of the cartilage [13–19]. Most of the studies are directed to determine how chemical composition and architecture influence cellular phenotype, differentiation, integration and extracellular matrix secretion during in vitro [20–25] and in vivo [26–32] assays.

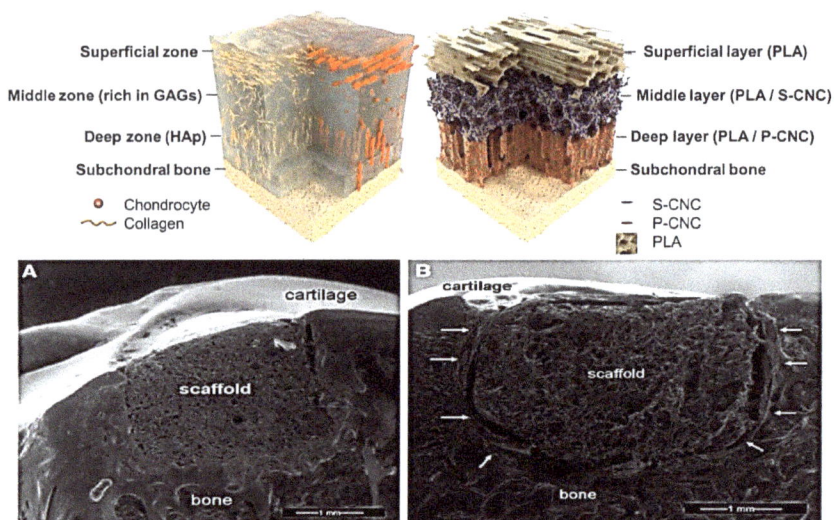

Figure 2. Top. How artificial scaffolds try to mimic the anisotropic characteristic in cartilage tissue [33,34]. Reprinted with permission of [33] Sandra Camarero-Espinosa et al. *Biomaterials*, 74: 42–52. **Bottom.** Matrix regeneration within a macroporous non-degradable implant for osteochondral defects is not enhanced with partial enzymatic digestion of the surrounding tissue. (**A**) Environmental scanning electron microscopy of a longitudinal slice taken through the cartilage-bone-implant. The specimen was retrieved at the 1 month post-operative time point. Good integration between the implant and the surrounding bone and articular cartilage was observed. (**B**) Environmental scanning electron microscopy image of a longitudinal slice taken through the cartilage-bone-implant construct at 3 months. Fibrous encapsulation of the implant is highlighted by arrows. Reprinted with permission of [34] Aaron J. Krych et al. *J. Mater. Sci. Mater. Med.*, 24: 2429–2437.

Several natural and synthetic polymers are being used to create novel materials, as collected in Table 1, in attempts to produce scaffolds for tissue engineering and regenerative medicine having clinical application [35].

Table 1. Scaffolds in clinical and preclinical use for cartilage regeneration.

Application	Material	Problem-result	Ref.
Nose (dorsal augmentation material in rhinoplasty)	Tissue-engineered chondrocyte PCS (Porcine Cartilage-derived Substance) scaffold construct.	Preliminary animal study: Excellent biocompatibility, neocartilage formation starts. However, it was not confirmed that the constructs contributed to the formation of neocartilage.	[36]

Polymers **2017**, *9*, 671

Table 1. *Cont.*

Application	Material	Problem-result	Ref.
Knee (subchondral bone)	Osteochondral biomimetic nanostructured scaffold Maioregen®	Better results in healing complex lesions in comparison with the implantation of a purely chondral scaffold.	[37]
	Cell-free biphasic scaffold: collagen-hydroxyapatite osteochondral scaffold	Statistically significant improvement in clinical scores. At 5 years, between 60.9% and 78.3% of the cases showed complete filling of the cartilage, complete integration of the graft, intact repaired tissue surface and homogeneous repaired tissue.	[38]
	Nanostructured biomimetic three-phasic collagen-hydroxiapatite construct	The implantations to treat chondral and osteochondral knee defects were effective in terms of clinical outcome, although MRI detected abnormal findings.	[39]
Knee (chondral defects)	Autologous ovine MNC Cell-seeded and cell-free DL-poly-lactide–*co*–glycolide (PLGA) scaffolds	The engineered tissue had not local or systemic adverse effects. However, only a poor integration of the tissue engineering product into adjacent tissue was reached and the formed ECM was not mature enough for long-lasting weight-loading resistance.	[40]
	Type I collagen-hydroxyapatite (Maioregen®) nanostructural biomimetic osteochondral scaffold	The use of the Maioregen® scaffold is a good procedure for the treatment of large osteochondral defects; however, the lesion site seems to influence the results. Patient affected in the medial femoral condyle showed better results.	[41]
	DeNovo (RevaFlex) engineered tissue graft	Preliminary evidence suggests that DeNovo ET implant is capable of spontaneous matrix formation with no immune response, improving function and recreating hyaline-like cartilage.	[42]
Knee (femoral condyles)	Biphasic cylindrical osteochondral composite construct of DL-poly-lactide–*co*–glycolide (PLGA). Its lower body is impregnated with β-tricalcium phosphate (TCP)	The regenerated osteochondral tissue was evaluated as a tissue of acceptable quality. Regenerated cartilage was defined as being hyaline when the ground substance was homogeneous without fibrous texture.	[43]
Tibial plateau (osteochondral scaffold)	Osteochondral biomimetic collagen-hydroxiapatite scaffold (Maioregen®, Fin-ceramica, Faenza, Italy)	MRI abnormalities. Clinical outcome with stable results up to a mid-term follow-up.	[44]

Table 1. *Cont.*

Application	Material	Problem-result	Ref.
Microfractured defect (for filling microfractures)	BioCartilageTM, product containing dehydrated, micronized allogeneic cartilage, implanted with the addition of platelet rich plasma	No human clinical outcomes data available. Data regarding results are limited to expert opinion.	[45]
	Chondroitin sulfate adhesive-Poly(ethylene glycol) diacrylate (PEGDA) hydrogel system combined with standard microfracture surgery	Significant increase in tissue fillers with defects in a short-term follow-up.	[46]
Knee (for donor site filling)	Artificial TruFit cylinders made of fully synthetic material called PolyGraft®-Material: 50% copolymer (PDLG), composed of 85% poly(D,L-lactide) and 15% glycolide; 40% calcium sulfphae, 10% PGA fibers	No clinical improvement could be found. The regeneration of the filled defects took more than 2 years, even though TruFit Plugs are supposed to stimulate cartilage and bone cell migration from the surrounding tissue to the synthetic cylinders.	[47]
	Porous poly(ethylene oxide)terephthalate/ butylene terephthalate) (PEOT/PBT) implants	Treated defects did not cause postoperative bleeding. Well integration. Surface stiffness was minimally improved compared to controls. Considerable biodegradation after 9 months. Congruent fibrocartilaginous surface repair with interspersed fibrous tissue formation in implanted sites. Donor site: fibrocartilaginous surface repair.	[48]
Shoulder	Engineered hyaluronic acid membrane, Hyalograft®	Using the hyaluronic membrane had no effect on the final outcome. No difference was observed between the fibrocartilage tissue formed after implementing microfractures and the fibrocartilage tissue grown on the hyaluronic acid membrane scaffold.	[49]

However, the complexity of the physical structure and properties of cartilage, including mechanical [50–54], anisotropic [55–57], nonlinear [58–60], inhomogeneous [61–63] and viscoelastic properties [64–66], are thought to be directly related to the failing of most of the attempts made to fabricate artificial substitutes for cartilage [67–71]. As a consequence, there are not yet biomaterials for cartilage regeneration in clinical use with satisfactory results. Scaffolds based on tissue-engineered constructs, osteochondral biomimetic scaffolds, cell-free biphasic or three-phasic scaffolds, autologous scaffolds, engineered-tissue grafts, porous implants have not demonstrated to be a satisfactory solution in clinical application. Therefore, it seems evident that there is a need to designi more suitable scaffolds and to develop new types of materials which can be used for cartilage regeneration.

In the case of articular cartilage repair, the required materials must provide successful mechanical properties, biological delivery, fixation of the device in situ and stability to the joint. Besides, the assays done to these materials need to be based on the intended biological effect and potential risks which have to be evaluated, such as toxicity, dedifferentiation, immunogenicity and contamination. Some preclinical trials with little animal models, such as rats and white New Zealand rabbits, are necessary to predict how biomaterials may behave during clinical assays. Moreover, some preliminary studies can be done in cadaveric human bodies; however extensive trials with large animal preclinical models are mandatory to obtain market approval. Therefore, just a few of the novel materials or tissue protocols are allowed to have clinical application or even to be commercialized. When trying to compare different studies of novel materials once they are introduced into clinical practice, there are other problems: (1) lack of homogeneity due to the different studied population; (2) short- and mid-follow-ups; (3) use of different evaluation systems; (4) new scaffold-based strategies for cartilage regeneration, either cell seeded or cell-free biomaterials; (5) procedures which differ in scaffold fixation methods, surgical approaches and postoperative rehabilitation phases. Therefore, there is the need among scientists, clinicians, industry and regulatory experts to improve communication and collaboration in order to overcome all the barriers in tissue engineering and to establish a defined road map to reach clinical application.

2. Hydrogels in Cartilage Regeneration

Hydrogels are emergent candidates for applications in cartilage regeneration. Hydrogels are three-dimensional hydrophilic polymer networks made up of water-soluble polymers, crosslinked by either covalent or physical methods [72] (Figure 3) to form a water-insoluble hydrogel [73,74]. Hydrogels can be composed of natural polysaccharides [75–77], proteins [78–82] or synthetic polymers [83–85]. Hydrogels are able to swell and retain great portions of water, from 20% to 99% by weight, when placed in aqueous solutions [86–88]. Hydrogels, tested as matrices to build up scaffolds, provide highly desirable 3D environments for cell growth, holding a great promise for the regeneration of cartilaginous tissue as in vitro and in vivo studies showed [89–94]. Several studies use cells to catalyze tissue formation while being distributed in 3D hydrogel matrices. Cell matrix adhesion to hydrogel is an important interaction which regulates stem cell survival, self-renewal, and differentiation. Using 3D culture systems (hydrogels) may provide an appropriate niche, scaffolding and environmental bioactive signals for cells. Depending on their physical structure and chemical composition, hydrogels can preserve a compositional and mechanical similarity with the native extracellular matrix of cartilage [95–98]. These properties are necessary for controlling cell response, differentiation and functional tissue regeneration [99–106].

One important reason for the choice of hydrogels in cartilage applications is the possibility of making them injectable which offers advantages over solid scaffolds such as the possibility of using a non-invasive approach. Injectable hydrogels can fill any shape defect and they may provide a homogeneous cell distribution within any defect size or shape prior to gelation [107–111]. Over recent years, a variety of naturally [112–119] and synthetically [120–122] derived materials such as silk [123,124], resilin [125], chitosan [126], chondroitin sulfate [127], hyaluronic acid [128–131], gelatin [132], agarose [133], alginate [134] poly(vinyl alcohol) (PVA) [95,135] poly(acrylic acid) [136], acrylamide [137] and many others have been used to form injectable hydrogels for cartilage repair.

Collagen II and glycosaminoglycans (GAGs) are cartilage-specific extracellular matrix components; they play a crucial role in regulating the expression of chondrocytic phenotype and in supporting chondrogenesis. They have been used for in vitro and in vivo assays [138]; many attempts have been made with different GAGs precursors of hydrogels to provide an appropriate biochemical and biomechanical environment for cells [139–146]. Unfortunately, hydrogels derived from GAGs degrade really fast, thus different chemical modifications have to be introduced. Several crosslinking degrees are necessary in attempt to modulate their degradation kinetics. However, their biological response is also modified [147–151]. In general, the studies found on literature can be divided into two main types: (1) cell-free hydrogel scaffolds; and (2) cell-seeded hydrogel scaffolds.

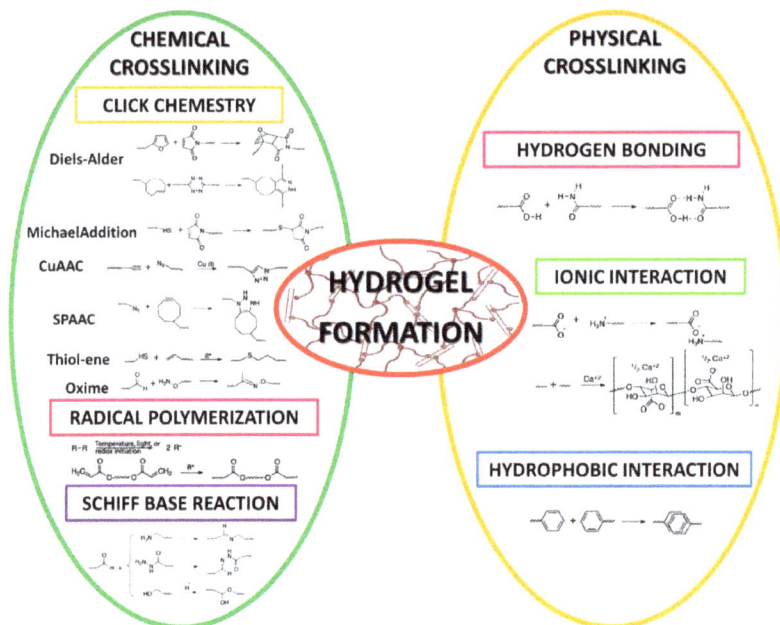

Figure 3. Chemical and physical crosslinking methods used to form hydrogels.

2.1. Cell-Free Hydrogel Scaffolds

Investigations based on cell-free hydrogel scaffolds focus on their physico-chemical characterization and mechanical properties [152]. These studies allow a full understanding of physical and chemical interactions within materials and how these interactions may affect biological and mechanical properties of scaffolds.

Poly(ethylene glycol) (PEG) and derived hydrogels are ones of the most widely used synthetic polymer for tissue engineering. The modulus for bovine articular cartilage, measured in compression mode, is 950 KPa, which is close to the value of the fully hydrated polyethylene glycol diacrylate (PEG-DA) hydrogel [87]. Polyvinyl alcohol is another synthetic polymer widely used to form hydrogels due to their excellent biocompatibility, high permeability to fluids (showing an equilibrium water content of $32 \pm 5\%$) and low friction coefficients (μ) in the range of 0.02 to 0.05 against smooth and wet substances. Some studies on PVA-based scaffolds aim to demonstrate that under tribological loading, friction and wear characteristics compatible to natural articular cartilage can be achieved [153]. As low friction coefficients are required for engineered cartilage, polyvinyl alcohol (PVA)/polyvinylpyrrolidone (PVP) hydrogels were synthesized with different polymerization degrees: 1700, 2400 and 2600 for the PVA; and different polymer concentrations: 10% w/w, 15% w/w and 20% w/w of PVA/PVP. It was found that the inner structures of hydrogels tend to be denser when polymer concentration and polymerization degree of PVA increase. While the friction coefficient increases (from 0.037 to 0.044) with an increment in the polymerization degree of PVA (average increase rate is approximate 3%), the friction coefficient decreases (from 0.033 to 0.03 for a 2.5 N load; from 0.049 to 0.045 for a 7.5 N load) with an increment in the polymer concentration (from 10% to 20%) in the low load region and under liquid lubrication. Thus, there is the need to keep friction coefficients stable under lubricated conditions [154].

Another study using PVA-based hydrogels, crosslinked with trimetaphosphate (STMP), revealed that fully hydrated hydrogels were covalently crosslinked systems when mechanically tested, with a rheological behavior (the G' changed from 0.01 MPa (0.01 Hz) to 0.02 MPa (15 Hz)) similar to that of tibia cartilage ($G' = 0.03$ for tissue surface and $G' = 0.11$ for overall tissue) [86]. As previously said, it is an important challenge to develop scaffolds which possess mechanical properties mimicking those of cartilage tissue, since cartilage is a complex nonlinear, viscoelastic and anisotropic extracellular matrix structure. T. Chen et al. [155] reported that hydrodynamic conditions, simulating the motion-induced flow fields between the articular surfaces within the synovial joint, induce the formation of a distinct superficial layer on tissue engineered cartilage hydrogels. These hydrodynamic conditions enhance, on the superficial layers, the production of cartilage matrix proteoglycan, type II collagen and a highly aligned fibrillary matrix which resembles the alignment pattern in native tissue surface zone.

Since many materials do not exhibit a low friction coefficient or withstand several loading cycles, some of them are infiltrated with an interpenetrating network hydrogel to form functional scaffolds which provide load-bearing and tribological properties, similar to native cartilage ones. For example: (1) a porous three-dimensionally woven poly(ε-caprolactone) fiber scaffold was infused with a "tough-gel" made of alginate and polyacrylamide [83]; (2) a boundary lubricant functionalized PVA-based hydrogel was developed to be used as a synthetic replacement for focal defects in articular cartilage [156]. Other techniques to develop hydrogels with high mechanical strength are: by using a double network or two-step polymerization [88,157]; or by functionalizing hydrogels with different organic or inorganic molecules [158].

2.2. Cell-Seeded Hydrogel Scaffolds

Cell-based hydrogel scaffold therapy is one of the main strategies being investigated in cartilage regeneration. Several scaffolds and materials are being evaluated. These studies focus on whether or not the hydrogels provide an appropriate biochemical and biomechanical environment for a long-lasting hyaline-type cartilage regeneration. Decellularized extracellular matrices from natural tissues like dermis or adipose are being studied as functional biologic scaffolds (Figure 4). It is possible to ensure the bioactivity of a substrate when scaffolds are seeded with a specific cell type, either chondrocytes or stem cells, and if these cells are able to proliferate regardless of "natural" conditions. As an example, Giavaresi G. et al. [159] evaluated in vitro the biological influence of a decellularized human dermal extracellular matrix on human chondrocytes (NHAC-kn) and mesenchymal stromal cells (hMSC). The study showed that at 24 h after seeding, cells adhered consistently to dermal membranes (NHAC-kn = 93% and hMSC = 98%); at 7 days, cell viability index was 98% for both cell cultures seeded on dermal membranes; and after 14 days of culture, the indexes increased significantly for both cell cultures ($p < 0.0005$; NHAC-kn = 136% and hMSC = 263%). Furthermore, a biohybrid composite scaffold, composed by combining a decellularized Wharton's jelly extracellular matrix with the polyvinyl alcohol (PVA)-based hydrogel, demonstrated its ability in promoting chondrocyte adhesion and scaffold colonization [160]. Other studies are examining extracellular matrices developed from porcine articular cartilage [161]. Although these substrates worked as proper scaffolds for the growth of cells, their therapeutic and functional efficiency in cartilage regeneration still need to be proved. Other cell therapies include implanting chondrogenic lines differentiated from mesenchymal stem cells (MSC) into different polysaccharides or synthetic hydrogels. Collagen hydrogels have proved to provide an appropriate 3D environment for MSC chondrogenesis, isolated from Wharton's jelly of human umbilical cord, and to be cytocompatible matrices with great potential for cartilage engineering [162].

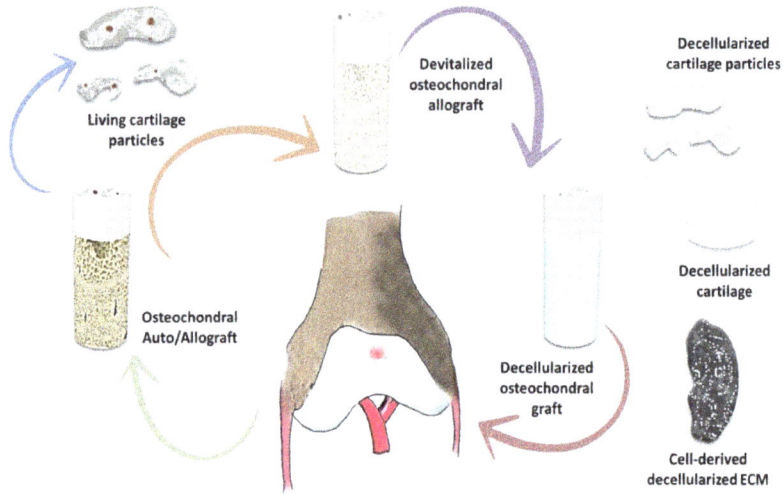

Figure 4. Matrix-based scaffold approaches for cartilage regeneration. Original illustration designed and provided by the authors.

One of the problems being reported when using cell therapies is the dedifferentiation of chondrocytes when cultured in two-dimensional cultures, making them less functional for cartilage repair. Wu L. et al. [163] hypothesized that functional exclusion of dedifferentiated chondrocytes can be achieved by detecting domains formation of collagen molecules deposited by chondrogenic cells into 3D environments. They reported a method which allows separation of functionally active chondrogenic cells, which produce high levels of collagen II, from functionally inferior dedifferentiated cells, which produce collagen X. To avoid dedifferentiation of cells once they are forming constructs, Lam J. et al. [164] investigated the ability of cell-laden bilayer hydrogels, by encapsulating chondrogenically and osteogenically pre-differentiated mesenchymal stem cells, by varying the period of chondrogenic pre-differentiation prior to implantation. Therefore, cell phenotype could be optimized in order to achieve ideal tissue repair. Furthermore, since regeneration of human articular cartilage is limited, various cellular sources have been studied, including adult and juvenile chondrocytes. Some studies have compared the formation of cartilage tissue, produced by juvenile, adult and osteoarthritic chondrocytes, inside 3D biomimetic hydrogels composed of poly(ethylene glycol) and chondroitin sulfate. It was found that after the cultured time, juvenile chondrocytes showed a greater upregulation of chondrogenic gene expression than adult chondrocytes, while OA chondrocytes showed a downregulation [101]. Another strategy being studied is the evaluation of therapeutic effects of intra-articular injections of hydrogels containing drugs used to treat osteoarthritis symptoms [165].

Some other studies are meant to analyze how the structure and fabrication methods of the scaffolds influence cells behavior. Due to their intimate contact with chondrocytes, scaffolds are important components of cell niche. Investigations into micro-architecture of scaffolds have revealed that mean pore size is cell-type specific and influences cellular shape, differentiation and extracellular matrix secretion (Figure 5). Studies in collagen-hyaluronic acid scaffolds, fabricated with different mean pore size, showed that scaffolds with the largest mean pore size (300 µm) stimulated significantly a higher cell proliferation, chondrogenic gene expression and cartilage-like matrix deposition [166]. When using synthetic materials to produce hydrogels, it is necessary to determine their in vitro pore size and mechanical stiffness after being rehydrated, in order to predict their in vivo behavior. Hui J. H. et al. [167] found that freeze-dried oligo[poly(ethylene glycol)fumarate] (OPF) hydrogels with a pore size ranged from 20 to 433 μm in diameter and a mechanical stiffness of 1 MPa when rehydrated, enhance the formation of hyaline-fibrocartilaginous mixed tissue. However, these hydrogels, implanted alone into cartilage defects, are insufficient to generate a homogenously hyaline cartilage repair tissue. Kwon H. et al. [168] demonstrated that scaffolds, with different pore size and fabrication methods, influence the microenvironment of chondrocytes and their response to proinflammatory substances. Having high levels of proinflammatory cytokines can cause cartilage destruction and instability of the engineered cartilage tissue. These authors found that silk scaffolds with larger pore sizes support higher levels of cartilage matrix and leach more efficiently proinflammatory cytokines into the medium, influencing cartilage gene expression.

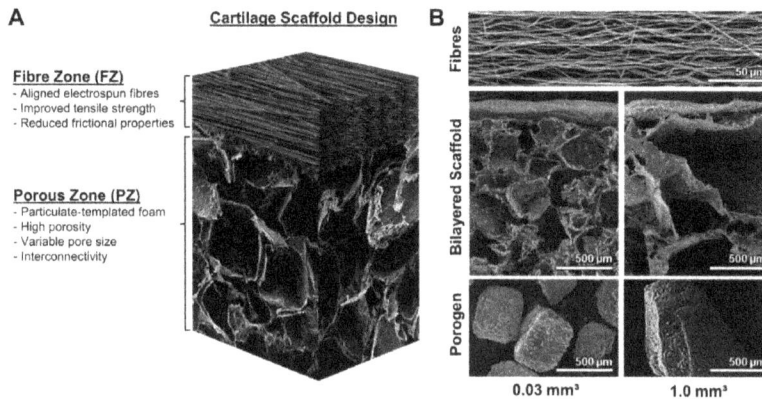

Figure 5. Bilayered cartilage scaffold (**A**) schematized by a diagram illustrating the electrospun fiber zone (FZ) deposited on a particulate-templated foam (PZ). The combination of the two distinct zones is designed to yield an anisotropic scaffold with a smooth articular surface and a more porous region for ECM deposition. (**B**) Electron microscopy images (top) of the aligned fiber zone that is shared between both scaffold varieties, (middle) the complete bilayered scaffolds with 0.3 mm³ (left) and 1.0 mm³ (right) pores, and (bottom) the sodium chloride porogens used to produce their respective scaffolds. Reprinted with permission from [169] J.A.M. Steele et al. Combinational scaffold morphologies for zonal articular cartilage engineering. *Acta Biomaterialia*. 10: 2065–2075.

Furthermore, each zone of cartilage tissue varies in regard to biochemical content, morphology and biomechanical function. Deeper cartilage zones present higher stiffness, higher proteoglycan concentration but lower cellular density and same collagen concentration along cartilage tissue. In a general structural perspective, cartilage can be simplified into two main regions: (1) the superficial zone which exhibits a high tensile strength and low friction coefficient to keep a smooth articulation; (2) a dense ECM region rich in proteoglycan molecules which give the tissue adequate compressive mechanical properties by producing a high osmotic pressure within the tissue. Therefore, when fabricating a bilayer or three-layer scaffold, pore size and fabrication method of each layer influence the microenvironment of chondrocytes. As shown by Steele J.A.M. et al. [169], tissue engineering scaffolds can be designed to vary in morphology and function, offering a template: (1) to mimic the structural organization and functional interface of cartilage superficial zone; (2) to increase the extracellular matrix production; (3) to enhance the anisotropic mechanical properties. These authors fabricated a multi-zone cartilage scaffold by electrostatic deposition of polymer microfibers onto particulate-templated scaffolds with 0.03 mm³ and 1.0 mm³ porogens. They demonstrated that bilayered scaffolds can closely mimic some of the structural characteristics of native cartilage due to: (1) the addition of aligned fiber membranes enhances the mechanical and surface properties of scaffolds; (2) zonal analysis of scaffolds showed region-specific variations in chondrocyte number, sulfated

GAG-rich extracellular matrix and chondrocytic gene expression; (3) smaller porogens (0.03 mm^3) yield higher GAGs accumulation and aggrecan gene expression.

It is important to understand the multi-scale biomechanical behavior of cartilage tissue, in order to realize the connection among joint kinematics, tissue-level mechanics, cell mechanics and mechanotransduction, matrix mechanics and the nanoscale mechanics of matrix macromolecules. Therefore, understanding mechanical behavior at each scale helps to correlate cell biology, matrix biochemistry and tissue structure/function of cartilage (Figure 6).

Moreover, the combination of cell-based therapies with growth factor delivery, which can locally signal cells promoting their function, is also being investigated. Since morphogenetic protein (BMP-2) and transforming growth factor (TGF-β) are cytokines proposed as stimulants for cartilage repair, it is necessary to undertake a detailed comparative analysis of their biological effects on chondrocytes [170]. As an example, when chondrocytes are encapsulated in PEG hydrogels functionalized with transforming growth TGF-β1, proliferation and matrix production increase, in comparison with cells in hydrogels where TGF-β1 is dosed in the culture medium or untreated TGF-β1 hydrogels [171]. In another study, chondroitin sulfate-bone marrow adhesive hydrogel was used to localize and carry BMP-2 protein, which enhance articular tissue formation. It was demonstrated that these hydrogels were capable of supporting articular chondrocytes, viability and phenotype retention, stimulating cells to produce hyaline-like extracellular matrix [172]. Although expensive growth factors in cultures are used, there is still a production of cartilage with inferior mechanical and structural properties compared with the natural tissue. However, recent evidence suggests that GAGs incorporated into tissue engineering scaffolds can isolate and/or activate growth factors, mimicking better the natural extracellular matrix [173]. For example, in the presence of TGF-β3 releasing microspheres, gellan gum hydrogels facilitate a greater cell proliferation than fibrin or agarose hydrogels. Histological and biochemical analysis of these hydrogels indicated that fibrin hydrogel was the least chondro-inductive, while agarose and gellan gum hydrogels supported more robust cartilage formation because of a greater GAGs accumulation within the constructs. Unfortunately, gellan gum hydrogels stained more intensely for collagen type II and collagen type I, suggesting a fibrocartilaginous tissue phenotype [174]. There are other studies focused on calcified cartilage zone, which provides mechanical integration between articular cartilage and subchondral bone. Lee W.D. et al. [175] developed tissue-engineered osteochondral-like constructs with bone marrow stromal cells (BMSC), as single cell source. Cartilage tissue and a porous bone substitute substrate were formed with an interfacial zone of calcified cartilage. The authors found that the presence of calcified cartilage increased the shear load that the construct may withstand at the interface. However, preclinical studies are needed to determine if these osteochondral-like constructs could repair joint defects in vivo.

Evaluating cell-free and cell-based hydrogels reviewed above, only a few of these biomaterials have been used in clinical applications [45,46,74,176–179] because of four main unsolved problems in tissue engineering: (1) toxicity of some crosslinking agents [180]; (2) lack of mechanical integrity [156,181–187]; (3) poor control of gelation kinetics [188,189]; (4) unsuitable degradation kinetics [190–194].

Since some of the reactions used to synthesize hydrogels are limited due to their complexity, the use of cytotoxic reagents, instability of some functional groups, possible side reactions and low coupling efficiency, there is the need to explore and exploit simple and highly efficient methods which may be applicable to a great variety of biodegradable polymeric precursors.

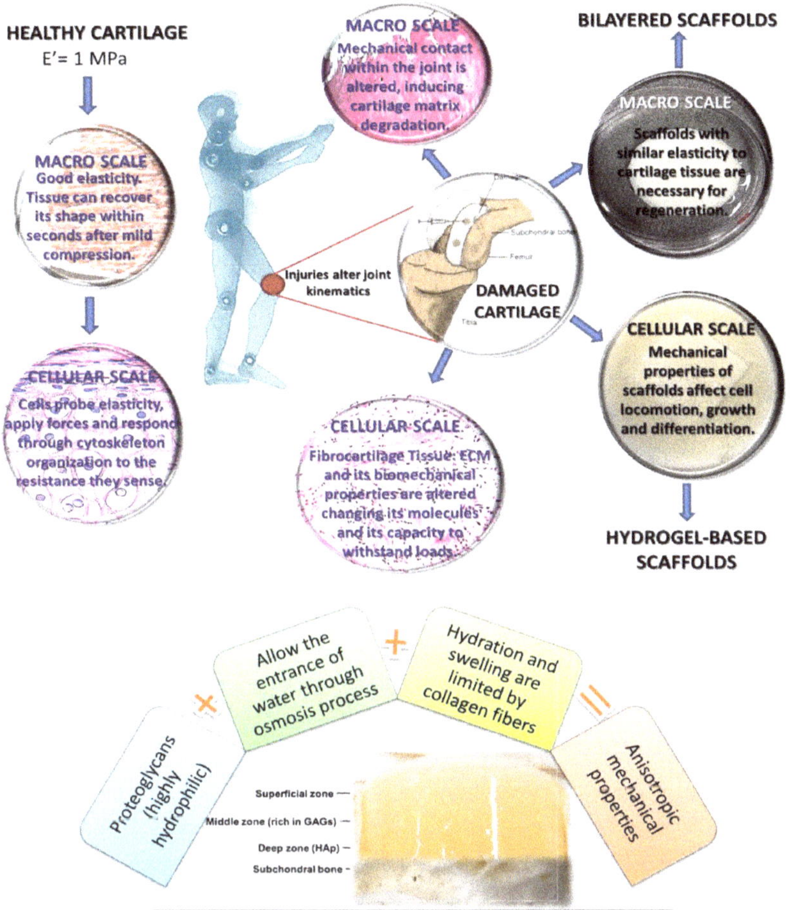

Polymers **2017**, *9*, 671

Figure 6. Correlation among the multi-scale biomechanical behavior of cartilage tissue. Original illustration designed and provided by the authors.

The tables embedded within Figure 6:

Ref.	Scaffold Material	Mechanical Properties
209	PCL control	Tensile Young's modulus E (MPa) = 2.09±0.21 Ultimate tensile strength (MPa) = 15.15±1.3 Compressive modulus (MPa)= 6.37±0.88
	PCL scaffold + hyaluronic acid hydrogel	Tensile Young's modulus E (MPa) = 2.08±0.24 Ultimate tensile strength (MPa) = 14.25±0.99 Compressive modulus (MPa)= 6.54±0.75
214	Si-HPMC hydrogel	Compressive storage moduli E′ (kPa) =5±3
	Si-HPMC/Nanofibers (3% wt)	Compressive elastic modulus E′ (kPa) = 27±2

Ref.	Scaffold type	Porogens size (mm^{-1})	Tensile modulus (MPa)	Tensile stress at failure (MPa)	Compressive modulus (kPa)	Compressive stress (kPa) at 10% strain	Interfacial stiffness (aligned) (N/mm)	Interface peak load (aligned) (N)	Surface roughness (R_a; μm)
169	Particulate-template PCL	0.03	0.04±0.02	0.05±0.01	40±5	5±1	N/A	N/A	16±10
	PCL	1.0	0.10±0.05	0.03±0.01	20±5	2±1	N/A	N/A	26±20
	Bilayered PCL	0.03	1.1±0.3	0.4±0.1	44±21	4.7±2.0	0.24±0.04	0.68±0.03	3.0±0.8
	PCL	1.0	1.6±0.5	0.4±0.2	14±4	2.0±0.4	0.23±0.002	1.15±0.03	2.1±0.5
	Fibre zone PCL	-	28±2	12±3	140±20	70±4	N/A	N/A	3±1

3. Polysaccharides Versus Synthetic Hydrogels

The degradation rate and mechanical properties of manufactured hydrogels must be compatible with the growth of new tissue [109]. As mentioned before, hydrogels can be made of either natural or synthetic polymers, each of them with advantages and disadvantages, or even a combination of both, whether or not a reduction of disadvantages from the individual components can be obtained. When using natural precursors, good biocompatibility and bioactivity are ensured in the hydrogel scaffolds. However, there will be a high degradation rate, in contact with body fluids or medium, and a limited mechanical behavior, since natural polymer components are extracted from tissues and subsequently reconstructed to form hydrogels [173,195]. Nevertheless, the strength of these natural hydrogels can be increased by making the polymer matrix denser, using a chemical crosslinking or making chemical modifications. On the other hand, using synthetic precursors may provide appropriate physical and chemical properties for hydrogel scaffolds; however a good cell biological response and an adequate degradation rate may not occur [196,197]. The strength of the synthetic hydrogels can be raised by changing the molecular weight of the starting polymers, increasing polymer concentration or even the degree of functionalization, using reactive groups during the crosslinking reaction.

3.1. Manufacturing Techniques and Their Influence on Hydrogel Properties

Depending on the polymer precursors, hydrogels can be synthesized in different ways. In the first step a polymer is modified with a functional group; then the polymer is crosslinked, either physically or chemically, to form a three-dimensional structure. While chemical hydrogels are covalently crosslinked, physical hydrogels are not. Crosslinking can take place at the same time or after the copolymerization [198]. In situ crosslinked cytocompatible injectable hydrogels can be formed using: (1) non-toxic chemical crosslinkers, as in the Michael Addition Reaction, Click Chemistry, Schiff Base Reaction, and photo-crosslinking reactions; (2) enzymes for a biological crosslinking; (3) physical interactions, such as ionic and hydrophobic ones; (4) supramolecular chemistry utilizing self-assembly molecules [199].

The morphology and physico-chemical structure of hydrogels also depend on processing conditions applied during their formation, for example using electrospinning or cryogelation techniques. Hydrogel morphologies may range from fibrils, characteristic of protein-based hydrogels such as collagen and fibrin, to amorphous, characteristic of synthetic hydrogels such as PEG. When using the electrospinning technique, it is possible to obtain hydrogels with aligned fibrils morphologies. Mirahmadi F. et al. [200] added degummed chopped silk fibers and electrospun silk fibers to thermosensitive chitosan/glycerophosphate hydrogels, to reinforce scaffolds for hyaline cartilage regeneration. The results showed that mechanical properties of hydrogels were significantly enhanced; besides the composition of the scaffolds supported the chondrogenic phenotype. Nevertheless, when using cryogelation, because of ice crystals, a controlled porosity can be induced into hydrogels, helping them recovering their shape [201]. Another method to fabricate three-dimensional porous hybrid scaffolds for articular cartilage repair is combining freeze-dried [192,202,203] natural components with synthetic polymers, provide scaffolds with mechanical strength and an environment similar to natural ECM, to let chondrocytes proliferate. Lyophilization or freeze-drying technique produces highly porous structures with open pores throughout the scaffolds. The pores are introduced into the scaffolds, first by ice crystal formation, then by freeze-drying them. For this reason pore architecture reflects the ice crystal morphology [203]. Novel collagen/polylactide (PLA), chitosan/PLA, and collagen/chitosan/PLA hybrid scaffolds were fabricated by freeze-drying technique [204]. It was observed that collagen binds water inside the scaffold structure and it helps cells to penetrate into the hybrid scaffolds. To enhance anisotropic properties of cartilage scaffolds, aligned unidirectional pores can be formed depending on the real alignment of cells and the type of extracellular matrix that has to be repaired. Collagen-hybrid scaffolds, constructed by directional freezing, were studied. When varying freezing rates and suspension media, it is possible to obtain collagen-hybrid scaffolds with unidirectional pores, tunable pore sizes and pore morphologies [203]. The results demonstrated that directed horizontal ice dendrite formation and vertical ice crystal nucleation are responsible for aligned unidirectional ice crystal growth and, consequently, for aligned unidirectional pore structure of the collagen-hybrid scaffolds.

Since conventional fabrication techniques may not provide a precisely control of pore size, interconnectivity or pore geometry for scaffolds, solid freeform fabrication (SFF) techniques are now being used to produce 3D scaffolds with an organized interconnected pore structure which ensures good functionality and good mechanical strength, necessary to maintain new cartilage formation.

Bioprinting and plotting are being used as freeform fabrication techniques. These emerging techniques are used to fabricate viable 3D tissue constructs through a precise deposition of cells in hydrogels [205]. However, scaffolds, fabricated by these printing systems, often lack of flexibility and adequate mechanical properties [138,206]. Xu T. et al. [207] described the construction of a hybrid inkjet printing/electrospinning system that can be used to fabricate viable cartilage engineered tissue. They fabricated a five-layer construct, 1 mm thick, made of electrospun polycaprolactone fibers alternated with inkjet printing of rabbit elastic chondrocytes suspended in a fibrin-collagen hydrogel. One week after printing, evidence showed more than 80% of cell viability, cell proliferation and formation of cartilage-like tissue in the five-layer construct, both in vitro and in vivo assays, and demonstrated an improvement of mechanical properties, in comparison with printed alginate or fibrin-collagen hydrogels.

Novel techniques of tissue scaffold fabrication, as ultrafast pulse DLW lithography, are attractive due to their 3D structuring capability, spatial resolution, scaling flexibility and diversity of working materials.

3.2. Degradation Kinetics, Physical Properties (Applicability) and Biological Effects

Hydrogels degradation kinetics should be compatible with new tissue formation kinetics, in order to ensure a good integration of the construct. If hydrogels degrade very fast, it will trigger occurrence of defects in the formed tissue, such as cysts. On the other hand, if a very slow degradation occurs, hydrogels will inhibit the formation of new biological material and their integration with the surrounding tissue. Degradation of hydrogels can take place by either hydrolytic or enzymatic mechanisms. The hydrolytic degradation happens when hydrogel is kept in contact with fluids by breaking the polymer chains or the crosslinked network. This type of degradation mechanism can be controlled by limiting the amount of degradable precursor, used to synthesize hydrogels. The enzymatic degradation is caused by cells when they begin to invade the hydrogel or when encapsulated cells within the hydrogel start to proliferate or migrate throughout it. Many natural origin proteins have sites of cleavage in the protease, allowing hydrogels to degrade during the replacement of the ECM. This type of degradation may also depend on the degree and the type of chemical crosslinking in the hydrogels used as scaffolds [201].

3.3. Specificities of Polysaccharide-Based Hydrogels

When trying to manufacture biomimetic scaffolds for cartilage tissue regeneration, naturally-derived hydrogels are widely used, due to their macromolecular properties and because the employed biopolymers are part of the natural tissue that needs to be healed. Most of the studied naturally-derived hydrogels are based on biopolymers such as collagen, gelatin, chitosan, hyaluronic acid, chondroitin sulfate, agarose, alginate and fibrin [109,208].

3.4. Biological Response of Polysaccharide-Based Hydrogels

Since hyaluronic acid is a fundamental component of natural cartilage matrix, some studies have shown its importance and its good qualities as an excellent naturally derived polymer. Injections of hyaluronan into osteoarthritic joints have proved to restore viscoelasticity, augment joint fluid flow, normalize endogenous hyaluronan synthesis, and provide joint function [109,112]. Some other studies [198] have demonstrated that hyaluronic acid is favorable for cell response, by maintaining chondrogenic phenotype and increasing collagen type II production and angiogenesis during in vivo assays. Another example of an excellent natural polymer is chitosan. Chitosan can easily form polyelectrolyte complexes with hyaluronan and chondroitin sulfate [112,209]. Hu X. et al. [198] tried to mimic the natural cartilage extracellular matrix by synthesizing a biological hydrogel made of hyaluronic acid, chondroitin sulfate modified with 11-azido-3,6,9-tri-oxaundeca-1-amin and gelatin modified with propiolic acid, via click chemistry. Even though the molecular modifications made to the biopolymers let biological hydrogels have good response (making chondrocytes adhere and proliferate on them during in vitro assays), degradation process was too fast. They showed a loss of 45% w in 4 weeks, and a release of 20% w of gelatin and 10% w of chondroitin sulfate during the first two weeks, leading to macroscopic shrinkage of hydrogels.

Furthermore, when combining both naturally-derived and synthetically-derived polymers, adequate degradation kinetics and biological response can be achieved. Park H. et al. [129] created, by photocrosslinking, injectable hydrogels consisting of methacrylated glycol chitosan (MeGC) and hyaluronic acid. The photopolymerized hydrogels were cytocompatible. The incorporation of hyaluronic acid increased cell proliferation while encapsulated chondrocytes improved the cartilaginous extracellular matrix production. Following the same guideline, based on fabricating a hybrid scaffold containing both biological and synthetic components, B.R. Mintz et al. [209] studied a hybrid scaffold, made of a hyaluronic acid-based hydrogel combined with a porous poly(ε-caprolactone) material. They tried to understand better the interface and the potential for integration between tissue engineered cartilage scaffolds and surrounding native tissue. They noticed that precursors provide a microenvironment which supports chondrocyte infiltration and proliferation, while maintaining seeded phenotype and structural integrity over a 6-week culture period.

4. Future Trends: From Combination of Materials to Hybrid Hydrogels

Identified problems hindering the application of hydrogels in cartilage regeneration, as described along this paper, include: mechanical properties [210] and mechanical instability [209,211]; dedifferentiation of chondrocytes [163]; toxicity of some of the used crosslinking agents [180]; poor control of gelation kinetics [188,189]; unsuitable degradation kinetics [190–194]. Figure 7 illustrates the combining requirements needed to create materials with biomimetic features.

Mechanical instability inhibits the integration of hydrogels with the surrounding native cartilage tissue when they are implanted [209,211]. In order to synthesize mechanically stable hydrogels and improve their mechanical properties, several options have been proposed. One of the most promising options relays on the principle that materials combination must show the ability to support matrix formation [210], as demonstrated by Boere K.W.M et al. [80]. In this research, it was determined that, when grafting two materials covalently (a 3-D-fabricated poly (hidroxymethylglycolide–*co*–ε–caprolactone)/poly(ε-caprolactone) thermoplastic polymer scaffold, functionalized with methacrylate groups and covalently linked to a chondrocyte-laden gelatin methacrylamide hydrogel), the binding strength between the materials improved significantly, resulting in the enhanced mechanical integrity of the reinforced hydrogel. Embedded chondrocytes in hydrogel scaffolds also showed significant cartilage-specific matrix deposition, both in vitro and in vivo assays.

Another promising option to enhance mechanical stability is by regenerating cartilage and bone tissue simultaneously using a two-phased scaffold, since ceramic-to-bone interface has a better and faster integration compared to hydrogel-to-cartilage interface [80]. Additionally it has been observed that bone integration is much faster than cartilage integration, occurring during 2 and 24 weeks after transplantation, respectively.

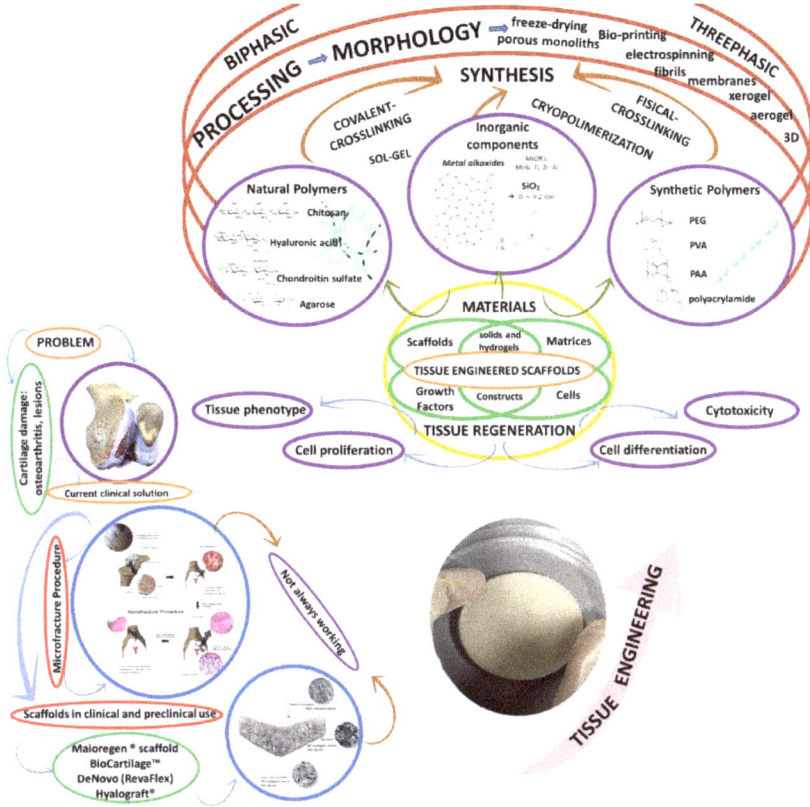

Figure 7. Correlation between current cell-based scaffolds used for cartilage regeneration. Original illustration designed and provided by the authors.

It is possible to have a stable fixation of a cartilage scaffold by exploring a fixation technique with the subchondral bone [206]. One way to accomplish this stable fixation is fabricating an osteochondral scaffold which facilitates fixation and integration with the surrounding cartilage tissue, accelerating the repair of defected articular cartilage when implanted. The general idea is that bone scaffolds act as anchors, providing mechanical stability for cartilage tissue regeneration, besides, the join between the bone component and the cartilage component should be strong enough to prevent dislocation or delamination on in vivo environment. In order to follow the theory mentioned above, Seol Y.-J. et al. [206] reinforced osteochondral scaffolds by developing combined scaffolds, made of hydrogel scaffolds anchoring to cartilage tissue and ceramic scaffolds anchoring to bone tissue. For in vivo assay, the combined scaffolds were press-fitted into osteochondral tissue defects, in rabbit knee joints. Hydrogel scaffolds and combined scaffolds were compared. After 12 weeks, in vivo experiments demonstrated that regeneration of osteochondral tissue, especially articular cartilage tissue regeneration, was better with combined scaffolds than with hydrogel scaffolds. Hydrogel scaffolds could not keep their initial position, suggesting that ceramic scaffolds in combined scaffolds provided mechanical stability for hydrogel scaffolds. Moreover, G. Camci-Unal et al. [202] realized that combined hydrogels can be biologically and physically tuned to yield within a range of different cell responses and, according to these responses, combined hydrogels may show potential therapeutic possibilities to treat either chondral or osteochondral lesions.

Following this trend, Yang S.S. et al. [138] developed a 3D plotting system to manufacture a biphasic graft which consists of cartilage and subchondral bone for application to osteochondral defects. A combined material (PLGH/alginate) was fabricated as supporting structure to induce a mature osteochondral graft. Cartilage-derived ECM or hydroxyapatite substances were blended with alginate and plotted together with human fetal cartilage-derived progenitor cells, either in the cartilage layer or in the subchondral bone one. The plotted biphasic osteochondral graft showed good integration between layers because no structural separation was observed, while there was dominant cartilage and bone tissue formation during differentiation assay. One of the limitations of using the osteochondral approach in combination with bone marrow derived MSCs is their terminal differentiation, as they seem to follow an endochondral ossification which can arrest differentiation at a stable cartilage hyaline-like phenotype during the chondrogenic process. For this reason, a chondrogenic stimulator, such as the recently described kartogenin which regulates Runx1 expression [212], has to be incorporated with a known inducer of chondrogenic differentiation and a suppressor of hypertrophy.

Another alternative to fulfill the inadequate mechanical strength of hydrogel is constructing a solid-supported thermogel, comprising hydrogel systems or demineralized bone matrix. Huang H. et al. [213] combined chitosan thermogel with demineralized bone matrix to produce solid-supported hydrogel scaffolds. This type of scaffolds provided sufficient strength for cartilage regeneration. They retained homogeneously more bone-derived mesenchymal stem cells (BMSCs) and they proved to have superior matrix production and chondrogenic differentiation in comparison with pure hydrogels and demineralized matrix by their own.

Using fibers of different natural materials to reinforce hydrogels is another way to offer mechanical strength to hybrid scaffolds [205,211]. Mechanical characteristics can be improved or modified using different strategies: (1) varying the number of fiber layers in the laminate; (2) combining different kinds of fibers and nanofiber sheets; (3) modifying crosslinking degree of hydrogels and fibers; (4) changing fibers content and surface treatment; (5) shifting fiber orientation. Fibers anisotropy is an excellent property to reach strong mechanical reinforcement at low charge levels. For example, Buchtová N. et al. [214] developed an injectable hydrogel based on a silanized cellulose derivative: hydroxypropyl methylcellulose interlinked with silica fibers. They proved that the compressive modulus of the hydrogel could be tunable, depending on the covalent bonding between biopolymer and silica fibers. In other approaches, the incorporation of bioactive species, such as cells, growth factors, peptides and proteins into the materials, is proposed to improve hydrogel scaffolds properties [73,215].

It can be deduced, from the reviewed studies, that combination of components yield up reinforced mechanical properties. However, to overcome the other requirements mentioned above, something different has to be done from the already investigated methods to fabricate materials. In authors' opinion, there is a barely explored alternative in material science: studying the potential of hybrid hydrogels based on natural polymers and inorganic components. Hybrids are considered to be materials formed by two components bonded at a molecular level. Commonly one of these components is organic and the other one is inorganic. The new hybrid materials (Figure 8) may show superior characteristics in comparison with the two component phases. This possibility may offer a great potential to design new materials with the complex properties required for cartilage regeneration.

Since degradation rate and mechanical properties can be fine-tuned through chemical and/or physical modifications on either naturally or synthetically derived scaffolds, an excellent opportunity is given to hybrids to emerge as a promising solution for cartilage regeneration. When fine-tuned, hybrids can acquire amorphous, semicrystalline, hydrogen-bonded or supramolecular physical structures [109,210].

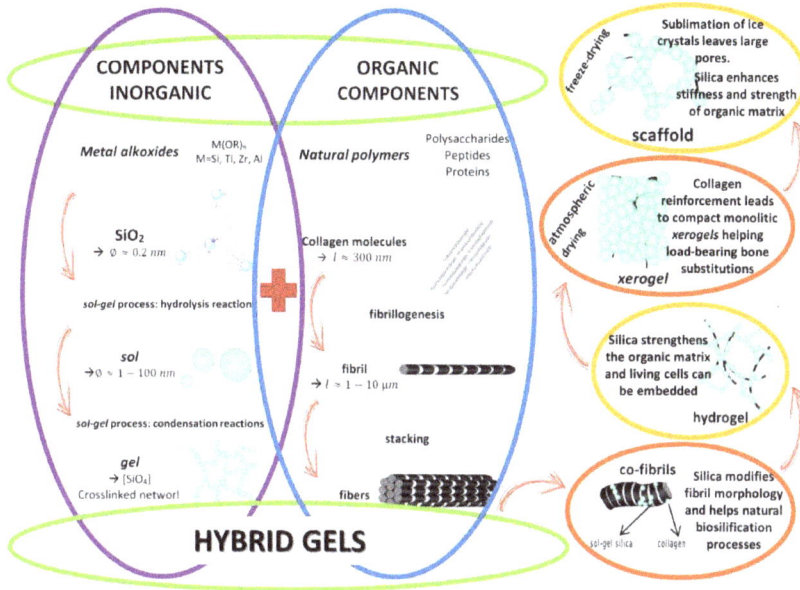

Figure 8. Advantages of hybrid *gels*. From soft mineralized hydrogels to hard compact *xerogels*. Original illustration designed and provided by the authors.

Hydrogels made from purely organic precursors are being used as viable materials for cartilage repair due to their ability to retain grate portions of water, swollen, distend and exhibit large changes of dimensions (volume changes of several- to 10-fold are common) [216,217]. These characteristics provide hydrogels with a low interfacial tension with water and other fluids, allowing them to reduce mechanical friction between tissues during implantation. The hydrophilic characteristics of hydrogels are caused by the presence of special hydrophilic molecules (–OH, –CONH, –CONH$_2$, and –SO$_3$H) found in the polymeric components. These molecules give hydrogels different absorption potential [216] and the ability to respond to a range of different stimuli, including temperature, pH, salt, specific (bio)chemical signals, and electric fields [217]. Nevertheless, hydrogels, based on a certain type of precursors, undergo uniform volumetric expansion and contraction, in response to several stimuli; therefore, the great added value of hybrid hydrogels is combining their precursors potential and restrictions. For example, to determine their degradation kinetics and enhanced biological and mechanical properties, it is possible to use: stiffer components like silica-based materials, which restrict swelling in hydrogels in certain positions or directions; or natural polysaccharide-based polymers hybridized with inorganic materials (hydroxyapatite, SiO$_2$, or demineralized bone matrix), which restrict the number and type of hydrophilic molecules.

A further step may be the integration of nanotubes with different chemical composition into hybrid scaffolds which may provide them with bioactive and mechanical properties. This is the case of three-dimensional porous collagen sponges where single-walled carbon nanotubes are incorporated into [218]. The incorporation of single-walled carbon nanotubes improved cell proliferation and GAGs production in the in vivo microenvironment, because nanotubes were internalized by cells, with benefit for controlled and localized delivery of biological factors. Another study is the design of a biomimetic nanostructured composite cartilage scaffold, via biologically-inspired rosette nanotubes (RNTs) and biocompatible non-woven poly (L-lactic acid) (PLLA) [219]. It was concluded that, RNTs have a similar morphology with native collagen fibers when self-assembled in aqueous conditions, and besides they increase glycosaminoglycan, collagen and protein production; their nanotopography and surface chemistry enhance chondrogenic differentiation. Another study tested the biocompatibility of 3D artificial hexagonal-pore shaped hybrid organic-inorganic microstructured scaffolds in a rabbit model [220].

The association of silica and polysaccharides within composites or hybrids has demonstrated therapeutic benefits in a wide range of bio-inspired silica-collagen materials [221]. Although this kind of materials has been prepared over nearly 15 years, their application in cartilage regeneration treatments has not been exploited. The great value of hybrid materials (Figure 7) is that it can be synthesized a large variety of structures and properties, from soft mineralized hydrogels to hard compact xerogel, depending on the soft (cartilage) or hard (osteochondral) tissue wanted to be repaired or regenerated. Moreover, to fully comprehend these potential materials and to raise their value for the development of innovative biomedical devices for cartilage regeneration, it is important studying the interplay between the organic-inorganic precursors, which is to follow carefully the polymer self-assembly process and the inorganic condensation mechanisms. Therefore, biological, mechanical and degradation properties can be modulated, guaranteeing bioactivity, cytocompatibility, and an eventual biocompatibility.

5. Concluding Remarks

Currently, there are no clinical satisfactory solutions for cartilage tissue regeneration. From this problem, it raises the need for developing new types of materials and designing more suitable scaffolds which can be used for cartilage regeneration. Investigations based on cell-free hydrogel scaffolds have focused on the optimization of physical-chemical and mechanical properties of matrices. Whereas, cell-seeded hydrogel scaffold studies focus on whether or not they provide an appropriate biochemical and biomechanical environment for regenerating a long-lasting hyaline-type cartilage. Since many materials neither exhibit a low friction coefficient nor withstand several loading cycles, many combinations of polysaccharides and synthetic hydrogels have been assayed to obtain load-bearing and tribological properties similar to native cartilage tissue ones. However, some unsolved problems hinder the application of these materials to the clinic: (1) toxicity of some crosslinking agents; (2) lack of mechanical integrity; (3) poor control of gelation kinetics; (4) unsuitable degradation kinetics.

Among the most promising options, to synthesize mechanically stable hydrogels which support matrix formation, there is the combination of materials to regenerate cartilage and bone tissue simultaneously, using a two/several-phased scaffold. These combinations can be biologically and physically tuned to yield, within a range of different cell responses and according to these responses, combined hydrogels which may show potential therapeutic possibilities to treat either chondral or osteochondral lesions. Osteochondral treatment introduces a new problem and requires the use of a chondrogenic stimulator, such as the recently described kartogenin, to induce chondrogenic differentiation and suppress hypertrophy. From materials science perspective, hybrid hydrogels based on natural polymers and inorganic components may offer a fine tuning of mechanical and biological response, required to reproduce the complexity of the cartilage tissue environment. Silica, hydroxyapatite or demineralized bone matrix/polysaccharide-based hybrids restrict swelling in hydrogels in certain positions or directions by reducing the number and the type of hydrophilic molecules. The association of silica and polysaccharides within composites or hybrids has demonstrated therapeutic benefits in a wide range of bio-inspired silica-collagen materials. To fully comprehend these potential materials and to raise their value for the development of innovative biomedical devices for cartilage regeneration, it is important studying the interplay between the organic-inorganic precursors, which is to follow carefully the polymer self-assembly process and the inorganic condensation mechanisms. Therefore, biological, mechanical and degradation properties can be modulated, bioactivity and cytocompatibility guaranteed and biocompatibility eventually achieved.

Acknowledgments: This work was supported by SIP-IPN 20170510 Project (Mexico), DGICYT Project, MAT2014-51918-C2-I-R (Spain), (CONACYT/CSIC)-EMHE program through project No. MHE200011 (Mexico-Spain). DA Sánchez-Téllez also acknowledges CONACYT (Mexico) for the scholarship given. The authors acknowledge Stephanie Cisneros-Téllez for helping them in the digitalization of Figures 1, 4 and 8.

Author Contributions: Daniela Anahí Sánchez-Téllez, Lucía Téllez-Jurado and Luís María Rodríguez-Lorenzo gave substantial contributions to the conception and design of the review content. Daniela Anahí Sánchez-Téllez wrote the paper. Daniela Anahí Sánchez-Téllez and Luís María Rodríguez-Lorenzo critically revised the final manuscript, making improvements in the drafting of the work. Daniela Anahí Sánchez-Téllez, Daniela Anahí Sánchez-Téllez and Luís María Rodríguez-Lorenzo gave final approval of the version to be published. The three authors, Daniela Anahí Sánchez-Téllez, Lucía Téllez-Jurado and Luís María Rodríguez-Lorenzo agreed in all aspects of the final written work, ensuring that any question related to the accuracy or integrity of any part of the work was appropriately investigated and resolved.

Conflicts of Interest: The authors declare no conflicts of interest.

References

1. Fickert, S.; Schattenberg, T.; Niks, M.; Weiss, C.; Their, S. Feasibility of arthroscopic 3-dimensional, purely autologous chondrocyte transplantation for chondral defects of the hip: A case series. *Arch. Orthop. Trauma Surg.* **2014**, *134*, 971–978. [CrossRef] [PubMed]

2. Baynat, C.; Andro, C.; Vincent, J.P.; Schiele, P.; Buisson, P.; Dubrana, F.; Gunepin, F.X. Actifit® synthetic meniscal substitute: Experience with 18 patients in Brest, France. *Orthop. Traumatol. Surg. Res.* **2014**, *100* (Suppl. 8), S385–S389. [CrossRef] [PubMed]

3. Verdonk, P.; Dhollander, A.; Almqvist, K.F.; Verdonk, R.; Victor, J. Treatment of osteochondral lesions in the knee using a cell-free scaffold. *Bone Jt. J.* **2015**, *97B*, 318–323. [CrossRef] [PubMed]

4. Case, J.M.; Scopp, J.M. Treatment of articular cartilage defects of the knee with microfracture and enhanced microfacture techniques. *Sports Med. Arthrosc.* **2016**, *24*, 63–68. [CrossRef] [PubMed]

5. Erggelet, C. Enhanced marrow stimulation techniques for cartilage repair. *Oper. Tech. Orthop.* **2014**, *24*, 2–13. [CrossRef]

6. Gille, J.; Behrens, P.; Volpi, P.; de Girolamo, L.; Reiss, E.; Zoch, W.; Anders, S. Outcome of autologous matrix induced chondrogenesis (AMIC) in cartilage knee surgery: Data of the AMIC registry. *Arch. Orthop. Trauma Surg.* **2013**, *133*, 87–93. [CrossRef] [PubMed]

7. Turajane, T.; Thitiset, T.; Honsawek, S.; Chaveewanakorn, U.; Aojanepong, J.; Papadopoulos, K.I. Assessment of chondrogenic differentiation potential of autologous activated peripheral blood stem cells on human early osteoarthritic cancellous tibial bone scaffold. *Musculoskelet. Surg.* **2014**, *98*, 35–43. [CrossRef] [PubMed]

8. Gaharwar, A.K.; Schexnailder, P.J.; Shmidt, G. Chapter 24: Nanocomposite polymer biomaterials for tissue repair of bone and cartilage: A material science perspective. In *Nanomaterials Handbook*; Taylor and Francis Group, LLC: Boca Raton, FL, USA, 2011; pp. 1–20.

9. Chung, C.; Burdick, J.A. Engineering cartilage tissue. *Adv. Drug Deliv. Rev.* **2008**, *60*, 243–262. [CrossRef] [PubMed]

10. Chang, G.; Xia, C.; Sherman, O.; Strauss, E.; Jazrawi, L.; Recht, M.P.; Regatte, R.R. High resolution morphologic imaging and T2 mapping of cartilage at 7 tesla: Comparison of cartilage repair patients and healthy controls. *Magn. Reson. Mater. Phys.* **2013**, *26*, 539–548. [CrossRef] [PubMed]

11. Mathieu, C.; Chevrier, A.; Lascau-Coman, V.; Rivard, G.E.; Hoemann, C.D. Stereological analysis of subchondral angiogenesis induced by chitosan and coagulation factors in microdrilled articular cartilage defects. *Osteoarthr. Cartil.* **2013**, *21*, 849–859. [CrossRef] [PubMed]

12. Siclari, A.; Mascaro, G.; Gentili, C.; Kaps, C.; Cancedda, R.; Boux, E. Cartilage repair in the knee with subchondral drilling augmented with a platelet-rich plasma-immersed polymer-based implant. *Knee Surg. Sports Traumatol. Arthrosc.* **2014**, *22*, 1225–1234. [CrossRef] [PubMed]

13. Accardi, M.A.; McCullen, S.D.; Callanan, A.; Chung, S.; Cann, P.M.; Stevens, M.M.; Dini, D. Effects of fiber orientation on the frictional properties and damage of regenerative articular cartilage surfaces. *Tissue Eng. Part A* **2013**, *19*, 2300–2310. [CrossRef] [PubMed]

14. Lebourg, M.; Rochina, J.R.; Sousa, T.; Mano, J.; Ribelles, J.L. Different hyaluronic acid morphology modulates primary articular chondrocyte behavior in hyaluronic acid-coated polycaprolactone scaffolds. *J. Biomed. Mater. Res. Part A* **2013**, *101A*, 518–527. [CrossRef] [PubMed]

15. Levorson, E.J.; Hu, O.; Mountziaris, P.M.; Kasper, F.K.; Mikos, A.G. Cell-derived polymer/extracellular matrix composite scaffolds for cartilage regeneration, Part 2: Construct devitalization and determination of chondroinductive capacity. *Tissue Eng. Part C Methods* **2014**, *20*, 358–372. [CrossRef] [PubMed]

16. Levingstone, T.J.; Matsiko, A.; Dickson, G.R.; O'Brien, F.J.; Gleeson, J.P. A biomimetic multi-layered collagen-based scaffold for osteochondral repair. *Acta Biomater.* **2014**, *10*, 1996–2004. [CrossRef] [PubMed]

17. Sutherland, A.J.; Beck, E.C.; Dennis, S.C.; Converse, G.L.; Hopkins, R.A.; Berkland, C.J.; Detamore, M.S. Decellularized cartilage may be a chondroinductive material for osteochondral tissue engineering. *PLoS ONE* **2015**, *10*, e0121966. [CrossRef] [PubMed]

18. Sancho-Tello, M.; Forriol, F.; Gastaldi, P.; Ruiz-Saurí, A.; Martín de Llano, J.J.; Novella-Maestre, E.; Antolinos-Turpín, C.M.; Gómez-Tejedor, J.A.; Gómez Ribelles, J.L.; Carda, C. Time evolution of in vivo articular cartilage repair induced by bone marrow stimulation and scaffold implantation in rabbits. *Int. J. Artif. Organs* **2015**, *38*, 210–223. [CrossRef] [PubMed]

19. Wang, J.; Yang, Q.; Cheng, N.; Tao, X.; Zhang, Z.; Sun, X.; Zhang, Q. Collagen/silk fibroin composite scaffold incorporated with PLGA microsphere for cartilage repair. *Mater. Sci. Eng. C* **2016**, *61*, 705–711. [CrossRef] [PubMed]

20. Shao, Z.; Zhang, X.; Pi, Y.; Yin, L.; Li, L.; Chen, H.; Zhou, C.; Ao, Y. Surface modification on polycaprolactone electrospun mesh and human decalcified bone scaffold with synovium-derived mesenchymal stem cells-affinity peptide for tissue engineering. *J. Biomed. Mater. Res. Part A* **2015**, *103A*, 318–329. [CrossRef] [PubMed]

21. Zhang, Y.; Tang, C.L.; Chen, W.J.; Zhang, Q.; Wang, S.L. Dynamic compression combined with exogenous SOX-9 promotes chondrogenesis of adipose-derived mesenchymal stem cells in PLGA scaffold. *Eur. Rev. Med. Pharmacol. Sci.* **2015**, *19*, 2671–2678. [PubMed]

22. Lehmann, M.; Martin, F.; Mannigel, K.; Kaltschmidt, K.; Sack, U.; Anderer, U. Three-dimensional scaffold-free fusion culture: The way to enhanced chondrogenesis of in vitro propagated human articular chondrocytes. *Eur. J. Histochem.* **2013**, *57*, e31. [CrossRef] [PubMed]

23. Kwon, H.; Sun, L.; Cairns, D.M.; Rainbow, R.S.; Preda, R.C.; Kaplan, D.L.; Zeng, L. The influence of scaffold material on chondrocytes under inflammatory conditions. *Acta Biomater.* **2013**, *9*, 6563–6575. [CrossRef] [PubMed]

24. Pretzel, D.; Linss, S.; Ahrem, H.; Endres, M.; Kaps, C.; Klemm, D.; Kinne, R.W. A novel in vitro bovine cartilage punch model for assessing the regeneration of focal cartilage defects with biocompatible bacterial nanocellulose. *Arthritis Res. Ther.* **2013**, *15*, R59. [CrossRef] [PubMed]

25. Chomchalao, P.; Pongcharoen, S.; Sutheerawattananonda, M.; Tiyaboonchai, W. Fibroin and fibroin blended three-dimensional scaffolds for rat chondrocyte culture. *Biomed. Eng. Online* **2013**, *12*, 28. [CrossRef] [PubMed]

26. Lee, S.U.; Lee, J.Y.; Joo, S.Y.; Lee, Y.S.; Jeong, C. Transplantation of a scaffold-free cartilage tissue analogue for the treatment of physeal cartilage injury of the proximal tibia in rabbits. *Yonsei Med. J.* **2016**, *57*, 441–448. [CrossRef] [PubMed]

27. O'Sullivan, N.A.; Kobayashi, S.; Ranka, M.P.; Zaleski, K.L.; Yaremchuk, M.J.; Bonassar, L.J.; Randolph, M.A. Adhesion and integration of tissue engineered cartilage to porous polyethylene for composite ear reconstruction. *J. Biomed. Mater. Res. Part B* **2015**, *103B*, 983–991. [CrossRef] [PubMed]

28. Schleicher, I.; Lips, K.S.; Sommer, U.; Schappat, I.; Martin, A.P.; Szalay, G.; Schnettler, R. Allogenous bone with collagen for repair of deep osteochondral defects. *J. Surg. Res.* **2013**, *185*, 667–675. [CrossRef] [PubMed]

29. Chang, N.J.; Lam, C.F.; Lin, C.C.; Chen, W.L.; Li, C.F.; Lin, Y.T.; Yeh, M.L. Transplantation of autologous endothelial progenitor cells in porous PLGA scaffolds create a microenvironment for the regeneration of hyaline cartilage in rabbits. *Osteoarthr. Cartil.* **2013**, *21*, 1613–1622. [CrossRef] [PubMed]

30. Sharma, S.; Lee, A.; Choi, K.; Kim, K.; Youn, I.; Trippel, S.B.; Panitch, A. Biomimetic aggrecan reduces cartilage extracellular matrix from degradation and lowers catabolic activity in ex vivo and in vivo models. *Macromol. Biosci.* **2013**, *13*, 1228–1237. [CrossRef] [PubMed]

31. Ding, C.; Qiao, Z.; Jiang, W.; Li, H.; Wei, J.; Zhou, G.; Dai, K. Regeneration of a goat femoral head using a tissue-specific, biphasic scaffold fabricated with CAD/CAM technology. *Biomaterials* **2013**, *34*, 6706–6716. [CrossRef] [PubMed]

32. Waldorff, E.I.; Roessler, B.J.; Zachos, T.A.; Miller, B.S.; McHugh, J.; Goldstein, S.A. Preclinical evaluation of a novel implant for treatment of a full-thickness distal femoral focal cartilage defect. *J. Arthroplast.* **2013**, *28*, 1421–1429. [CrossRef] [PubMed]

33. Camarero-Espinosa, S.; Rothen-Rutishauser, B.; Weder, C.; Foster, E.J. Directed cell growth in multi-zonal scaffolds for cartilage tissue engineering. *Biomaterials* **2016**, *74*, 42–52. [CrossRef] [PubMed]

34. Krych, A.J.; Wanivenhaus, F.; Ng, K.W.; Doty, S.; Warren, R.F.; Maher, S.A. Matrix generation within a macroporous non-degradable implant for osteochondral defects is not enhanced with partial enzymatic digestion of the surrounding tissue: Evaluation in an in vivo rabbit model. *J. Mater. Sci. Mater. Med.* **2013**, *24*, 2429–2437. [CrossRef] [PubMed]

35. Irion, V.H.; Flanigan, D.C. New and emerging techniques in cartilage repair: Other scaffold-based cartilage treatment options. *Oper. Tech. Sports Med.* **2013**, *21*, 125–137. [CrossRef]

36. Kim, Y.S.; Park, D.Y.; Cho, Y.H.; Chang, J.W.; Choi, J.W.; Park, J.K.; Min, B.H.; Shin, Y.S.; Kim, C.H. Cultured chondrocyte and porcine cartilagederived substance (PCS) construct as a possible dorsal augmentation material in rhinoplasty: A preliminary animal study. *J. Plast. Reconstr. Aesthet. Surg.* **2015**, *68*, 659–666. [CrossRef] [PubMed]

37. Filardo, G.; Kon, E.; Perdisa, F.; Di Matteo, B.; Di Martino, A.; Iacono, F.; Zaffagnini, S.; Balboni, F.; Vaccari, V.; Marcacci, M. Osteochondral scaffold reconstruction for complex knee lesions: A comparative evaluation. *Knee* **2013**, *20*, 570–576. [CrossRef] [PubMed]

38. Kon, E.; Filardo, G.; Di Martino, A.; Busacca, M.; Moio, A.; Perdisa, F.; Marcacci, M. Clinical results and mri evolution of a nano-composite multilayered biomaterial for osteochondral regeneration at 5 years. *Am. J. Sports Med.* **2014**, *42*, 158–165. [CrossRef] [PubMed]

39. Kon, E.; Filardo, G.; Perdisa, F.; Di Martino, A.; Busacca, M.; Balboni, F.; Sessa, A.; Marcacci, M. A one-step treatment for chondral and osteochondral knee defects: Clinical results of a biomimetic scaffold implantation at 2 years of follow-up. *J. Mater. Sci. Mater. Med.* **2014**, *25*, 2437–2444. [CrossRef] [PubMed]

40. Caminal, M.; Moll, X.; Codina, D.; Rabanal, R.M.; Morist, A.; Barrachina, J.; Garcia, F.; Pla, A.; Vives, J. Transitory improvement of articular cartilage characteristics after implantation of polylactide: Polyglycolic acid (PLGA) scaffolds seeded with autologous mesenchymal stromal cells in a sheep model of critical-sized chondral defect. *Biotechnol. Lett.* **2014**, *36*, 2143–2153. [CrossRef] [PubMed]

41. Delcogliano, M.; de Caro, F.; Scaravella, E.; Ziveri, G.; De Biase, C.F.; Marotta, D.; Marenghi, P.; Delcogliano, A. Use of innovative biomimetic scaffold in the treatment for large osteochondral lesions of the knee. *Knee Surg. Sports Traumatol. Arthrosc.* **2014**, *22*, 1260–1269. [CrossRef] [PubMed]

42. McCormick, F.; Cole, B.J.; Nwachukwu, B.; Harris, J.D.; Adkisson, H.D.; Farr, J. Treatment of focal cartilage defects with a juvenile allogeneic 3-dimensional articular cartilage graft. *Oper. Techniq. Sports Med.* **2013**, *21*, 95–99. [CrossRef]

43. Chiang, H.; Liao, C.J.; Hsieh, C.H.; Shen, C.Y.; Huang, Y.Y.; Jiang, C.C. Clinical feasibility of a novel biphasic osteochondral compositefor matrix-associated autologous chondrocyte implantation. *Osteoarthr. Cartil.* **2013**, *21*, 589–598. [CrossRef] [PubMed]

44. Kon, E.; Filardo, G.; Venieri, G.; Perdisa, F.; Marcacci, M. Tibial plateau lesions. Surface reconstruction with a biomimetic osteochondral scaffold: Results at 2 years of follow-up. *Inj. Int. J. Care Inj.* **2014**, *45S*, S121–S125. [CrossRef] [PubMed]

45. Abrams, G.A.; Mall, N.A.; Fortier, L.A.; Roller, B.L.; Cole, B.J. BioCartilage: Background and operative technique. *Oper. Tech. Sports Med.* **2013**, *21*, 116–124. [CrossRef]

46. Sharma, B.; Fermanian, S.; Gibson, M.; Unterman, S.; Herzka, D.A.; Cascio, B.; Coburn, J.; Hui, A.Y.; Marcus, N.; Gold, G.E.; et al. Human cartilage repair with a photoreactive adhesive-hydrogel composite. *Sci. Transl. Med.* **2013**, *5*, 167ra6. [CrossRef] [PubMed]

47. Quarch, V.M.; Enderle, E.; Lotz, J.; Frosch, K.H. Fate of large donor site defects in osteochondral transfer procedures in the knee joint with and without TruFit plugs. *Arch. Orthop. Trauma Surg.* **2014**, *134*, 657–666. [CrossRef] [PubMed]

48. Bartha, L.; Hamann, D.; Pieper, J.; Péters, F.; Riesle, J.; Vajda, A.; Novak, P.K.; Hangody, L.R.; Vasarhelyi, G.; Bodó, L.; et al. A clinical feasibility study to evaluate the safety and efficacy of PEOT/PBT implants for human donor site filling during mosaicplasty. *Eur. J. Orthop. Surg. Traumatol.* **2013**, *23*, 81–91. [CrossRef] [PubMed]

49. Porcellini, G.; Merolla, G.; Campi, F.; Pellegrini, A.; Bodanki, C.S.; Paladini, P. Arthroscopic treatment of early glenohumeral arthritis. *J. Orthop. Traumatol.* **2013**, *14*, 23–29. [CrossRef] [PubMed]

50. Krase, A.; Abedian, R.; Steck, E.; Hurschler, C.; Richter, W. BMP activation and WNT-signalling affect biochemistry and functional biomechanical properties of cartilage tissue engineering constructs. *Osteoarthr. Cartil.* **2014**, *22*, 284–292. [CrossRef] [PubMed]

51. Panadero, J.A.; Vikingsson, L.V.; Gomez Ribelles, J.L.; Lanceros-Mendez, S.; Sencadas, V. In vitro mechanical fatigue behavior of poly-ε-caprolactone macroporous scaffolds for cartilage tissue engineering: Influence of pore filling by a poly(vinyl alcohol) gel. *J. Biomed. Mater. Res. Part B Appl. Biomater.* **2014**, *103*, 1037–1043. [CrossRef] [PubMed]

52. Motavalli, M.; Whitney, G.A.; Dennis, J.E.; Mansour, J.M. Investigating a continuous shear strain function for depth-dependent properties of native and tissue engineering cartilage using pixel-size data. *J. Mech. Behav. Biomed. Mater.* **2013**, *28*, 62–70. [CrossRef] [PubMed]

53. Hendriks, J.A.; Moroni, L.; Riesle, J.; de Wijn, J.R.; van Blitterswijk, C.A. The effect of scaffold-cell entrapment capacity and physico-chemical properties on cartilage regeneration. *Biomaterials* **2013**, *34*, 4259–4265. [CrossRef] [PubMed]

54. Griebel, A.J.; Khoshgoftar, M.; Novak, T.; van Donkelaar, C.C.; Neu, C.P. Direct noninvasive measurement and numerical modeling of depth-dependent strains in layered agarose constructs. *J. Biomech.* **2014**, *47*, 2149–2156. [CrossRef] [PubMed]

55. Ahn, H.; Kim, K.J.; Park, S.Y.; Huh, J.E.; Kim, H.J.; Yu, W.R. 3D braid scaffolds for regeneration of articular cartilage. *J. Mech. Behav. Biomed. Mater.* **2014**, *34*, 37–46. [CrossRef] [PubMed]

56. MacBarb, R.F.; Chen, A.L.; Hu, J.C.; Athanasiou, K.A. Engineering functional anisotropy in fibrocartilage neotissues. *Biomaterials* **2013**, *34*, 9980–9989. [CrossRef] [PubMed]

57. Moradi, A.; Pramanik, S.P.; Ataollahi, F.; Khalil, A.A.; Kamarul, T.; Pingguan-Murphy, B. A comparison study of different physical treatments on cartilage matrix derived porous scaffolds for tissue engineering applications. *Sci. Technol. Adv. Mater.* **2014**, *15*, 065001. [CrossRef] [PubMed]

58. Peng, G.; McNary, S.M.; Athanasiou, K.A.; Reddi, A.H. Surface zone articular chondrocytes modulate the bulk and surface mechanical properties of the tissue-engineered cartilage. *Tissue Eng. Part A* **2014**, *20*, 3332–3341. [CrossRef] [PubMed]

59. Causin, P.; Sacco, R.; Verri, M. A multiscale approach in the computational modeling of the biophysical environment in artificial cartilage tissue regeneration. *Biomech. Model. Mechanobiol.* **2013**, *12*, 763–780. [CrossRef] [PubMed]

60. Da, H.; Jia, S.J.; Meng, G.L.; Cheng, J.H.; Zhou, W.; Xiong, Z.; Mu, Y.J.; Liu, J. The impact of compact layer in biphasic scaffold on osteochondral tissue engineering. *PLoS ONE* **2013**, *8*, e54838. [CrossRef] [PubMed]

61. Shimomura, K.; Moriguchi, Y.; Ando, W.; Nansai, R.; Fujie, H.; Hart, D.A.; Gobbi, A.; Kita, K.; Horibe, S.; Shino, K.; et al. Osteochondral repair using a scaffold-free tissue-engineered construct derived from synovial mesenchymal stem cells and a hydroxyapatite-based artificial bone. *Tissue Eng. Part A* **2014**, *20*, 2291–2304. [CrossRef] [PubMed]

62. Dabiri, Y.; Li, L.P. Influences of the depth-dependent material inhomogeneity of articular cartilage on the fluid pressurization in the human knee. *Med. Eng. Phys.* **2013**, *35*, 1591–1598. [CrossRef] [PubMed]

63. Pierce, D.M.; Ricken, T.; Holzapfel, G.A. Modeling sample/patient-specific structural and diffusional responses of cartilage using DT-MRI. *Int. J. Numer. Methods Biomed. Eng.* **2013**, *29*, 807–821. [CrossRef] [PubMed]

64. Chang, D.P.; Guilak, F.; Jay, G.D.; Zauscher, S. Interaction of lubricin with type II collagen surfaces: Adsorption, friction, and normal forces. *J. Biomech.* **2014**, *47*, 659–666. [CrossRef] [PubMed]

65. Greene, G.W.; Olszewska, A.; Osterberg, M.; Zhu, H.; Horn, R. A cartilage-inspired lubrication system. *Soft Matter* **2014**, *10*, 374–382. [CrossRef] [PubMed]

66. Laurenti, K.C.; de Albuquerque Haach, L.C.; dos Santos, A.R., Jr.; de Almeida Rollo, J.M.D.; Bezerra de Menezes Reiff, R.; Minarelli Gaspar, A.M.; de Moraes Purquerio, B.; Fortulan, C.A. Cartilage reconstruction using self-anchoring implant with functional gradient. *Mater. Res.* **2014**, *17*, 638–649. [CrossRef]

67. Utzschneider, S.; Lorber, V.; Dedic, M.; Paulus, A.C.; Schröder, C.; Gottschalk, O.; Schmitt-Sody, M.; Jansson, V. Biological activity and migration of wear particles in the knee joint: An in vivo comparison of six different polyethylene materials. *J. Mater. Sci. Mater. Med.* **2014**, *25*, 1599–1612. [CrossRef] [PubMed]

68. Nürnberger, S.; Meyer, C.; Ponomarev, I.; Barnewitz, D.; Resinger, C.; Klepal, W.; Albrecht, C.; Marlovits, S. Equine articular chondrocytes on MACT scaffolds for cartilage defect treatment. *Anat. Histol. Embryol.* **2013**, *42*, 332–343. [CrossRef] [PubMed]

69. Vahdati, A.; Wagner, D.R. Implant size and mechanical properties influence the failure of the adhesive bond between cartilage implants and native tissue in a finite element analysis. *J. Biomech.* **2013**, *46*, 1554–1560. [CrossRef] [PubMed]

70. Bulman, S.E.; Coleman, C.M.; Murphy, J.M.; Medcalf, N.; Ryan, A.E.; Barry, F. Pullulan: A new cytoadhesive for cell-mediated cartilage repair. *Stem Cell Res. Ther.* **2015**, *6*, 34. [CrossRef] [PubMed]

71. Vikingsson, L.; Gómez-Tejedor, J.A.; Gallego Ferrer, G.; Gómez Ribelles, J.L. An experimental fatigue study of a porous scaffold for the regeneration of articular cartilage. *J. Biomech.* **2015**, *48*, 1310–1317. [CrossRef] [PubMed]

72. Kharkar, P.M.; Kiick, K.L.; Kloxin, A.M. Designing degradable hydrogels for orthogonal control of cell microenvironments. *Chem. Soc. Rev.* **2013**, *42*, 7335–7372. [CrossRef] [PubMed]

73. Delighkaris, K.; Tadele, T.S.; Olthuis, W.; van den Berg, A. Hydrogel-based devices for biomedical applications. *Sens. Actuators B Chem.* **2010**, *147*, 765–774. [CrossRef]

74. Kim, I.L.; Mauck, R.L.; Burdick, J.A. Hydrogel design for cartilage tissue engineering: A case study with hyaluronic acid. *Biomaterials* **2011**, *32*, 8771–8782. [CrossRef] [PubMed]

75. Martins, E.A.N.; Michelacci, Y.M.; Baccarin, R.Y.A.; Cogliati, B.; Silva, L.S.L.C. Evaluation of chitosan-GP hydrogel biocompatibility in osteochondral defects: An experimental approach. *BMC Vet. Res.* **2014**, *10*, 197. [CrossRef] [PubMed]

76. Kazusa, H.; Nakasa, T.; Shibuya, H.; Ohkawa, S.; Kamei, G.; Adachi, N.; Deie, M.; Nakajima, N.; Hyon, S.H.; Ochi, M. Strong adhesiveness of a new biodegradable hydrogel glue, LYDEX, for use on articular cartilage. *J. Appl. Biomater. Funct. Mater.* **2013**, *11*, 180–186. [CrossRef] [PubMed]

77. Mesallati, T.; Buckley, C.T.; Kelly, D.J. Engineering articular cartilage-like grafts by self-assembly of infrapatellar fat pad-derived stem cells. *Biotechnol. Bioeng.* **2014**, *111*, 1686–1698. [CrossRef] [PubMed]

78. Bhat, S.; Lidgren, L.; Kumar, A. In vitro neo-cartilage formation on a three-dimensional composite polymeric cryogel matrix. *Macromol. Biosci.* **2013**, *13*, 827–837. [CrossRef] [PubMed]

79. Foss, C.; Merzari, E.; Migliaresi, C.; Motta, A. Silk fibroin/hyaluronic acid 3D matrices for cartilage tissue engineering. *Biomacromolecules* **2013**, *14*, 38–47. [CrossRef] [PubMed]

80. Boere, K.W.; Visser, J.; Seyednejad, H.; Rahimian, S.; Gawlitta, D.; van Steenbergen, M.J.; Dhert, W.J.; Hennink, W.E.; Vermonden, T.; Malda, J. Covalent attachment of a three-dimensionally printed thermoplast to a gelatin hydrogel for mechanically enhanced cartilage constructs. *Acta Biomater.* **2014**, *10*, 2602–2611. [CrossRef] [PubMed]

81. Parmar, P.A.; Chow, L.W.; St-Pierre, J.P.; Horejs, C.M.; Peng, Y.Y.; Werkmeister, J.A.; Ramshaw, J.A.; Stevens, M.M.; Paresh, A.; Parmar, L.W.C.; et al. Collagen-mimetic peptide-modifiable hydrogels for articular cartilage regeneration. *Biomaterials* **2015**, *54*, 213–225. [CrossRef] [PubMed]

82. Omobono, M.A.; Zhao, X.; Furlong, M.A.; Kwon, C.H.; Gill, T.J.; Randolph, M.A.; Redmond, R.W. Enhancing the stiffness of collagen hydrogels for delivery of encapsulated chondrocytes to articular lesions for cartilage regeneration. *J. Biomed. Mater. Res. Part A* **2015**, *103A*, 1332–1338. [CrossRef] [PubMed]

83. Liao, I.C.; Moutos, F.T.; Estes, B.T.; Zhao, X.; Guilak, F. Composite three-dimensional woven scaffolds with interpenetrating network hydrogels to create functional synthetic articular cartilage. *Adv. Funct. Mater.* **2013**, *23*, 5833–5839. [CrossRef] [PubMed]

84. Antonioli, E.; Lobo, A.O.; Ferretti, M.; Cohen, M.; Marciano, F.R.; Corat, E.J.; Trava-Airoldi, V.J. An evaluation of chondrocyte morphology and gene expression on superhydrophilic vertically-aligned multi-walled carbon nanotube films. *Mater. Sci. Eng. C* **2013**, *33*, 641–647. [CrossRef] [PubMed]

85. Liu, M.; Ishida, Y.; Ebina, Y.; Sasaki, T.; Hikima, T.; Takata, M.; Aida, T. An anisotropic hydrogel with electrostatic repulsion between cofacially aligned nanosheets. *Nature* **2015**, *517*, 68–72. [CrossRef] [PubMed]

86. Leone, G.; Bidini, A.B.; Lamponi, S.; Magnani, A. States of water, surface and rheological characterisation of a new biohydrogel as articular cartilage substitute. *Polym. Adv. Technol.* **2013**, *24*, 824–833. [CrossRef]

87. Drira, Z.; Yadavalli, V.K. Nanomechanical measurements of polyethyleneglycol hydrogels using atomic force microscopy. *J. Mech. Behav. Biomed. Mater.* **2013**, *18*, 20–28. [CrossRef] [PubMed]

88. Ronken, S.; Wirz, D.; Daniels, A.U.; Kurokawa, T.; Gong, J.P.; Arnold, M.P. Double-network acrylamide hydrogel compositions adapted to achieve cartilage-like dynamic stiffness. *Biomech. Model. Mechanobiol.* **2013**, *12*, 243–248. [CrossRef] [PubMed]

89. Snyder, T.N.; Madhavan, K.; Intrator, M.; Dregalla, R.C.; Park, D. A fibrin/hyaluronic acid hydrogel for the delivery of mesenchymal stem cells and potential for articular cartilage repair. *J. Biol. Eng.* **2014**, *8*, 10. [CrossRef] [PubMed]

90. Ahearne, M.; Liu, Y.; Kelly, D.J. Combining freshly isolated chondroprogenitor cells from the infrapatellar fat pad with a growth factor delivery hydrogel as a putative single stage therapy for articular cartilage repair. *Tissue Eng. Part A* **2014**, *20*, 930–939. [CrossRef] [PubMed]

91. Mesallati, T.; Buckley, C.T.; Kelly, D.J. A comparison of self-assembly and hydrogel encapsulation as a means to engineer functional cartilaginous grafts using culture expanded chondrocytes. *Tissue Eng. Part C* **2014**, *20*, 52–63. [CrossRef] [PubMed]

92. Ramesh, S.; Rajagopal, K.; Vaikkath, D.; Nair, P.D.; Madhuri, V. Enhanced encapsulation of chondrocytes within a chitosan/ hyaluronic acid hydrogel: A new technique. *Biotechnol. Lett.* **2014**, *36*, 1107–1111. [CrossRef] [PubMed]

93. Ha, C.W.; Park, Y.B.; Chung, J.Y.; Park, Y.G. Cartilage repair using composites of human umbilical cord blood-derived mesenchymal stem cells and hyaluronic acid hydrogel in a minipig model. *Stem Cells Transl. Med.* **2015**, *4*, 1–8. [CrossRef] [PubMed]

94. Zeng, L.; Chen, X.; Zhang, Q.; Yu, F.; Li, Y.; Yao, Y. Redifferentiation of dedifferentiated chondrocytes in a novel three-dimensional microcavitary hydrogel. *J. Biomed. Mater. Res. Part A* **2015**, *103A*, 1693–1702. [CrossRef] [PubMed]

95. Murakami, T.; Sakai, N.; Yamaguchi, T.; Yarimitsu, S.; Nakashima, K.; Sawae, Y.; Suzuki, A. Evaluation of a superior lubrication mechanism with biphasic hydrogels for artificial cartilage. *Tribol. Int.* **2015**, *89*, 19–26. [CrossRef]

96. Chen, K.; Zhang, D.; Cui, X.; Wang, Q. Research on swing friction lubrication mechanisms and the fluid load support characteristics of PVA–HA composite hydrogel. *Tribol. Int.* **2015**, *90*, 412–419. [CrossRef]

97. Baykal, D.; Underwood, R.J.; Mansmann, K.; Marcolongo, M.; Kurtz, S.M. Evaluation of friction properties of hydrogels based on a biphasic cartilage model. *J. Mech. Behav. Biomed. Mater.* **2013**, *28*, 263–273. [CrossRef] [PubMed]

98. Guo, Y.; Guo, J.; Bai, D.; Wang, H.; Zheng, X.; Guo, W.; Tian, W. Hemiarthroplasty of the shoulder joint using a custom-designed high-density nano-hydroxyapatite/polyamide prosthesiswith a polyvinyl alcohol hydrogel humeral head surface in rabbits. *Artif. Organs* **2014**, *38*, 580–586. [CrossRef] [PubMed]

99. Chiang, C.-W.; Chen, W.-C.; Liu, H.-W.; Chen, C.-H. Application of synovial fluid mesenchymal stem cells: Platelet-rich plasma hydrogel for focal cartilage defect. *J. Exp. Clin. Med.* **2014**, *6*, 118–124. [CrossRef]

100. Sridhar, B.V.; Brock, J.L.; Silver, J.S.; Leight, J.L.; Randolph, M.A.; Anseth, K.S. Development of a cellularly degradable PEG hydrogel to promote articular cartilage extracellular matrix deposition. *Adv. Healthc. Mater.* **2015**, *4*, 702–713. [CrossRef] [PubMed]

101. Smeriglio, P.; Lai, J.H.; Dhulipala, L.; Behn, A.W.; Goodman, S.B.; Smith, R.L.; Maloney, W.J.; Yang, F.; Bhutani, N. Comparative potential of juvenile and adult human articular chondrocytes for cartilage tissue formation in three-dimensional biomimetic hydrogels. *Tissue Eng. Part A* **2015**, *21*, 147–155. [CrossRef] [PubMed]

102. Zhao, F.; He, W.; Yan, Y.; Zhang, H.; Zhang, G.; Tian, D.; Gao, H. The application of polysaccharide biocomposites to repair cartilage defects. *Int. J. Polym. Sci.* **2014**, *2014*, 654597. [CrossRef]

103. Zeng, L.; Yao, Y.; Wang, D.A.; Chen, X. Effect of microcavity alginate hydrogel with different pore sizes on chondrocyte culture for cartilage tissue engineering. *Mater. Sci. Eng. C* **2014**, *34*, 168–175. [CrossRef] [PubMed]

104. Rackwitz, L.; Djouad, F.; Janjanin, S.; Nöth, U.; Tuan, R.S. Functional cartilage repair capacity of de-differentiated, chondrocyteand mesenchymal stem cell-laden hydrogels in vitro. *Osteoarthr. Cartil.* **2014**, *22*, 1148–1157. [CrossRef] [PubMed]

105. Ponnurangam, S.; O'Connell, G.D.; Chernyshova, I.V.; Wood, K.; Hung, C.T.; Somasundaran, P. Beneficial effects of cerium oxide nanoparticles in development of chondrocyte-seeded hydrogel constructs and cellular response to interleukin insults. *Tissue Eng. Part A* **2014**, *20*, 2908–2919. [CrossRef] [PubMed]

106. Dashtdar, H.; Murali, M.R.; Abbas, A.A.; Suhaeb, A.M.; Selvaratnam, L.; Tay, L.X.; Kamarul, T. PVA-chitosan composite hydrogel versus alginate beads as a potential mesenchymal stem cell carrier for the treatment of focal cartilage defects. *Knee Surg. Sports Traumatol. Arthrosc.* **2015**, *23*, 1368–1377. [CrossRef] [PubMed]

107. Choi, B.; Kim, S.; Fan, J.; Kowalski, T.; Petrigliano, F.; Evseenko, D.; Lee, M. Covalently conjugated transforming growth factor-β1 in modular chitosan hydrogels for the effective treatment of articular cartilage defects. *Biomater. Sci.* **2015**, *3*, 742. [CrossRef] [PubMed]

108. Puppi, D.; Chiellini, F.; Piras, A.M.; Chiellini, E. Polymeric materials for bone and cartilage repair. *Prog. Polym. Sci.* **2010**, *35*, 403–440. [CrossRef]

109. Tan, H.; Marra, K.G. Injectable, biodegradable hydrogels for tissue engineering applications. *Materials* **2010**, *3*, 1746–1767. [CrossRef]

110. Seol, D.; Magnetta, M.J.; Ramakrishnan, P.S.; Kurriger, G.L.; Choe, H.; Jang, K.; Martin, J.A.; Lim, T.H. Biocompatibility and preclinical feasibility tests of a temperature-sensitive hydrogel for the purpose of surgical wound pain control and cartilage repair. *J. Biomed. Mater. Res. Part B* **2013**, *101B*, 1508–1515. [CrossRef] [PubMed]

111. Petit, A.; Sandker, M.; Müller, B.; Meyboom, R.; van Midwoud, P.; Bruin, P.; Redout, E.M.; Versluijs-Helder, M.; van der Lest, C.H.; Buwalda, S.J.; et al. Release behavior and intra-articular biocompatibility of celecoxib-loaded acetyl-capped PCLA-PEG-PCLA thermogels. *Biomaterials* **2014**, *35*, 7919–7928. [CrossRef] [PubMed]

112. Muzzarelli, R.A.A.; Greco, F.; Busilacchi, A.; Sollazzo, V.; Gigante, A. Chitosan, hyaluronan and chondroitin sulfate in tissue engineering for cartilage regeneration: A review. *Carbohydr. Polym.* **2012**, *89*, 723–739. [CrossRef] [PubMed]

113. Tamaddon, M.; Walton, R.S.; Brand, D.D.; Czernuszka, J.T. Characterisation of freeze-dried type ii collagen and chondroitin sulfate scaffolds. *J. Mater. Sci. Mater. Med.* **2013**, *24*, 1153–1165. [CrossRef] [PubMed]

114. Park, H.; Lee, K.Y. Cartilage regeneration using biodegradable oxidized alginate/hyaluronate hydrogels. *J. Biomed. Mater. Res. Part A* **2014**, *102A*, 4519–4525. [CrossRef] [PubMed]

115. Walker, K.J.; Madihally, S.V. Anisotropic temperature sensitive chitosan-based injectable hydrogels mimicking cartilage matrix. *J. Biomed. Mater. Res. Part B* **2015**, *103B*, 1149–1160. [CrossRef] [PubMed]

116. Sheehy, E.J.; Mesallati, T.; Vinardell, T.; Kelly, D.J. Engineering cartilage or endochondral bone: A comparison of different naturally derived hydrogels. *Acta Biomater.* **2015**, *13*, 245–253. [CrossRef] [PubMed]

117. Kaderli, S.; Viguier, E.; Watrelot-Virieux, D.; Roger, T.; Gurny, R.; Scapozza, L.; Möller, M.; Boulocher, C.; Jordan, O. Efficacy study of two novel hyaluronic acid-based formulations for viscosupplementation therapy in an early osteoarthrosic rabbit model. *Eur. J. Pharm. Biopharm.* **2015**, *96*, 388–395. [CrossRef] [PubMed]

118. Ren, C.D.; Gao, S.; Kurisawa, M.; Ying, J.Y. Cartilage synthesis in hyaluronic acid–tyramine constructs. *J. Mater. Chem. B* **2015**, *3*, 1942. [CrossRef]

119. Kim, J.; Lin, B.; Kim, S.; Choi, B.; Evseenko, D.; Lee, M. TGF-β1 conjugated chitosan collagen hydrogels induce chondrogenic differentiation of human synovium-derived stem cells. *J. Biol. Eng.* **2015**, *9*, 1. [CrossRef] [PubMed]

120. Jovanović, Z.; Radosavljević, A.; Kačarević-Popović, Z.; Stojkovska, J.; Perić-Grujić, A.; Ristić, M.; Matić, I.Z.; Juranić, Z.D.; Obradovic, B.; Mišković-Stanković, V. Bioreactor validation and biocompatibility of Ag/poly(n-vinyl-2-pyrrolidone) hydrogel nanocomposites. *Colloids Surf. B Biointerfaces* **2013**, *105*, 230–235. [CrossRef] [PubMed]

121. Dua, R.; Centeno, J.; Ramaswamy, S. Augmentation of engineered cartilage to bone integration using hydroxyapatite. *J. Biomed. Mater. Res. Part B* **2014**, *102 B*, 922–932. [CrossRef] [PubMed]

122. Pan, Y.; Shen, Q.; Pan, C.; Wang, J. Compressive mechanical characteristics of multi-layered gradient hydroxyapatite reinforced poly (vinyl alcohol) gel biomaterial. *J. Mater. Sci. Technol.* **2013**, *29*, 551–556. [CrossRef]

123. Yodmuang, S.; McNamara, S.L.; Nover, A.B.; Mandal, B.B.; Agarwal, M.; Kelly, T.A.; Chao, P.H.; Hung, C.; Kaplan, D.L.; Vunjak-Novakovic, G. Silk microfiber-reinforced silk hydrogel composites for functional cartilage tissue repair. *Acta Biomater.* **2015**, *11*, 27–36. [CrossRef] [PubMed]

124. Parkes, M.; Myant, C.; Dini, D.; Cann, P. Tribology-optimised silk protein hydrogels for articular cartilage repair. *Tribol. Int.* **2015**, *89*, 9–18. [CrossRef]

125. Su, R.S.; Kim, Y.; Liu, J.C. Resilin: Protein-based elastomeric biomaterials. *Acta Biomater.* **2014**, *10*, 1601–1611. [CrossRef] [PubMed]

126. Kuo, C.Y.; Chen, C.H.; Hsiao, C.Y.; Chen, J.P. Incorporation of chitosan in biomimetic gelatin/chondroitin-6-sulfate/hyaluronan cryogel for cartilage tissue engineering. *Carbohydr. Polym.* **2015**, *117*, 722–730. [CrossRef] [PubMed]

127. Zhao, L.; Gwon, H.J.; Lim, Y.M.; Nho, Y.C.; Kim, S.Y. Hyaluronic acid/chondroitin sulfate-based hydrogel prepared by gamma irradiation technique. *Carbohydr. Polym.* **2014**, *102*, 598–605. [CrossRef] [PubMed]

128. Ni, Y.; Tang, Z.; Cao, W.; Lin, H.; Fan, Y.; Guo, L.; Zhang, X. Tough and elastic hydrogel of hyaluronic acid and chondroitin sulfate as potential cell scaffold materials. *Int. J. Biol. Macromol.* **2015**, *74*, 367–375. [CrossRef] [PubMed]

129. Park, H.; Choi, B.; Hu, J.; Lee, M. Injectable chitosan hyaluronic acid hydrogels for cartilage tissue engineering. *Acta Biomater.* **2013**, *9*, 4779–4789. [CrossRef] [PubMed]

130. Yang, R.; Tan, L.; Cen, L.; Zhang, Z. An injectable scaffold based on crosslinked hyaluronic acid gel for tissue regeneration. *RSC Adv.* **2016**, *6*, 16838. [CrossRef]

131. Fiorica, C.; Palumbo, F.S.; Pitarresi, G.; Gulino, A.; Agnello, S.; Giammona, G. Injectable in situ forming hydrogels based on natural and synthetic polymers for potential application in cartilage repair. *RSC Adv.* **2015**, *5*, 19715. [CrossRef]

132. Matsuzaki, T.; Matsushita, T.; Tabata, Y.; Saito, T.; Matsumoto, T.; Nagai, K.; Kuroda, R.; Kurosaka, M. Intra-articular administration of gelatin hydrogels incorporating rapamycinemicelles reduces the development of experimental osteoarthritis in a murine model. *Biomaterials* **2014**, *35*, 9904–9911. [CrossRef] [PubMed]

133. Nims, R.J.; Cigan, A.D.; Albro, M.B.; Hung, C.T.; Ateshian, G.A. Synthesis rates and binding kinetics of matrix products in engineered cartilage constructs using chondrocyte-seeded agarose gels. *J. Biomech.* **2014**, *47*, 2165–2172. [CrossRef] [PubMed]

134. Sun, J.; Tan, H. Alginate-based biomaterials for regenerative medicine applications. *Materials* **2013**, *6*, 1285–1309. [CrossRef] [PubMed]

135. Shi, Y.; Xiong, D.S.; Peng, Y.; Wang, N. Effects of polymerization degree on recovery behavior of PVA/PVP hydrogels as potential articular cartilage prosthesis after fatigue test. *eXPRESS Polym. Lett.* **2016**, *10*, 125–138. [CrossRef]

136. Bichara, D.A.; Bodugoz-Sentruk, H.; Ling, D.; Malchau, E.; Bragdon, C.R.; Muratoglu, O.K. Osteochondral defect repair using a polyvinyl alcohol-polyacrylic acid (PVA-PAAc) hydrogel. *Biomed. Mater.* **2014**, *9*. [CrossRef] [PubMed]

137. Yang, H.Y.; van Ee, R.J.; Timmer, K.; Craenmehr, E.G.M.; Huang, J.H.; Öner, F.C.; Dhert, W.J.A.; Kragten, A.H.M.; Willems, N.; Grinwis, G.C.M.; et al. A novel injectable thermoresponsive and cytocompatible gel of poly(N-isopropylacrylamide) with layered double hydroxides facilitates siRNA delivery into chondrocytes in 3D culture. *Acta Biomater.* **2015**, *23*, 214–228. [CrossRef] [PubMed]

138. Yang, S.S.; Choi, W.H.; Song, B.R.; Jin, H.; Lee, S.J.; Lee, S.H.; Lee, J.; Kim, Y.J.; Park, S.R.; Park, S.-H.; et al. Fabrication of an osteochondral graft with using a solid freeform fabrication system. *Tissue Eng. Regen. Med.* **2015**, *12*, 239–248. [CrossRef]

139. Kim, J.E.; Kim, S.H.; Jung, Y. In situ chondrogenic differentiation of bone marrow stromal cells in bioactive self-assembled peptide gels. *J. Biosci. Bioeng.* **2015**, *120*, 91–98. [CrossRef] [PubMed]

140. Park, Y.B.; Song, M.; Lee, C.H.; Kim, J.A.; Ha, C.W. Cartilage repair by human umbilical cord blood-derived mesenchymal stem cells with different hydrogels in a rat model. *J. Orthop. Res.* **2015**, 1580–1586. [CrossRef] [PubMed]

141. Mao, H.; Kawazoe, N.; Chen, G. Cellular uptake of single-walled carbon nanotubes in 3D extracellular matrix-mimetic composite collagen hydrogels. *J. Nanosci. Nanotechnol.* **2014**, *14*, 2487–2492. [CrossRef] [PubMed]

142. Aberle, T.; Franke, K.; Rist, E.; Benz, K.; Schlosshauer, B. Cell-type specific four-component hydrogel. *PLoS ONE* **2014**, *9*, e86740. [CrossRef] [PubMed]

143. Chung, J.Y.; Song, M.; Ha, C.W.; Kim, J.A.; Lee, C.H.; Park, Y.B. Comparison of articular cartilage repair with different hydrogel-human umbilical cord blood-derived mesenchymal stem cell composites in a rat model. *Stem Cell Res. Ther.* **2014**, *5*, 39. [CrossRef] [PubMed]

144. Salamon, A.; van Vlierberghe, S.; van Nieuwenhove, I.; Baudisch, F.; Graulus, G.-J.; Benecke, V.; Alberti, K.; Neumann, H.-G.; Rychly, J.; Martins, J.C.; et al. Gelatin-based hydrogels promote chondrogenic differentiation of human adipose tissue-derived mesenchymal stem cells in vitro. *Materials* **2014**, *7*, 1342–1359. [CrossRef] [PubMed]

145. Fisher, M.B.; Henning, E.A.; Söegaard, N.B.; Dodge, G.R.; Steinberg, D.R.; Mauck, R.L. Maximizing cartilage formation and integration via a trajectory-based tissue engineering approach. *Biomaterials* **2014**, *35*, 2140–2148. [CrossRef] [PubMed]

146. Song, K.; Li, L.; Li, W.; Zhu, Y.; Jiao, Z.; Lim, M.; Fang, M.; Shi, F.; Wang, L.; Liu, T. Three-dimensional dynamic fabrication of engineered cartilage based on chitosan/gelatin hybrid hydrogel scaffold in a spinner flask with a special designed steel frame. *Mater. Sci. Eng. C* **2015**, *55*, 384–392. [CrossRef] [PubMed]

147. Mhanna, R.; Kashyap, A.; Palazzolo, G.; Vallmajo-Martin, Q.; Becher, J.; Möller, S.; Schnabelrauch, M.; Zenobi-Wong, M. Chondrocyte culture in three dimensional alginate sulfate hydrogels promotes proliferation while maintaining expression of chondrogenic markers. *Tissue Eng. A* **2014**, *20*, 1454–1464. [CrossRef] [PubMed]

148. Kesti, M.; Müller, M.; Becher, J.; Schnabelrauch, M.; D'Este, M.; Eglin, D.; Zenobi-Wong, M. A versatile bioink for three-dimensional printing of cellular scaffolds based on thermally and photo-triggered tandem gelation. *Acta Biomater.* **2015**, *11*, 162–172. [CrossRef] [PubMed]

149. Ishikawa, M.; Yoshioka, K.; Urano, K.; Tanaka, Y.; Hatanaka, T.; Nii, A. Biocompatibility of cross-linked hyaluronate (Gel-200) for the treatment of knee osteoarthritis. *Osteoarthr. Cartil.* **2014**, *22*, 1902–1909. [CrossRef] [PubMed]

150. Mazaki, T.; Shiozaki, Y.; Yamane, K.; Yoshida, A.; Nakamura, M.; Yoshida, Y.; Zhou, D.; Kitajima, T.; Tanaka, M.; Ito, Y.; et al. A novel, visible light-induced, rapidly cross-linkable gelatin scaffold for osteochondral tissue engineering. *Sci. Rep.* **2014**, *4*, 4457. [CrossRef] [PubMed]

151. Zhao, M.; Chen, Z.; Liu, K.; Wan, Y.Q.; Li, X.D.; Luo, X.W.; Bai, Y.G.; Yang, Z.L.; Feng, G. Repair of articular cartilage defects in rabbits through tissue-engineered cartilage constructed with chitosan hydrogel and chondrocytes. *J. Zhejiang Univ.-Sci. B (Biomed. Biotechnol.)* **2015**, *16*, 914–923. [CrossRef] [PubMed]

152. Kyomoto, M.; Moro, T.; Yamane, S.; Hashimoto, M.; Takatori, Y.; Ishihara, K. Effect of UV-irradiation intensity on graft polymerization of 2-methacryloyloxyethyl phosphorylcholine on orthopedic bearing substrate. *Mater. Res. Part A* **2014**, *102A*, 3012–3023. [CrossRef] [PubMed]

153. Sardinha, V.M.; Lima, L.L.; Belangero, W.D.; Zavaglia, C.A.; Bavaresco, V.P.; Gomes, J.R. Tribological characterization of polyvinyl alcohol hydrogel as substitute of articular cartilage. *Wear* **2013**, *301*, 218–225. [CrossRef]

154. Shi, Y.; Xiong, D. Microstructure and friction properties of PVA/PVP hydrogels for articular cartilage repair as function of polymerization degree and polymer concentration. *Wear* **2013**, *35*, 280–285. [CrossRef]

155. Chen, T.; Hilton, M.J.; Brown, E.B.; Zuscik, M.J.; Awad, H.A. Engineering superficial zone features in tissue engineered cartilage. *Biotechnol. Bioeng.* **2013**, *110*, 1476–1486. [CrossRef] [PubMed]

156. Blum, M.M.; Ovaert, T.C. Low friction hydrogel for articular cartilage repair: Evaluation of mechanical and tribological properties in comparison with natural cartilage tissue. *Mater. Sci. Eng. C* **2013**, *33*, 4377–4383. [CrossRef] [PubMed]

157. Fan, C.; Liao, L.; Zhang, C.; Liu, L. A tough double network hydrogel for cartilage tissue engineering. *J. Mater. Chem. B* **2013**, *1*, 4251–4258. [CrossRef]

158. Blum, M.M.; Ovaert, T.C. Investigation of friction and surface degradation of innovative boundary lubricant functionalized hydrogel material for use as artificial articular cartilage. *Wear* **2013**, *301*, 201–209. [CrossRef]

159. Giavaresi, G.; Bondioli, E.; Melandri, D.; Giardino, R.; Tschon, M.; Torricelli, P.; Cenacchi, G.; Rotini, R.; Castagna, A.; Veronesi, F.; et al. Response of human chondrocytes and mesenchymal stromal cells to a decellularized human dermis. *BMC Musculoskelet. Disord.* **2013**, *14*, 12. [CrossRef] [PubMed]

160. Stocco, E.; Barbon, S.; Dalzoppo, D.; Lora, S.; Sartore, L.; Folin, M.; Parnigotto, P.P.; Grandi, C. Tailored PVA/ECM scaffolds for cartilage regeneration. *Biomed. Res. Int.* **2014**, *2014*, 762189. [CrossRef] [PubMed]

161. Baek, J.H.; Kim, K.; Yang, S.S.; Park, S.H.; Song, B.R.; Yun, H.W.; Jeong, S.I.; Kim, Y.J.; Min, B.H.; Kim, M.S. Preparation of extracellular matrix developed using porcine articular cartilage and in vitro feasibility study of porcine articular cartilage as an anti-adhesive film. *Materials* **2016**, *9*, 49. [CrossRef] [PubMed]

162. Chen, X.; Zhang, F.; He, X.; Xu, Y.; Yang, Z.; Chen, L.; Zhou, S.; Yang, Y.; Zhou, Z.; Sheng, W.; et al. Chondrogenic differentiation of umbilical cord-derived mesenchymal stem cells in type I collagen-hydrogel for cartilage engineering. *Inj. Int. J. Care Inj.* **2013**, *44*, 540–549. [CrossRef] [PubMed]

163. Wu, L.; Gonzalez, S.; Shah, S.; Kyupelyan, L.; Petrigliano, F.A.; McAllister, D.R.; Adams, J.S.; Karperien, M.; Tuan, T.L.; Benya, P.D.; et al. Extracellular matrix domain formation as an indicator of chondrocyte dedifferentiation and hypertrophy. *Tissue Eng. Part C* **2014**, *20*, 160–168. [CrossRef] [PubMed]

164. Lam, J.; Lu, S.; Lee, E.J.; Trachtenberg, J.E.; Meretoja, V.V.; Dahlin, R.L.; van den Beucken, J.J.; Tabata, Y.; Wong, M.E.; Jansen, J.A.; et al. Osteochondral defect repair using bilayered hydrogels encapsulating both chondrogenically and osteogenically pre-differentiated mesenchymal stem cells in a rabbit model. *Osteoarthr. Cartil.* **2014**, *22*, 1291–1300. [CrossRef] [PubMed]

165. Lu, H.-T.; Sheu, M.-T.; Lin, Y.-F.; Lan, J.; Chin, Y.-P.; Hsieh, M.-S.; Cheng, C.-W.; Chen, C.-H. Injectable hyaluronic-acid-doxycycline hydrogel therapy in experimental rabbit osteoarthritis. *Vet. Res.* **2013**, *9*, 68. [CrossRef] [PubMed]

166. Matsiko, A.; Gleeson, J.P.; O'Brien, F.J. Scaffold mean pore size influences mesenchymal stem cell chondrogenic differentiation and matrix deposition. *Tissue Eng. Part A* **2015**, *21*, 486–497. [CrossRef] [PubMed]

167. Hui, J.H.; Ren, X.; Afizah, M.H.; Chian, K.S.; Mikos, A.G. Oligo[poly(ethylene glycol)fumarate] hydrogel enhances osteochondral repair in porcine femoral condyle defects. *Clin. Orthop. Relat. Res.* **2013**, *471*, 1174–1185. [CrossRef] [PubMed]

168. Kwon, H.; Rainbow, R.S.; Sun, L.; Hui, C.K.; Cairns, D.M.; Preda, R.C.; Kaplan, D.L.; Zeng, L. Scaffold structure and fabrication method affect proinflammatory milieu in three-dimensional-cultured chondrocytes. *J. Biomed. Mater. Res. Part A* **2015**, *103A*, 534–544. [CrossRef] [PubMed]

169. Steele, J.A.; McCullen, S.D.; Callanan, A.; Autefage, H.; Accardi, M.A.; Dini, D.; Stevens, M.M. Combinatorial scaffold morphologies for zonal articular cartilage engineering. *Acta Biomater.* **2014**, *10*, 2065–2075. [CrossRef] [PubMed]

170. Perrier-Groult, E.; Pasdeloup, M.; Malbouyres, M.; Galéra, P.; Mallein-Gerin, F. Control of collagen production in mouse chondrocytes by using a combination of bone morphogenetic protein-2 and small interfering RNA targeting Col1a1 for hydrogel-based tissue-engineered cartilage. *Tissue Eng. Part C* **2013**, *19*, 652–664. [CrossRef] [PubMed]

171. Sridhar, B.V.; Doyle, N.R.; Randolph, M.A.; Anseth, K.S. Covalently tethered TGF-β1 with encapsulated chondrocytes in a PEG hydrogel system enhances extracellular matrix production. *J. Biomed. Mater. Res. Part A* **2014**, *102A*, 4464–4472. [CrossRef] [PubMed]

172. Simson, J.A.; Strehin, I.A.; Lu, Q.; Uy, M.O.; Elisseeff, J.H. An adhesive bone marrow scaffold and bone morphogenetic-2 protein carrier for cartilage tissue engineering. *Biomacromolecules* **2013**, *14*, 637–643. [CrossRef] [PubMed]

173. Ayerst, B.I.; Day, A.J.; Nurcombe, V.; Cool, S.M.; Merry, C.L. New strategies for cartilage regeneration exploiting selected glycosaminoglycans to enhance cell fate determination. *Biochem. Soc. Trans.* **2014**, *42*, 703–709. [CrossRef] [PubMed]

174. Ahearne, M.; Kelly, D.J. A comparison of fibrin, agarose and gellan gum hydrogels as carriers of stem cells and growth factor delivery microspheres for cartilage regeneration. *Biomed. Mater.* **2013**, *8*, 035004. [CrossRef] [PubMed]

175. Lee, W.D.; Hurtig, M.B.; Pilliar, R.M.; Stanford, W.L.; Kandel, R.A. Engineering of hyaline cartilage with a calcified zone using bone marrow stromal cells. *Osteoarthr. Cartil.* **2015**, *23*, 1307–1315. [CrossRef] [PubMed]

176. Kaderli, S.; Boulocher, C.; Pillet, E.; Watrelot-Virieux, D.; Rougemont, A.L.; Roger, T.; Viguier, E.; Gurny, R.; Scapozza, L.; Jordan, O. A novel biocompatible hyaluronic acid–chitosan hybrid hydrogel for osteoarthrosis therapy. *Int. J. Pharm.* **2015**, *483*, 158–168. [CrossRef] [PubMed]

177. Medved, F.; Gonser, P.; Lotter, O.; Albrecht, D.; Amr, A.; Schaller, H.E. Severe posttraumatic radiocarpal cartilage damage: First report of autologous chondrocyte implantation. *Arch. Orthop. Trauma Surg.* **2013**, *133*, 1469–1475. [CrossRef] [PubMed]

178. Niemietz, T.; Zass, G.; Hagmann, S.; Diederichs, S.; Gotterbarm, T.; Richter, W. Xenogeneic transplantation of articular chondrocytes into full-thickness articular cartilage defects in minipigs: Fate of cells and the role of macrophages. *Cell Tissue Res.* **2014**, *358*, 749–761. [CrossRef] [PubMed]

179. Adachi, N.; Ochi, M.; Deie, M.; Nakamae, A.; Kamei, G.; Uchio, Y.; Iwasa, J. Implantation of tissue-engineered cartilage-like tissue for the treatment for full-thickness cartilage defects of the knee. *Knee Surg. Sports Traumatol. Arthrosc.* **2014**, *22*, 1241–1248. [CrossRef] [PubMed]

180. Buwalda, S.J.; Boere, K.W.; Dijkstra, P.J.; Feijen, J.; Vermonden, T.; Hennink, W.E. Hydrogels in a historical perspective: From simple networks to smart materials. *J. Control. Release* **2014**, *190*, 254–273. [CrossRef] [PubMed]

181. Vikingsson, L.; Gallego Ferrer, G.; Gómez-Tejedor, J.A.; Gómez Ribelles, J.L. An "in vitro" experimental model to predict the mechanical behavior of macroporous scaffolds implanted in articular cartilage. *J. Mech. Behav. Biomed. Mater.* **2014**, *32*, 125–131. [CrossRef] [PubMed]

182. Yusong, P.; Qianqian, S.; Chengling, P.; Jing, W. Prediction of mechanical properties of multilayer gradient hydroxyapatite reinforced poly(vinyl alcohol) gel biomaterial. *J. Biomed. Mater. Res. Part B* **2013**, *101B*, 729–735. [CrossRef] [PubMed]

183. Pan, Y.; Xiong, D. Stress-relaxation models of nano-HA/PVA gel biocomposites. *Mech. Time-Depend. Mater.* **2013**, *17*, 195–204. [CrossRef]

184. Du, G.; Gao, G.; Hou, R.; Cheng, Y.; Chen, T.; Fu, J. Tough and fatigue resistant biomimetic hydrogels of interlaced self-assembled conjugated polymer belts with a polyelectrolyte network. *Chem. Mater.* **2014**, *26*, 3522–3529. [CrossRef]

185. Cao, Yi.; Xiong, D.; Niu, Y.; Mei, Y.; Yin, Z.; Gui, J. Compressive properties and creep resistance of a novel, porous, semidegradable poly(vinyl alcohol)/poly(lactic-*co*-glycolic acid) scaffold for articular cartilage repair. *J. Appl. Polym. Sci.* **2014**, *131*, 40311. [CrossRef]

186. Gonzalez, J.S.; Alvarez, V.A. Mechanical properties of polyvinyl alcohol/hydroxyapatite cryogel as potential artificial cartilage. *J. Mech. Behav. Biomed. Mater.* **2014**, *34*, 47–56. [CrossRef] [PubMed]

187. Chen, K.; Zhang, D.; Dai, Z.; Wang, S.; Ge, S. Research on the interstitial fluid load support characteristics and start-up friction mechanisms of PVA-HA-silk composite hydrogel. *J. Bionic Eng.* **2014**, *11*, 378–388. [CrossRef]

188. Mansour, J.M.; Gu, D.W.; Chung, C.Y.; Heebner, J.; Althans, J.; Abdalian, S.; Schluchter, M.D.; Liu, Y.; Welter, J.F. Towards the feasibility of using ultrasound to determine mechanical properties of tissues in a bioreactor. *Ann. Biomed. Eng.* **2014**, *40*, 2190–2202. [CrossRef] [PubMed]

189. Yun, A.; Lee, S.-H.; Kim, J. A phase-field model for articular cartilage regeneration in degradable scaffolds. *Bull. Math. Biol.* **2013**, *75*, 2389–2409. [CrossRef] [PubMed]

190. Sandker, M.J.; Petit, A.; Redout, E.M.; Siebelt, M.; Müller, B.; Bruin, P.; Meyboom, R.; Vermonden, T.; Hennink, W.E.; Weinans, H. In situ forming acyl-capped PCLA-PEG-PCLA triblock copolymer based hydrogels. *Biomaterials* **2013**, *34*, 8002–8011. [CrossRef] [PubMed]

191. Fukui, T.; Kitamura, N.; Kurokawa, T.; Yokota, M.; Kondo, E.; Gong, J.P.; Yasuda, K. Intra-articular administration of hyaluronic acid increases the volume of the hyaline cartilage regenerated in a large osteochondral defect by implantation of a double-network gel. *J. Mater. Sci. Mater. Med.* **2014**, *25*, 1173–1182. [CrossRef] [PubMed]

192. Haaparanta, A.M.; Järvinen, E.; Cengiz, I.F.; Ellä, V.; Kokkonen, H.T.; Kiviranta, I.; Kellomäki, M. Preparation and characterization of collagen/PLA, chitosan/PLA, and collagen/chitosan/PLA hybrid scaffolds for cartilage tissue engineering. *J. Mater. Sci. Mater. Med.* **2014**, *25*, 1129–1136. [CrossRef] [PubMed]

193. Lin, H.; Cheng, A.W.; Alexander, P.G.; Beck, A.M.; Tuan, R.S. Cartilage tissue engineering application of injectable gelatin hydrogel with in situ visible-light-activated gelation capability in both air and aqueous solution. *Tissue Eng. Part A* **2014**, *2014*, 2402–2411. [CrossRef] [PubMed]

194. Li, X.; Li, Y.; Zuo, Y.; Qu, D.; Liu, Y.; Chen, T.; Jiang, N.; Li, H.; Li, J. Osteogenesis and chondrogenesis of biomimetic integrated porous PVA/gel/V-n-HA/pa6 scaffolds and BMSCs construct in repair of articular osteochondral defect. *J. Biomed. Mater. Res. Part A* **2015**, *103A*, 3226–3236. [CrossRef] [PubMed]

195. Hoch, E.; Tovar, G.E.; Borchers, K. Biopolymer-based hydrogels for cartilage tissue engineering. *Bioinspir. Biomim. Nanobiomater.* **2016**, *5*, 51–66. [CrossRef]

196. Tuan, R.S.; Chen, A.F.; Klatt, B.A. Cartilage regeneration. *J. Am. Acad Orthop. Surg.* **2013**, *21*, 303–311. [CrossRef] [PubMed]

197. Sivashanmugam, A.; Kumar, R.A.; Priya, M.V.; Nair, S.V.; Jayakumar, R. An overview of injectable polymeric hydrogels for tissue engineering. *Eur. Polym. J.* **2015**, *72*, 543–565. [CrossRef]

198. Hu, X.; Li, D.; Zhou, F.; Gao, C. Biological hydrogel synthesized from hyaluronic acid, gelatin and chondroitin sulfate by click chemistry. *Acta Biomater.* **2011**, *7*, 1618–1626. [CrossRef] [PubMed]

199. Yang, J.-A.; Yeom, J.; Hwang, B.W.; Hoffman, A.S.; Hahn, S.K. In situ-forming injectable hydrogels for regenerative medicine. *Prog. Polym. Sci.* **2014**, *39*, 1973–1986. [CrossRef]

200. Mirahmadi, F.; Tafazzoli-Shadpour, M.; Shokrgozar, M.A.; Bonakdar, S. Enhanced mechanical properties of thermosensitive chitosan hydrogel by silk fibers for cartilage tissue engineering. *Mater. Sci. Eng. C* **2013**, *33*, 4786–4794. [CrossRef] [PubMed]

201. Moreira Teixeira, L.S.; Patterson, J.; Luyten, F.P. Skeletal tissue regeneration: Where can hydrogels play a role? *Int. Orthop.* **2014**, *38*, 1861–1876. [CrossRef] [PubMed]

202. Camci-Unal, G.; Cuttica, D.; Annabi, N.; Demarchi, D.; Khademhosseini, A. Synthesis and characterization of hybrid hyaluronic acid-gelatin hydrogels. *Biomacromolecules* **2013**, *14*, 1085–1092. [CrossRef] [PubMed]

203. Pot, M.W.; Faraj, K.A.; Adawy, A.; van Enckevort, W.J.; van Moerkerk, H.T.; Vlieg, E.; Daamen, W.F.; van Kuppevelt, T.H. Versatile wedge-based system for the construction of unidirectional collagen scaffolds by directional freezing: Practical and theoretical considerations. *Appl. Mater. Interfaces* **2015**, *7*, 8495–8505. [CrossRef] [PubMed]

204. Mao, H.; Kawazoe, N.; Chen, G. Cell response to single-walled carbon nanotubes in hybrid porous collagen sponges. *Colloids Surf. B Biointerfaces* **2015**, *126*, 63–69. [CrossRef] [PubMed]

205. Daly, A.C.; Critchley, S.E.; Rencsok, E.M.; Kelly, D.J. A comparison of different bioinks for 3D bioprinting of fibrocartilage and hyaline cartilage. *Biofabrication* **2016**, *8*, 045002. [CrossRef] [PubMed]

206. Seol, Y.J.; Park, J.Y.; Jeong, W.; Kim, T.H.; Kim, S.Y.; Cho, D.W. Development of hybrid scaffolds using ceramic and hydrogel for articular cartilage tissue regeneration. *J. Biomed. Mater. Res. Part A* **2015**, *103A*, 1404–1413. [CrossRef] [PubMed]

207. Xu, T.; Binder, K.W.; Albanna, M.Z.; Dice, D.; Zhao, W.; Yoo, J.J.; Atala, A. Hybrid printing of mechanically and biologically improved constructs for cartilage tissue engineering applications. *Biofabrication* **2013**, *5*, 015001. [CrossRef] [PubMed]

208. Vázquez-Portalatín, N.; Kilmer, C.E.; Panitch, A.; Liu, J.C. Characterization of collagen type I and II blended hydrogels for articular cartilage tissue engineering. *Biomacromolecules* **2016**, *17*, 3145–3152. [CrossRef] [PubMed]

209. Mintz, B.R.; Cooper, J.A., Jr. Hybrid hyaluronic acid hydrogel/poly(ε-caprolactone) scaffold provides mechanically favorable platform for cartilage tissue engineering studies. *J. Biomed. Mater. Res. Part A* **2014**, *102A*, 2918–2926. [CrossRef] [PubMed]

210. Sun, W.; Xue, B.; Li, Y.; Qin, M.; Wu, J.; Lu, K.; Wu, J.; Cao, Y.; Jiang, Q.; Wang, W. Polymer-supramolecular polymer double-network hydrogel. *Adv. Funct. Mater.* **2016**, *26*, 9044–9052. [CrossRef]

211. Formica, F.A.; Öztürk, E.; Hess, S.C.; Stark, W.J.; Maniura-Weber, K.; Rottmar, M.; Zenobi-Wong, M. A bioinspired ultraporous nanofiber-hydrogel mimic of the cartilage extracellular matrix. *Adv. Funct. Mater.* **2016**, *5*, 3129–3138. [CrossRef] [PubMed]

212. Kang, M.L.; Jeong, S.Y.; Im, G.I. Hyaluronic acid hydrogel functionalized with self-assembled micelles of amphiphilic PEGylated kartogenin for the treatment of osteoarthritis. *Tissue Eng. Part A* **2017**, *23*, 630–639. [CrossRef] [PubMed]

213. Huang, H.; Zhang, X.; Hu, X.; Dai, L.; Zhu, J.; Man, Z.; Chen, H.; Zhou, C.; Ao, Y. Directing chondrogenic differentiation of mesenchymal stem cells with a solid-supported chitosan thermogel for cartilage tissue engineering. *Biomed. Mater.* **2014**, *9*, 035008. [CrossRef] [PubMed]

214. Buchtová, N.; Réthoré, G.; Boyer, C.; Guicheux, J.; Rambaud, F.; Vallé, K.; Belleville, P.; Sanchez, C.; Chauvet, O.; Weiss, P.; et al. Nanocomposite hydrogels for cartilage tissue engineering: Mesoporous silica nanofibers interlinked with siloxane derived polysaccharide. *J. Mater. Sci. Mater. Med.* **2013**, *24*, 1875–1884. [CrossRef] [PubMed]

215. Wang, T.; Lai, J.H.; Yang, F. Effects of hydrogel stiffness and extracellular compositions on modulating regeneration by mixed populations of stem cells and chondrocytes in vivo. *Tissue Eng. Part A* **2016**, *22*, 1348–1356. [CrossRef] [PubMed]

216. Gyles, D.A.; Castro, L.D.; Silva, J.O.C., Jr.; Ribiero-Costa, R.M. A review of the designs and prominent biomedical advances of natural and synthetic hydrogel formulations. *Eur. Polym. J.* **2017**, *88*, 373–392. [CrossRef]

217. Jeon, S.-J.; Hauser, A.W.; Hayward, R.C. Shape-morphing materials from stimuli-responsive hydrogel hybrids. *Acc. Chem. Res.* **2017**, *50*, 161–169. [CrossRef] [PubMed]
218. Jeon, J.E.; Vaquette, C.; Theodoropoulos, C.; Klein, T.J.; Hutmacher, D.W. Multiphasic construct studied in an ectopic osteochondral defect model. *J. R. Soc. Interface* **2014**, *11*, 20140184. [CrossRef] [PubMed]
219. Childs, A.; Hemraz, U.D.; Castro, N.J.; Fenniri, H.; Zhang, L.G. Novel biologically-inspired rosette nanotube PLLA scaffolds for improving human mesenchymal stem cell chondrogenic differentiation. *Biomed. Mater.* **2013**, *8*, 065003. [CrossRef] [PubMed]
220. Mačiulaitis, J.; Deveikytė, M.; Rekštytė, S.; Bratchikov, M.; Darinskas, A.; Šimbelytė, A.; Daunoras, G.; Laurinavičienė, A.; Laurinavičius, A.; Gudas, R.; et al. Preclinical study of SZ20180 material 3D microstructured scaffolds for cartilage tissue engineering made by femtosecond direct laser writing lithography. *Biofabrication* **2015**, *7*, 015015. [CrossRef] [PubMed]
221. Heinemann, S.; Coradin, T.; Desimone, M.F. Bio-inspired silica-collagen materials: Applications and perspectives in the medical field. *Biomater. Sci.* **2013**, *1*, 688–702. [CrossRef]

polymers

MDPI

Article

Microsphere-Based Hierarchically Juxtapositioned Biphasic Scaffolds Prepared from Poly(Lactic-*co*-Glycolic Acid) and Nanohydroxyapatite for Osteochondral Tissue Engineering

K. T. Shalumon [1,*], Chialin Sheu [1], Yi Teng Fong [2], Han-Tsung Liao [2] and Jyh-Ping Chen [1,2,3,4,*]

[1] Department of Chemical and Materials Engineering, Chang Gung University, Kwei-San, Taoyuan 33302, Taiwan; chialinsheu@gmail.com
[2] Department of Plastic and Reconstructive Surgery and Craniofacial Research Center, Chang Gung Memorial Hospital, Kwei-San, Taoyuan 33305, Taiwan; evausatw@gmail.com (Y.T.F.); lia01211@gmail.com (H.-T.L.)
[3] Graduate Institute of Health Industry and Technology, Research Center for Industry of Human Ecology, Chang Gung University of Science and Technology, Kwei-San, Taoyuan 33302, Taiwan
[4] Department of Materials Engineering, Ming Chi University of Technology, Tai-Shan, New Taipei City 24301, Taiwan
*   Correspondence: shalumon@gmail.com (K.T.S.); jpchen@mail.cgu.edu.tw (J.-P.C.); Tel.: +886-3-211-8800 (ext. 3741) (K.T.S.); +886-3-211-8800 (ext. 5298) (J.-P.C.)

Academic Editor: Insung S. Choi
Received: 12 October 2016; Accepted: 5 December 2016; Published: 10 December 2016

Abstract: This study aims to prepare biphasic osteochondral scaffolds based on seamless joining of sintered polymer and polymer/ceramic microspheres for co-culture of chondrocytes and bone marrow stem cells (BMSCs). Poly(lactide-*co*-glycolide) (PLGA) microspheres and 10% nanohydroxyapatite (nHAP)-incorporated PLGA (PGA/nHAP) microspheres were prepared through the oil-in-water precipitation method. Virgin (V) and composite (C) scaffolds were prepared from 250–500 μm PLGA and PLGA/nHAP microspheres, respectively, while osteochondral (OC) scaffolds were fabricated through the combination of V and C scaffolds. Physico-chemical properties of scaffolds were characterized through microscopic-spectroscopic evaluations. The effect of nHAP in scaffolds was investigated through thermogravimetric analysis and mechanical testing, while surface hydrophobicity was tested through contact angle measurements. Rabbit chondrocytes and BMSCs were used for cell culture, and cell morphology and proliferation were determined from SEM and DNA assays. Alizarin red and Alcian blue stains were used to identify the in vitro bone and cartilage tissue-specific regeneration, while cetylpyridinium chloride was used to quantitatively estimate calcium in mineralized bone. For co-culture in OC scaffolds, BMSCs were first seeded in the bone part of the scaffold and cultured in osteogenic medium, followed by seeding chondrocytes in the cartilage part, and cultured in chondrocyte medium. High cell viability was confirmed from the Live/Dead assays. Actin cytoskeleton organization obtained by DAPI-phalloidin staining revealed proper organization of chondrocytes and BMSCs in OC scaffolds. Immunofluorescent staining of bone (type I collagen and osteocalcin (OCN)) and cartilage marker proteins (type II collagen (COL II)) confirmed cellular behavior of osteoblasts and chondrocytes in vitro. Using an ectopic osteochondral defect model by subcutaneous implantation of co-cultured OC scaffolds in nude mice confirmed cell proliferation and tissue development from gross view and SEM observation. IF staining of OCN and COL II in the bone and cartilage parts of OC scaffolds and tissue-specific histological analysis exhibited a time-dependent tissue re-modeling and confirmed the potential application of the biphasic scaffold in osteochondral tissue engineering.

Polymers **2016**, *8*, 429

Keywords: microspheres; poly(lactide-*co*-glycolide); nanohydroxyapatite; scaffold; tissue engineering; osteochondral; biphasic

---

1. Introduction

Osteochondral defects are major tissue defects caused by traumatic injury or disease, due to the damage of both articular cartilage and underlying subchondral bone [1]. Osteochondral grafting, prosthetic joint replacement, and chondroplasty are the available methods for osteochondral repair. However, donor site morbidity, abrasion, surface loosening of implants, and risk of infection put osteochondral tissue engineering as a high demand task [2]. In osteochondral tissue engineering, it provides a favorable alternative therapy to engineer a scaffold with the same structural and mechanical properties as a native cartilage–bone plug [3–5]. A suitably designed scaffold architecture is important for the individual formation of bone [6–8], cartilage [9–11] or both, in osteochondral defects, with the combinatory approach being of greatest interest owing to the difficulty in achieving the goal [12,13]. The fabrication of such biphasic scaffolds, which can control the formation of a composite bone and cartilage architecture, remains a significant challenge.

Osteochondral defect repair or regeneration requires three components; bone, cartilage, and bone–cartilage interface. Though several studies have undertaken for the better understanding of bone and cartilage tissue engineering over the years, still, osteochondral tissue repair is unlike these two separate areas. The true challenge in osteochondral tissue repair lies in the effectiveness of osteochondral interface formation, its mechanical strength, structure, and biology. Considering the structure–morphology relations, a bi-phasic or bi-layer scaffold design is more appropriate for facing such challenges. Each layer of the scaffold should be engineered to exhibit tissue-specific biophysical conditions and environments suitable to support the regeneration of two tissue types: articular cartilage and subchondral bone. Only few reports are available related to integration of bi-layered scaffolds for osteochondral tissue engineering using natural or synthetic polymers [14–18]. Bi-layered scaffolds can be made in two ways: separately engineering the bone and cartilage tissues and combining them together or engineering both tissue types in an integrated scaffold. Compared to the former one, integrated bi-layered scaffolds provide a gradual transition of bone and the cartilage within a single scaffold, thus avoiding the procedure to joint both parts before implantation. Various methods have been reported for the fabrication of osteochondral scaffolds in which a three-dimensional (3D) printing/salt leaching technique for the production of cartilage and bone regions using poly(lactic-*co*-glycolic acid)/poly(lactic acid) (PLGA/PLA) and PLGA/tricalcium phosphate was reported [19]. A two-stage crosslinking procedure composing oligo(poly(ethylene glycol) fumarate) (OPF) and OPF/gelatin microparticles in an integrated bi-layered scaffold was reported [20]. Chen et al. reported the development of a bi-layered scaffold through a combined solvent casting, salt leaching, and freeze drying method for making bi-layered scaffolds of collagen and PLGA/collagen for cartilage and bone layers, respectively [21]. However, most of the reports do not meet the requirements, such as mechanical strength, porosity, moldability, and pore interconnectivity [22], which are essential in developing the best osteochondral scaffold for tissue engineering.

Numerous biomaterials such as proteins or carbohydrate-based polymers, including collagen, hyaluronan, and chitosan, and synthetic materials such as bioactive ceramics and synthetic polymers are being investigated for interface tissue engineering [23–26]. Though natural materials are superior in their biological interaction with host tissues, their relative mechanical inferiority and instability compared with native cartilage makes them inappropriate for clinical applications. At the same time, having relatively low biocompatibility with natural materials, synthetic materials have high controllability in their chemical, mechanical, and structural properties and thereby could satisfy the increasing clinical demand. Considering their excellent osteoconductive/osteoinductive capabilities [17,27,28], synthetic bioactive ceramics such as nanohydroxyapatite (nHAP), tricalcium phosphate, silicate bioactive glass, *etc.* could be used in the bone part of the bi-layered scaffold for osteochondral tissue engineering.

The cell type used for osteochondral repair has a great influence on its overall outcome. A controlled environment is required to prevent cell differentiation into another phenotype. That is, to prevent chondrocytes from differentiating into fibroblasts or to arrest stem cell differentiation at the chondrocyte stage without further proceeding to ossification [29]. Similarly, cells in a bone compartment should be able to maintain their own phenotype, without getting disturbed by surroundings or related factors. Hence, incorporation of growth factors was put forward, to enhance the cell performance in vivo. The function of growth factors is to establish an inducing environment for cells to differentiate into a specific phenotype. Either single growth factor or a combination of growth factors was used in osteochondral tissue development. Various reports are available on the use of single or multiple growth factors in osteochondral regeneration [30–32]. However, growth factors have not been validated. Besides, due to the inability to control the release of growth factors precisely to designated parts in osteochondral scaffolds, both cell types will experience inductive effects from both growth factors and will eventually regenerate a single cell phenotype. It should be noted that incorporating growth factors or platelet-rich plasma (PRP) in the osteochondral scaffold could support angiogenesis/vascularization [33].

In this work, we are approaching with a microsphere-based PLGA-nHAP/PLGA integrated scaffold with hierarchically designed PLGA/nHAP microspheres in the bone part while with PLGA microspheres in the cartilage part. The osteo-conductive nHAP was incorporated in situ to achieve uniform distribution within the bone part of the scaffold to induce differentiation. We postulate that since nHAP could induce osteogenic differentiation of bone marrow stem cells (BMSCs) that are in direct physical contact with the bone part of the scaffold, a stem cell-based approach could be used here to first create the bone part of an osteochondral zone. Pre-differentiation of BMSCs to the osteogenic lineage was achieved through the combined effect of osteo-conductive PLGA/nHAP in osteogenic medium (OM) for successful creation of the new bone tissue. Subsequently, chondrocytes were seeded in the cartilage part, maintained in the chondrocyte medium and evaluated for osteochondral tissue regeneration in vitro and in vivo.

2. Materials and Methods

2.1. Materials

PLGA with a lactide-to-glycolide ratio of 85:15 (intrinsic viscosity 0.96 dL/g) was purchased from Green Chemical Inc., Taipei, Taiwan, while calcium hydrogen phosphate ($Ca_2HPO_4 \cdot 2H_2O$) was purchased from Showa Co., Gyoda, Japan. Calcium carbonate ($CaCO_3$), sodium hydroxide (NaOH), cetylpyridinium chloride and poly(vinyl alcohol) were obtained from Sigma-Aldrich (St. Louis, MO, USA). Dichloromethane (CH_2Cl_2) was procured from Alfa Aesar, Ward Hill, MA, USA, while Dulbecco's Modified Eagle's Medium (DMEM) and fetal bovine serum (FBS) were purchased from Invitrogen (Carlsbad, CA, USA) and Thermo Fisher Scientific (Waltham, MA, USA), respectively. Dulbecco's Modified Eagle's Medium/Nutrient Mixture F-12 (DMEM/F-12) was purchased from Sigma-Aldrich (St. Louis, MO, USA). All chemicals were used as received without further purification.

2.2. Preparation of Nanohydroxyapatite (nHAP)

The chemical precipitation method using $Ca_2HPO_4 \cdot 2H_2O$ and $CaCO_3$ was used to prepare nHAP particles, as reported in our previous studies [34]. Briefly, 0.86 g of $CaHPO_4 \cdot 2H_2O$ and 0.335 g of $CaCO_3$ were gently mixed in 2.5 M NaOH solution at 75 °C for 1 h, followed by termination of the reaction by keeping the mixture in an ice bath. The resultant solution was centrifuged, multiple washed with double distilled water, and dried at 70 °C for 24 h to obtain nHAP.

2.3. Preparation of PLGA and PLGA/nHAP Microspheres

Both PLGA and PLGA/nHAP microspheres were prepared by the oil-in-water emulsion/solvent evaporation method, as reported by Yang et al. [35]. For PLGA microspheres, 4.2 g PLGA was dissolved in 30 mL dichloromethane to get 14% PLGA solution, followed by slowly pouring into 1.6 L polyvinyl alcohol (PVA) solution (0.5%) and stirred overnight at 360 rpm. For PLGA/nHAP microspheres with 15% of nHAP, 4.2 g PLGA was mixed in 30 mL dichloromethane with 630 mg nHAP and allowed to undergo microsphere formation through overnight stirring in 0.5% PVA solution. Formed microspheres were washed with distilled water, dried under vacuum, and sorted using metallic sieves (Figure 1, step 1).

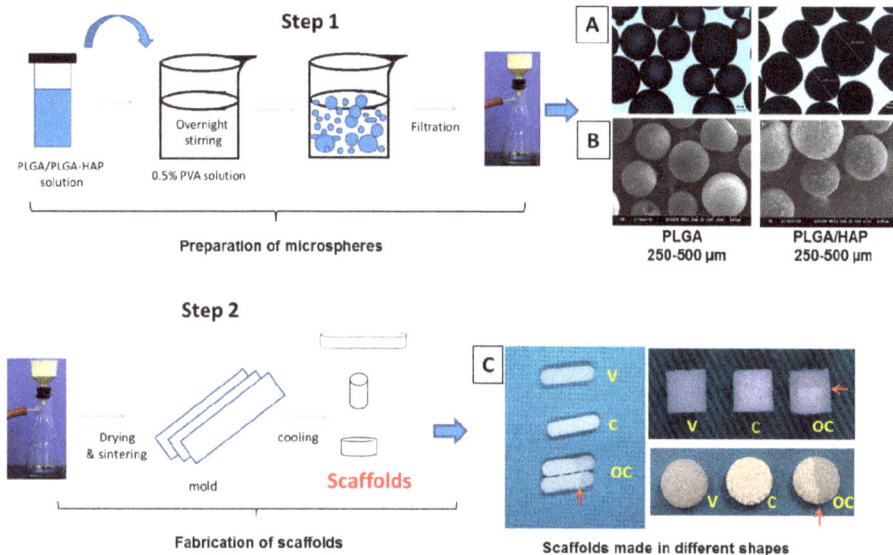

Figure 1. Schematic representation of preparation of PLGA and PLGA/nHAP (Poly(lactide-*co*-glycolide)/nanohydroxyapatite) microspheres by the oil-in-water emulsion/solvent evaporation method (Step 1) and the corresponding light microscopic (**A**) and SEM images (**B**) of microspheres in the range of 250–500 μm. Step 2 represents the sintering of microspheres to make virgin (V), composite (C) and osteochondral (OC) scaffolds of various shapes with the gross views shown in (**C**). Red arrows in (**C**) point out the interface regions in the curve-edged rectangular, disc, and cylindrical-shaped OC scaffolds with distinguishable separating zones for cartilage and bone parts, respectively.

2.4. Fabrication of Scaffolds

Selected microspheres with size ranging from 250 to 500 μm were sintered in a pre-designed stainless steel mold at 85 °C for 90 min to produce three types of scaffolds: virgin (V) scaffolds using PLGA microspheres, composite (C) scaffolds using PLGA/nHAP microspheres, and osteochondral (OC) scaffolds with hierarchically juxtapositioned V and C scaffolds in a single biphasic scaffold using PLGA and PLGA/nHAP microspheres. The scaffolds could be made into various shapes such as cylinders (4 mm diameter × 4 mm height), discs (6 mm diameter × 2 mm height), and curve-edged rectangles (18 mm length × 4 mm width × 2 mm height) (Figure 1, Step 2) to be used in various experiments.

2.5. Characterization of Scaffolds

The size and morphology of the prepared nHAP was determined using a transmission electron microscope (TEM) (JEM-2000EXII, JEOL, Tokyo, Japan) at 75 kV, whereas the crystallographic phases of nHAP were examined through an X-ray diffractometer (XRD) (D5005, Siemens AG, Munich, Germany) with a CuKa source (wavelength = 1.54056 Å), a goniometric plate, and a quartz monochromater having a scanning speed of $2°/min$ from $20°$ to $60°$. A Horiba FT-730 spectrometer (Horiba Ltd., Kyoto, Japan) was used to record the Fourier transformed infrared spectroscopy (FTIR) spectrum of nHAP at a wavenumber range of 400–4000 cm^{-1} with a resolution of 2 cm^{-1}. The shape, size, and morphology of the microspheres and sintered scaffolds were observed through an inverted microscope (IX-71, Olympus Co., Tokyo, Japan) and a scanning electron microscope (SEM) (S-3000N, Hitachi Ltd., Tokyo, Japan) at 15 kV. The FTIR spectra of the V scaffold was compared with nHAP and the OC scaffold using a Horiba FT-730 spectrometer. Further confirmation on the incorporation of nHAP in OC scaffolds was determined through XRD analysis in comparison with V scaffolds and nHAP. In addition, presence of nHAP in C and OC scaffolds was tested through elemental analysis by an energy dispersive X-ray (EDX) analyzer (EX-250, Horiba Ltd., Kyoto, Japan).

Thermal characteristics of V and C scaffolds as well as the quantitative estimation of nHAP in C scaffolds were measured through thermogravimetric analysis (TGA) using TGA 2050 from TA instruments (New Castle, DE, USA). The samples were heated at a controlled rate of 10 °C/min from 25 to 700 °C and the resultant decomposition curve was plotted as residual weight (%) vs. temperature (°C).

Mechanical stability of the prepared V, C, and OC scaffolds was tested using an ElectroForce® 5200 BioDynamic® Test Instrument from Bose (Eden Prairie, MN, USA) equipped with a 250 N load cell. All samples were prepared in cylindrical shapes with diameter and height 5 mm each. A cross head speed of 5 mm/min was applied and the result was plotted as stress (MPa) vs. strain (%).

Surface characteristics of the V, C, and OC scaffolds were analyzed using a sessile drop contact angle goniometer (FTA125, First Ten Ångstroms, Portsmouth, VA, USA). Disc-shaped scaffolds with 1 cm diameter and 2 mm thickness were fixed on the sample holder. The wettability of the scaffold was evaluated by gently dripping a double-distilled water droplet on the surface. Respective images were recorded 2 s and 30 s and the contact angle was calculated by fitting a mathematical expression to the shape of the liquid drop and then calculating the slope of the tangent to the liquid drop at the liquid–solid–vapor interface line.

2.6. In Vitro Cell Culture Studies

2.6.1. Cell Isolation and Expansion

Bone marrow stem cells (BMSCs) were isolated from New Zealand white rabbits following the procedures described previously [36]. 5×10^5 BMSCs were re-suspended in 10 mL cell culture medium (DMEM medium containing 10% FBS and 1% penicillin-streptomycin) in a T75 culture flask and maintained at 37 °C in 5% CO_2 atmosphere. The medium was changed every 2 days until reaching 80%–90% confluence. The supernatant medium was removed along with residual impurities followed by triplicate washing with phosphate-buffered saline (PBS). The adhered cells were detached using 0.05% trypsin–EDTA solution and transferred to a centrifuge tube. The tube was centrifuged for 5 min at $1000 \times g$ and 4 °C and the cell pellet was collected by removing the supernatant. The collected cell pellet was re-suspended in 10 mL cell culture medium and split into four T75 flasks. Cells at passage 3–4 were used for further studies. Chondrocytes were harvested from the knee articular cartilage of New Zealand white rabbits, as reported earlier [37]. The cells were sub-cultured in chondrocyte medium (DMEM/F12 supplemented with 10% FBS and 1% antibiotic/antimycotic) and passage two was selected for cell seeding.

2.6.2. Cell Attachment and Morphology

Biocompatibility of the scaffolds was first tested through cell attachment using BMSCs. Cells were cultured on cylindrical shaped V, C, and OC scaffolds up to 28 days. Scaffolds sterilized by UV light were taken in a 24-well culture plate and pre-wet with DMEM followed by culturing with BMSCs in cell culture medium at a seeding density of 3×10^4 cells/scaffold. Cells were seeded to scaffolds in both vertical and horizontal positions and cultured in 5% CO_2 environment at 37 °C with regular replacement of fresh medium every two days. Morphology of the cells on the scaffold surface and interior was analyzed using a SEM (S-3000N, Hitachi Ltd., Tokyo, Japan).

2.6.3. Mono-Culture with BMSCs and Chondrocytes

Qualitative assessment on the viability of adhered BMSCs and chondrocytes in C and V scaffolds were evaluated through a Live/Dead viability/cytotoxicity assay kit (Molecular Probes, Eugene, OR, USA). BMSCs were seeded in pre-wet cylindrical C scaffolds at a seeding density of 3×10^4 cells/scaffold, whereas chondrocytes were seeded in V scaffolds at the same seeding density and cultured up to 28 days. C scaffolds were cultured in osteogenic medium (OM) (DMEM with 50 μM L-ascorbic acid phosphate, 0.1 μM dexamethasone, 10 mM glycerol 2-phosphate, 1% antibiotic-antimycotic, and 10% FBS), whereas V scaffolds were cultured in chondrocyte medium (DMEM/F12 supplemented with 10% FBS and 1% antibiotic-antimycotic). The culture medium was replaced with fresh medium every 2 days and washed with PBS prior to staining. The Live/Dead staining solution was prepared by mixing 2 μM calcein AM (excitation 494 nm and emission 517 nm for live cells) with 5 μM of ethidium homodimer-1 (EthD-1) (excitation 528 nm and emission 617 nm for dead cells) in culture medium. Samples were incubated with the staining solution at 37 °C for 30 min and imaged under a Zeiss LSM 510 Meta confocal laser scanning microscope (Carl Zeiss AG, Jena, Germany).

2.6.4. Cell Proliferation

Cylindrical-shaped V, C, and OC scaffolds were sterilized by UV light for 4 h and placed in a 24-well culture plate. All scaffolds were pre-wet with DMEM followed by seeding with BMSCs at a density of 1×10^4 cells/scaffold and the cells were allowed to adhere at 37 °C for 4 h. After the incubation period, the scaffolds were transferred to a new culture plate containing 1 mL OM and placed in a 37 °C humidified 5% CO_2 incubator. The cell number was determined by DNA assay using Hoechst 33258 [38].

2.6.5. Co-Culture with BMSCs and Chondrocytes

Cylindrical-shaped OC scaffolds with nHAP-containing bone part were designed to be cultured with BMSCs followed by culturing chondrocytes in the cartilage part. Briefly, the bone part of OC scaffolds was rinsed in OM followed by seeding with BMSCs (2×10^4 cells/scaffold), and maintained in OM for 7, 14, and 21 days. At each time period, the OM was removed and the cartilage part of the OC scaffold was seeded with chondrocytes (1×10^4 cells/scaffold) and maintained for another 7, 14, and 21 days in chondrocyte medium. Thus, the OC scaffold was immersed in OM and chondrocyte medium respectively before and after chondrocyte seeding in the cartilage part. The respective morphology of BMSCs and chondrocytes in the bone and cartilage parts of OC scaffolds were monitored through SEM. Morphology of BMSCs in the bone part of OC scaffolds after each co-culture time point was further evaluated through SEM observation to verify the cell behavior in bone part with additional culture in chondrocyte medium.

2.6.6. Alizarin Red and Alcian Blue Staining

The effect of various stimulating factors in tissue-specific cell differentiation was analyzed through Alizarin red (AR) staining of the bone part and Alcian blue (AB) staining of the cartilage part. AR staining was done for both mono- and co-cultured samples, while AB staining was performed only for co-culture. In mono-culture, disc-shaped scaffolds (V, C, and OC) were seeded with BMSCs (1×10^4 cells/scaffold) and cultured in normal medium (NM, DMEM with 10% FBS and 1% antibiotic-antimycotic) and OM. Acellular scaffolds were considered as controls for comparative evaluation towards bone formation. After 14 days culture in both NM and OM, samples were rinsed thrice with PBS and fixed with glutaraldehyde solution (2.5%) for 2 h. 0.5 g Alizarin red S (ARS) was dissolved in 25 mL deionized water adjusted to pH 4.1~4.3 with ammonium hydroxide to get the AR staining solution and each scaffold was immersed in 1 mL of the same, followed by incubation for 1 h at room temperature. Excess dye was washed off after incubation using deionized water, and the presence of mineral deposition was qualitatively evaluated by observing intensity of the red color on the scaffold surface. In co-culture, OC scaffolds were seeded with 1×10^4 BMSCs in the bone part and cultured for 14 days (OM), followed by co-culturing with chondrocytes (1×10^4 cells) in the cartilage part (chondrocyte medium) for 7 and 14 days. Scaffolds were separately analyzed through AR and AB staining by immersing the whole scaffold in 1 mL staining solution, as mentioned before. After incubation and washing, the intensity of the red and blue color on the bone and cartilage part was analyzed from photographs.

2.6.7. Calcium Quantification

Calcium quantification from mineralization was measured through cetylpyridinium chloride (CPC) treatment. Scaffolds utilized for qualitative measurement of AR staining were washed with distilled deionized water and treated with 1 mL of 10% CPC solution for 1 h to chelate calcium ions. Both mono- and co-cultured samples were analyzed and the absorbance of the solution was read at 540 nm in an ELISA reader (OD_{540}) (Synergy HTX, BioTek, Winooski, VT, USA) and normalized with cell number.

2.6.8. Viability and Cytoskeletal Expression of Cells in Co-Culture

Viability of co-cultured BMSCs and chondrocytes in the bone and cartilage part of OC scaffolds were observed through confocal laser scanning microscope. Cell viability was tested through a Live/Dead viability/cytotoxicity assay kit. OC scaffolds pre-wet with DMEM was seeded with BMSCs at a seeding density of 2×10^4 cells in the bone part and maintained in OM in a 24-well plate for two different time durations, viz: 14 and 21. Further, chondrocytes were seeded in the cartilage part (1×10^4 cells) and maintained in chondrocyte medium for 7, 14, and 21 days. The scaffolds were washed with PBS, stained with the Live/Dead staining solution, and imaged for both bone and cartilage parts (excitation/emission 494/517 nm for live cells and 528/617 nm for dead cells) under a Zeiss LSM 510 Meta confocal laser scanning microscope. Cytoskeletal expression of BMSCs and chondrocytes in OC scaffolds were evaluated using phalloidin–tetramethylrhodamine B isothiocyanate (phalloidin–TRITC) (Sigma-Aldrich, St. Louis, MO, USA). The experimental design and time points for analysis were similar to Live/Dead staining. For cytoskeleton staining, cell-seeded OC scaffolds were treated with 4% formaldehyde in PBS for 20 min, followed by dehydration with ethanol and permeabilization with 0.1% Triton X-100 for 5 min. Cytoskeleton was stained with 50 µg/mL phalloidin–TRITC solution for 30 min at room temperature and then washed with PBS to remove unreacted phalloidin conjugates. Nuclear staining was done with (4′,6-diamidino-2-phenylindole) (DAPI) (KPL, Gaithersburg, MD, USA). Fluorescent images were recorded under a LSM 510 Meta confocal laser scanning microscope at an excitation/emission wavelength of 540/570 nm for phalloidin and 360/460 nm for DAPI.

2.6.9. Immunofluorescent Staining of COL I, OCN, and COL II

Immunofluorescent staining was done for co-cultured OC scaffolds to observe and compare the bone and cartilage-specific protein expression. Cylindrical OC scaffolds pre-wet with DMEM were seeded with BMSCs at a seeding density of 2×10^4 cells in the bone part and maintained in OM for different culture durations from 7 to 28 days. Furthermore, OC scaffolds with 14-day-cultured BMSCs (in bone part) were co-cultured with chondrocytes in cartilage part at a seeding density of 1×10^4 cells and maintained the whole scaffold in chondrocyte medium for another 21 days. Cells in the scaffold were fixed by immersing in 4% formaldehyde in PBS for 30 min followed by washing with PBST (PBS with 0.1% Tween 20) for 3 times, 15 min each. Nonspecific labeling was blocked by incubating with 1 mL of HyBlock 1-min Blocking Buffer® (Goal Bio, Taipei, Taiwan) followed by washing with PBST. OC scaffolds were incubated in either type I collagen (COL I) primary antibody (1:200 in PBS, mouse monoclonal anti-collagen I, Abcam, Cambridge, MA, USA) or osteocalcin (OCN) primary antibody (1:200 in PBS, guinea pig polyclonal anti-osteocalcin, Cloud-Clone Co., Houston, TX, USA) for 1 h to analyze the bone formation in the BMSCs-seeded part. Type II collagen (COL II) primary antibody (1:200 in PBS, guinea pig polyclonal anti-collagen II, Cloud-Clone Co., USA) was used to identify the cartilage-specific protein generation. Each sample was rinsed in PBS for 20 min followed by incubating COL I-treated scaffolds in Cy3-conjugated goat anti-mouse IgG secondary antibody (Jacksons Laboratories, Bar Harbor, ME, USA), OCN-treated scaffolds in FITC-conjugated goat anti-guinea pig IgG secondary antibody (Jacksons Laboratories, USA), or COL II-treated scaffolds in FITC-conjugated goat anti-guinea pig IgG secondary antibody (Jacksons Laboratories, USA) for 1 h. Another PBST washing was given prior to the addition of DAPI for nuclear staining. The immunofluorescent-stained OC scaffolds were observed for COL I at day 7, 14, and 28, while OCN formation was evaluated at day 14, 21, and 28. COL II formation was tested at day 14 and 21. Samples were rinsed once more before imaging under a laser scanning confocal microscope.

2.7. In Vivo Animal Study

2.7.1. Animal Implantation

All animal procedures were approved by the Institutional Animal Care and Use Committee of Chang Gung University (IACUC Approval No. CGU16-043). 6–8 weeks old female nude mice weighing 20–30 g purchased from the National Laboratory Animal Center (Taipei, Taiwan) were used for the in vivo implantation. Pre-sterilized disc-shaped OC scaffolds were rinsed with DMEM followed by seeding with BMSCs at a seeding density of 2×10^4 cells in the bone part and maintained in OM for 14 days in a 24-well plate with replacement of fresh medium every two days. OC scaffolds were further seeded with chondrocytes in the cartilage part with a cell seeding density of 1×10^4 cells and maintained in chondrocyte medium for another 4 h for cell attachment. Acellular cylindrical OC scaffolds were used as control. Prior to surgery, all the animals were anesthetized using a mixture of 7 mg/kg xylazine (Rompun®, Bayer, Leverkusen, Germany) and 140 mg/kg ketamine (Ketalar®, Hoffman-La Roche Ltd., Basel, Switzerland). The back sides of the animals were sterilized with 75% alcohol solution, two subcutaneous pockets on each side of the back of the animal were made and one acellular and three co-cultured OC scaffolds were implanted intra-muscularly. The animals were sacrificed 4, 8, and 12 weeks post-implantation using overdosed pentobarbital and implants were harvested for further analysis.

2.7.2. Gross and Microscopic Observation

Samples harvested at week 4, 8, and 12 were gross evaluated in comparison with controls. Effects of implantation on scaffold morphology, cell adhesion, and cell spreading were further verified through SEM analysis. Harvested samples were cut through the center of bone–cartilage intersections, washed with PBS, and fixed in 10% formaldehyde for 4 h. After PBS washing, alcohol gradient was used to remove the water content from the sample and vacuum dried overnight. Scaffolds were fixed on the carbon tape-pasted aluminum stub with the cross-sectional area facing outwards. Morphology of the cell-seeded scaffolds was recorded in comparison with acellular ones using SEM.

2.7.3. Immunofluorescent Staining

Samples harvested at 4, 8, and 12 weeks were immersed in 4% formaldehyde for 4 h followed by washing thrice with PBST. Surface adhered fibrous tissue layer was removed and scaffolds were cut through the bone–cartilage interface for separate immunofluorescent (IF) staining. Nonspecific labeling was blocked followed by washing with PBST. The bone part of the scaffold was incubated in OCN primary antibody while the cartilage part was treated with COL II primary antibody for 1 h. Both samples were rinsed in PBS for 20 min followed by incubation in FITC-conjugated goat anti-guinea pig IgG secondary antibody for 1 h. DAPI was used for nuclear staining. Both IF-stained scaffold parts were observed under a laser scanning confocal microscope for OCN and COL II formation in vivo.

2.7.4. Histological Analysis

The harvested OC scaffolds were immersed in 10% formaldehyde at room temperature and dehydrated using alcohol gradient, followed by embedding in paraffin wax and sectioned into 10 μm thick tissue sections in glass slides. The slides were further heated in a 70 °C oven followed by immersing in xylene for 5 min. The process was repeated thrice. Samples were re-hydrated using alcohol gradient and finally with water. After 5 min treatment with Tris-buffered saline containing 0.1% Tween 20, the bone part of the tissue sections was subjected to hematoxylin and eosin (H&E), AR, Masson's trichrome stains, while the cartilage part was stained with H&E and AB. The images were recorded under an inverted optical microscope (IX-71, Olympus Co., Tokyo, Japan).

2.8. Statistical Analysis

All data were reported as mean $\pm$ standard deviation (SD) and the one-way ANOVA LSD test was used to determine the significant differences. A value of $p < 0.05$ was considered statistically significant.

3. Results and Discussion

3.1. Microsphere Preparation and Scaffold Fabrication

PLGA microspheres with or without nHAP were prepared through the standard oil-in-water emulsion/solvent evaporation method, as reported earlier [35]. The experimental procedures are shown in Step 1 of Figure 1. The size of spheres was distributed in the range of 75–800 μm, among which 250–500 μm was selected as the ideal size range. Specifically, microspheres in this size range were chosen, as this range may provide optimal pore sizes to allow cellular infiltration and cell-to-cell interaction [39]. The maximum yield within this size range was obtained by parameter optimization during the preparation step. Figure 1A,B is the respective light microscopic and SEM images of the prepared PLGA microspheres with size range 250–500 μm without (left) and with (right) nHAP. The effective incorporation of nHAP can be verified through the radio-opaque nature of PLGA/nHAP microspheres (right) compared to the transparent nature of virgin PLGA microspheres (left) in Figure 1A. The surface morphology of the same was further characterized through SEM and shown in Figure 1B. PLGA/nHAP microspheres had relatively rough morphology (right) compared to PLGA microspheres (left), due to the presence of nHAP particles. The scaffold was fabricated from the microspheres, as depicted in Step 2 of Figure 1. Figure 1C indicates the gross views of scaffolds prepared in various shapes with respect to different required experimental procedures in vitro and in vivo. Due to the osteo-conductive nature, the incorporation on nHAP in the scaffold will enhance its osteogenic differentiation effect toward BMSCs, and thereby is expected to be a key scaffold material for bone formation. V and C denote the scaffolds made from PLGA microspheres and PLGA/nHAP microspheres, respectively. OC represents a biphasic integrated structure of cartilage and bone parts sintered together in a single scaffold. Red arrows in Figure 1C point out the interface region in the curve-edged rectangular, disc, and cylindrical-shaped OC scaffolds, with distinguishable separating zones for cartilage and bone parts, respectively. The bright white appearance of the bone part is due to the presence of nHAP in PLGA/nHAP microspheres while the transparent nature of the cartilage part is attributed to the absence of the same in PLGA microspheres. The capability to generate OC scaffolds in various shapes also demonstrates the flexibility of the microsphere-based biphasic scaffold fabrication method to meet the need in repairing complex osteochondral defects.

3.2. Scaffold Characterization

3.2.1. Morphology of Spheres and Scaffolds

The prepared nHAP was characterized by TEM. The particles had a smooth round morphology with an average diameter of 15 ± 4 nm (Figure 2A). The morphology of sintered microspheres in V, C, and OC scaffolds is shown through cross-sectional SEM observation (Figure 2B). It is evident from the images that all microspheres showed uniform surface sintered to each other, making the scaffolds mechanically stable and macro-porous. The white dots in the C scaffolds further confirm the existence of nHAP in PLGA/nHAP microspheres after sintering [40]. The interface of OC scaffolds exhibits a rough surface textured with white-dotted PLGA/nHAP microspheres on one side of the image and positioned next to the relatively smoother PLGA microspheres on the other side. All scaffolds had a pore size ranging from 50 to 250 μm, making them ideal for cell penetration throughout the scaffolds [41].

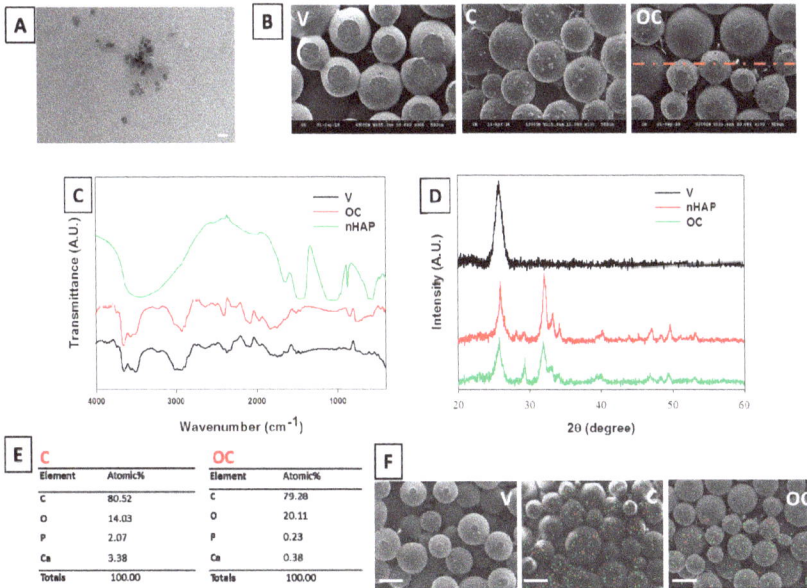

Figure 2. Physico-chemical properties of scaffolds. (**A**) TEM image of nHAP (bar = 20 nm); (**B**) SEM images of V, C, and OC scaffolds; (**C,D**) FTIR and XRD of nHAP and V and OC scaffolds; (**E**) EDX elemental atomic compositions of C and OC scaffolds; (**F**) elemental mapping of V, C, and OC scaffolds with Ca and P as green and red dots, respectively (bar = 300 μm). The red line in (**B**) indicates the interface region in the OC scaffold.

3.2.2. FTIR Analysis

nHAP and V and OC scaffolds were further characterized through FTIR (Figure 2C). The characteristic phosphate stretching vibrations were observed at 576, 612, and 930–1040 cm^{-1}, whereas the stretching bands at 1510–1541 cm^{-1} confirm the presence of the hydroxyl group. The broad band observed at 3460–3560 cm^{-1} corresponds to the stretching mode of the hydroxyl group, which is characteristic of calcium phosphates such as nHAP [17]. PLGA exhibited characteristic alkyl group frequencies in the broad range 2800–2950 cm^{-1}, hydroxyl vibrations at 3690, and 633 cm^{-1}, C=O stretching at 1710 cm^{-1}. There are also stretching bands due to asymmetric and symmetric C–C(=O)–O vibrations between 1105 and 1300 cm^{-1} [42]. The bands in these regions are useful in the characterization of esters. The phosphate bands at 578, 610, and 928 cm^{-1} in the OC scaffolds further confirm the presence of nHAP in the scaffold formed by sintered PLGA/nHAP.

3.2.3. XRD Measurements

Crystallinity of nHAP was confirmed by XRD and compared with JCPDS-090432 (Figure 2D). Characteristic 2θ values of nHAP, V, and OC scaffolds were recorded and compared. nHAP showed prominent 2θ peaks at 32.2° (211 plane) and 25.9° (002 plane) with minor 2θ values at 29.3°, 33.2°, 34.3, 40.2°, 46.9°, 49.7°, and 53.3° [43]. V scaffolds had a single peak of PLGA at 25.7°, which is overlapping with the prominent nHAP peak. OC scaffolds exhibited the presence of the respective typical nHAP crystalline peaks with lowered peak intensities.

3.2.4. Stoichiometric Analysis of nHAP by EDX

Apart from FTIR and XRD, nHAP content in C and OC scaffolds were cross-confirmed with EDX analysis. Figure 2E represents the EDX elemental compositions of the C and OC scaffolds. The presence of Ca and P peaks in both the C and OC scaffolds is evident, with the peak intensity of the OC being lower than that of the C, attributing to the lower overall nHAP content in the former. The scaffolds show a Ca/P stoichiometric ratio of 1.63 and 1.65 for the C and OC, respectively, in close relation with the theoretical value 1.67 for HAP [44]. Presence of nHAP in the C and OC scaffolds was further represented through elemental mapping, in which calcium was denoted as green dots and phosphate as red dots (Figure 2F). The V scaffolds did not show mapping colors contrast due to the absence of nHAP, while the C scaffolds showed color dots representing Ca and P throughout the scaffold. An interface mapping of the OC scaffolds revealed one part (lower half) of the scaffold is with Ca and P dots while the other without, confirming the distinguishable borderline between the bone and cartilage part of the OC scaffold.

3.2.5. Thermal and Mechanical Analysis

The thermal properties of the V and C scaffolds were analyzed with TGA. As can be seen from Figure 3A, both scaffolds were thermally stable until ~274 °C from room temperature, without having any significant loss on the scaffold weight. The onset of weight loss was found at ~286 °C for both the V and C scaffolds. Further heating resulted in 50% weight loss at ~359 °C. The decomposition curve for both scaffolds was similar until reaching 376 °C, with a remaining weight of 12.3%. Further heating resulted in complete decomposition of the V scaffold at ~449 °C while the C scaffold had ~10% weight remaining. The residual weight percentage of nHAP-containing C scaffolds refers to the amount of nHAP present in the scaffolds due to the stability of ceramic particles, even at elevated temperatures. Most of the biodegradable polymeric materials are more prone to decomposition at higher temperatures, leading to complete charring prior to reaching 500 °C [34]. However, the ceramic counterparts of the same in the scaffold are much more stable, allowing it to retain residual weight during the thermal decomposition cycle at higher temperatures, thus accounting for the quantitative evaluation of nHAP in the C scaffolds.

Polymers **2016**, *8*, 429

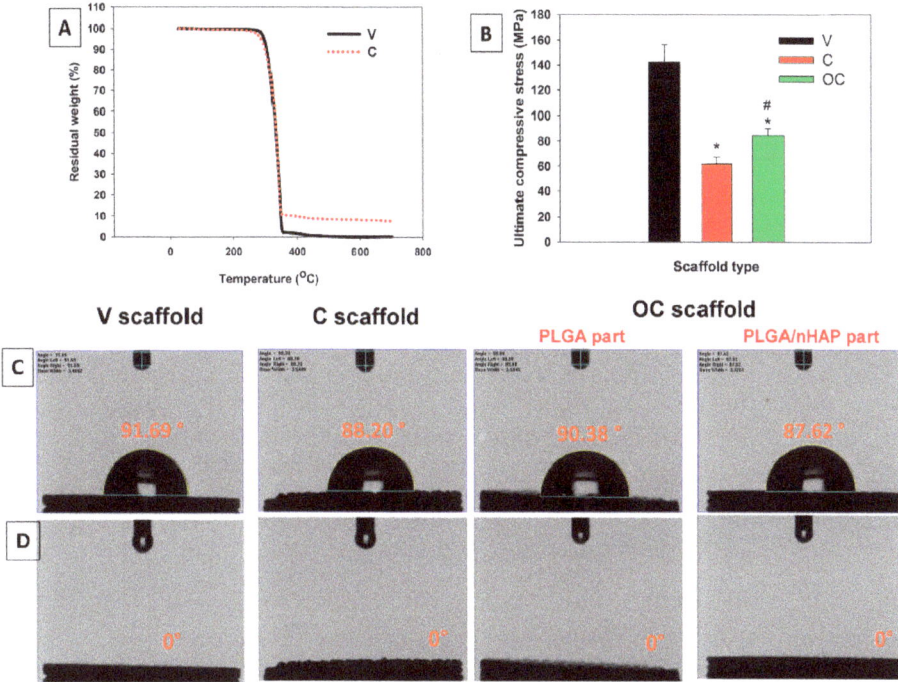

Figure 3. Thermal decomposition behavior (**A**) and mechanical testing (**B**) of V and C, and OC scaffolds. Surface hydrophilicity of V, C, and OC scaffolds by measuring water contact angles after 2 s (**C**) and 30 s (**D**); * $p < 0.05$ compared with V; # $p < 0.05$ compared with C.

Vigilant selection of materials in osteochondral tissue engineering is crucial, since supporting the substrate's mechanical strength, structural support, and favorable culturing environment is critical during the early stages of the regenerative process. Optimal mechanical properties and suitable structures will ultimately promote bone–cartilage formation and implant healing. The V scaffolds had the maximum ultimate compressive strength of 142 ± 14 MPa, which is significantly higher than that of the C (62 ± 6 MPa) and OC (85 ± 5 MPa) scaffolds (Figure 3B). The median compressive strength value for the OC scaffolds should be due to the combination of PLGA and PLGA/nHAP microspheres in a single construct, accounting for the combinatory effect of both V and C scaffolds, which are made from PLGA and PLGA/nHAP microspheres separately. Generally, incorporation of an optimum percentage nHAP should increase the mechanical strength of the C scaffolds [45]. The addition of nHAP particles to PLGA microspheres may result in a reduction in mechanical strength of C compared to V scaffolds. Though the percentage of nHAP in PLGA/nHAP microspheres is ~10%, the irony of reduction in compressive strength refers to the sintering aspects of scaffolds. The sintering mechanism of scaffolds is through the surface melting of compressed microspheres in the heating cycle and permanent adhering of the same during the cooling cycle. Since PLGA/nHAP microspheres had uniformly distributed nHAP on the surface, the thermal resistibility of the same would have interfered with the surface melting–adhering mechanism during the heat sintering process, thereby reducing the mechanical strength of the C scaffolds. More studies on the time–temperature dependence of the scaffold's mechanical strength are necessary to elucidate this aspect.

3.2.6. Contact Angle Measurements

Contact angle is a measure of hydrophobicity or surface wettability of a material. The higher the contact angle values, the higher the hydrophobicity of the material is. The goniometer was used here to measure the hydrophobicity of the scaffolds based on pictures of a water droplet on the scaffold surface. Here, the contact angle was measured at two time points, 2 s (Figure 3C) and 30 s (Figure 3D) after placing the drop on the scaffold surface. The V and C scaffolds displayed a contact angle of $91.69°$ and $88.20°$, respectively, showing their less hydrophilic nature, similar to previous reports [46]. The OC parts fabricated from PLGA and PLGA/nHAP microspheres also showed similar values at $90.38°$ and $87.62°$, respectively, revealing that different microspheres in different parts of the OC scaffold maintain its hydrophilicity after the physical sintering process. While considering the shape of the water droplet after 30 s, the contact angle was zero, or none of the scaffolds showed the existence of a water droplet on top of the scaffold surface. This could be attributed to the porosity of microspheres rather than wettability. Since all scaffolds have large pore sizes between microspheres (Figure 2B), drops could easily be absorbed into the scaffold interior through gravitational force and thereby rapidly disappear from the scaffold surface. Contact angle measurements confirm that though the surface characteristic is less hydrophilic in nature, faster absorption of water drops after 30 s indicates the high porosity of scaffolds, which is one of the essential requirements for an ideal scaffold for tissue regeneration.

3.3. In Vitro Cell Culture Studies

3.3.1. Cell Morphology

The use of mesenchymal stem cells on biodegradable implants has been shown to be effective for repairing osteochondral defects in vivo [47]. Also, many results report the added advantages of using two different cell types to regenerate osteochondral defects [40]. In this work, we first analyzed the ability of the OC scaffolds to successfully support BMSCs attachment and proliferation in vitro, in comparison with V and C scaffolds. When treating osteochondral defects, the "bone-like" layer should aim at the repair of the cortical-subchondral bone, while the "cartilage-like layer" must provide a good environment for cartilage regeneration. The effect of microsphere surface characteristics and interior pore gap on cell adhesion, spreading, and cell–cell interconnectivity was analyzed through surface (Figure 4A) and cross-sectional (Figure 4B) SEM observations. Though the intended use was for osteochondral repair, the cell behavior was studied using only BMSCs to verify the general cell characteristics such as adhesion and spreading and also to evaluate the ability of the sintered microsphere structure to support cell growth and spreading without obstructing the scaffold pores, which are the key factors when considering waste and nutrient transport in the scaffold. At day 1, cells attached well to both surface and interior sintered microspheres of all scaffolds, albeit with low cell density and spreading. Cell morphology at day 7 was much altered, where cells started spreading more on the sphere-to-sphere intersections by outspreading filopodial extensions regardless of scaffold type. Much robust cell behavior was observed at day 14, where cells extended their growth on the wall of the microsphere, giving a thin cellular coating on the sphere surface. The cell density in the interstitial space was much higher than at previous time points, as observed from both surface and cross-sectional images. The cell density was significantly higher than earlier culture duration at day 21 and 28. At these points, cells became more flattened, cellular wall density on microsphere surfaces became much denser, and cells connected adjacent spheres through filopodial extension, as evidenced from the images. The cross-sectional image indicates that cells were able to penetrate deeper into the scaffold pores, which is an important observation, since a good cell distribution within the whole scaffold prominently affects the overall performance of the construct, without limiting the diffusion of nutrients and oxygen. Though all scaffolds exhibit similar characteristics in terms of cell interactions, we speculate that this result is quite promising since it validates that the microsphere-based scaffolds possess an adequate pore size and pore distribution to effectively allow cells to adhere and maintain their functions.

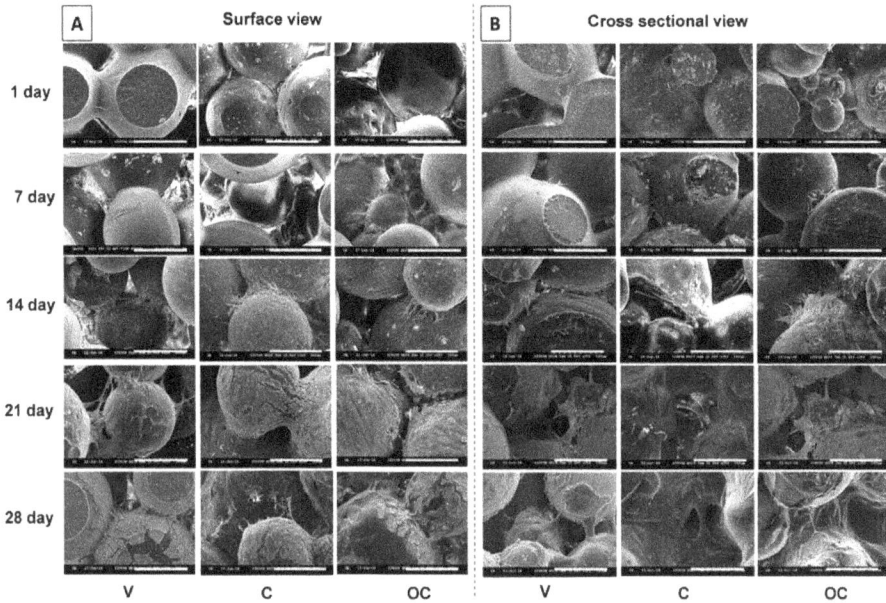

Figure 4. Cell infiltration and cell morphology of BMSCs (bone marrow stem cells) in V, C, and OC scaffolds. The increased cell density and spreading from day 1 to day 28 from SEM images of scaffold surface (**A**) and cross section (**B**) validates the biocompatibility and pore interconnectivity in all scaffolds. Bar = 200 μm

3.3.2. Viability of BMSCs and Chondrocytes in Mono-Culture

The viability of cells in scaffolds was evaluated using a Live/Dead assay kit that shows cell morphology and distribution in the scaffold interior after imaged by confocal microscopy (Figure 5). As mentioned before, the behavior of cells in the OC scaffolds was evaluated using chondrocytes and BMSCs for the V and C scaffolds separately, using chondrocyte and osteogenic medium. Due to the inability of a laser scanning confocal microscope to have high depth scanning arrays for the scaffolds, scaffolds were scanned both axially (from the top or the bottom of scaffolds) and radially to determine cell distribution in 3D within the scaffold. Uniform cell distribution and high cell viability in both the V and C scaffolds could be confirmed from the green fluorescence signal of live cells and few red signals from dead cells. The cell density is lower at day 7 than day 14, irrespective of the scaffold type or the scan direction. Cell morphology at day 21 and 28 were almost similar to that at day 14, which can be correlated with the flattened cell morphology on the spherical surface of the microsphere, as shown from the SEM images in Figure 4. Unlike other types of scaffolds, cells in a microsphere-based scaffold adhere first, spread on the microsphere wall, and then deposit as a monolayer over the surface of the microsphere in the early stages, followed by thickening of the cellular layer as reported previously [48]. A few dark empty spaces inside cellular circles are due to the out-of-focus plane of different microspheres inside the scaffold. However, no significant difference in cell morphology was observed among chondrocytes-seeded V scaffolds or BMSCs-seeded C scaffolds at any point of the culture period. Nevertheless, cells grew well and proliferated around the microspheres and throughout the scaffold, cross-confirming the advantage of the macroporous structure of the scaffolds for tissue regeneration in vitro.

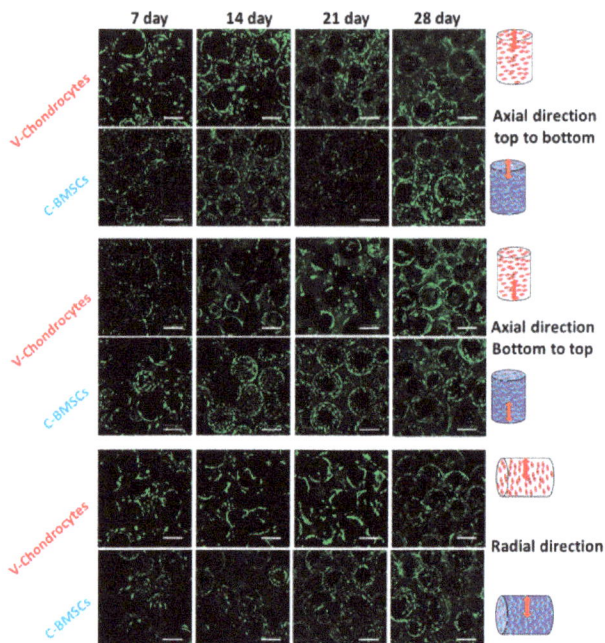

Figure 5. Live/Dead assays of mono-cultured chondrocytes in V scaffolds and BMSCs in C scaffolds. Confocal laser scanning microscope images from axial and radial directions at day 7, 14, 21, and 28 indicate the spread nature and uniform distribution of both cells. Bar = 300 μm.

3.3.3. Cell Proliferation

Cell proliferation in scaffolds was evaluated through DNA contents (Figure 6A). The cell number after 4 h of attachment (day 0) was taken as the reference point to be compared with later periods of cell proliferation of BMSCs in V, C, and OC scaffolds. All scaffolds exhibited an increase in cell number after day 7, compared to day 0 and then reached a maximum at day 14. There was no significant increase in cell number following day 14, ascribed to the differentiation of BMSCs in OM to osteoblasts with reduced cell proliferation rate. Our previous studies also confirmed the same trend of stem cells differentiation in osteoconductive-inductive environments [34]. The rationale behind the choice of BMSCs alone for cell proliferation rather than using both cell types is based on two factors: in vitro co-culture and in vivo aspects. In the former case, since V and C scaffolds are separately compared here with OC scaffolds, mono-culture strategy is practically more significant to evaluate the cell proliferation and differentiation potential. Though the V part is assigned for chondrocytes and the C part for BMSCs, culturing two different cell types in OC scaffolds at two different time points and further with cell number evaluation is practically irrelevant. Moreover, most of the in vitro co-culture experiments performed throughout this study was by seeding BMSCs in the bone part for pre-optimized times, followed by seeding chondrocytes in the cartilage part and maintaining the culture for another two or three weeks, which only aimed at examining the potential of both cell types to maintain their phenotype in co-culture conditions. Considering the in vivo aspect, OC scaffolds were designed to culture with BMSCs in the bone part for the first 14 days, followed by seeding with chondrocytes in the cartilage part for 4 h prior to implantation, denoting the non-requisite strategy of co-culture in cell proliferation studies. Further, the logic behind the culture of V scaffolds with BMSCs rather than chondrocytes is to validate the effect of osteo-conductive nHAP in C and OC scaffolds towards stem cell differentiation.

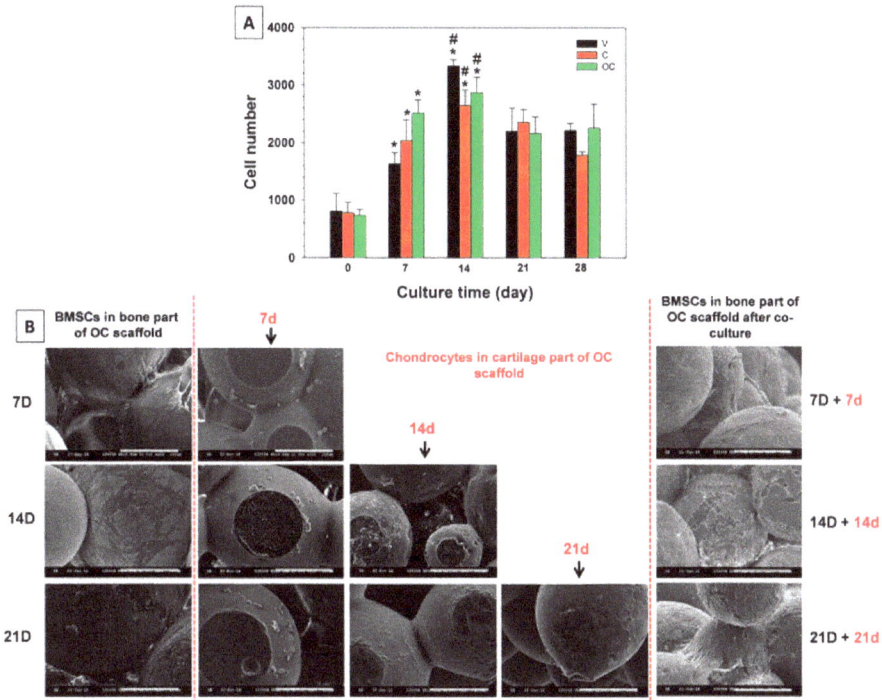

Figure 6. (**A**) Cell numbers of BMSCs from DNA assays when cultured in in V, C, and OC scaffolds up to 28 days. * $p < 0.05$ compared with day 0, # $p < 0.05$ compared with day 7 for each scaffold; (**B**) SEM images of the morphologies of BMSCs in the bone part of OC scaffolds before and after co-culture with chondrocytes. 7 D, 14 D, and 21 D represents the morphology of mono-cultured BMSCs in the bone part for 7, 14, and 21 days, while 7 d, 14 d, and 21 d is the morphology of co-cultured chondrocytes in cartilage part of OC scaffolds after 7, 14, and 21 days. 7 D + 7 d, 14 D + 14 d, and 21 D + 21 d denote the morphologies of BMSCs in the bone part of OC scaffolds after immersing in chondrocyte medium for another 7, 14, and 21 days during co-culture. Bar = 200 μm.

3.3.4. Co-Culture with BMSCs and Chondrocytes

Bi-phasic tissue engineering always has the complexity of recognizing proper cues toward preferred cell differentiation and thereby develops separate tissues. Co-culture offers the benefits of distinguished regeneration and faster development to multiple tissue types compared with mono-culture. Figure 6B shows the SEM images of morphologies of the BMSCs-seeded bone part and chondrocytes-seeded cartilage part of the OC scaffolds at various co-culture durations. The left panel represents the BMSCs morphology in the bone part in OM at day 7, 14, and 21. Scaffolds seeded with BMSCs in the bone part for 7 d had similar cell adhesion patterns as observed before in Figure 4, while the morphology of the chondrocytes-grown cartilage part for another 7 d had a relatively more round morphology, as shown in the middle panel (red letters). The extent of cell spreading and flattening increased with time for both cell types. The morphology of chondrocytes at each co-culture time point was similar, irrespective of BMSCs culture durations in the bone part. Similar to BMSCs, chondrocytes spread well and flattened throughout the scaffolds with respect to co-culture durations, with the maximum cell density at day 21, which could be correlated with the fluorescently stained chondrocyte morphology shown in Figure 5. However, after co-culture, it is necessary to prove the existence of BMSCs with no alteration in cell morphology. The right panel is the SEM images of

BMSCs in the bone part after co-culture with chondrocytes. The 7 D + 7 d represents the morphology of BMSCs cultured in the bone part in OM 7 days, followed by immersing in chondrocyte medium for another 7 days while co-culturing. In a similar aspect, 14 D + 14 d and 21 D + 21 d SEM images respectively represent the morphology of BMSCs after 28 days (i.e., 14 days +14 days) and 42 days (i.e., 21 days + 21 days) BMSCs/chondrocyte co-culture durations. Though BMSCs in the bone part have undergone more culture time in chondrocyte medium during co-culture, the cell morphology of BMSCs confirms cell survival/proliferation was not affected by co-culture. Only a few reports are available on the successful regeneration of biphasic tissues from mono-culture using a single cell source such as BMSCs, due to the difficulty of maintaining distinguished environments in a single scaffold. Studies are not satisfactory to establish an ideal scaffold design with separate zones for loading specific growth factors/proteins, which could lead to separate tissues such as bone and cartilage. The restriction of specific growth factors/proteins eluted from the scaffold to be within a specific region of the environment that BMSCs are exposed to during co-culture is considerably challenging [49]. Our scaffold design without eluting growth factors/proteins but with osteo-conductive nHAP in the bone part and co-cultured chondrocytes as the cartilage part is very promising.

3.3.5. Qualitative Evaluation on Bone–Cartilage Formation in Vitro

Figure 7A,B demonstrate positive staining for Alizarin red (AR), while Figure 7C depicts the positive staining for Alcian blue (AB). The AR stain evaluates calcium deposits in the mineralized extracellular matrix (ECM) of differentiated BMSCs when calcium forms an Alizarin red S-calcium complex in a chelation process and the degree of mineralization will be directly proportional to the intensity of the stain. Considering the cell proliferation, the duration of BMSCs mono-culture in the bone part of OC scaffolds was selected at 14 days. The effect of nHAP alone in C scaffolds and the bone part of the OC scaffolds towards BMSCs differentiation was tested through mono-culture in normal medium (NM) and osteo-inductive medium (OM). Both the C scaffolds and bone part of the OC scaffolds displayed a significant increase in AR stain intensity due to the mineralization of BMSCs. These results are supportive of the fact that nHAP particles in the C and OC scaffolds, due to its osteo-conductive nature [50,51], which triggered BMSCs differentiation along an osteogenic pathway, even in the absence of soluble growth factors in NM. However, nHAP alone is not enough to guarantee osteoblast differentiation. Therefore, the combined effects of osteo-conductive nHAP and osteogenic inducing factors were evaluated through BMSCs cultured in OM. That the maximum stain intensity was observed in the C and OC scaffolds announces the synergistic effect of nHAP and inducing factors on commitment of BMSCs toward the osteogenic lineage in OM [52,53]. To eliminate the possibility that positive staining was due to the calcium ions in nHAP within the scaffolds, AR staining was also performed on acellular control scaffolds, which showed much less stain intensity compared to BMSCs-seeded scaffolds. The results demonstrate that BMSCs mono-culture in the presence of osteogenic factors could lead to a dramatic increase in mineral deposition, which further leads to bone formation. Based on mono-culture results, OC scaffolds were seeded with BMSCs in the bone part and cultured in OM for 14 days, followed by chondrocytes seeding in the cartilage part for 7 and 14 days (Figure 7B). As expected, the 14-day (14 D + 14 d) co-cultured scaffolds had maximum AR stain intensity in the bone part compared to the 7-day (14 D + 7 d) co-cultured samples, indicating continued mineral deposition of BMSCs even in chondrocyte medium, due to the presence of osteo-conductive nHAP.

Alcian blue is any member of a family of polyvalent basic dyes and is used to stain acidic polysaccharides such as glycosaminoglycans (GAGs) in cartilages. Therefore, AB stain could demonstrate that chondrocytes were able to deposit a GAGs-rich ECM in the cartilage part of OC scaffolds. Similar to AR staining, OC scaffold having pre-differentiated BMSCs in the bone part was co-cultured with chondrocytes in the cartilage part for 7 days (14 D + 7 d) and 14 days (14 D + 14 d), followed by AB staining of the whole scaffold. That positive AB staining was only observed in the cartilage part of OC scaffolds at both time points confirms the maintenance of chondrogenic phenotype of seeded chondrocytes during co-culture and cartilage formation in vitro (Figure 7C). False positive AB staining could also be discounted through AB staining on acellular control scaffolds. Thus, the qualitative measurements of bone–cartilage formation in OC scaffolds through AR–AB stains endorse the potential of the same in osteochondral regeneration.

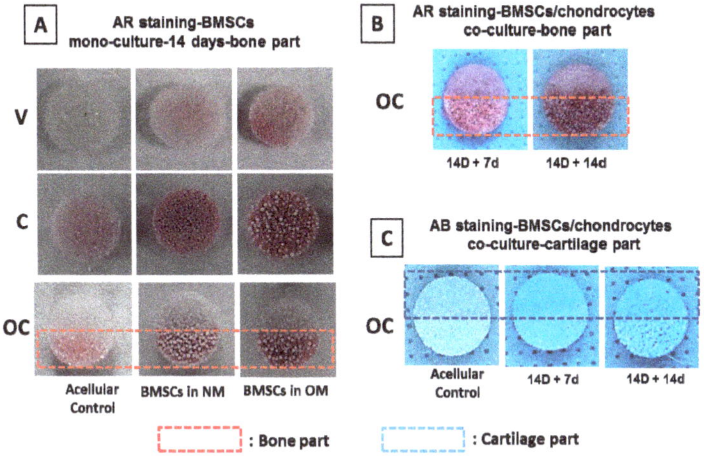

Figure 7. (**A**) The Alizarin red (AR) staining images of acellular control and BMSCs-seeded V, C, and OC scaffolds after mono-culture in normal medium (NM) and osteogenic medium (OM) for 14 days; (**B,C**) are the Alizarin red (AR) and Alcian blue (AB) staining images of BMSCs in the bone part and chondrocytes in the cartilage part of OC scaffolds. 14 D + 7 d and 14 D + 14 d denote the culture of BMSCs in the bone part of OC scaffolds in OM for 14 days followed by immersing in chondrocyte medium for another 7 and 14 days during co-culture.

3.3.6. Calcium Quantification

Positive AR staining from gross view observation (Figure 7A,B) is the evidence for BMSCs mineralization. However, calcium quantification with cetylpyridinium chloride (CPC) could estimate the extent of calcium-based mineral deposits by BMSCs. When scaffolds react with CPC, it extracts the ARS bound to Ca^{2+} on the cell surface and the absorbance of the extracted solution at 540 nm will provide direct estimation of the extent of BMSCs mineralization. Figure 8A represents the calcium quantification of BMSCs mono-cultured in V, C, and OC scaffolds for 14 days whereas Figure 8B denotes those co-cultured in OC scaffolds. AR-stained V, C, and OC scaffolds mono-cultured with BMSCs in NM is denoted as VN, CN, and OCN while the same construct cultured in OM was termed VO, CO, and OCO. Considering the same scaffold, the absorbance values in NM show the least reading while in OM it was much higher. However, C scaffolds in NM had a relatively higher absorbance value than V in OM, probably due to the binding of ARS to nHAP in the scaffold. That OC scaffolds had a higher absorbance value than V scaffolds but lower than C scaffolds in OM indicates the synergistic effect of nHAP and osteogenic inductive factors in OM in enhancing calcium deposition, in accordance with the qualitative AR staining results of BMSCs mono-culture shown in Figure 7A.

Besides mono-culture, OC scaffolds co-cultured with BMSCs and chondrocytes were further investigated for calcium content. OC scaffolds with 14 days pre-differentiated BMSCs in the bone part, co-cultured with chondrocytes in the cartilage part for 7 and 14 days are denoted as 14 D + 7 d and 14 D + 14 d, respectively, in Figure 8B. The absorbance values observed for 14 D + 7 d was lower compared to 14 D + 14 d, indicating continued differentiation of BMSCs in chondrocyte medium after the commitment of BMSCs toward the osteogenic lineage during the initial culture in OM. Furthermore, the calcium content of the 14 D + 7 d co-culture group is much higher than that of any of the BMSCs mono-culture groups, due to the prolonged culture duration of 21 d. This culture duration is longer than the culture time (14 days) required for the maximum cell number (Figure 6A), which is critical in continued differentiation and mineralization of BMSCs. Since the calcium deposition of stem cells starts at later stages of cell proliferation, duration of culture period is very critical in the extent of calcium deposition [34,54]. Indeed, these results can be correlated with the cell proliferation data shown in Figure 6A, claiming the maximum cell number at day 14, followed by a steady value on the remaining days. This confirms the differentiation of BMSC to osteoblasts in the later stages of cell proliferation and thereby accounts for the higher calcium deposition for co-culture of BMSCs and chondrocytes in OC scaffolds. The results are also convincing for the fact that BMSCs are not influenced by the presence of the chondrocyte medium used for co-culture, but continue their mineralization and maintain their phenotype throughout the in vitro culture period.

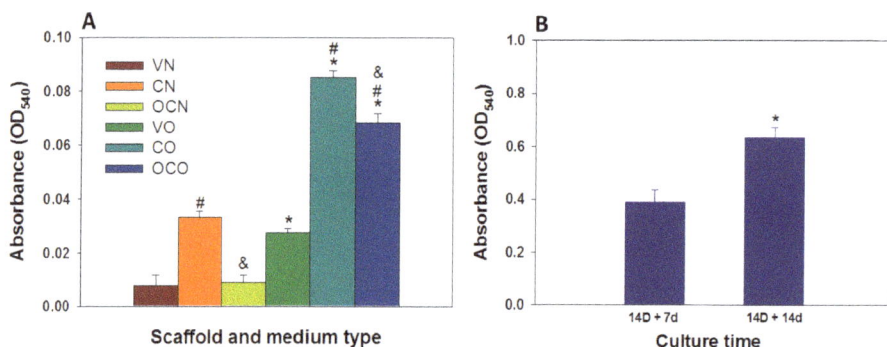

Figure 8. (**A**) Quantitative estimation of calcium deposition in V, C, and OC scaffolds after mono-culture of BMSCs in different scaffolds in normal medium (NM) and osteogenic medium (OM) for 14 days. VN, CN, and OCN denote V, C, and OC scaffolds in NM whereas VO, CO, and OCO denote V, C, and OC scaffolds in OM. * $p < 0.05$ compared with NM for each scaffold, # $p < 0.05$ compared with V for each medium, & $p < 0.05$ compared with C for each medium; (**B**) Total calcium deposition in the bone part of co-cultured BMSCs and chondrocytes in OC scaffolds. 14 D + 7 d denotes 14 days in osteogenic medium and 7 days in chondrocyte medium; 14 D + 14 d denotes 14 days in osteogenic medium and 14 days in chondrocyte medium. * $p < 0.05$.

3.3.7. Viability and Morphology of BMSCs and Chondrocytes during Co-Culture

Apart from mono-culture, viability of BMSCs and chondrocytes during co-culture was evaluated through Live/Dead assays (Figure 9A). As in mono-culture, the absence of red signals demonstrates the high viability of both BMSCs and chondrocytes during co-culture. BMSCs cultured for the initial 14 and 21 days had no distinguishable difference in cell morphology as observed for mono-culture in Figure 5. Chondrocytes exhibited comparable cell morphology in the cartilage part of OC scaffolds at each time point, irrespective of BMSCs culture duration. Similar to BMSCs, density of viable cells was the maximum for chondrocytes at day 21 of co-culture, indicating the high survival rate of the same. The densely populated live chondrocytes at day 21 indicates the cartilage part of OC scaffolds provides a favorable environment for cartilage development, while comparable cell densities at the same time point for the BMSCs-seeded bone part endorses the favorable cellular response of the OC scaffolds to maintain cell viability and support growth of BMSCs during co-culture for bone development. SEM images of the morphology of BMSCs in the bone part of OC scaffolds (Figure 6B) after various durations of chondrocyte co-culture cross-confirm the ability of BMSC to maintain the spread morphology.

Polymers 2016, 8, 429

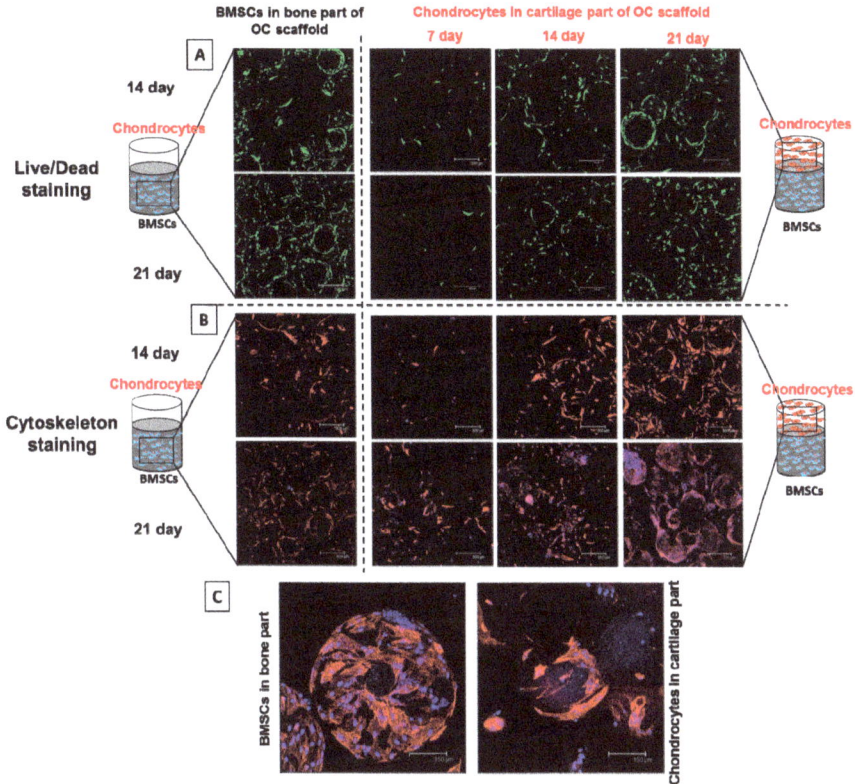

Figure 9. Live/Dead cell assays (**A**) and cytoskeletal arrangement determined by phalloidin/DAPI staining (**B**) of co-cultured BMSCs and chondrocytes in OC scaffolds. Bar = 300 μm; (**C**) The cytoskeletal arrangement of BMSCs and chondrocytes on a single microsphere surface in the bone and cartilage part at day 14. Bar = 150 μm.

The morphology of co-cultured cells on microspheres of OC scaffolds was validated through staining of the actin cytoskeleton of adhered cells. The cytoskeleton plays important roles in cell morphology, adhesion, growth, and signaling [55]. Changes in the cytoskeleton of the cell allow the cell to migrate, divide, and maintain its shape. The cytoskeleton consists of three components: actin filaments, intermediate filaments, and microtubules. The backbone of the cytoskeleton is F-actin, which clusters to form actin filaments. As explained before, unlike other scaffolds, cells on microspheres adhere first, spread on the microsphere wall, and make a monolayer over the microsphere in the early stages of cell adhesion. However, in the later stages, cells preferentially grow around the gaps among the microspheres, fill the gaps, and connect each adjacent sphere through a cellular bridging [56]. Being an abundant protein in cells, actin staining could help to visualize the migration and 3D proliferation of the cells within the scaffold structure. The versatility of the confocal microscope to scan various planes and their corresponding maximum projection would provide a better understanding of cellular morphology throughout the matrix. Figure 9B represents the confocal images of rhodamine-phalloidin stained cytoskeletal actin filaments of BMSCs and chondrocytes in bone and cartilage parts of OC scaffolds, respectively. Similar to Live/Dead staining, the intensity of actin staining of differentiated BMSCs in the bone part of OC scaffolds was almost similar at day 14 and 21, confirming the possible transformation of BMSCs to osteoblast differentiation. Considering the cartilage part at different co-culture times, chondrocytes had minimum stain intensity at day 7, enhanced at day 14, and reached the maximum at day 21, confirming the same trend of cell proliferation. Both cells almost covered the microsphere gaps and surface at later stages of culture. The cell spreading on a single microsphere surface is shown in the enlarged microscopic images in Figure 9C. Since both cells attained the shapes of the microspheres they adhered, actin staining was not a conclusive tool to declare the morphological difference between cells in bone and cartilage parts. More detailed biochemical assays are mandatory to reveal the efficiency of osteo-conductive/inductive factors and the BMSCs-chondrocytes approach towards osteochondral regeneration.

3.3.8. Immunofluorescent Staining of Bone- and Cartilage-Specific Proteins

Figure 10A,B reveals time-lapsed IF staining of bone-specific proteins COL I and OCN in the bone part during co-culture. Figure 10C is the corresponding IF staining of cartilage-specific protein COL II in the cartilage part during co-culture. COL I production in the bone part of OC scaffolds was on day 7, 14, and 21 in contrast to day 14, 21, and 28 for OCN, as COL I and OCN are respectively the early and late markers of osteogenic differentiation. The blue fluorescent images represent DAPI-stained nucleus, red-colored Cy3 shows COL I in the ECM, and merged images are the combination of DAPI and Cy3 (Figure 10A). The production of COL I was minimum at day 7 compared to day 14, judging from the staining intensity, while day 21 had almost similar COL II protein expression to day 14. The relatively similar COL I protein production at day 14 and 21 for BMSCs cultured in the bone part of OC scaffolds is validated through the early stage COL I expression pattern of BMSCs differentiation [54,57]. The well spread appearance of Cy3-conjugated COL I at later stages of observation also confirms the uniform protein distribution within the ECM.

Bone formation in OC scaffolds was further confirmed through OCN protein expression, as shown in Figure 10B. Unlike COL I, OCN had almost no expression at day 14 (shown in green), though the DAPI-stained cell nucleus was present. However, the expression of FITC-conjugated OCN present in the ECM was higher at day 21, which further enhanced to a maximum at day 28. Being a late expression protein marker during bone formation, OCN displayed the same trend as reported in our previous studies [34,54]. Figure 10C demonstrates the protein expression of COL II in the cartilage part of OC scaffolds during co-culture. Similar to OCN, positive COL II staining using FITC-conjugated antibody confirmed cartilage formation in the cartilage part of OC scaffolds [58]. DAPI-stained cell nucleus in the well distributed COL II confirms the proliferation of chondrocytes during the regeneration of cartilage in the cartilage part of OC scaffolds, leading to the formation of osteochondral tissue. The green signal from FITC-conjugated secondary antibody indicates COL II distribution is more spherical in shape along the contour of a microsphere at day 14, whereas it is well distributed throughout the scaffold at day 21, confirming the expansion of chondrocyte ECM towards the pores within the scaffold. Actin-stained chondrocyte morphology in co-cultured OC scaffolds (Figure 9B) could be compared with the current observations in Figure 10C, confirming the attachment of chondrocytes over the microsphere surface at day 14 but extending cell deposition to the pores at day 21, similar to the COL II expression. The results are conclusive of the fact that the BMSCs-seeded bone part and chondrocyte-seeded cartilage part of OC scaffolds showed a great extent of bone and cartilage formation in vitro.

Other than difference in material composition in OC scaffolds employed here, it is also feasible to develop PLGA microsphere-based scaffolds with continuous gradients of bioactive cues [59]. The OC scaffold can be designed with opposing gradients of chondrogenic microspheres (encapsulating transforming growth factor-β1) and osteogenic microspheres (encapsulating bone morphogenetic protein-2) to direct the differentiation of BMSCs regionally toward osteogenesis and chondrogenesis [60].

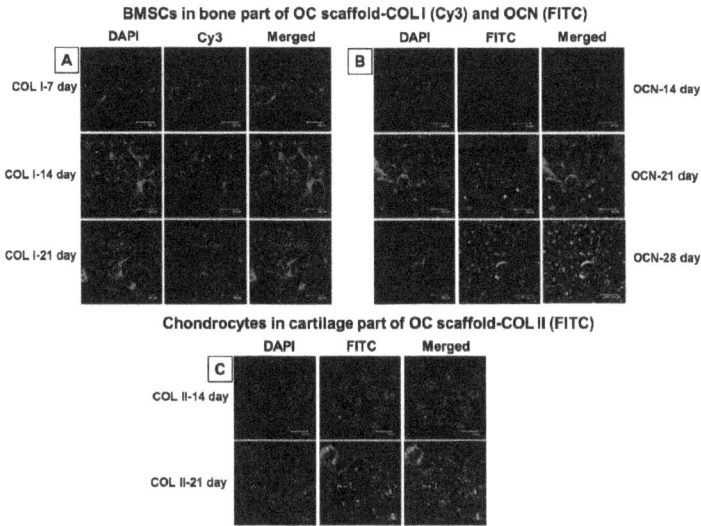

Figure 10. Immunofluorescent staining images using DAPI for cell nucleus and Cy3-conjugated antibody for type I collagen (COL I) (**A**); DAPI for cell nucleus and FITC-conjugated antibody for osteocalcin (OCN) (**B**); DAPI for cell nucleus and FITC-conjugated antibody for type II collagen (COL II) (**C**) at various durations of BMSCs-chondrocytes co-culture in OC scaffolds. Bar = 300 μm.

3.4. In Vivo Animal Studies

3.4.1. Gross and Microscopic Evaluation

For in vivo studies, subcutaneous implantation in nude mice was used to evaluate the biocompatibility of acellular OC scaffolds (control) and cell-seeded OC scaffolds constructs (sample) and their potential in ectopic osteochondral tissue formation. Figure 11A is the gross view photographic observations of harvested controls and samples 4, 8, and 12 weeks post-implantation. Both bone and cartilage parts had blood vessel formation on the surface and interiors of the scaffold (black arrows) at all time points with more intense vascularization after week 12. The shape of the scaffolds was found to be slightly deforming with harvesting times, due to in vivo degradation. Moreover, mice fibrous tissue layer adhesions on scaffold peripherals were observed, which might be due to the high porosity of OC scaffolds. The cross-sectional SEM imaging was performed to observe the cells inside the scaffold at each time point (Figure 11B). A very dense population of sphere-bound cells was observed in both bone and cartilage parts of the OC scaffolds after 4 weeks. In both cases, most of the cells were bridging the gap between adjacent microspheres and thereby filling the porous intersections. Control samples had few cells adhered to microspheres due to fibroblasts infiltration from the host tissue. Both control and sample groups showed unchanged microsphere morphology, confirming their physical stability for short in vivo durations. A dramatic variation in scaffold morphology was observed at week 8, where bone and cartilage parts were surrounded with cellular layers of differentiated BMSCs and chondrocytes, respectively. A few spheres had completely disappeared or covered with newly-formed tissue. The sphere shape deformation was more at 8 weeks compared to week 4, suggesting initiation of bio-degradation. Control samples still had some host cell infiltration similar to 4 weeks, but negligible in comparison with bone and cartilage parts of cell-seeded scaffolds. A complete cellular regeneration was observed on both parts of the OC scaffold after 12 weeks of implantation. Microspheres were partially or completely degraded and cellular multilayers replaced the porous intersections of the scaffold. The narrow cellular bridging observed at 4 and 8 weeks re-formed to thick intense tissue channels at week 12, confirming the faster regeneration of osteochondral tissue. Unlike previous observations, control samples at 12 week displayed a complete collapse of spherical integrity within the scaffold, showing no signs of regeneration. Thus, the microscopic observations of implanted OC scaffolds assure the regeneration potential of co-cultured OC scaffolds in vivo.

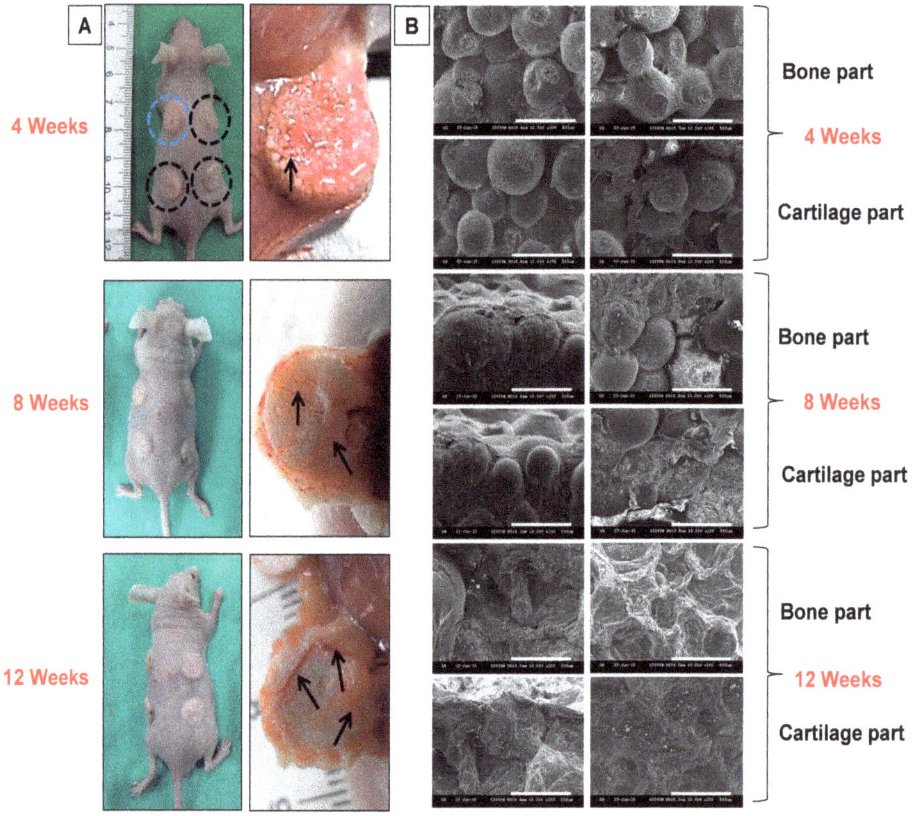

Figure 11. The gross views (**A**) and SEM images (**B**) of acellular OC scaffolds (control) and cell-seeded OC scaffolds (sample) harvested from nude mice 4, 8, and 12 weeks post-implantation. The blue and black circles indicate the control and the sample, respectively. Black arrows indicate blood vessel formation on scaffolds in (**A**); BMSCs and chondrocytes were co-cultured in OC scaffolds and bone and cartilage parts of scaffolds are observed by SEM in (**B**). Bar = 500 μm.

3.4.2. Immunofluorescent Staining for Bone and Cartilage Formation *in Vivo*

Figure 12A,B are images of FITC-labelled staining of OCN and COL II at 4, 8, and 12 weeks, with separate images of DAPI staining of cell nucleus. The 4-week samples showed a complete distribution of OCN deposition over the microspheres with concomitant extension to the surrounding pore space. The mono-dispersed DAPI-stained nucleus observed alongside OCN asserts the multi-layered osteoblasts distribution inside the scaffolds. Unchanged spherical morphology of scaffolds further cross-confirms the relatively higher physical integrity of the scaffolds even after 4 weeks implantation, as observed in SEM images of the 4-week bone part in Figure 11B. At week 8, OCN extended more to the surroundings from deformed sphere surface, revealing enhanced OCN deposition into pores (white arrows). This could be due to the degradation of microspheres along with faster differentiation of BMSCs. Nevertheless, the OCN signal of 12-week harvested samples was dramatically higher in comparison with previous time points. Spherical morphology had completely disappeared and thick OCN distribution was found throughout the scaffold, which could also be interrelated with the scaffold morphology observed for the 12-week bone part samples in Figure 11B. A similar trend in COL II deposition was observed for cartilage regeneration (Figure 12B). As in the bone part, the intensity of cell secreted COL II in the cartilage part of OC scaffolds increased with time, reaching the maximum at week 12. The wide dispersal of COL II at this time point coincides with the 12-week OCN deposition in the bone part and affirms the effective regeneration of both tissues simultaneously. The cell-seeded scaffolds were compared with acellular controls in Figure 12C to eliminate the false positive staining of bone and cartilage marker proteins. DAPI nuclear stains were observed due to the host cell infiltration to the interior of the scaffold, however, there was no positive OCN and COL II staining observed for the control, confirming the effective regeneration of osteochondral tissue regeneration in vivo.

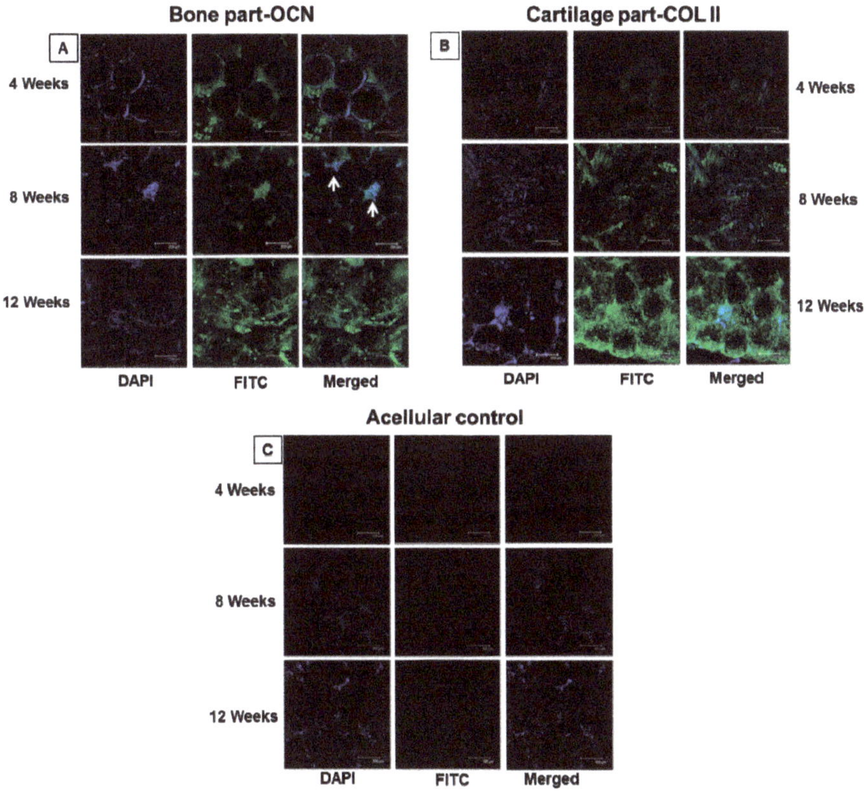

Figure 12. Immunofluorescent staining images using DAPI for cell nucleus and FITC-conjugated antibody for osteocalcin (OCN) in the bone part (**A**) and DAPI for cell nucleus and FITC-conjugated antibody for type II collagen (COL II) in the cartilage part (**B**) of cell-seeded OC scaffolds. The samples were harvested from nude mice 4, 8, and 12 weeks post-implantation. BMSCs and chondrocytes were co-cultured in the OC scaffold and bone and cartilage parts of the scaffolds were observed by confocal microscope. White arrows indicate enhanced OCN deposition into pores. The same staining results were shown for acellular controls in (**C**). Bar = 300 μm.

3.4.3. Histological Analysis

The osteochondral regeneration was finally confirmed through the histological staining of bone and cartilage parts of OC scaffolds through H&E, AR, AB, and Masson's trichrome staining.

Figure 13A,B respectively represents the histological staining images of the BMSCs-seeded bone part and chondrocytes-seeded cartilage part of the OC scaffolds, harvested 4, 8, and 12 weeks post-implantation. H&E, AR, and Masson's trichrome were used for the bone part while H&E and AB were selected for the cartilage part. All staining images of samples were compared with acellular controls. Considering the H&E staining, both bone and cartilage parts revealed significant cell growth inside the scaffold at all implantation durations. H&E is the prime clue for osteoid formation with osteoblasts in the bone matrix, which can be very well observed as the purple background at 4, 8, and 12 weeks. It confirmed cell migration and proliferation, with drastic change in cell number in comparison with acellular controls. The purple color observed at 4 weeks in both bone and cartilage parts is the surrounding tissue entrapped in the tissue slice, which can be confirmed through the absence of microspheres around that area. The intensity of H&E stain at week 8 and 12 of samples for bone and cartilage parts was significantly higher than for controls as well as for the 4-week samples. However, deformed sphere shapes observed at both parts of the scaffolds confirms the biodegradation of the scaffold at 12 weeks, as observed from SEM images in Figure 11B. The blurring of the purple color intensity toward the center of degraded microspheres at 12 weeks displays the uniformly wide dispersion of regenerated tissues in both regions.

This trend was followed in AR and Masson's trichrome stains for bone and AB stain for cartilage. Positive AR staining can reassure the capability of OC scaffolds to continuously support the differentiation of BMSCs into the osteogenic lineage, by displaying bright red spots on the tissue matrix. Basically, AR stain can attach Ca^{2+} ions present in the mineralized bone part, which stems from the differentiation of BMSCs to osteoblasts. Indeed, the mineralized bone part was positively stained with ARS dye and exhibited a linear dependence of stain intensity with duration of implantation. Similar to H&E staining, AR dye had a larger area of staining at week 8 and showed the maximum at a later stage of week 12, endorsing the time-dependent mineralization of BMSCs in bone parts of OC scaffolds. Together with the negative AR staining results of controls at any time point post-implantation, we affirm bone regeneration in the bone part of the co-cultured OC scaffolds in vivo. Masson's trichrome staining was used to detect the presence of collagen in regenerated bone tissue. It could reveal the presence of osteoids as blue stains in the sections of cell-seeded OC scaffolds when BMSCs are differentiated into osteoblasts. The intensity of blue color at the sphere interfaces was much inferior at 4 weeks, but intensified at 8 and 12 weeks. At week 8, the collagen deposition was higher in larger scaffold pores but fewer where sphere–sphere bonding density was lower. At the later stages of implantation, at 12 weeks, the entire sphere interfaces were uniformly stained with Masson's trichrome, with much higher intensity than week 4 and 8, which reveals the potential of co-cultured OC scaffolds to regenerate bone in this region.

AB stain highlights the chondrocyte excreted GAGs in the cartilage part of implanted scaffolds (Figure 13B). The AB blue staining intensity of the sample had a slight increment compared to the control at week 4. However, both the control and sample had stable microsphere morphology as observed for the bone part at the same time point. After 8 weeks, degradation of spheres and restoration of cartilage was observed in the sample, compared to the control. The GAGs are expected to be secreted from the newly formed chondrocytes. Though the density of lacuna was lesser, the available staining area at the microsphere interface was considerably sufficient to emphasis the presence of uniformly regenerated cartilage tissue in the OC scaffolds. However, reduction in intensity of the AB-stained cartilage part at week 12 might be due to the larger number of empty spaces created by partially degraded microspheres in the scaffolds. The possibility of a deceleration in cartilage formation at week 12 was ruled out by the existence of high density COL II deposition found at the same time, from IF staining of COL II in Figure 12B. Thus, the histology results specify the efficacy of co-cultured OC scaffolds towards osteochondral tissue formation, which is conferred through the synergistic effect of pre-differentiated BMSCs in the bone part and chondrocytes in the cartilage part. Using PLGA microsphere-based scaffolds containing a continuous gradient in both material composition (nHAP) and encapsulated growth factors, an in vivo study showed that scaffolds with gradients in both material composition and bioactive signals led to faster osteochondral regeneration, and resulted in restoration of the overlying cartilage with quality integration between the cartilage and bone [61].

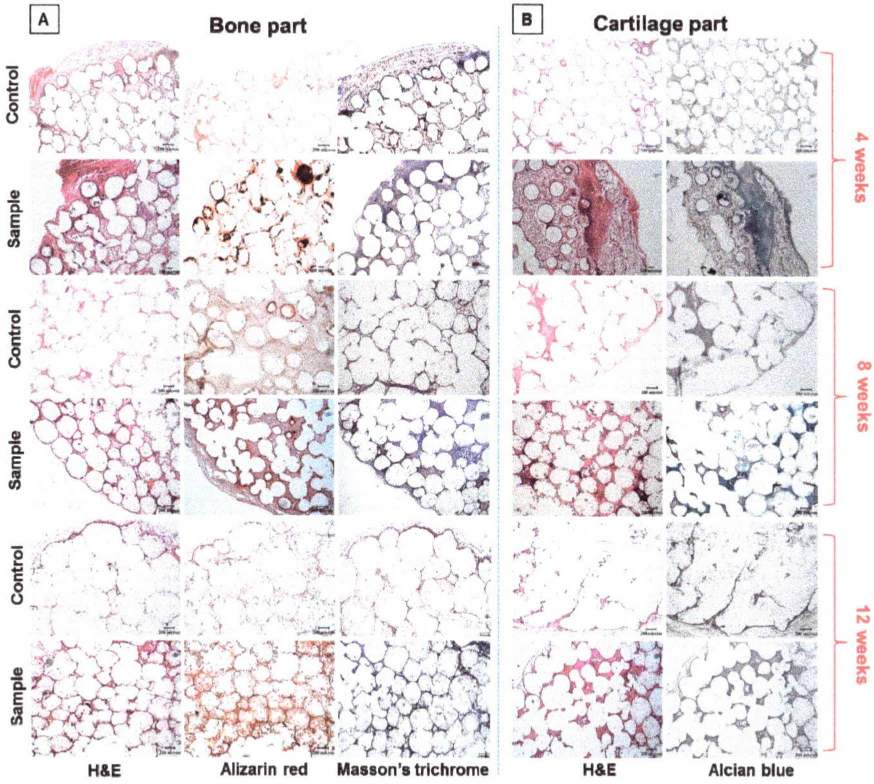

Figure 13. In vivo histological staining of bone (**A**) and cartilage (**B**) parts of co-cultured BMSCs and chondrocytes in OC scaffolds. H&E, Alizarin red, and Masson's trichrome stains of the bone part and H&E and Alcian blue stains of the cartilage part are shown for the cell-seeded sample and the acellular control 4, 8, and 12 weeks post-implantation in nude mice. Bar = 200 μm.

4. Conclusions

This osteochondral tissue engineering is an exciting research area that develops functional strategies for the osteochondral-related clinical issues. In this study, PLGA-based osteochondral scaffolds have been thoroughly investigated for its osteochondral re-modelling efficiency in vitro and in vivo. Three types of scaffolds, i.e., V from PLGA microspheres, C from PLGA/nHAP microspheres, and OC with a combination of PLGA and PLGA/nHAP microspheres, were fabricated through heat sintering using microspheres of size 250–500 μm. TEM observations confirmed the uniform morphology of nHAP, while SEM revealed the smooth and rough morphology of PLGA and PLGA/nHAP microspheres. Both FTIR and XRD data ensured the presence of nHAP in C and OC scaffolds, while elemental analysis confirmed ideal calcium–phosphate ratio of entrapped nHAP. Loading of nHAP in PLGA/nHAP microspheres was estimated to be ~10% by TGA, whereas elemental mapping portrayed the pre-designed arrangement of PLGA/nHAP and PLGA microspheres in OC scaffolds. Mechanical testing of the scaffolds displayed maximum strength for V and minimum for C while OC scaffolds had the value in between both. Live/Dead cell assays of V and C scaffolds single-cultured with chondrocytes, and BMSCs showed similar cell attachment and viability, irrespective of scaffold type. Cell proliferation studies using BMSCs in OM pointed toward the possible mineralization of BMSCs after two weeks of culture and thereby validated the BMSCs culture duration in the bone part of OC scaffolds, prior to chondrocyte seeding. Morphology of pre-differentiated BMSCs in the bone part of OC scaffolds was found unaltered even after co-culture, and based on these results, calcium deposition in the mineralized bone part and GAGs-deposited cartilage part was cross-confirmed through AR staining/CPC treatment and AB staining, respectively. Co-cultured cells in OC scaffolds were found to be viable through Live/Dead assays, while cells with round morphology were observed from cytoskeletal F-actin staining. The presence of COL I/OCN in the bone part and COL II in the cartilage part through IF staining showed the potential of OC scaffolds for biphasic tissue development through co-culture. Most importantly, the positive IF staining and histology results of the newly-formed tissues from the explanted constructs from nude mice endorsed the advantage of hierarchically designed microsphere-based biphasic scaffolds towards effective regeneration of osteochondral tissue.

We concluded that microsphere-based bi-phasic tissue development has a high potential in osteochondral tissue repair. Further studies with graded nHAP concentration and barriers at the bone–cartilage interface zones are promising towards effective remodeling of osteochondral defects in a clinical scenario. Further modifications of the scaffold design with an osteo-conductive gradient in a mechanically tuned bone part and hydrogel-based cartilage part may even have more impacts in the regeneration potential of complex osteochondral defects in vivo.

Acknowledgments: This work was supported by the Ministry of Science and Technology, Taiwan, ROC (MOST-105-2314-B-182-009) and Chang Gung Memorial Hospital (BMRP 249, CMRPD3D0201 and CMRPD3D0202). The Microscope Core Laboratory, Chang Gung Memorial Hospital, Linkou is acknowledged for the expert technical assistance.

Author Contributions: K. T. Shalumon, Han-Tsung Liao and Jyh-Ping Chen conceived and designed the experiments; K. T. Shalumon and Yi Teng Fong performed the experiments and analyzed the data; K. T. Shalumon and Jyh-Ping Chen wrote the paper.

Conflicts of Interest: The authors declare no conflict of interest.

References

1. O'Driscoll, S.W. The healing and regeneration of articular cartilage. *J. Bone Jt. Surg. Am.* **1998**, *80*, 1795–1812. [CrossRef]

2. Swieszkowski, W.; Tuan, B.H.; Kurzydlowski, K.J.; Hutmacher, D.W. Repair and regeneration of osteochondral defects in the articular joints. *Biomol. Eng.* **2007**, *24*, 489–495. [CrossRef] [PubMed]

3. Yang, P.J.; Temenoff, J.S. Engineering orthopedic tissue interfaces. *Tissue Eng. B* **2009**, *15*, 127–141. [CrossRef] [PubMed]

4. Mano, J.F.; Reis, R.L. Osteochondral defects: Present situation and tissue engineering approaches. *J. Tissue Eng. Regen. Med.* **2007**, *1*, 261–273. [CrossRef] [PubMed]

5. Guarino, V.; Gloria, A.; Raucci, M.G.; Ambrosio, L. Hydrogel-based platforms for the regeneration of osteochondral tissue and intervertebral disc. *Polymers* **2012**, *4*, 1590–1612. [CrossRef]

6. Shi, S.; Gronthos, S.; Chen, S.; Reddi, A.; Counter, C.M.; Robey, P.G.; Wang, C.Y. Bone formation by human postnatal bone marrow stromal stem cells is enhanced by telomerase expression. *Nat. Biotechnol.* **2002**, *20*, 587–591. [CrossRef] [PubMed]

7. Yousefi, A.-M.; James, P.F.; Akbarzadeh, R.; Subramanian, A.; Flavin, C.; Oudadesse, H. Prospect of stem cells in bone tissue engineering: A review. *Stem Cells Int.* **2016**, *2016*, 13. [CrossRef] [PubMed]

8. Zhang, N.; Wang, Y.; Xu, W.; Hu, Y.; Ding, J. Poly(lactide-co-glycolide)/hydroxyapatite porous scaffold with microchannels for bone regeneration. *Polymers* **2016**, *8*, 218. [CrossRef]

9. Lee, W.D.; Hurtig, M.B.; Pilliar, R.M.; Stanford, W.L.; Kandel, R.A. Engineering of hyaline cartilage with a calcified zone using bone marrow stromal cells. *Osteoarthr. Cartil.* **2015**, *23*, 1307–1315. [CrossRef] [PubMed]

10. Vinatier, C.; Guicheux, J. Cartilage tissue engineering: From biomaterials and stem cells to osteoarthritis treatments. *Ann. Phys. Rehabil. Med.* **2016**, *59*, 139–144. [CrossRef] [PubMed]

11. Wang, S.-J.; Zhang, Z.-Z.; Jiang, D.; Qi, Y.-S.; Wang, H.-J.; Zhang, J.-Y.; Ding, J.-X.; Yu, J.-K. Thermogel-coated poly(ε-caprolactone) composite scaffold for enhanced cartilage tissue engineering. *Polymers* **2016**, *8*, 200. [CrossRef]

12. Yan, L.-P.; Oliveira, J.M.; Oliveira, A.L.; Reis, R.L. Current concepts and challenges in osteochondral tissue engineering and regenerative medicine. *ACS Biomater. Sci. Eng.* **2015**, *1*, 183–200. [CrossRef]

13. Nukavarapu, S.P.; Dorcemus, D.L. Osteochondral tissue engineering: Current strategies and challenges. *Biotechnol. Adv.* **2013**, *31*, 706–721. [CrossRef] [PubMed]

14. Hutmacher, D.W. Scaffolds in tissue engineering bone and cartilage. *Biomaterials* **2000**, *21*, 2529–2543. [CrossRef]

15. Angele, P.; Kujat, R.; Nerlich, M.; Yoo, J.; Goldberg, V.; Johnstone, B. Engineering of osteochondral tissue with bone marrow mesenchymal progenitor cells in a derivatized hyaluronan-gelatin composite sponge. *Tissue Eng.* **1999**, *5*, 545–554. [CrossRef] [PubMed]

16. Gao, J.; Dennis, J.E.; Solchaga, L.A.; Awadallah, A.S.; Goldberg, V.M.; Caplan, A.I. Tissue-engineered fabrication of an osteochondral composite graft using rat bone marrow-derived mesenchymal stem cells. *Tissue Eng.* **2001**, *7*, 363–371. [CrossRef] [PubMed]

17. Oliveira, J.M.; Rodrigues, M.T.; Silva, S.S.; Malafaya, P.B.; Gomes, M.E.; Viegas, C.A.; Dias, I.R.; Azevedo, J.T.; Mano, J.F.; Reis, R.L. Novel hydroxyapatite/chitosan bilayered scaffold for osteochondral tissue-engineering applications: Scaffold design and its performance when seeded with goat bone marrow stromal cells. *Biomaterials* **2006**, *27*, 6123–6137. [CrossRef] [PubMed]

18. Wang, X.; Grogan, S.P.; Rieser, F.; Winkelmann, V.; Maquet, V.; La Berge, M.; Mainil-Varlet, P. Tissue engineering of biphasic cartilage constructs using various biodegradable scaffolds: An in vitro study. *Biomaterials* **2004**, *25*, 3681–3688. [CrossRef] [PubMed]

19. Sherwood, J.K.; Riley, S.L.; Palazzolo, R.; Brown, S.C.; Monkhouse, D.C.; Coates, M.; Griffith, L.G.; Landeen, L.K.; Ratcliffe, A. A three-dimensional osteochondral composite scaffold for articular cartilage repair. *Biomaterials* **2002**, *23*, 4739–4751. [CrossRef]

20. Holland, T.A.; Bodde, E.W.; Baggett, L.S.; Tabata, Y.; Mikos, A.G.; Jansen, J.A. Osteochondral repair in the rabbit model utilizing bilayered, degradable oligo(poly(ethylene glycol) fumarate) hydrogel scaffolds. *J. Biomed. Mater. Res. A* **2005**, *75*, 156–167. [CrossRef] [PubMed]

21. Chen, G.; Tanaka, J.; Tateishi, T. Osteochondral tissue engineering using a PLGA–collagen hybrid mesh. *Mater. Sci. Eng. C* **2006**, *26*, 124–129. [CrossRef]

22. Seo, S.-J.; Mahapatra, C.; Singh, R.K.; Knowles, J.C.; Kim, H.-W. Strategies for osteochondral repair: Focus on scaffolds. *J. Tissue Eng.* **2014**, *5*. [CrossRef] [PubMed]

23. Keeney, M.; Pandit, A. The osteochondral junction and its repair via bi-phasic tissue engineering scaffolds. *Tissue Eng. B* **2009**, *15*, 55–73. [CrossRef] [PubMed]

24. Sharma, B.; Elisseeff, J.H. Engineering structurally organized cartilage and bone tissues. *Ann. Biomed. Eng.* **2004**, *32*, 148–159. [CrossRef] [PubMed]

25. Altamura, D.; Pastore, S.G.; Raucci, M.G.; Siliqi, D.; De Pascalis, F.; Nacucchi, M.; Ambrosio, L.; Giannini, C. Scanning small- and wide-angle X-ray scattering microscopy selectively probes ha content in gelatin/hydroxyapatite scaffolds for osteochondral defect repair. *ACS Appl. Mater. Interfaces* **2016**, *8*, 8728–8736. [CrossRef] [PubMed]

26. Mano, J.F.; Silva, G.A.; Azevedo, H.S.; Malafaya, P.B.; Sousa, R.A.; Silva, S.S.; Boesel, L.F.; Oliveira, J.M.; Santos, T.C.; Marques, A.P.; et al. Natural origin biodegradable systems in tissue engineering and regenerative medicine: Present status and some moving trends. *J. R. Soc. Interface* **2007**, *4*, 999–1030. [CrossRef] [PubMed]

27. Kokubo, T. Bioactive glass ceramics: Properties and applications. *Biomaterials* **1991**, *12*, 155–163. [CrossRef]

28. Gentile, P.; Wilcock, C.; Miller, C.; Moorehead, R.; Hatton, P. Process optimisation to control the physico-chemical characteristics of biomimetic nanoscale hydroxyapatites prepared using wet chemical precipitation. *Materials* **2015**, *8*, 2297–2310. [CrossRef]

29. Nagai, T.; Sato, M.; Kutsuna, T.; Kokubo, M.; Ebihara, G.; Ohta, N.; Mochida, J. Intravenous administration of anti-vascular endothelial growth factor humanized monoclonal antibody bevacizumab improves articular cartilage repair. *Arthritis Res. Ther.* **2010**, *12*, R178. [CrossRef] [PubMed]

30. Wang, X.; Wenk, E.; Zhang, X.; Meinel, L.; Vunjak-Novakovic, G.; Kaplan, D.L. Growth factor gradients via microsphere delivery in biopolymer scaffolds for osteochondral tissue engineering. *J. Control. Release* **2009**, *134*, 81–90. [CrossRef] [PubMed]

31. Babensee, J.E.; McIntire, L.V.; Mikos, A.G. Growth factor delivery for tissue engineering. *Pharm. Res.* **2000**, *17*, 497–504. [CrossRef] [PubMed]

32. Chen, F.-M.; Zhang, M.; Wu, Z.-F. Toward delivery of multiple growth factors in tissue engineering. *Biomaterials* **2010**, *31*, 6279–6308. [CrossRef] [PubMed]

33. Yousefi, A.M.; Hoque, M.E.; Prasad, R.G.; Uth, N. Current strategies in multiphasic scaffold design for osteochondral tissue engineering: A review. *J. Biomed. Mater. Res. A* **2015**, *103*, 2460–2481. [CrossRef] [PubMed]

34. Lai, G.J.; Shalumon, K.T.; Chen, J.P. Response of human mesenchymal stem cells to intrafibrillar nanohydroxyapatite content and extrafibrillar nanohydroxyapatite in biomimetic chitosan/silk fibroin/ nanohydroxyapatite nanofibrous membrane scaffolds. *Int. J. Nanomed.* **2015**, *10*, 567–584.

35. Yang, Y.-Y.; Chia, H.-H.; Chung, T.-S. Effect of preparation temperature on the characteristics and release profiles of PLGA microspheres containing protein fabricated by double-emulsion solvent extraction/evaporation method. *J. Control. Release* **2000**, *69*, 81–96. [CrossRef]

36. Jeng, L.B.; Chung, H.Y.; Lin, T.M.; Chen, J.P.; Chen, Y.L.; Lu, Y.L.; Wang, Y.J.; Chang, S.C. Characterization and osteogenic effects of mesenchymal stem cells on microbeads composed of hydroxyapatite nanoparticles/reconstituted collagen. *J. Biomed. Mater. Res. A* **2009**, *91*, 886–893. [CrossRef] [PubMed]

37. Chen, J.P.; Cheng, T.H. Thermo-responsive chitosan-*graft*-poly(*n*-isopropylacrylamide) injectable hydrogel for cultivation of chondrocytes and meniscus cells. *Macromol. Biosci.* **2006**, *6*, 1026–1039. [CrossRef] [PubMed]

38. Kim, Y.J.; Sah, R.L.; Doong, J.Y.; Grodzinsky, A.J. Fluorometric assay of DNA in cartilage explants using hoechst 33258. *Anal. Biochem.* **1988**, *174*, 168–176. [CrossRef]

39. Huang, W.; Li, X.; Shi, X.; Lai, C. Microsphere based scaffolds for bone regenerative applications. *Biomater. Sci.* **2014**, *2*, 1145–1153. [CrossRef]

40. Galperin, A.; Oldinski, R.A.; Florczyk, S.J.; Bryers, J.D.; Zhang, M.; Ratner, B.D. Integrated bi-layered scaffold for osteochondral tissue engineering. *Adv. Healthc. Mater.* **2013**, *2*, 872–883. [CrossRef] [PubMed]

41. Loh, Q.L.; Choong, C. Three-dimensional scaffolds for tissue engineering applications: Role of porosity and pore size. *Tissue Eng. B* **2013**, *19*, 485–502. [CrossRef] [PubMed]

42. Erbetta, C.D.A.C.; Alves, R.J.; Resende, J.M.E.; Freitas, R.F.D.S.; Sousa, R.G.D. Synthesis and characterization of poly(D,L-lactide-*co*-glycolide) copolymer. *J. Biomater. Nanobiotechnol.* **2012**, *3*, 208–225. [CrossRef]

43. Chen, J.-P.; Tsai, M.-J.; Liao, H.-T. Incorporation of biphasic calcium phosphate microparticles in injectable thermoresponsive hydrogel modulates bone cell proliferation and differentiation. *Colloid Surf. B* **2013**, *110*, 120–129. [CrossRef] [PubMed]

44. Daculsi, G.; Legeros, R.Z.; Nery, E.; Lynch, K.; Kerebel, B. Transformation of biphasic calcium phosphate ceramics in vivo: Ultrastructural and physicochemical characterization. *J. Biomed. Mater. Res.* **1989**, *23*, 883–894. [CrossRef] [PubMed]

45. Lu, H.H.; El-Amin, S.F.; Scott, K.D.; Laurencin, C.T. Three-dimensional, bioactive, biodegradable, polymer–bioactive glass composite scaffolds with improved mechanical properties support collagen synthesis and mineralization of human osteoblast-like cells in vitro. *J. Biomed. Mater. Res. A* **2003**, *64*, 465–474. [CrossRef] [PubMed]

46. Shi, X.; Wang, Y.; Ren, L.; Lai, C.; Gong, Y.; Wang, D.-A. A novel hydrophilic poly(lactide-*co*-glycolide)/ lecithin hybrid microspheres sintered scaffold for bone repair. *J. Biomed. Mater. Res. A* **2010**, *92*, 963–972. [CrossRef] [PubMed]

47. Uematsu, K.; Hattori, K.; Ishimoto, Y.; Yamauchi, J.; Habata, T.; Takakura, Y.; Ohgushi, H.; Fukuchi, T.; Sato, M. Cartilage regeneration using mesenchymal stem cells and a three-dimensional poly-lactic-glycolic acid (PLGA) scaffold. *Biomaterials* **2005**, *26*, 4273–4279. [CrossRef] [PubMed]

48. Aravamudhan, A.; Ramos, D.M.; Jenkins, N.A.; Dyment, N.A.; Sanders, M.M.; Rowe, D.W.; Kumbar, S.G. Collagen nanofibril self-assembly on a natural polymeric material for the osteoinduction of stem cells in vitro and biocompatibility in vivo. *RSC Adv.* **2016**, *6*, 80851–80866. [CrossRef]

49. Nooeaid, P.; Salih, V.; Beier, J.P.; Boccaccini, A.R. Osteochondral tissue engineering: Scaffolds, stem cells and applications. *J. Cell. Mol. Med.* **2012**, *16*, 2247–2270. [CrossRef] [PubMed]

50. Lin, L.; Chow, K.L.; Leng, Y. Study of hydroxyapatite osteoinductivity with an osteogenic differentiation of mesenchymal stem cells. *J. Biomed. Mater. Res. A* **2009**, *89*, 326–335. [CrossRef] [PubMed]

51. Ahmadzadeh, E.; Talebnia, F.; Tabatabaei, M.; Ahmadzadeh, H.; Mostaghaci, B. Osteoconductive composite graft based on bacterial synthesized hydroxyapatite nanoparticles doped with different ions: From synthesis to in vivo studies. *Nanomedicine* **2016**, *12*, 1387–1395. [CrossRef] [PubMed]
52. Meinel, L.; Karageorgiou, V.; Fajardo, R.; Snyder, B.; Shinde-Patil, V.; Zichner, L.; Kaplan, D.; Langer, R.; Vunjak-Novakovic, G. Bone tissue engineering using human mesenchymal stem cells: Effects of scaffold material and medium flow. *Ann. Biomed. Eng.* **2004**, *32*, 112–122. [CrossRef] [PubMed]
53. Nishimura, I.; Hisanaga, R.; Sato, T.; Arano, T.; Nomoto, S.; Ikada, Y.; Yoshinari, M. Effect of osteogenic differentiation medium on proliferation and differentiation of human mesenchymal stem cells in three-dimensional culture with radial flow bioreactor. *Regen. Ther.* **2015**, *2*, 24–31. [CrossRef]
54. Shalumon, K.T.; Lai, G.J.; Chen, C.H.; Chen, J.P. Modulation of bone-specific tissue regeneration by incorporating bone morphogenetic protein and controlling the shell thickness of silk fibroin/chitosan/nanohydroxyapatite core-shell nanofibrous membranes. *ACS Appl. Mater. Interfaces* **2015**, *7*, 21170–21181. [CrossRef] [PubMed]
55. Stossel, T.P. On the crawling of animal cells. *Science* **1993**, *260*, 1086–1094. [CrossRef] [PubMed]
56. Borden, M.; El-Amin, S.F.; Attawia, M.; Laurencin, C.T. Structural and human cellular assessment of a novel microsphere-based tissue engineered scaffold for bone repair. *Biomaterials* **2003**, *24*, 597–609. [CrossRef]
57. Black, C.R.M.; Goriainov, V.; Gibbs, D.; Kanczler, J.; Tare, R.S.; Oreffo, R.O. Bone tissue engineering. *Curr. Mol. Biol. Rep.* **2015**, *1*, 132–140. [CrossRef] [PubMed]
58. Kuo, C.-Y.; Chen, C.-H.; Hsiao, C.-Y.; Chen, J.-P. Incorporation of chitosan in biomimetic gelatin/chondroitin-6-sulfate/hyaluronan cryogel for cartilage tissue engineering. *Carbohydr. Polym.* **2015**, *117*, 722–730. [CrossRef] [PubMed]
59. Dormer, N.H.; Singh, M.; Wang, L.; Berkland, C.J.; Detamore, M.S. Osteochondral interface tissue engineering using macroscopic gradients of bioactive signals. *Ann. Biomed. Eng.* **2010**, *38*, 2167–2182. [CrossRef] [PubMed]
60. Dormer, N.H.; Singh, M.; Zhao, L.; Mohan, N.; Berkland, C.J.; Detamore, M.S. Osteochondral interface regeneration of the rabbit knee with macroscopic gradients of bioactive signals. *J. Biomed. Mater. Res. A* **2012**, *100*, 162–170. [CrossRef] [PubMed]
61. Mohan, N.; Dormer, N.H.; Caldwell, K.L.; Key, V.H.; Berkland, C.J.; Detamore, M.S. Continuous gradients of material composition and growth factors for effective regeneration of the osteochondral interface. *Tissue Eng. A* **2011**, *17*, 2845–2855. [CrossRef] [PubMed]

polymers

MDPI

Article

A Biodegradable Microneedle Cuff for Comparison of Drug Effects through Perivascular Delivery to Balloon-Injured Arteries

Dae-Hyun Kim [1], Eui Hwa Jang [1], Kang Ju Lee [2], Ji Yong Lee [2], Seung Hyun Park [2], Il Ho Seo [2], Kang Woog Lee [1], Seung Hyun Lee [1], WonHyoung Ryu [2,*] and Young-Nam Youn [1,*]

[1] Division of Cardiovascular Surgery, Severance Cardiovascular Hospital, Yonsei University College of Medicine, 50-1 Yonsei-ro, Sedaemun-gu, Seoul 03722, Korea; vet1982@hanmail.net (D.-H.K.); nurjih83@yuhs.ac (E.H.J.); sysebg@yuhs.ac (K.W.L.); henry75@yuhs.ac (S.H.L.)
[2] School of Mechanical Engineering, Yonsei University, 50 Yonsei-ro, Sedaemun-gu, Seoul 03722, Korea; knjulee@gmail.com (K.J.L.); bamemo@naver.com (J.Y.L.); dwitong@naver.com (S.H.P.); kiayora@gmail.com (I.H.S.)
* Correspondence: whryu@yonsei.ac.kr (W.R.); ynyoun@yuhs.ac (Y.-N.Y.); Tel.: +82-2-2123-5821 (W.R.); +82-2-2228-8487 (Y.-N.Y.); Fax: +82-2-312-2159 (W.R.); +82-2-313-2992 (Y.-N.Y.)

Academic Editors: João F. Mano and Insung S. Choi
Received: 5 January 2017; Accepted: 3 February 2017; Published: 8 February 2017

Abstract: Restenosis at a vascular anastomosis site is a major cause of graft failure and is difficult to prevent by conventional treatment. Perivascular drug delivery has advantages as drugs can be diffused to tunica media and subintima while minimizing the direct effect on endothelium. This in vivo study investigated the comparative effectiveness of paclitaxel, sirolimus, and sunitinib using a perivascular biodegradable microneedle cuff. A total of 31 New Zealand white rabbits were used. Rhodamine was used to visualize drug distribution ($n = 3$). Sirolimus- ($n = 7$), sunitinib- ($n = 7$), and paclitaxel-loaded ($n = 7$) microneedle cuffs were placed at balloon-injured abdominal aortae and compared to drug-free cuffs ($n = 7$). Basic histological structures were not affected by microneedle devices, and vascular wall thickness of the device-only group was similar to that of normal artery. Quantitative analysis revealed significantly decreased neointima formation in all drug-treated groups ($p < 0.001$). However, the tunica media layer of the paclitaxel-treated group was significantly thinner than that of other groups and also showed the highest apoptotic ratio ($p < 0.001$). Proliferating cell nuclear antigen (PCNA)-positive cells were significantly reduced in all drug-treated groups. Sirolimus or sunitinib appeared to be more appropriate for microneedle devices capable of slow drug release because vascular wall thickness was minimally affected.

Keywords: anastomosis; drug delivery; microneedle; restenosis; peripheral vascular disease

1. Introduction

Application of biodegradable polymers for medical purposes—such as drug delivery, nanoparticle imaging technology, and soft tissue reconstruction—has been widely reported [1–4]. Recently, a lot of polymeric devices such as films/wraps, meshes, rings, or micro/nano particles were also introduced for drug delivery to various cardiovascular lesions [5–10].

Cardiovascular disease is a major cause of death worldwide [11]. In particular, obstructive diseases of small-sized vessels like myocardial infarction have the highest mortality rates among various cardiovascular diseases [11]. Although continuous progress has been made, successful implementation of non-surgical treatments such as balloon angioplasty and endovascular stenting remains challenging due to restenosis derived from neointimal hyperplasia [12,13].

When vascular endothelium is damaged by any cause, smooth muscle cells (SMCs) migrate to an intima layer of the vessel and excessively proliferate, leading to neointimal hyperplasia [14]. Therefore, various attempts have been made to locally deliver anti-proliferative agents, such as paclitaxel and sirolimus, via the use of drug-eluting stent (DES) or drug-eluting balloons [15–18]. Although promising-inhibitory effects on neointimal hyperplasia have been reported in various preclinical and clinical studies, these methods are limited in enabling sustained, controlled drug delivery to the tunica media layer, or as effectively as desired [19,20]. In addition, loss of vascular patency due to late stent thrombosis or restenosis was reported in up to 20% of patients in long-term follow-up studies on DES [12,13].

When restenosis and intimal hyperplasia cannot be limited by intraluminal therapy, arterial replacement or bypass surgery is currently accepted as a general surgical treatment method [11]. However, the occurrence rate of restenosis due to neointimal hyperplasia at vascular anastomosis sites is also relatively high in small-sized vessels less than 6 mm in diameter [11]. In several studies, extraluminal drug delivery using a perivascular biodegradable material was reported to reduce the recurrent rate [12,21]. Such an approach is advantageous as drugs can be diffused to not only the tunica adventitia but also tunica media and subintima while minimizing the direct effect on endothelium [12].

To explore this advantage further, we developed a perivascular cuff containing an array of microneedles (MN) that has been shown to drastically increase drug delivery efficiency compared with devices without MN [20,22]. In addition, the device itself did not affect normal vascular structures and demonstrated a capability to effectively inhibit neointima formation by delivering paclitaxel to the vascular tissue layer [20]. However, thinning of the tunica media layer was observed in paclitaxel-loading groups on histopathological examination.

In the present study, we thus aim to compare the effects other anti-proliferating drugs with paclitaxel on inhibition of neointima formation as well as thinning of tunica media when delivered through a MN cuff. Specifically, sirolimus is widely used in DES and has been shown to inhibit neointimal hyperplasia effectively through cytostatic mechanisms while sunitinib malate salt (sunitinib) was recently reported to show an inhibitory effect on neointimal hyperplasia. Therefore, these two drugs were compared with paclitaxel.

2. Methods

2.1. MN Device Fabrication

The MN cuff was fabricated through the following steps, as previously reported: MN array fabrication by thermal drawing, dip coating of drug onto MN end tips, and post-annealing for cuff shape form (Figure 1) [20,22]. Briefly, a biodegradable polymer film of 100 μm thickness was fabricated with poly(lactic-*co*-glycolic acid) (PLGA) 90/10 (M_w = 268,000, Samyang Biopharmaceuticals, Gyeonggi-do, Korea). Subsequently, a 3 × 3 array of micro-pillars attached to a heating cartridge was lowered toward the PLGA film on a hot chuck. After making contact between the micro-pillars and PLGA film, the micro-pillar array was lifted up with an automatic micro-stage and this motion resulted in thermal drawing of the heated PLGA. Detailed shapes of the MNs were adjusted by carefully modulating pillar and film temperature, drawing speeds, drawing steps, and drawing distance for each step.

Paclitaxel, sirolimus, and sunitinib (P9600, R5000 and S-8803, LC Laboratories, Woburn, MA, USA) were used as anti-proliferative drugs in this study (Figure 2A,B). The drug formulations for sustained perivascular drug delivery were mixed at a 3:1:0.33 weight ratio of DMSO, PLGA50/50 (PDLG5010, PURAC, Gorinchem, The Netherlands), and anti-proliferative drug. As previously reported, 1.2 μg of each drug was loaded into MN end tips by dip coating with the drug formulations [23]. Then MN devices were annealed in a stainless cylinder 3 mm in diameter to form a cuff shape.

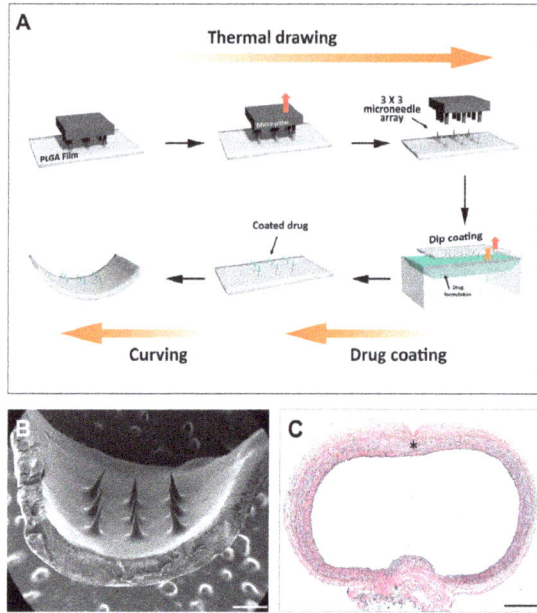

Figure 1. (**A**) A schematic diagram of microneedle cuff fabrication; (**B**) A SEM image of microneedle cuff (the scale bar indicates 500 µm); (**C**) A Microscopic image of an abdominal aorta with the microneedle insertion. The asterisk indicates the insertion mark by the microneedle cuff device (hematoxylin and eosin staining; the scale bar indicates 500 µm).

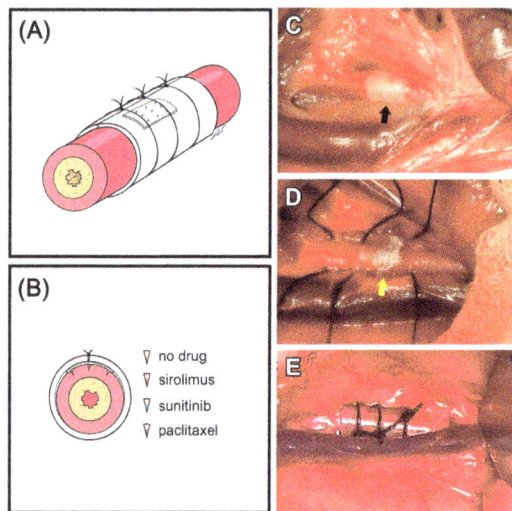

Figure 2. (**A**) A schematic view of microneedle device installation on an artery post induction of neointimal hyperplasia; (**B**) Cross-sectional view of the device installed on the artery and scheme of small box); each experimental group; (**C**) Balloon injury before MN device installation (black arrow: saline-inflated balloon); (**D**) MN device (yellow arrow) over the abdominal aorta; (**E**) Tightly-fixed MN device with Tygon tube and silks.

2.2. Animal Experiment

A total of 31 healthy male New Zealand white rabbits weighing 3.3 ± 0.2 kg (range: 2.9–3.6 kg) were used in the present study. Animal care and experimental procedures were conducted in accordance with the guidelines approved by the Institutional Animal Care and Use Committee (IACUC) at Yonsei University Health System, Seoul, Korea (IACUC approval No. 2015-0020, 2015).

The following experimental groups were compared in terms of the delivery effectiveness and the anti-neointimal formation effect of each drug after arterial balloon injury: rhodamine group (*n* = 3), fresh artery + rhodamine B-loaded MN device; I + D group (*n* = 7), balloon injury + MN device only; I + Sir group (*n* = 7), balloon injury + sirolimus-loaded MN device; I + Sun group (*n* = 7), balloon injury + sunitinib-loaded MN device; and I + Ptx group (*n* = 7), balloon injury + paclitaxel-loaded MN device. Samples of normal aorta were obtained from three New Zealand white rabbits used in tracheal transplantation-associated research, which was unrelated to vascular pathology (IACUC approval No. 2014-0172-2, 2014).

Each animal was intramuscularly injected with 5 mg/kg of xylazine and 15 mg/kg of Zoletil® every 15 min as a premedication. After intubation with a 3.0 or 3.5 mm endotracheal tube, 1.5%–2.0% isoflurane was used to maintain inhalation anesthesia. All the animals received crystalloid solution (10 mL/kg/h) throughout the surgical procedure.

After each animal was anesthetized, an abdominal midline incision was performed. The bowel was retracted to expose the abdominal aorta, and blunt dissection was performed to separate surrounding connective tissue. Then the right inguinal area was incised to expose the right femoral artery. Heparin (100 U/kg) was administered intravenously. A 3-Fr Fogarty embolectomy catheter was inserted through the right femoral artery and positioned in the abdominal aorta. The abdominal aorta was de-endothelialized with three passes of a balloon inflated with 0.05 ml saline (Figure 2C). The MN device was then positioned and fixed with a Tygon® tube and three 4–0 silks (Figure 2D,E). After removing the balloon catheter, the right femoral artery was ligated, and the abdominal cavity, subcutaneous tissue, and skin were closed with general procedures. As analgesic, 1 mg/kg of ketolorac tromethamine was administered intramuscularly, two times per day for one week. For antibiotics, 5 mg/kg of enrofloxacin was administered subcutaneously, two times per day for one week. Aspirin (10 mg/kg) was administered orally once per day for one week.

2.3. Fluorescent Microscopic Analysis

To visualize post-delivery drug distribution within vascular tissue, a fluorescent model drug, rhodamine B, was used for a two-week in vivo study. For sampling, the animal was induced into anesthesia using the same method as described above, and 100 U/kg of heparin was administered intravenously. Vascular clamps were applied at the cranial and caudal portions of the artery around the device location. After collecting samples by en-bloc resection, animals were euthanized by bolus injection of high-dose KCl.

The collected samples were trimmed to appropriate sizes after removing the MN devices. They were then put into molds, embedded into optimal cutting temperature (OCT) compound, frozen in a −80 °C deep freezer, and finally cryosectioned to 5-μm thickness. Fluorescent images were taken using a computer-assisted image analysis program (DP Controller, Olympus), with exposure time maintained as 0.6 seconds at ISO 100. Under the same conditions, fluorescent images of normal vessels were taken as a reference control.

2.4. Histopathological Examinations

For histopathological analysis, animals were euthanized in the same method described above, and samples were collected. Each sample was processed using standard methods and embedded in paraffin, then sectioned to 4-μm thickness and stained.

Hematoxylin (Sigma-Aldrich, St. Louis, MO, USA) and eosin (BBC Biochemical, Mt Vernon, WA, USA) staining was used to identify basic pathological changes such as vessel wall thickness, narrowing of vessel lumen, and neointimal formation. Movat's pentachrome staining (Movat pentachrome stain kit, Empire Genomics, Buffalo, NY, USA) was used to identify detailed vessel components.

For immunohistochemistry, the sections were deparaffinized using standard protocols, and antigen retrieval was performed with proteinase K. Then, sections were incubated for 10 min in 3% hydrogen peroxide (Duksan Hydrogen peroxide 3059, Gyeonggi-do, Korea) to inactivate endogenous peroxidase, and blocked by incubating for 1 h in 5% bovine serum albumin. Subsequently, the sections were labelled with the mouse anti-PCNA antibody (AbD Serotec, MCA1558). The antibody-labelled sections were then incubated with Dako EnVision+ System-HRP-labelled polymer antimouse kit solution (Dako4001, Denmark). DAP (K-3468, Dako, Glostrup, Denmark) staining was performed for 3 min at room temperature for tissue visualization. Sections were then counterstained with hematoxylin. Microscopic assessment was conducted by an independent pathologist (from a different institution) who was blinded to the treatment allocation.

Quantitative analysis was performed using Image J software (National Institutes of Health, Bethesda, MD, USA). A layer inside the internal elastic lamina was measured as the area of the neointima, and another layer between the internal and external elastic lamina was measured as the area of the tunica media. The extent of stenosis was determined as follows: % neointimal formation = neointimal area $\times$ 100/(neointimal + luminal area). The average tunica media thickness from eight sites in each vessel was compared.

2.5. TUNEL (Terminal Deoxynucleotidyl Transferase dUTP Nick end Labeling) Assay

To compare the percentage of apoptotic vascular SMCs in the tunica media, TUNEL immunohistochemical staining (In Situ Cell Death Detection kit, Roche, Mannheim, Germany) was performed, and nuclear fast red staining was performed as counterstaining. Microscopic images at $400\times$ magnification from eight sites of each vessel were used for apoptotic cell counting. Apoptotic ratio (%) was calculated as the ratio of apoptotic nuclei to total cell nuclei.

2.6. Statistical Analysis

All data are expressed as mean $\pm$ standard deviation. Statistical analyses were performed using GraphPad Prism 5.0 software (GraphPad Software, Inc., San Diego, CA, USA). Normal data distribution was determined using Shapiro-Wilk test. One-way ANOVA and post hoc Bonferroni tests were used to compare mean tunica media thickness, TUNEL-positive cells, and PCNA-positive cells between groups. Kruskal-Wallis test and Mann Whitney test were applied for testing overall/pair-wise group mean differences in other quantitative analysis results between groups. A p-value < 0.05 was considered statistically significant.

3. Results

3.1. Drug Distribution in Vascular Tissue by MN Device

Fluorescent imaging of tissue sections confirmed rhodamine B was distributed almost in the entire vascular tissue. Fluorescence signal of rhodamine was detected in not only tunica adventitia, but also tunica media layer and subintima (Figure 3).

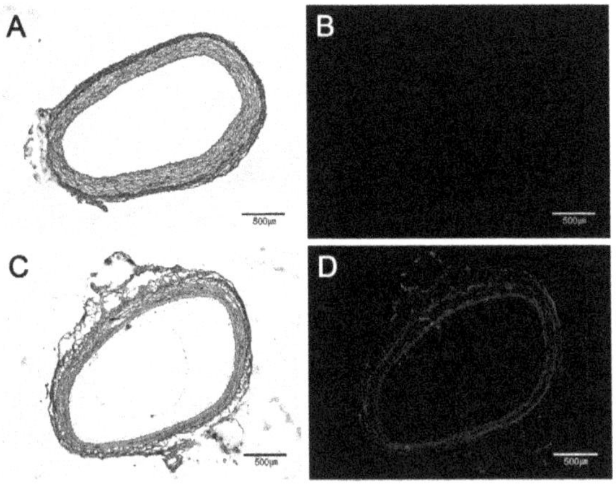

Figure 3. Fluorescent images of drug distribution. Compared to that in normal abdominal aorta (**B**), well-distributed rhodamine B delivered by the MN device is shown as red color in vascular tissue (**D**). (**A**) A light microscopic image of normal abdominal aorta; (**B**) A fluorescent image of normal abdominal aorta; (**C**) A light microscopic image of rhodamine B-delivered abdominal aorta; (**D**) A fluorescent image of rhodamine B-delivered abdominal aorta. The bar indicates 500 μm in all micrographs.

3.2. Quantitative Analysis

The quantitative analysis results are presented in Table 1 and Figure 4. The neointimal area was significantly decreased in all drug treated groups (Figure 4A). The measured area of the tunica media showed no statistical difference between groups (Figure 4B). The measured area of the vessel lumen of sirolimus- (I + Sir) or sunitinib- (I + Sun) treated group did not statistically differ from the area of normal vessel or drug-free (I + D) group (Figure 4C). However, the paclitaxel- (I + Ptx) treated group showed significant dilatation compared to both the normal vessel and the groups treated with other anti-proliferative drugs. Neointimal formation (%) was significantly decreased in all drug-loaded groups with no statistical significance between drug-treated groups (Figure 4D). The ratio of tunica media area to neointimal area was also significantly decreased in all drug-loaded groups (Figure 4E). The measured thickness of the tunica media layer of normal vessel, I + D group, I + Sir group, and I + Sun group showed no statistical difference between the groups (Figure 4F). However, tunica media thickness in the I + Ptx group was significantly thinner than other groups.

Table 1. Summary of quantitative analysis results. The measured neointimal area and % of neointimal formation were significantly decreased in all drug-treated groups. However, the tunica media layer of the paclitaxel-treated group was significantly thinner than that of other groups and also showed the highest apoptotic ratio (I + D: balloon injury + MN device only group; I + Sir: balloon injury + sirolimus-loaded group; I + Sun: balloon injury + sunitinib-loaded group; balloon injury + paclitaxel-loaded group; -: not assessed; *** $p < 0.001$ when compared with other groups; ** $p < 0.01$ when compared with I + Ptx group; * $p < 0.05$ when compared with I + D group; †† $p < 0.01$ when compared with I + Sir group and I + Sun group).

Parameters	Normal artery	I + D	I + Sir	I + Sun	I + Ptx
Neointimal area (mm²)	-	0.56 ± 0.13 ***	0.14 ± 0.08	0.15 ± 0.04	0.26 ± 0.10
Tunica media area (mm²)	-	1.09 ± 0.18	1.24 ± 0.31	1.23 ± 0.37	0.82 ± 0.23
Luminal area (mm²)	1.67 ± 0.31	1.05 ± 0.64	1.94 ± 0.59	1.42 ± 0.47	4.90 ± 0.90 ***
Neointimal formation (%)	-	33.59 ± 19.85 ***	7.42 ± 5.59	9.02 ± 4.08	4.70 ± 2.31
Neointimal area/tunica media area (%)	-	52.57 ± 13.92 ***	11.26 ± 6.80 **	12.86 ± 4.07 **	33.43 ± 11.62 *
Tunica media thickness (mm)	0.20 ± 0.05	0.18 ± 0.04	0.19 ± 0.05	0.17 ± 0.04	0.07 ± 0.02 ***
PCNA-positive cells (%)	3.03 ± 1.37	70.82 ± 5.11 ***	33.59 ± 7.26	32.59 ± 2.49	14.82 ± 0.46 ††
TUNEL-positive cells (%)	6.58 ± 2.27 ***	38.33 ± 3.30	37.99 ± 8.78	34.71 ± 4.48	85.60 ± 7.14 ***

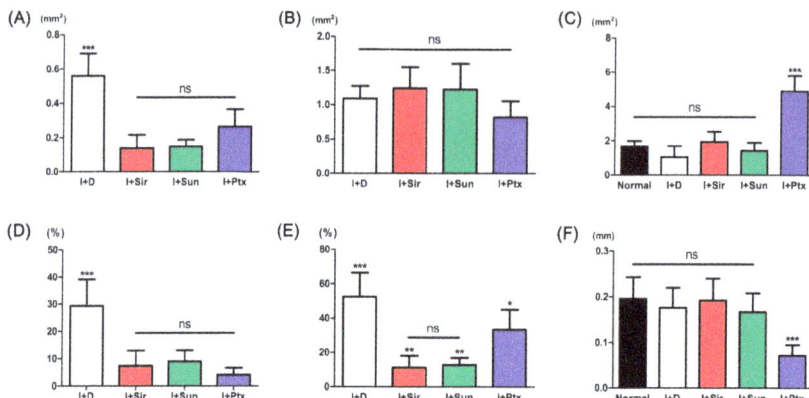

Figure 4. Quantitative analysis of neointimal area (**A**); tunica media area (**B**); luminal area (**C**); neointimal formation (**D**); ratios of neointima and tunica media area (**E**); and tunica media layer thickness (**F**). * $p < 0.05$ when compared with I + D group, ** $p < 0.01$ when compared with I + Ptx group, *** $p < 0.001$ when compared with other groups, ns = no significant difference.

3.3. Histopathological Findings

In the I + D group, excessive formation of neointima was apparent in all animals (Figure 5A,E). Under high magnification, the neointima showed typical vascular scar tissue comprised of abundant SMCs of reddish-purple color and collagenous matrix of blue color on Movat's pentachrome staining (Figure 6A). In addition, some breakdown of the elastic lamina was observed at the tunica media layer (Figure 6A).

The I + Sir group (Figure 5B,F) and I + Sun group (Figure 5C,G) showed similar histological features. They displayed limited formation of neointima compared to the I + D group, and neointima consisting of few SMCs and bluish collagen matrix was revealed with pentachrome stain (Figure 6B,C). Abundant SMCs and elastic lamina were also observed at the tunica media layer in both groups (Figure 6B,C).

In the I + Ptx group, a lesser degree of neointima formation was observed compared to the I + D group (Figure 5D,H). However, the tunica media layer was distinctively thinner (Figure 6D). It appeared mostly blue in color on pentachrome staining, because the cellularity of the tunica media layer was significantly lower and thus the staining was predominantly of the collagen matrix (Figure 6D).

Figure 5. Low magnification images at four weeks after MN device installation. Note that distinctively thin tunica media layers with vascular dilatation were observed in the paclitaxel-treated group (**D,H**). (**A,E**) I + D group (animal No. 4); (**B,F**) I + Sir group (animal No. 11); (**C,G**) I + Sun group (animal No. 12); (**D,H**) I + Ptx group (animal No. 17). ((**A–D**): hematoxylin and eosin staining, (**E–H**): Movat's pentachrome staining; the bar indicates 500 µm in all micrographs).

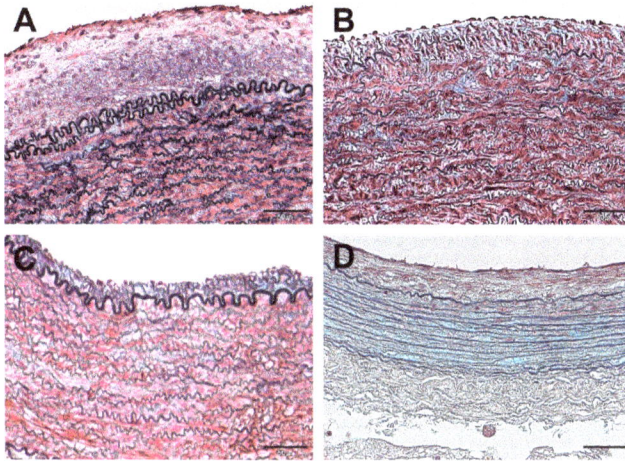

Figure 6. High magnification images at four weeks after MN device installation (Movat's pentachrome staining; the bar indicates 50 μm in all micrographs). All drug-treated groups (**B–D**) displayed limited formation of neointima compared to the device-only group (**A**). The smooth muscle cells were rarely seen in the paclitaxel-treated group (**D**). (**A**) I + D group (animal No. 7); (**B**) I + Sir group (animal No. 8); (**C**) I + Sun group (animal No. 13); (**D**) I + Ptx group (animal No. 16).

When compared to normal vessel, the ratio of PCNA-positive cells in all MN device groups was statistically significantly higher ($p < 0.001$) (Figure 7). However, the ratio of PCNA-positive cells of all drug-treated groups were significantly lower than those of the I + D group ($p < 0.001$). In addition, the I + Sun group and I + Sir group did not statistically differ, but I + Ptx group was significantly lower than both groups ($p < 0.01$).

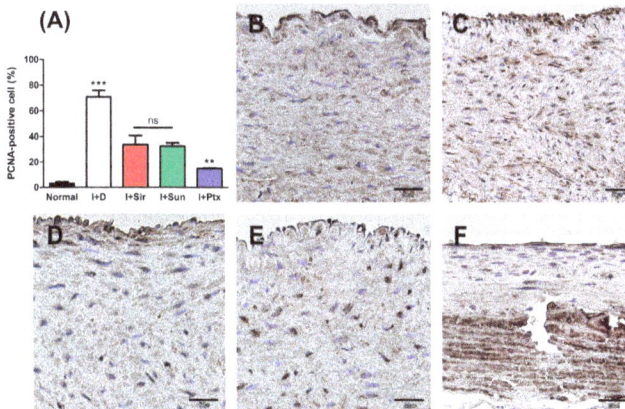

Figure 7. Anti-PCNA immunohistochemical results at four weeks after MN device installation. PCNA-positive cells were significantly reduced in all drug-treated groups. (**A**) A graph of the ratios of PCNA-positive cells to all nucleated cells. ** $p < 0.01$ when compared with I + Sir group and I + Sun group. *** $p < 0.001$ when compared with other groups, ns = no significant difference; (**B**) normal abdominal aorta; (**C**) I + D group (animal No. 5); (**D**) I + Sir group (animal No. 9); (**E**) I + Sun group (animal No. 13); (**F**) I + Ptx group (animal No. 16). The bar indicates 50 μm in all micrographs.

3.4. Apoptosis of Vascular Smooth Muscle Cells

When compared to a normal vessel, the apoptotic ratio in all MN device groups was statistically significantly higher ($p < 0.001$) (Figure 8). About 85% SMCs in the I + Ptx group displayed apoptosis, and the apoptotic nuclei ratio was significantly higher ($p < 0.001$) compared to all other groups. The I + Sir group, I + Sun group, and I+D group did not statistically differ.

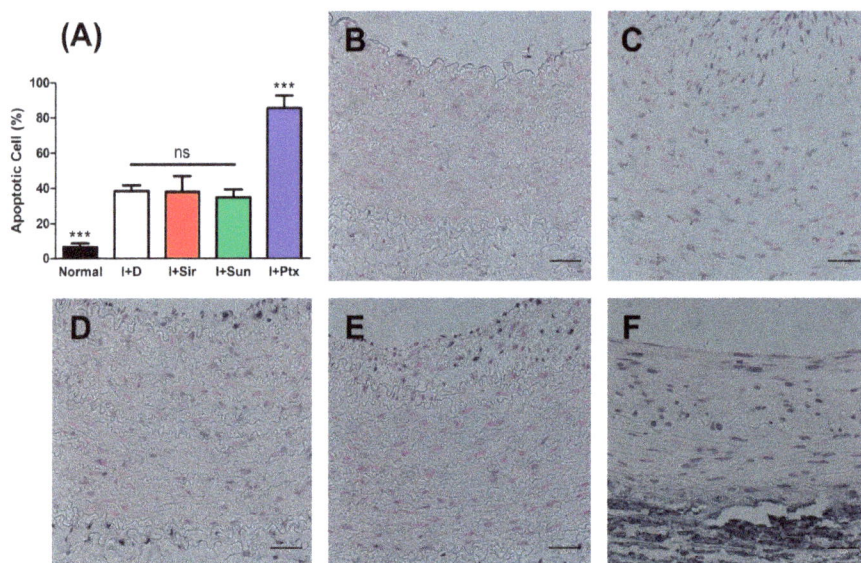

Figure 8. TUNEL staining results at four weeks after MN device installation. The paclitaxel-treated group (**F**) showed the highest apoptotic ratio. (**A**) A graph of apoptotic ratio (%). *** $p < 0.001$ when compared with other groups, ns = no significant difference; (**B**) normal abdominal aorta; (**C**) I + D group (animal No. 7); (**D**) I + Sir group (animal No. 10); (**E**) I + Sun group (animal No. 15); (**F**) I + Ptx group (animal No. 17). The bar indicates 50 μm in all micrographs.

4. Discussion

In this study, the effectiveness of an MN-based drug delivery system was demonstrated by showing its inhibitory effect on neointimal formation in balloon-injured arteries using paclitaxel-, sirolimus-, and sunitinib-loaded MN device.

Perivascular wraps fabricated from a biodegradable polymer such as poly(ε-caprolactone) (PCL) or PLGA have been introduced as a form of extraluminal drug-delivery device [24,25]. As mentioned above, one of the advantages of an extraluminal device is that the drug not only remains at the tunica externa layer but also gradually diffuses to the tunica media and subintima [12]. However, despite several successful reports in animal studies, application to human clinical trials has been relatively slow thus far, and related reports are lacking [12]. In addition, due to the possibility of drug loss to the non-contacted part of the vessel, delivering the desired amount of drug to the blood vessel wall remains a challenge [20]. To increase the efficacy of drug delivery, an intraluminal device employing a needle for the direct injection of drug into the adventitia has also been introduced [26]. However, this method is inappropriate for sustained delivery and still has the disadvantage of possible damage to vascular endothelium [12]. The perivascular MN device used in this study enables sustained delivery for a few weeks to months, with local drug delivery directly to the vascular tissue and high delivery efficiency.

We previously performed an animal study of the MN device using paclitaxel as a model drug [20]. Although neointimal formation was effectively inhibited, more than 90% of SMCs were identified as

TUNEL-positive cells in an additional apoptosis study, as was similarly reproduced in the present study. Unlike sirolimus, paclitaxel is known to induce apoptosis of SMCs in S phase and G2/M phase in injured arteries [27]. The higher percentage of apoptotic cells observed with paclitaxel than with the other drugs in the present study is thought to be due to this mechanism. The very high ratio of TUNEL-positive cells is ascribed to the sustained drug release from the device. In addition, apoptosis induced by paclitaxel could have occurred in other cells related to the vascular remodeling such as fibroblasts in the tunica adventitia. Therefore, it is likely that an abnormal remodeling process after balloon injury contributed to the thinning of the arterial wall observed in this study.

Sirolimus is also a widely-used drug for the prevention of restenosis. A number of studies have reported different effects on the vessel due to different mechanisms of action between sirolimus and paclitaxel [27–30]. Pires et al. reported that high-dose paclitaxel increased the number of apoptotic cells and decreased SMCs and collagen content of the tunica media, while the arteries were not significantly affected by sirolimus treatment [29]. In addition, according to the study of Parry et al., the perivascular application of both drugs to the rat carotid artery injury model led to effective inhibition of neointimal hyperplasia; however, paclitaxel induced apoptotic cell death while sirolimus mainly inhibited neointimal formation by a cytostatic mechanism [27]. These results are consistent with our observation of a lower ratio of apoptotic cells in the sirolimus group than in the paclitaxel group.

There is some evidence that platelet-derived growth factor (PDGF) is associated with SMCs migration and proliferation during neointimal formation after vascular injury [31,32]. According to Ferns et al., expression of PDGF and PDGF receptors was upregulated upon vessel injury, and a polyclonal antibody to PDGF showed an inhibitory effect on neointimal hyperplasia in balloon-injured artery [32]. PDGF receptor tyrosine kinase inhibitors such as imatinib also showed an inhibitory effect on neointimal hyperplasia [33,34]. Sunitinib is a multi-target inhibitor of receptor tyrosine kinases including vascular endothelial growth factor (VEGF) receptor and PDGF receptor subtypes, so it could be a potential drug candidate for inhibition of neointimal hyperplasia [35,36]. Recently, Ishii et al. reported that orally administered sunitinib significantly inhibited neointimal hyperplasia in balloon-injured rat carotid artery by reducing cell proliferation [36]. Sanders et al. introduced a PLGA perivascular bilayer wrap device for which sunitinib was used as a model drug [35]. They confirmed that high amounts of the drug remained in the vascular segment for up to four weeks using a porcine model [35]. In our study, sunitinib also showed a similar inhibitory effect on neointimal formation to the other tested drugs. Similar to sirolimus-treated group, the PCNA-positive cell level of the sunitinib group was lower than I + D group. In addition, apoptotic ratio was lower than that of the paclitaxel groups and similar to that of the MN device-only group. For these reasons, local delivery of sunitinib appears promising for prevention of neointimal hyperplasia in various vascular obstructive diseases. In particular, since VEGF is also known to be involved in neointimal formation at the surgical anastomosis site of the vessel [35], sunitinib is expected to be more effective for prevention of restenosis after vascular anastomosis, such as coronary artery bypass grafting.

The relatively high ratio of apoptotic cells observed in the I + D group is another important discussion point. According to the study of Roque et al., the ratio of apoptotic cells increased until the first week after balloon injury [37]. However, apoptotic cells gradually decreased and were rarely seen in the vessel by the fourth week [37]. The first possible reason for this discrepancy is the possibility of overestimation of apoptotic cells in the present study. Generally, it is accepted that TUNEL positivity is not synonymous with apoptosis [38]. The TUNEL assay can stain pre-apoptotic cells that may not proceed to apoptosis and do not have apoptotic morphology [38]. The staining pattern and nuclear morphology in other groups was quite different from the apoptotic nuclei in the paclitaxel group, which were believed to be certain apoptotic cells. Therefore, it is possible that TUNEL-positive results in other groups might have included pre-apoptotic cells. In particular, the finding of few differences in wall thickness and structures between the device-introduced artery and a normal artery could be evidence for this hypothesis. As another possible reason, the relatively rigid base of the device and outer Tygon tube were likely to have negative effects on the vessel. Continuous shear stress to the blood vessel wall has been reported to induce arterial dilatation [39]. However, whether mechanical stress from the device is directly related to cell apoptosis cannot be determined with this study design. Further comparison study using a flexible microneedle device is required.

In conclusion, this study demonstrated effective drug delivery by a perivascular MN device. Inhibitory effects of neointimal formation by paclitaxel, sirolimus, and sunitinib were also identified. However, considering its capability for slow drug release, sirolimus and sunitinib appear to be more appropriate drugs for this device because they showed effective suppression of neointima formation with lower effect on the vascular wall thickness. An assessment of long-term safety study is planned for a follow-up study.

Acknowledgments: The authors would like to thank Dong-Su Jang, MFA, (Medical Illustrator, Medical Research Support Section, Yonsei University College of Medicine, Seoul, Korea) for his help with the illustrations. This study was supported by a faculty research grant of Yonsei University College of Medicine (6-2014-0116). This research was also financially supported by research grants of the National Research Foundation of Korea (NRF) funded by the Korean Government (MSIP) (NRF-2016R1A2B4010487) and the Korea Health Technology R&D Project through the Korea Health Industry Development Institute (KHIDI), funded by the Ministry of Health & Welfare, Republic of Korea (HI08C2149).

Author Contributions: Young-Nam Youn and WonHyoung Ryu conceived and designed the experiments; Dae-Hyun Kim, Eui Hwa Jang, Kang Ju Lee, Ji Yong Lee, Seung Hyun Park, Il Ho Seo, and Kang Woog Lee performed the experiments; Dae-Hyun Kim and Eui Hwa Jang analyzed the data; Seung Hyun Lee contributed reagents/materials/analysis tools; Dae-Hyun Kim wrote the paper.

Conflicts of Interest: The authors declare no conflict of interest.

References

1. Brannigan, R.P.; Dove, A.P. Synthesis, properties and biomedical applications of hydrolytically degradable materials based on aliphatic polyesters and polycarbonates. *Biomater. Sci.* **2016**, *5*, 9–21. [CrossRef] [PubMed]
2. Daimon, Y.; Kamei, N.; Kawakami, K.; Takeda-Morishita, M.; Izawa, H.; Takechi-Haraya, Y.; Saito, H.; Sakai, H.; Abe, M.; Ariga, K. Dependence of Intestinal Absorption Profile of Insulin on Carrier Morphology Composed of β-Cyclodextrin-Grafted Chitosan. *Mol. Pharm.* **2016**, *13*, 4034–4042. [CrossRef] [PubMed]
3. Tolstik, E.; Osminkina, L.A.; Akimov, D.; Gongalsky, M.B.; Kudryavtsev, A.A.; Timoshenko, V.Y.; Heintzmann, R.; Sivakov, V.; Popp, J. Linear and Non-Linear Optical Imaging of Cancer Cells with Silicon Nanoparticles. *Int. J. Mol. Sci.* **2016**, *17*, 1536. [CrossRef] [PubMed]
4. Takanari, K.; Hashizume, R.; Hong, Y.; Amoroso, N.J.; Yoshizumi, T.; Gharaibeh, B.; Yoshida, O.; Nonaka, K.; Sato, H.; Huard, J.; et al. Skeletal muscle derived stem cells microintegrated into a biodegradable elastomer for reconstruction of the abdominal wall. *Biomaterials* **2017**, *113*, 31–41. [CrossRef] [PubMed]
5. Chaudhary, M.A.; Guo, L.W.; Shi, X.; Chen, G.; Gong, S.; Liu, B.; Kent, K.C. Periadventitial drug delivery for the prevention of intimal hyperplasia following open surgery. *J. Control. Release* **2016**, *233*, 174–180. [CrossRef] [PubMed]
6. Yu, X.; Takayama, T.; Goel, S.A.; Shi, X.; Zhou, Y.; Kent, K.C.; Murphy, W.L.; Guo, L.W. A rapamycin-releasing perivascular polymeric sheath produces highly effective inhibition of intimal hyperplasia. *J. Control. Release* **2014**, *191*, 47–53. [CrossRef] [PubMed]
7. Skalský, I.; Szárszoi, O.; Filová, E.; Pařízek, M.; Lytvynets, A.; Malušková, J.; Lodererová, A.; Brynda, E.; Lisá, V.; Burdíková, Z.; et al. A perivascular system releasing sirolimus prevented intimal hyperplasia in a rabbit model in a medium-term study. *Int. J. Pharm.* **2012**, *427*, 311–319. [CrossRef] [PubMed]
8. Kanjickal, D.; Lopina, S.; Evancho-Chapman, M.M.; Schmidt, S.; Donovan, D. Sustained local drug delivery from a novel polymeric ring to inhibit intimal hyperplasia. *J. Biomed. Mater. Res. A* **2010**, *93*, 656–665. [CrossRef] [PubMed]
9. Rajathurai, T.; Rizvi, S.I.; Lin, H.; Angelini, G.D.; Newby, A.C.; Murphy, G.J. Periadventitial rapamycin-eluting microbeads promote vein graft disease in long-term pig vein-into-artery interposition grafts. *Circ. Cardiovasc. Interv.* **2010**, *3*, 157–165. [CrossRef] [PubMed]
10. Shi, X.; Chen, G.; Guo, L.W.; Si, Y.; Zhu, M.; Pilla, S.; Liu, B.; Gong, S.; Kent, K.C. Periadventitial application of rapamycin-loaded nanoparticles produces sustained inhibition of vascular restenosis. *PLoS ONE* **2014**, *9*, e89227. [CrossRef] [PubMed]

11. Wang, X.; Lin, P.; Yao, Q.; Chen, C. Development of small-diameter vascular grafts. *World J. Surg.* **2007**, *31*, 682–689. [CrossRef] [PubMed]

12. Seedial, S.M.; Ghosh, S.; Saunders, R.S.; Suwanabol, P.A.; Shi, X.; Liu, B.; Kent, K.C. Local drug delivery to prevent restenosis. *J. Vasc. Surg.* **2013**, *57*, 1403–1414. [CrossRef] [PubMed]

13. Riede, F.N.; Pfisterer, M.; Jeger, R. Long-term safety of drug-eluting stents. *Expert. Rev. Cardiovasc. Ther.* **2013**, *11*, 1359–1378. [CrossRef] [PubMed]

14. Lee, T.; Roy-Chaudhury, P. Advances and new frontiers in the pathophysiology of venous neointimal hyperplasia and dialysis access stenosis. *Adv. Chronic Kidney Dis.* **2009**, *16*, 329–338. [CrossRef] [PubMed]

15. Schwartz, R.S.; Chronos, N.A.; Virmani, R. Preclinical restenosis models and drug-eluting stents: Still important, still much to learn. *J. Am. Coll. Cardiol.* **2004**, *44*, 1373–1385. [CrossRef] [PubMed]

16. Finn, A.V.; Kolodgie, F.D.; Harnek, J.; Guerrero, L.J.; Acampado, E.; Tefera, K.; Skorija, K.; Weber, D.K.; Gold, H.K.; Virmani, R. Differential response of delayed healing and persistent inflammation at sites of overlapping sirolimus- or paclitaxel-eluting stents. *Circulation* **2005**, *112*, 270–278. [CrossRef] [PubMed]

17. Nakazawa, G.; Finn, A.V.; John, M.C.; Kolodgie, F.D.; Virmani, R. The significance of preclinical evaluation of sirolimus-, paclitaxel-, and zotarolimus-eluting stents. *Am. J. Cardiol.* **2007**, *100*, 36M–44M. [CrossRef] [PubMed]

18. Buechel, R.; Stirnimann, A.; Zimmer, R.; Keo, H.; Groechenig, E. Drug-eluting stents and drug-coated balloons in peripheral artery disease. *Vasa* **2012**, *41*, 248–261. [CrossRef] [PubMed]

19. Venkatraman, S.; Boey, F. Release profiles in drug-eluting stents: Issues and uncertainties. *J. Control. Release* **2007**, *120*, 149–160. [CrossRef] [PubMed]

20. Lee, K.J.; Park, S.H.; Lee, J.Y.; Joo, H.C.; Jang, E.H.; Youn, Y.N.; Ryu, W. Perivascular biodegradable microneedle cuff for reduction of neointima formation after vascular injury. *J. Control. Release* **2014**, *192*, 174–181. [CrossRef] [PubMed]

21. Le, V.; Johnson, C.G.; Lee, J.D.; Baker, A.B. Murine model of femoral artery wire injury with implantation of a perivascular drug delivery patch. *J. Vis. Exp.* **2015**, *96*, e52403. [CrossRef] [PubMed]

22. Choi, C.K.; Kim, J.B.; Jang, E.H.; Youn, Y.N.; Ryu, W.H. Curved biodegradable microneedles for vascular drug delivery. *Small* **2012**, *8*, 2483–2488. [CrossRef] [PubMed]

23. Choi, C.K.; Lee, K.J.; Youn, Y.N.; Jang, E.H.; Kim, W.; Min, B.K.; Ryu, W. Spatially discrete thermal drawing of biodegradable microneedles for vascular drug delivery. *Eur. J. Pharm. Biopharm.* **2013**, *83*, 224–233. [CrossRef] [PubMed]

24. Signore, P.E.; Machan, L.S.; Jackson, J.K.; Burt, H.; Bromley, P.; Wilson, J.E.; McManus, B.M. Complete inhibition of intimal hyperplasia by perivascular delivery of paclitaxel in balloon-injured rat carotid arteries. *J. Vasc. Interv. Radiol.* **2001**, *12*, 79–88. [CrossRef]

25. Kohler, T.R.; Toleikis, P.M.; Gravett, D.M.; Avelar, R.L. Inhibition of neointimal hyperplasia in a sheep model of dialysis access failure with the bioabsorbable vascular wrap paclitaxel-eluting mesh. *J. Vasc. Surg.* **2007**, *45*, 1029–1038. [CrossRef] [PubMed]

26. Ikeno, F.; Lyons, J.; Kaneda, H.; Baluom, M.; Benet, L.Z.; Rezaee, M. Novel percutaneous adventitial drug delivery system for regional vascular treatment. *Catheter. Cardiovasc. Interv.* **2004**, *63*, 222–230. [CrossRef] [PubMed]

27. Parry, T.J.; Brosius, R.; Thyagarajan, R.; Carter, D.; Argentieri, D.; Falotico, R.; Siekierka, J. Drug-eluting stents: Sirolimus and paclitaxel differentially affect cultured cells and injured arteries. *Eur. J. Pharmacol.* **2005**, *524*, 19–29. [CrossRef] [PubMed]

28. Wessely, R.; Blaich, B.; Belaiba, R.S.; Merl, S.; Görlach, A.; Kastrati, A.; Schömig, A. Comparative characterization of cellular and molecular anti-restenotic profiles of paclitaxel and sirolimus. Implications for local drug delivery. *Thromb. Haemost.* **2007**, *97*, 1003–1012. [CrossRef] [PubMed]

29. Pires, N.M.; Eefting, D.; de Vries, M.R.; Quax, P.H.; Jukema, J.W. Sirolimus and paclitaxel provoke different vascular pathological responses after local delivery in a murine model for restenosis on underlying atherosclerotic arteries. *Heart* **2007**, *93*, 922–927. [CrossRef] [PubMed]

30. Silva, G.V.; Fernandes, M.R.; Madonna, R.; Clubb, F.; Oliveira, E.; Jimenez-Quevedo, P.; Branco, R.; Lopez, J.; Angeli, F.S.; Sanz-Ruiz, R.; et al. Comparative healing response after sirolimus- and paclitaxel-eluting stent implantation in a pig model of restenosis. *Catheter. Cardiovasc. Interv.* **2009**, *73*, 801–808. [CrossRef] [PubMed]

31. Majesky, M.W.; Reidy, M.A.; Bowen-Pope, D.F.; Hart, C.E.; Wilcox, J.N.; Schwartz, S.M. PDGF ligand and receptor gene expression during repair of arterial injury. *J. Cell Biol.* **1990**, *111*, 2149–2158. [CrossRef] [PubMed]
32. Ferns, G.A.; Raines, E.W.; Sprugel, K.H.; Motani, A.S.; Reidy, M.A.; Ross, R. Inhibition of neointimal smooth muscle accumulation after angioplasty by an antibody to PDGF. *Science* **1991**, *253*, 1129–1132. [CrossRef] [PubMed]
33. Makiyama, Y.; Toba, K.; Kato, K.; Hirono, S.; Ozawa, T.; Saigawa, T.; Minagawa, S.; Isoda, M.; Asami, F.; Ikarashi, N.; et al. Imatinib mesilate inhibits neointimal hyperplasia via growth inhibition of vascular smooth muscle cells in a rat model of balloon injury. *Tohoku J. Exp. Med.* **2008**, *215*, 299–306. [CrossRef] [PubMed]
34. Vamvakopoulos, J.E.; Petrov, L.; Aavik, S.; Lehti, S.; Aavik, E.; Hayry, P. Synergistic suppression of rat neointimal hyperplasia by rapamycin and imatinib mesylate: Implications for the prevention of accelerated arteriosclerosis. *J. Vasc. Res.* **2006**, *43*, 184–192. [CrossRef] [PubMed]
35. Sanders, W.G.; Hogrebe, P.C.; Grainger, D.W.; Cheung, A.K.; Terry, C.M. A biodegradable perivascular wrap for controlled, local and directed drug delivery. *J Control. Release* **2012**, *161*, 81–89. [CrossRef] [PubMed]
36. Ishii, S.; Okamoto, Y.; Katsumata, H.; Egawa, S.; Yamanaka, D.; Fukushima, M.; Minami, S. Sunitinib, a small-molecule receptor tyrosine kinase inhibitor, suppresses neointimal hyperplasia in balloon-injured rat carotid artery. *J. Cardiovasc. Pharmacol. Ther.* **2013**, *18*, 359–366. [CrossRef] [PubMed]
37. Roque, M.; Cordon-Cardo, C.; Fuster, V.; Reis, E.D.; Drobnjak, M.; Badimon, J.J. Modulation of apoptosis, proliferation, and p27 expression in a porcine coronary angioplasty model. *Atherosclerosis* **2000**, *153*, 315–322. [CrossRef]
38. Angelini, A.; Visonà, A.; Calabrese, F.; Pettenazzo, E.; Yacoub, A.; Valente, M.; Bonandini, E.M.; Jori, G.; Pagnan, A.; Thiene, G. Time Course of Apoptosis and Proliferation in vascular Smooth Muscle Cells after Balloon Angioplasty. *Basic Appl. Myol.* **2002**, *12*, 33–42.
39. Markos, F.; Ruane O'Hora, T.; Noble, M.I. What is the mechanism of flow-mediated arterial dilatation. *Clin. Exp. Pharmacol. Physiol.* **2013**, *40*, 489–494. [CrossRef] [PubMed]

polymers

MDPI

Article

Cell Proliferation on Polyethylene Terephthalate Treated in Plasma Created in SO_2/O_2 Mixtures

Nina Recek [1,*], Matic Resnik [2], Rok Zaplotnik [1], Miran Mozetic [1], Helena Motaln [3], Tamara Lah-Turnsek [3] and Alenka Vesel [1]

[1] Jozef Stefan Institute, Jamova cesta 39, 1000 Ljubljana, Slovenia; rok.zaplotnik@ijs.si (R.Z.);
     miran.mozetic@ijs.si (M.M.); alenka.vesel@guest.arnes.si (A.V.)
[2] Jozef Stefan International Postgraduate School, Jamova cesta 39, 1000 Ljubljana, Slovenia; matic.resnik@ijs.si
[3] National Institute of Biology, Vecna pot 111, 1000 Ljubljana, Slovenia; helena.motaln@nib.si (H.M.);
     tamara.lah@nib.si (T.L.-T.)
*   Correspondence: nina.recek@ijs.si; Tel.: +386-1-477-3398

Academic Editor: Insung S. Choi
Received: 3 January 2017; Accepted: 22 February 2017; Published: 25 February 2017

Abstract: Samples of polymer polyethylene terephthalate were exposed to a weakly ionized gaseous plasma to modify the polymer surface properties for better cell cultivation. The gases used for treatment were sulfur dioxide and oxygen of various partial pressures. Plasma was created by an electrodeless radio frequency discharge at a total pressure of 60 Pa. X-ray photoelectron spectroscopy showed weak functionalization of the samples' surfaces with the sulfur, with a concentration around 2.5 at %, whereas the oxygen concentration remained at the level of untreated samples, except when the gas mixture with oxygen concentration above 90% was used. Atomic force microscopy revealed highly altered morphology of plasma-treated samples; however, at high oxygen partial pressures this morphology vanished. The samples were then incubated with human umbilical vein endothelial cells. Biological tests to determine endothelialization and possible toxicity of the plasma treated polyethylene terephthalate samples were performed. Cell metabolic activity (MTT) and in vitro toxic effects of unknown compounds (TOX) were assayed to determine the biocompatibility of the treated substrates. The biocompatibility demonstrated a well-pronounced maximum versus gas composition which correlated well with development of the surface morphology.

Keywords: polymer; polyethylene terephthalate (PET); SO_2 plasma treatment; surface modification; wettability; X-ray photoelectron spectroscopy (XPS); atomic force microscopy (AFM); human umbilical endothelial cell (HUVEC) proliferation; MTT; toxicity; scanning electron microscopy (SEM)

1. Introduction

Polymer materials often require surface modification to achieve the best results in particular applications [1–7]. Different methods can be used for modification of the polymer surface properties, including chemical treatments, irradiation with photons, irradiation with ion or electron beams and treatment by gaseous plasmas created by electric discharges (corona, dielectric barrier discharge, glow discharge, etc.). Among all of these methods, plasma treatments remain the most prominent techniques for surface modification [8–14]. By appropriate selection of the type of discharge, treatment conditions, and working gases, it is possible to graft different surface functional groups, to change surface morphology, and even to create nanostructures on the treated surface. This way the surface wettability can be changed from superhydrophilic to superhydrophobic finishes [15].

In biomedical applications, and many others like surface cleaning, sterilization, improving surface wettability, and adhesive properties, the best results are most commonly achieved by the use of oxygen-containing plasma. Especially in biomedical applications, oxygen plasma was found to improve hemocompatibility of polymer grafts by making the polymer surface antithrombogenic because of reduced platelet adhesion [16]. Furthermore, oxygen plasma also improves cell adhesion and proliferation [17,18]. Therefore, oxygen plasma is a good alternative to the other methods used to make surface antithrombogenic, such as chemical grafting or various coatings like gelatin, heparin, fucoidan, etc. [19,20].

It has been demonstrated that sulfonic functional groups can also act in the antithrombogenic manner [21,22]. Commonly, such surface finish was accomplished by chemical synthesis of polymers containing sulfonic groups or by using special sulfated coatings like fucoidan [23–25]. In only a few attempts the SO_2 plasma treatment was utilized as an alternative technique. Yet, biocompatibility tests of oxidized sulfur groups revealed some contradictory results. Some authors have observed increased platelet activation (i.e., increased surface thrombogenic effect) [26,27], whereas the others have reported on decreased platelet adhesion and activation (i.e., increased surface anti-thrombogenic effect) [21,22,28]. Despite the scarce literature, it appears that SO_2 plasma-treated surfaces do not always display the most optimal antithrombogenic properties, as the best results were obtained for sulfonated surfaces prepared by chemical synthesis.

Indeed, very few papers were published regarding investigated cell adhesion on polymer surfaces treated with SO_2 plasma. The first report was provided by Klee et al. [29], who studied adhesion of human umbilical vein endothelial cells (HUVEC), as well as fibronectin adsorption to medical grade polyvinyl chloride treated with SO_2 plasma. They observed better fibrinogen adsorption on SO_2 plasma-treated surface. Cell proliferation was in direct correlation with the fibronectin adsorption. [29]. Another reason for good cell proliferation was also a water contact angle of 67° presenting moderate hydrophilicity of the surface which is regarded as optimal for cell proliferation [30]. In another paper Wang et al. investigated the adhesion of dog vascular smooth muscle cells to SO_2 plasma-treated polybutylene succinate surface and again found improved cell adhesion and growth on plasma-treated surface compared to the untreated one [31]. Likewise, Gugala was investigating the proliferation of rat osteoblast cells grown on polylactide surface treated with plasma containing SO_2/H_2 mixture [32]. In his study, SO_2/H_2 plasma treatment proved to negatively affect the cell growth. These rare results require further research of biocompatibility properties of surfaces with grafted oxidized sulfur groups.

More studies exis regarding the determination of chemical properties of SO_2 plasma-treated surfaces. SO_2 plasma was used for surface modification of different materials including polyethylene (PE) [26,27,33], polyethylene terephthalate (PET) [33], polymethyl methacrylate [34], polyvinylchloride (PVC) [29], polytetrafluoroethylene (PTFE) [35], polypropylene (PP) [33,36], polyesterpoly-uretane [36], heptylamine [37], octadiene [37], clay (laponite) [38], and highly-oriented pyrolytic graphite (HOPG) [36]. In these studies even the polytetrafluoroethylene (PTFE), which is quite difficult to activate by plasma, showed promising results [35]. Among the papers reporting results on SO_2 plasma treatment, it is worth mentioning the works performed by Holländer et al. [33], Siow et al. [37], and Fatyeyeva et al. [38].

Fatyeyeva et al. [38] have studied the influence of discharge parameters (plasma power, gas flow rate, and treatment time) on kinetics of grafting sulfur moieties onto the treated clay surface. They found that increasing the gas flow leads to a significant decrease of sulfur content on the substrate surfaces; however, the sulfur content increased by increasing power and treatment time. Siow et al. [37] performed a detailed XPS characterization of plasma-treated heptylamine and 1,7-octadiene polymers and their post-treatment aging. They reported on groups with a higher oxidation state to be more influenced by aging in air. Holländer et al. [33] has compared treatment of polymers in pure SO_2 plasma with those treated in a mixture of SO_2 with oxygen and hydrogen. They found formation of highly oxidized sulfur groups in $SO_2 + O_2$ mixture, whereas in $SO_2 + H_2$ mixture sulfides prevailed. In pure SO_2 plasma, both sulfides and highly-oxidized groups were observed. The effects of treatment conditions (discharge power and gas flow rate) were studied as well. At higher powers and flow rates more sulfur was detected on the surface. Furthermore, low plasma power lead to producing more sulfur in a low oxidation state, whereas a high flow of SO_2 and a high power favored the formation of highly-oxidized species.

Although there are discrepancies between different reports it can be summarized that SO_2 plasma causes functionalization of polymers with sulfur groups, preferentially in the form of sulfites, however, the influence of such surface finish on proliferation of biological cells is still not understood well. In this study we systematically investigated the influence of SO_2 plasma, as well as SO_2/O_2 mixtures on PET surface modification. PET is a material commonly used for artificial body implants like vascular grafts. To make PET surface biocompatible it should exhibit good endothelialization. The final goal was thus to create a biocompatible surface using the mixture of SO_2/O_2. To the best of our knowledge, no other group has done such work. SO_2 plasma was chosen to mimic heparin, which is a commonly used coating on commercial vascular grafts for improving biocompatibility. Therefore, surface chemistry, morphology, and wettability of plasma-treated PET surfaces, as well as cell adhesion and surface toxicity were investigated.

2. Materials and Methods

2.1. Plasma Treatment

Biaxially-oriented polyethylene terephthalate (PET) from Goodfellow Ltd. (Huntingdon, UK) was used as the substrate. The polymer foil with a thickness of 0.125 mm was cut to small square pieces with a size of 1×1 cm^2. The samples were treated in a quartz-glass discharge tube presented schematically in Figure 1. The tube was 80 cm long and 4 cm in diameter. The discharge tube was pumped with a rotary pump operating at a nominal pumping speed of 80 m^3 h^{-1}. The base pressure was below 1 Pa. A mixture of $SO_2 + O_2$ was leaked to the experimental system on the other side as shown in Figure 1. The total pressure was set to 60 Pa where a maximum dissociation occurred, as measured by a catalytic probe. The ratio of SO_2/O_2 was varied. The purity of gases (which were supplied by Messer, Bad Soden, Germany) was 99.999 vol % and 99.98 vol % for O_2 and SO_2, respectively. A coil of six turns was mounted in the center of the Pyrex tube as shown in Figure 1. Plasma was created by a radio frequency (RF) generator coupled to the coil via a matching network. The generator operated at the standard frequency of 13.56 MHz and its nominal power was set to 150 W. Under such discharge conditions the plasma was sustained in the E-mode. Samples were treated in plasma for 30 s to allow for surface saturation. After the treatments they were characterized using atomic force microscopy (AFM), X-ray photoelectron spectroscopy (XPS), and water contact angle measurements (WCA).

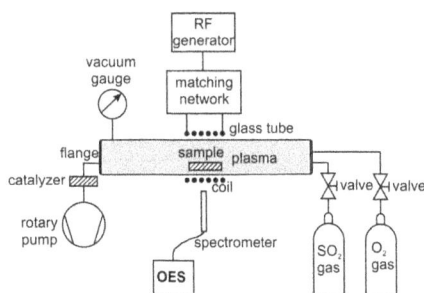

Figure 1. Schematic diagram of the experimental system.

2.2. Plasma Characterisation

Plasma was characterized using optical emission spectroscopy (OES) [39–41]. OES measurements were performed in a quartz tube with a 16-bit Avantes AvaSpec 3648 fiber optic spectrometer (Avantes Inc., Louisville, CO, USA). A nominal spectral resolution was 0.8 nm and spectra were recorded in the range from 200 to 1100 nm. The combined deuterium tungsten reference light source was used to determine the spectral response of the spectrometer. The measured OES spectra were calibrated with this spectral response. The integration time used to record OES spectra was 2 s.

2.3. Atomic Force Microscopy (AFM) Measurements

An AFM (Solver PRO, NT-MDT, Moscow, Russia) was used to characterize the topology of the samples. All measurements were performed in tapping mode using ATEC-NC-20 tips (Nano and More GmbH, Limerick, Ireland) with a resonance frequency of 210–490 kHz and force constant of 12–110 Nm^{-1}. The surface roughness was calculated from AFM images taken over an area of 2×2 μm^2 and 5×5 μm^2 using the program Spip 5.1.3 (Image Metrology A/S, Hørsholm, Denmark). Surface roughness was expressed in terms of average roughness (Ra).

2.4. Characterisation by X-Ray Photoelectron Spectroscopy (XPS)

XPS characterization of polymer samples was performed to determine their chemical composition after plasma treatment using an XPS (a model TFA XPS Physical Electronics, Chanhassen, MN, US). The samples were excited with monochromatic Al K$\alpha_{1,2}$ radiation at 1486.6 eV over an area with a diameter of 400 μm. Photoelectrons were detected with a hemispherical analyzer positioned at an angle of 45° with respect to the normal of the sample surface. XPS survey spectra were measured at a pass-energy of 187 eV using an energy step of 0.4 eV, whereas high-resolution spectra were measured at a pass-energy of 23.5 eV using an energy step of 0.1 eV. An additional electron gun was used for surface neutralization during XPS measurements. All spectra were referenced to the main C1s peak of the carbon atoms, which was assigned a value of 284.8 eV. The measured spectra were analyzed using MultiPak v8.1c software (Ulvac-Phi Inc., Kanagawa, Japan, 2006) from Physical Electronics, which was supplied with the spectrometer.

2.5. Contact Angle Measurements

The surface wettability was measured immediately after plasma treatment by determining the water contact angle (WCA) with a demineralized water droplet of volume 2 μL. Contact angles were measured by See System (Advex Instruments, Brno, Czech Republic). For each sample, five measurements were taken to minimize the statistical error. The contact angles were determined by the software supplied by the producer.

2.6. Cell Adhesion and Morphology Studies

Human umbilical endothelial cells (HUVEC; purchased from ATCC, Manassas, VA, USA) were cultured in a minimum essential medium (MEM; Sigma-Aldrich, Taufkirchen, Germany) supplemented with 10% fetal bovine serum (FBS; Sigma-Aldrich, Taufkirchen, Germany), 100 U penicillin, 1000 U streptomycin, 2 mM L-glutamine, and plated at density of 3000 cells/cm^2. For the investigation of cell adhesion and morphology, the cells were seeded at a density of 2×10^4 cells in 100 μL drop of medium on the upper side of the polymers (concentration: 2.55×10^4 cells/cm^2) and left for 1, 2, 3, 5, and 24 h to attach at 37 °C in a humidified atmosphere of 5% CO$_2$. Cells were seeded onto modified polymer in duplicates for each time and plasma treatment condition.

Cell adhesion and morphology was assessed after 24 h of incubation (time allowed for cells to firmly attach on the surface) by scanning electron microscopy (SEM). Briefly, the polymer samples with the attached cells were fixed in 2% glutaraldehyde (Sigma-Aldrich, Taufkirchen, Germany) in phosphate-buffered solution for 5 min, followed by dehydration through an increasing gradient of ethanol and then vacuum dried by the critical point method. Finally, the samples were covered by a thin layer of gold and analyzed by SEM. For gold evaporation a PECS instrument (Model 682) from Gatan GmbH (München, Germany) was used. SEM analyses were performed using a JEOL JSM-840 Scanning Electron Microscope (JEOL, Tokyo, Japan).

2.6.1. MTT (3-(4,5-dimethylthiazol-2-yl)-2,5-diphenyltetrazolium bromide) Assay

HUVECs were seeded and cultured in the same manner as for the cell adhesion and morphology investigation by SEM. The MTT-related colorimetric assay (EZ4U; Biomedica, Wien, Austria) was used to determine cell growth and viability, according to the manufacturer's instructions. The method is based on the fact that living cells are capable of reducing less-colored tetrazolium salts into intensely-colored formazan derivatives. This reduction process requires functional mitochondria, which are inactivated within a few minutes after cell death. Briefly, after 1, 2, 3, 5, and 24 h of HUVEC cell incubation on the differently modified polymer surfaces the medium was removed and the polymer samples were rinsed with phosphate buffer saline to remove for all non-attached cells. Then 200 μL of fresh Hanks' Balanced Salt Solution (HBSS) (Sigma-Aldrich, Taufkirchen, Germany) mixed with the tetrazolium agent were added to each well with the polymer sample of the 24-well plate. After 1, 2, 3, 5, and 24 h of incubation, supernatants were transferred into 96-well plates and the absorbance was measured at OD 570/690 nm with SynergyTM HT Microplate Reader (Bio-TeK Instruments, Inc., Winooski, VT, USA).

2.6.2. In Vitro Toxicology Assay (TOX)

The sulforhodamine B assay measures total biomass staining cellular proteins with sulforhodamine B. Cells were seeded on the samples at density of 3000 cells/cm^2, tests were performed in duplicate, and each test included a blank containing complete medium without cells.

Fifty percent of (*w/v*) trichloroacetic acid (TCA, Sigma-Aldrich, Taufkirchen, Germany) solution and wash solution were prepared according to manufacturing instructions. Samples were removed from incubator into a laminar flow hood and cells were fixed by gently layering cold 50% (*w/v*) TCA solution on top of the growth medium. 96-well plates were incubated for 1 h at 4 °C and then rinsed with water several times to remove TCA solution, serum proteins, etc. Additionally, 0.4% sulforhodamine B solution was added in an amount sufficient to cover the culture surface area (~50% of the culture medium volume). Cells were then allowed to stain for 20–30 min. At the end of the staining period, the stain was removed and the cells were rinsed quickly with wash solution (1% acetic acid), until unincorporated dye was removed. Wash times were kept to a minimum to reduce desorption of protein-bound dye. After being rinsed, the cultures were air dried until no moisture was visible. The incorporated dye was then solubilized in a volume of sulforhodamine B assay solubilization solution (10 mM base solution, Sigma-Aldrich, Taufkirchen, Germany) equal to the original volume of culture medium and liberated from the cells. Cultures were allowed to stand for 5 min at room temperature, while pipetting up and down to enhance mixing of the dye. Absorbance was measured spectrophotometrically at a wavelength of 565 nm. Blank background optical density was measured in wells incubated with growth medium without cells. The background absorbance of multiwell plates was measured at 690 nm and subtracted from the measurement at 565 nm. An increase or decrease in the number of cells (total biomass) resulted in a concomitant change in the amount of dye incorporated by the cells in the culture. This indicated the degree of cytotoxicity caused by the test material.

2.6.3. Statistical Analysis

All the above experiments were performed in duplicate and independently repeated at least three times, unless otherwise stated. The results obtained are shown as the mean ± SE (standard error of the mean) for duplicates of cultures. Student's *t*-test was used to test the effect different plasma modifications of PET have on the adhesion and metabolic activity of HUVECs and a value of $p < 0.05$ was considered significant.

3. Results and Discussion

3.1. Plasma Characterisation

Gaseous plasma was characterized by optical emission spectroscopy and the resultant spectra for selected gas mixtures are presented in Figure 2. A large continuum in the ultraviolet (UV) range is attributed to the radiative relaxation of the SO_2 molecule [42,43]. The large continuum is explained by the fact that the final state is the ground one, thus even electrons of moderate energy are capable of exciting SO_2 molecules to the radiative states. The molecules are excited by electron impact and radiate in the broad range from UV A to UV C. Other spectral features are attributed to emissions from excited oxygen atoms. As usual, the most intensive is the line at 777 nm followed by the line at 845 nm. The radiation from the O-atoms is much weaker than the continuum because of the high threshold for excitation of the oxygen radiative states (approximately 11 eV).

The intensity of radiation arising from SO_2 molecules and O atoms depends on the concentration of gases in the gas mixture. Figure 2 reveals a gradual decrease of the continuum and increase of the atomic lines as the concentration of oxygen in the gas mixture increases. The integral radiation in the UV range as a function of the oxygen concentration in the $SO_2 + O_2$ mixture is plotted in Figure 3. The dots in this figure are measured points and the curve is the best fit. The integral intensity remains fairly unchanged up to the oxygen concentration of approximately 30 vol % and then decreases with the increasing concentration. Such behavior is explained by decreasing electrons' density or their temperature because of the addition of substantial oxygen amounts. Namely, oxygen molecules represent additional channels for loss of electron energy and/or density because of dissociation and attachment onto oxygen atoms.

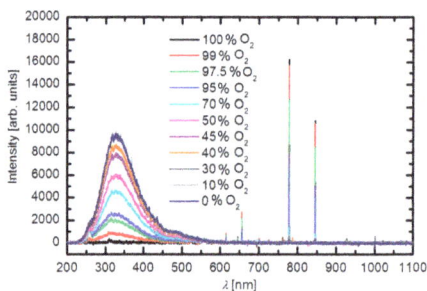

Figure 2. Spectra of plasma sustained in various gas mixtures. Parameter is the oxygen concentration in the $SO_2 + O_2$ mixture.

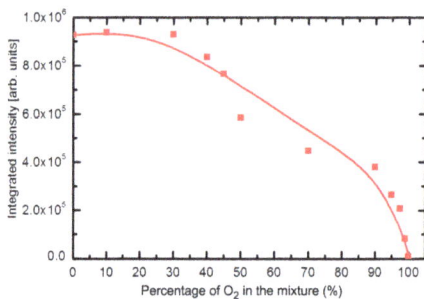

Figure 3. Integral radiation in the UV range versus oxygen concentration in $SO_2 + O_2$ mixture. Zero percent of O_2 in the mixture refers to pure SO_2 plasma, whereas 100% of O_2 refers to pure oxygen plasma.

3.2. Surface Characterisation of Plasma-Treated Samples

Plasma-treated samples were characterized by XPS just after the treatment. The concentration of oxygen and sulfur on the polymer surface as extracted from the XPS survey spectra is presented in Figure 4. The concentration of sulfur is approximately 2.5 at % irrespective from plasma gas mixture. Obviously, the surface of the samples is saturated with sulfur groups upon a half-minute of plasma treatment what is sound with the results reported by other authors [33]. The reason for the saturation is the very high dissociation rate in our plasma and rather long treatment time. Additionally, the oxygen concentration on the polymer surface is not affected by adding oxygen to the gas mixture, because it remained close to the value for the sample treated in pure SO_2 plasma (38 at %), except for the sample treated in a mixture with the highest oxygen content (10% SO_2 + 90% O_2) where it is somehow increased. The oxygen concentration obtained upon treatment in pure oxygen plasma is added to Figure 4 for comparison. For untreated PET sample, the oxygen concentration was 25 at % (not shown in the figure).

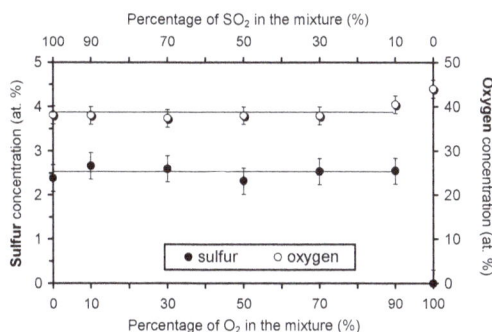

Figure 4. Concentration of sulfur and oxygen on the PET polymer treated in various SO_2 + O_2 mixtures. Zero percent of O_2 in the mixture refers to pure SO_2 plasma, whereas 100% of O_2 refers to pure oxygen plasma.

The high-resolution XPS spectra of C1s and S2p are presented in Figures 5 and 6, respectively. Curves for all gas mixtures overlap so one can conclude that the surface functionalization does not depend on the concentration of gases. The S2p peak is observed at the binding energy of approximately 169 eV what is typical for sulfites (SO_3^{2-}) [37]. Such functional groups have been observed before, even for the case of oxygen-plasma treatment of sulfur-containing polymers [44,45].

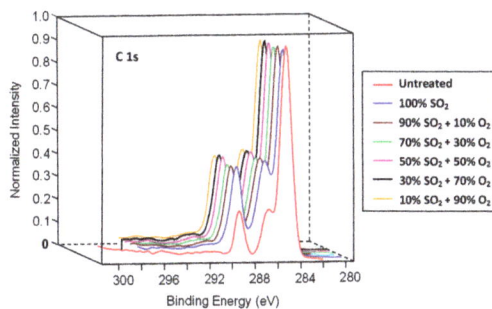

Figure 5. High-resolution C1s peak for samples treated at various SO_2 + O_2 mixtures.

Figure 6. High-resolution S2p peak for samples treated at various $SO_2 + O_2$ mixtures.

More interesting are the results on the evolution of the surface morphology. Typical three-dimensional AFM images of samples treated by plasma of various gas mixtures are presented in Figure 7. The morphology of the untreated sample is presented in Figure 7a. The material is rather smooth on the micrometer scale, but the morphology of the samples treated by plasma undergoes interesting modifications.

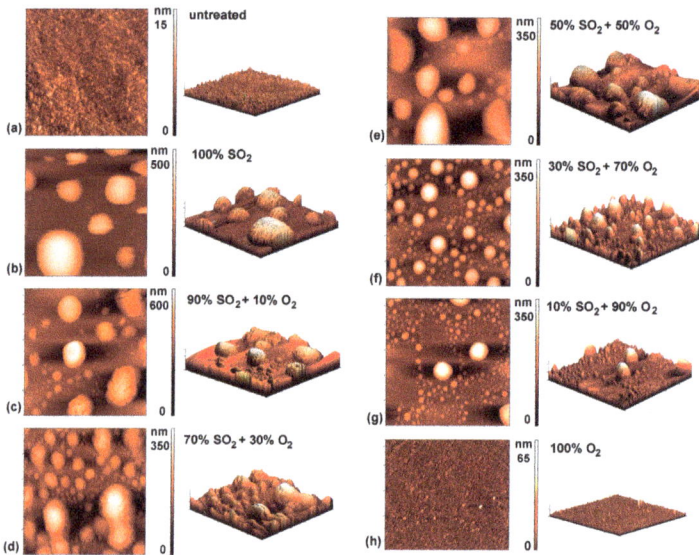

Figure 7. AFM images (5 × 5 μm^2) of the samples treated in various $SO_2 + O_2$ mixtures: (**a**) untreated; (**b**) 100% SO_2; (**c**) 10% O_2 + 90% SO_2; (**d**) 30% O_2 + 70% SO_2; (**e**) 50% O_2 + 50% SO_2; (**f**) 70% O_2 + 30% SO_2; (**g**) 90% O_2 + 10% SO_2; and (**h**) 100% O_2.

Figure 7b represents an AFM image of the sample treated in pure SO_2 plasma. The scale on the vertical axis is nearly 100 times larger than for the untreated sample. Nearly spherical features appeared on the polymer surface upon plasma treatment. The features are distributed randomly on the surface and the lateral dimensions are roughly a micrometer, whereas the height is somehow smaller. It appears as if droplets were formed on the surface of originally smooth material. The formation of such droplets is not typical for plasma-treated polymers. Although numerous authors reported nanostructuring of polymers upon plasma treatment [46–54] such a surface finish was rarely reported. The formation of such droplets cannot be explained by deposition of a third material because no other material but the polymer sample was introduced into the plasma reactor. The features were obviously formed by transformation of the polymer surface film of thickness of the order of 100 nm. Etching of polymer in plasma used in our experiments cannot cause formation of such droplets because the etching rate is just a few nm/s [55]. The formation of the droplets is rather explained by modification of the polymer structure in the surface film of thickness typical for the penetration depth of the UV radiation. As explained in the Subsection 3.1, SO_2 plasma is a rich source of the UV radiation in the bread range from 190 to 400 nm (Figure 2). According to Kim, Ahn, and Sancaktar [56] the penetration depth of the UV radiation at wavelength of 248 and 193 nm in PET polymer is 62 and 34 nm, respectively. The penetration depth increases with increasing wavelength (decreasing photon energy). The UV radiation is absorbed in the polymer and caused a bond scission. The resulting low-mass molecular fragments rearrange according to the thermodynamic laws [57]. Since the surface energy of materials treated by oxygen-containing plasma is increased, the surface tension favorites formation of spherical features as observed in Figure 7b.

The UV radiation from plasma decreases with increasing oxygen content in the gas mixture and completely vanishes for the case of pure oxygen (Figure 3). Pure oxygen plasma does not emit UV radiation except in the VUV range where the penetration depth is minimal [58]. The morphology of the samples treated with plasma of various concentrations of gases follows the intensity of UV radiation: the droplets height decrease with the increasing oxygen content. The lateral dimension also decreases with increase of the oxygen content. The sample treated at 90% SO_2 (Figure 7g) where the integral UV intensity is only a third of the original value (Figure 3) reveals only a few droplets of microscopic dimensions and such features vanish completely in the case of treatment with pure oxygen plasma. The sample in Figure 7h assumes morphology typical for treatment of PET in pure oxygen plasma [59]. The key modification of the PET polymers treated in plasma created in SO_2 gases is therefore rich surface morphology what is explained by degradation of the polymer chains because of extensive UV radiation.

Surface functionalization and a rich surface topography may have a large effect on the surface wettability. In Figure 8 are shown water contact angles for the PET polymer treated in various mixtures. A water contact angle decreased from initial ~78° to approximately 5° after the treatment in pure SO_2 plasma. When adding oxygen to the discharge, a slight increase of the contact angle is observed to approximately 20° when the mixture of 90% O_2 + 10% SO_2 was used. When pure oxygen plasma was used, the contact value decreased and was approximately similar to the one obtained in pure SO_2 plasma. Such variation of the contact angles can be explained by a different surface morphology, because the chemical state of the surface was practically similar for all the samples (Figures 4–6). The influence of surface morphology on the contact angle of polymers with virtually the same surface functional groups has been already elaborated in the scientific literature [60]. There are also many other papers about tuning the wettability of materials by changing the surface roughness [61–63].

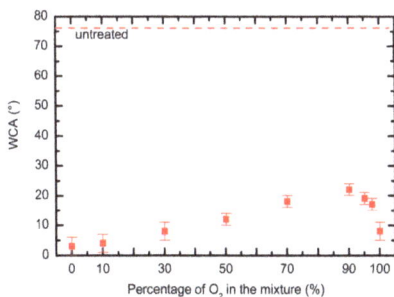

Figure 8. Water contact angles of the PET polymer treated in various SO_2 + O_2 mixtures. Zero percent of O_2 in the mixture refers to pure SO_2 plasma, whereas 100% of O_2 refers to pure oxygen plasma.

3.3. Cell Adhesion on Plasma-Treated Surfaces

The polymer surface morphology and wettability may influence adhesion and proliferation of mammalian cells [64–68]. To reveal the influence of the morphology and wettability of PET samples functionalized with sulfate groups on the biological response, we performed systematic measurements of the cell adhesion efficiency using HUVEC cells. Biological tests were adapted to study the cell adhesion to plasma modified surfaces and their toxic effects via the metabolic activity of cells adhered to the plasma modified substrates. Figure 9 represents the results of the cellular viability using the MTT assay and Figure 10 shows the behavior of cells using the TOX test. Experiments were performed using various incubation times of 1, 2, 3, 5, and 24 h. The value of 100% was attributed to the untreated samples that served us as a control. Both figures indicate improved biocompatibility of the PET samples treated by plasma compared with the untreated ones. Differences were observed as early as one hour after cells seeding and persisted in all incubation times. After 24 h of incubation, a slight decrease in the cell viability was observed as compared to the viability after 5 h of incubation (Figure 9). At the beginning, the cells first speeded up the metabolism (ATP (adenosine triphosphate) production for synthesis of heat shock proteins and surface adhesive proteins), which was observed after 3 and 5 h of incubation as an increased cell viability. Between 5 and 24 h some cells undergo apoptosis, whereas the other cells continue growing, but slow down the metabolism to save the energy and nutrients available. Another reason for slowing down the metabolism may be toxicity of the plasma-treated samples on the adhered HUVEC cells.

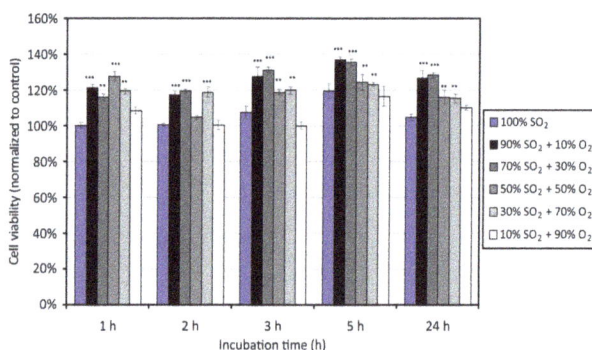

Figure 9. Results of the MTT assay for HUVEC cell proliferation on PET surfaces treated in various $SO_2 + O_2$ mixtures. Symbols * represent statistical significance (** represents statistical significance at $p < 0.01$ compared with the control. *** represents statistical significance at $p < 0.001$ compared with the control). Mean values ($\pm$ SE) for the respective triplicates are given.

The results obtained by the MTT and TOX assays are complementary—based on different experimental techniques. The MTT assay revealed metabolic activity of HUVEC adhered on the sample surface, which is directly related to cell viability. Whereas, the TOX test determined the total biomass of cells (live vs. dead) adhered on the samples, based on the intracellular proteins, detected by the test. Information on the mass of adhered cells was obtained, from what was possible to conclude on the toxic effects of the samples surface.

Since the results obtained by the MTT and TOX tests are scattered because of the reasons explained above, the sum of the deviations from the untreated samples is shown in Figure 11. This figure represents the sum of deviation for each treatment setting the value for the untreated sample to zero. For all treated samples there is a well-pronounced maximum in the cell proliferation versus the gas mixture. The values at various gas mixtures for particular incubation times are fitted with a parabola.

The maxima for each incubation time appear between 40% and 50% of oxygen concentration. These maxima appear at moderately rough surfaces (Figure 7).

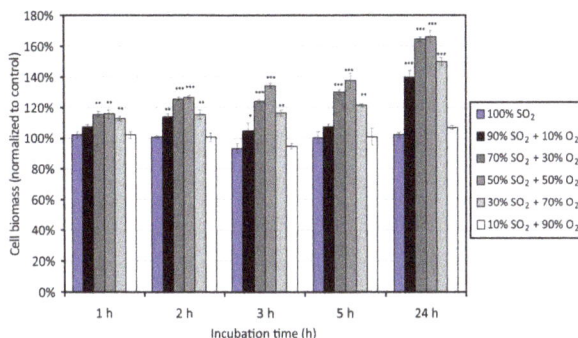

Figure 10. Surface cytotoxicity of PET surfaces treated in various $SO_2 + O_2$ mixtures. * represents statistical significance (* statistically significant at $p < 0.05$ compared with the control. ** represents statistical significance at $p < 0.01$ compared with the control. *** represents statistical significance at $p < 0.001$ compared with the control). Mean values ($\pm$ SE) for the respective triplicates are given.

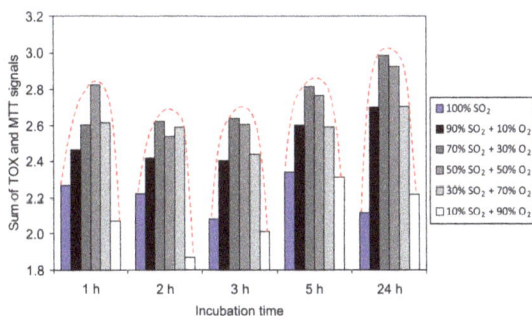

Figure 11. The sum of deviations from untreated samples observed for the MTT and TOX tests.

Both MTT and TOX tests are quantitative and reveal metabolic activity and total biomass of cells seeded onto substrates. The metabolic activity and total biomass should be reflected in the morphology of cells, so we performed also SEM characterization of selected samples. This technique is qualitative because the cells are usually randomly distributed on a substrate. The most representative SEM images are shown in Figure 12. Figure 12a represents a PET sample treated with 100% SO_2 and the cells adhered to it. As expected, the cells were clustered unevenly on the surface and have a rather circular morphology what is typical for the cells not able to fully adhere to the surface, and/or the cells responding to cytotoxic effect, indicating the apoptosis. A 10% amount of oxygen in the gas mixture allows for a surface finish which is more suitable for cell proliferation because the cells have longitudinal shape rich in long protrusions (Figure 12b). The image in Figure 12c (oxygen concentration of 40% where the MTT and TOX tests show the best results) reveals well-proliferating cells. Their spread and elongated morphology indicates good adhesion on the polymer surface. Finally, the image in Figure 12d (90% of oxygen) reveals cells of morphology similar to those on samples treated by pure SO_2 plasma. The SEM images are therefore sound with the quantitative results obtained by the MTT and TOX tests. Rounded cluttered cells on Figure 12d are metabolically less active than spread cells on the Figure 12c, but the higher number of cells in both Figure 12c and d indicates increased biomass detected with the TOX test after 24 h of incubation.

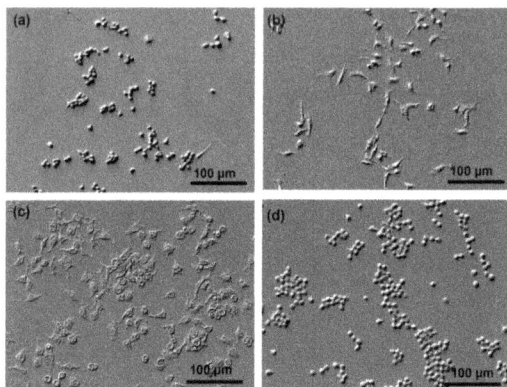

Figure 12. SEM images of cells on PET surfaces treated in various $SO_2 + O_2$ mixtures after 24 h of incubation: (**a**) 100% SO_2; (**b**) 90% SO_2 + 10% O_2; (**c**) 60% SO_2 + 40% O_2; and (**d**) 10% SO_2 + 90% O_2.

Given that the PET samples treated in various gas mixtures differ only in their surface morphology but not in their chemical composition, our results clearly show that the surface morphology of plasma-treated polymers has an important influence on cell adhesion and proliferation. Regarding the surface wettability of the plasma-treated samples, it seems that it does not have an important influence on the cell adhesion in our case, because changes in the surface wettability are not so pronounced to explain changes in the cell adhesion and proliferation. Furthermore, a trend of monotonous increase of the contact angles with the increasing oxygen content in the mixture is not observed in the case of cell proliferation.

4. Conclusions

The combination of various experimental biological techniques allowed for studying the behavior of HUVEC cells seeded on PET substrates which had been previously modified by gaseous plasma created in various mixtures of sulfur dioxide and oxygen gases. The best proliferation was observed in the case of nearly the same amounts of both mentioned gasses. The key reason for better proliferation of this type of human cell is likely due to the appropriate structuring of the substrate morphology of the sulfurized polymer surface. The best improvement of the polymer biocompatibility was obtained in the gas mixtures of approximately 60% SO_2 where the surface of the sulfonated polymer was covered with nearly spherical structures of lateral dimension of approximately 1 micrometer. The rich morphology of the samples treated in gaseous plasma with substantial concentration of sulfur dioxide was explained by modification of the polymer within the surface film of several 100 nm. The modification was due to the bond scission of this type of polymer caused by the ultraviolet radiation.

Acknowledgments: This research was financially supported by the Slovenian Research Agency ARRS (project grant number L7-6782).

Author Contributions: Nina Recek conceived and designed the experiments; Matic Resnik and Rok Zaplotnik performed plasma treatment and plasma characterization; Alenka Vesel performed surface characterization of the samples; Nina Recek and Helena Motaln performed and interpreted biological experiments under supervision of Tamara Lah-Turnsek; Alenka Vesel and Miran Mozetic analyzed the data and wrote the paper.

References

1. Mozetic, M.; Primc, G.; Vesel, A.; Zaplotnik, R.; Modic, M.; Junkar, I.; Recek, N.; Klanjsek-Gunde, M.; Guhy, L.; Sunkara, M.K.; et al. Application of extremely non-equilibrium plasmas in the processing of nano and biomedical materials. *Plasma Sources Sci. Technol.* **2015**, *24*, 015026. [CrossRef]

2. Labay, C.; Canal, J.M.; Modic, M.; Cvelbar, U.; Quiles, M.; Armengol, M.; Arbos, M.A.; Gil, F.J.; Canal, C. Antibiotic-loaded polypropylene surgical meshes with suitable biological behaviour by plasma functionalization and polymerization. *Biomaterials* **2015**, *71*, 132–144. [CrossRef] [PubMed]

3. Puac, N.; Petrovic, Z.L.; Radetic, M.; Djordjevic, A. Low pressure RF capacitively coupled plasma reactor for modification of seeds, polymers and textile fabrics. *Mater. Sci. Forum* **2005**, *494*, 291–296. [CrossRef]

4. Lopez-Garcia, J.; Lehocky, M.; Humpolicek, P.; Novak, I. On the correlation of surface charge and energy in non-thermal plasma-treated polyethylene. *Surf. Interface Anal.* **2014**, *46*, 625–629. [CrossRef]

5. Vandrovcova, M.; Grinevich, A.; Drabik, M.; Kylian, O.; Hanus, J.; Stankova, L.; Lisa, V.; Choukourov, A.; Slavinska, D.; Biederman, H.; Bacakova, L. Effect of various concentrations of Ti in hydrocarbon plasma polymer films on the adhesion, proliferation and differentiation of human osteoblast-like MG-63 cells. *Appl. Surf. Sci.* **2015**, *357*, 459–472. [CrossRef]

6. Labay, C.; Canal, C.; Garcia-Celma, M.J. Influence of corona plasma treatment on polypropylene and polyamide 6.6 on the release of a model drug. *Plasma Chem. Plasma Process.* **2010**, *30*, 885–896. [CrossRef]

7. Rodriguez-Emmenegger, C.; Kylian, O.; Houska, M.; Brynda, E.; Artemenko, A.; Kousal, J.; Alles, A.B.; Biederman, H. Substrate-independent approach for the generation of functional protein resistant surfaces. *Biomacromolecules* **2011**, *12*, 1058–1066. [CrossRef] [PubMed]

8. Petrovic, Z.L.; Puac, N.; Malovic, G.; Lazovic, S.; Maletic, D.; Miletic, M.; Mojsilovic, S.; Milenkovic, P.; Bugarski, D. Application of non-equilibrium plasmas in medicine. *J. Serb. Chem. Soc.* **2012**, *77*, 1689–1699. [CrossRef]

9. Popelka, A.; Novak, I.; Lehocky, M.; Bilek, F.; Kleinova, A.; Mozetic, M.; Spirkova, M.; Chodak, I. Antibacterial treatment of LDPE with halogen derivatives via cold plasma. *Express. Polym. Lett.* **2015**, *9*, 402–411. [CrossRef]

10. Labay, C.; Canal, J.M.; Canal, C. Relevance of surface modification of polyamide 6.6 fibers by air plasma treatment on the release of caffeine. *Plasma Process. Polym.* **2012**, *9*, 165–173. [CrossRef]

11. Kuzminova, A.; Shelemin, A.; Kylian, O.; Choukourov, A.; Valentova, H.; Krakovsky, I.; Nedbal, J.; Slavinska, D.; Biederman, H. Study of the effect of atmospheric pressure air dielectric barrier discharge on Nylon 6,6 foils. *Polym. Degrad. Stabil.* **2014**, *110*, 378–388. [CrossRef]

12. Kregar, Z.; Biscan, M.; Milosevic, S.; Mozetic, M.; Vesel, A. Interaction of argon, hydrogen and oxygen plasma early afterglow with polyvinyl chloride (PVC) materials. *Plasma Process. Polym.* **2012**, *9*, 1020–1027. [CrossRef]

13. Novak, I.; Popelka, A.; Valentin, M.; Chodak, I.; Spirkova, M.; Toth, A.; Kleinova, A.; Sedliacik, J.; Lehocky, M.; Maronek, M. Surface behavior of polyamide 6 modified by barrier plasma in oxygen and nitrogen. *Int. J. Polym. Anal. Ch.* **2014**, *19*, 31–38. [CrossRef]

14. Labay, C.; Canal, J.M.; Navarro, A.; Canal, C. Corona plasma modification of polyamide 66 for the design of textile delivery systems for cosmetic therapy. *Appl. Surf. Sci.* **2014**, *316*, 251–258. [CrossRef]

15. Lopez-Garcia, J.; Primc, G.; Junkar, I.; Lehocky, M.; Mozetic, M. On the hydrophilicity and water resistance effect of styrene-acrylonitrile copolymer treated by CF_4 and O_2 plasmas. *Plasma Process. Polym.* **2015**, *12*, 1075–1084. [CrossRef]

16. Modic, M.; Junkar, I.; Vesel, A.; Mozetic, M. Aging of plasma treated surfaces and their effects on platelet adhesion and activation. *Surf. Coat. Technol.* **2012**, *213*, 98–104. [CrossRef]

17. Recek, N.; Resnik, M.; Motaln, H.; Lah-Turnsek, T.; Augustine, R.; Kalarikkal, N.; Thomas, S.; Mozetic, M. Cell adhesion on polycaprolactone modified by plasma treatment. *Int. J. Polym. Sci.* **2016**, *2016*, 354–366. [CrossRef]

18. Kuzminova, A.; Vandrovcova, M.; Shelemin, A.; Kylian, O.; Choukourov, A.; Hanus, J.; Bacakova, L.; Slavinska, D.; Biederman, H. Treatment of poly(ethylene terephthalate) foils by atmospheric pressure air dielectric barrier discharge and its influence on cell growth. *Appl. Surf. Sci.* **2015**, *357*, 689–695. [CrossRef]

19. Li, B.; Lu, F.; Wei, X.; Zhao, R. Fucoidan: Structure and bioactivity. *Molecules* **2008**, *13*, 1671–1695. [CrossRef] [PubMed]

20. Klement, P.; Du, Y. J.; Berry, L.; Andrew, M.; Chan, A.K.C. Blood-compatible biomaterials by surface coating with a novel antithrombin-heparin covalent complex. *Biomaterials* **2002**, *23*, 527–535. [CrossRef]
21. Grasel, T.G.; Cooper, S.L. Properties and biological inteactions of polyurethane anionomers: Effect of sulfonate incorporation. *J. Biomed. Mater. Res.* **1989**, *23*, 311–338. [CrossRef] [PubMed]
22. Okkema, A.Z.; Visser, S.A.; Cooper, S.I. Physical and blood-contacting properties of polyurethanes based on a sulfonic acid-containing diol chain extender. *J. Biomed. Mater. Res.* **1991**, *25*, 1371–1395. [CrossRef] [PubMed]
23. Vesel, A.; Mozetic, M.; Strnad, S. Improvement of adhesion of fucoidan on polyethylene terephthalate surface using gas plasma treatments. *Vacuum* **2011**, *85*, 1083–1086. [CrossRef]
24. Ozaltin, K.; Lehocky, M.; Humpolicek, P.; Pelkova, J.; Saha, P. A new route of fucoidan immobilization on low density polyethylene and its blood compatibility and anticoagulation activity. *Int. J. Mol. Sci.* **2016**, *17*, 908–911. [CrossRef] [PubMed]
25. Popelka, A.; Novak, I.; Lehocky, M.; Junkar, I.; Mozetic, M.; Kleinova, A.; Janigova, I.; Slouf, M.; Bilek, F.; Chodak, I. A new route for chitosan immobilization onto polyethylene surface. *Carbohyd. Polym.* **2012**, *90*, 1501–1508. [CrossRef] [PubMed]
26. Jui-Che, L.; Cooper, S.L. Surface characterization and ex vivo blood compatibility study of plasmamodified small diameter tubing: effect of sulphur dioxide and hexamethyl-disiloxane plasmas. *Biomaterials* **1995**, *16*, 1017–1023. [CrossRef]
27. Tze-Man, K.; Jui-Che, L.; Cooper, S.L. Surface characterization and platelet adhesion studies of plasma-sulphonated polyethylene. *Biomaterials* **1993**, *14*, 657–664. [CrossRef]
28. Inagaki, N.; Tasaka, S.; Miyazaki, H. Sulfonic acid group-containing thin films prepared by plasma polymerization. *J. Appl. Polym. Sci.* **1989**, *38*, 1829–1838. [CrossRef]
29. Klee, D.; Villari, R.V.; Höcker, H.; Dekker, B.; Mittermayer, C. Surface modification of a new flexible polymer with improved cell adhesion. *J. Mater. Sci. Mater. Med.* **1994**, *5*, 592–595. [CrossRef]
30. Shin, Y.N.; Kim, B.S.; Ahn, H.H.; Lee, J.H.; Kim, K.S.; Lee, J.Y.; Kim, M.S.; Khang, G.; Lee, H.B. Adhesion comparison of human bone marrow stem cells on a gradient wettable surface prepared by corona treatment. *Appl. Surf. Sci.* **2008**, *255*, 293–296. [CrossRef]
31. Wang, L.C.; Chen, J.W.; Liu, H.I.; Chen, Z.Q.; Zhang, Y.; Wang, C.Y.; Feng, Z.G. Synthesis and evaluation of biodegradable segmented multiblock poly(ether ester) copolymers for biomaterial applications. *Polym. Int.* **2004**, *53*, 2145–2154. [CrossRef]
32. Gugala, Z.; Gogolewski, S. Attachment, growth, and activity of rat osteoblasts on polylactide membranes treated with various lowtemperature radiofrequency plasmas. *J. Biomed. Mater. Res. A* **2006**, *76A*, 288–299. [CrossRef] [PubMed]
33. Holländer, A.; Kröpke, S. Polymer surface treatment with SO$_2$-containing plasmas. *Plasma Process. Polym.* **2010**, *7*, 390–402. [CrossRef]
34. Hiratsuka, A.; Fukui, H.; Suzuki, Y.; Muguruma, H.; Sakairi, K.; Matsushima, T.; Maruo, Y.; Yokoyama, K. Sulphur dioxide plasma modification on poly(methyl methacrylate) for fluidic devices. *Curr. Appl. Phys.* **2008**, *8*, 198–205. [CrossRef]
35. Caro, J.C.; Lappan, U.; Simon, F.; Pleul, D.; Lunkwitz, K. On the low-pressure plasma treatment of PTFE (polytetrafluoroethylene) with SO$_2$ as process gas. *Eur. Polym. J.* **1999**, *35*, 1149–1152. [CrossRef]
36. Coen, M.C.; Keller, B.; Groening, P.; Schlapbach, L. Functionalization of graphite, glassy carbon, and polymer surfaces with highly oxidized sulfur species by plasma treatments. *J. Appl. Phys.* **2002**, *92*, 5077–5083. [CrossRef]
37. Siow, K.S.; Britcher, L.; Kumar, S.; Griesser, H.J. Sulfonated surfaces by sulfur dioxide plasma surface treatment of plasma polymer films. *Plasma Process. Polym.* **2009**, *6*, 583–592. [CrossRef]
38. Fatyeyeva, K.; Poncin-Epaillard, F. Sulfur dioxide plasma treatment of the clay (laponite) particles. *Plasma Chem. Plasma Process.* **2011**, *31*, 449–464. [CrossRef]
39. Kregar, Z.; Biscan, M.; Milosevic, S.; Elersic, K.; Zaplotnik, R.; Primc, G.; Cvelbar, U. Optical emission characterization of extremely reactive oxygen plasma during treatment of graphite samples. *Mater. Tehnol.* **2012**, *46*, 25–30.
40. Kregar, Z.; Biscan, M.; Milosevic, S.; Vesel, A. Monitoring oxygen plasma treatment of polypropylene with optical emission spectroscopy. *IEEE Trans. Plasma Sci.* **2011**, *39*, 1239–1246. [CrossRef]

41. Krstulovic, N.; Labazan, I.; Milosevic, S.; Cvelbar, U.; Vesel, A.; Mozetic, M. Optical emission spectroscopy characterization of oxygen plasma during treatment of a PET foil. *J. Phys. D Appl. Phys.* **2006**, *39*, 3799–3804. [CrossRef]

42. Xie, C.; Hu, X.; Zhou, L.; Xie, D.; Guo, H. Ab initio determination of potential energy surfaces for the first two UV absorption bands of SO_2. *J. Chem. Phys.* **2013**, *139*, 14–30. [CrossRef] [PubMed]

43. Zaplotnik, R.; Vesel, A.; Mozetič, M. Characteristics of gaseous plasma created in SO_2 by inductively coupled RF discharge in E and H modes. *J. Appl. Phys.* **2016**, *120*, 163–169. [CrossRef]

44. Vesel, A.; Zaplotnik, R.; Modic, M.; Mozetic, M. Hemocompatibility properties of a polymer surface treated in plasma containing sulfur. *Surf. Interface Anal.* **2016**, *48*, 601–605. [CrossRef]

45. Cvelbar, U.; Mozetič, M.; Junkar, I.; Vesel, A.; Kovač, J.; Drenik, A.; Vrlinič, T.; Hauptman, N.; Klanjšek-Gunde, M.; Markoli, B.; Krstulović, N.; Milošević, S.; Gaboriau, F.; Belmonte, T. Oxygen plasma functionalization of poly(p-phenilene sulphide). *Appl. Surf. Sci.* **2007**, *253*, 8669–8673. [CrossRef]

46. Tsougeni, K.; Vourdas, N.; Tserepi, A.; Gogolides, E.; Cardinaud, C. Mechanisms of oxygen plasma nanotexturing of organic polymer surfaces: from stable super hydrophilic to super hydrophobic surfaces. *Langmuir* **2009**, *25*, 11748–11759. [CrossRef] [PubMed]

47. Wohlfart, E.; Fernández-Blázquez, J.P.; Knoche, E.; Bello, A.; Pérez, E.; Arzt, E.; del Campo, A. Nanofibrillar Patterns by Plasma Etching: The influence of polymer crystallinity and orientation in surface morphology. *Macromolecules* **2010**, *43*, 9908–9917. [CrossRef]

48. Slepicka, P.; Kasalkova, N.S.; Stranska, E.; Bacakova, L.; Svorcik, V. Surface characterization of plasma treated polymers for applications as biocompatible carriers. *Express Polym. Lett.* **2013**, *7*, 535–545. [CrossRef]

49. Kontziampasis, D.; Constantoudis, V.; Gogolides, E. Plasma Directed organization of nanodots on polymers: Effects of polymer type and etching time on morphology and order. *Plasma Process. Polym.* **2012**, *9*, 866–872. [CrossRef]

50. Li, Y.P.; Shi, W.; Li, S.Y.; Lei, M.K. Transition of water adhesion on superhydrophobic surface during aging of polypropylene modified by oxygen capacitively coupled radio frequency plasma. *Surf. Coat. Technol.* **2012**, *213*, 139–144. [CrossRef]

51. Li, Y.P.; Li, S.Y.; Shi, W.; Lei, M.K. Hydrophobic over-recovery during aging of polyethylene modified by oxygen capacitively coupled radio frequency plasma: A new approach for stable superhydrophobic surface with high water adhesion. *Surf. Coat. Technol.* **2012**, *206*, 4952–4958. [CrossRef]

52. Cortese, B.; Morgan, H. Controlling the wettability of hierarchically structured thermoplastics. *Langmuir* **2012**, *28*, 896–904. [CrossRef] [PubMed]

53. Fernández-Blázquez, J.P.; Fell, D.; Bonaccurso, E.; Campo, A.D. Superhydrophilic and superhydrophobic nanostructured surfaces via plasma treatment. *J. Colloid Interface Sci.* **2011**, *357*, 234–238. [CrossRef] [PubMed]

54. Palumbo, F.; Di Mundo, R.; Cappelluti, D.; d'Agostino, R. Superhydrophobic and superhydrophilic polycarbonate by tailoring chemistry and nano-texture with plasma processing. *Plasma Process. Polym.* **2011**, *8*, 118–126. [CrossRef]

55. Doliska, A.; Vesel, A.; Kolar, M.; Stana-Kleinschek, K.; Mozetic, M. Interaction between model poly(ethylene terephthalate) thin films and weakly ionised oxygen plasma. *Surf. Interface Anal.* **2012**, *44*, 56–61. [CrossRef]

56. Kim, J.; Ahn, D.U.; Sancaktar, E. The effects of excimer laser irradiation on surface morphology development in stretched poly(ethylene therephthalate), poly(butylene therephthalate) and polystyrene films. In *Polymer Surface Modification, Relevance to Adhesion*, 1st ed.; Mittal, K.L., Ed.; CRC Press: Boca Raton, FL, USA, 2007; Volume 4, pp. 33–86.

57. Kostov, K.G.; Nishime, T.M.C.; Hein, L.R.O.; Toth, A. Study of polypropylene surface modification by air dielectric barrier discharge operated at two different frequencies. *Surf. Coat. Technol.* **2013**, *234*, 60–66. [CrossRef]

58. Cismaru, C.; Shohet, J.L. Plasma vacuum ultraviolet emission in an electron cyclotron resonance etcher. *Appl. Phys. Lett.* **1999**, *74*, 2599–2601. [CrossRef]

59. Vesel, A.; Junkar, I.; Cvelbar, U.; Kovac, J.; Mozetic, M. Surface modification of polyester by oxygen- and nitrogen-plasma treatment. *Surf. Interface Anal.* **2008**, *40*, 1444–1453. [CrossRef]

60. Izdebska, J.; Sabu, T. *Printing on polymers: Fundamentals and applications*, 1st ed.; Elsevier Science Publishing: Waltham, USA, 2016; pp. 116–118.

61. Ramiasa-MacGregor, M.; Mierczynska, A.; Sedev, R.; Vasilev, K. Tuning and predicting the wetting of nanoengineered material surface. *Nanoscale* **2016**, *8*, 4635–4642. [CrossRef] [PubMed]

62. Bico, J.; Tordeux, C.; Quere, D. Rough wetting. *Europhys. Lett.* **2001**, *55*, 214–220. [CrossRef]
63. Encinas, N.; Pantoja, M.; Abenojar, J.; Martínez, M.A. Control of wettability of polymers by surface roughness modification. *J. Adhes. Sci. Technol.* **2010**, *24*, 1869–1883. [CrossRef]
64. Flemming, R.G.; Murphy, C.J.; Abrams, G.A.; Goodman, S.L.; Nealey, P.F. Effects of synthetic micro- and nano-structured surfaces on cell behavior. *Biomaterials* **1999**, *20*, 573–588. [CrossRef]
65. Anselme, K.; Davidson, P.; Popa, A.M.; Giazzon, M.; Liley, M.; Ploux, L. The interaction of cells and bacteria with surfaces structured at the nanometre scale. *Acta Biomater.* **2010**, *6*, 3824–3846. [CrossRef] [PubMed]
66. Martínez, E.; Engel, E.; Planell, J.A.; Samitier, J. Effects of artificial micro- and nano-structured surfaces on cell behaviour. *Ann. Anat.* **2009**, *191*, 126–135. [CrossRef] [PubMed]
67. Lord, M.S.; Foss, M.; Besenbacher, F. Influence of nanoscale surface topography on protein adsorption and cellular response. *Nano Today* **2010**, *5*, 66–78. [CrossRef]
68. Mendes, P.M. Cellular nanotechnology: Making biological interfaces smarter. *Chem. Soc. Rev.* **2013**, *42*, 9207–9218. [CrossRef] [PubMed]

polymers

MDPI

Article

Artificial Spores: Immunoprotective Nanocoating of Red Blood Cells with Supramolecular Ferric Ion-Tannic Acid Complex

Taegyun Park [1], Ji Yup Kim [1], Hyeoncheol Cho [1], Hee Chul Moon [1], Beom Jin Kim [1],
Ji Hun Park [1], Daewha Hong [2], Joonhong Park [3] and Insung S. Choi [1,*]

[1] Center for Cell-Encapsulation Research, Department of Chemistry, Korea Advanced Institute of Science and
    Technology (KAIST), Daejeon 34141, Korea; xorbs7467@kaist.ac.kr (T.P.); y123812@kaist.ac.kr (J.Y.K.);
    harry0305@kaist.ac.kr (H.C.); arbalest@kaist.ac.kr (H.C.M.); kimbj20@kaist.ac.kr (B.J.K.);
    pjh1987@kaist.ac.kr (J.H.P.)
[2] Department of Chemistry and Chemistry Institute of Functional Materials, Pusan National University,
    Busan 46241, Korea; dwhong17@pusan.ac.kr
[3] Department of Laboratory Medicine, College of Medicine, The Catholic University of Korea,
    St. Mary's Hospital, Daejeon 34943, Korea; miziro@catholic.ac.kr
* Correspondence: ischoi@kaist.ac.kr; Tel.: +82-42-350-2840

Academic Editor: Ruth Freitag
Received: 20 March 2017; Accepted: 12 April 2017; Published: 13 April 2017

Abstract: The blood-type-mismatch problem, in addition to shortage of blood donation, in blood transfusion has prompted the researchers to develop universal blood that does not require blood typing. In this work, the "cell-in-shell" (i.e., artificial spore) approach is utilized to shield the immune-provoking epitopes on the surface of red blood cells (RBCs). Individual RBCs are successfully coated with supramolecular metal-organic coordination complex of ferric ion (Fe^{III}) and tannic acid (TA). The use of isotonic saline (0.85% NaCl) is found to be critical in the formation of stable, reasonably thick (20 nm) shells on RBCs without any aggregation and hemolysis. The formed "RBC-in-shell" structures maintain their original shapes, and effectively attenuate the antibody-mediated agglutination. Moreover, the oxygen-carrying capability of RBCs is not deteriorated after shell formation. This work suggests a simple but fast method for generating immune-camouflaged RBCs, which would contribute to the development of universal blood.

Keywords: artificial spores; cell-surface engineering; immunoprotection; nanocoating; red blood cells; supramolecular complex

1. Introduction

The "cell-in-shell" structures (a.k.a., artificial spores [1–3] or micrometric Iron Men [4]) are the emerging cell hybrid entities in biomedical and nanomedicinal fields, where individual live cells are encapsulated within nanometric (<100 nm) shells. Microbial and mammalian cells have been coated with silica, silica-titania, polymers, and metal-organic frameworks, and the cells inside are protected from the harmful, and often lethal, attack of enzymes, nanoparticles, heat, or UV light [5–12]. Live cells also have been interfaced or three-dimensionally confined with liposomes [13], carbon nanotubes [14], and graphene [15,16]. The shell formation and degradation are further controlled chemically, which allows for temporal cytoprotection of therapeutically functional cells during in vitro manipulation and storage, inspired by sporulation and germination processes found in nature [12,17–20]. Among the functional cells employed so far in the field of artificial spores, red blood cells (RBCs) would be one of the simplest but highly important cells in cell therapy and related fields [21]. RBCs are the anucleate, non-dividing cells, the main role of which is oxygen delivery in the body. Approximately 85 million units of RBCs are transfused annually worldwide [22], but blood supply, mostly supported by donation, falls short of the demand. Blood-type mismatch poses a worse problem in transfusion medicine, causing antibody-mediated immune responses, such as cross-type agglutination and hemolysis, which lead to life-threatening situations [23]. To tackle these problems in blood transfusion, immune-camouflaged "RBC-in-shell" structures have been proposed as a universal blood, where the immune-provoking epitopes on RBC surfaces are shielded by the encasing shells. The approach of artificial spores would be much simpler and more cost-effective than other strategies, such as enzymatic cleavage of the antigens [24] and biological production of O (Rh$^+$) RBCs from hematopoietic stem cells [25,26]. Moreover, the approach of enzymatic cleavage can be applied only to A and B antigens, not to the protein-based D (Rh) antigen, because glycosidases are mainly used for the cleavage reactions [27]; the in vitro RBC production is difficult to scale up [28]. RBCs have been coated with polyelectrolyte multilayers (PEMs) by multi-step layer-by-layer (LbL) assembly for the attenuation of antigen-antibody recognition [29]. One-step coating of polydopamine [30] or plant-derived pyrogallol [31] also has recently been applied for immunoprotection of RBCs. However, all the reported methods require hours of coating steps, which precludes the clinical applications. In this paper, we report a simple but rapid method for fabricating immune-camouflaged RBCs, based on the supramolecular metal-organic coordination complex of ferric ion (Fe^{III}) and tannic acid (TA) (Figure 1).

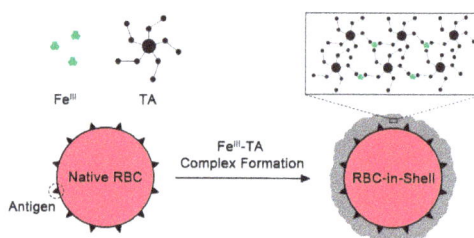

Figure 1. Schematic representation for supramolecular shell formation of ferric ion (Fe^{III}) and tannic acid (TA) on individual red blood cells (RBCs).

2. Results and Discussion

TA is a natural polyphenol, which has several galloyl groups attached to a glucose molecule. Because the galloyl groups of TA form multivalent coordination bonds with Fe^{III}, TA molecules are cross-linked by Fe^{III} to form a supramolecular metal-organic Fe^{III}-TA complex (Fe^{III}-TA-MOC). The universal surface-binding affinity of TA (and its assemblies) allows for the Fe^{III}-TA nanofilm formation on various substrates including planar or particulate substrates having different surface properties [32,33]. We and others have previously demonstrated that the material-independent Fe^{III}-TA nanocoating method could be utilized to coat individual living cells, such as yeast, *Escherichia coli*, HeLa, NIH 3T3, PC-12, and Jurkat cells, without significant loss of cell viability [17,18,20]. However, this method cannot be directly applied to RBCs, because RBCs are extremely sensitive to the tonicity of the reaction solutions and chemicals. In this work, we optimized the coating conditions of the supramolecular Fe^{III}-TA-MOC for RBC coating and successfully generated "RBC-in-shell" structures.

We used the isotonic 0.85% NaCl solution during the entire coating processes including washing. RBCs were found to be swollen and hemolyzed rapidly in deionized (DI) water (data not shown). We also optimized the concentration of TA and found that 0.05 mg mL^{-1} was optimal for RBCs; above that concentration were clumped RBCs (Figure S1). The mass ratio of TA to FeCl$_3$·6H$_2$O was fixed to 4:1 for uniform coating according to the previous studies [17,32]. Briefly, to the RBC suspension in the 0.85% NaCl solution were added the TA stock solution and the FeCl$_3$·6H$_2$O stock solution (final concentration: [TA] = 0.05 mg mL^{-1}, [Fe^{III}] = 0.0125 mg mL^{-1}) sequentially, with 10-second mixing after each addition. 3-(*N*-morpholino)propanesulfonic acid (MOPS)-buffered saline (pH 7.4) was then added for pH stabilization, which made Fe^{III}-TA-MOC stable. The whole coating processes were conducted in one pot for less than one minute, and repeated four times to generate the Fe^{III}-TA-coated RBC (RBC@[Fe^{III}-TA]). We found that the addition sequence of TA and Fe^{III} was critical in the stable formation of Fe^{III}-TA coats. When the Fe^{III} stock solution (>0.001 mg mL^{-1}) was added to the RBC suspension, RBCs aggregated uncontrollably (Figure S2), presumably because of multivalent interactions of metallic cations with RBCs [34]. However, when the TA stock solution was added first to the RBC suspension, individual RBCs were successfully coated with Fe^{III}-TA-MOC without agglutination. We think that the pre-deposited TA molecules rapidly formed Fe^{III}-TA-MOC, when Fe^{III} was added; the Fe^{III}-TA-MOC formation on RBC surfaces would decrease effective Fe^{III} concentration in solution, and, additionally, the Fe^{III}-TA-MOC further bound with free TA in solution.

The isotonic saline solution of RBC@[Fe^{III}-TA] was changed from red to dark purple in color, indicating that the Fe^{III}-TA coat was successfully formed (Figure 2a). We also visualized the Fe^{III}-TA coat by adsorbing Alexa Fluor® 647-conjugated bovine serum albumin (BSA-Alexa Fluor® 647) onto RBC@[Fe^{III}-TA] (Figure 2c). The confocal laser-scanning microscopy (CLSM) images clearly showed red-fluorescent rings only for RBC@[Fe^{III}-TA]. When we treated RBCs only with TA, we did not observe any fluorescence (Figure S3). We characterized RBC@[Fe^{III}-TA] by Raman spectroscopy, scanning electron microscopy (SEM), transmission electron microscopy (TEM), and atomic force microscopy (AFM). The Raman spectrum of RBC@[Fe^{III}-TA] showed intense bands at 1349 and 1487 cm^{-1}, corresponding to C-C ring vibration and C-H bending of TA (Figure 2b). In the SEM micrographs, the surface became rougher and more grainy after coating than the surface of native RBCs (Figure 2d). Although RBCs were crumpled after Fe^{III}-TA coating, any defects or fractures were not observed. To further characterize the cell membranes in the SEM images, we immersed RBCs (native or coated) in DI water for hypotonic lysis and air-dried them on a flat silicon wafer. Compared with the smooth surface of lysed native RBCs, the surfaces of lysed RBC@[Fe^{III}-TA] cells were grainy, indicating that the Fe^{III}-TA coat was not affected greatly by hypotonic lysis (Figure 2e). On a closer view, many wrinkles were clearly observed only on the membrane of lysed RBC@[Fe^{III}-TA] cells. These results suggested that the Fe^{III}-TA coat reinforced the integrity of RBC membranes; it has been well reported that hollow capsules are folded when they collapse, unless their capsule shells have defects [32,35].

The uniform Fe^{III}-TA coat was clearly observed in the TEM images of microtomed RBC@[Fe^{III}-TA], and the average thickness was measured to be around 20 nm (Figure 3a). The thickness of the Fe^{III}-TA coat was also estimated by the AFM line-profile analysis. The minimum heights of the collapsed membranes were measured to be about 10 and 50 nm for native RBC and RBC@[Fe^{III}-TA], respectively (Figure 3b). Therefore, the film thickness was calculated to be about 20 nm, which was in agreement with the results from the TEM analysis. It is to note that the relatively thick films (5 nm per coating) were formed with the use of low concentrations of TA and Fe^{III} (final concentration: [TA] = 0.05 mg mL^{-1}; [Fe^{III}] = 0.0125 mg mL^{-1}). For comparison, the previous reports with polystyrene particles showed that the 10-nm-thick film was formed with 0.4 mg mL^{-1} of TA and 0.1 mg mL^{-1} of Fe^{III} [32]. The thick film formation in this work was attributed to the use of isotonic saline instead of DI water, because we found that NaCl increased the film thickness with gold substrates as a model (Figure S4). In the model studies, the gold substrates were coated with Fe^{III}-TA-MOC in DI water or isotonic saline ([TA] = 0.05 mg mL^{-1}; [Fe^{III}] = 0.0125 mg mL^{-1}), and the film thickness was measured by ellipsometry. The ellipsometric measurements indicated that the films formed in isotonic saline were 3.4 times thicker than those in DI water. The effects of ions on film thickness in the Fe^{III}-TA coating is currently being investigated in detail.

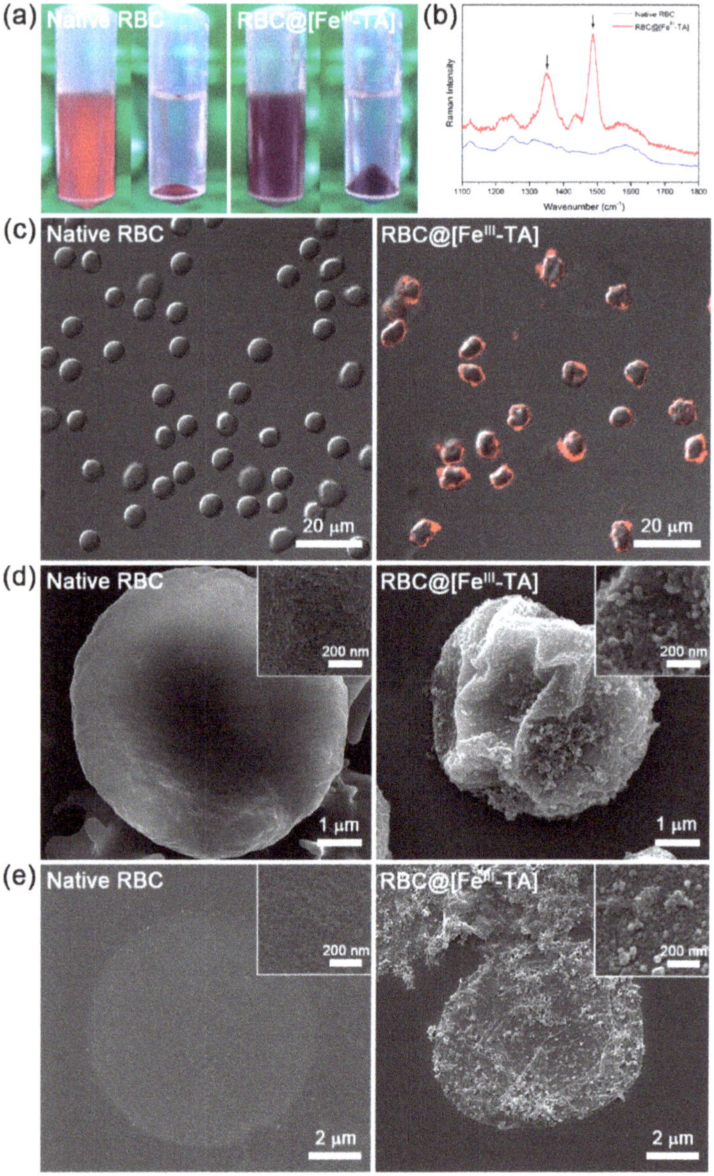

Figure 2. Characterizations of native RBC and RBC@[Fe^{III}-TA]. (**a**) Photographs of RBC suspension and pellet before and after shell formation. (**b**) Raman spectra of (blue) native RBC and (red) RBC@[Fe^{III}-TA]. Black arrows indicate the strong bands attributed to the ring structures of TA. (**c**) CLSM images of native RBC and RBC@[Fe^{III}-TA] after incubation with BSA-Alexa Fluor® 647. (**d**) SEM micrographs of native RBC and RBC@[Fe^{III}-TA]. (**e**) SEM micrographs of the membrane of native RBC and RBC@[Fe^{III}-TA] after hypotonic lysis.

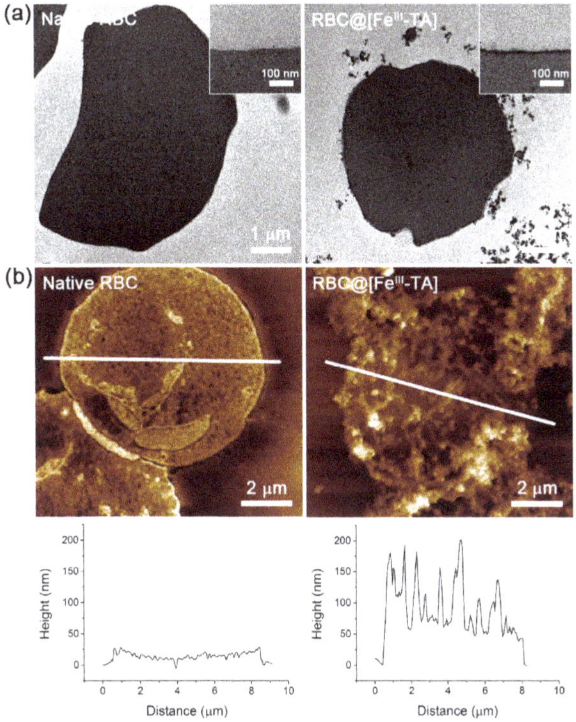

Figure 3. (**a**) TEM micrographs of native RBC and RBC@[Fe^{III}-TA]. (**b**) (**top**) AFM micrographs and (**bottom**) line-profile graphs of native RBC and RBC@[Fe^{III}-TA]. White lines in AFM micrographs indicate the path of line-profile analysis.

To assess the immunoprotective effect of Fe^{III}-TA coats, we performed the antibody-mediated agglutination assay. Because ABO/D blood typing should be conducted before blood transfusion, RBCs of type A, B, and D (Rh) were used for the assay. Each type of RBCs is determined by the presence of specific surface antigens, which causes agglutination with the anti-type antibodies. When treated with their anti-type sera (anti-A, anti-B, or anti-D (Rh)), RBC@[Fe^{III}-TA] remained unaffected, while native RBCs harshly agglutinated (Figure 4a). The results indicated that the Fe^{III}-TA coat was uniformly formed on all types of RBCs and successfully prevented the access of antibodies to the RBC surface-antigens, acting as an immunoprotective barrier. Oxygen transport is a vital function of RBCs in the body, and, therefore, many immunoprotective strategies have tried to maintain the oxygen-carrying capacity of native RBCs after modification for the development of universal RBCs. To investigate the oxygen-carrying property of "RBC-in-shell" structures, native or coated RBCs were added to an O$_2$-purged (initial oxygen concentration: ~39%) phosphate-buffered saline (PBS, 10 mM, pH 7.4), and the dissolved oxygen concentration was monitored over time with an oxygen probe (Figure 4b). Two time-lapse graphs for native RBC and RBC@[Fe^{III}-TA] had similar shapes to each other, and there was no significant difference in the amount of oxygen consumption after saturation (8.9% for native RBC and 9.4% for RBC@[Fe^{III}-TA]), indicating that the Fe^{III}-TA coat did not inhibit oxygen diffusion and hemoglobin function.

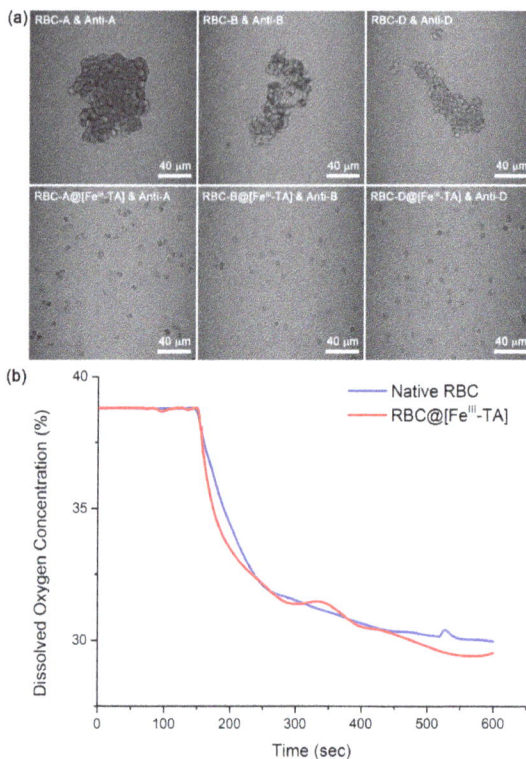

Figure 4. (**a**) Antibody-mediated agglutination assay. Native RBCs or RBC@[Fe^{III}-TA] cells were mixed with their anti-type sera, and the optical images were taken after one hour. (**b**) Oxygen consumption graphs. The dissolved oxygen concentration (%) in PBS (pH 7.4) was plotted as a function of time. The initial oxygen concentration dissolved in the O$_2$-purged PBS was about 39%, and the oxygen concentration was recorded by the oxygen probe connected with LabQuest$^®$.

3. Experimental Section

3.1. Materials

Tannic acid (TA, Sigma), iron(III) chloride hexahydrate (FeCl$_3$·6H$_2$O, Sigma, St. Louis, MO, USA), 3-(N-morpholino)propanesulfonic acid (MOPS, Sigma, St. Louis, MO, USA), sodium chloride (NaCl, Daejung, Siheung, Korea), Alexa Fluor$^®$ 647-conjugated albumin from bovine serum (BSA-Alexa Fluor$^®$ 647, Life Technologies, Carlsbad, CA, USA), phosphate-buffered saline (PBS, pH 7.4, Sigma, St. Louis, MO, USA) were used as received. Anti-A, anti-B, and anti-D (Rh) antisera were purchased from Asan Pharmaceutical (Seoul, Korea). Ultrapure water (18.3 MΩ·cm) from the Human Ultrapure System (Human Corp., Seoul, Korea) was used.

3.2. Red Blood Cell (RBC) Samples

Blood samples were obtained after completion of clinical testing. All studied samples were collected from human subjects who provided the written informed consent, and the study protocol was approved by the Institutional Review Board of the Catholic University of Korea. The collected blood was centrifuged at 600 g for 5 min at room temperature. The plasma and the buffy coat at the top were removed by a pipette, and the remaining red blood cells were washed twice with 0.85% (*w/v*) NaCl.

3.3. RBC Coating with Fe^{III}-TA Complex

Washed RBCs were re-suspended in 490 μL of 0.85% NaCl to be 1% hematocrit. The 5 μL of TA stock solution (5 mg mL^{-1} in 0.85% NaCl) and the 5 μL of FeCl$_3$·6H$_2$O stock solution (1.25 mg mL^{-1} in 0.85% NaCl) were added sequentially to the RBC suspension with 10 s mixing between the additions (final concentration: [TA] = 0.05 mg mL^{-1}; [Fe^{III}] = 0.0125 mg mL^{-1}). The resulting suspension was mixed for 10 s, and, after addition of 500 μL of MOPS-buffered saline (20 mM MOPS, 0.85% NaCl, pH 7.4) for pH stabilization, washed three times with 0.85% NaCl to remove any remaining TA and FeCl$_3$, leading to the formation of single-coated RBCs. The coating processes, from addition of TA to washing, were repeated four times to generate quadruple-coated RBC@[Fe^{III}-TA]. To visualize the Fe^{III}-TA coat, native RBCs or RBC@[Fe^{III}-TA] cells were suspended in 200 μL of the BSA-Alexa Fluor® 647 solution (0.4 mg mL^{-1} in 0.85% NaCl), and the resulting suspension was incubated for 15 min at room temperature. After washing with 0.85% NaCl, the samples were characterized by confocal laser-scanning microscopy (LSM 700, Carl Zeiss, Oberkochen, Germany). To measure the film thickness, native RBCs or RBC@[Fe^{III}-TA] cells were lysed by immersing them in DI water. The suspension was centrifuged at 3000 g for 30 s, and the supernatant was removed. After washing twice with DI water, the membranes of lysed native RBC or RBC@[Fe^{III}-TA] were re-suspended in DI water and dropped onto a piece of silicon wafer (5 mm × 5 mm). After drying the samples in the air, they were analyzed by atomic force microscopy (Innova, Bruker, Billerica, MA, USA). Typical scans were conducted in tapping mode with OTESPA silicon cantilevers (Bruker, Billerica, MA, USA), and the coat thickness was analyzed by using line-profile data.

3.4. Effect of NaCl on Film Thickness: Gold Substrates

Gold-coated silicon wafers were cut into approximately 1 cm × 1 cm slides, and cleaned by sonication in acetone and ethanol, prior to use. The cleaned slides were dried under Ar gas and soaked in DI water or isotonic saline solution in a 12-well plate. The stock solutions of TA and FeCl$_3$·6H$_2$O were sequentially added to yield final concentrations of 0.05 and 0.0125 mg mL^{-1}, respectively. The pH of the solution was then raised to ca. 8 by adding 1 M NaOH solution. After 10 s of incubation, the coated slides were washed with DI water and dried under Ar gas. The coating process was repeated four times, and the film thickness was measured with an L116s ellipsometer (Gaertner Scientific Corporation, Skokie, IL, USA).

3.5. Characterizations

Raman spectra were obtained with a LabRAM ARAMIS spectrometer (HORIBA Jobin Yvon, Edison, NJ, USA). The 633-nm line of an air-cooled He/Ne laser was used as an excitation source. Field-emission scanning electron microscopy (FE-SEM) imaging was performed with a Philips XN30FEG microscope (FEI-Philips Co., Hillsboro, OR, USA) with an accelerating voltage of 10 kV, after sputter-coating with platinum. Transmission electron microscopy (TEM) imaging was performed with an LEO 912AB microscope (Carl Zeiss, Oberkochen, Germany). Specimens were fixed with glutaraldehyde and paraformaldehyde, and then dehydrated in ethanol. The fixed samples were embedded in Epon 812 resin. Thin sections were cut by using ULTRACUT UCT ultramicrotome (Leica, Wetzlar, Germany) and stained with uranyl acetate and lead citrate.

3.6. Antibody-Mediated Agglutination and Oxygen Consumption

The immunoprotective capability of Fe^{III}-TA coating was evaluated by investigating the attenuation of antibody-mediated agglutination of RBCs. The native RBC or RBC@[Fe^{III}-TA] (1% hematocrit in PBS) were mixed with their anti-type sera (10:1, *v/v*), and the images were taken by optical microscopy (LSM 700, Carl Zeiss, Oberkochen, Germany) after one hour. The oxygen-carrying capacity of native RBC and RBC@[Fe^{III}-TA] was examined by measuring the amount of oxygen uptake. The initial oxygen concentration dissolved in the O$_2$-purged PBS was about 39%, and the oxygen concentration was recorded by an oxygen probe connected with LabQuest® (Vernier, Beaverton, OR, USA). The probe was calibrated with the sodium sulfite calibration solution (Vernier, Beaverton, OR, USA), prior to measurement, and immersed in 2 mL of O$_2$-purged PBS. Once the signal was stabilized, a suspension of native RBC or RBC@[Fe^{III}-TA] (100 µL in PBS, 1.5 × 10^8 cells) was transfused into the O$_2$-purged PBS. The oxygen concentration was recorded over the whole process until the signal was stabilized.

4. Conclusions

In summary, we demonstrated that the supramolecular Fe^{III}-TA coating on RBC surfaces effectively attenuated the immune response while maintaining the oxygen-carrying capacity. Coordination-driven supramolecular complexation between Fe^{III} and TA was rapid, and made uniform films on individual RBC surfaces, which was confirmed by various characterizations. The coating conditions were optimized for RBCs that are very sensitive to the tonicity of the solution and chemicals. Especially, we found that the use of isotonic saline enhanced the coating efficiency as well as maintained membrane integrity. The chemical motif in supramolecular complex formation of Fe^{III} and TA could be utilized for other materials that contain multivalent catechol groups, such as poly(ethylene glycol) (PEG)-catechol and hyaluronic acid-catechol conjugates [36,37]. Especially, the use of PEG-catechol conjugates for RBC coating would be beneficial in the reduction of potential recognition by immune cells in the body. Considering that the Fe^{III}-catechol coating is a rapid, easy, and inexpensive process, we also believe that the formation of RBC-in-shell structures would have a potential in the manufacture of universal RBCs for medical transfusion.

Supplementary Materials: The following are available online at www.mdpi.com/2073-4360/9/4/140/s1. Figure S1: Optical images of RBCs after 10-min treatment of tannic acid (TA): (a) 0.1 mg mL^{-1}, (b) 0.08 mg mL^{-1}, and (c) 0.05 mg mL^{-1}, Figure S2: Optical images of RBCs after 10-min treatment of FeCl$_3$·6H$_2$O: (a) 0.01 mg mL^{-1}, (b) 0.005 mg mL^{-1}, and (c) 0.001 mg mL^{-1}, Figure S3: CLSM image of TA-treated native RBCs after incubation with BSA-Alexa Fluor® 647, Figure S4: Ellipsometric thickness of Fe^{III}-TA films on a gold substrate. The films were formed either in water or in isotonic saline (mean ± S.D., $N = 3$).

Acknowledgments: This work was supported by the End Run Project of KAIST (N01150623). Daewha Hong thanks the financial support by Pusan National University Research Grant, 2017. The authors thank Myeong-seon Jeong at Korea Basic Science Institute (KBSI) for the TEM analysis.

Author Contributions: Taegyun Park, Ji Yup Kim, Hyeoncheol Cho, Hee Chul Moon, Beom Jin Kim, Ji Hun Park, Daewha Hong, Joonhong Park, and Insung S. Choi contributed to the experiments and discussed the data. The manuscript was written by Taegyun Park and Insung S. Choi. All authors approved the final version of the manuscript.

Conflicts of Interest: The authors declare no conflict of interest.

References

1. Yang, S.H.; Hong, D.; Lee, J.; Ko, E.H.; Choi, I.S. Artificial Spores: Cytocompatible encapsulation of individual living cells within thin, tough artificial shells. *Small* **2013**, *9*, 178–186. [CrossRef] [PubMed]

2. Hong, D.; Park, M.; Yang, S.H.; Lee, J.; Kim, Y.-G.; Choi, I.S. Artificial spores: Cytoprotective nanoencapsulation of living cells. *Trends Biotechnol.* **2013**, *31*, 442–447. [CrossRef] [PubMed]

3. Park, J.H.; Yang, S.H.; Lee, J.; Ko, E.H.; Hong, D.; Choi, I.S. Nanocoating of single cells: From maintenance of cell viability to manipulation of cellular activities. *Adv. Mater.* **2014**, *26*, 2001–2010. [CrossRef] [PubMed]

4. Park, J.H.; Hong, D.; Lee, J.; Choi, I.S. Cell-in-shell hybrids: Chemical nanoencapsulation of individual cells. *Acc. Chem. Res.* **2016**, *49*, 792–800. [CrossRef] [PubMed]

5. Yang, S.H.; Lee, K.-B.; Kong, B.; Kim, J.-H.; Kim, H.-S.; Choi, I.S. Biomimetic encapsulation of individual cells with silica. *Angew. Chem. Int. Ed.* **2009**, *48*, 9160–9163. [CrossRef] [PubMed]

6. Yang, S.H.; Kang, S.M.; Lee, K.-B.; Chung, T.D.; Lee, H.; Choi, I.S. Mussel-inspired encapsulation and functionalization of individual yeast cells. *J. Am. Chem. Soc.* **2011**, *133*, 2795–2797. [CrossRef] [PubMed]

7. Yang, S.H.; Ko, E.H.; Jung, Y.H.; Choi, I.S. Bio-inspired functionalization of silica-encapsulated yeast cells. *Angew. Chem. Int. Ed.* **2011**, *50*, 6115–6118. [CrossRef] [PubMed]

8. Ko, E.H.; Yoon, Y.; Park, J.H.; Yang, S.H.; Hong, D.; Lee, K.-B.; Shon, H.K.; Lee, T.G.; Choi, I.S. Bioinspired, cytocompatible mineralization of silica-titania composites: Thermo-protective nanoshell formation for individual chlorella cells. *Angew. Chem. Int. Ed.* **2013**, *52*, 12279–12282. [CrossRef] [PubMed]

9. Lee, J.; Choi, J.; Park, J.H.; Kim, M.-H.; Hong, D.; Cho, H.; Yang, S.H.; Choi, I.S. Cytoprotective silica coating of individual mammalian cells through bioinspired silicification. *Angew. Chem. Int. Ed.* **2014**, *53*, 8056–8059. [CrossRef] [PubMed]

10. Kim, B.J.; Park, T.; Moon, H.C.; Park, S.-Y.; Hong, D.; Ko, E.H.; Kim, J.Y.; Hong, J.W.; Han, S.W.; Kim, Y.-G.; et al. Cytoprotective alginate/polydopamine core/shell microcapsules in microbial encapsulation. *Angew. Chem. Int. Ed.* **2014**, *53*, 14443–14446. [CrossRef] [PubMed]

11. Hong, D.; Lee, H.; Ko, E.H.; Lee, J.; Cho, H.; Park, M.; Yang, S.H.; Choi, I.S. Organic/inorganic double-layered shells for multiple cytoprotection of individual living cells. *Chem. Sci.* **2015**, *6*, 203–208. [CrossRef]

12. Liang, K.; Richardson, J.J.; Cui, J.; Caruso, F.; Doonan, C.J.; Falcaro, P. Metal-organic framework coatings as cytoprotective exoskeletons for living cells. *Adv. Mater.* **2016**, *28*, 7910–7914. [CrossRef] [PubMed]

13. Chowdhuri, S.; Cole, C.M.; Devaraj, N.K. Encapsulation of living cells within giant phospholipid liposomes formed by the inverse-emulsion technique. *ChemBioChem* **2016**, *17*, 886–889. [CrossRef] [PubMed]

14. Chen, X.; Tam, U.C.; Czlapinski, J.L.; Lee, G.S.; Rabuka, D.; Zettl, A.; Bertozzi, C.R. Interfacing carbon nanotubes with living cells. *J. Am. Chem. Soc.* **2006**, *128*, 6292–6293. [CrossRef] [PubMed]

15. Wojcik, M.; Hauser, M.; Li, W.; Moon, S.; Xu, K. Graphene-enabled electron microscopy and correlated super-resolution microscopy of wet cells. *Nat. Commun.* **2015**, *6*, 7384. [CrossRef] [PubMed]

16. Yang, S.H.; Lee, T.; Seo, E.; Ko, E.H.; Choi, I.S.; Kim, B.-S. Interfacing living yeast cells with graphene oxide nanosheaths. *Macromol. Biosci.* **2012**, *12*, 61–66. [CrossRef] [PubMed]

17. Park, J.H.; Kim, K.; Lee, J.; Choi, J.Y.; Hong, D.; Yang, S.H.; Caruso, F.; Lee, Y.; Choi, I.S. A cytoprotective and degradable metal-polyphenol nanoshell for single-cell encapsulation. *Angew. Chem. Int. Ed.* **2014**, *53*, 12420–12425. [CrossRef]

18. Li, W.; Bing, W.; Huang, S.; Ren, J.; Qu, X. Mussel byssus-like reversible metal-chelated supramolecular complex used for dynamic cellular surface engineering and imaging. *Adv. Funct. Mater.* **2015**, *24*, 3775–3784. [CrossRef]

19. Yang, S.H.; Choi, J.; Palanikumar, L.; Choi, E.S.; Lee, J.; Kim, J.; Choi, I.S.; Ryu, J.-H. Cytocompatible in situ cross-linking of degradable lbl films based on thiol-exchange reaction. *Chem. Sci.* **2015**, *6*, 4698–4703. [CrossRef]

20. Lee, J.; Cho, H.; Choi, J.; Kim, D.; Hong, D.; Park, J.H.; Yang, S.H.; Choi, I.S. Chemical sporulation and germination: Cytoprotective nanocoating of individual mammalian cells with a degradable tannic acid-Fe^{III} complex. *Nanoscale* **2015**, *7*, 18918–18922. [CrossRef] [PubMed]

21. Villa, C.H.; Anselmo, A.C.; Mitragotri, S.; Muzykantov, V. Red blood cells: Supercarriers for drugs, biologicals, and nanoparticles and inspiration for advanced delivery systems. *Adv. Drug Deliv. Rev.* **2016**, *106*, 88–103. [CrossRef] [PubMed]

22. Carson, J.L.; Grossman, B.J.; Kleinman, S.; Tinmouth, A.T.; Marques, M.B.; Fung, M.K.; Holcomb, J.B.; Illoh, O.; Kaplan, L.J.; Katz, L.M.; et al. Red blood cell transfusion: A clinical practice guideline from the Aabb. *Ann. Intern. Med.* **2012**, *157*, 49–58. [CrossRef] [PubMed]

23. Mujahid, A.; Dickert, F.L. Blood group typing: From classical strategies to the application of synthetic antibodies generated by molecular imprinting. *Sensors* **2016**, *16*, 51. [CrossRef] [PubMed]

24. Liu, Q.P.; Sulzenbacher, G.; Yuan, H.; Bennett, E.P.; Pietz, G.; Saunders, K.; Spence, J.; Nudelman, E.; Levery, S.B.; White, T.; et al. Bacterial glycosidases for the production of universal red blood cells. *Nat. Biotechnol.* **2007**, *25*, 454–464. [CrossRef] [PubMed]

25. Giarratana, M.-C.; Rouard, H.; Dumont, A.; Kiger, L.; Safeukui, I.; Le Pennec, P.-Y.; François, S.; Trugnan, G.; Peyrard, T.; Marie, T.; et al. Proof of principle for transfusion of in vitro-generated red blood cells. *Blood* **2011**, *118*, 5071–5079. [CrossRef] [PubMed]

26. Douay, L.; Andreu, G. Ex vivo production of human red blood cells from hematopoietic stem cells: What is the future in transfusion? *Transfus. Med. Rev.* **2007**, *21*, 91–100. [CrossRef] [PubMed]

27. Reid, M.E.; Mohandas, N. Red blood cell blood group antigens: Structure and function. *Semin. Hematol.* **2004**, *41*, 93–117. [CrossRef] [PubMed]

28. Timmins, N.E.; Nielsen, L.K. Blood cell manufacture: Current methods and future challenges. *Trends Biotechnol.* **2009**, *27*, 415–422. [CrossRef] [PubMed]

29. Mansouri, S.; Merhi, Y.; Winnik, F.M.; Tabrizian, M. Investigation of layer-by-layer assembly of polyelectrolytes on fully functional human red blood cells in suspension for attenuated immune response. *Biomacromolecules* **2011**, *12*, 585–592. [CrossRef] [PubMed]

30. Wang, B.; Wang, G.; Zhao, B.; Chen, J.; Zhang, X.; Tang, R. Antigenically shielded universal red blood cells by polydopamine-based cell surface engineering. *Chem. Sci.* **2014**, *5*, 3463–3468. [CrossRef]

31. Kim, J.Y.; Lee, H.; Park, T.; Park, J.; Kim, M.-H.; Cho, H.; Youn, W.; Kang, S.M.; Choi, I.S. Artificial spores: Cytocompatible coating of living cells with plant-derived pyrogallol. *Chem. Asian J.* **2016**, *11*, 3183–3187. [CrossRef] [PubMed]

32. Ejima, H.; Richardson, J.J.; Liang, K.; Best, J.P.; van Koeverden, M.P.; Such, G.K.; Cui, J.; Caruso, F. One-step assembly of coordination complexes for versatile film and particle engineering. *Science* **2013**, *341*, 154–157. [CrossRef] [PubMed]

33. Kim, S.; Kim, D.S.; Kang, S.M. Reversible layer-by-layer deposition on solid substrates inspired by mussel byssus cuticle. *Chem. Asian J.* **2014**, *9*, 63–66. [CrossRef] [PubMed]

34. Jandl, J.H.; Simmons, R.L. The agglutination and sensitization of red blood cells by metallic cations: Interactions between multivalent metals and the red-cell membrane. *Br. J. Haematol.* **1957**, *3*, 19–38. [CrossRef] [PubMed]

35. Kozlovskaya, V.; Kharlampieva, E.; Drachuk, I.; Cheng, D.; Tsukruk, V.V. Responsive microcapsule reactors based on hydrogen-bonded tannic acid layer-by-layer assemblies. *Soft Matter* **2010**, *6*, 3596–3608. [CrossRef]

36. Ju, Y.; Cui, J.; Müllner, M.; Suma, T.; Hu, M.; Caruso, F. Engineering low-fouling and pH-degradable capsules through the assembly of metal-phenolic networks. *Biomacromolecules* **2015**, *16*, 807–814. [CrossRef] [PubMed]

37. Ju, Y.; Cui, J.; Sun, H.; Müllner, M.; Dai, Y.; Guo, J.; Bertleff-Zieschang, N.; Suma, T.; Richardson, J.J.; Caruso, F. Engineered metal-phenolic capsules show tunable targeted delivery to cancer cells. *Biomacromolecules* **2016**, *17*, 2268–2276. [CrossRef] [PubMed]

![polymers logo] *polymers*

MDPI

Article

Calcium Silicate Improved Bioactivity and Mechanical Properties of Poly(3-hydroxybutyrate-co-3-hydroxyvalerate) Scaffolds

Cijun Shuai [1,2], Wang Guo [1], Chengde Gao [1], Youwen Yang [1], Yong Xu [1], Long Liu [1], Tian Qin [1], Hang Sun [1], Sheng Yang [3], Pei Feng [1,*] and Ping Wu [4,*]

[1] State Key Laboratory of High Performance Complex Manufacturing, Central South University, Changsha 410083, China; shuai@csu.edu.cn (C.S.); guowang@csu.edu.cn (W.G.); gaochengde@csu.edu.cn (C.G.); yangyouwen@csu.edu.cn (Y.Y.); xuyong2927@csu.edu.cn (Y.X.); liulong@csu.edu.cn (L.L.); qintian@csu.edu.cn (T.Q.); shsunhang@csu.edu.cn (H.S.)
[2] Key Laboratory of Organ Injury, Aging and Regenerative Medicine of Hunan Province, Changsha 410008, China
[3] Human Reproduction Center, Shenzhen Hospital of Hongkong University, Shenzhen 518000, China; yangs8@hku-szh.org
[4] College of Chemistry, Xiangtan University, Xiangtan 411105, China
* Correspondence: fengpei@csu.edu.cn (P.F.); pingwu@xtu.edu.cn (P.W.); Tel.: +86-731-8480-5412 (P.F.); +86-731-5829-2251 (P.W.)

Academic Editor: Insung S. Choi
Received: 2 April 2017; Accepted: 12 May 2017; Published: 14 May 2017

Abstract: The poor bioactivity and mechanical properties have restricted its biomedical application, although poly(3-hydroxybutyrate-*co*-3-hydroxyvalerate) (PHBV) had good biocompatibility and biodegradability. In this study, calcium silicate (CS) was incorporated into PHBV for improving its bioactivity and mechanical properties, and the porous PHBV/CS composite scaffolds were fabricated via selective laser sintering (SLS). Simulated body fluid (SBF) immersion tests indicated the composite scaffolds had good apatite-forming ability, which could be mainly attributed to the electrostatic attraction of negatively charged silanol groups derived from CS degradation to positively charged calcium ions in SBF. Moreover, the compressive properties of the composite scaffolds increased at first, and then decreased with increasing the CS content, which was ascribed to the fact that CS of a proper content could homogeneously disperse in PHBV matrix, while excessive CS would form continuous phase. The compressive strength and modulus of composite scaffolds with optimal CS content of 10 wt % were 3.55 MPa and 36.54 MPa, respectively, which were increased by 41.43% and 28.61%, respectively, as compared with PHBV scaffolds. Additionally, 3-(4,5-dimethylthiazol-2-yl)-2,5-diphenyltetrazolium bromide (MTT) assay indicated MG63 cells had a higher proliferation rate on PHBV/CS composite scaffolds than that on PHBV. Alkaline phosphatase (ALP) staining assay demonstrated the incorporation of CS significantly promoted osteogenic differentiation of MG63 cells on the scaffolds. These results suggest that the PHBV/CS composite scaffolds have the potential in serving as a substitute in bone tissue engineering.

Keywords: poly(3-hydroxybutyrate-*co*-3-hydroxyvalerate) (PHBV); calcium silicate; composite scaffolds; bioactivity; mechanical properties; cytocompatibility

---

1. Introduction

A scaffold, which acts as a temporary extracellular matrix, plays a significant role in regulating cell functions and tissue regeneration [1–3]. Therefore, it should have good biological properties, proper mechanical properties, and interconnected porous structures, as well as customized shape [4,5].

Poly (3-hydroxybutyrate-co-3-hydroxyvalerate) (PHBV) has been an attractive scaffold material due to its favorable biocompatibility, biodegradability, and processibility [6–8]. However, there still exist two significant problems that restrict its biomedical application, especially in bone tissue engineering. One is its lack of bioactivity, which leads to a difficulty in obtaining good integration, i.e., forming a direct contact and good interfacial bonding between the implanted scaffolds and the surrounding tissues [9]. The other is its poor mechanical properties, which is unable to provide sufficient mechanical support during tissue regeneration. Calcium silicate (CS), as a silicon-calcium based bioceramic, possesses excellent bioactivity [10–12] as its degradation products, including silanol groups and calcium ions, can accelerate the formation and deposition of apatite [13]. Moreover, CS can act as a rigid filler to reinforce polymer materials by microcracking and crack pinning [14,15]. In addition, it was reported that silicon ions and calcium ions released from CS could stimulate human mesenchymal stem cells proliferation and osteogenic differentiation [16,17].

Incorporating bioactive reinforcement phases into polymer scaffolds for improving bioactivity and mechanical properties has attracted much attention. Jack [18] prepared PHBV/hydroxyapatite (HA) composite scaffolds using a modified thermally induced phase-separation technique and found the introduction of HA greatly increased the stiffness and strength, and improved the in vitro bioactivity of the scaffolds. Li [15] prepared PHBV/bioglass (BG) composite scaffolds by a salt particulate leaching method and their results showed that the compressive strength and apatite-forming ability of the scaffolds were significantly improved by adding BG particles. Zhu [19] prepared silk fibroin/CS composite scaffolds by a freeze-drying method and concluded that incorporating CS into the silk fibroin can not only enhance the mechanical strength, but also improve the bioactivity and cytocompatibility of the scaffolds. However, these conventional techniques had poor control of the geometries and porous structures of scaffolds [3,20]. Selective laser sintering (SLS), an additive manufacturing technology, is quite capable of fabricating scaffolds with arbitrary geometries that matched with defected tissues. Furthermore, it is able to control the internal porous structures of scaffolds [21,22].

Based on the above consideration, in this study we incorporated CS into PHBV for the purpose of improving its bioactivity and mechanical properties, and fabricated three-dimensional porous PHBV/CS composite scaffolds via SLS in view of the potential for use as bone scaffolds. The effects of CS contents and soaking time on the apatite-forming ability of composite scaffolds were studied in simulated body fluid (SBF). Moreover, the compressive strength and modulus of composite scaffolds with different CS content were studied by compression tests. The dispersion state of CS in PHBV matrix on the surface and cross section was analyzed by scanning electron microscope (SEM). Additionally, the proliferation and osteogenic differentiation of MG63 cells cultured on PHBV/CS composite scaffolds was evaluated using MTT assay and alkaline phosphatase (ALP) staining, respectively.

2. Materials and Methods

2.1. Materials

PHBV with poly (3-hydroxyvalerate) content of 3 mol%, molecular weight of 280 kDa, and a density of 1.25 g/cm^3 was purchased from Tianan Biologic Material Co. (Ningbo, China). Calcium silicate (CS) powder was obtained from Kunshan Huaqiao New Materials Co. (Suzhou, China).

The following procedures [23] were mainly involved in preparing PHBV/CS composite powders: (a) ultrasonically dispersing PHBV powder of a certain amount in anhydrous ethanol for 30 min with an ultrasonic cleaner JP-040ST (Skymen Cleaning Equipment Shenzhen Co.,Ltd., Shenzhen, China); (b) adding CS powder of a proportional amount into the PHBV solution and then ultrasonically dispersing for another 30 min; (c) magnetically stirring the above PHBV/CS solution with a magnetic stirrer JB-5 (Jintan Ronghua Instrument Manufacturing Co., Ltd., Changzhou, China); (d) filtering the PHBV/CS solution to obtain the powder and then drying at 60 °C for 12 h in an electrothermal blowing dry box (101-00S, Guangzhou Daxiang Electronic Machinery Equipment Co., Ltd., Guangzhou, China); and (e) mechanically milling the dried powder using a planetary ball mill (DECO-PBM-V-0.4L,

Changsha Deco Equipment Co., Ltd., Changsha, China). Five kinds of powders of different compositions were prepared by incorporating 0, 5, 10, 15, and 20 wt % CS into PHBV and labeled as PHBV, PHBV/5%CS, PHBV/10%CS, PHBV/15%CS, and PHBV/20%CS, respectively.

2.2. Fabrication of the Scaffold

Three-dimensional PHBV/CS composite scaffolds were fabricated using a selective laser sintering (SLS) machine, which was developed by our team. The SLS machine mainly consists of a carbon dioxide laser (SR 10i, Rofin-Sinar Laser GmbH, Hamburg, Germany) and a galvanometer-based scanning system (3D scanhead-300-15D, Beijing Century Sunny Technology Co., Beijing, China). The machine controls laser by scanning system to selectively sinter powder layers according to the cross-section profiles to form three-dimensional parts layer-by-layer. Briefly, one layer of powder was deposited on the worktable and the laser beam under the control of a computer selectively sintered the deposited powder layer according to the cross-section profiles of the designed scaffold. Then, the worktable descended by a height of a layer thickness, and subsequent powder layers were deposited and sintered directly on the top of the previously sintered layers. Afterwards, the above-mentioned layer-by-layer sintering procedure was cycled until the whole scaffold was formed. Finally, the designed scaffold was obtained after removing the unsintered powder by blowing high-pressure air. The main processing parameters of SLS were set up as follows: laser power 2 W, scanning speed 200 mm/s, scanning spacing 0.1 mm, and layer thickness 0.1 mm.

2.3. Microstructure Characterization

Microtopography of powders, surface, and cross section morphologies of scaffolds, and the morphology of MG63 cells cultured on scaffolds, were characterized using a scanning electron microscope (SEM) (FEI Quanta-200, FEI Co., Hillsboro, OR, USA) under high vacuum conditions at 20 kV. Element composition of the specimen surface was examined by energy-dispersive spectroscopy (EDS) (Neptune XM4, EDAX Inc., Mahwah, NJ, USA) which was installed on the SEM. All SEM specimens were oven dried, mounted on stubs, and sputtered with gold before observation. The phase composition of powders and scaffolds was identified by X-ray diffraction (XRD) (Bruker D8 Advance Diffractometer, German Bruker Co., Karlsruhe, Germany) using Cu Kα radiation (λ = 1.5406 Å) with scattering angles (2θ) range 5–70° and a scanning rate of 8°/min.

2.4. Compression Tests

The compression tests of the scaffolds were carried out on a universal tester (WD-01, Shanghai Zhuoji Instruments Co., Shanghai, China) at room temperature. The loading speed was 1 mm/min. Five specimens per group were used for each composition of the scaffolds. The compressive strength and compressive modulus were determined from the compression stress-strain curve.

2.5. Immersion Tests

The bioactivity of PHBV/CS composite scaffolds was assessed by immersing the scaffold specimens (with diameters of 8 mm and thicknesses of 2 mm) in simulated body fluid (SBF) which has similar ion concentrations to human blood plasma [24,25]. The SBF was changed every other day. After a predetermined incubation time, the specimens were sacrificed (three specimens per composition per time point), and then gently washed with distilled water, followed by drying at 37 °C in a electrothermal blowing dry box (101-00S). Finally, the surface morphologies and element compositions of the soaked specimens were characterized by SEM and EDS, respectively.

2.6. Cytocompatibility

Cell proliferation on the PHBV/CS composite scaffolds was evaluated by 3-[4,5-dimethylthiazol-2-yl]-2,5-diphenyltetrasodium bromide (MTT) assay, with the PHBV scaffolds without CS serving as the control. MG63 cells (American Type Culture Collection, Manassas, VA, USA) were cultured in low glucose Dulbecco's Modified Eagle's Medium supplemented with 10% fetal bovine serum and 1% antibiotic-antimycotic solution under 37 °C and 5% CO_2. Before cell seeding, the scaffold specimens (with diameters of 8 mm and thicknesses of 2 mm) were sterilized by ultraviolet radiation and washed with phosphate-buffered solution (PBS), followed by being kept in the culture medium for 24 h. Then, the MG63 cells at a density of $2 \times 10^5/dm^2$ were seeded onto the scaffold specimens and cultured for different time. During the last 4 h of the predetermined culture time, the scaffold specimens were incubated with MTT. Subsequently, the precipitated formazan salts were dissolved in dimethylsulphoxide. Finally, the absorbance at 570 nm was measured by an enzyme immunosorbent assay reader.

The morphology of MG63 cells was observed by SEM. Briefly, the scaffold/cell constructs were washed with PBS and fixed with 2.5% glutaraldehyde in PBS after four days. Then, the cells on the scaffolds were dehydrated with gradient ethanol solutions, followed by drying in a drying box at 37 °C overnight. Finally, the specimens were mounted on stubs and sputtered with gold for SEM observation. Osteogenic differentiation of MG63 cells was evaluated by alkaline phosphatase (ALP) staining. Briefly, the cell/scaffold constructs were rinsed with PBS after a predetermined culture time, followed by treatment with 0.1% triton solution to obtain cell lysates. Then, the cell lysates were hydrolyzed by p-nitrophenyl phosphate in alkaline buffer solution. Finally, a light microscope was used to view the cells staining positive for ALP.

2.7. Statistical Analysis

The data were expressed as the mean $\pm$ standard deviation ($n = 4$). The statistical comparison was performed using Student's t-test and differences were considered significant when $p < 0.05$.

3. Results and Discussion

The SEM morphologies of PHBV, CS, and PHBV/20%CS powders are shown in Figure 1. The PHBV powder was spherical, with a particle size was about 1 μm (Figure 1a). The CS powder was irregular in shape and its particle size ranged mainly from 0.5 to 5 μm (Figure 1b). For the PHBV/20%CS composite powders, the CS particles were distributed randomly on the PHBV particles (Figure 1c). The corresponding XRD patterns are shown in Figure 1d. PHBV presented two strong diffraction peaks at about 2θ = 13.4 and 16.8° which were assigned to (020) and (110) planes, respectively. Additionally, other peaks at about 2θ = 20.1, 21.4, 22.6, 25.5, and 27.1° were assigned to (021), (101), (111), (121), and (040) planes, respectively [26,27]. CS presented a strong diffraction peak at about 2θ = 23.1°, which was assigned to the (400) plane. In addition, other peaks at about 2θ = 25.3, 26.9, 28.9, 30.0, 36.2, 38.2, 39.2, 41.3, 49.8, and 53.3° were assigned to (002), (20-2), (202), (320), (122), (520), (20-3), (521), (040), and (72-2) planes, respectively [28–30]. For the PHBV/CS composite powder, the diffraction peaks of both PHBV and CS existed. Furthermore, there existed no new phases, which indicated that no chemical reaction occurred between PHBV and CS during the preparation of the composite powders.

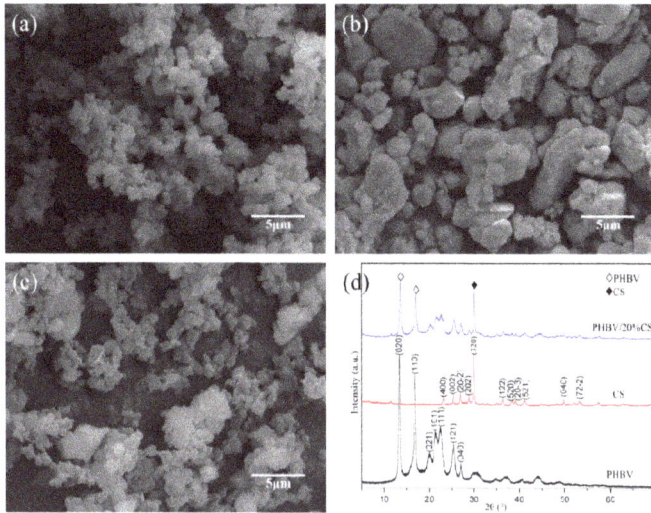

Figure 1. SEM (scanning electron microscope) morphologies of (**a**) PHBV; (**b**) CS; and (**c**) PHBV/20%CS composite powders and (**d**) the corresponding XRD (X-ray diffraction) patterns.

A representative three-dimensional rectangular PHBV/10%CS porous scaffold (15 mm × 15 mm × 5 mm) prepared by SLS is shown in Figure 2. It presented a well ordered and interconnected porous structure. The average pore size was approximately 500 μm (Figure 2d). This kind of interconnected porous structure may be beneficial for transport of nutrients and excretion of metabolites [31,32] and, thus, may play an important role in regulating cell functions and tissue ingrowth. Bose et al. [33] suggested that the pore size should be at least 100 μm for successful diffusion of oxygen and nutrients for cell survivability. Guo et al. [34] reported that the macropores 400–500 μm in size benefited cell infiltration and bone ingrowth. Tarafder [35] implanted tricalcium phosphate scaffolds with a 350 μm pore size into rat femurs and founded new bone formed after two weeks.

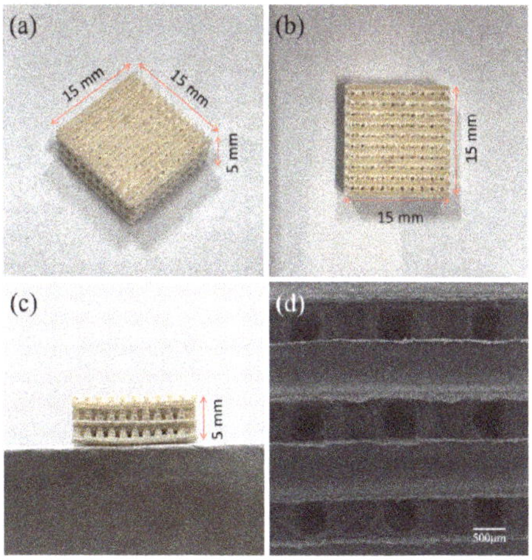

Figure 2. (**a–c**) Optical images and (**d**) SEM morphology of a representative three-dimensional rectangular PHBV/10%CS porous scaffold fabricated via SLS (selective laser sintering).

The XRD patterns of the SLS-fabricated scaffolds of different formulations were shown in Figure 3. With the increase of CS content, its diffraction peak intensity increased while PHBV decreased, which indicated that CS was blended into the composite scaffolds and did not recrystallize since its melting point (about 952 °C) [36] was far higher than that of PHBV (about 164 °C) [37]. The diffraction peak positions of PHBV and CS remained unchanged and no other peaks were detected after laser sintering, which indicated that CS was compatible with PHBV. It is noteworthy that adding only 5% CS remarkably decreased the diffraction peak of PHBV in the composite scaffold, suggesting that the incorporation of CS significantly decreased the crystallinity of PHBV. This may be due to CS particles of larger size than PHBV hindered the growth of PHBV grains during recrystallization. The crystallinity was one of the important factors that influenced the degradation rate of the polymers [38–40]. The decrease of crystallinity of PHBV could accelerate the degradation of the scaffolds, which may be beneficial for promoting tissue repair.

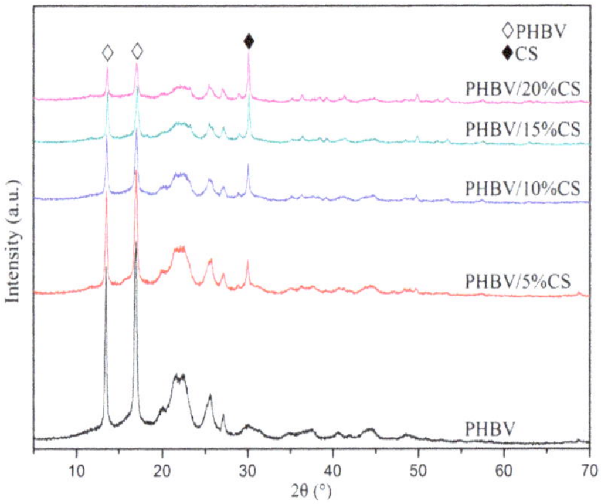

Figure 3. XRD patterns of the SLS-fabricated scaffolds of different formulations.

The surface morphologies of PHBV and PHBV/CS composite scaffolds with different CS content were shown in Figure 4. The surface of the PHBV scaffold was flat and smooth. With the addition of 5 wt % CS, the surface became slightly rougher and several CS particles are exposed. The amount of exposed particles increases with the increase of CS, leading to a rougher surface. Figure 4f showed the EDS spectra of zones A and B. The strongest peaks corresponding to zones A and B belong to elements Si and C, respectively, which implied that they consisted mainly of CS and PHBV, respectively. Moreover, when its content is relatively low (no more than 10 wt %), the CS particles can disperse relatively uniformly in the PHBV matrix. However, the individual CS particles became continuous when its content was excessive (larger than 10 wt %). The cross section morphologies of the PHBV scaffold and PHBV/CS composite scaffolds with different CS content were shown in Figure 5. In general, the dispersion state of CS particles in the PHBV matrix on the cross-section underwent a similar change trend to that on the surface with the CS content increasing. The optimal dispersion state of CS in the PHBV matrix on the cross section also resulted from 10 wt % CS, where the particles could disperse relatively uniformly in the PHBV matrix.

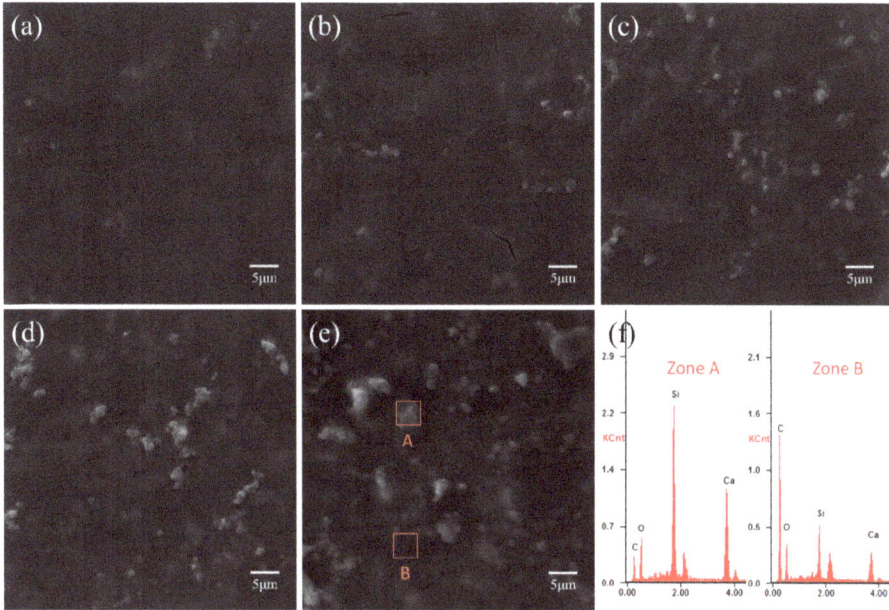

Figure 4. SEM surface morphologies of (**a**) PHBV scaffold; (**b**) PHBV/5%CS scaffold; (**c**) PHBV/10%CS scaffold; (**d**) PHBV/15%CS scaffold, and (**e**) PHBV/20%CS scaffold; and (**f**) the EDS (energy-dispersive spectroscopy) spectra of Zones A and B.

Figure 5. SEM cross section morphologies of (**a**) PHBV scaffold, (**b**) PHBV/5%CS scaffold, (**c**) PHBV/10%CS scaffold, (**d**) PHBV/15%CS scaffold, and (**e**) PHBV/20%CS scaffold.

Figure 6 shows the compressive strength and compressive modulus of the PHBV/CS scaffolds as a function of CS content. The compressive strength and compressive modulus of the PHBV scaffold was 2.51 MPa and 28.41 MPa, respectively. After adding 5 wt % CS, they increased by 25.10% and 17.04%, respectively. With increasing the CS content to 10 wt %, the compressive strength and modulus reached the optimal values of 3.55 MPa and 36.54 MPa, respectively, reflecting an increase of 41.43% and 28.61%, respectively. However, the compressive strength and modulus decreased when the CS content further increased to more than 10 wt %. Nevertheless, the compressive strength and modulus of all of the PHBV/CS composite scaffolds were still higher than the PHBV scaffold.

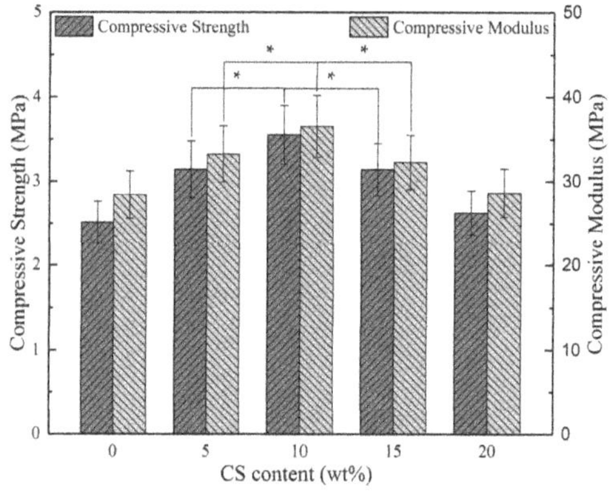

Figure 6. The compressive strength and compressive modulus of PHBV/CS scaffolds as a function of CS content; * represents $p < 0.05$.

The significant increases in compressive properties of the PHBV/CS scaffolds were attributed to the strong reinforcement effects imparted by CS particles due to its high modulus and strength. Moreover, the distribution of the reinforcement phases played a significant role in influencing the compressive properties of polymer/inorganic composite scaffolds. As shown in Figures 4 and 5, when the content was no more than 10%, the CS particles could disperse uniformly in the matrix so that the compressive properties increased with its content increasing. However, when the content of CS was excessive, namely more than 10%, the CS phases became continuous and, thus, decreased the compressive properties of the PHBV/CS scaffolds. It was well known that the continuous phases of stiff fillers in polymer matrixes can decrease the mechanical properties of the composites as they deteriorate the interface bonding between the fillers and matrices [41].

The SEM surface morphologies of the PHBV/CS scaffolds of different formulations after immersion in SBF for 14 days are shown in Figure 7. The PHBV scaffold exhibited a smooth surface (Figure 7a), whereas some particles deposited on all of the composite scaffolds. Moreover, the amount of the deposits increased with the CS content increasing. It is noticeable that large numbers of the deposits agglomerated on the surface of the PHBV/20%CS scaffold. The elemental compositions of the deposits (Zone C) were analyzed by EDS (Figure 7f). Not only were the elements Ca and O distinctly detected, but so was P. It is noteworthy that both PHBV and CS did not contain the element P, while hydroxyapatite ($Ca_{10}(PO_4)_6(OH)_2$) does. Moreover, the morphology and size of the particles in the deposits (in Figure 7) were similar to that of the hydroxyapatite deposits induced on wollastonite (CS) ceramic samples in SBF reported by Liu [42]. Therefore, it was implied that the deposits rich in P consisted mainly of hydroxyapatite.

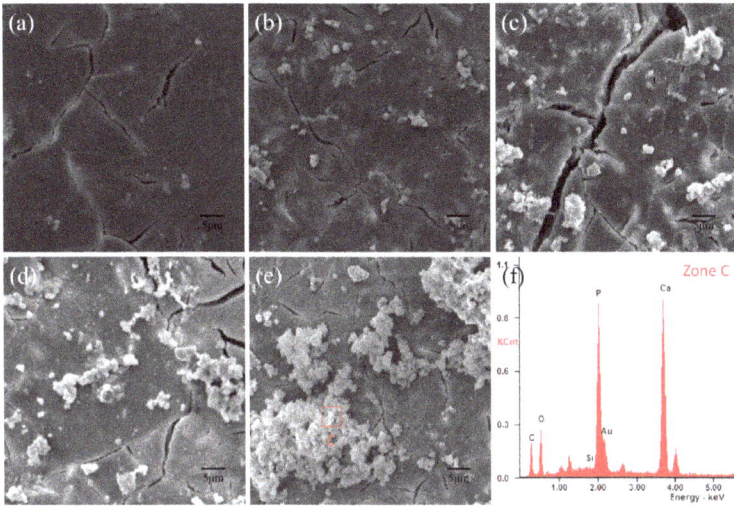

Figure 7. SEM surface morphologies of the (**a**) PHBV; (**b**) PHBV/5%CS; (**c**) PHBV/10%CS; (**d**) PHBV/15%CS and (**e**) PHBV/20%CS scaffolds after being immersed in SBF for 14 days; (**f**) The EDS spectrum of deposits in Zone C.

As the optimal content of CS was 10% for the mechanical properties, the PHBV/10%CS scaffolds were selected to further study their bioactivity for different soaking times. The surface morphologies of the PHBV/10%CS scaffolds after immersion in SBF for 7, 14, 21, and 28 days were observed using SEM (Figure 8). After seven days of immersion, there were only small amounts of hydroxyapatite depositing on the PHBV/10%CS composite scaffolds. With an increase in the immersion, the amounts of hydroxyapatite increased. After 28 days of immersion, large amounts of hydroxyapatite deposited and aggregated on the surface. The results indicated that the PHBV/10%CS composite scaffolds had favorable bioactivity.

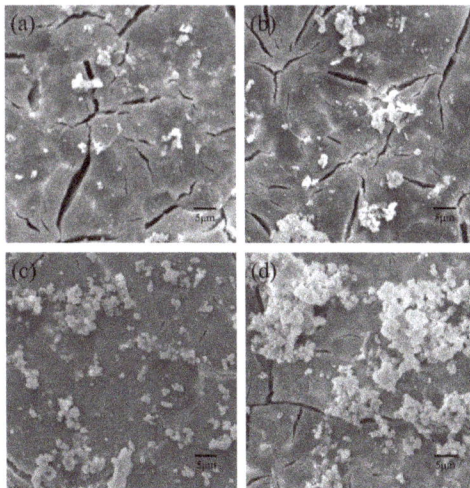

Figure 8. Surface morphologies of the PHBV/10%CS composite scaffolds after immersing in SBF for (**a**) 7, (**b**) 14, (**c**) 21, and (**d**) 28 days.

It was well known that a negatively charged surface would absorb cations in solution [43–45]. The improved bioactivity of the scaffolds may be attributed to the electrostatic attraction between positively charged ions and negatively charged silanol groups derived from CS degradation. The formation processes of hydroxyapatite are briefly discussed as follows. When the PHBV/CS composite scaffold was immersed in the SBF solution, Ca^{2+} release from CS in the surface layer of the scaffold dissolved into the SBF solution. At the same time, protons from the SBF penetrated into the surface layer of the scaffold. As a result, the ion exchange (Equation (1)) occurred due to the different chemical potentials of the ions; thus, the silanol group ($\equiv$Si–OH) was created:

$$\equiv\text{Si–O–Ca–O–Si}\equiv + 2H^+ \rightarrow 2\equiv\text{Si–OH} + Ca^{2+} \tag{1}$$

With the CS continuously degrading, large amounts of silanol formed and deposited on the surface of the scaffold, and the pH of the solution increased. Subsequently, the ion exchange (Equation (2)) occurred on the surface of scaffold:

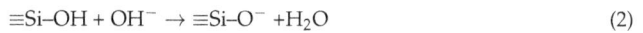

$$\equiv\text{Si–OH} + OH^- \rightarrow \equiv\text{Si–O}^- + H_2O \tag{2}$$

As a consequence, a negatively charged surface, which was abundant in negatively charged functional groups ($\equiv$Si–O$^-$), was created. Then, the positively charged Ca^{2+} in the solution was attracted to the surface of the scaffold by the negatively charged functional group ($\equiv$Si–O$^-$). Afterwards, PO_4^{2-} was also attracted to the scaffold surface by Ca^{2+}. When the ionic activity product of hydroxyapatite was high enough on the surface, the reaction, shown by Equation (3), occurred and hydroxyapatite began to nucleate:

$$10\,Ca^{2+} + 6\,PO_4^{2-} + OH^- \rightarrow Ca_{10}(PO4)_6(OH)_2 \tag{3}$$

The hydroxyapatite crystal nuclei spontaneously grew through continually adsorbing Ca^{2+}, OH^-, HPO_4^-, and PO_4^{2-} from the SBF solution. Finally, large numbers of hydroxyapatite granules formed and aggregated on the surface of the scaffold.

As the PHBV/10%CS composite scaffolds had not only favorable bioactivity, but also the optimal mechanical properties, they were further studied by cell experiments, with the PHBV scaffolds serving as a control. The proliferation of MG63 cells on the scaffolds was assessed by MTT assay (Figure 9a). The absorbance values were positively correlated to the viable cell numbers. After a one-day culture, the absorbance corresponding to PHBV/10%CS scaffold was higher than that corresponding to the PHBV scaffold. As the culture time was prolonged, the proliferation level on the scaffolds both increased, which indicated MG63 cells were cytocompatible with both types of scaffolds. It is noteworthy that a remarkable difference in absorbance between PHBV/10%CS and PHBV scaffolds was observed on day 7, where the absorbance corresponding to the former was almost double that of the latter. The MTT results indicated the incorporation of CS increased the cytocompatibility of PHBV scaffolds and promoted MG63 cell proliferation. The morphologies of MG63 cells cultured on the PHBV and PHBV/10% composite scaffolds on day 4 are shown in Figure 9b,c. It could be seen that MG63 cells adhered and spread well on both of the scaffold surfaces with filopodium and lamellipodium on day 4, and there were no remarkable differences in the morphologies of MG63 cells cultured on PHBV/10%CS composite and PHBV scaffolds.

Osteogenic differentiation of MG63 cells cultured on the PHBV and PHBV/10% composite scaffolds was assessed by ALP staining for 1, 4, and 7 days (Figure 10). It was obvious that the ALP activity increased with increasing culture time for both types of scaffolds. More importantly, the ALP activity corresponding to PHBV/10%CS composite scaffolds was much higher than that corresponding to PHBV scaffolds at the same time point. The ALP assay results indicated that the incorporation of CS significantly improved the osteogenic differentiation of MG63 cells.

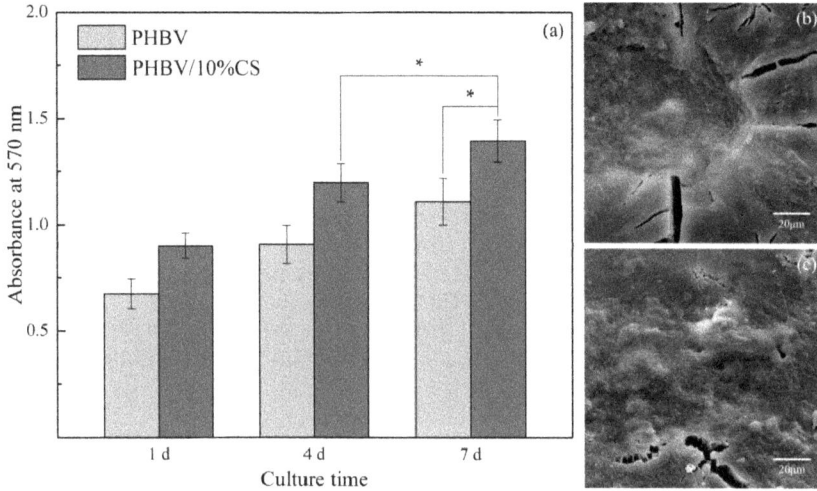

Figure 9. (a) The proliferation of MG63 cells cultured on the PHBV and PHBV/10% composite scaffolds (* represents $p < 0.05$); the SEM morphologies of MG63 cells cultured on the (b) PHBV and (c) PHBV/10% composite scaffolds on day 4.

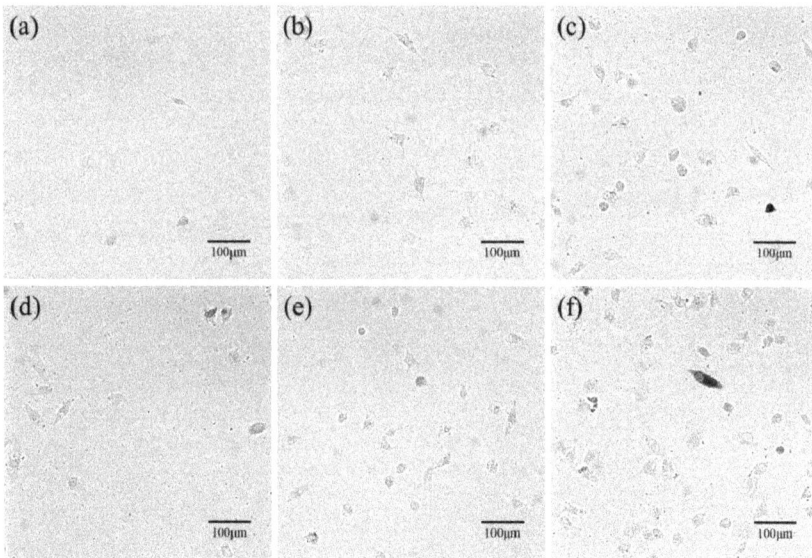

Figure 10. Osteogenic differentiation of MG63 cells cultured on the (a–c) PHBV scaffolds and (d–f) PHBV/10% composite scaffolds for (a,d) 1, (b,e) 4, and (c,f) 7 days. ALP (alkaline phosphatase) activity increased with the increase in culture time and the ALP activity for PHBV/10% composite scaffolds was higher than that for PHBV scaffolds at the same time point.

It was known that the surface characteristics of a material can influence the cell response and behavior [46]. Postiglione [47] reported that bone tissue showed better interactions with titanium implants of a rough surface compared with a relatively smooth one. Deligianni et al. [48] founded the cell adhesion, proliferation, differentiation, and detachment strength of human bone marrow cells on hydroxyapatite discs increased as the roughness of the discs increased. As shown in Figure 4, the incorporation of CS changed the flat and smooth surface morphology of the PHBV scaffold to a relatively rough surface. Such a rough surface with exposed CS particles may provide more target spots for cell adhesion and thereby facilitate cell proliferation. Furthermore, the cell-material interactions were also influenced by the ion species and level released from the material. Xynos et al. [49] cultured osteoblasts on bioactive glass and found that the ion products released from the glass, especially Si and Ca, could stimulate cell proliferation. Shie et al. [50] found that the silicon ions of an appropriate concentration could effectively support the proliferation of osteoblast-like cells and positively stimulate biological responses in MG63 cells through producing bone-specific proteins. Moreover, Ca ions could activate Ca-sensing receptors in osteoblast cells, thus increasing the expression of growth factors, e.g., IGF-I or IGF-II [51,52], and favored osteoblast proliferation, differentiation, and extracellular matrix (ECM) mineralization [53]. Furthermore, Si ions were reported to likely be the key factor for mediating the mineralization and nodule formation [54]. Therefore, it was expected that the Si and Ca ions released from the CS component in the composite scaffolds may play a significant role in promoting MG63 cell proliferation and osteogenic differentiation.

4. Conclusions

Three-dimensional PHBV/CS composite scaffolds were fabricated via SLS and exhibited a well ordered and interconnected porous structure. The incorporation of CS remarkably improved the bioactivity of the scaffolds, which could be attributed to the fact that the degradation products of CS including silanol groups and calcium ions could accelerate the formation and deposition of apatite through electrostatic attraction. Moreover, the compressive strength and modulus were increased by introducing CS, and the enhancement efficiency was dependent on the amount and dispersion state of CS particles in the PHBV matrix. The optimal CS content was 10 wt %, where the compressive strength and compressive modulus were 3.55 and 36.54 MPa, respectively, which was an increase of 41.43% and 28.61%, respectively, compared with the PHBV scaffolds. Additionally, the incorporation of CS significantly promoted the proliferation and osteogenic differentiation of MG63 cells on the scaffolds. This study indicated the biological and mechanical enhanced PHBV scaffolds by incorporating CS may be a promising substitution for bone tissue engineering.

Acknowledgments: This work was supported by the following funds: (1) The Natural Science Foundation of China (51575537, 81572577); (2) Overseas, Hong Kong, and Macao Scholars Collaborated Researching Fund of National Natural Science Foundation of China (81428018); (3) Hunan Provincial Natural Science Foundation of China (14JJ1006, 2016JJ1027); (4) The Project of Innovation-driven Plan of Central South University (2015CXS008, 2016CX023); (5) The Open-End Fund for the Valuable and Precision Instruments of Central South University; (6) The fund of the State Key Laboratory of Solidification Processing at NWPU (SKLSP201605); (7) The fund of the State Key Laboratory for Powder Metallurgy; and (8) The Fundamental Research Funds for the Central Universities of Central South University.

Author Contributions: Pei Feng and Ping Wu conceived and designed the experiments; Cijun Shuai and Wang Guo prepared the scaffold specimens and performed the compression tests and the immersion tests; Chengde Gao and Youwen Yang performed cell culture experiments; Yong Xu, Long Liu, and Tian Qin performed the SEM and XRD characterization; Hang Sun and Sheng Yang analyzed the data; and Cijun Shuai and Wang Guo wrote the paper. All authors discussed the results and approved the final manuscript.

Conflicts of Interest: The authors declare no conflict of interest.

References

1. Hollister, S.J. Porous scaffold design for tissue engineering. *Nat. Mater.* **2005**, *4*, 518. [CrossRef] [PubMed]
2. Hutmacher, D.W. Scaffold design and fabrication technologies for engineering tissues—State of the art and future perspectives. *J. Biomat. Sci.-Polym. Ed.* **2001**, *12*, 107–124. [CrossRef]
3. Yeong, W.-Y.; Chua, C.-K.; Leong, K.-F.; Chandrasekaran, M. Rapid prototyping in tissue engineering: Challenges and potential. *Trends Biotechnol.* **2004**, *22*, 643–652. [CrossRef] [PubMed]
4. Raghunath, J.; Rollo, J.; Sales, K.M.; Butler, P.E.; Seifalian, A.M. Biomaterials and scaffold design: Key to tissue-engineering cartilage. *Biotechnol. Appl. Biochem.* **2007**, *46*, 73–84.

5. Burg, K.J.; Porter, S.; Kellam, J.F. Biomaterial developments for bone tissue engineering. *Biomaterials* **2000**, *21*, 2347–2359. [CrossRef]

6. Chen, G.Q.; Wu, Q. The application of polyhydroxyalkanoates as tissue engineering materials. *Biomaterials* **2005**, *26*, 6565–6578. [CrossRef] [PubMed]

7. Zhao, K.; Deng, Y.; Chen, J.C.; Chen, G.-Q. Polyhydroxyalkanoate (PHA) scaffolds with good mechanical properties and biocompatibility. *Biomaterials* **2003**, *24*, 1041–1045. [CrossRef]

8. Sabir, M.I.; Xu, X.; Li, L. A review on biodegradable polymeric materials for bone tissue engineering applications. *J. Mater. Sci.* **2009**, *44*, 5713–5724. [CrossRef]

9. Mistry, A.S.; Pham, Q.P.; Schouten, C.; Yeh, T.; Christenson, E.M.; Mikos, A.G.; Jansen, J.A. In vivo bone biocompatibility and degradation of porous fumarate-based polymer/alumoxane nanocomposites for bone tissue engineering. *J. Biomed. Mater. Res. A* **2010**, *92*, 451–462. [PubMed]

10. Xu, S.; Lin, K.; Wang, Z.; Chang, J.; Wang, L.; Lu, J.; Ning, C. Reconstruction of calvarial defect of rabbits using porous calcium silicate bioactive ceramics. *Biomaterials* **2008**, *29*, 2588–2596. [CrossRef] [PubMed]

11. Xue, W.; Liu, X.; Zheng, X.; Ding, C. In vivo evaluation of plasma-sprayed wollastonite coating. *Biomaterials* **2005**, *26*, 3455–3460. [CrossRef] [PubMed]

12. Liu, X.; Morra, M.; Carpi, A.; Li, B. Bioactive calcium silicate ceramics and coatings. *Biomed. Pharmacother.* **2008**, *62*, 526–529. [CrossRef] [PubMed]

13. Liu, X.; Ding, C.; Chu, P.K. Mechanism of apatite formation on wollastonite coatings in simulated body fluids. *Biomaterials* **2004**, *25*, 1755–1761. [CrossRef] [PubMed]

14. Unal, H.; Mimaroglu, A.; Alkan, M. Mechanical properties and morphology of nylon-6 hybrid composites. *Polym. Int.* **2003**, *53*, 56–60. [CrossRef]

15. Li, H.; Du, R.; Chang, J. Fabrication, characterization, and in vitro degradation of composite scaffolds based on PHBV and bioactive glass. *J. Biomater. Appl.* **2005**, *20*, 137–155. [CrossRef] [PubMed]

16. Fei, L.; Wang, C.; Xue, Y.; Lin, K.; Chang, J.; Sun, J. Osteogenic differentiation of osteoblasts induced by calcium silicate and calcium silicate/β-tricalcium phosphate composite bioceramics. *J. Biomed. Mater. Res. B* **2012**, *100B*, 1237–1244. [CrossRef] [PubMed]

17. Saravanan, S.; Vimalraj, S.; Vairamani, M.; Selvamurugan, N. Role of mesoporous wollastonite (calcium silicate) in mesenchymal stem cell proliferation and osteoblast differentiation: A cellular and molecular study. *J. Biomed. Nanotechnol.* **2015**, *11*, 1124–1138. [CrossRef] [PubMed]

18. Jack, K.S.; Velayudhan, S.; Luckman, P.; Trau, M.; Grøndahl, L.; Cooper-White, J. The fabrication and characterization of biodegradable HA/PHBV nanoparticle–polymer composite scaffolds. *Acta Biomater.* **2009**, *5*, 2657–2667. [CrossRef] [PubMed]

19. Zhu, H.; Shen, J.; Feng, X.; Zhang, H.; Guo, Y.; Chen, J. Fabrication and characterization of bioactive silk fibroin/wollastonite composite scaffolds. *Mat. Sci. Eng. C* **2010**, *30*, 132–140. [CrossRef]

20. Gao, C.; Deng, Y.; Feng, P.; Mao, Z.; Li, P.; Yang, B.; Deng, J.; Cao, Y.; Shuai, C.; Peng, S. Current progress in bioactive ceramic scaffolds for bone repair and regeneration. *Int. J. Mol. Sci.* **2014**, *15*, 4714–4732. [CrossRef] [PubMed]

21. Williams, J.M.; Adewunmi, A.; Schek, R.M.; Flanagan, C.L.; Krebsbach, P.H.; Feinberg, S.E.; Hollister, S.J.; Das, S. Bone tissue engineering using polycaprolactone scaffolds fabricated via selective laser sintering. *Biomaterials* **2005**, *26*, 4817–4827. [CrossRef] [PubMed]

22. Tan, K.; Chua, C.; Leong, K.; Cheah, C.; Cheang, P.; Bakar, M.A.; Cha, S. Scaffold development using selective laser sintering of polyetheretherketone–hydroxyapatite biocomposite blends. *Biomaterials* **2003**, *24*, 3115–3123. [CrossRef]

23. Pei, F.; Peng, S.; Ping, W.; Gao, C.; Wei, H.; Deng, Y.; Shuai, C. A space network structure constructed by tetraneedlelike ZnO whiskers supporting boron nitride nanosheets to enhance comprehensive properties of poly(L-lacti acid) scaffolds. *Sci. Rep.* **2016**, *6*, 33385.

24. Kokubo, T.; Takadama, H. How useful is SBF in predicting in vivo bone bioactivity? *Biomaterials* **2006**, *27*, 2907–2915. [CrossRef] [PubMed]

25. Shuai, C.; Feng, P.; Wu, P.; Liu, Y.; Liu, X.; Lai, D.; Gao, C.; Peng, S. A combined nanostructure constructed by graphene and boron nitride nanotubes reinforces ceramic scaffolds. *Chem. Eng. J.* **2017**, *313*, 487–497. [CrossRef]

26. Lei, C.; Zhu, H.; Li, J.; Li, J.; Feng, X.; Chen, J. Preparation and characterization of polyhydroxybutyrate-co-hydroxyvalerate/silk fibroin nanofibrous scaffolds for skin tissue engineering. *Polym. Eng. Sci.* **2014**, *55*, 365–371. [CrossRef]

27. Ten, E.; Jiang, L.; Wolcott, M.P. Crystallization kinetics of poly(3-hydroxybutyrate-co-3-hydroxyvalerate)/cellulose nanowhiskers composites. *Carbohyd. Polym.* **2012**, *90*, 541–550. [CrossRef] [PubMed]

28. Pei, L.Z.; Yang, L.J.; Yang, Y.; Fan, C.G.; Yin, W.Y.; Chen, J.; Zhang, Q.F. A green and facile route to synthesize calcium silicate nanowires. *Mater. Charact.* **2010**, *61*, 1281–1285. [CrossRef]

29. Tiimob, B.J.; Rangari, V.K.; Jeelani, S. Effect of reinforcement of sustainable β-casio3 nanoparticles in bio-based epoxy resin system. *J. Appl. Polym. Sci.* **2014**, *131*, 19.

30. Gao, C.; Liu, T.; Shuai, C.; Peng, S. Enhancement mechanisms of graphene in nano-58S bioactive glass scaffold: Mechanical and biological performance. *Sci. Rep.* **2014**, *4*, 4712. [CrossRef] [PubMed]

31. Feng, P.; Niu, M.; Gao, C.; Peng, S.; Shuai, C. A novel two-step sintering for nano-hydroxyapatite scaffolds for bone tissue engineering. *Sci. Rep.* **2014**, *4*, 5599. [CrossRef] [PubMed]

32. Zhang, N.; Wang, Y.; Xu, W.; Hu, Y.; Ding, J. Poly(lactide-co-glycolide)/hydroxyapatite porous scaffold with microchannels for bone regeneration. *Polymers* **2016**, *8*, 218. [CrossRef]

33. Bose, S.; Roy, M.; Bandyopadhyay, A. Recent advances in bone tissue engineering scaffolds. *Trends Biotechnol.* **2012**, *30*, 546. [CrossRef] [PubMed]

34. Guo, H.; Su, J.; Wei, J.; Kong, H.; Liu, C. Biocompatibility and osteogenicity of degradable Ca-deficient hydroxyapatite scaffolds from calcium phosphate cement for bone tissue engineering. *Acta Biomater.* **2009**, *5*, 268–278. [CrossRef] [PubMed]

35. Tarafder, S.; Balla, V.K.; Davies, N.M.; Bandyopadhyay, A.; Bose, S. Microwave sintered 3D printed tricalcium phosphate scaffolds for bone tissue engineering. *J. Tissue Eng. Regen. M.* **2013**, *7*, 631–641. [CrossRef] [PubMed]

36. Cai, Y.Q.; Lin, Y.Y.; Li, X.; Huang, J.T.; Aoki, T. In Calcium behaviors in MnZn ferrite at different temperatures. *Key Eng. Mat.* **2012**, *512–515*, 1412–1415. [CrossRef]

37. Jenkins, M.; Cao, Y.; Howell, L.; Leeke, G. Miscibility in blends of poly(3-hydroxybutyrate-co-3-hydroxyvalerate) and poly(ε-caprolactone) induced by melt blending in the presence of supercritical CO_2. *Polymer* **2007**, *48*, 6304–6310. [CrossRef]

38. Tsuji, H.; Ikada, Y. Properties and morphology of poly(L-lactide) 4. Effects of structural parameters on long-term hydrolysis of poly(L-lactide) in phosphate-buffered solution. *Polym. Degrad. Stabil.* **2000**, *67*, 179–189. [CrossRef]

39. Loo, S.C.; Ooi, C.P.; Wee, S.H.; Boey, Y.C. Effect of isothermal annealing on the hydrolytic degradation rate of poly(lactide-co-glycolide) (PLGA). *Biomaterials* **2005**, *26*, 2827–2833. [PubMed]

40. Hurrell, S.; Cameron, R.E. The effect of initial polymer morphology on the degradation and drug release from polyglycolide. *Biomaterials* **2002**, *23*, 2401. [CrossRef]

41. Fu, S.-Y.; Feng, X.-Q.; Lauke, B.; Mai, Y.-W. Effects of particle size, particle/matrix interface adhesion and particle loading on mechanical properties of particulate–polymer composites. *Compos. Part B-Eng.* **2008**, *39*, 933–961. [CrossRef]

42. Liu, X.; Ding, C.; Wang, Z. Apatite formed on the surface of plasma-sprayed wollastonite coating immersed in simulated body fluid. *Biomaterials* **2001**, *22*, 2007–2012. [CrossRef]

43. Ohgaki, M.; Kizuki, T.; Katsura, M.; Yamashita, K. Manipulation of selective cell adhesion and growth by surface charges of electrically polarized hydroxyapatite. *J. Biomed. Mater. Res.* **2001**, *57*, 366–373. [CrossRef]

44. Kabaso, D.; Gongadze, E.; Perutková, S.; Matschegewski, C.; Kralj-Iglic, V.; Beck, U.; Van, R.U.; Iglic, A. Mechanics and electrostatics of the interactions between osteoblasts and titanium surface. *Comput. Method Biomech.* **2011**, *14*, 469–482. [CrossRef] [PubMed]

45. Liu, X.; Zhao, X.; Fu, R.K.; Ho, J.P.; Ding, C.; Chu, P.K. Plasma-treated nanostructured TiO_2 surface supporting biomimetic growth of apatite. *Biomaterials* **2005**, *26*, 6143–6150. [CrossRef] [PubMed]

46. Anselme, K. Osteoblast adhesion on biomaterials. *Biomaterials* **2000**, *21*, 667–681. [CrossRef]

47. Postiglione, L.; Di Domenico, G.; Ramaglia, L.; Di Lauro, A.; Di Meglio, F.; Montagnani, S. Different titanium surfaces modulate the bone phenotype of SaOS-2 osteoblast-like cells. *Eur. J. Histochem.* **2004**, *48*, 213. [PubMed]

48. Deligianni, D.D.; Katsala, N.D.; Koutsoukos, P.G.; Missirlis, Y.F. Effect of surface roughness of hydroxyapatite on human bone marrow cell adhesion, proliferation, differentiation and detachment strength. *Biomaterials* **2001**, *22*, 87–96. [CrossRef]

49. Xynos, I.D.; Edgar, A.J.; Buttery, L.D.; Hench, L.L.; Polak, J.M. Ionic products of bioactive glass dissolution increase proliferation of human osteoblasts and induce insulin-like growth factor ii mRNA expression and protein synthesis. *Biochem. Bioph. Res. Commun.* **2000**, *276*, 461–465. [CrossRef] [PubMed]

50. Shie, M.Y.; Ding, S.J.; Chang, H.C. The role of silicon in osteoblast-like cell proliferation and apoptosis. *Acta Biomater.* **2011**, *7*, 2604. [CrossRef] [PubMed]

51. Marie, P.J. The calcium-sensing receptor in bone cells: A potential therapeutic target in osteoporosis. *Bone* **2010**, *46*, 571–576. [CrossRef] [PubMed]

52. Valerio, P.; Pereira, M.; Goes, A.; Leite, M.F. Effects of extracellular calcium concentration on the glutamate release by bioactive glass (BG60S) preincubated osteoblasts. *Biomed Mater* **2009**, *4*, 045011. [CrossRef] [PubMed]

53. Maeno, S.; Niki, Y.; Matsumoto, H.; Morioka, H.; Yatabe, T.; Funayama, A.; Toyama, Y.; Taguchi, T.; Tanaka, J. The effect of calcium ion concentration on osteoblast viability, proliferation and differentiation in monolayer and 3D culture. *Biomaterials* **2005**, *26*, 4847–4855. [CrossRef] [PubMed]

54. Gough, J.E.; Jones, J.R.; Hench, L.L. Nodule formation and mineralisation of human primary osteoblasts cultured on a porous bioactive glass scaffold. *Biomaterials* **2004**, *25*, 2039–2046. [CrossRef] [PubMed]

polymers

MDPI

Article

Osteochondral Regeneration Induced by TGF-β Loaded Photo Cross-Linked Hyaluronic Acid Hydrogel Infiltrated in Fused Deposition-Manufactured Composite Scaffold of Hydroxyapatite and Poly (Ethylene Glycol)-Block-Poly(ε-Caprolactone)

Yi-Ho Hsieh [1,2], Ming-Fa Hsieh [1], Chih-Hsiang Fang [1], Cho-Pei Jiang [3], Bojain Lin [1,4] and Hung-Maan Lee [1,5,*]

[1] Department of Biomedical Engineering, Chung Yuan Christian University, 200 Chung Pei Road, Chung-Li District, Taoyuan City 320, Taiwan; dilantin11@gmail.com (Y.-H.H.); mfhsieh@cycu.edu.tw (M.-F.H.); danny07291991@hotmail.com (C.-H.F.); linbojain@gmail.com (B.L.)
[2] Department of Orthopedics, Min-Sheng General Hospital, 168, ChingKuo Rd, Taoyuan 330, Taiwan
[3] Department of Power Mechanical Engineering, National Formosa University, Yunlin County 632, Taiwan; kasu23@gmail.com
[4] Department of Orthopedics, Taoyuan Armed Forces General Hospital, No. 168, Zhongxing Road, Longtan District, Taoyuan City 325, Taiwan
[5] Department of Orthopedics, Hualien Tzu Chi General Hospital, No. 707, Sec. 3, Chung Yang Rd, Hualien 970, Taiwan
* Correspondence: hungmaan@ms12.hinet.net; Tel.: +886-972-766-158

Academic Editor: Insung S. Choi
Received: 6 March 2017; Accepted: 14 May 2017; Published: 20 May 2017

Abstract: The aim of this study was to report the fabrication of porous scaffolds with pre-designed internal pores using a fused deposition modeling (FDM) method. Polycaprolactone (PCL) is a suitable material for the FDM method due to the fact it can be melted and has adequate flexural modulus and strength to be formed into a filament. In our study, the filaments of methoxy poly(ethylene glycol)-block-poly(ε-caprolactone) having terminal groups of carboxylic acid were deposited layer by layer. Raw materials having a weight ratio of hydroxyapatite (HAp) to polymer of 1:2 was used for FDM. To promote cell adhesion, amino groups of the Arg-Gly-Asp(RGD) peptide were condensed with the carboxylic groups on the surface of the fabricated scaffold. Then the scaffold was infiltrated with hydrogel of glycidyl methacrylate hyaluronic acid loading with 10 ng/mL of TGF-β1 and photo cross-linked on the top of the scaffolds. Serious tests of mechanical and biological properties were performed in vitro. HAp was found to significantly increase the compressive strength of the porous scaffolds. Among three orientations of the filaments, the lay down pattern 0°/90° scaffolds exhibited the highest compressive strength. Fluorescent staining of the cytoskeleton found that the osteoblast-like cells and stem cells well spread on RGD-modified PEG-PCL film indicating a favorable surface for the proliferation of cells. An in vivo test was performed on rabbit knee. The histological sections indicated that the bone and cartilage defects produced in the knees were fully healed 12 weeks after the implantation of the TGF-β1 loaded hydrogel and scaffolds, and regenerated cartilage was hyaline cartilage as indicated by alcian blue and periodic acid-schiff double staining.

Keywords: poly(ε-caprolactone); osteoarthritis; cartilage; scaffold; hydrogel; fused deposition manufacturing

Polymers **2017**, *9*, 182

1. Introduction

Articular cartilage is hyaline cartilage covering the articular surface of bones. Due to its avascular characteristic, once injured, articular cartilage has a poor self-repair ability. Treatment includes conservative treatment and surgical approaches, but there is currently no effective treatment for this disease. Bone marrow stimulation (microfracture surgery) could produce cartilage formation, but the newly-formed cartilage is fibrocartilage instead of hyaline cartilage. Several methods of chondrocyte implantation were used to treat cartilage defects in humans, though implanted grafts do not provide mechanical stability and various side effects lead to procedure failure [1,2]. Osteochondral plug transplantation (mosaicplasty) has been developed but the limitations of this technique include donor site morbidity and limited availability of grafts that can be harvested. The long-term outcome is constrained by the age and gender of the patient and the size of the wound [3–5].

Various tissue-engineering approaches were developed using natural or synthetic biomaterials as scaffolds for cell growth. The porous scaffolds are traditionally fabricated by the particulate leaching method, adding pore-forming agents and their combinations [6]. However, the pore size and interconnection of pores could not be precisely controlled throughout the entire scaffold. Recently, rapid prototyping (RP) technology has emerged to produce custom-made scaffolds and provide a new method to control the internal architecture of the scaffolds for cell growth [7,8]. Among the natural biopolymers and synthetic polymers, poly(ε-caprolactone) (PCL) has been widely studied for 3D printing applications because of the longer biodegradation time in the human body as compared to poly(lactide-co-glycolide) [9–12]. Woodfield et al. employed FDM to fabricate articulating scaffolds having a natural curvature of the surface for fitting into the individual patient's knee [13]. They showed that chondrocytes grew into scaffolds and secreted the ECM of cartilage [14]. Boere et al. showed that chondrocytes could be embedded into 3D-fabricated polymeric scaffolds, showing significant deposition of ECM in their constructs, both in vitro and in vivo [15]. Shao et al. implanted PCL scaffolds into cancellous bone while fibrin glue was applied on top of the cartilage surface. Therefore, PCL scaffolds are a promising matrix for bone regeneration and fibrin glue for cartilage regeneration [16].

The successful interaction between the cell and the surface of the biomaterial is important to the scaffolds. The PCL scaffold has good biocompatibility and biodegradability, but lacks biological factors on its surface, leading cells to not easily attach and grow. RGD (Arg-Gly-Asp) is a peptide for biopolymer functionalization and can trigger cell lines to adhere and proliferate on the polymer's surface. Characterizing the chemical modification and, therefore, the extent of RGD (Arg-Gly-Asp) at the polymer's surface can significantly promote cell adhesion and proliferation [17,18].

Hydrogels can simulate the microenvironments of natural tissues, encapsulate cells homogeneously, and deliver drugs, which makes them attractive for cartilage tissue engineering [19,20]. Hydrogels loaded with growth factor could ideally fit into irregular cartilage wounds while releasing the growth factor [21], and three-dimensional networks of hydrogels are known to proliferate and differentiate mesenchymal stem cells [22]. For example, Guo et al. Used a bi-layered hydrogel of oligo(poly(ethylene glycol) fumarate) to encapsulate TGF-β loaded gelatin microparticles for chondrogenic differentiation and osteochondral regeneration [23,24]. However, the osteochondral defects were not healed using their biphasic scaffolds in vivo. For delivery of the growth factors, hyaluronic acid (HA), among other hydrogels, has been grafted with glycidyl methacrylate to bestow a HA photo cross-linkable property [25,26]. The light-cured HA hydrogel displays various properties, such as sealant of irregular defects of cartilage [27], the carrier of growth factors [28], ECM to enhance MSC differentiation [29], or to maintain the differentiated phenotype [30].

Recently, tissue engineering for articular cartilage repair has focused on biphasic scaffolds [31]. Cartilage damage involves not only the articular surface of the knee, but also the subchondral bone. Currently, the performance of scaffolds combining with hydrogels reported so far with respect to damaged cartilage is not satisfactory. Therefore, we aimed to fabricate a biphasic scaffold with two layers for osteochondral regeneration. The bottom layer is a fused-deposited methoxy poly(ethylene glycol)-block-poly(β-caprolactone) scaffold (mPEG-PCL), which was first modified by adding hydroxyapatite (HAp) powder and grafting RGD peptide to the surface of the scaffold. The surface layer of the mPEG-PCL scaffold was infiltrated with glycidyl methacrylate-hyaluronic acid (GMHA) hydrogel containing transforming growth factor TGF-β1 for cartilage healing.

2. Materials and Methods

2.1. Materials

Poly(ethylene glycol) (PEG, number-average molecular weight (M_n = 4000 g·mol^{-1}) was obtained from Showa (Tokyo, Japan); ε-caprolactone (ε-CL), stannous octoate (Sn(Oct)$_2$), 1,6-Diphenyl-1,3,5-hexatriene (DPH), dimethyl sulfoxide (DMSO), 3-(4,5-Dimethylthiazol-2-yl)-2,5-diphenyltetrazolium bromide (MTT), phosphotungstic acid (PTA), 2-Hydroxy-2-methylpropiophenone, Arg-Gly-Asp (RGD peptide), Hydroxyapatite, Noble agar, potassium phosphate monobasic, potassium bromide (KBr), sodium chloride, and sodium phosphate were all obtained from Sigma-Aldrich Chem. Inc. (St. Louis, MO, USA). 1,4-dioxane, Acetone, acetonitrile (ACN), Triethylamine, D-chloroform (CDCl3), dichloromethane, ethyl ether, ethanol, hexane, and tetrahydrofuran were provided by ECHO Chemicals (Miaoli, Taiwan).

2.2. Synthesis of Biopolymers

In this study, biopolymer synthesis is divided into two parts. The first part synthesized the diblock copolymers (mPEG-PCL), and the second part modified the diblock copolymers to terminal function groups (mPEG-PCL-COOH) (Figure 1), as in our previous reports [9]. In brief, the mPEG-PCL diblock copolymers were synthesized by ring-opening polymerization of ε-CL using mPEG as the macro-initiator and Sn(Oct)$_2$ serving as a catalyst. The hydroxyl group of mPEG-PCL was modified to a carboxylic acid group and was connected with mPEG-PCL diblock copolymers. The theoretical molecular weight of mPEG-PCL is defined as 9450 Da whereas the molecular weight of mPEG is 550 Da.

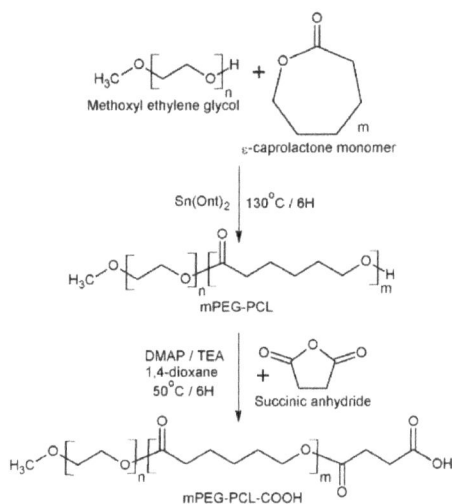

Figure 1. Synthesis scheme of the mPEG-PCL-COOH diblock copolymer.

2.3. Characterization of Biopolymers

Thermal properties were measured by differential scanning calorimetry (Jade DSC, PerkinElmer, Waltham, MA, USA). Samples with weights ranging from 5 to 10 mg were put into the aluminum pans with two series of heating and cooling. The samples were heated up to 150 °C and cooled down to 20 °C. All operations were conducted at the rate of 10 °C/min. The vibrational spectra of Fourier transform infrared spectroscopy were carried out on a FTIR 410 (JASCO, Tokyo, Japan) ranging from 4000 to 400 cm^{-1}. The powdery polymers were mixed with KBr powder and compressed into a disk for FTIR measurements. Two-hundred fifty-six (256) scans were performed in all specimens, and the spectrum was recorded. The average molecular weight (M_w) and the polydispersity (PDI, M_w/M_n) of the polymers were determined by Gel Permeation Chromatography (GPC 270, Viscotek, Malvern, UK) connected with a refractive index detector. Tetrahydrofuran (THF) was used as an eluent. The molecular weight was calculated using standard polystyrene samples as references. The molecular structure of the polymer was determined by nuclear magnetic resonance spectrophotometry (500 MHz, Bruker, MA, USA). The ^{1}H NMR spectra of the block biopolymers were recorded, using D-chloroform (CDCl$_3$) as the solvent. Further, the average molecular weight and grafting percentage of the biopolymers were calculated based on the spectra.

2.4. Fabrication of Porous Scaffolds of mPEG-PCL-COOH/HAp

The powders of polymer and HAp were mixed at a weight ratio of 2:1 in a glass container. To ensure thorough mixing of the powders, the mixing was performed using a homemade roller. The fused deposition modeling of the mixed powder was performed in a homemade air pressure-aided deposition system. It consists of a sample compartment (stainless steel cylinder) surrounded by a heating system (heater, controller, and nozzle), the air pressure pump, three-axes motors, and self-developed software (NI LabView).

The mixed powder was heated up to 60 °C for complete melting. Afterward, a pressure of 15 psi was applied to extrude the molten sample from an 18-gauge nozzle onto a computer-controlled *x-y-z* table. The extruded filament was laid down layer-by-layer in three different layer orientations: 0°/90°, staggered 0°/90°, and 0°/90°/+45°/−45° at a speed of 4 mm/s. The structure of scaffolds was computer-designed (Figure 2). The average pore size measurements of the scaffolds can be found in our previous study [9] In our study, scaffold surfaces were modified by RGD peptide grafting. RGD peptide is known to enhance cell adhesion by binding the intergen αvβ3 in the cell membrane [32]. In brief, polymers created COOH functional groups, and were prepared by our home-made RP machine. Scaffolds were immersed in a solution of dimethylaminopropyl-3-ethylcarbodiimide hydrochloride (EDC, 0.2 M) + *N*-hydroxysuccinimide (NHS, 0.1 M) in (2-(*N*-morpholino)-ethanesulfonic acid (MES buffer, 0.1 M in d^{2}H$_2$O) for 30 min to activate the hydroxyl groups and then rinsed in d^{2}H$_2$O. After that, immobilization of RGD peptides was achieved with a solution of RGD peptides (10^{-3} M) by phosphate buffered solution (PBS) for 24 h at 4 °C. After grafting, the scaffolds were rinsed with d^{2}H$_2$O (100 mL) for 10 min in order to remove non-grafted peptides.

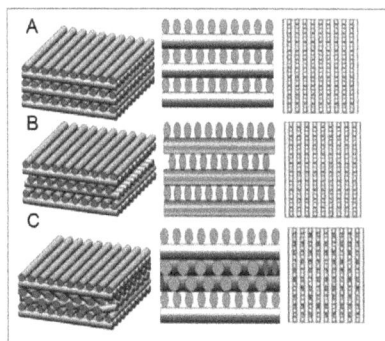

Figure 2. The structure of (**A**) lay-down patterns 0°/90° (90ECR/90ECR); (**B**) staggered 0°/90° (SECR/SECR); and (**C**) 0°/90°/+45°/−45° (45ECR/45ECR) scaffold designs.

2.5. Determination of Inorganic Components in the Scaffolds

Thermogravimetric analysis was used to determine the mass change of a scaffold as a function of temperature or time. This analysis confirmed that the composite contained the correct amounts of inorganic components, e.g., hydroxyapatite. Samples with weights ranging from 10 to 20 mg were put into the furnace with one series of heating. The sample was heated from 25 °C to 900 °C. All operations were conducted at the rate of 10 °C/min.

2.6. Mechanical Testing of the Scaffolds

Compression tests were conducted on all structures of the fabricated scaffolds. Compressive properties of the scaffolds were measured by using the material compression test system (Chun Yen Testing Machines CY-6040A4). Specimen dimensions were 10 mm in length, 10 mm in width, and 4 mm in height. Conducted with a cross-head displacement speed of 1 mm/min, compressive strengths were obtained from the testing system and Young's modulus were calculated.

2.7. In Vitro Degradation

Rapid-prototyped scaffolds ($10 \times 10 \times 4$ mm) were prepared into rectangle shape for degradation testing. In vitro degradation study was carried out in PBS solution according to the standard protocol ASTM F1635 [33]. The RP scaffolds grafting RGD peptide were sterilized under UV light irradiation for 2 h, placed in 100 mL glass container containing 50 mL of PBS and incubated at 37 °C shaking with 50 rpm for three months. Scaffolds of each degradation period were taken out at different periods of time, washed with distilled water, and vacuum dried for characterization analysis.

2.8. Synthesis of UV-Curable GMHA

First, 1 g of hyaluronic acid (Mw = 4.4×10^5) was dissolved in 100 mL PBS and covered overnight under continuous stirring. After it was fully dissolved, 100 mL *N,N*-dimethylformamide (DMF), 18.04 mL triethylamine (TEA) and 35.11 mL glycidyl methacrylate (GM) were added separately, in that order, and stirred for 10 days. After that, the solution was then precipitated as a white solid in acetone of 10-fold volume, centrifuged to remove the acetone at 5000 rpm for 10 min. The precipitate was then dissolved in deionized water, and subjected to dialysis (MWCO = 8 kDa) in deionized water for 24 h. Finally, the samples were frozen in -80 °C and freeze-dried to obtain the product.

2.9. Isolation and Identification of Mesenchymal Stem Cells

The mesenchymal stem cells (MSCs) were isolated from BALB/c mice. Bone marrow was collected from six-week old BALB/c mice that were sacrificed by carbon dioxide. Their femurs were carefully cleaned of adherent soft tissue and metaphyseal ends of the bones were removed to expose the bone marrow cavity, and bone marrow was harvested through a 0.22 μm filter by flushing with Dulbecco's modified Eagle's medium-low glucose (DMEM-LG) supplemented with 10% fetal bovine serum (FBS) and 1% penicillin/streptomycin. MSCs were characterized by positive localization of the multipotent mesenchymal stem cell markers, like CD105, and negative localization of CD34. MSCs should characteristically show positive localization of CD44, CD29, CD105, and CD90, but no localization of hematopoietic markers CD45 or CD34. Flow cytometery could be used to determine the characteristic of MSCs. The cultured MSCs were retrieved by trypsin digestion. Cells aliquots (1×10^5) were washed with fluorescence-activated cell sorting buffer, called stain buffer (2% FBS, 0.1% NaN$_3$ in PBS), incubated on ice for 30 min, and stained with fluorescein isothiocyanate (FITC)- and phycoerythrin (PE)-conjugated monoclonal antibodies against mouse CD34 and CD105 (eBioScience). After washing twice with FACS buffer, the cells were fixed with 1% paraformaldehyde (Sigma) in PBS. At least 10,000 events were collected for further analysis using a FACSVantage cytometer and CellQuest software (BD, San Jose, CA, USA).

2.10. Cytotoxicity Assay

MSCs were seeded onto scaffolds (length = 10 mm, width = 10 mm and height = 4 mm) and GMHA gel (concentration of 0.25% to 1%) in the 24-well plate at a density of 2×10^4 cells/mL, respectively. Each well was cultured with low-glucose Dulbecco's Modified Eagle's Medium (DMEM-LG) with 10% fetal bovine serum (FBS, Gibco) and 1% penicillin/streptomycin from 24 h to 72 h, and cell activity was determined by the MTT (3-(4,5-dimethylthiazol-2-yl)-2,5,-diphenyl-tetrazolium bromide) assay. MTT reagent of 5 mg/mL was added 100 µL to each well, followed by incubation for four hours in an incubator at 37 °C with 5% CO_2. After removing the medium, 1 mL DMSO was then added to each well to dissolve formazan, and transferred to a 96-well microtiter shaking plate for 15 min, and OD at 570 nm were determined with an ELISA plate reader.

2.11. Cell Adhesion and Viability

The MSCs were pre-seeded in each of the scaffolds with 1×10^5 cells/mL in a 24-well culture plate, followed by incubation for four hours in an incubator at 37 °C with 5% CO_2. After cell seeding, all of scaffolds were induced to undergo growth medium for one month. For staining of cytoskeletons, all of the culture media was removed, adding 4% buffered paraformaldehyde for 30 min, then removed, permeabilized in 0.1% Triton X-100, and incubated with Invitrogen Alexa Fluor® 488 Phalloidin to stain the cytoskeleton protein for 30 min. After incubation, the samples were washed with PBS. Nuclei were stained with 4′6-diamidino-2-phenylindole (DAPI) for 1 min before imaging through the Nikon ECLIPSE Ti-s fluorescence microscope. Cell viability was assessed using an Invitrogen LIVE/DEAD Viability/Cytotoxicity Kit containing approximately 2 µM calcein-AM and 4 µM EthD-1 as a working solution.

2.12. Chondrogenic Differentiation

MSCs were pre-seeded in each scaffold with 1×10^7 cells/mL in a 24-well culture plate, followed by incubation for four hours in an incubator at 37 °C, 5% CO_2. After that, all of the scaffolds were induced to undergo chondrogenic differentiation. Specifically, the constructs were cultured in basic medium as the control consisted of DMEM-high glucose supplemented with 100 nM dexamethasone, 50 µg/mL ascorbic acid, 100 µg/mL sodium pyruvate, 40 µg/mL proline, and 50 mg/mL ITS + 1 liquid media supplement. The chondrogenic differentiation medium was the basic medium with 10 ng/mL recombinant human transforming growth factor-beta 1 (TGF-beta 1; PeproTech Inc., Rocky Hill, NJ, USA). At days 7, 14, 21, and 28, the samples were harvested for subsequent characterization.

2.13. Real-Time Polymerase Chain Reaction (qPCR) Analysis of Gene Expression

Total RNA was extracted using the mirVana™ miRNA Isolation Kit (Invitrogen, Carlsbad, CA, USA). Isolated RNA was reverse-transcribed with a High-Capacity cDNA Reverse Transcription Kit (Invitrogen, Carlsbad, CA, USA), and real-time PCR analysis was performed using the ABI 7300 Real-Time PCR System (Applied Biosystems, Foster City, CA, USA) with Power SYBR® Green PCR Master Mix (Invitrogen, Paisley, UK). Real-time PCR conditions were as follows: 95 °C for 8.5 min, followed by 40 cycles of 95 °C for 30 s, 58 °C for 30 s, and 72 °C for 30 s. The relative expression ratio between the target genes and the control group were calculated. The PCR primers were as follows: Sox9: forward, 5′-CGGCTCCAGCAAGAACAAG-3′ and reverse, 3′-TTGTGCAGATGCGGGTACTG-5′; Aggrecan: forward, 5′-AGATGGCACCCTCCGATAC-3′ and reverse, 3′-ACACACCTCGGAAGCAGAAG-5′; Col2a1: forward, 5′-GGAGGGAACGGTCCACGAT-3′ and reverse, 3′-AGTCCGCGTATCCACAA-5′; GAPDH: forward, 5′-GCATTGTGGAAGGGCTCA-3′ and reverse, 3′-GGGTAGGAACACGGAAGG-5′. Transcription levels were normalized to GAPDH.

2.14. Biochemical Analysis

Sulfated glycosaminoglycan (sGAG) content was spectrophotometrically determined with a Biocolor[TM] Blyscan assay kit by 9-dimethylmethylene blue chloride (DMMB) methods. Cell number was determined via extraction of total DNA with a genomic Geno Plus[TM] DNA extraction miniprep system and quantification of DNA content was determined using a Quant-iT PicoGreen dsDNA Assay Kit. Quantitative total sGAG and total DNA were normalized for differences in each sample.

2.15. Animal Study

Six male New Zealand white rabbits (eight weeks old) were chosen as the animal model. Rabbits were generally anaesthetized, and had their knees shaved and disinfected. Both knees of each animal were operated during the same surgery, with arthrotomy made through a longitudinal medial parapatellar incision and lateral dislocation of the patella. A 3 mm diameter circle was drilled in the center of the medial femoral condyle and holes were left to bleed. Then, scaffolds and light-curing gel containing TGF-beta 1 were implanted into the defect, cross-linking with 365 nm ultraviolet light for 5 min (Figure 3). Joints were harvested after three months from implantation and evaluated histologically as described below. Joints were fixed in 4% buffered paraformaldehyde for one to two weeks (replaced with fresh 4% buffered paraformaldehyde per three days). After that, joints were demineralized by 10% ethylenediaminetetraacetic acid (EDTA) solution for one month (replaced with fresh 10% EDTA solution per week). When demineralization finished, all of the specimens were dehydrated in increasing grades of ethanol, defatted and cleared with xylene, and embedded in wax.

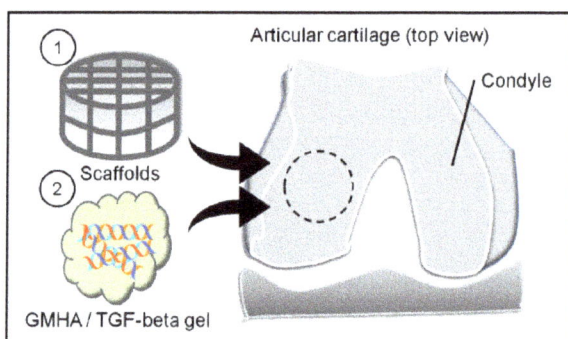

Figure 3. The illustration of animal model. Control group: create a bony defect 3 mm in diameter and 8 mm in depth without scaffold implantation. Experimental group: scaffolds and light-curing gel containing TGF-beta 1 were implanted into the defect, and cross-linked with 365 nm ultraviolet light.

2.16. Statistical Analysis

The results of all experimental data were expressed as means ± standard deviation and compared using Student's *t*-test. The difference was considered significant when $p < 0.05$.

3. Results

3.1. Characterization of Diblock Biopolymers

Characteristics were observed between the DSC diagrams of mPEG-PCL and mPEG-PCL-COOH polymers. The melting temperatures of mPEG-PCL and mPEG-PCL-COOH were identified at 60.63 °C and 59.06 °C, respectively. The freezing temperatures were identified at 32.92 °C and 36.34 °C, respectively. These results are in good agreement with the previous study in DSC [34], and no significant differences were observed between the mPEG-PCL and mPEG-PCL-COOH. Therefore, the temperature of the rapid prototyping machine heat module should be higher than 60 °C to prepare the scaffolds. In the FTIR spectra of the biopolymers, a strong absorption band associated with the carbonyl group at 1731.76–1727.91 cm^{-1} was observed. Another peak at 1180.22 cm^{-1} indicated the C–O–C stretching. The FTIR spectrum of mPEG-PCL-COOH showed the presence of hydroxyl (OH–) stretching modes in the carboxyl group at 3444.24 cm^{-1} (Figure 4). Molecular weight and distribution of biopolymers were also characterized by GPC and ^{1}H NMR, respectively, and summarized in Table 1. The results of GPC are shown in Table 1. As can be seen, the total molecular weight and polydispersity (PDI, M_w/M_n) of mPEG-PCL-COOH biopolymer is 9514.33 ± 389.70 and 1.19 ± 0.01, respectively. The typical signals of the ^{1}H NMR spectra of both mPEG-PCL and mPEG-PCL-COOH were detected as shown in Figure 5, and CDCl$_3$ was used as a solvent, thus confirming that the ring opening polymerization and coupling reaction were carried out. As seen in the spectrum of mPEG-PCL, the characteristic peaks at 1.37, 1.62, 2.29, and 4.06 ppm are attributed to the methylene protons of –(CH$_2$)$_3$–, –OCCH$_2$– and –CH$_2$OOC– in the PCL unit. The PEG unit shows its characteristic peaks at 3.38 and 3.64 ppm, which are the signals of CH$_3$– and –CH$_2$CHO–. The spectra of mPEG-PCL-COOH show succinic anhydride ring opening modifications of the carboxyl group at 2.66 ppm. The average molecular weight (M_n) was calculated by using ^{1}H NMR spectroscopy, in accordance with the previous study [35].

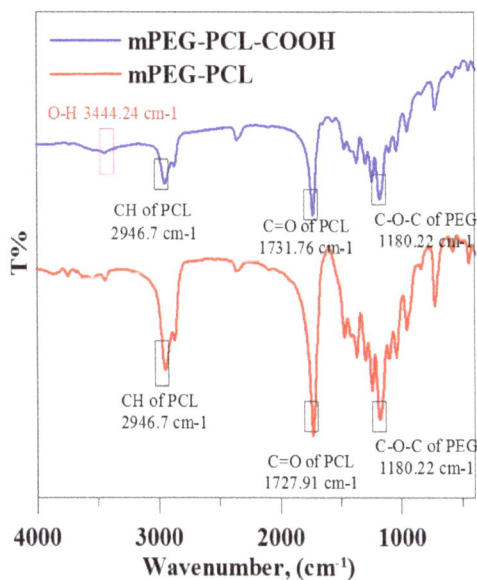

Figure 4. The FTIR spectra of mPEG-PCL and mPEG-PCL-COOH, respectively.

Table 1. M_n, M_w, and PDI (M_n/M_w) of mPEG-PCL determined by GPC and ^{1}H NMR.

Number average molecular weight (M_n) of NMR	12546.07 Da
Number average molecular weight (M_n) of GPC	7969.67 ± 289.0144 Da
Weight average molecular weight (M_w) of GPC	9514.33 ± 389.6977 Da
Polydispersity (PDI) of GPC	1.19 ± 0.0061

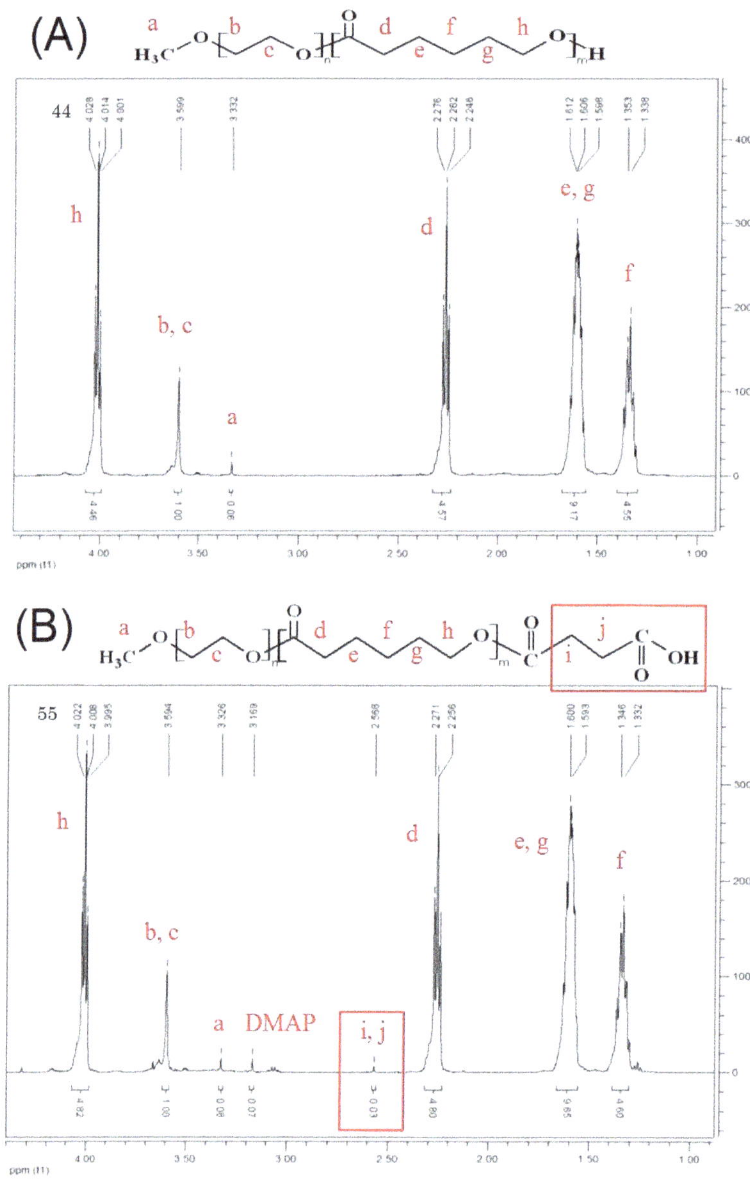

Figure 5. ^{1}H NMR spectrum of (**A**) mPEG-PCL and (**B**) mPEG-PCL-COOH.

3.2. The Inorganic Component in the Scaffolds

Specimens taken from random spots and weighing approximately 10 mg, were heated from 25 to 900 °C at 10 °C per min. From thermogravemetric analysis (TGA), it was observed that the onset of thermal degradation started at about 300 °C, and at 400 °C the entire amount of PCL had degraded. However, there was no thermal degradation observed in pure HAp, as heating was performed only up to 900 °C, Thermal stability of the HAp was evident from the TGA curve obtained up to 900 °C. In fact, the thermal degradation temperature of HAp is in the range between 1360 °C to 1400 °C [36].

The total weight loss of the HAp sample of about 2 wt % can be attributed to the loss of physically and chemically absorbed water and CO_2 elimination as a result of the decarbonation process in the range 400 °C to 900 °C. As seen in the TGA of the scaffolds, the weight percentage that was left in the pan indicated the composition of HAp, which was 33 wt %. This is a good indication that the physically blended biocomposite was homogeneously mixed.

3.3. Mechanical Properties and Degradation of Scaffolds

To investigate the influence of HAp incorporated into the scaffolds on the mechanical properties, compression testing was performed on three different structures (90ECR, 45ECR, and SECR) and HAp impregnation. The compressive strength of three different structures of scaffolds, adding HAp or not, is shown in Figure 6. The compressive strength of 90ECRH (with a lay-down patterns of 0°/90°) structure is the highest, which was similar to articular cartilage [37]. Therefore, 90ECRH was chosen for the following experiments. In the degradation test, scaffolds were immersed in PBS at 37 °C, shaking at 50 rpm to imitate the human articular environment for three months. It is well-known that proper degradation in a physiological environment is one of the most important characteristics of a scaffold in tissue engineering. The degradation behavior of the scaffolds was evaluated from the weight loss and morphological changes. The incorporation of time increases the degradation of the scaffolds. After 12 weeks of degradation, sample weight lost 85 wt %, compressive and Young's modulus decreased from 7.15 MPa to 2.47 MPa and from 120.92 MPa to 75.76 MPa, respectively. the half-life of the scaffold is over 60 days.

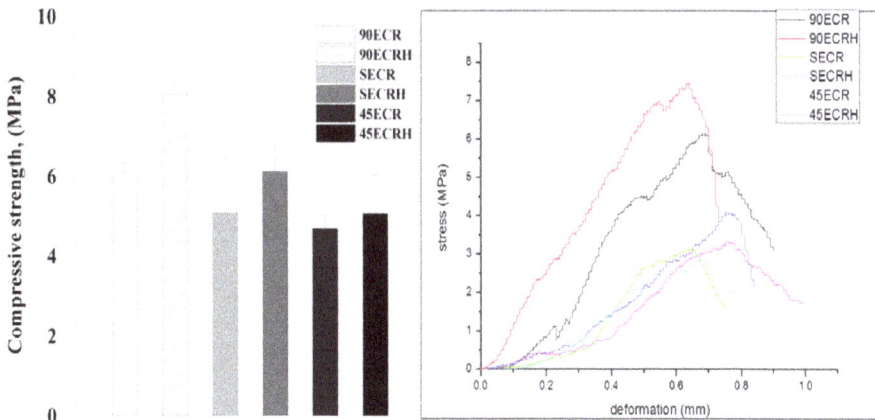

Figure 6. Compressive strength of each structure of scaffolds (three groups of samples were tested).

3.4. Scanning Electron Microscope Imaging

The SEM image of materials and scaffold before degradation and after degradation were scanned by Hitachi TM-1000. In the image of scaffolds before degradation, there is a smooth surface of fiber. After 12 weeks of immersion in PBS, scaffolds exhibited the rough surface with HAp exposure. The image of the scaffolds clearly illustrates the white dots on the fracture and erosion surface, and homogeneously distributed inside the scaffolds. According to the fracture and erosion surfaces, the scaffold underwent surface erosion.

3.5. Extraction and Identification of MSCs

In this study, all of the mesenchymal stem cells were isolated from four-weeks old BALB/c mice from BioLASCO. The MSCs positively expressed CD29, CD44, CD105, and Sca-1. Nevertheless, they were negative for CD3, CD19, CD11b, CD45, CD117, CD34, TER-119, CD86, H-2kb, and I-Ab [38].

We chose FITC-conjugated CD34 and PE-conjugated CD105 to confirm cell surface markers by BD FACSCalibur. Over 80% of the cells showed red fluorescence (CD105-PE) and only 2% of cells showed green fluorescence (CD34-FITC). Therefore, the cells were identified to be MSCs isolated from the femurs of BALB/c mice.

3.6. Cytotoxicity

Cytotoxicity of the scaffolds and synthesized gels were assessed by MTT assay. The cells have been seeded on the scaffold or on the gel to co-culture for three days. Figure 7A shows the proliferation behavior of MSCs on the surface of scaffolds after cells were seeded and cultured from day 1 to day 3 under standard conditions. Cell viability increased as culturing time increased. After three days culturing, the O.D. value increased one-fold in comparison to day 1. In Figure 7B the cell viability of GMHA gel is shown. There is a slight cytotoxicity in higher concentrations of GMHA; nevertheless, the O.D. value increased as the culturing time increased. Therefore, for the in vivo study, we chose 1% GMHA gel to encapsulate the growth factor.

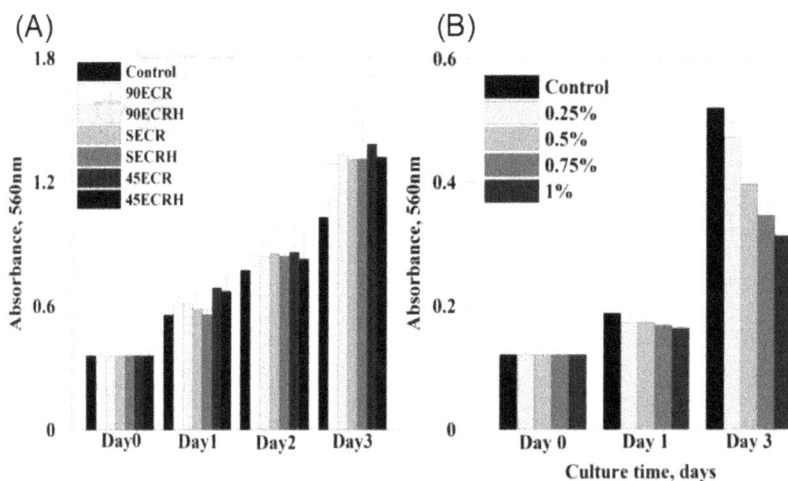

Figure 7. Cytotoxicity of (**A**) scaffolds and (**B**) GMHA by MTT assay. Error bars represent means with standard deviation.

3.7. Cell Adhesion and Cell Viability

The morphology and cytoskeleton organization of the cells on the constructs was analyzed after phalloidin staining of the F-actin. For cells adhesion on 2D slides coated with materials, RGD peptide grafting could enhance the formation of F-actin which appeared as a green fiber inside the cells. The DNA content of cells also showed higher quantification in comparison to un-grafted and control groups (Figure 8). In 3D culturing, obviously, cells increased significantly with the increase of the culture time on the constructs, and cell viability was good according to the LIVE/DEAD staining, as shown in Figure 9. The DNA content of cells showed the highest quantification on day 21.

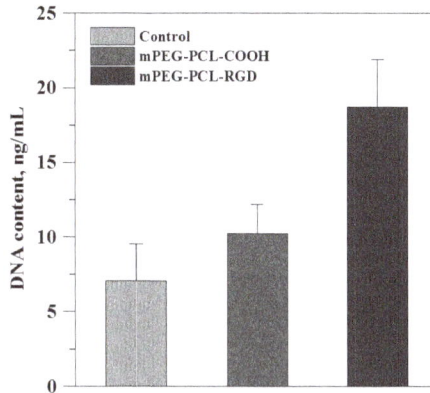

Figure 8. The DNA content of cell adhesion on 2D materials. Three days after cell culture, DNA detection showed that grafting RGD peptide in the material can effectively enhance cell attachment and proliferation.

Figure 9. The phalloidin staining of the F-actin and LIVE/DEAD staining of cells on 3D scaffolds after 7, 14, 21, and 28 days of culturing.

3.8. Gene Expression

To evaluate the chondrogenic and hypertrophic differentiation of MSCs, real-time polymerase chain reaction (qPCR) was performed for selecting chondrogenic markers (type II collagen, aggrecan, and Sox9). Gene expressions were normalized to GAPDH and to cells culturing with growth medium of selected chondrogenic markers. Figure 10 shows the relative gene expression of Col2a1 (Figure 10A), aggrecan (Figure 10B), and Sox9 (Figure 10C), respectively. MSCs seeded on scaffolds, culturing with chondrogenic differentiation medium showed higher aggrecan gene expression on day 28, higher Col2a1 gene expression on day 21, and higher Sox9 gene expression on day 14, respectively.

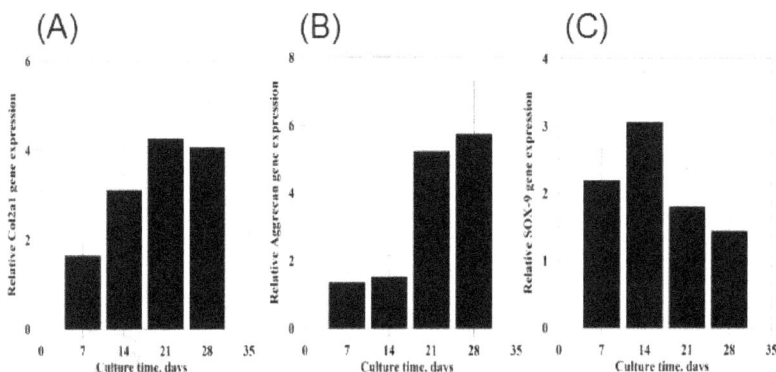

Figure 10. Chondrogenic differentiation of MSCs in aggregate culture following transduction. Relative expression of mRNA of (**A**) type II collagen; (**B**) aggrecan; and (**C**) Sox9 of MSCs cultured in vitro for up to 28 days. mRNA expression levels were normalized to the housekeeping gene GAPDH and control group.

3.9. sGAG Content

MSCs could differentiate into chondrocytes in the presence of transforming growth factor -beta and dexamethasone. In our study, chondrogenic differentiation of MSCs was induced by chondrogenic differentiation medium containing transforming growth factor-beta, dexamethasone, and 1% ITS + 1 supplement for four weeks. Sulfated glycosaminoglycan (sGAG) quantification analysis was performed by using the Blyscan™ Glycosaminoglycan Assay kit. After four weeks of culture, Figure 11 showed the content of sGAG normalized to the content of DNA, culturing in the chondrogenic differentiation medium has a higher level compared with that cultured in the growth medium. The sGAG/DNA ratio increased significantly as the culture time increased.

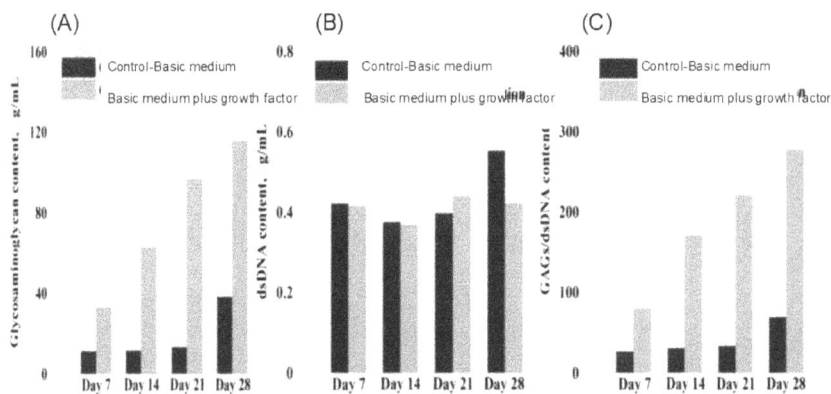

Figure 11. Biochemical analysis of MSCs differentiation. (**A**) sGAG contents of the MSCs after 7, 14, 21, and 28 days of differentiation; (**B**) DNA contents of the MSCs after 7, 14, 21, and 28 days of differentiation and (**C**) normalized s-GAG contents with respect to DNA contents.

3.10. Histological Analysis

In histological analysis, rabbit articular cartilage sections were stained with Alcian blue and periodic acid-schiff (PAS) double-staining with a Polyscience Alcian Blue/PAS kit. Alcian blue staining indicated sGAG accumulation, imparting a blue color to the acidic mucins and other carboxylated or weakly sulphated acid mucosubstances. Periodic Acid Schiff reaction was then used to stain the basement membranes, glycogen, and neutral mucosubstances pink to red. Mixtures of neutral and acidic mucosubstances will appear purple due to positive reactions with both Alcian Blue and PAS. Figure 12 shows sections stained by Alcian Blue and PAS double staining, after 12 weeks scaffolds and GMHA gel containing TGF-beta 1 were implanted. Articular cartilage and subchondral bone were healed in comparison to the control group, which showed only bone formation.

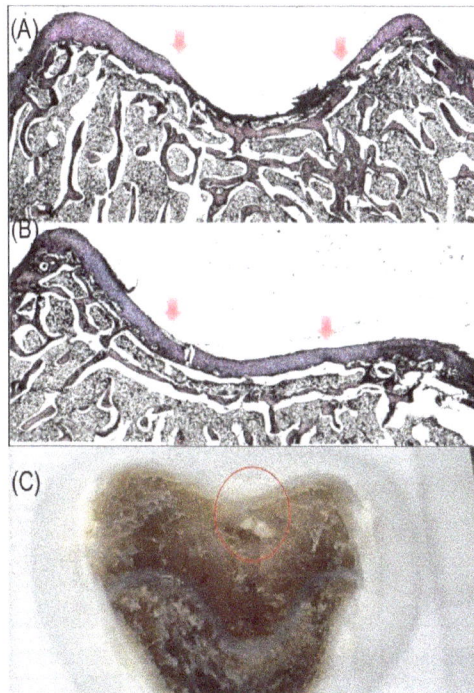

Figure 12. The histological staining (Alcian blue/PAS staining) and specimen of cartilage sections of rabbit after 12 weeks after implantation. (**A**) The control group with only bone formation; and (**B**) the experimental group showed bone and hyalin cartilage regeneration (arrow means defect); (**C**) The specimen of the experimental group showed good cartilage and bone regeneration on the surface layer and residual un-degraded scaffold in the deep layer.

4. Discussion

Tissue engineering and modelling of native articular cartilage in vivo requires an appropriate density and function of chondrocytes in a three-dimensional environment. Many studies have reported how one or more materials affect osteoarthritis, however, in those cases, their research should isolate primary cells from volunteers or patient-self to pre-culture with materials before implanting [3,39,40]. In our study, we proposed MSCs mixing with the implant by capillarity so that we do not need to pre-culture the chondrocytes with the implant before using. Therefore, scaffolds play an important role in directing 3D tissue regeneration [41]. According to Woodfield et al. [13] and

Wang et al. [7], traditional processing could not control the intra-structure of the scaffold. Using RP, it is possible to control the critical factors affecting tissue regeneration within the scaffold, like pore size and interconnection. Pore size of 300 μm is better for MSCs proliferation and chondrogenic differentiation [42], and a high interconnection of the scaffold could result in capillarity and homogeneous distribution of cells. Adding HAp to the prototyped scaffold could increase the mechanical properties. The array type of the filament could influence the structure properties, as well. In our experiments, compositions of polymers and HAp with a weight ratio 2:1, and laying down patterns $0°/90°$ significantly increased the compressive strength up to 30% (Figure 6). The compressive strength of the scaffold is 7.15 Mpa, which is consistent with the values reported by De Santis et al. [43] for rapid-prototyped hydroxyapatite-reinforced PCL/PEG scaffolds, ranging between 5.7 MPa and 13 MPa. The Young's modulus of our RP HAp-reinforced scaffold is 120.92 Mpa, also similar to the value reported by De Santis et al., ranging between 41 MPa and 125 MPa. The values are higher than those without HAp reinforcement, as reported by Wang et al. [31].

In in vitro studies, MSCs could adhere, proliferate, and differentiate on the scaffold biomanufactured through RP. TGF-beta will activate TGF-β receptor II on the cell surface and then bind with TGF-β receptor I, inducing TGF-β receptor II and TGF-β receptor I to combine in a heteromeric complex, activating the downstream Smad pathway [44,45]. In gene expression of chondrogenic differentiation, Sox (Sry-type high mobility group box) acts as a DNA-binding protein involved in the regulation of sex determination, bone development, and neurogenesis formation in the animal development process [46]. Sox9 is one of the transcription factors in the Sox family involved in the regulation of bone development. It has a transcriptional activation domain in the high-mobility-group-box DNA-binding domain, combining with a special sequence of minor DNA grooves. In the process of chondrogenic differentiation, Sox9 will bind chondrocyte-specific enhancers of Col2a1, activating the target gene of chondrogenic differentiation [47,48]. MSCs cultured with appropriate conditions, affecting growth factor TGF-beta, the Sox9 gene will express in the early stage of chondrogenic differentiation (Figure 10C) [49]. The Sox9 gene can also enhance aggrecan promoter activity [50], that is, aggrecan gene expression is slower than Sox9 gene expression (Figure 10B).

In in vivo studies, we reported the successful use of RP scaffolds infiltrated with GMAH gel containing TGF-beta for regeneration of articular cartilage defects in rabbits. We implanted RP scaffolds infiltrated with GMHA gel containing TGF-beta into osteochondral defects in rabbits, without exogenous cell seeding. These scaffolds could recruit endogenous bone marrow MSCs for chondrogenesis and facilitate hyaline cartilage regeneration. There are three types of cartilage (hyaline cartilage, fibrocartilage, and elastic cartilage) [51]. Compared to other types of cartilage, hyaline cartilage has fewer fibers in its matrix, and microscopy investigations show that this type of cartilage looks very smooth. After 12 weeks from implantation, sections were stained by Alcian blue/PAS, and we found that cartilage and subchondral bone were healed in the implanted group; nonetheless, we found no cartilage formation in the control group (Figure 12). RP scaffolds can promote bone healing as is known from our previous study [10]; nevertheless, this is our first time to promote cartilage regeneration. The new formation of cartilage is hyaline cartilage, and the fibrocartilage was composed of a fibrous layer, proliferative layer, hypertrophic layer, and mineralized layer. The hyaline cartilage had a homogeneous structure, and consisted simply of a cartilaginous matrix and chondrocytes [52].

Unlike other studies in cartilage regeneration [13,17,25,31], our study integrated required factors for cartilage growth. We used FDM to fabricate a biodegradable porous PCL scaffold, modified by RGD peptide grafting for cell adhesion, and light-curing HA gel containing TGF-beta 1 on the surface for cartilage regeneration. This biphasic scaffold can offer both mechanical support and a biological environment for bone and cartilage regeneration.

5. Conclusions

In our study, we showed that a rapid-prototyped scaffold infiltrated with UV light-cured hyaluronic acid hydrogel containing growth factor TGF-β1 could enhance the healing of the

Polymers **2017**, *9*, 182

osteochondral defect in the knees of rabbits. This approach showed good integration of scaffolds coupled to hyaline cartilage repair tissue without damaging the adjacent native articular cartilage. The osteochondral defect was filled with the bio-composite and biodegradable implant, blood with mesenchymal stem cells was added to the implant by capillarity, and then differentiated into chondrocytes on it. Compared to current treatments of osteoarthritis, our study proposed a new clinical option for a surgical treatment of osteoarthritis.

Acknowledgments: The authors would like to thank the Ministry of Science and Technology of the Republic of China under Contract No. 102-2632-M-033-001-MY3, MOST-105-2221-E-033-018, and MOST-104-2221-E-033-068 and Yeezen general hospital for financial support.

Author Contributions: Ming-Fa Hsieh and Hung-Maan Lee conceived and designed the experiments; Chih-Hsiang Fang and Cho-Pei Jiang performed the experiments; Yi-Ho Hsieh and Chih-Hsiang Fang analyzed the data; Yi-Ho Hsieh, Chih-Hsiang Fang and Ming-Fa Hsieh wrote the paper.

Conflicts of Interest: The authors declare no conflict of interest.

References

1. Matricali, G.A.; Dereymaeker, G.P.; Luyten, F.P. Donor site morbidity after articular cartilage repair procedures: A review. *Acta orthop. Belg.* **2010**, *76*, 669–674. [PubMed]
2. Darling, E.M.; Athanasiou, K.A. Rapid phenotypic changes in passaged articular chondrocyte subpopulations. *J. Orthop. Res.* **2005**, *23*, 425–432. [CrossRef] [PubMed]
3. Jiang, C.C.; Chiang, H.; Liao, C.J.; Lin, Y.J.; Kuo, T.F.; Shieh, C.S.; Huang, Y.Y.; Tuan, R.S. Repair of porcine articular cartilage defect with a biphasic osteochondral composite. *J. Orthop. Res.* **2007**, *25*, 1277–1290. [CrossRef] [PubMed]
4. Filardo, G.; Kon, E.; Perdisa, F.; Tetta, C.; Di Martino, A.; Marcacci, M. Arthroscopic mosaicplasty: Long-term outcome and joint degeneration progression. *Knee* **2015**, *22*, 36–40. [CrossRef] [PubMed]
5. Solheim, E.; Hegna, J.; Oyen, J.; Harlem, T.; Strand, T. Results at 10 to 14 years after osteochondral autografting (mosaicplasty) in articular cartilage defects in the knee. *Knee* **2013**, *20*, 287–290. [CrossRef] [PubMed]
6. Mou, Z.L.; Duan, L.M.; Qi, X.N.; Zhang, Z.Q. Preparation of silk fibroin/collagen/hydroxyapatite composite scaffold by particulate leaching method. *Mater. Lett.* **2013**, *105*, 189–191. [CrossRef]
7. Wang, X.; Yan, Y.; Zhang, R. Recent trends and challenges in complex organ manufacturing. *Tissue Eng. B Rev.* **2010**, *16*, 189–197. [CrossRef] [PubMed]
8. Hu, C.Z.; Tercero, C.; Ikeda, S.; Nakajima, M.; Tajima, H.; Shen, Y.J.; Fukuda, T.; Arai, F. Biodegradable porous sheet-like scaffolds for soft-tissue engineering using a combined particulate leaching of salt particles and magnetic sugar particles. *J. Biosci. Bioeng.* **2013**, *116*, 126–131. [CrossRef] [PubMed]
9. Jiang, C.P.; Chen, Y.Y.; Hsieh, M.F. Biofabrication and in vitro study of hydroxyapatite/mPEG-PCL-mPEG scaffolds for bone tissue engineering using air pressure-aided deposition technology. *Mater. Sci. Eng. C Mater. Boil. Appl.* **2013**, *33*, 680–690. [CrossRef] [PubMed]
10. Liao, H.T.; Chen, Y.Y.; Lai, Y.T.; Hsieh, M.F.; Jiang, C.P. The osteogenesis of bone marrow stem cells on mPEG-PCL-mPEG/hydroxyapatite composite scaffold via solid freeform fabrication. *BioMed Res. Int.* **2014**, *2014*, 321549. [CrossRef] [PubMed]
11. Kreklau, B.; Sittinger, M.; Mensing, M.B.; Voigt, C.; Berger, G.; Burmester, G.R.; Rahmanzadeh, R.; Gross, U. Tissue engineering of biphasic joint cartilage transplants. *Biomaterials* **1999**, *20*, 1743–1749. [CrossRef]
12. Schaefer, D.; Martin, I.; Shastri, P.; Padera, R.F.; Langer, R.; Freed, L.E.; Vunjak-Novakovic, G. In Vitro generation of osteochondral composites. *Biomaterials* **2000**, *21*, 2599–2606. [CrossRef]
13. Woodfield, T.B.; Guggenheim, M.; von Rechenberg, B.; Riesle, J.; van Blitterswijk, C.A.; Wedler, V. Rapid prototyping of anatomically shaped, tissue-engineered implants for restoring congruent articulating surfaces in small joints. *Cell Prolif.* **2009**, *42*, 485–497. [CrossRef] [PubMed]
14. Woodfield, T.B.; Malda, J.; de Wijn, J.; Peters, F.; Riesle, J.; van Blitterswijk, C.A. Design of porous scaffolds for cartilage tissue engineering using a three-dimensional fiber-deposition technique. *Biomaterials* **2004**, *25*, 4149–4461. [CrossRef] [PubMed]

15. Boere, K.W.; Visser, J.; Seyednejad, H.; Rahimian, S.; Gawlitta, D.; van Steenbergen, M.J.; Dhert, W.J.; Hennink, W.E.; Vermonden, T.; Malda, J. Covalent attachment of a three-dimensionally printed thermoplast to a gelatin hydrogel for mechanically enhanced cartilage constructs. *Acta Biomater.* **2014**, *10*, 2602–2611. [CrossRef] [PubMed]

16. Shao, X.X.; Hutmacher, D.W.; Ho, S.T.; Goh, J.C.; Lee, E.H. Evaluation of a hybrid scaffold/cell construct in repair of high-load-bearing osteochondral defects in rabbits. *Biomaterials* **2006**, *27*, 1071–1080. [CrossRef] [PubMed]

17. Gentile, P.; Ferreira, A.M.; Callaghan, J.T.; Miller, C.A.; Atkinson, J.; Freeman, C.; Hatton, P.V. Multilayer Nanoscale Encapsulation of Biofunctional Peptides to Enhance Bone Tissue Regeneration In Vivo. *Adv. Healthc. Mater.* **2017**, *6*. [CrossRef] [PubMed]

18. Gentile, P.; Ghione, C.; Tonda-Turo, C.; Kalaskar, D.M. Peptide functionalisation of nanocomposite polymer for bone tissue engineering using plasma surface polymerization. *RSC Adv.* **2015**, *5*, 80039–80047. [CrossRef]

19. Li, X.; Chen, S.; Li, J.; Wang, X.; Zhang, J.; Kawazoe, N.; Chen, G. 3D culture of chondrocytes in gelatin hydrogels with different stiffness. *Polymers* **2016**, *8*, 269. [CrossRef]

20. Iwasaki, N.; Kasahara, Y.; Yamane, S.; Igarashi, T.; Minami, A.; Nisimura, S.I. Chitosan-based hyaluronic acid hybrid polymer fibers as a scaffold biomaterial for cartilage tissue engineering. *Polymers* **2010**, *3*, 100–113. [CrossRef]

21. Engler, A.J.; Sen, S.; Sweeney, H.L.; Discher, D.E. Matrix elasticity directs stem cell lineage specification. *Cell* **2006**, *126*, 677–689. [CrossRef] [PubMed]

22. Temenoff, J.S.; Mikos, A.G. Injectable biodegradable materials for orthopedic tissue engineering. *Biomaterials* **2000**, *21*, 2405–2412. [CrossRef]

23. Guo, X.; Liao, J.; Park, H.; Saraf, A.; Raphael, R.M.; Tabata, Y.; Kasper, F.K.; Mikos, A.G. Effects of TGF-beta3 and preculture period of osteogenic cells on the chondrogenic differentiation of rabbit marrow mesenchymal stem cells encapsulated in a bilayered hydrogel composite. *Acta Biomater.* **2010**, *6*, 2920–2931. [CrossRef] [PubMed]

24. Guo, X.; Park, H.; Young, S.; Kretlow, J.D.; van den Beucken, J.J.; Baggett, L.S.; Tabata, Y.; Kasper, F.K.; Mikos, A.G.; Jansen, J.A. Repair of osteochondral defects with biodegradable hydrogel composites encapsulating marrow mesenchymal stem cells in a rabbit model. *Acta Biomater.* **2010**, *6*, 39–47. [CrossRef] [PubMed]

25. Schmidt, C.E.; Leach, J.B. Neural tissue engineering: Strategies for repair and regeneration. *Ann. Rev. Biomed. Eng.* **2003**, *5*, 293–347. [CrossRef] [PubMed]

26. Smeds, K.A.; Pfister-Serres, A.; Miki, D.; Dastgheib, K.; Inoue, M.; Hatchell, D.L.; Grinstaff, M.W. Photocrosslinkable polysaccharides for in situ hydrogel formation. *J. Biomed. Mater. Res.* **2001**, *54*, 115–121. [CrossRef]

27. Figueroa, D.; Espinosa, M.; Calvo, R.; Scheu, M.; Valderrama, J.J.; Gallegos, M.; Conget, P. Treatment of acute full-thickness chondral defects with high molecular weight hyaluronic acid; an experimental model. *Rev. Esp. Cir. Ortop. Traumatol.* **2014**, *58*, 261–266. [CrossRef] [PubMed]

28. Bian, L.; Zhai, D.Y.; Tous, E.; Rai, R.; Mauck, R.L.; Burdick, J.A. Enhanced MSC chondrogenesis following delivery of TGF-β3 from alginate microspheres within hyaluronic acid hydrogels in vitro and in vivo. *Biomaterials* **2011**, *32*, 6425–6434. [CrossRef] [PubMed]

29. Kim, I.L.; Mauck, R.L.; Burdick, J.A. Hydrogel design for cartilage tissue engineering: A case study with hyaluronic acid. *Biomaterials* **2011**, *32*, 8771–8782. [CrossRef] [PubMed]

30. Spiller, K.L.; Maher, S.A.; Lowman, A.M. Hydrogels for the repair of articular cartilage defects. *Tissue Eng. B Rev.* **2011**, *17*, 281–299. [CrossRef] [PubMed]

31. Wang, S.J.; Zhang, Z.Z.; Jiang, D.; Qi, Y.S.; Wang, H.J.; Zhang, J.Y.; Ding, J.X.; Yu, J.K. Thermogel-coated poly(ε-caprolactone) composite scaffold for enhanced cartilage tissue engineering. *Polymers* **2016**, *8*, 200. [CrossRef]

32. Anselme, K. Osteoblast adhesion on biomaterials. *Biomaterials* **2000**, *21*, 667–681. [CrossRef]

33. ASTM F1635. *Standard Test Method for In Vitro Degradation Testing of Hydrolytically Degradable Polymer Resins and Fabricated Forms for Surgical Implants*; ASTM: West Conshohocken, PA, USA, 2016.

34. Kweon, H.; Yoo, M.K.; Park, I.K.; Kim, T.H.; Lee, H.C.; Lee, H.S.; Oh, J.S.; Akaike, T.; Cho, C.S. A novel degradable polycaprolactone networks for tissue engineering. *Biomaterials* **2003**, *24*, 801–808. [CrossRef]

35. Choi, C.Y.; Chae, S.Y.; Kim, T.H.; Jang, M.K.; Cho, C.S.; Nah, J.W. Preparation and characterizations of poly(ethylene glycol)-poly(epsilon-caprolactone) block copolymer nanoparticles. *Bull. Korean Chem. Soc.* **2005**, *26*, 523–528.

36. Liao, C.J.; Lin, F.H.; Chen, K.S.; Sun, J.S. Thermal decomposition and reconstitution of hydroxyapatite in air atmosphere. *Biomaterials* **1999**, *20*, 1807–1813. [CrossRef]

37. Kerin, A.J.; Wisnom, M.R.; Adams, M.A. The compressive strength of articular cartilage. *Proc. Inst. Mech. Eng. H* **1998**, *212*, 273–280. [CrossRef] [PubMed]

38. Sung, J.H.; Yang, H.M.; Park, J.B.; Choi, G.S.; Joh, J.W.; Kwon, C.H.; Chun, J.M.; Lee, S.K.; Kim, S.J. Isolation and characterization of mouse mesenchymal stem cells. *Transplant. Proc.* **2008**, *40*, 2649–2654. [CrossRef] [PubMed]

39. Ko, J.Y.; Kim, K.I.; Park, S.; Im, G.I. In Vitro chondrogenesis and in vivo repair of osteochondral defect with human induced pluripotent stem cells. *Biomaterials* **2014**, *35*, 3571–3581. [CrossRef] [PubMed]

40. Lee, J.M.; Im, G.I. SOX trio-co-transduced adipose stem cells in fibrin gel to enhance cartilage repair and delay the progression of osteoarthritis in the rat. *Biomaterials* **2012**, *33*, 2016–2024. [CrossRef] [PubMed]

41. Raghunath, J.; Rollo, J.; Sales, K.M.; Butler, P.E.; Seifalian, A.M. Biomaterials and scaffold design: Key to tissue-engineering cartilage. *Biotechnol. Appl. Biochem.* **2007**, *46*, 73–84. [PubMed]

42. Matsiko, A.; Gleeson, J.P.; O'Brien, F.J. Scaffold mean pore size influences mesenchymal stem cell chondrogenic differentiation and matrix deposition. *Tissue Eng. A* **2015**, *21*, 486–497. [CrossRef] [PubMed]

43. De Santis, R.; D'amora, U.; Russo, T.; Ronca, A.; Gloria, A.; Ambrosio, L. 3D fibre deposition and stereolithography techniques for the design of multifunctional nanocomposite magnetic scaffolds. *J. Mater. Sci. Mater. Med.* **2015**, *26*, 250. [CrossRef] [PubMed]

44. Derynck, R.; Feng, X.H. TGF-beta receptor signaling. *Biochim. Biophys. Acta* **1997**, *1333*, F105–F150. [PubMed]

45. Massague, J. TGF-beta signal transduction. *Ann. Rev. Biochem.* **1998**, *67*, 753–791. [CrossRef] [PubMed]

46. Wright, E.M.; Snopek, B.; Koopman, P. Seven new members of the Sox gene family expressed during mouse development. *Nucleic Acids Res.* **1993**, *21*, 744. [CrossRef] [PubMed]

47. Ikeda, T.; Kawaguchi, H.; Kamekura, S.; Ogata, N.; Mori, Y.; Nakamura, K.; Ikegawa, S.; Chung, U. Distinct roles of Sox5, Sox6, and Sox9 in different stages of chondrogenic differentiation. *J. Bone Min. Metab.* **2005**, *23*, 337–340. [CrossRef] [PubMed]

48. Bi, W.; Deng, J.M.; Zhang, Z.; Behringer, R.R.; de Crombrugghe, B. Sox9 is required for cartilage formation. *Nat. Genet.* **1999**, *22*, 85–89. [PubMed]

49. Kulyk, W.M.; Franklin, J.L.; Hoffman, L.M. Sox9 expression during chondrogenesis in micromass cultures of embryonic limb mesenchyme. *Exp. Cell Res.* **2000**, *255*, 327–332. [CrossRef] [PubMed]

50. Sekiya, I.; Tsuji, K.; Koopman, P.; Watanabe, H.; Yamada, Y.; Shinomiya, K.; Nifuji, A.; Noda, M. SOX9 enhances aggrecan gene promoter/enhancer activity and is up-regulated by retinoic acid in a cartilage-derived cell line, TC6. *J. Biol. Chem.* **2000**, *275*, 10738–10744. [CrossRef] [PubMed]

51. Holden, P.K.; Li, C.; Da Costa, V.; Sun, C.H.; Bryant, S.V.; Gardiner, D.M.; Wong, B.J.F. The effects of laser irradiation of cartilage on chondrocyte gene expression and the collagen matrix. *Lasers Surg. Med.* **2009**, *41*, 487–491. [CrossRef] [PubMed]

52. Huang, L.; Li, M.; Li, H.; Yang, C.; Cai, X. Study of differential properties of fibrochondrocytes and hyaline chondrocytes in growing rabbits. *Br. J. Oral Maxillofac. Surg.* **2015**, *53*, 187–193. [CrossRef] [PubMed]

polymers

MDPI

Article

In Vitro Biocompatibility, Radiopacity, and Physical Property Tests of Nano-Fe$_3$O$_4$ Incorporated Poly-L-lactide Bone Screws

Hsin-Ta Wang [1,†], Pao-Chang Chiang [2,†], Jy-Jiunn Tzeng [3], Ting-Lin Wu [3], Yu-Hwa Pan [4,5], Wei-Jen Chang [6,7,*] and Haw-Ming Huang [6,8,9,*]

[1]   School of Organic and Polymeric, National Taipei University of Technology, Taipei 10608, Taiwan; htwang@mail.ntut.edu.tw
[2]   Dental Department, Wan Fang Hospital, Taipei Medical University, Taipei 11696, Taiwan; jumbo117@gmail.com
[3]   Graduate Institute of Biomedical Materials and Tissue Engineering, College of Biomedical Engineering, Taipei Medical University, Taipei 11031, Taiwan; fungus0429@yahoo.com.tw (J.-J.T.); goku273031@hotmail.com (T.-L.W.)
[4]   Department of General Dentistry, Chang Gung Memorial Hospital, Taipei 10507, Taiwan; shalom.dc@msa.hinet.net
[5]   Chang Gung University, Taoyuan 33371, Taiwan
[6]   School of Dentistry, College of Oral Medicine, Taipei Medical University, Taipei 11031, Taiwan
[7]   Dental Department, Taipei Medical University Shuang-Ho Hospital, New Taipei City 23561, Taiwan
[8]   Graduate Institute of Biomedical Optomechatronics, College of Biomedical Engineering, Taipei 11031, Taiwan
[9]   Ph.D Program in Biotechnology Research and Development, College of Pharmacy, Taipei Medical University, Taipei 11031, Taiwan
*   Correspondence: cweijen1@tmu.edu.tw (W.-J.C.); hhm@tmu.edu.tw (H.-M.H.); Tel.: +886-291-937-9783 (W.-J.C & H.-M.H.)
†   Hsin-Ta Wang and Pao-Chang Chiang contributed equally to this work.

Academic Editors: Insung S. Choi and João F. Mano
Received: 18 April 2017; Accepted: 24 May 2017; Published: 26 May 2017

Abstract: The aim of this study was to fabricate biodegradable poly-L-lactic acid (PLLA) bone screws containing iron oxide (Fe$_3$O$_4$) nanoparticles, which are radiopaque and 3D-printable. The PLLA composites were fabricated by loading 20%, 30%, and 40% Fe$_3$O$_4$ nanoparticles into the PLLA. The physical properties, including elastic modulus, thermal properties, and biocompatibility of the composites were tested. The 20% nano-Fe$_3$O$_4$/PLLA composite was used as the material for fabricating the 3D-printed bone screws. The mechanical performance of the nano-Fe$_3$O$_4$/PLLA bone screws was evaluated by anti-bending and anti-torque strength tests. The tissue response and radiopacity of the nano-Fe$_3$O$_4$/PLLA bone screws were assessed by histologic and CT imaging studies using an animal model. The addition of nano-Fe$_3$O$_4$ increased the crystallization of the PLLA composites. Furthermore, the 20% nano-Fe$_3$O$_4$/PLLA composite exhibited the highest thermal stability compared to the other Fe$_3$O$_4$ proportions. The 3D-printed bone screws using the 20% nano-Fe$_3$O$_4$/PLLA composite provided excellent local tissue response. In addition, the radiopacity of the 20% nano-Fe$_3$O$_4$/PLLA screw was significantly better compared with the neat PLLA screw.

Keywords: radiopaque polymer; 3D printed bone screw; iron oxide nanoparticles; poly-L-lactic acid

1. Introduction

Metallic bone screws are widely used in the healing of bony defects. Nonetheless, their use requires a second surgical procedure to remove them after healing [1,2]. Additionally, the high elastic modulus of metallic implants results in stress shielding effect, which leads to a decrease in bone quality and delayed bone healing [3]. Recently, bone screws fabricated with biodegradable polymers were introduced to overcome these problems. Among these biodegradable polymers, polylactic acid (PLA) is commonly used in orthopedic devices, dental rehabilitation, drug delivery, and injectable tissue engineering [4]. One of the advantages of using PLA to fabricate bone implants is that it is easy to shape the bone graft to fit the defect.

Recently, additive manufacturing technologies have been used to manufacture customized scaffolds. Among these technologies, fused deposition modelling (FDM) has been extensively applied for the fabrication of three-dimensional (3D) bio-scaffolds in the biomedical field [5–9]. Previous reports indicate that PLA can be used as an FDM printing material for bone tissue engineering [3,10]. However, using PLA for fabricating bone implanta has two disadvantages. First, due to the low mechanical properties, the application of PLA must be limited only at non-weight bearing sites. Second, due to its low specific gravity and electron density, PLA cannot be detected by X-ray radiography [11]. This drawback makes evaluating the degradation and position of PLA devices during placement and healing difficult.

To overcome these problems, researchers began incorporating inorganic materials into the PLA to make composite materials. Fe_3O_4 nanoparticles exhibited excellent osteogenic effects [12–14]. In 2011, De Santis et al. incorporated iron oxide (Fe_3O_4) nanoparticles into poly(ε-caprolactone) (PCL) matrix to fabricate magnetic scaffolds. They found that the addition of the Fe_3O_4 nanoparticles increased the elastic modulus of the PCL by 10% [8]. Recently, several researchers added Fe_3O_4 nanoparticles into PLA to fabricate multifunctional biomaterials [15–17]. Although Fe_3O_4 nanoparticles show strong contrast in magnetic resonance imaging (MRI) studies [18], whether or not Fe_3O_4 nanoparticle–PLA composites demonstrate radiopacity is not well studied. In 2015, Huang's group developed an X-ray imaging-enhanced PLA bone screws using Fe_3O_4 nanoparticle–PLA composites [19]. Their results showed that nano-Fe_3O_4/PLLA screws provided better osteogenic properties compared to that of neat PLLA screws. In addition, this material can be used as a 3D printing material. However, the in vitro biocompatibility and thermal properties of the material were not discussed in their report.

The aim of this study was to fabricate Fe_3O_4 nanoparticle-PLA bone screws using FDM technology. The local tissue response and in vitro calorimetric and mechanical properties of the fabricated screws were evaluated.

2. Materials and Methods

Before the fabrication procedure of the nano-Fe_3O_4/PLLA composite began, the Fe_3O_4 nanoparticles (99.9%, 50 nm, Long Ton, Inc., Taipei, Taiwan) and PLLA powder (molecular weight of 100 kDa, Wei Mon Industry Co., Taipei, Taiwan) were dried at 80 °C overnight. Then, a twin-screw extruder was used to mix the two materials at 150 °C. The extruded nano-Fe_3O_4/PLLA was cooled in a water bath at 25 °C. A pelletizer was used to cut the nano-Fe_3O_4/PLLA composite into small granules. Four nano-Fe_3O_4/PLLA composites containing 0, 20%, 30% and 40% (*w/w*) Fe_3O_4 were manufactured to test for the optimal nano-Fe_3O_4/PLLA ratio for fabricating the bone screws. Three nano-Fe_3O_4/PLLA samples were prepared by injection molding at an injection temperature of 210 °C to test the thermal and biological properties of the composite material. For cell culture tests, nano-Fe_3O_4/PLLA discs with a diameter of 1 cm were fabricated. In addition, 1.65-mm nano-Fe_3O_4/PLLA rods 20 cm in length were also produced to serve as a feeding material for FDM printing.

2.1. Thermal Properties of the Fe_3O_4/PLLA Composites

The thermal properties of the nano-Fe_3O_4/PLLA composites were analyzed using differential scanning calorimetry (DSC) (Q100, TA Instruments, Inc., California, CA, USA). A dried sample (5 mg) under nitrogen flow (50 mL/min) was used. During DSC tests, the samples were heated from 25 to 200 °C at a rate of 10 °C/min. Then, the samples were maintained at 200 °C for 2 min.

During the cooling stage, the samples were cooled from 200 to 30 °C at 10 °C/min and remained at this temperature for 3 min. The heating–cooling scans were performed twice. To reduce the error due to thermal resistance between the composite and the bottom of the DSC crucible and irregular molecular structure in the composite, the glass transition temperature (T_g), cold crystallization temperature (T_c), and melting temperature (T_m) of the samples were determined from the second scan. The thermal stability of the samples was detected using a thermogravimeter (TGA, TG 209 F3 Tarsus, Netzsch, Gerätebau GmbH, Bavarian, Germany). During the tests, 3-mg samples were heated from room temperature to 900 °C at a rate of 20 °C/min. The thermal decomposition temperatures (T_{onset}) and the degradation peak temperature (T_{peak}) were reduced.

2.2. In Vitro Biocompatibility Tests of the Fe₃O₄/PLLA Composites

2.2. In Vitro Biocompatibility Tests of the Fe_3O_4/PLLA Composites

To test the in vitro biocompatability of Fe_3O_4/PLLA composites, MG63 osteoblast-like cells were cultured on injection molded Fe_3O_4/PLLA discs that were placed in 24-well culture plates at a density of 1×10^4 cells/well. The cells were cultured in Dulbecco's Modified Eagle's Medium (DMEM; HyClone, Logan, UT, USA) with with 10% fetal bovine serum, 4 mM L-glutamine, and 1% penicillin–streptomycin. The discs were cultured in 5% CO_2 at 37 °C and 100% humidity for 1–4 days. The cell viability was determined by staining with MTT (3-[4,5-dimethyl-thiazol-2-yl]-2,5-diphenyltetrazolium bromide) reagent (MTT kit; Roche Applied Science, Mannheim, Germany). The optical absorbance measured at the wavelengths of 570/690 nm in a microplate reader (Model 2020, Anthos Labtec Instruments, Eugendorf, Wals, Austria) was used to determine the cell viability.

2.3. Fabrication of Nano-Fe_3O_4/PLLA Bone Screws

The screws designed and 3D printed in this study were 3.1 mm in diameter and 12 mm in length with a self-tapping tip. The thread width was 0.3 mm. The geometry of the screw model was established using commercial software (Solidworks, Inc., Waltham, MA, USA). The meshed model was imported into 3D printing software (ReplicatorG 0037, Makerbot, NY, USA) and transferred to the G-code format (Skenforge, Makerbot, NY, USA). As shown in Figure 1, a fused deposition modeling (FDM) machine (Born One, Wanwall, Honk Kong, China) was used to fabricate nano-Fe_3O_4/PLLA bone screws. The printing thickness was 0.15 mm for each layer. The injection molded nano-Fe_3O_4/PLLA rods were fed into the extraction nozzle at an extrusion temperature of 185 °C. During the physical property tests, we found that the nano-Fe_3O_4/PLLA composite with mix ratios greater than 20% significantly decreased the thermal stability of the PLLA. Thus, only 20% nano-Fe_3O_4/PLLA bone screws were fabricated and used in the following experiments. For all the tests, neat PLLA screws were used as the control group.

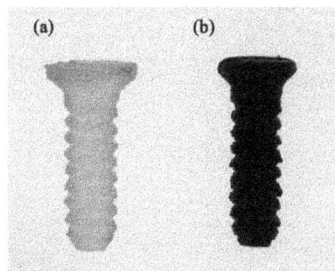

Figure 1. FDM fabricated bone screws using (**a**) neat PLLA and (**b**) 20% nano-Fe_3O_4/PLLA composites.

2.4. Mechanical Tests of the Bone Screws

To test the mechanical strength of the FDM-made nano-Fe_3O_4/PLLA bone screws, the ultimate anti-torque strength and bending strength of the screws were evaluated. According to previous studies [20,21], solid-rigid polyurethane form bone blocks (Sawbones Pacific Research Laboratories Inc., Washington, WA, USA) were used as simulated bone. The simulated bone blocks were cut into cubic units (1 cm × 2 cm × 2 cm). For each bone block specimen, pilot holes (2.5 mm in diameter) were prepared using a drill with the same diameter as the bone screw. After cavity preparation, the fabricated bone screws were screwed into the simulated bone blocks. Before testing, the samples were clamped onto a clamp stand. The peak anti-torque strength of the screws was averaged from the measurements of five samples using a digital torque meter (TQ-8800, Lulton Electronic Enterprise, Taipei, Taiwan).

For the bending strength test, the screw/simulated bone samples were mounted on a custom-designed holding stage and rigidly fixed on the base plate of the universal testing machine (AGS-1000D, Shimadzu, Tokyo, Japan). As shown in Figure 2, the test screws were 90° off the axis to produce a bending force on the samples. A vertical pre-load of 0.5 mm was directly applied to the screw by a force head before testing. Then, the force was applied downward at a rate of 0.05 mm/s until the samples fractured. Five specimens were tested, and for each, the maximum applied load was recorded. The average value of the maximum applied load causing each tested sample to fail was defined as the "bending strength". For all the mechanical tests, neat PLLA screws served as controls.

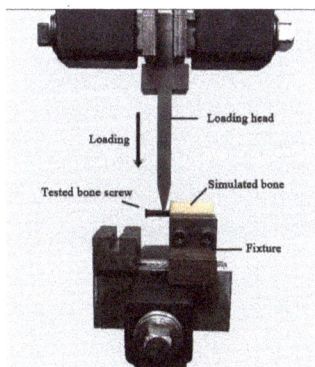

Figure 2. Schematic setup of the bending strength test.

2.5. Animal Experiments

The Institutional Animal Care and Use Committee approved the protocols for all of the animal testing performed in this study (LAC-2014-0123). Three New Zealand white rabbits aged eight months and weighing 3.0–3.5 kg were used as subjects. For each rabbit, one 20% nano-Fe_3O_4/PLLA bone screw was implanted in the left femoral condyle while a neat PLLA screw (control) was implanted in the right leg of the rabbits. Before the bone screw operations, the animal received intramuscular injection (15 mg/kg tiletamine-zolazepam) (Zoletil 50, Virbac, Carros Cedex, France) for general anesthesia. In addition, 2% epinephrine (1.8 mL) was injected at each femoral condyle for local anesthesia. After the skin incision and muscle dissection, a drilled cavity (2.5 mm in diameter) was prepared at both implant sites.

After the bone screws had been implanted, the muscle and skin were closed with absorbable sutures (Vicryl® 4.0, Ethicon, Somerville, NJ, USA). To control pain and reduce the risk of infection, antibiotics were intramuscularly administered for three consecutive days. The animals were killed four weeks post-surgery using a humane method.

The femoral condyles with the bone screws were excised and fixed in 10% formalin. Micro-computed tomographic (micro-CT) images of the bone samples were taken at an energy level of 55 kV at 181 A (Skyscan 1076, Skyscan, Antwerp, Belgium) to determine the radiopacity of the nano-Fe_3O_4/PLLA screws. After micro-CT examination, histomorphometric observation was performed using the same bone blocks. Demineralization was done according to the method of Ayukawa et al. 1998. The bone blocks were cut with a diamond blade saw into 2 mm slices [22]. After dehydration, the slices were embedded in paraffin wax, and 10 μm sections were prepared using an ultramicrotome (Bright 5040, Bright Instrument, Cambs, UK). For histological examination, the specimens were stained with hematoxylin and eosin and observed using a light microscope (CH2, Nikon, Tokyo, Japan) equipped with a digital camera (Coolpix 950, Nikon, Tokyo, Japan) [23].

2.6. Statistical Analysis

Differences in elastic modulus and cell viability between the nano-Fe_3O_4/PLLA composites with various mix ratios were tested using one-way analysis of variance. For the mechanical tests of the fabricated bone screws, Student's *t*-test was used to examine the differences in anti-torque strength and bending strength between the neat PLLA screws and the 20% nano-Fe_3O_4/PLLA screws. Probability values less than 0.05 were considered significant.

3. Results

Figure 3 shows the DSC thermograms recorded for the nano-Fe_3O_4/PLLA composites. The thermograms of the samples showed no significant glass transition. Table 1 shows that the crystallization temperature (T_c) of the PLLA composites mixed with various amounts of nano-Fe_3O_4 particles (103–109 °C) was lower than that of neat PLA (117.1 °C). That is, the nano-Fe_3O_4/PLLA composite started to crystallize before the neat PLLA. Additionally, thermograms of the neat PLLA display marked double melting peaks at 162.3 and 167.9 °C for the first and second melting peaks, respectively. As the nano-Fe_3O_4/PLLA mix ratio increased, the double melting peak changed to a single melting peak. The high-temperature melting peak of nano-Fe_3O_4/PLLA composites did not shift with increasing mix ratio.

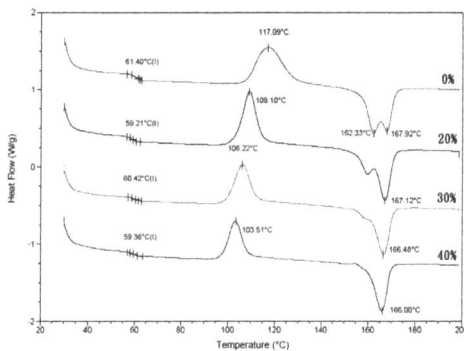

Figure 3. DSC thermograms of the four nano-Fe_3O_4/PLLA composites with various nano-Fe_3O_4/PLLA ratios.

Table 1. Thermal properties of nano-Fe_3O_4/PLLA composites.

Fe_3O_4 (%)	T_g	T_c	T_m	T_{onset}	T_{peak}
0%	61.4	117.1	167.9	309.0	363.9
20%	59.2	109.1	167.1	317.7	338.2
30%	60.4	106.2	166.5	314.6	325.0
40%	59.4	103.5	166	311.9	320.4

Figure 4 illustrates the thermography of the nano-Fe$_3$O$_4$/PLLA composites. The addition of 20% nano-Fe$_3$O$_4$ particles to the PLLA improved the thermal stability of the polymer (Table 1). The initial decomposition temperature (T_{onset}) of the neat PLLA was 309.0 °C, which rose to 317.7 °C when 20% nano-Fe$_3$O$_4$ particles were added. However, it decreased to 314.6 and 311.9 °C when the mix ratios were 30% and 40%, respectively. The presence of nano-Fe$_3$O$_4$ particles reduced the degradation peak temperature (T_{peak}). For the neat PLLA, the T_{peak} was recorded as 363.9 °C. T_{peak} was 338.2 °C for the 20% nano-Fe$_3$O$_4$/PLLA composite, 325.0 °C for the 30% composite, and 320.4 °C for the 40% nano-Fe$_3$O$_4$/PLLA composite.

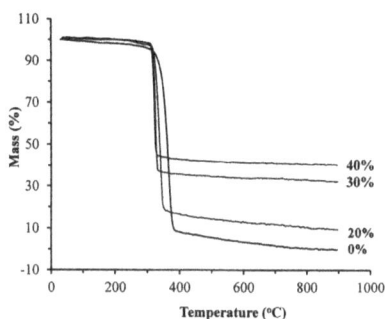

Figure 4. TGA patterns of the four nano-Fe$_3$O$_4$/PLLA composites with various nano-Fe$_3$O$_4$/PLLA ratios.

MG63 osteoblast-like cells were cultured on the Fe$_3$O$_4$/PLLA composites for four days to test the biocompatibility of the Fe$_3$O$_4$/PLLA composites. Figure 5 illustrates the growth of the MG63 cells seeded on the nano-Fe$_3$O$_4$/PLLA composites. The cells demonstrated normal growth curves on all the composites. The cell numbers were increased in both the Fe$_3$O$_4$/PLLA composite groups and the neat PLLA group throughout the entire culture period. When comparing cells grown on the nano-Fe$_3$O$_4$/PLLA composites with the various nano-Fe$_3$O$_4$/PLLA ratios, no statistically significant differences in cell numbers were observed for any counting interval over the four-day period.

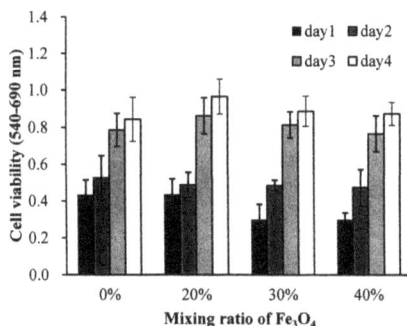

Figure 5. Cell growth assay for MG63 cells seeded onto the nano-Fe$_3$O$_4$/PLLA composites and incubated for four days.

In Figure 6a, the bending strength of the neat PLLA bone screw is 20.6 $\pm$ 5.9 N. The addition of 20% nano-Fe$_3$O$_4$ to the PLLA had no significant effect on the bending strength of the screw. A similar result was obtained in the anti-torque strength experiment. The maximum anti-torque strength of the neat PLLA was 5.9 $\pm$ 1.0 N-cm (Figure 6b). The statistical analysis showed no difference between the neat PLLA and 20% nano-Fe$_3$O$_4$/PLLA screws.

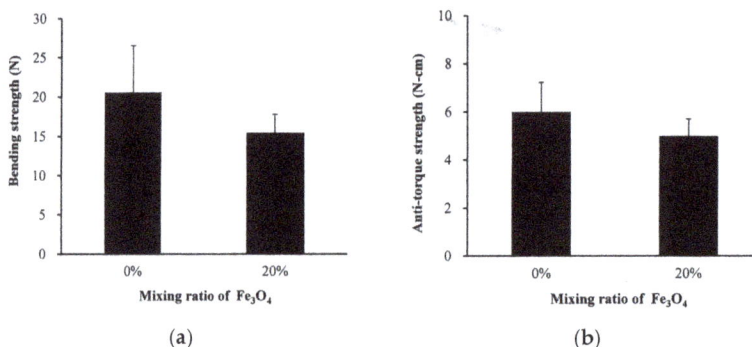

(a)   (b)

Figure 6. Comparison of (**a**) bending strength and (**b**) maximum anti-torque strength between bone screws manufactured with neat PLLA and 20% nano-Fe$_3$O$_4$/PLLA composite.

When the bone screws were placed in the rabbits for four weeks, new bone formation at the implant/bone interface was observed both in nano-Fe$_3$O$_4$/PLLA group (Figure 7a) and neat PLLA group (Figure 7b). Furthermore, along with the degradation process, bone tissue grew into the nano-Fe$_3$O$_4$/PLLA screw at the bone interface between threads. Leached nano-Fe$_3$O$_4$/PLLA debris was surrounded by bone tissue without an observable inflammatory response or side effect (Figure 7a). Figure 8 shows typical micro-CT images of the implanted bone screws. The neat PLLA screws showed no radiopacity and could not be distinguished from the surrounding tissue (Figure 8a). However, the addition of 20% nano-Fe$_3$O$_4$ particles significantly improved the radiopacity of the PLLA screw. The boundary and location of the 20% nano-Fe$_3$O$_4$/PLLA screw were easily identified without any blooming artifact (Figure 8b).

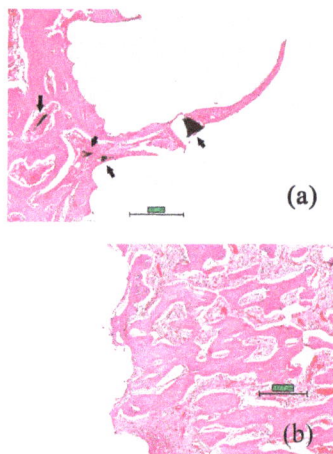

Figure 7. Histology of the bone tissue at the screw/bone interface for (**a**) 20% nano-Fe$_3$O$_4$/PLLA group and (**b**) neat PLLA group after four weeks of healing. The black arrow indicates debris leached from the nano-Fe$_3$O$_4$/PLLA composite, which is surrounded by bone tissue. There is no observable local toxicity in the bone tissue. Black arrows indicate the leached nano-Fe$_3$O$_4$/PLLA debris was surrounded by bone tissue. Scale bar: (**a**) 0.3 mm (**b**) 0.5 mm.

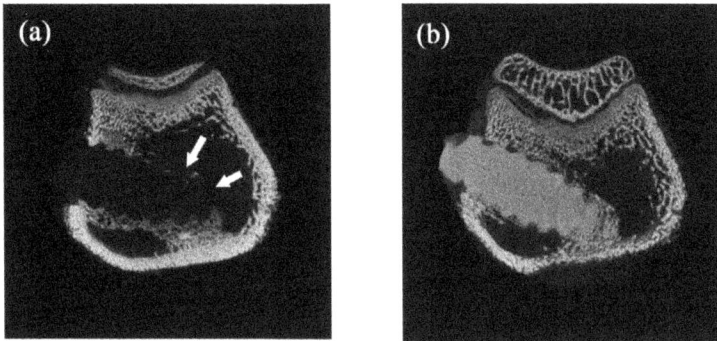

Figure 8. Typical micro-CT images of (**a**) neat PLLA screw and (**b**) 20% nano-Fe$_3$O$_4$/PLLA screw implanted in rabbit bone. The addition of 20% Fe$_3$O$_4$ nanoparticles significantly improved the radiopacity of the PLLA screws.

4. Discussion

We applied an FDM method for 3D printing of nano-Fe$_3$O$_4$/PLLA composite bone screws. The screws were produced by extruding the PLLA composite via a heated extrusion nozzle. Thus, the nano-Fe$_3$O$_4$/PLLA used in this study was a thermoplastic material with the appropriate viscosity when melted so that it could be extruded from the extrusion nozzle. FDM technology may be used to print poly(ε-caprolactone)/Fe$_3$O$_4$ [8] and neat PLLA [3] devices directly for tissue engineering. Nonetheless, whether the addition of nano-Fe$_3$O$_4$ particles to the PLLA affects their viscosity and 3D printability has not been systematically investigated. Although body temperature is only 36–37 °C, the material may be subjected to a temperature of 185 °C when used for FDM fabrication. Thus the thermal stability of the nano-Fe$_3$O$_4$/PLLA composite at such a high temperature was tested in this study. The glass transition temperature is related to the physical properties of the PLLA. The DSC thermograms of the samples (Figure 3) showed no significant shift in glass temperature transition. Thus, the fabricated nano-Fe$_3$O$_4$/PLLA composites may be used to manufacture the bone screws using the FDM method (Figure 1).

The crystallization temperature (T_c) was affected by the addition of nano-Fe$_3$O$_4$ particles. It is well known that a lower crystallization temperature results in faster crystallization [24]. Thus, the addition of nano-Fe$_3$O$_4$ particles into the PLLA reduced the energy required to achieve crystallinity. The melting behavior of PLLA is complex. As shown in Figure 3, the neat PLLA exhibited double melting peaks. This is because PLLA is a semi-crystalline polymer made up of different types of crystals [24,25]. Additionally, the presence of the filler significantly modified the overall crystallinity. The change in the ratio of the first and the second melting peaks of the samples indicates that the nano-Fe$_3$O$_4$ particles influence the size of the crystals. The low-temperature melting peak of the PLLA composites increased with the increasing nano-Fe$_3$O$_4$/PLLA ratio. It is because the increase in nano-Fe$_3$O$_4$ makes the crystallization of PLLA more complete, and the number of imperfect crystals decreases [26]. Overall, the changes of T_c and T_m (Figure 3 and Table 1) indicate that the nano-Fe$_3$O$_4$ particles act as a nucleating agent to enhance the crystallization rate [24,27]. This improves the perfection of the crystals in the PLLA [24,27].

Researchers have added various contrast agents to enhance the radiopacity of polymeric devices. Most investigators reported that these compounds reduced the thermal stability of the polymer [11,28,29]. However, our results showed a different trend. The addition to Fe$_3$O$_4$ nanoparticles improving not only the radiopacity of the PLLA (Figure 8b) but also increasing the thermal stability of the polymer. Figure 4 and Table 1 show that the highest decomposition temperatures occurred when 20% Fe$_3$O$_4$ nanoparticles were incorporated into the PLLA. This result is consistent with the report by Rakmae et al. (2011). They found that the addition of filler to the PLLA increased the thermal decomposition temperatures (T_{onset}) of the composite compared with that of the neat polymer [30]. They reported that the thermal stability of the PLLA composite was due to the thermally stable filler, which decreased the decomposition of the polymer. These additives may act as barriers to prevent heat transfer [30]. This improvement also contributes to the interface interaction between the nanoparticles and the PLLA [31]. However, the composites containing 30% or 40% Fe_3O_4 have lower thermal decomposition temperatures than that of 20% samples. It may be due to the decrement of molecular weight during mixing with a higher ratio of inorganic filler [32]. The addition of 20% nano-Fe_3O_4 particles provided the best dispersion and distribution of nano-Fe_3O_4 particles in the PLLA.

The CT image shown in Figure 8b demonstrated that the addition of nano-Fe_3O_4 particles at 20% provided excellent X-ray visibility. These findings are consistent with previous reports that the addition of 20% contrast agent to polymers provided optimal X-ray visibility without altering the physical properties of the polymer [29,33]. Thus, it is reasonable to suggest that the 20% nano-Fe_3O_4/PLLA has optimal thermal stability as well as radiopacity. However, the bending strength of this composite is still not large enough for use in weight-bearing areas. It only can be used in non-weight-bearing areas.

Although PLLA is a non-toxic, biodegradable material, the biocompatibility of the added contrast agents is still a concern when it leaches into the surrounding tissue during the degradation process. For example, $BaSO_4$ was introduced as the contrast agent for increasing the radiopacity of bone cement. However, it was reported that $BaSO_4$ induced osteolysis [28]. Figure 7 shows the osteoblastic cells cultured on the surfaces of the nano-Fe_3O_4/PLLA discs. The cells demonstrated normal morphology and growth curves. When the material was 3D printed to make the screw and implanted into the tibia of the rabbits, bone tissue growth into the screw was observed (Figure 7). The leached nano-Fe_3O_4/PLLA was found in the bone marrow as well as at the screw-bone interface. No adverse effects on the rabbit bone tissue were found. Thus, the nano-Fe_3O_4/PLLA composite was safe for use as an implant. However, the animal study was performed for only four weeks, the long-term systemic toxicity and inflammation response due to hydrolosis of this PLLA screw cannot be determined. This is the limitation of the current study. In addition, to reduce the toxicity and inflammation response, lower amounts of iron oxide nanoparticles and iron oxide nanoparticles coated with a biocompatible polymer shell [8] can also be considered to incorporate into PLLA polymer.

5. Conclusions

In conclusion, the 20% nano-Fe_3O_4/PLLA bone screw fabricated in this study demonstrated good biocompatibility, physical, and radiopacity properties. It can be a potential material for fabricating 3D-printed bone screws with radiopacity. These results may serve as a useful reference for future advanced studies of such composites.

Acknowledgments: This study was supported by grants from Taipei Medical University and the National Taipei University of Technology, Taipei, Taiwan (USTP-NTUT-TMU-104-10).

Author Contributions: Wei-Jen Chang and Haw-Ming Huang conceived and designed the experiments; Jy-Jiunn Tzeng and Ting-Lin Wu performed the experiments; Hsin-Ta Wang and Pao-Chang Chiang analyzed the data and wrote the paper; Yu-Hwa Pan contributed reagents, materials, and analysis tools.

Conflicts of Interest: The authors declare no conflict of interest.

References

1. Temenoff, J.S.; Mikos, A.G. Injectable biodegradable materials for orthopaedic tissue engineering. *Biomaterials* **2000**, *21*, 2405–2412. [CrossRef]
2. Jackson, D.W.; Simon, T.M. Tissue engineering principles in orthopaedic surgery. *Clin. Orthop. Relat. Res.* **1999**, *367*, S31–S45. [CrossRef]
3. Giordano, R.A.; Wu, B.M.; Borland, S.W.; Cima, L.G.; Sachs, E.M.; Cima, M.J. Mechanical properties of dense polylactic acid structures fabricated by three dimensional printing. *J. Biomater. Sci. Polym. Ed.* **1996**, *8*, 63–75. [CrossRef] [PubMed]

4.   Lei, K.; Shen, W.; Cao, L.; Yu, L.; Ding, J. An injectable thermogel with high radiopacity. *Chem. Commun.* **2015**, *51*, 6080–6083. [CrossRef] [PubMed]

5.   Puppi, D.; Mota, C.; Gazzarri, M.; Dinucci, D.; Gloria, A.; Myrzabekova, M.; Ambrosio, L.; Chiellini, F. Additive manufacturing of wet-spun polymeric scaffolds for bone tissue engineering. *Biomed. Microdevices* **2012**, *14*, 1115–1127. [CrossRef] [PubMed]

6.   Mota, C.; Puppi, D.; Dinucci, D.; Errico, C.; Bártolo, P.; Chiellini, F. Dual-scale polymeric constructs as scaffolds for tissue engineering. *Materials* **2011**, *4*, 527–542. [CrossRef]

7.   Khalil, S.; Nam, J.; Sun, W. Multi-nozzle deposition for construction of 3D biopolymer tissue scaffolds. *Rapid Prototyp. J.* **2005**, *11*, 9–17. [CrossRef]

8.   De Santis, R.; Gloria, A.; Russo, T.; D'Amora, U.; Zeppetelli, S.; Dionigi, C.; Sytcheva, A.; Herrmannsdörfer, T.; Dediu, V.; Ambrosio, L. A basic approach toward the development of nanocomposite magnetic scaffolds for advanced bone tissue engineering. *J. Appl. Polym. Sci.* **2011**, *122*, 3599–3605. [CrossRef]

9.   De Santis, R.; Gloria, A.; Russo, T.; D'Amora, U.; D'Antò, V.; Bollino, F.; Catauro, M.; Mollica, F.; Rengo, S.; Ambrosio, L. Advanced composites for hard-tissue engineering based on PCL/organic–inorganic hybrid fillers: from the design of 2D substrates to 3D rapid prototyped scaffolds. *Polym. Compos.* **2013**, *34*, 1413–1417. [CrossRef]

10.   Bose, S.; Vahabzadeh, S.; Bandyopadhyay, A. Bone tissue engineering using 3D printing. *Mater. Today* **2013**, *16*, 496–504. [CrossRef]

11.   James, N.R.; Philip, J.; Jayakrishnan, A. Polyurethanes with radiopaque properties. *Biomaterials* **2006**, *27*, 160–166. [CrossRef] [PubMed]

12.   Wu, Y.; Jiang, W.; Wen, X.; He, B.; Zeng, X.; Wang, G.; Gu, Z. A novel calcium phosphate ceramic-magnetic nanoparticle composite as a potential bone substitute. *Biomed. Mater.* **2010**, *5*, 15001. [CrossRef] [PubMed]

13.   Meng, J.; Zhang, Y.; Qi, X.; Kong, H.; Wang, C.; Xu, Z.; Xie, S.; Gu, N.; Xu, H. Paramagnetic nanofibrous composite films enhance the osteogenic responses of pre-osteoblast cells. *Nanoscale* **2010**, *2*, 2565–2569. [CrossRef] [PubMed]

14.   Shan, D.; Shi, Y.; Duan, S.; Wei, Y.; Cai, Q.; Yang, X. Electrospun magnetic poly(L-lactide) (PLLA) nanofibers by incorporating PLLA-stabilized Fe$_3$O$_4$ nanoparticles. *Mater. Sci. Eng. C* **2013**, *33*, 3498–3505. [CrossRef] [PubMed]

15.   Xu, B.; Dou, H.; Tao, K.; Sun, K.; Ding, J.; Shi, W.; Guo, X.; Li, J.; Zhang, D.; Sun, K. "Two-in-One" fabrication of Fe$_3$O$_4$/MePEG–PLA composite nanocapsules as a potential ultrasonic/MRI dual contrast agent. *Langmuir* **2011**, *7*, 12134–12142. [CrossRef] [PubMed]

16.   Wei, Q.; Li, T.; Wang, G.; Li, H.; Qian, Z.; Yang, M. Fe$_3$O$_4$ nanoparticles-loaded PEG–PLA polymeric vesicles as labels for ultrasensitive immunosensors. *Biomaterials* **2010**, *31*, 7332–7339. [CrossRef] [PubMed]

17.   Shen, L.K.; Fan, K.H.; Wu, T.L.; Huang, H.M.; Leung, T.K.; Chen, C.J.; Chang, W.J. Fabrication and magnetic testing of a poly-L-lactide biocomposite incorporating magnetite nanoparticles. *J. Polym. Eng.* **2014**, *34*, 237–240. [CrossRef]

18.   Baumgartner, J.; Bertinetti, L.; Widdrat, M.; Hirt, A.M.; Faivre, D. Formation of magnetite nanoparticles at low temperature: From superparamagnetic to stable single domain particles. *PLoS ONE* **2013**, *8*, e5707. [CrossRef] [PubMed]

19.   Chang, W.J.; Pan, Y.H.; Tzeng, J.J.; Wu, T.L.; Fong, T.H.; Feng, S.W.; Huang, H.M. Development and testing of X-ray imaging-enhanced poly-L-lactide bone screws. *PLoS ONE* **2015**, *10*, e0140354. [CrossRef] [PubMed]

20.   Hong, J.; Lim, Y.J.; Park, S.O. Quantitative biomechanical analysis of the influence of the cortical bone and implant length on primary stability. *Clin. Oral. Implants Res.* **2012**, *23*, 1193–1197. [CrossRef] [PubMed]

21.   Hsu, J.T.; Fuh, L.J.; Tu, M.G.; Li, Y.F.; Chen, K.T.; Huang, H.L. The effects of cortical bone thickness and trabecular bone strength on noninvasive measures of the implant primary stability using synthetic bone models. *Clin. Implant Dent. Relat. Res.* **2013**, *15*, 251–261. [CrossRef] [PubMed]

22.   Ayukawa, Y.; Takeshita, F.; Inoue, T.; Yoshinari, M.; Shimono, M.; Suetsugu, T.; Tanaka, T. An immunoelectron microscopic localization of noncollagenous bone proteins (osteocalcin and osteopontin) at the bone-titanium interface of rat tibiae. *J. Biomed. Mater. Res.* **1998**, *41*, 111–119. [CrossRef]

23.   Xiang, W.; Baolin, L.; Yan, J.; Yang, X. The effect of bone morphogenetic protein on osseointegration of titanium implants. *J. Oral. Maxillofac. Surg.* **1993**, *51*, 647–651. [CrossRef]

24. Cipriano, T.F.; da Silva, A.L.N.; da Silva, A.H.M.; da Fonseca Thomé de Sousa, A.M.F.; da Silva, G.M.; Rocha, M.G. Thermal, rheological and morphological properties of poly (Lactic Acid) (PLA) and talc composites. *Polímeros* **2014**, *24*, 276–282.

25. Gregorova, A.; Sedlarik, V.; Pastorek, M.; Jachandra, H.; Stelzer, F. Effect of compatibilizing agent on the properties of highly crystalline composites based on poly(lactic acid) and wood flour and/or mica. *J. Polym. Environ.* **2011**, *19*, 372–381. [CrossRef]

26. Cai, T.H. Crystallization and melting behavior of biodegradable poly(L-lactic acid)/talc composites. *J. Chem.* **2012**, *9*, 1569–1574. [CrossRef]

27. Suryanegara, L.; Nakagaito, A.N.; Yano, H. The effect of crystallization of PLA on the thermal and mechanical properties of microfibrillated cellulose-reinforced PLA composites. *Compos. Sci. Technol.* **2009**, *69*, 1187–1192. [CrossRef]

28. Aldenhoff, Y.B.; Kruft, M.A.; Pijpers, A.P.; van der Veen, F.H.; Bulstra, S.K.; Kuijer, R.; Koole, L.H. Stability of radiopaque iodine-containing biomaterials. *Biomaterials* **2002**, *23*, 881–886. [CrossRef]

29. Kiran, S.; James, N.R.; Jayakrishnan, A.; Joseph, R. Polyurethane thermoplastic elastomers with inherent radiopacity for biomedical applications. *J. Biomed. Mater. Res. A* **2012**, *100*, 3472–3479. [CrossRef] [PubMed]

30. Rakmae, S.; Ruksakulpiwat, Y.; Sutapun, W.; Suppakarn, N. Effects of mixing technique and filler content on physical properties of bovine bone-based CHA/PLA composites. *J. Appl. Polym. Sci.* **2011**, *122*, 2433–2441. [CrossRef]

31. Albano, C.; Gonzalez, G.; Palacios, J.; Karam, A.; Covis, M. PLLA–HA vs. PLGA–HA characterization and comparative analysis. *Polym. Compos.* **2013**, *34*, 1433–1442. [CrossRef]

32. Liu, X.; Wang, T.; Chow, L.C.; Yang, M.; Mitchell, J.W. Effects of inorganic fillers on the thermal and mechanical properties of poly(lactic acid). *Int. J. Polym. Sci.* **2014**, *2014*, 827028. [CrossRef] [PubMed]

33. Coutu, J.M.; Fatimi, A.; Berrahmoune, S.; Soulez, G.; Lerouge, S. A new radiopaque embolizing agent for the treatment of endoleaks after endovascular repair: Influence of contrast agent on chitosan thermogel properties. *J. Biomed. Mater. Res. B* **2013**, *101*, 153–161. [CrossRef] [PubMed]

MDPI

Article

Preparation of Pendant Group-Functionalized Diblock Copolymers with Adjustable Thermogelling Behavior

Bo Keun Lee [1], Ji Hoon Park [1], Seung Hun Park [1], Jae Ho Kim [1], Se Heang Oh [2], Sang Jin Lee [3], Bun Yeoul Lee [1] and Moon Suk Kim [1,*]

[1] Department of Molecular Science and Technology, Ajou University, Suwon 443-749, Korea; acousticjazz@ajou.ac.kr (B.K.L.); jhp@ajou.ac.kr (J.H.P.); hpt88@ajou.ac.kr (S.H.P.); jhkim@ajou.ac.kr (J.H.K.); bunyeoul@ajou.ac.kr (B.Y.L.)

[2] Department of Nanobiomedical Science, Dankook University, Cheonan 330-714, Korea; seheangoh@dankook.ac.kr

[3] Wake Forest Institute for Regenerative Medicine, Winston-Salem, NC 27157, USA; sjlee@wfubmc.edu

* Correspondence: moonskim@ajou.ac.kr; Tel.: +82-31-219-2608

Academic Editor: Insung S. Choi
Received: 9 May 2017; Accepted: 18 June 2017; Published: 20 June 2017

Abstract: Recently, several thermogelling materials have been developed for biomedical applications. In this study, we prepared methoxy polyethylene glycol (MPEG)-b-(poly(ε-caprolactone)-ran-poly(2-chloride-ε-caprolactone) (PCL-ran-PfCL)) (MP-Cl) diblock copolymers at room temperature via the ring-opening polymerization of caprolactone (CL) and 2-chloride-ε-caprolactone (fCL) monomers, using the terminal alcohol of MPEG as the initiator in the presence of HCl. MPEG-b-(poly(ε-caprolactone)-ran-poly(2-azide-ε-caprolactone) (PCL-ran-PCL-N$_3$)) (MP-N$_3$) was prepared by the reaction of MP-Cl with sodium azide. MPEG-b-(poly(ε-caprolactone)-ran-poly(2-amine-ε-caprolactone) (PCL-ran-PCL-NH$_2$)) (MP-NH$_2$) was subsequently prepared by Staudinger reaction. MP-Cl and MP-N$_3$ showed negative zeta potentials, but MP-NH$_2$ had a positive zeta potential. MP-Cl, MP-N$_3$, and MP-NH$_2$ solutions formed opaque emulsions at room temperature. The solutions exhibited a solution-to-hydrogel phase transition as a function of the temperature and were affected by variation of the chloride, azide, and the amine pendant group, as well as the amount of pendant groups present in their structure. Additionally, the phase transition of MP-Cl, MP-N$_3$, and MP-NH$_2$ copolymers was altered by pendant groups. The solution-to-hydrogel phase transition was adjusted by tailoring the crystallinity and hydrophobicity of the copolymers in aqueous solutions. Collectively, MP-Cl, MP-N$_3$, and MP-NH$_2$ with various pendant-group contents in the PCL segment showed a solution-to-hydrogel phase transition that depended on both the type of pendant groups and their content.

Keywords: thermogelling; diblock copolymer; pendant group; crystallinity; phase transition

---

1. Introduction

Over the past few decades, several thermogelling materials have been investigated extensively for a broad range of biomedical applications, such as in drug- or cell-delivery carriers [1]. Thermogelling materials can absorb water and swell in aqueous environments, which allow them to retain large amounts of fluids [2–4]. Fully hydrated thermogelling material solutions can incorporate various therapeutic agents such as drugs, growth factors, gene products, or cells by simple mixing. Thermogelling material solutions can be injected and solidified at the specific injected body site [5–7].

Various thermogelling materials have been developed. Among them, block copolymers consisting of (methoxy)polyethylene glycol (MPEG), poly(propylene oxide), and biodegradable polyesters, such as poly(L-lactic acid) (PLLA), poly(glycolic acid), their copolyesters, or poly(ε-caprolactone) (PCL), have been reported as potential thermogelling candidates [8–12].

Previously, we developed MPEG-*b*-PCL (MP) and MPEG-*b*-polyester diblock copolymers, (MPEG-*b*-(PCL-*ran*-PLLA), MPEG-*b*-(PCL-*ran*-poly(trimethylene carbonate)), and MPEG-*b*-(PCL-*ran*-poly(1,4-dioxan-2-one)), as thermogelling materials [13–16].

Among these, the MP material exhibited favorable thermogelling properties, such as long-term gel persistence. Furthermore, MP was synthesized by ring-opening polymerization of ε-caprolactone (CL) in the presence of HCl·Et$_2$O as monomer activator [13]. This ring-opening polymerization is especially attractive for biomedical application because HCl·Et$_2$O can be easily removed from MP.

Recently, we examined the thermogelling properties of MP diblock copolymers derivatized with a carboxylic acid group, an amine group, and a zwitterionic group at the end of an MP chain [17,18]. The functional groups on MP diblock copolymers engaged in intra- and inter-molecular interactions to stabilize or destabilize diblock copolymer chain aggregates. The end-group derivatization of MP diblock copolymers can lead to macroscopic gelation through the formation of intra- and inter-molecular interactions.

In a later study, we prepared thermogelling diblock copolymers using benzyl, hydroxyl or carboxylic acid groups in the pendant-position derivatization of diblock copolymers [19]. We found that intra- and inter-molecular interactions among hydrophobic PCL segments can stabilize or destabilize aggregation depending on functional pendant groups.

Based on these results, we concluded that the formation or destruction of a structured network of PCL hydrophobic blocks depends on the functional groups in the chain end or the pendant positions and thus induces different thermal phase-transition behaviors.

In this work, on the basis of previous studies, we introduced different pendant groups into MP to alter the thermogelling behavior of the MP copolymer in aqueous solutions, as shown in Figure 1. We firstly describe the ring-opening polymerization of CL and 2-chloride-ε-caprolactone (fCL) monomers by metal-free method in the presence of HCl·Et$_2$O. The ROP of CL and fCL were randomly copolymerized with varying feed ratios to prepare MPEG-*b*-(poly(ε-caprolactone)-*ran*-poly(2-chloride-ε-caprolactone)) (MPEG-*b*-(PCL-*ran*-PfCL)) (MP-Cl) diblock copolymers.

Next, MPEG-*b*-(poly(ε-caprolactone)-*ran*-poly(2-azide-ε-caprolactone)) (MPEG-*b*-(PCL-*ran*-PCL-N$_3$)) (MP-N$_3$) was prepared by the reaction of MP-Cl with sodium azide. Subsequent Staudinger reaction yielded MPEG-*b*-(poly(ε-caprolactone)-*ran*-poly(2-amine-ε-caprolactone)) (MPEG-*b*-(PCL-*ran*-PCL-NH$_2$)) (MP-NH$_2$). The effects of chloride, azide and amine pendant groups on intra- and inter-molecular interactions were tested. Then, we examined the solution-to-hydrogel phase transitions of MP-Cl, MP-N$_3$, and MP-NH$_2$ as a function of temperature to understand how the type and amount of pendant groups and the solubility and crystallinity of MP-Cl, MP-N$_3$, and MP-NH$_2$ stabilize or destabilize aggregation.

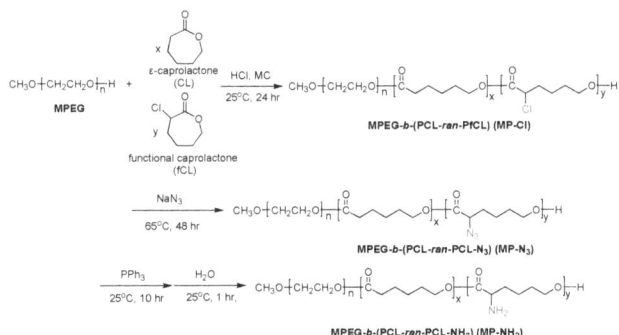

Figure 1. Synthesis of MP-Cl, MP-N$_3$ and MP-NH$_2$ diblock copolymers.

2. Materials and Methods

2.1. Materials

MPEG (number-average molecular weight M_n = 750, Sigma, St. Louis, MO, USA) and HCl (Aldrich; 1.0 M solution in diethyl ether, Sigma, St. Louis, MO, USA) were used as received. ε-Caprolactone (CL) (TCI, Tokyo, Japan) was distilled over CaH_2 at a reduced pressure. 2-Chlorocyclohexanone (TCI, Tokyo, Japan), 3-chloroperoxybenzoic acid (Sigma, St. Louis, MO, USA), sodium azide (Sigma, St. Louis, MO, USA), and triphenylphosphine (Sigma, St. Louis, MO, USA) were used as received. CH_2Cl_2 was distilled sequentially from $CaCl_2$ and CaH_2 in a nitrogen atmosphere before use.

2.2. Characterization

^{1}H-NMR and ^{13}C-NMR spectra were measured using a Varian Mercury Plus 400 (Oxford Instruments, Abingdon, UK) with $CDCl_3$ in the presence of Tetramethylsilian (TMS) as an internal standard. The molecular-weight distributions of MP-Cl, MP-N$_3$, and MP-NH$_2$ were measured at 40 °C using a YL-Clarity GPC system (YL 9170 RI detector) equipped with three columns (Shodex K-802, K-803 and K-804 polystyrene gel columns). For this measurement, $CHCl_3$ was used as the eluent, at a flow rate of 1.0 mL/min, and polystyrene was used for calibration. FT-IR spectra were measured by using a Thermo Nicolet 6700 spectrometer (Thermo Electron Corporation, Runcorn, UK). The melting temperature (T_m) and the heat of fusion (ΔH_m) of MP-Cl, MP-N$_3$ and MP-NH$_2$ were determined using differential scanning calorimetry (DSC; Q1000, TA Instruments, Eschborn, Germany) performed from −80 to 200 °C at a heating rate of 5 °C/min in a nitrogen atmosphere. DSC of MP-Cl, MP-N$_3$, and MP-NH$_2$ in aqueous solutions (20 wt %) was performed from 10 to 60 °C at a heating rate of 5 °C/min in a nitrogen atmosphere. The heat of fusion per gram of MP-Cl, MP-N$_3$ and MP-NH$_2$ was calculated according to the area under the peak. The crystallinity of MP-Cl, MP-N$_3$ and MP-NH$_2$ in aqueous solutions (20 wt %) was measured by X-ray diffraction (XRD; D/MAX-2500V/PC, Rigaku, Tokyo, Japan). A Ni filter at 35 kV and 15 mA generated the radiation. The samples were placed in a quartz sample holder and scanned from 10 to 60° at a scanning rate of 5°/min. The electrophoretic mobility of 0.1 wt % MP-Cl, MP-N3 and MP-NH2 copolymer solutions was measured by electrophoretic light scattering photometer (ELSZ-1000; Otsuka Electronics, Osaka, Japan) at 37 °C. The measured mobility was converted to zeta potential of copolymer solutions.

2.3. Synthesis of MPEG-b-PCL Diblock Copolymer (MP)

MP diblock copolymers with PCL molecular weights (2410 g/mol) using MPEG (750 g/mol) were prepared via a previously reported block copolymerization method [8].

2.4. Preparation of 2-Chloro-ε-caprolactone (fCL)

All glasses were heated in a vacuum and flushed with dry nitrogen stream for drying. All reactions were conducted in dry nitrogen stream. 3-Chloroperoxybenzoic acid (5.2 g, 30 mmol) was placed in a 100 mL flask and dissolved at CH_2Cl_2 (20 mL). 2-Chlorocyclohexanone (2 g, 15 mmol) was added to the solution, which was maintained at room temperature. After 48 h, the resulting solution was washed with a saturated aqueous NaHSO$_3$ solution and an aqueous NaHCO$_3$ solution several times, followed by washing with a brine solution. The organic phase was dried over anhydrous magnesium sulfate and concentrated by evaporation, yielding fCL, i.e., a light yellowish viscous liquid. The obtained mixture was purified by silica-gel column chromatography using a solution of *n*-hexane and ethyl acetate (v/v = 70/30, R_f = 0.6) as an eluent, yielding 1.07 g (7.2 mmol, 48%) of clear liquid. ^{1}H-NMR (CDCl$_3$): 4.81 (*s*, 1H, –CH(Cl)–) 4.61 (*t*, 2H, –OCH$_2$–), 4.22 (*t*, 2H, –OCH$_2$–), 2.05 (*m*, 4H, –CH$_2$–, –CH$_2$–), 1.82 (*m*, 4H, –CH$_2$–). ^{13}C-NMR (CDCl$_3$): 169.6, 69.2, 58.8, 34.0, 29.1, 26.1. Anal. Calcd. for C$_6$H$_9$ClO$_2$, C: 48.50, H: 6.11. Found: C: 48.27, H: 6.02.

2.5. Synthesis of MPEG-b-(poly(ε-caprolactone)-ran-poly(2-chloride-ε-caprolactone))
(MPEG-b-(PCL-ran-PfCL)) (MP-Cl) Diblock Copolymers

All glasses were heated in a vacuum and dried via flushing with dry nitrogen stream. We used the typical polymerization process for producing MP-Cl-97/3 with a CL/fCL ratio of 97/3 using MPEG (750 g/mol) as an initiator, as follows. MPEG (1.6 g, 2.1 mmol) and toluene (55 mL) were added into a flask. Azeotropic distillation was performed to remove water from the MPEG and toluene. Then toluene was removed via distillation. CL (4.92 g, 43 mmol) and fCL (0.2 g, 1.3 mmol) were introduced to the MPEG at room temperature under dry nitrogen stream. The polymerization was started by the addition of a 1.0 M solution of HCl·Et$_2$O (3.2 mL, 3.2 mmol) at room temperature. After 24 h, the mixture was poured into a mixture of *n*-hexane to precipitate a polymer. The precipitated polymers were obtained from the supernatant by decantation, dissolved in CH$_2$Cl$_2$, and then filtered. The resulting polymer solution was concentrated by rotary evaporation and dried in a vacuum, yielding a colorless copolymer. In the same way, MP-Cl copolymers (MP-Cl-97/3, MP-Cl-95/5, MP-Cl-90/10 and MP-Cl-85/15) were prepared. The molecular weights and ratios of the PCL and PCL-Cl segments in the copolymers were determined by comparing the intensity of the carbonyl carbons in PCL and PCL-Cl segments at δ = 172.7 and 169.6 ppm, respectively, at ^{13}C-NMR peaks.

2.6. Synthesis of MPEG-b-(poly(ε-caprolactone)-ran-poly(2-azide-ε-caprolactone))
(MPEG-b-(PCL-ran-PCL-N$_3$)) (MP-N$_3$) Diblock Copolymers

MP-Cl (5 g, 1.6 mmol, with 0.05 mmol concentration of Cl group) and DMF (23 mL) were inserted into a flask. Sodium azide (0.01 g, 0.1 mmol) was added to the MP-Cl solution at room temperature under nitrogen, and the mixture was stirred at 65 °C for 48 h. The reaction mixture was then poured into a mixture of *n*-hexane and ethyl ether (v/v = 4/1) to precipitate the polymer, which was separated from the supernatant by decantation. The obtained polymer was redissolved in CH$_2$Cl$_2$. The resulting solution was washed with a brine solution. The organic phase was concentrated using a rotary evaporator and dried in vacuum to obtain a yellowish copolymer.

2.7. Synthesis of MPEG-b-(poly(ε-caprolactone)-ran-poly(2-amine-ε-caprolactone))
(MPEG-b-(PCL-ran-PCL-NH$_2$)) (MP-NH$_2$) Diblock Copolymers

MP-N$_3$ (4 g, 1.3 mmol, concentration of azide group: 0.04 mmol) and DMF (20 mL) were introduced into a flask. Triphenylphosphine (0.02 g, 0.08 mmol) was added to the MP-N$_3$ solution at room temperature under nitrogen, and the mixture was stirred at room temperature. After 10 h, H$_2$O (1 mL) was poured into the reaction mixture, and the mixture was stirred for 1 h. The reaction mixture was then poured into a mixture of *n*-hexane and ethyl ether (v/v = 4/1) to precipitate the polymer, which was separated from the supernatant by decantation. The obtained polymer was redissolved in CH$_2$Cl$_2$. The resulting solution was washed with a saturated aqueous NaHCO$_3$ solution several times. The organic phase was dried over anhydrous magnesium sulfate and concentrated by evaporation to obtain an opaque copolymer.

2.8. Determination of Sol-to-Gel Phase-Transition Times via Tilting Experiment

To examine the gelation time of the MP-Cl, MP-N$_3$ and MP-NH$_2$ copolymer solutions, the copolymers were dissolved at 20 wt % in 4 mL vials at 80 °C in deionized water (DW). The MP-Cl, MP-N$_3$ and MP-NH$_2$ copolymers were obtained as homogeneous opaque emulsions and were stored at 4 °C. After 48 h, the homogeneous opaque emulsions in the vials were gently stirred at room temperature, and the vials were immediately immersed in a 37 °C water bath. While the vials were maintained at 37 °C, the time taken by the emulsions to stop exhibiting flow was determined and defined as the gelation time.

2.9. Determination of Sol-to-Gel Phase-Transition Times via Rheological Measurement

The gelation time of the MP-Cl, MP-N$_3$ and MP-NH$_2$ copolymers was measured using a rheometer (MCR 102, Anton Paar, Ostfildern, Germany) with peltier temperature-control system for bottom plates and a 25.0 mm stainless-steel parallel-plate measuring system. The storage modulus G′ and loss modulus G″ were measured under a 1.0% strain level at 37 °C and calculated using the software of the instrument. The gelation time was determined at the crossover point of the G′ and G″ curves.

2.10. Viscosity Measurements

MP-Cl, MP-N$_3$, and MP-NH$_2$ copolymers (0.5 g, 0.15 mmol) was dissolved in a 4 mL vial at 80 °C to obtain a concentration of 20 wt % by using DW and then stored at 4 °C. After 48 h, the viscosities of the copolymer solutions were measured using a Brookfield Viscometer DV-III Ultra, which was equipped with a programmable rheometer and circulating baths featuring a programmable controller (TC-502P). The viscosity of the MP-Cl, MP-N$_3$ and MP-NH$_2$ copolymer solutions was determined using a T-F spindle at 0.1 rpm from 10 to 60 °C in increments of 1 °C.

3. Results and Discussion

3.1. Preparation and Characterization of MP-Cl, MP-N$_3$, and MP-NH$_2$

Firstly, MP-Cl was synthesized at room temperature via the ring-opening polymerization of the monomer CL and fCL using the terminal alcohol of MPEG as the initiator in the presence of HCl·Et$_2$O. The colorless MP-Cl diblock copolymers were obtained in almost quantitative yield.

^{1}H- and ^{13}C-NMR of the MP-Cl diblock copolymers exhibited characteristic peaks of MPEG, PCL, and PCL-Cl (Figures 2 and 3). The methoxy, methylene, and methine protons 1, 3 and 9 were observed at δ = 3.48, 4.21, and 4.18 ppm at ^{1}H-NMR peaks. The carbonyl carbons (–C(=O)) of PCL and PCL-Cl segments were observed at δ = 172.7 and 169.6 ppm at ^{13}C-NMR peaks, respectively. ^{13}C-NMR of the MP-Cl diblock copolymer also showed a peak that could be assigned to –C(H)(Cl)– at δ = 62.15 ppm. The ratios of the PCL and PCL-Cl segments in the copolymers were determined according to the carbon-integration ratios of carbonyl in the PCL and PCL-Cl segments, which agreed well with the expected values. Additionally, the molecular weights of MP-Cl determined by NMR spectroscopy were close to theoretical values calculated from different ratios of CL and fCL. This indicates that the polymerization procedure yielded targeted MP-Cl diblock copolymers with a PCL-Cl content of 3–15 mol % in the PCL segment. The synthesis results for MP-Cl with Cl pendant group contents of 3–15 mol % in the PCL segment was summarized in Supplementary Materials Table S1. This result clearly showed that we successfully prepared the MP-Cl by metal-free ring-opening polymerization of the monomer CL and fCL.

Next, MP-N$_3$ was obtained as light yellowish diblock copolymers through the modification of MP-Cl using sodium azide. The synthesis results for MP-N$_3$ with azide pendant group contents of 3–15 mol % in the PCL segment was summarized in Supplementary Materials Table S1. ^{1}H- and ^{13}C-NMR of the MP-N$_3$ diblock copolymer also exhibited characteristic peaks corresponding to PCL, MPEG, and MP-N$_3$ (Figures 2 and 3). Specially, the methine proton 15 of the –C(H)(N$_3$)– was observed at δ = 3.85 ppm. The MP-N$_3$ diblock copolymer showed a new ^{13}C-NMR peak owing to –C(H)(N$_3$)– at δ = 61.82 ppm but complete disappearance of the signal at δ = 169.6 ppm, assigned to –C(=O) of the PCL-Cl segments. The Fourier transform infrared (FT-IR) spectra of the MP-N$_3$ diblock copolymer exhibited characteristic peaks around 2105 cm^{-1}, which are assigned to the stretching of the azide group (Supplementary Materials Figure S1). This indicates that the chloride group on MP-Cl was quantitatively changed into the azide group.

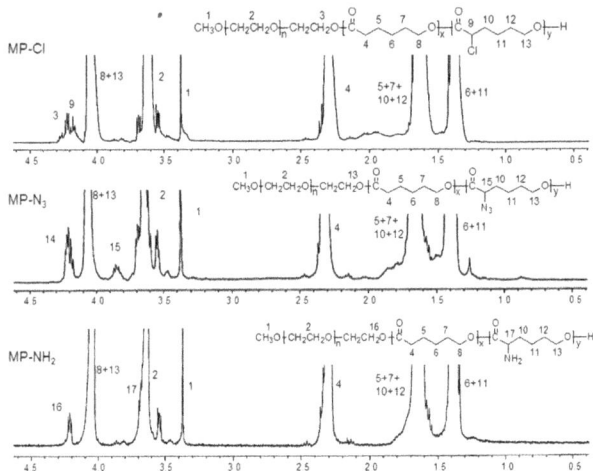

Figure 2. ^{1}H-NMR spectra of MP-Cl, MP-N$_3$ and MP-NH$_2$ diblock copolymers.

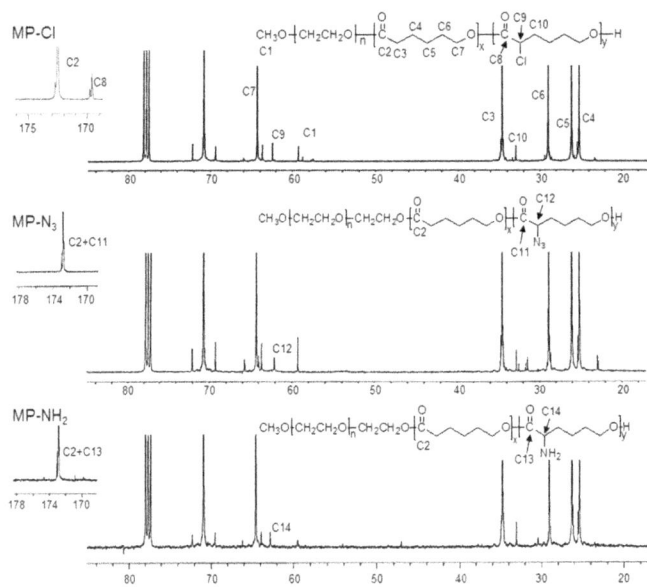

Figure 3. ^{13}C-NMR spectra of MP-Cl, MP-N$_3$ and MP-NH$_2$ diblock copolymers.

Finally, an MP-NH$_2$ diblock copolymer was obtained by the subsequent modification of MP-N$_3$ using the Staudinger reaction. The MP-NH$_2$ diblock copolymer had the appearance of –C(H)(NH$_2$) at δ = 62.97 ppm of ^{13}C-NMR. In addition, the azide group was completely disappeared in the FT-IR spectra of the MP-NH$_2$ diblock copolymer.

The obtained MP-N$_3$ and MP-NH$_2$ diblock copolymers with different composition ratios were compared with a corresponding diblock copolymer via elemental analysis (Supplementary Materials Table S2). The elemental analysis of the MP-N$_3$ and MP-NH$_2$ diblock copolymers showed values almost similar to the expected values at all composition ratios. This result indicates that the pendant azide and amine groups were stoichiometrically modified into MP diblock copolymers and demonstrates that the procedure described here generated the azide and amine-modified MP diblock copolymers.

To examine the electrostatic properties according to the variation in the composition ratios of chloride, azide, and amine pendant groups, the zeta potential was measured in solutions of MP, MP-Cl, MP-N$_3$ and MP-NH$_2$ diblock copolymers (Figure 4). The zeta potential of the MP solution was -2.7 mV. The MP-Cl and MP-N$_3$ exhibited negative zeta potentials at all composition ratios because of the negative properties of the chloride and azide groups on MP-Cl and MP-N$_3$. The negative zeta potentials of the chloride- and azide-group on alkyl or polymer chain were reported [20–23]. The zeta potentials of MP-N$_3$ were lower than those of MP-Cl at the same composition ratios. The zeta potentials of the MP-Cl and MP-N$_3$ solutions decreased almost linearly from -3.6 to -13.2 mV and -12.3 to -22.5 mV, respectively, for the chloride- and azide-group contents of 3–15 mol %. MP-NH$_2$ had a positive zeta potential of 4.8 mV and increased almost linearly to 14 mV as the amine content increased.

These results indicate that the electrostatic properties of the MP-Cl, MP-N$_3$ and MP-NH$_2$ copolymers were affected by the variation of the chloride, azide and amine pendant groups, as well as the amount of pendant groups in the structure.

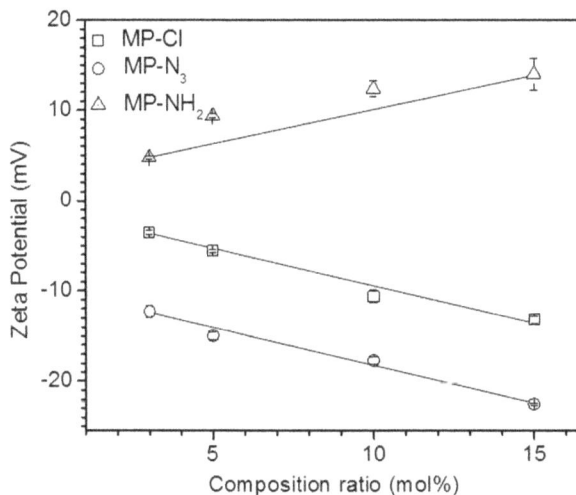

Figure 4. Zeta potential versus block composition ratios for MP-Cl, MP-N$_3$ and MP-NH$_2$ diblock copolymers with 3–15 mol % of chloride, azide and amine pendant groups.

3.2. Solution Properties of MP-Cl, MP-N$_3$ and MP-NH$_2$

Aqueous solutions of the MP-Cl, MP-N$_3$ and MP-NH$_2$ diblock copolymers were prepared by dissolving them in DW at 80 °C. The aqueous MP-Cl, MP-N$_3$ and MP-NH$_2$ solutions formed opaque emulsions at room temperature (Figure 5). The solution properties of all the copolymers were determined according to the emulsion formation time, the phase transition, and the gelation time (Supplementary Materials Table S3). The MP has an emulsion formation time of 20 min and a gelation time below 10 s.

Then, MP, MP-Cl, MP-N$_3$ and MP-NH$_2$ diblock copolymers in the emulsion-sol state were individually incubated at 37 °C. The time taken by incubation at 37 °C to stop exhibiting a flow of the emulsions was determined as the gelation time, as summarized in Supplementary Materials Table S3.

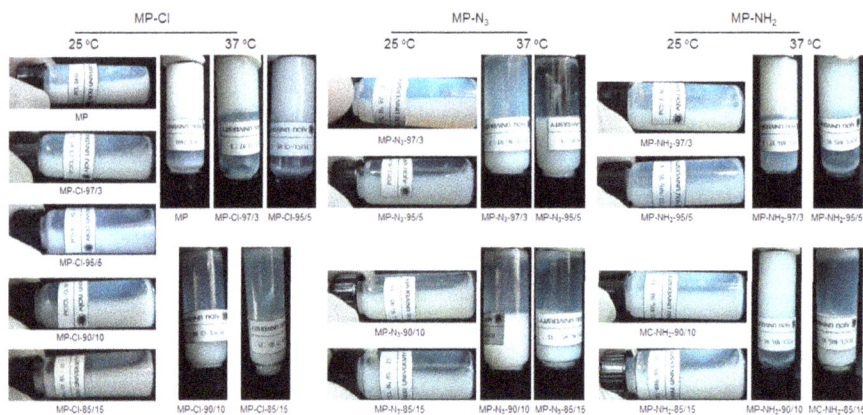

Figure 5. Images of MP-Cl, MP-N$_3$, and MP-NH$_2$ diblock copolymer solutions with 3–15 mol % chloride, azide and amine pendant groups at 25 and 37 °C.

The emulsion formation time of MP-Cl decreased as the chloride-group content increased. The emulsion formation time of MP-Cl with a chloride group content of 3 mol % was ~15 min, and the time sharply decreased to 3 min as the chloride group content increased. The phase transition of MP-Cl with chloride group contents of 3 and 5 mol % was observed at 37 °C, and the gelation time was ~30 s. MP-Cl with chloride group contents of 10 and 15 mol % had a short emulsion formation time and no phase transition, probably owing to easy formation of emulsion of the MP-Cl diblock copolymers in the aqueous solution.

The emulsion formation time of MP-N$_3$ decreased as the azide group content increased. The emulsion formation of MP-N$_3$ was less than 10 min. In addition, for MP-N$_3$, at all azide group contents, there was no phase transition, owing to easy formation of emulsion of MP-N$_3$ diblock copolymers in the aqueous solution.

MP-NH$_2$ with amine group contents of 3 and 5 mol % had solubilization times of 10 and 5 min, respectively, and a solution-to-hydrogel phase transition at 37 °C. The gelation time increased from 20 to 50 s as the amine group content increased. MP-NH$_2$ with an amine group content of 15 mol % formed emulsion in the aqueous solution and thus showed no phase transition.

These results indicate that the phase transition and the gelation time of the MP-Cl, MP-N$_3$ and MP-NH$_2$ copolymers were affected by the variation of the chloride, azide and amine pendant groups as well as the amount of pendant groups present in their structure.

3.3. Rheological Properties of MP-Cl, MP-N$_3$ and MP-NH$_2$

The viscosities of a series of MP-Cl, MP-N$_3$ and MP-NH$_2$ copolymers containing various pendant groups were measured as a function of temperature. Figure 6 shows the viscosity-temperature curves of MP-Cl, MP-N$_3$ and MP-NH$_2$ with 3–15 mol % of chloride, azide and amine pendant groups. Below 28 °C, the viscosities of the MP-Cl, MP-N$_3$ and MP-NH$_2$ solutions at all pendant group ratios were in the range of 1–1000 cP, demonstrating that this fluid-like zone corresponded to an emulsion of MP-Cl, MP-N$_3$ and MP-NH$_2$.

At 28–50 °C, the viscosities of the MP-Cl, MP-N$_3$ and MP-NH$_2$ solutions increased, indicating that in this zone, the copolymers became gel-like and that the interactions between the hydrophobic blocks in MP-Cl, MP-N$_3$ and MP-NH$_2$ were strengthened in this temperature range. As the content of pendant groups increased, the onset temperatures increased, and maximum viscosities decreased.

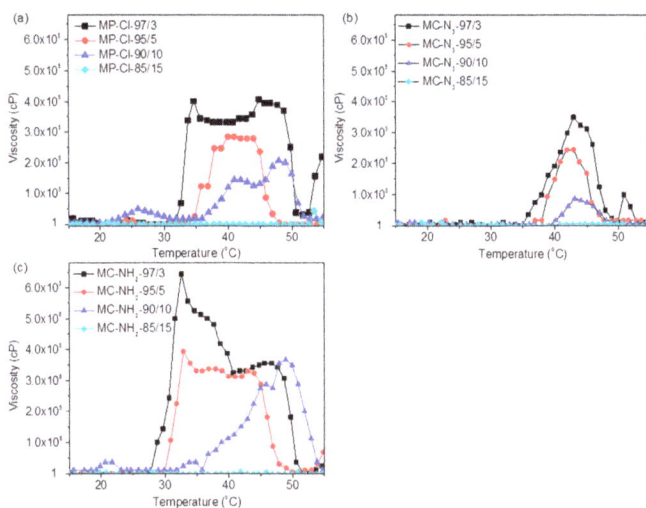

Figure 6. Viscosity-versus-temperature curves of (**a**) MP-Cl with 3–15 mol % of chloride pendant groups, (**b**) MP-N$_3$ with 3–15 mol % of azide pendant groups, and (**c**) MP-NH$_2$ with 3–15 mol % of amine pendant groups.

For MP-Cl copolymers (Figure 7a), the onset temperatures of gelation increased as the amount of chloride pendant groups increased in the MP-Cl copolymer, except in the case of MP-Cl with 15 mol % of chloride. MP-Cl with 3, 5 and 10 mol % of chloride exhibited onset temperatures of 33, 35 and 38 °C, respectively. The order of maximum viscosities of MP-Cl at 37 °C was 3 > 5 > 10 mol % of the chloride group in the MP-Cl copolymer. MP-Cl with 15 mol % of the chloride group was dissolved in water, yielding a viscosity of 1 cP, and no phase-transition behavior was observed.

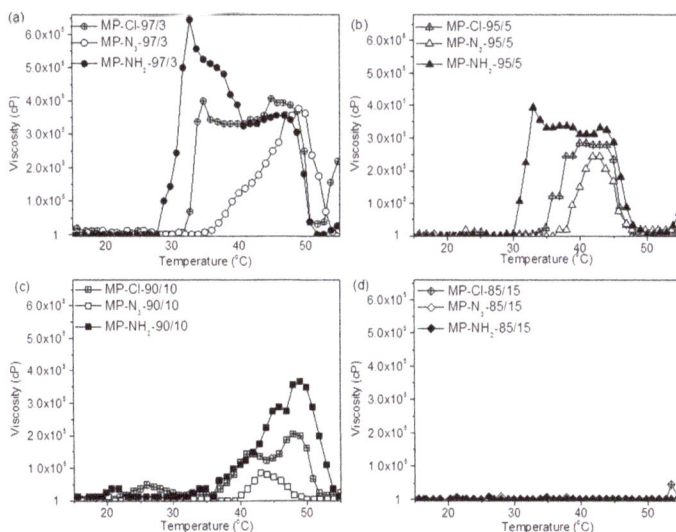

Figure 7. Viscosity-versus-temperature curves of MP-Cl, MP-N$_3$, and MP-NH$_2$ diblock copolymer solutions with (**a**) 3, (**b**) 5, (**c**) 10, and (**d**) 15 mol % of pendant groups.

MP-N$_3$ and MP-NH$_2$ exhibited similar onset temperatures and orders of viscosity at 37 °C to MP-Cl copolymers (Figure 7b,c). A solution-to-hydrogel phase transition was observed upon introduction of chloride, azide and amine up to concentrations of 10 mol %. MP-Cl, MP-N$_3$ and MP-NH$_2$ with 15 mol % of chloride, azide and amine showed a viscosity of 1 cP, indicating no phase transition.

The copolymers in order of increasing onset temperatures was MP-NH$_2$ < MP-Cl < MP-N$_3$ up to a concentration of 10 mol %. MP-Cl, MP-N$_3$ and MP-NH$_2$ copolymers with 15 mol % of pendant groups exhibited a viscosity of 1 cP, indicating no phase transition. The maximum viscosities decreased in the order of MP-NH$_2$ > MP-Cl > MP-N$_3$.

The order of the onset temperatures and maximum viscosities were similar to the order of the zeta potentials for the MP-Cl, MP-N$_3$ and MP-NH$_2$ copolymers. For MP-NH$_2$ copolymers, a cationic pendant amine group on the PCL can stabilize aggregation between amine groups on the hydrophobic PCL segments via hydrogen bonding, but higher concentrations of amine pendant groups in aqueous solutions induce plenty of hydrogen bonding with surrounding large water.

A negative pendant chloride group on the PCL slightly stabilized the aggregation between the hydrophobic PCL segments at low chloride contents but induced the destabilization of aggregation between PCL segments at high chloride contents. Additionally, the large negative potential of the azide group on the PCL showed less stabilization than the chloride group at all pendant contents.

Collectively, the pendant chloride, azide and amine groups of the MP-Cl, MP-N$_3$ and MP-NH$_2$ copolymers altered the packing or aggregation in the diblock copolymer solution, which was mediated by intra- and inter-molecular hydrophobic interactions or hydrogen bonding between pendant groups. These results suggest that pendant groups or the compositional ratio of MP-Cl, MP-N$_3$ and MP-NH$_2$ can be used to adjust the temperature-responsive window.

3.4. Crystallization of MP-Cl, MP-N$_3$ and MP-NH$_2$

As described in the previous section, the phase transition of the MP-Cl, MP-N$_3$ and MP-NH$_2$ copolymers was altered by pendant groups through intra- and inter-molecular aggregation of the PCL blocks. To understand the intra- and inter-molecular aggregation between the hydrophobic domains of MP-Cl, MP-N$_3$ and MP-NH$_2$, the crystalline properties of MP-Cl, MP-N$_3$ and MP-NH$_2$ with 3–15 mol % of chloride, azide and amine pendant groups were examined according to the thermal properties via DSC and XRD in the bulk and aqueous solutions.

The thermal properties of the copolymers were measured as summarized in Supplementary Materials Table S4. The DSC thermograms of the diblock copolymer showed the peak assignable to melting of the hydrophobic domains, which could be attributed to the crystallization enthalpy of the hydrophobic domains. The thermal analysis indicated that the MP diblock copolymer exhibited a melting temperature of 56 °C and that the crystallization enthalpy of the MP diblock copolymers was 70 J/g. The solutions of the MP diblock copolymer exhibited a crystallization enthalpy of 11 J/g at a melting temperature of 47.6 °C. The degrees of crystallinity in the aqueous-solution copolymers were estimated to be 37% via XRD.

MP-Cl diblock copolymers with 3–15 mol % of chloride pendant groups on the PCL segment showed a decreasing crystallization enthalpy—from 53 to 18 J/g—with increasing number of chloride pendant groups. The solutions of MP-Cl diblock copolymers also showed a decrease of the crystallization enthalpy, which reached zero at a chloride group content of 15 mol %, probably because of the complete phase mixing of the MP-Cl diblock copolymers in the aqueous solution.

The degrees of crystallinity in the aqueous-solution state of the MP-Cl diblock copolymers decreased from 32% to 17% with increasing chloride pendant groups. These results indicate that the chloride group on MP prohibited aggregation between the hydrophobic PCL segments. In the bulk and aqueous phases, MP-N$_3$ with 3–15 mol % of azide groups in the PCL segment showed almost the same behavior as the crystallization enthalpy and the degrees of crystallinity of the MP-Cl diblock copolymers. This result indicates that the azide group in the PCL segment prohibited aggregation between the hydrophobic segments.

Polymers **2017**, *9*, 239

The thermal properties of MP-NH$_2$ copolymers containing 3, 5 and 10 mol % of amine groups showed increases in the crystallization enthalpy—from 52 to 77 J/g—and the degrees of crystallinity—from 31% to 36%—with increasing amine pendant groups, but with 15 mol % of amine groups, the crystallization enthalpy decreased to 33 J/g, and the degree of crystallinity decreased to 22%.

According to the XRD analysis of the MP-Cl diblock copolymer, the total crystallinity decreased from 32% to 17% as the amount of the chloride pendant groups increased. For the MP-N$_3$ diblock copolymer, the total crystallinity decreased from 30% to 23% as the amount of the azide pendant groups increased.

For the MP-NH$_2$ diblock copolymer, the total crystallinity increased from 31% to 36% as the amount of the amine group increased, except in the case of MP-NH$_2$-85/15. DSC and XRD indicate that the crystallinity increased with the amount of amine pendant groups. This indicates that the crystallinity of the MP-Cl, MP-N$_3$ and MP-NH$_2$ copolymers was affected by the variation of the chloride, azide and amine pendant groups, as well as the amount of pendant groups in their structure.

The plausible mechanism for thermogelling of the amphiphilic diblock copolymers is proposed by the formation of the microphase-separated micelles and the interconnection between micelles [24]. Thus, large crystallinity of hydrophobic domains can enhance and stabilize packing of interconnected micelles. The presence of crystallinity in aqueous medium in this work strongly supports the idea that the MP-Cl, MP-N$_3$ and MP-NH$_2$ diblock copolymer solutions stabilize the aggregation of hydrophobic domains via intra- and inter-molecular interactions. However, the crystallinity of hydrophobic domains of the MP-Cl, MP-N$_3$ and MP-NH$_2$ diblock copolymer solutions could rise or decline because the functional pendant groups on MP diblock copolymers could stabilize or destabilize diblock copolymer chain aggregation.

Collectively, the pendant chloride and azide groups induced weakened and/or prohibited aggregation between the hydrophobic segments. The amine groups enhanced aggregation between the hydrophobic segments via intra- and inter-molecular hydrogen-bonding interactions but weakened the aggregation between the hydrophobic segments via complete phase mixing at an amine-group content of 15 mol %.

4. Conclusions

We successfully prepared the MP-Cl by metal-free ring-opening polymerization of the monomer CL and fCL. Then, MP-N$_3$ and MP-NH$_2$ with azide and amine pendant group contents of 3–15 mol % in the PCL segment showed a solution-to-hydrogel phase transition that depended on both the type of pendant groups and their content. The solution-to-hydrogel phase transition temperature was adjusted by tailoring the crystallinity and hydrophobicity of the copolymers in aqueous solutions. Thus, we consider that the proposed MP-Cl, MP-N$_3$ and MP-NH$_2$ copolymers with various pendant groups can be utilized as injectable drug and cell carriers. Meanwhile, further experiments will be needed to investigate the pH-responsible thermogelling behavior and their solution properties of pendant group-functionalized diblock copolymers.

Supplementary Materials: The following are available online at www.mdpi.com/2073-4360/9/6/239/s1, Figure S1: FT-IR spectra of MP-Cl, MP-N$_3$, and MP-NH$_2$ diblock copolymers, Table S1: Synthesis of MP-Cl, MP-N$_3$, and MP-NH$_2$ diblock copolymers, Table S2: Elementary analysis of MP-N3 and MP-NH2 diblock copolymers, Table S3: Thermogelling properties of MP-Cl, MP-N$_3$, and MP-NH$_2$ diblock copolymers, Table S4: Thermal properties of MP-Cl, MP-N3, and MP-NH2 diblock copolymers.

Acknowledgments: This study was supported by a grant from a Basic Science Research Program (2016R1A2B3007448) through the National Research Foundation of Korea (NRF) funded by the Ministry of Education. Partial data published in the abstract 207 of the 2015 TechConnect World Innovation Conference.

Author Contributions: Bo Keun Lee, Ji Hoon Park and Seung Hun Park performed the experiments; Bo Keun Lee, Ji Hoon Park, Jae Ho Kim, Bun Yeoul Lee, Se Heang Oh and Sang Jin Lee analyzed the data; Jae Ho Kim, Bun Yeoul Lee and Se Heang Oh contributed analysis tools (zeta potential, pH measurement and FT-IR); Moon Suk Kim designed the experiments and wrote the paper.

Conflicts of Interest: The authors declare no conflict of interest.

References

1. Jing, Y.; Quan, C.; Liu, B.; Jiang, Q.; Zhang, C. A Mini Review on the Functional Biomaterials Based on Poly(lactic acid) Stereocomplex. *Polym. Rev.* **2016**, *56*, 262–286. [CrossRef]
2. Yue, K.; Trujillo-de Santiago, G.; Alvarez, M.M.; Tamayol, A.; Annabi, N.; Khademhosseini, A. Synthesis, properties, and biomedical applications of gelatin methacryloyl (GelMA) hydrogels. *Biomaterials* **2015**, *73*, 254–271. [CrossRef] [PubMed]
3. Kotsuchibashi, Y.; Ebara, M.; Aoyagi, T.; Narain, R. Recent Advances in Dual Temperature Responsive Block Copolymers and Their Potential as Biomedical Applications. *Polymers* **2016**, *8*, 380. [CrossRef]
4. Mahapatra, C.; Jin, G.-Z.; Kim, H.-W. Alginate-hyaluronic acid-collagen composite hydrogel favorable for the culture of chondrocytes and their phenotype maintenance. *Tissue Eng. Regen. Med.* **2016**, *13*, 538–546. [CrossRef]
5. Wu, Y.L.; Wang, H.; Qiu, Y.-K.; Liow, S.S.; Li, Z.; Loh, X.J. PHB-Based Gels as Delivery Agents of Chemotherapeutics for the Effective Shrinkage of Tumors. *Adv. Healthc. Mater.* **2016**, *5*, 2679–2685. [CrossRef] [PubMed]
6. Wei, Z.; Zhao, J.; Chen, Y.M.; Zhang, P.; Zhang, Q. Self-healing polysaccharide-based hydrogels as injectable carriers for neural stem cells. *Sci. Rep.* **2016**, *6*, 37841. [CrossRef] [PubMed]
7. Shimojo, A.A.M.; Galdames, S.E.M.; Perez, A.G.M.; Ito, T.H.; Luzo, A.C.M.; Santana, M.H.A. In Vitro Performance of Injectable Chitosan-Tripolyphosphate Scaffolds Combined with Platelet-Rich Plasma. *Tissue Eng. Regen. Med.* **2016**, *13*, 21–30. [CrossRef]
8. Kim, D.Y.; Kwon, D.Y.; Kwon, J.S.; Kim, J.H.; Min, B.H.; Kim, M.S. Injectable in situ-forming hydrogels for regenerative medicines. *Polym. Rev.* **2015**, *55*, 407–452. [CrossRef]
9. Jang, J.Y.; Park, S.H.; Park, J.H.; Lee, B.K.; Yun, J.-H.; Lee, B.; Kim, J.H.; Min, B.H.; Kim, M.S. In vivo osteogenic differentiation of human dental pulp stem cells embedded in an injectable in vivo-forming hydrogel. *Macromol. Biosci.* **2016**, *16*, 1158–1169. [CrossRef] [PubMed]
10. Zhao, X.; Guo, B.; Ma, P.X. Single component thermo-gelling electroactive hydrogels from poly(caprolactone)-poly(ethylene glycol)-poly(caprolactone)-graft-aniline tetramer amphiphilic copolymers. *J. Mater. Chem. B* **2015**, *3*, 8459–8468. [CrossRef]
11. Lim, A.C.; Alvarez-Lorenzo, C.; Mano, J.F. Design Advances in Particulate Systems for Biomedical Applications. *Adv. Healthc. Mater.* **2016**, *5*, 1687–1723. [CrossRef] [PubMed]
12. Hsu, S.; Chen, C.; Hung, K.; Tsai, Y.; Suming, L. Thermo-Responsive Polyurethane Hydrogels Based on Poly(ε-caprolactone) Diol and Amphiphilic Polylactide-Poly(Ethylene Glycol) Block Copolymers. *Polymers* **2016**, *8*, 252. [CrossRef]
13. Kim, M.S.; Hyun, H.; Khang, G.; Lee, H.B. Preparation of thermosensitive diblock copolymers consisting of MPEG and polyesters. *Macromolecules* **2006**, *39*, 3099–3102. [CrossRef]
14. Kang, Y.M.; Lee, S.H.; Lee, J.Y.; Son, J.S.; Kim, B.S.; Lee, B.; Chun, H.J.; Min, B.H.; Kim, J.H.; Kim, M.S. A Biodegradable, Injectable, Gel System Based on MPEG-b-(PCL-ran-PLLA) Diblock Copolymers with an Adjustable Therapeutic Window. *Biomaterials* **2010**, *31*, 2453–2460. [CrossRef] [PubMed]
15. Hyun, H.; Park, S.H.; Kwon, D.Y.; Khang, G.; Lee, H.B.; Kim, M.S. Thermo-responsive injectable MPEG-polyester diblock copolymers for sustained drug release. *Polymers* **2014**, *6*, 2670–2683. [CrossRef]
16. Kim, M.S.; Hyun, H.; Seo, K.S.; Cho, Y.H.; Khang, G.; Lee, H.B. Preparation and characterization of MPEG-PCL diblock copolymers with thermo-responsive sol-gel-sol behavior. *J. Polym. Sci. Part A Polym. Chem.* **2006**, *44*, 5413–5423. [CrossRef]
17. Kim, J.I.; Lee, S.H.; Kang, H.J.; Kwon, D.Y.; Kim, D.Y.; Kang, W.S.; Kim, J.H.; Kim, M.S. Examination of phase transition behavior of ion group functionalized MPEG-b-PCL diblock copolymers. *Soft Matter* **2011**, *7*, 8650–8656. [CrossRef]
18. Shim, S.W.; Kwon, D.Y.; Park, J.H.; Kim, J.H.; Chun, H.J.; Koh, Y.J.; Kim, M.S. Preparation of zwitterionic sulfobetaine end-functionalized poly(ethylene glycol)-b-poly(caprolactone) diblock copolymers and examination of their thermogelling properties. *J. Polym. Sci. Part A Polym. Chem.* **2014**, *52*, 2185–2191. [CrossRef]

Polymers **2017**, *9*, 239

19. Kim, J.I.; Kim, D.Y.; Kwon, D.Y.; Kang, H.J.; Kim, J.H.; Min, B.H.; Kim, M.S. An Injectable Biodegradable Temperature-Responsive Gel with an Adjustable Persistence Window. *Biomaterials* **2012**, *33*, 2823–2834. [CrossRef] [PubMed]

20. Lameiras, F.S.; de Souza, A.L.; de Melo, V.A.R.; Nunes, E.H.M.; Brag, I.D. Measurement of the Zeta Potential of Planar Surfaces With a Rotating Disk. *Mater. Res.* **2008**, *11*, 217–219. [CrossRef]

21. Kandhakumar, G.; Kalaiarasi, C.; Kumaradhas, P. Structure and charge density distribution of amine azide based hypergolic propellant molecules: A theoretical study. *Can. J. Chem.* **2016**, *94*, 126–136. [CrossRef]

22. Hu, X.; Li, D.; Zhou, F.; Gao, C. Biological hydrogel synthesized from hyaluronic acid, gelatin and chondroitin sulfate by click chemistry. *Acta Biomater.* **2011**, *7*, 1618–1626. [CrossRef] [PubMed]

23. Guney, A.; Ozdilek, C.M.; Kangal, O.; Burat, F. Flotation characterization of PET and PVC in the presence of different Plasticizers. *Sep. Purif. Technol.* **2015**, *151*, 47–56. [CrossRef]

24. Singh, N.K.; Lee, D.S. In situ gelling pH- and temperature-sensitive biodegradable block copolymer hydrogels for drug delivery. *J. Control. Release* **2014**, *193*, 214–227. [CrossRef] [PubMed]

Article

Biphasic Calcium Phosphate (BCP)-Immobilized Porous Poly (D,L-Lactic-*co*-Glycolic Acid) Microspheres Enhance Osteogenic Activities of Osteoblasts

Kyu-Sik Shim [1,2,†], Sung Eun Kim [2,†], Young-Pil Yun [2], Somang Choi [1,2], Hak-Jun Kim [2], Kyeongsoon Park [3,*]and Hae-Ryong Song [2,*]

[1]   Department of Biomedical Science, College of Medicine, Korea University, Anam-dong, Seongbuk-gu, Seoul 02841, Korea; breakdown88@nate.com (K.-S.S.); chlthakd1029@naver.com (S.C.)

[2]   Department of Orthopedic Surgery and Rare Diseases Institute, Korea University Medical College, Guro Hospital, #80, Guro-dong, Guro-gu, Seoul 08308, Korea; sekim10@korea.ac.kr (S.E.K.); ofeel0479@korea.ac.kr (Y.-P.Y.); dakjul@korea.ac.kr (H.-J.K.)

[3]   Department of Systems Biotechnology, College of Biotechnology and Natural Resources, Chung-Ang University, 4726 Seodong-daero, Daedeok-myeon, Anseong-si, Gyeonggi-do 17546, Korea

*   Correspondence: kspark1223@cau.ac.kr (K.P.); songhae@korea.ac.kr (H.-R.S.); Tel.: +82-31-670-3357 (K.P.); +82-2-2626-2481 (H.-R.S.)

†   These authors contributed equally to this work.

Received: 27 June 2017; Accepted: 18 July 2017; Published: 21 July 2017

Abstract: The purpose of this study was to evaluate the potential of porous poly (D,L-lactic-*co*-glycolic acid) (PLGA) microspheres (PMSs) immobilized on biphasic calcium phosphate nanoparticles (BCP NPs) (BCP-IM-PMSs) to enhance osteogenic activity. PMSs were fabricated using a fluidic device, and their surfaces were modified with L-lysine (aminated-PMSs), whereas the BCP NPs were modified with heparin–dopamine (Hep-DOPA) to obtain heparinized–BCP (Hep-BCP) NPs. BCP-IM-PMSs were fabricated via electrostatic interactions between the Hep-BCP NPs and aminated-PMSs. The fabricated BCP-IM-PMSs showed an interconnected pore structure. In vitro studies showed that MG-63 cells cultured on BCP-IM-PMSs had increased alkaline phosphatase activity, calcium content, and mRNA expression of osteocalcin (OCN) and osteopontin (OPN) compared with cells cultured on PMSs. These data suggest that BCP NP-immobilized PMSs have the potential to enhance osteogenic activity.

Keywords: porous microspheres (PMSs); biphasic calcium phosphate (BCP); MG-63 cells; osteogenic activity

1. Introduction

To regenerate bone defects, bone grafts such as autografts and allografts are widely used. These approaches lead to strong autograft osteoinduction, osteoconduction, and osteogenesis as well as allograft osteoinduction and osteoconduction [1]. However, bone grafts have some disadvantages in terms of pain, donor site morbidity, rejection, and disease transmission [2,3]. Given these problems, bone tissue engineering involving a three-dimensional (3D) porous scaffold, stem cells, and cellular stimulation (e.g., via growth factors or drugs) has become a recent focus of interest. Such engineering could potentially help with regeneration or repair of bone defects [4]. 3D porous scaffolds provide the structural support for cell adhesion, proliferation, and differentiation. They also act as substrates for induction of new bone formation at the defect or injury site through surgical procedures [5,6].

However, it is difficult to regenerate or repair bone defects in non-invasive or minimally invasive surgical procedures using 3D porous scaffolds.

From a clinical perspective, injectable scaffolds are a desirable alternative substrate for tissue regeneration of irregular-shaped bone defects since only minimally invasive surgical procedures are required. Moreover, injectable scaffolds have the advantages of small scars, short operation times, and lower risk of infection since only 10–16 gauge needles are used [7,8]. Many injectable scaffolds have been investigated, among which microspheres are generally believed to have the most utility as vehicles for sustained drug or protein delivery due to their inherently small size, small volume, large specific surface area, and high drug loading efficiency [9–11]. Porous microspheres have been determined to be suitable carriers for cells, drugs, and proteins due to their interconnected structures extending between external surface pores and internal core pores [12,13]. However, naked microspheres or porous microspheres (i.e., without cells, proteins, drugs, or osteoconductive material) have potentially limited ability to replace or regenerate bone tissue.

Bioceramics such as hydroxyapatite (HAp), calcium carbonate, calcium sulfate, and β-tricalcium phosphate (β-TCP) have been used for the treatment of bone defects for over 30 years [14]. Biphasic calcium phosphate (BCP) ceramics are related to a group of bone substitute materials consisting of mixtures of HAp and β-TCP in fixed ratios [15,16]. These materials have been shown to possess excellent osteoconductivity, bioactivity, biocompatibility, and biodegradability [17–19]. Although BCP ceramics have many advantages, their intrinsic brittleness makes them unsuitable in calcium phosphate-based scaffolds with complex structures [20–22]. To overcome these drawbacks, composite materials have been pursued. Compared with ceramic and polymers, these composite materials have been shown to have better mechanical properties and osteogenic activity [22,23]. Polymer/ceramic composite scaffolds have been fabricated by chemical foaming, solvent casting, particle/salt leaching, and thermally-induced phase separation [24–28]. However, polymer/ceramic composite scaffolds also have several limitations, including the presence of large aggregates and limited exposure of ceramics on the scaffold surface. As an alternative approach, ceramics have been coated on the surfaces of polymer scaffolds using simulated body fluid (SBF) [23,29,30]. The SBF coating method requires a time-consuming incubation and maintenance of SBF pH, which limits the effectiveness of this approach to generate bioactive bone-like apatite layers on the scaffold surface. A previous study conducted by Kim et al. demonstrated that nano-hydroxyapatite (N-HAp) grafted with anionic poly(ethylene glycol methacrylate phosphate, PolyEGMP) was uniformly immobilized on a chitosan scaffold surface [31]. In addition, these scaffolds were shown to have strong osteogenic potential based on evaluation of alkaline phosphate (ALP) activity in vitro and animal studies. More recently, HAp nanoparticles (NPs) modified with heparin-dopamine (Hep-DOPA) and lactoferrin (LF) were shown to significantly increase alkaline phosphatase (ALP) activity, calcium content, and mRNA expression of osteocalcin (OCN) and osteopontin (OPN) compared with HAp and Hep–HAp NPs [32].

Based on these previous studies, we fabricated porous microspheres (PMSs) using a fluidic device and then introduced positively charged amine groups onto the surfaces of the PMSs through modification with L-lysine. Additionally, Hep-DOPA was anchored on the surface of BCP NPs to introduce negative charge. Finally, Hep-BCP NPs modified with Hep-DOPA were immobilized on the surface of the aminated-PMSs modified with L-lysine in the MES buffer solution, thereby fabricating BCP-immobilized PMSs (BCP-IM-PMSs). The objective of this study was to investigate the potential of BCP-IM-PMSs to enhance osteogenic activity. Hence, we investigated alkaline phosphatase (ALP) activity, calcium content, and gene expression in vitro. Also, we compared it to in vitro osteogenic activities of unmodified PMSs and BCP-mixed PMSs (BCP-IM-PMSs).

2. Materials and Methods

2.1. Materials

Poly (D,L-lactic-*co*-glycolic acid) (PLGA, 50:50, molecular weight: 30,000–60,000), poly vinyl alcohol (PVA, molecular weight: 13,000–23,000, 98% hydrolyzed), dichloromethane (DCM), gelatin from porcine skin, ascorbic acid, dexamethasone, and β-glycerophosphate were purchased from Sigma-Aldrich (St. Louis, MO, USA). Dulbecco's modified Eagle's medium (DMEM), fetal bovine serum (FBS), phosphate-buffered saline (PBS), and penicillin-streptomycin were obtained from Gibco BRL (Rockville, MD, USA). Cell counting kit-8 (CCK-8) reagents were obtained from Dojindo, Inc. (Kumamoto, Japan). Biphasic calcium phosphate (BCP; Hydroxyapatite = 60%, Tricalcium phosphate = 40%) nanoparticles were kindly donated by Ossgen Corporation (Gyeongbuk, Korea). MG-63 cells (human osteosarcoma cell line) were obtained from the Korea Cell Line Bank (KCLB No. 21427, Seoul, Korea). All chemical reagents were of the purest analytical grade available.

2.2. Fabrication of Porous PLGA Microspheres (PMSs), BCP-Mixed PMSs, and BCP-Immobilized PMSs

To fabricate the porous PLGA microspheres (PMSs), BCP-mixed PMSs (BCP-M-PMSs), and BCP-immobilized PMSs (BCP-IM-PMSs), a fluidic device was used as described in our previous work [12,13]. Briefly, to fabricate the BCP-mixed PMSs, BCP NPs (37.8 mg) at a given weight ratio of PLGA to BCP (27 wt %) were dispersed in DCM. PLGA (140 mg) was added to DCM with BCP NPs. Gelatin (7.5 wt %) and PVA (2 wt %) were also added to the BCP/PLGA solution. A homogenizer (Ultra-Turrax T-25 Basic, IKA, Staufen im Breisgau, Germany) was used for 1 min to create a water-in-oil (W/O) polymer emulsion. The W/O emulsion is hereafter referred to as the discontinuous phase, while the PVA solution is hereafter referred to as the continuous phase. The discontinuous and continuous phases had flow rates of 0.06 and 0.65 mL/min, respectively. The BCP-M-PMSs were immersed in warm water at 40 °C under gentle stirring for 1 h to remove residual gelatin, after which the BCP-M-PMSs were washed with PBS and lyophilized for 3 days. To fabricate the BCP-IM-PMSs, the surfaces of the BCP NPs were first modified with Hep-DOPA to anchor the negatively charged groups. In brief, BCP NPs were immersed in 10 mM Tris·HCl (pH 8.0) and then dissolved in Hep-DOPA. This reaction was allowed to proceed for 24 h in the dark. After the reaction, Hep-anchored BCP (Hep-BCP) NPs were rinsed with distilled water (DW) and lyophilized for 3 days. The particle size and zeta potential value of Hep-BCP NPs were 333.17 ± 17.79 nm and -12.72 ± 1.24 mV, respectively. Hep-BCP NPs were well dispersed in aqueous solution due to surface immobilization of negatively charged heparin on BCP NPs. The surface of the PMSs was also modified by L-lysine to anchor the positively-charged groups. L-lysine (2 mg/mL) was dissolved in 0.1 M MES buffer (pH 5.6). Then, PMSs were added to the aforementioned L-lysine solutions and shaken gently overnight. The L-lysine-anchored PMSs are hereafter referred to as aminated-PMSs. To immobilize the Hep-BCP NPs on the surfaces and inside the PMSs, Hep-BCP NPs (37.8 mg) [weight ratio PLGA:BCP (27 wt %)] were dispersed in 0.1 M MES buffer (pH 5.6) using a bath-type ultra sonicator (Powersonic 405, Hwashin Inc., Soul, Korea) for 3 h at 350 W and 4 °C. The aminated-PMSs were then immersed in the Hep-BCP solution. The BCP-immobilized PMSs were washed with DW and lyophilized for 3 days. The BCP-immobilized PMSs are hereafter referred to as BCP-IM-PMSs (Figure 1).

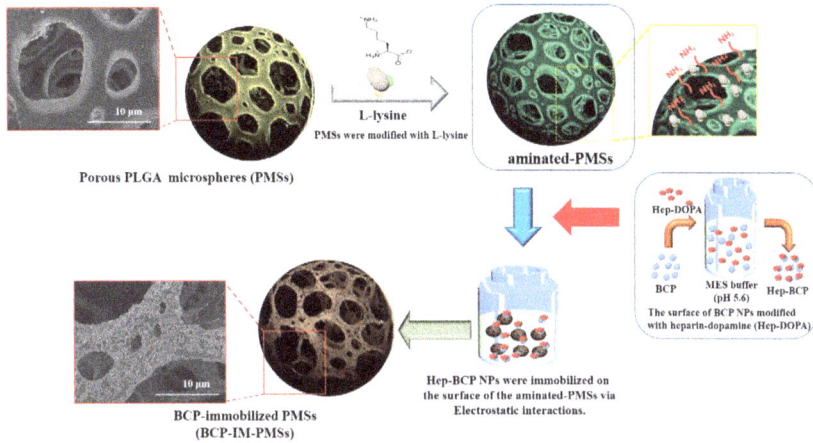

Figure 1. Schematic diagram of the fabrication of biphasic calcium phosphate (BCP)-immobilized porous microspheres (BCP-IM-PMSs). PMSs were fabricated using a fluidic device. Next, the fabricated PMSs were treated with L-lysine, yielding aminated-PMSs. BCP nanoparticles (NPs) were modified with heparin-dopamine (Hep-DOPA) to obtain Hep-BCP NPs. Finally, BCP-immobilized PMSs (BCP-IM-PMSs) were fabricated by immobilization of Hep-BCP NPs on the surface of the aminated-PMSs via electrostatic interactions.

2.3. Characterization of the PMSs, Aminated-PMSs, BCP-M-PMSs, and BCP-IM-PMSs

The morphologies and elemental compositions of the PMSs, aminated-PMSs, BCP-M-PMSs, and BCP-IM-PMSs were determined using a field emission scanning electron microscope (FE-SEM, S-2300, Hitachi, Tokyo, Japan) with energy dispersive X-ray spectroscopy (EDX). To this end, the specimens were coated with gold using a sputter-coater (Eiko IB, Tokyo, Japan), and the FE-SEM was operated at 3 kV. To evaluate the average pore size of each group, selected ($n = 50$) porous PMSs, aminated-PMSs, BCP-M-PMSs, and BCP-IM-PMSs were randomly determined using Image J analysis software (Softonic International, R.B., Barcelona, Spain) based on FE-SEM images. The surface chemical compositions of the PMSs, aminated-PMSs, BCP-M-PMSs, and BCP-IM-PMSs were analyzed by an X-ray photoelectron spectroscopy (XPS) K-Alpha spectrometer (Thermo Electron, Rockford, IL, USA) at the Korea Basic Science Institute Busan Center (Busan, Korea).

2.4. Cell Proliferation

CCK-8 reagents were used to assess cell proliferation as previously described [12,13,32]. Briefly, MG-63 cells were carefully seeded at a concentration of 1×10^5 cells/mL on 10 mg of PMSs, aminated-PMSs, BCP-M-PMSs, and BCP-IM-PMSs in 24-well tissue culture plates. Cells were maintained in DMEM supplemented with 10% FBS and 1% antibiotics (100 U/mM penicillin, 0.1 mg/mL streptomycin). CCK-8 reagents were added to each sample and incubated for 1 h after 3 and 7 days of incubation. Cell proliferation was determined by measuring the absorbance of each well at 450 nm using a Flash Multimode Reader (Varioskan™, Thermo, 3001–2019, Grand Island, NY, USA).

2.5. Alkaline Phosphatase (ALP) Activity

MG-63 cells were obtained from Korea Cell Line Bank (No. 21427, Seoul, Korea). In this study, we used MG-63 cells because they are commonly used to evaluate osteogenic evaluation of bioceramics, bioactive drugs, peptides, and proteins in/on various scaffolds.

MG-63 cells were carefully seeded at a density of 1×10^5 cells/mL on 10 mg of PMSs, aminated-PMSs, BCP-M-PMSs, or BCP-IM-PMSs in 24-well tissue culture plates. Cells were cultured in DMEM supplemented with 10% FBS, 50 µg/mL ascorbic acid, 10 nM dexamethasone, and 10 mM β-glycerophosphate with 100 U/mL penicillin and 100 µg/mL streptomycin for 10 days. At predetermined time intervals (3, 7, and 10 days), each sample was rinsed with PBS, and 1× RIPA buffer was added for cell lysis. To remove cell debris, cell lysates were centrifuged at 13,500 rpm for 1 min. After centrifugation, *p*-nitrophenyl phosphate solution was added to the supernatants. The reactions were incubated for 30 min at 37 °C and then stopped with 1 M NaOH. Based on the standard of *p*-nitrophenyl phosphate solution, the ALP activity was determined by converting *p*-nitrophenyl phosphate to *p*-nitrophenol. ALP activity was evaluated using a Flash Multimode Reader to measure the absorbance at 405 nm.

2.6. Calcium Deposition

MG-63 cells (1×10^5 cells/well) were carefully seeded on 10 mg of PMSs, aminated-PMSs, BCP-M-PMSs, or BCP-IM-PMSs in 24-well tissue culture plates. At predetermined time intervals (days 7 and 21), 0.5 M HCl (400 µL) was added to the cell/substrate mixtures. After centrifugation, calcium deposition in the supernatants was measured using a QuantiChrom Calcium Assay Kit (DICA-500, BioAssay Systems, Hayward, CA, USA) according to the manufacturer's instructions. Calcium chloride was used as a standard. The amount of deposited calcium was determined using a Flash Multimode Reader to measure the absorbance at 612 nm.

2.7. Gene Expression

To quantify the mRNA expression levels of the osteogenic differentiation markers osteocalcin (OCN) and osteopontin (OPN), quantitative real-time polymerase chain reaction (RT-PCR) was performed on days 7 and 21. MG-63 cells (1×10^5 cells/well) were seeded on 10 mg of PMSs, aminated-PMSs, BCP-M-PMSs, or BCP-IM-PMSs and cultured in 24-well tissue culture plates. Total RNA (1 µg) was used for cDNA synthesis with AccuPower RT PreMix (Bioneer, Daejeon, Korea). The following primers were used: osteocalcin (OCN): (F) 5′-TTG GTG CAC ACC TAG CAG AC-3′, (R) 5′-ACC TTA TTG CCC TCC TGC TT-3′; osteopontin (OPN): (F) 5′-GAG GGC TTG GTT GTC AGC-3′, (R) 5′-CAA TTC TCA TGG TAG TGA GTT TTC C-3′; and glyceraldehyde 3-phosphate dehydrogenase (GAPDH) (F) 5′-ACT TTG TCA AGC TCA TTT CC-3′, (R) 5′-TGC AGC GAA CTT TAT TGA TG-3′. PCR amplification and detection were carried out on an ABI7300 Real-Time Thermal Cycler (Applied Biosystems, Foster, CA, USA) using a DyNAmoTM SYBR® Green qPCR Kit (Finnzymes, Espoo, Finland). OCN and OPN mRNA expression levels were normalized to those of GAPDH and are expressed as relative values.

2.8. Statistical Analysis

Quantitative data are presented as mean ± standard deviation. Data were compared between groups via one-way ANOVA using SYSTAT software (Chicago, IL, USA). Differences were considered to be statistically significant at * $p < 0.05$ and ** $p < 0.01$.

3. Results

3.1. Characterization of PMSs, aminated–PMSs, BCP-M-PMSs, and BCP-IM-PMSs

Figure 2 shows the morphologies and sizes of the PMSs, aminated-PMSs, BCP-M-PMSs, and BCP-IM-PMSs as determined by FE-SEM. The fabricated PMSs, aminated-PMSs, BCP-M-PMSs, and BCP-IM-PMSs were round and exhibited interconnected porous structures in their interiors and on their surfaces. The BCP-M-PMSs and BCP-IM-PMSs both displayed coarse surfaces, whereas the PMSs and aminated-PMSs had smooth surfaces. Additionally, the surfaces of the BCP-IM-PMSs were rougher than those of the BCP-M-PMSs. The average pore sizes were 24 ± 7 µm for PMSs, 19 ± 6 µm for aminated-PMSs, 22 ± 11 µm for BCP-M-PMSs, and 23 ± 7 µm for BCP-IM-PMSs. EDX analysis was also used to confirm the elemental compositions of the PMSs, aminated-PMSs, BCP-M-PMSs, and BCP-IM-PMSs. Successful anchoring of L-lysine on the surface of and inside the PMSs was confirmed by the increased N content of the aminated-PMSs compared with that of the naked PMSs. The BCP-M-PMSs and BCP-IM-PMSs were characterized by increased P and Ca contents. In particular, the BCP-IM-PMSs had lower N content compared to aminated PMSs, and showed higher P and Ca contents than the BCP-M-PMSs (Table 1), indicating that the negatively charged Hep-BCP NPs were successfully immobilized on the aminated PMSs. Next, the surface chemical compositions of the PMSs, aminated-PMSs, BCP-M-PMSs, and BCP-IM-PMSs were evaluated by XPS (Table 2). In the L-lysine-anchored PMSs, the N1 content was increased compared to that of the PMSs. In addition, the P2p and Ca2p contents of both the BCP-M-PMSs and BCP-IM-PMSs were greater than those of the aminated-PMSs. Furthermore, the P2p and Ca2p contents of the BCP-IM-PMSs were 0.87% to 1.02% and 1.91% to 3.87% greater, respectively, than those of the BCP-M-PMSs.

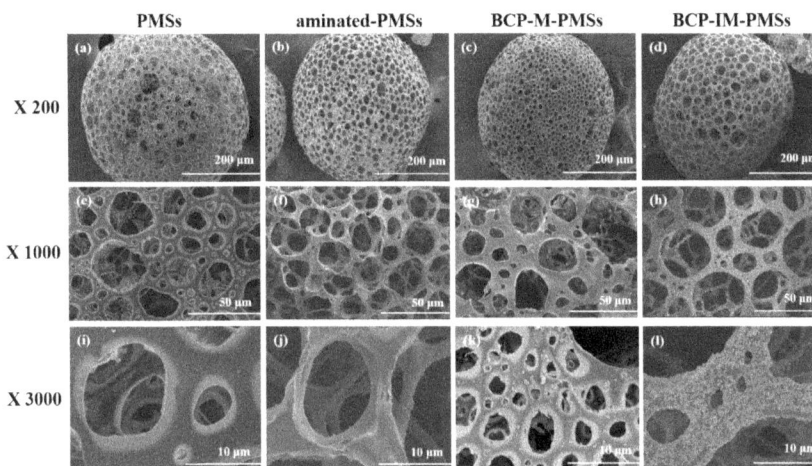

Figure 2. Scanning electron microscope (SEM) images of porous microspheres (PMSs) (a,e,i), aminated-PMSs (b,f,j), BCP-mixed PMSs (BCP-M-PMSs) (c,g,k), and BCP-IM-PMSs (d,h,l).

Table 1. Elemental compositions of PMSs, aminated PMSs, BCP-M-PMSs, and BCP-IM-PMSs.

Samples	C (%)	O (%)	N (%)	P (%)	Ca (%)	Total (%)
PMSs	57.31	42.69	-	-	-	100
aminated-PMSs	42.36	30.44	27.20	-	-	100
BCP-M-PMSs	54.14	43.51	-	0.99	1.36	100
BCP-IM-PMSs	44.52	38.96	11.84	1.90	2.78	100

Table 2. Surface elemental compositions of PMSs, aminated PMSs, BCP-M-PMSs, and BCP-IM-PMSs.

Samples	C1s (%)	O1s (%)	N1s (%)	P2p (%)	Ca2p (%)	Total (%)
PMSs	73.96	26.04	-	-	-	100
aminated-PMSs	56.48	34.80	8.72	-	-	100
BCP-M-PMSs	63.54	33.68	-	0.87	1.91	100
BCP-IM-PMSs	61.12	28.55	5.44	1.02	3.87	100

3.2. Cell Proliferaiton

After 3 and 7 days of incubation, the proliferation of MG-63 cells grown on PMSs, aminated-PMSs, BCP-M-PMSs, or BCP-IM-PMSs was investigated (Figure 3). In our previous study, we already confirmed with confocal laser scanning microscope that the carefully seeded cells were attached in/on the PMSs [33]. MG-63 cells grown on PMSs with BCP exhibited significantly greater proliferation than cells grown on PMSs or on aminated-PMSs at days 3 and 7 (** $p < 0.01$). Moreover, MG-63 cell proliferation on BCP-IM-PMSs was significantly different than that of cells on BCP-M-PMSs on days 3 and 7 (** $p < 0.01$).

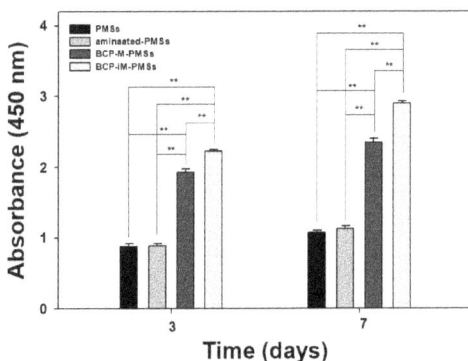

Figure 3. Proliferation of MG-63 cells cultured on PMSs, aminated-PMSs, BCP-M-PMSs, or BCP-IM-PMSs for 3 or 7 days ($n = 5$) (* $p < 0.05$ and ** $p < 0.01$).

3.3. ALP Activity

The ALP activity of MG-63 cells cultured on PMSs, aminated-PMSs, BCP-M-PMSs, or BCP-IM-PMSs was determined after 3, 7, and 10 days (Figure 4). At day 3, the ALP activity of MG-63 cells cultured on BCP-M-PMSs or BCP-IM-PMSs was significantly different from that of cells cultured on PMSs alone (** $p < 0.01$). At day 7, the ALP activity of MG-63 cells grown on BCP-M-PMSs or BCP-IM-PMSs was significantly increased compared to that of MG-63 cells cultured on PMSs (** $p < 0.01$). At 10 days, MG-63 cells grown on BCP-M-PMSs or BCP-IM-PMSs had significantly greater ALP activity than cells grown on PMSs (** $p < 0.01$). Additionally, MG-63 cells grown on BCP-IM-PMSs had significantly higher ALP activity than MG-63 cells grown on BCP-M-PMSs at days 3, 7, and 10 days (* $p < 0.05$ and ** $p < 0.01$, respectively). However, ALP activity was not markedly different on days 3, 7, and 10 for MG-63 cells grown on PMSs versus cells grown on aminated-PMSs.

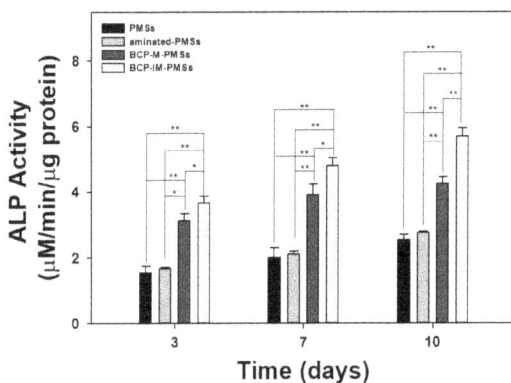

Figure 4. Alkaline phosphatase (ALP) activity of MG-63 cells cultivated on PMSs, aminated-PMSs, BCP-M-PMSs, or BCP-IM-PMSs after 3, 7, or 10 days of incubation ($n = 5$) (* $p < 0.05$ and ** $p < 0.01$).

3.4. Calcium Deposition

Figure 5 shows the amounts of calcium deposited by MG-63 cells cultivated on PMSs, aminated-PMSs, BCP-M-PMSs, or BCP-IM-PMSs after 7 or 21 days of incubation. On day 7, MG-63 cells cultured on BCP-IM-PMSs had deposited significantly greater calcium than MG-63 cells grown on PMSs (** $p < 0.01$). On day 21, MG-63 cells cultured on BCP-IM-PMSs had deposited markedly more calcium compared with cells cultured on PMSs (** $p < 0.01$). Moreover, MG-63 cells cultured on BCP-IM-PMSs had deposited significantly different amounts of calcium on days 7 and 21 compared to cells cultured on BCP-M-PMSs (** $p < 0.01$).

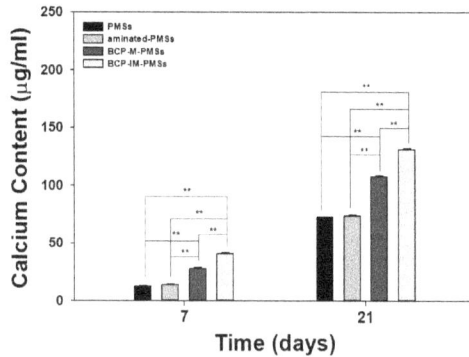

Figure 5. Amount of calcium deposited by MG-63 cells grown on PMSs, aminated-PMSs, BCP-M-PMSs, or BCP-IM-PMSs after 7, 14, or 21 days of incubation ($n = 5$) (* $p < 0.05$ and ** $p < 0.01$).

3.5. Gene Expression

After 7 and 21 days of incubation, real-time PCR was used to determine the mRNA expression levels of OCN and OPN in MG-63 cells cultured on PMSs, aminated-PMSs, BCP-M-PMSs, or BCP-IM-PMSs (Figure 6). As shown in Figure 6a,b, OCN and OPN mRNA expression levels were significantly different at days 7 and 21 between MG-63 cells grown on BCP-M-PMSs or BCP-IM-PMSs compared with those grown on PMSs (** $p < 0.01$). In addition, OCN and OPN expression levels were markedly higher on days 7 and 21 in MG-63 cells grown on BCP-IM-PMSs than in cells grown on BCP-M-PMSs (** $p < 0.01$). However, the OCN and OPN expression levels in MG-63 cells cultured on PMSs and MG-63 cells cultured on aminated-PMSs were not significantly different at any time-point.

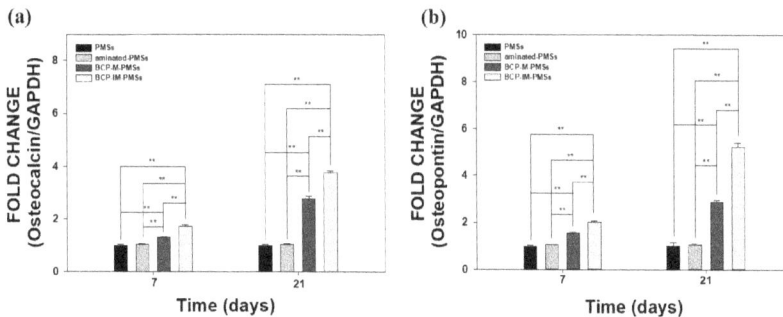

Figure 6. Real-time PCR analysis of (**a**) osteocalcin (OCN) and (**b**) osteopontin (OPN) expression in MG-63 cells after 7 or 21 days of incubation with PMSs, aminated-PMSs, BCP-M-PMSs, or BCP-IM-PMSs ($n = 5$) (* $p < 0.05$ and ** $p < 0.01$).

4. Discussion

The purpose of this study was to investigate the potential of BCP-IM-PMSs to enhance osteogenic activity through in vitro studies. PMSs, which have porous structures, are believed to be good environments for fostering cell attachment, proliferation, and differentiation [34,35]. In the conventional method, the porogen ammonium bicarbonate (NH_4HCO_3) is added to the appropriate solution for PMS preparation in a water-in-oil-in-water (W/O/W) emulsion system [35,36]. The porogen is removed by solvent casting/particulate leaching, thereby forming interconnective pores in both the surface and interior. However, porogen-derived PMSs have inherent restrictions. For example, the high-shear natural method of emulsification generates microspheres with heterogeneous sizes and morphologies [34,35]. In addition, it is time-consuming to completely eliminate the porogens [34,37]. In this study, we fabricated PMSs with or without BCP with interconnected pores using a fluidic device. The SEM images demonstrated that the fabricated PMSs with or without BCP had spherical structures and open surface and interior pores. Our previous studies demonstrated that PMSs made from fluidic devices are round, with interconnected surfaces and interior pores [12,13]. In addition, high magnification SEM images revealed that both PMSs and aminated-PMSs had relatively smooth surfaces. On the other hand, PMSs with BCP NPs had rough surfaces. Moreover, BCP-IM-PMSs had much rougher surfaces than BCP-M-PMSs because the BCP NPs were immobilized on the surface of the PMSs. Chitosan scaffolds with immobilized HAp NPs were previously shown to have rougher, more uniform surfaces compared with HAp NP-mixed chitosan scaffolds [31]. Notably, the SEM results of Kim et al. [31] are consistent with our results.

To fabricate BCP-IM-PMSs, Hep-DOPA was used to easily modify the BCP NP surface. This step utilizes the marine mussel's adhesive mechanism [38]. DOPA, which harbors ortho-dihydroxyphenyl (catechol) functional groups, readily changes different materials including organic, inorganic, and metallic materials in alkaline solutions (pH 8.0) [38]. Our previous studies reported that Hep-DOPA molecules were readily immobilized on the surface of PMS, hydroxyapatite (HAp), and titanium substrates in basic environments [12,13,32,39]. Thus, based on our previous studies, we chose to immobilize Hep-DOPA molecules on the surfaces of the BCP NPs. In the dispersion test, the BCP NPs modified with Hep-DOPA were well-dispersed in aqueous conditions, whereas pure BCP NPs were rapidly precipitated in aqueous conditions within 30 min (data not shown). Consistent with our previous report, HAp NPs with adherent Hep-DOPA molecules were well-diffused in aqueous conditions for up to 6 h and maintained turbidity for 24 h. However, pure HAp NPs were quickly precipitated in 1 h due to extensive aggregation [32]. These results suggest that heparin-immobilized BCP or HAp NPs prohibit the aggregation of BCP or HAp NPs, possibly due to the repulsion of heparin molecules, which have a high content of negatively-charged sulfate and carboxyl groups. Thus, we chose to fabricate BCP NP-immobilized PMSs because we hypothesized that they might remain well-dispersed during surface immobilization on aminated-PMSs.

The EDX and XPS results showed that PMSs with anchored L-lysine showed increased nitrogen (N) content compared with naked PMSs. Lee and colleagues [40] similarly confirmed that an L-lysine–grafted titanium (Ti) substrate had increased nitrogen content compared with the Ti substrate only. Phosphorus and calcium were found in both BCP-M-PMSs and BCP-IM-PMSs; however, BCP-IM-PMSs showed higher phosphorus and calcium levels than BCP-M-PMSs. This result suggests that the Hep-BCP NPs immobilized on the surface of the aminated-PMSs via electrostatic interactions are more surface-exposed than the PMSs with the BCP NPs via the mixing method. In one previous study, chitosan scaffolds with surface-immobilized HAp were reported to show increased phosphorus and calcium contents compared with HAp-mixed chitosan scaffolds [31]. The XPS result of this study is consistent with our results.

To evaluate the osteogenic capabilities of the PMSs, aminated-PMSs, BCP-M-PMSs, and BCP-IM-PMSs, ALP activity, calcium content, and osteocalcin (OCN) and osteopontin (OPN) expression were measured in MG-63 cells. At days 3 and 7, MG-63 cells grown on BCP-M-PMSs or BCP-IM-PMSs exhibited significantly increased proliferation compared to those grown on

PMSs (** $p < 0.01$). Moreover, MG-63 cells cultivated on BCP-IM-PMSs exhibited markedly greater proliferation on days 3 and 7 than those cultivated on BCP-M-PMSs (** $p < 0.01$). However, MG-63 cells grown on PMSs and aminated-PMSs exhibited similar proliferation during the 7-day culture period. Cells cultured on scaffolds with surface-immobilized nano-HAp exhibited significantly greater proliferation than cells cultured on scaffolds alone or N-HAp-mixed scaffolds [23,31]. These results imply that surface-placed nanoscale ceramics are key for optimal cell adhesion and proliferation.

As key components of bone matrix vesicles and mineralization, ALP activity and calcium content are often evaluated as markers of early and late osteogenic differentiation, respectively [12,13,39,40]. The ALP activity of MG-63 cells cultured on PMSs, aminated-PMSs, BCP-M-PMSs, or BCP-IM-PMSs was increased for culture times up to 10 days. Furthermore, ALP activity was significantly higher on days 3, 7, and 10 in cells cultured on PMSs with BCP NPs compared with cells cultured on PMSs alone. Moreover, significant differences in ALP activity were observed between MG-63 cells cultured on BCP-IM-PMSs versus those cultured on BCP-M-PMSs. Consistent with the ALP activity results, the calcium contents of MG-63 cells grown on BCP-M-PMSs or BCP-IM-PMSs were significantly greater than those of cells grown on PMSs. Additionally, MG-63 cells cultured on BCP-IM-PMSs had significantly different calcium content compared with cells cultured on BCP-M-PMSs. Kim et al. [29] demonstrated that HAp-coated PLGA scaffolds modified by incubation with 5× simulated body fluid (SBF) solution showed increased ALP activity and calcium deposition of osteoblast cells compared with unmodified PLGA scaffolds. Moreover, Kim et al. [31] confirmed that human adipose-derived stem cells (hADSCs) cultured on HAp-immobilized chitosan scaffolds showed significantly different ALP activity compared to those cultured on chitosan scaffolds alone. Taken together, these results demonstrate that BCP/HAp-immobilized and BCP/HAp-coated scaffolds directly affect both early and late differentiation of osteoblast cells and stem cells.

To evaluate osteogenic differentiation, we performed real-time PCR to measure the gene expression of OCN and OPN after 7 and 21 days of incubation. Expression of OCN and OPN is known to be upregulated in osteoblast-like cells after stimulation of osteogenic differentiation [12,39,40]. At days 7 and 21, MG-63 cells cultured on PMSs with BCP NPs exhibited significantly higher OCN and OPN expression than cells cultured on PMSs. Moreover, the OCN and OPN mRNA levels in MG-63 cells grown on BCP-IM-PMSs were markedly higher than those in cells grown on BCP-M-PMSs. These results are consistent with the ALP activity and calcium content findings. The mRNA levels of OCN, OPN, and Runt-related transcription factor2 (*Runx2*) in rat bone marrow stem cells (BMSCs) grown on hydroxyapatite-coated genipin-chitosan conjugation scaffolds (HGCCS) have been shown to be significantly higher than those in cells grown on a genipin cross-linked chitosan framework (GCF) [41]. These results indicate that BCP immobilization and HAp coating on the PMS or scaffold surface both directly promote osteogenic differentiation of osteoblast cells and stem cells.

Recently, interest has increased in using polymer/ceramic composite microspheres as injectable scaffolds in bone regeneration. These composite microspheres have many advantages, including the osteoconductive properties of ceramics and a favorable environment for cell adhesion, proliferation, and differentiation. In this study, we showed that PMSs and aminated-PMSs had negligible osteogenic effects and mineralization, and the surface modification of PMSs with BCP NPs could more greatly promote the osteogenic activities of MG-63 cells than mixing of BCP NPs within PMSs. These four types of microspheres have similar pore sizes. However, BCP-IM-PMSs showed the highest osteogenic activities compared to other PMSs because surface immobilized BCP NPs on PMSs was similar to the minerals of natural bone to favor cell differentiation, and the released calcium and phosphate ions after degradation of BCP NPs might facilitate the formation of the biological apatite on scaffolds [42]. This might indicate that surface immobilization of BCP NPs on PMSs leads to the enhanced osteogenesis effects rather than pore sizes of PMSs. Thus, we confirmed that that BCP-IM-PMSs markedly enhanced osteogenic activity compared to BCP-M-PMSs. The surface of porous microspheres can be easily modified with BCP NPs and also used as delivery carriers of many kinds of drugs (i.e., anti-cancer drugs, antibiotics, anti-inflammatory drugs, and bone stimulating medications). Thus, porous microspheres with multifunctional properties can be broadly used to treat fractures, tumors, and inflammatory diseases. For their clinical application, future studies must be performed in animal models with bone defects associated with fractures, tumors, and inflammation.

5. Conclusions

We fabricated PMSs using a fluidic device. The fabricated PMSs were modified with L-lysine and showed a positively-charged surface. The BCP NPs were also modified by Hep-DOPA to obtain Hep-BCP NPs, which had a negatively-charged surface. Next, Hep-BCP NPs were immobilized on the surface of the aminated-PMSs via electrostatic interactions to obtain BCP-IM-PMSs. We demonstrated that MG-63 cells cultured with BCP-IM-PMSs showed significantly higher ALP activity, calcium deposition, and expression of osteogenic differentiation genes (OCN and OPN) compared with cells cultured with PMSs or BCP-M-PMSs. Therefore, we suggest that BCP-immobilized porous microspheres can be used to enhance osteogenic activity.

Acknowledgments: This study was financially supported by a grant from the Korean Health Technology R&D Project, Ministry of Health & Welfare, Republic of Korea (HI11C0388), and by a Korea University Guro Hospital Grant (O1500231).

Author Contributions: Sung Eun Kim, Kyeongsoon Park, and Hae-Ryong Song conceived and designed the study; Kyu-Sik Shim and Somang Choi conducted the experiments; Young-Pil Yun and Hak-Jun Kim analyzed the data; Sung Eun Kim and Kyeongsoon Park wrote the paper and reviewed the final paper.

Conflicts of Interest: The authors declare no conflict of interest.

References

1. Oryan, A.; Alidadi, S.; Moshiri, A.; Maffulli, N. Bone regenerative medicine: Classic options, novel strategies, and future directions. *J. Orthop. Surg. Res.* **2014**, *9*, 18. [CrossRef] [PubMed]
2. Nandi, S.K.; Roy, S.; Mukherjee, P.; Kundu, B.; De, D.K.; Basu, D. Orthopaedic applications of bone graft & graft substitutes: A review. *Indian J. Med. Res.* **2010**, *132*, 15–30. [PubMed]
3. Athanasiou, V.T.; Papachristou, D.J.; Panagopoulos, A.; Saridis, A.; Scopa, C.D.; Megas, P. Histological comparison of autograft, allograft-dbm, xenograft, and synthetic grafts in a trabecular bone defect: An experimental study in rabbits. *Med. Sci. Monit.* **2010**, *16*, BR24–BR31. [PubMed]
4. Amini, A.R.; Laurencin, C.T.; Nukavarapu, S.P. Bone tissue engineering: Recent advances and challenges. *Crit. Rev. Biomed. Eng.* **2012**, *40*, 363–408. [CrossRef] [PubMed]
5. Antonov, E.N.; Bagratashvili, V.N.; Whitaker, M.J.; Barry, J.J.; Shakesheff, K.M.; Konovalov, A.N.; Popov, V.K.; Howdle, S.M. Three-dimensional bioactive and biodegradable scaffolds fabricated by surface-selective laser sintering. *Adv. Mater.* **2004**, *17*, 327–330. [CrossRef] [PubMed]
6. Bettinger, C.J.; Weinberg, E.J.; Kulig, K.M.; Vacanti, J.P.; Wang, Y.D.; Borenstein, J.T.; Langer, R. Three-dimensional microfluidic tissue-engineering scaffolds using a flexible biodegradable polymer. *Adv. Mater.* **2006**, *18*, 165–169. [CrossRef] [PubMed]
7. Dreifke, M.B.; Ebraheim, N.A.; Jayasuriya, A.C. Investigation of potential injectable polymeric biomaterials for bone regeneration. *J. Biomed. Mater. Res. A* **2013**, *101*, 2436–2447. [CrossRef] [PubMed]
8. Munarin, F.; Petrini, P.; Bozzini, S.; Tanzi, M.C. New perspectives in cell delivery systems for tissue regeneration: Natural-derived injectable hydrogels. *J. Appl. Biomater. Funct. Mater.* **2012**, *10*, 67–81. [CrossRef] [PubMed]
9. Freiberg, S.; Zhu, X.X. Polymer microspheres for controlled drug release. *Int. J. Pharm.* **2004**, *282*, 1–18. [CrossRef] [PubMed]
10. Biondi, M.; Ungaro, F.; Quaglia, F.; Netti, P.A. Controlled drug delivery in tissue engineering. *Adv. Drug Deliv. Rev.* **2008**, *60*, 229–242. [CrossRef] [PubMed]
11. Wang, H.A.; Leeuwenburgh, S.C.G.; Li, Y.B.; Jansen, J.A. The use of micro- and nanospheres as functional components for bone tissue regeneration. *Tissue Eng. B* **2012**, *18*, 24–39. [CrossRef] [PubMed]
12. Kim, S.E.; Yun, Y.P.; Shim, K.S.; Park, K.; Choi, S.W.; Suh, D.H. Effect of lactoferrin-impregnated porous poly(lactide-*co*-glycolide) (plga) microspheres on osteogenic differentiation of rabbit adipose-derived stem cells (radscs). *Colloids Surf. B* **2014**, *122*, 457–464. [CrossRef] [PubMed]

13. Kim, S.E.; Yun, Y.P.; Shim, K.S.; Park, K.; Choi, S.W.; Shin, D.H.; Suh, D.H. Fabrication of a bmp-2-immobilized porous microsphere modified by heparin for bone tissue engineering. *Colloids Surf. B* **2015**, *134*, 453–460. [CrossRef] [PubMed]

14. Hulbert, S.F.; Young, F.A.; Mathews, R.S.; Klawitter, J.J.; Talbert, C.D.; Stelling, F.H. Potential of ceramic materials as permanently implantable skeletal prostheses. *J. Biomed. Mater. Res.* **1970**, *4*, 433–456. [CrossRef] [PubMed]

15. LeGeros, R.Z.; Lin, S.; Rohanizadeh, R.; Mijares, D.; LeGeros, J.P. Biphasic calcium phosphate bioceramics: Preparation, properties and applications. *J. Mater. Sci. Mater. Med.* **2003**, *14*, 201–209. [CrossRef] [PubMed]

16. Suneelkumar, C.; Datta, K.; Srinivasan, M.R.; Kumar, S.T. Biphasic calcium phosphate in periapical surgery. *J. Conserv. Dent.* **2008**, *11*, 92–96. [CrossRef] [PubMed]

17. Yuan, H.; van Blitterswijk, C.A.; de Groot, K.; de Bruijn, J.D. Cross-species comparison of ectopic bone formation in biphasic calcium phosphate (BCP) and hydroxyapatite (HA) scaffolds. *Tissue Eng.* **2006**, *12*, 1607–1615. [CrossRef] [PubMed]

18. Stahli, C.; Bohner, M.; Bashoor-Zadeh, M.; Doebelin, N.; Baroud, G. Aqueous impregnation of porous beta-tricalcium phosphate scaffolds. *Acta Biomater.* **2010**, *6*, 2760–2772. [CrossRef] [PubMed]

19. Nie, L.; Chen, D.; Suo, J.; Zou, P.; Feng, S.; Yang, Q.; Yang, S.; Ye, S. Physicochemical characterization and biocompatibility in vitro of biphasic calcium phosphate/polyvinyl alcohol scaffolds prepared by freeze-drying method for bone tissue engineering applications. *Colloids Surf. B* **2012**, *100*, 169–176. [CrossRef] [PubMed]

20. Rezwan, K.; Chen, Q.Z.; Blaker, J.J.; Boccaccini, A.R. Biodegradable and bioactive porous polymer/inorganic composite scaffolds for bone tissue engineering. *Biomaterials* **2006**, *27*, 3413–3431. [CrossRef] [PubMed]

21. Sun, H.; Yang, H.L. Calcium phosphate scaffolds combined with bone morphogenetic proteins or mesenchymal stem cells in bone tissue engineering. *Chin. Med. J. (Engl.)* **2015**, *128*, 1121–1127. [PubMed]

22. Nie, L.; Suo, J.P.; Zou, P.; Feng, S.B. Preparation and properties of biphasic calcium phosphate scaffolds multiply coated with ha/plla nanocomposites for bone tissue engineering applications. *J. Nanomater.* **2012**, *2012*, 213549. [CrossRef]

23. Kang, S.W.; Yang, H.S.; Seo, S.W.; Han, D.K.; Kim, B.S. Apatite-coated poly(lactic-*co*-glycolic acid) microspheres as an injectable scaffold for bone tissue engineering. *J. Biomed. Mater. Res. A* **2008**, *85*, 747–756. [CrossRef] [PubMed]

24. Kim, S.; Kim, S.S.; Lee, S.H.; Eun Ahn, S.; Gwak, S.J.; Song, J.H.; Kim, B.S.; Chung, H.M. In vivo bone formation from human embryonic stem cell-derived osteogenic cells in poly(d,l-lactic-co-glycolic acid)/hydroxyapatite composite scaffolds. *Biomaterials* **2008**, *29*, 1043–1053. [CrossRef] [PubMed]

25. Zhang, P.; Hong, Z.; Yu, T.; Chen, X.; Jing, X. In vivo mineralization and osteogenesis of nanocomposite scaffold of poly(lactide-*co*-glycolide) and hydroxyapatite surface-grafted with poly(L-lactide). *Biomaterials* **2009**, *30*, 58–70. [CrossRef] [PubMed]

26. Luangphakdy, V.; Walker, E.; Shinohara, K.; Pan, H.; Hefferan, T.; Bauer, T.W.; Stockdale, L.; Saini, S.; Dadsetan, M.; Runge, M.B.; et al. Evaluation of osteoconductive scaffolds in the canine femoral multi-defect model. *Tissue Eng. A* **2013**, *19*, 634–648. [CrossRef] [PubMed]

27. Mekala, N.K.; Baadhe, R.R.; Parcha, S.R.; Yalavarthy, P.D. Physical and degradation properties of PLGA scaffolds fabricated by salt fusion technique. *J. Biomed. Res.* **2013**, *27*, 318–325. [PubMed]

28. Maquet, V.; Boccaccini, A.R.; Pravata, L.; Notingher, I.; Jerome, R. Porous poly(α-hydroxyacid)/bioglass composite scaffolds for bone tissue engineering. I: Preparation and in vitro characterisation. *Biomaterials* **2004**, *25*, 4185–4194. [CrossRef] [PubMed]

29. Kim, S.S.; Park, M.S.; Gwak, S.J.; Choi, C.Y.; Kim, B.S. Accelerated bonelike apatite growth on porous polymer/ceramic composite scaffolds in vitro. *Tissue Eng.* **2006**, *12*, 2997–3006. [CrossRef] [PubMed]

30. Zhang, R.Y.; Ma, P.X. Porous poly(L-lactic acid)/apatite composites created by biomimetic process. *J. Biomed. Mater. Res.* **1999**, *45*, 285–293. [CrossRef]

31. Kim, S.E.; Choi, H.W.; Lee, H.J.; Chang, J.H.; Choi, J.; Kim, K.J.; Lim, H.J.; Jun, Y.J.; Lee, S.C. Designing a highly bioactive 3D bone-regenerative scaffold by surface immobilization of nano-hydroxyapatite. *J. Mater. Chem.* **2008**, *18*, 4994–5001. [CrossRef]

32. Kim, S.E.; Lee, D.W.; Yun, Y.P.; Shim, K.S.; Jeon, D.I.; Rhee, J.K.; Kim, H.J.; Park, K. Heparin-immobilized hydroxyapatite nanoparticles as a lactoferrin delivery system for improving osteogenic differentiation of adipose-derived stem cells. *Biomed. Mater.* **2016**, *11*, 025004. [CrossRef] [PubMed]

Polymers **2017**, *9*, 297

33. Lee, J.Y.; Kim, S.E.; Yun, Y.P.; Choi, S.W.; Jeon, D.I.; Kim, H.J.; Park, K.; Song, H.R. Osteogenesis and new bone formation of alendronate-immobilized porous plga microspheres in a rat calvarial defect model. *J. Ind. Eng. Chem.* **2017**, *52*, 277–286. [CrossRef]

34. Cai, Y.P.; Chen, Y.H.; Hong, X.Y.; Liu, Z.G.; Yuan, W.E. Porous microsphere and its applications. *Int. J. Nanomed.* **2013**, *8*, 1111–1120.

35. Lee, T.J.; Kang, S.W.; Bhang, S.H.; Kang, J.M.; Kim, B.S. Apatite-coated porous poly(lactic-*co*-glycolic acid) microspheres as an injectable bone substitute. *J. Biomater. Sci. Polym. E* **2010**, *21*, 635–645. [CrossRef] [PubMed]

36. Chung, H.J.; Kim, I.K.; Kim, T.G.; Park, T.G. Highly open porous biodegradable microcarriers: In vitro cultivation of chondrocytes for injectable delivery. *Tissue Eng. A* **2008**, *14*, 607–615. [CrossRef] [PubMed]

37. Li, J.S.; Chen, Y.; Mak, A.F.T.; Tuan, R.S.; Li, L.; Li, Y. A one-step method to fabricate plla scaffolds with deposition of bioactive hydroxyapatite and collagen using ice-based microporogens. *Acta Biomater.* **2010**, *6*, 2013–2019. [CrossRef] [PubMed]

38. Lee, H.; Dellatore, S.M.; Miller, W.M.; Messersmith, P.B. Mussel-inspired surface chemistry for multifunctional coatings. *Science* **2007**, *318*, 426–430. [CrossRef] [PubMed]

39. Kim, S.E.; Yun, Y.P.; Lee, J.Y.; Park, K.; Suh, D.H. Osteoblast activity of mg-63 cells is enhanced by growth on a lactoferrin-immobilized titanium substrate. *Colloid Surf. B* **2014**, *123*, 191–198. [CrossRef] [PubMed]

40. Lee, S.Y.; Yun, Y.P.; Song, H.R.; Chun, H.J.; Yang, D.H.; Park, K.; Kim, S.E. The effect of titanium with heparin/bmp-2 complex for improving osteoblast activity. *Carbohydr. Polym.* **2013**, *98*, 546–554. [CrossRef] [PubMed]

41. Wang, G.C.; Zheng, L.; Zhao, H.S.; Miao, J.Y.; Sun, C.H.; Ren, N.; Wang, J.Y.; Liu, H.; Tao, X.T. In vitro assessment of the differentiation potential of bone marrow-derived mesenchymal stem cells on genipin-chitosan conjugation scaffold with surface hydroxyapatite nanostructure for bone tissue engineering. *Tissue Eng. A* **2011**, *17*, 1341–1349. [CrossRef] [PubMed]

42. Le Nihouannen, D.; Saffarzadeh, A.; Gauthier, O.; Moreau, F.; Pilet, P.; Spaethe, R.; Layrolle, P.; Daculsi, G. Bone tissue formation in sheep muscles induced by a biphasic calcium phosphate ceramic and fibrin glue composite. *J. Mater. Sci. Mater. Med.* **2008**, *19*, 667–675. [CrossRef] [PubMed]

polymers

MDPI

Article

In Vitro Evaluation of Essential Mechanical Properties and Cell Behaviors of a Novel Polylactic-*co*-Glycolic Acid (PLGA)-Based Tubular Scaffold for Small-Diameter Vascular Tissue Engineering

Nuoxin Wang [1,2], Wenfu Zheng [2], Shiyu Cheng [2], Wei Zhang [2,*], Shaoqin Liu [1,*] and Xingyu Jiang [1,2,3,*]

1   School of Life Science and Technology, Harbin Institute of Technology, 2 Yikuang Road, Nangang District, Harbin 150001, China; wangnx@nanoctr.cn
2   Beijing Engineering Research Center for BioNanotechnology & CAS Key Laboratory for Biological Effects of Nanomaterials and Nanosafety, CAS Center for Excellence in Nanoscience, National Center for NanoScience and Technology, 11 Beiyitiao, Zhongguancun, Haidian District, Beijing 100190, China; zhengwf@nanoctr.cn (W.Z.); chengsy@nanoctr.cn (S.C.)
3   The University of Chinese Academy of Sciences, 19 A Yuquan Road, Shijingshan District, Beijing 100049, China
*   Correspondence: zhangw@nanoctr.cn (W.Z.); shaoqinliu@hit.edu.cn (S.L.); xingyujiang@nanoctr.cn (X.J.)

Received: 12 June 2017; Accepted: 27 July 2017; Published: 30 July 2017

Abstract: In this paper, we investigate essential mechanical properties and cell behaviors of the scaffolds fabricated by rolling polylactic-*co*-glycolic acid (PLGA) electrospinning (ES) films for small-diameter vascular grafts (inner diameter < 6 mm). The newly developed strategy can be used to fabricate small diameter vascular grafts with or without pre-seeded cells, which are two main branches for small diameter vascular engineering. We demonstrate that the mechanical properties of our rolling-based scaffolds can be tuned flexibly by the number of layers. For cell-free scaffolds, with the increase of layer number, burst pressure and suture retention increase, elastic tensile modulus maintains unchanged statistically, but compliance and liquid leakage decrease. For cell-containing scaffolds, seeding cells will significantly decrease the liquid leakage, but there are no statistical differences for other mechanical properties; moreover, cells live and proliferate well in the scaffold after a 6-day culture.

Keywords: mechanical property; cell behavior; small diameter vascular graft; fibrin glue; rolling

1. Introduction

Cardiovascular diseases have become one of the leading threats to human lives at present [1]. In clinics, great success on large diameter vascular grafts has been achieved by using synthetic polymers (e.g., expanded polytetrafluoroethylene (ePTFE)) as the substitute material [2,3]. However, for small diameter grafts, these materials have suboptimal performance [2–4]. In the fields of tissue engineering and regenerative medicine, small diameter vascular grafts have posed a central challenge, which has attracted researchers' considerable attention. Researchers now find that biodegradable engineered scaffolds, either seeded with cells or not, may be a practical way to address this problem [5–9]. Scaffolds made of material/cell hybrids or material-only will be remodeled into tissue-like structures as the materials degrade and cells infiltrate. These studies have greatly advanced the development of this field and provided us various solutions or tools for small-diameter vascular regeneration.

In 2012, our group developed a stress-induced self-rolling technique to fabricate multi-layered tubular scaffolds by rolling polydimethylsiloxane (PDMS) membranes into three-dimensional (3D)

tubes [10]. This method might be a promising solution to small-diameter vascular engineering. However, PDMS is a bio-inert material that cannot degrade in vivo, and thus not an ideal material for blood vessel substitute [11]. Based on this work, recently, our group developed a novel strategy to construct a kind of rolling-based fully biodegradable scaffolds (i.e., polylactic-*co*-glycolic acid (PLGA) in this study) by a single step whose layers are bonded by fibrin glue [12]. This design synthesizes the advantages of several previous strategies. It can realize fabrication of cell-free scaffolds within 10 min and cell-laden scaffolds within 70 min manually with minimal ancillary equipment. Multiple parameters of the scaffolds, such as diameter, wall thickness, mechanical strength, and cell type and distribution in each layer, can be facilely modulated.

As the scaffolds would be used as blood vessel substitutes, we carry out several essential mechanical property tests and cell behavior assessments before animal implantation trials (Figure 1). For cell-free scaffolds, evaluation of the mechanical properties (including burst pressure, suture retention, compliance, and so forth) is a critical step to prove their feasibility for implantation; while for cell-laden scaffolds, besides its mechanical properties, cell behaviors (including cell viability, proliferation, and migration) are other vital aspects that should be considered. In this paper, we will report the tests of the mechanical properties of both scaffolds and the results of cell behaviors of cell-containing scaffolds under static culture condition. Our data demonstrate that the mechanical properties of the scaffolds can be modulated by the number of layers and cells survive and proliferate well in the cell-containing scaffolds. These results will guide future application of these scaffolds in animal trials.

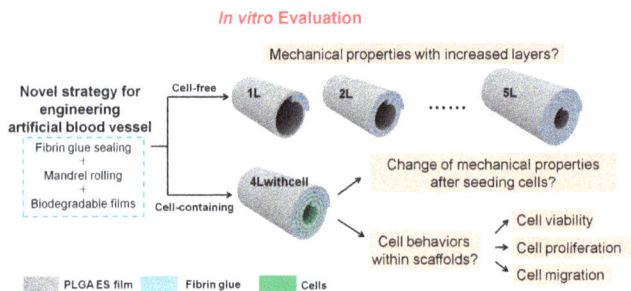

Figure 1. Schematic of in vitro evaluation of the novel artificial blood vessel. Cell-free and cell-containing scaffolds can be fabricated by this method. The substrate material is polylactic-*co*-glycolic acid (PLGA) electrospinning (ES) film. The model cell used is C2C12 mouse myoblast cell. The cell-free scaffolds containing 1.25, 2.25, 3.25, 4.25, and 5.25 layers are noted as 1L, 2L, 3L, 4L, and 5L, respectively. The 4L scaffolds containing cells in its innermost three layers are noted as 4Lwithcell. To fabricate the layered scaffolds without cells, the PLGA films were cut into rectangles with proper sizes, coated with two components of the fibrin glue on two sides of the film, and rolled around an ePTFE mandrel with proper outer diameters by hand. When rolling, the two components of the fibrin glue would react with each other and bond the layers. After the mandrel was gently extracted, the residual glue components in the scaffolds was washed with PBS and eventually only the reacted components that formed fibrin glue would be left in the scaffold. This process is illustrated in Supplementary Materials Figure S2. To fabricate the layered scaffolds with cells, cells were patterned by polydimethylsiloxane (PDMS) chambers. After cell attachment, the chambers were peeled off and the scaffolds were rolled up just the same as those without cells. This process is illustrated in Supplementary Materials Figure S3. In this paper, for cell-free scaffolds, the mechanical property changes with increased layers will be evaluated; for 4Lwithcell scaffolds, the change of mechanical property after seeding cells compared with 4L scaffolds and cell behaviors (cell viability, cell proliferation, and cell migration) within scaffolds will be evaluated.

2. Materials and Methods

2.1. Materials

PLGA75:25 (polylactic-*co*-glycolic acid, mass ratio of polylactic acid (PLA) and polyglycolic acid (PGA) is 75:25) polymers (pharmaceutical grade), fibronectin (FN), and the fibrin medical adhesive (Porcine Fibrin Sealant Kit), were purchased from Lakeshore Biomaterials Co., Ltd. (Eden Presley, MN, USA), Sigma Co., Ltd. (Shanghai, China), and Puji Medical Technology Development Co., Ltd. (Hangzhou, China), respectively. The fibrin medical adhesive contains 0.04 mg/mL fibrinectin in its component A, and 450 IU/mL thrombin in its component B. In the fabrication, the two components were both applied 10 μL/cm^2 on the films. Methyl cellulose solution (1.8%, average molecular weight ranging from 10,000 Da to 220,000 Da) was purchased from BioRoYee Co., Ltd. (Beijing, China). We diluted it into a concentration of 0.25% using distilled water when used for plasma mimics in liquid leakage tests and burst pressure tests according to a published report [13]. Other reagents were all of analytical grade bought from Beijing Chemical Factory.

2.2. PLGA ES Film Preparation

For PLGA electrospinning, PLGA particles were dissolved into the blend of ethyl acetate and dimethyl formamide (DMF) with a mass ratio of 4:1 at 30 (*w*/*w*) %. The temperature and humidity for electrospinning were ~20 °C and ~30%, respectively. The voltage of 12 kV was generated by a direct-current (DC) high-voltage generator (SL150, Spellman, New York, NY, USA). The collection distance was 15 cm and the collection time was around 40 min. All films to be seeded with cells were sterilized by a cobalt ray radiation of 10 kGy, incubated with fibronectin (FN) at a concentration of 20 μg/mL in phosphate buffer solution (PBS) at 37 °C for 1 h to enhance cell adhesion, and washed with PBS once before use.

2.3. PMMA Substrate and PDMS Chamber Fabrication

Polymethyl methacrylate (PMMA) substrates with pre-designed patterns were prepared by digitally controlled micromachining. PDMS pre-polymer (base) and catalyst (curing agent) were mixed at a ratio of 10:1, cured against the patterned PMMA substrate, and then incubated at 80 °C for 2 h for sodification. PDMS chambers were then gently peeled off the substrate surface. The fabrication process of PDMS chambers from PMMA substrates was shown in Supplementary Materials Figure S1.

2.4. Cell Culture, Staining and Seeding

C2C12 mouse myoblast cells (ATCC, Manassas, VA, USA) were cultured in Dulbecco's modified Eagle medium (DMEM, Invitrogen, Carlsbad, CA, USA) containing 10% fetal bovine serum (FBS, Invitrogen), 1% penicillin-streptomycin (PS, Invitrogen), 1% Gluta-max (Invitrogen), at 37 °C with 5% CO$_2$. Before seeding into PDMS chambers, cells were stained with dyes or not according to different experiments (for cell adhesion, cell viability, and cell proliferation tests, as well as cell migration observed by scanning electron microscope (SEM), the cells were not stained; for cell migration tests using a fluorescent assay, the cells were prelabeled by CellTracker dyes from Invitrogen Co. Ltd). We then collected cells and delivered them into the PDMS chambers at a density of 2×10^4 cm^{-2}. Before injecting the cell suspension into designated channels, PDMS chambers had been placed on PLGA films and sealed by fibrin glue to avoid liquid leakage.

2.5. Fabrication of Scaffolds with or without Cells

To fabricate the layered scaffolds without cells, the PLGA films were cut into rectangles with proper sizes, coated with two components of the fibrin glue on two sides of the film, and rolled around an ePTFE mandrel with proper outer diameters (in this experiment: 2 mm outer diameter of the mandrel, corresponding a scaffold with 2-mm inner diameter) by hand at a speed of 2–4 cm·min^{-1}.

When rolling, the two components of the fibrin glue would react with each other to form a sticky fibrin gel and bond the layers. The mandrel was gently extracted and the scaffold was ready. The residual glue components in the scaffolds were washed with PBS and eventually only the reacted components that formed fibrin glue would be left in the scaffold. To investigate the mechanical properties of the scaffolds with different layers, we fabricated the scaffolds without cells containing 1.25, 2.25, 3.25, 4.25, and 5.25 layers (noted as 1L, 2L, 3L, 4L, and 5L, respectively). The extra 0.25 layer was used to anchor the scaffolds and to facilitate the fabrication. This fabrication process is illustrated in Supplementary Materials Figure S2 (using 4L scaffolds as an example). To fabricate the layered scaffolds with cells, cells were seeded on the film with the aid of PDMS chambers. PDMS chambers were used to ensure the cells in the right position and density. The PDMS chamber was peeled off and the scaffolds were rolled up just the same as those without cells. In this experiment, we selected 4 layered scaffolds (4.25 layers) to seeded its innermost three layers with C2C12 cells (noted as 4Lwithcell) to investigate the mechanical properties differences from its 4L alternatives without cells, its cell viability and cell proliferation in each layer, and cell migration between layers. This fabrication process is illustrated in Supplementary Materials Figure S3 (using 4Lwithcell scaffolds as an example).

2.6. Measurement of Mechanical Properties

Mechanical properties were measured according to the following protocols. Both cell-free and cell-containing scaffolds were immersed in the culture medium or PBS after fabrication for 20 min at room temperature before measurements were conducted. All tests were applied on three grafts. The room temperature was around 20 °C for all tests.

Suture retention test: one end of a 1.5-cm long graft was clamped at onto a dynamic mechanical analysis machine (DMA, Q800, TA instrument, New Castle, DE, USA). A single bite suture (9-0 Niklon suture, Jiaxin, China) was placed 2 mm from the edge of the other end. A constant pulling rate of 1 mm/min was applied until the suture was pulled out. The maximum force of pulling was recorded as the suture retention.

Burst pressure test: one end of a 2-cm long graft was hermetically clamped, and the other end of the graft was hermetically connected to a syringe on a syringe pump (PHD ULTRA, Harvard, Boston, MA, USA). A constant rate of 50 mL/min was applied to fill 0.25% methyl cellulose solution (plasma mimics) into the graft. The peak pressure before the graft bursts was tested by a pressure gage (AZ 82100 and AZ 8205, Taiwan, China) as the burst pressure. The home-made setup used for burst pressure tests is shown in Supplementary Materials Figure S4.

Wall thickness test: the graft was pressed tightly together and the total thickness of the graft wall was measured by a digital meter. The wall thickness equals the half of the total thickness.

Liquid leakage test: a 2-cm long graft was placed on two vascular catheters, and hermetically sealed with two elastic threads. The graft was flushed with water or 0.25% methyl cellulose solution (plasma mimics) at a pressure of 16 kPa (corresponding to the average systolic pressure of 120 mmHg) for 3 min. The liquid that leaked through the grafts was collected and results were expressed in $mL \cdot min^{-1} \cdot cm^{-2}$. The home-made setup used for liquid leakage tests and the following compliance tests is shown in Supplementary Materials Figure S5.

Compliance test: compliance (C) was calculated according to the following equation [14].

$$C = \frac{\frac{D_{inner}(P2) - D_{inner}(P1)}{D_{inner}(P1)}}{P2 - P1} \times 10^4 \qquad (1)$$

where P1 and P2 are the lower (80 mmHg) and higher (120 mmHg) pressures, respectively.

$$D_{inner} = \sqrt{D_{outer}^2 - \frac{4(A_{wall})}{\pi}} \qquad (2)$$

where D_{inner} is inner diameter of the graft, D_{outer} is outer diameter of the graft, and A_{wall} is axial cross-sectional area of the vessel wall [15]. A_{wall} is calculated from the cross-sectional area $A_{wall,0}$ at zero pressure [16].

$$A_{wall} = A_{wall,0} = \pi \cdot h_0 \cdot (D_{outer,0} - h_0) \tag{3}$$

where h_0 is wall thickness measured at zero lumenal pressure, $D_{outer,0}$ represents outer diameter measured at zero lumenal pressure. To simplify the calculation, we employed average value of wall thickness for each type of graft as the value of h_0. We recorded the outer diameter under different lumenal pressures use a digital camera and analyzed using ImageJ 1.43m (NIH USA, 2008) and PhotoShop 6.0 (Adobe).

Tensile elastic modulus:

Tensile elastic modulus was calculated according to the following equations [15]:

Circumferential Stretch Ratio ($\lambda_{\theta\theta}$):

$$\lambda_{\theta\theta} = \frac{D_{inner}}{D_{inner,0}} \tag{4}$$

Circumferential Ring Strain ($T_{\theta\theta}$):

$$T_{\theta\theta} = \frac{1}{2}(\lambda_{\theta\theta}^2 - 1) \tag{5}$$

Cauchy Stress ($\sigma_{\theta\theta}$):

$$\sigma_{\theta\theta} = \frac{P \cdot D_{outer} \cdot \lambda_{\theta\theta}}{2h_0} - P \tag{6}$$

We calculated tensile elastic modulus (E) within the physiologic pressure range from the slope of stress-strain curves between 80 mm Hg and 120 mm Hg lumenal pressure:

$$E = \frac{\sigma_{\theta\theta}(P120) - \sigma_{\theta\theta}(P80)}{T_{\theta\theta}(P120) - T_{\theta\theta}(P80)} \tag{7}$$

2.7. Cell Viability Test

C2C12 cells (without staining) were seeded on the films with the aid of PDMS chambers at the density of 2×10^4 cm^{-2} (as shown in Supplementary Materials Figure S3). After fabrication and culture of the scaffolds in DMEM for 3 or 6 days, the scaffolds were unrolled. The cells in each layer were then washed with PBS and stained with the LIVE/DEAD kit (Invitrogen). The image was observed with confocal microscopy.

2.8. Cell Proliferation Test by a Fluorescent Assay

C2C12 cells (without staining) were seeded on the films with the aid of PDMS chambers at a density of 2×10^4 cm^{-2} (as shown in Supplementary Materials Figure S3). After fabrication and culture of the scaffolds in DMEM for 3 or 6 days, the scaffolds were unrolled. The cells in each layer were washed with PBS and then lysed (cell culture lysis reagent part #E153A, Promega, Madison, WI, USA) using 1 mL 1× Lysis Buffer. The cell density per layer was measured using a cell proliferation assay according to the manufacturer's protocol (CyQuant Cell Proliferation Assay Kit, Invitrogen). Briefly, 1/10 of the buffer containing lysed cells was collected in a 96-well plate. Add CyQUANT® GR dye (a proprietary dye that exhibits strong fluorescence enhancement when bound to nucleic acids, with a maximal excitation at 480 nm and a maximal emission at 520 nm) into the buffer to reach a final working concentration of 1× and a total volume of 100.5 μL in each well. After incubating at room temperature for 30 min, the 96-well plate was scanned in a microplate reader (EnSpire Multimode Plate Readers, PerkinElmer, Waltham, MA, USA). We used the excitation light at 480 nm, and collected the emission light at 520 nm. Three cell-seeded grafts were used in this test. We also plotted a standard curve of the relationship between the cell number and emission intensity at 520 nm. Briefly, the cells were collected and counted after trypsinization, and then the cell suspensions containing 100,000, 75,000, 50,000, 25,000, 12,500, and 6000 cells were pelleted using a centrifuge at 1200 rpm for 5 min. After discarding the supernatant, 100.5 μL CyQUANT® GR dye-containing cell lysis was added into the collected cells and the solution was transferred into 96-well plates. After incubation at room temperature for 30 min, we read the emission intensity at 520 nm. The background of each sample (the intensity of sample without containing cells) has been deducted. We converted the emission intensity value of tested samples into cell number per square centimeters.

2.9. Cell Migration Test by a Fluorescent Assay

Pre-labeled C2C12 cells were seeded onto the films with the aid of PDMS chambers at a density of 2×10^4 cm^{-2} (as shown in Supplementary Materials Figure S6). The cells on the first, second, and third layer were stained with CellTracker Green, CellTracker Orange, and CellTracker DeepRed, respectively. After fabrication and culture of the scaffolds in DMEM for 3 or 6 days, the scaffolds were unrolled. Random fields of each layer were scanned with all three excitation/emission (ex/em) waves: Cell Tracker Green- (ex/em) 488/517 nm, Cell Tracker Orange- (ex/em) 543/565 nm, and Celltracker Deep Red- (ex/em) 633/650 nm. The spectra windows for emission collection were set so that there was no cross-talk among the three dyes. If the stained color in some layers occurs to other layers, it demonstrates that the migration takes place from some layers to other layers.

2.10. Cell Adhesion, Proliferation, and Migration Tests by SEM

To investigate the cell adhesion, proliferation, and migration by SEM, we seeded cells without staining only on the second layer of the 4L scaffolds with the aid of PDMS chambers at the density of 2×10^4 cm^{-2} (as shown in Supplementary Materials Figure S7). Cell adhesion was checked after 12 h of incubation in a cell culture incubator. After culture of 3 days and 6 days, we unrolled the scaffolds. Cell proliferation on the second layer of the scaffold was checked by SEM. Cell migration from the second layer to the first, third, and fourth layer was checked by SEM. The flat film and the unrolled scaffolds were fixed in 4% paraformaldehyde for 2 h, washed with distilled water for 3 times, and dehydrated in gradient using 30%, 50%, 70%, 80%, and 90% each once and 100% ethanol for three times. Each dehydration was performed for 5–10 min. The samples were then air dried at room temperature, mounted on a metal stub, and sprayed with gold for 60 s before SEM observations.

2.11. Film Thickness, Fiber Size, Pore Size, and Cell Size Measurements

The thickness of the films was measured by an electronic digital micrometer (Guanglu, Guilin, China). The average fiber diameter was measured from SEM images; 200 fibers were manually measured and analyzed using ImageJ software (NIH USA, 2008). From SEM images, pore sizes and cell sizes were measured by manually fitting an ellipse in representative pores formed by fibers in the same plane and cells adhering on the film. The size of each pore or cell was the average between the long and short diameters of the fitted ellipse; 100 pores and 50 cells were manually measured and analyzed using ImageJ software. Results are given as mean ± standard deviation.

2.12. PLGA Film Degradation

To test the degradation property of ES PLGA films, the films were immersed into DMEM medium, and incubated at 37 °C with 5% CO$_2$ for 2 weeks. The films were washed with distilled water three times, lyophilized using a lyophilizer (FD-1A-50, Biocool, Beijing, China), and observed by SEM.

2.13. Statistics

Differences between two groups were examined via unpaired two-tailed Student's *t*-test using GraphPad Prism 6.0 software (GraphPad Software, Inc., La Jolla, CA, Country). A value of $p < 0.05$ was considered to be statistically significant.

3. Results

3.1. PLGA ES Film Characterization and Mechanical Properties for Scaffolds without Cells

The PLGA films possess a thickness of 41.3 ± 2.1 μm ($n = 10$). The fiber size is 0.59 ± 0.28 μm ($n = 200$) and the pore size of the ES film is 2.92 ± 1.76 μm ($n = 100$) (Supplementary Materials Figure S8). With the increase of the layer number (1 to 5), the wall thickness increases almost linearly (from 42 ± 1 μm to 200 ± 8 μm) (Figure 2). The burst pressure also increases from 0.015 ± 0.0017 MPa for a 1-layered scaffold to 0.142 ± 0.013 MPa for a 5-layered scaffold. The suture retention increases with layer numbers, from 0.16 ± 0.021 N for 1 layer to 0.77 ± 0.19 N for 5 layers. For the compliance, modulus, and water leakage tests, because scaffolds with 1 and 2 layers cannot bear the pressure of 120 mmHg, we could only obtain the data of these parameters for scaffolds with 3–5 layers. Compliance between 80 mmHg and 120 mmHg decreases with the layer number, from $6.88 \pm 1.94\%/100$ mmHg for 3 layers to $3.41 \pm 0.64\%/100$ mmHg for 5 layers. Under 120 mmHg pressure, we observe a similar decreasing trend on water leakage (3.83 ± 0.23, 2.22 ± 0.13, 1.55 ± 0.10 mL/(min·cm^2) for 3, 4, 5 layers, respectively and 0.25% methyl cellulose solution leakage (0.85 ± 0.17, 0.55 ± 0.076, 0.23 ± 0.045 mL/(min·cm^2) for 3, 4, 5 layers, respectively) (Figure 2 and Supplementary Materials Figure S9). Tensile elastic modulus has no significant changes with the increase of layer number, and values are 1929 ± 257, 2719 ± 693, and 2264 ± 181 kPa for 3, 4, and 5 layers, respectively. The pressure-inner radius curves show that the radius of scaffolds with 3–5 layers increases steadily with the increase of lumenal pressure (Supplementary Materials Figure S10). According to Supplementary Materials Table S1, there are no significant differences for circumferential ring strain for 3–5 layered scaffolds at 80 and 120 mmHg, but Cauthy stress significantly decreases from 3–5 layered scaffolds at 80 and 120 mmHg. In addition, the film shows slight degradation and apparent shrinkage after a 2-week incubation (Supplementary Materials Figure S8).

3.2. Mechanical Properties for Scaffolds with Cells

Because the typical structure of a blood vessel includes three cell layers [17], we then compared the mechanical properties of 4 layered scaffolds with and without 3 cell-containing layers (the scaffolds with cells contain cells in their 1st to 3rd layers, and the 4th layer as the outermost layer does not contain cells, in order to reinforce the whole structure). The cell size on the film is 30.5 ± 7.4 μm ($n = 50$), much larger than the pore size (Supplementary Materials Figures S8 and S10). Incorporating cells does not significantly increase the wall thickness, burst pressure, suture retention, compliance, and tensile elastic modulus of the scaffold (Figure 3). The compliance and tensile elastic modulus appear to have significant changes, but there are no statistical significances ($p = 0.0858$ and $p = 0.0711$, respectively). Liquid leakage of scaffolds containing cells has a sharp decrease compared with that of the cell-free ones (0.80 ± 0.15 vs. 2.22 ± 0.13 mL/(min·cm^2) for water, and 0.17 ± 0.036 vs. 0.55 ± 0.076 mL/(min·cm^2) for 0.25% methyl cellulose solution) (Figure 3 and Supplementary Materials Figure S9). The pressure-inner radius curves show that the radii of both 4L and 4Lwithcell scaffolds increase steadily with the increase of lumenal pressure (Supplementary Materials Figure S11). There are no significant differences on circumferential ring strain both at 80 and 120 mmHg and Cauthy stress for 4L and 4Lwithcell scaffolds at 120 mmHg. However, Cauthy stress of 4Lwithcell scaffolds has a significant decrease compared with that of 4L (Supplementary Materials Table S1).

Figure 2. Mechanical properties for cell-free scaffolds. (**A–C**) Wall thickness (**A**), burst pressure (**B**), and suture retention (**C**) of scaffolds with 1–5 layers. (**D–F**) Compliance (**D**), tensile elastic modulus (**E**), and water leakage (**F**) of scaffolds with 3–5 layers. Data of (**D,E,F**) do not have the results of 1 and 2 layered scaffolds, because the scaffolds with 1 and 2 layers cannot bear the pressure of 120 mmHg. All tests were biological triplicates. * indicates the *p* value smaller than 0.05.

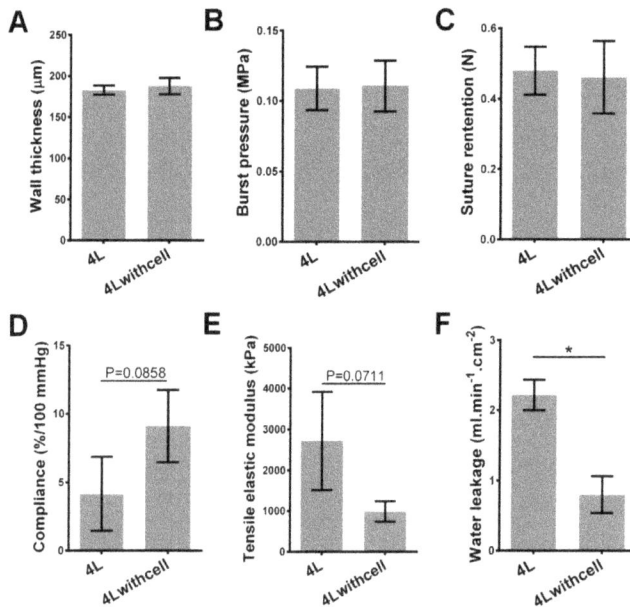

Figure 3. Comparison of mechanical properties of 4 layered scaffolds with and without cells. Wall thickness (**A**), burst pressure (**B**), suture retention (**C**), compliance (**D**), tensile elastic modulus (**E**), and water leakage (**F**). All tests were biological triplicates. * indicates the *p* value smaller than 0.05.

3.3. Cell Behaviors in Cell-Containing Scaffolds

After evaluation of the mechanical properties, we tested the cell behaviors in cell-containing scaffolds, including cell viability, cell proliferation, as well as cell migration.

3.3.1. Cell Viability in Each Layer

We tested the cell viability of the scaffolds either with or without the mandrel, namely, the ePTFE mandrels were pulled out or not when incubating the scaffolds in the culture medium. After static culture of 3 and 6 days, we stained the cells on the unrolled films in each layer using LIVE/DEAD Kit. From the confocal images, we can rarely see the PI-positive (red, indicating dead cells); almost all the cells are Calcein-positive live cells (green) (Figure 4). There are no obvious differences of the cell viability between the scaffolds with and without mandrel and also between the tested scaffolds and the flat film serving as the control. Moreover, this phenomenon is not affected by culturing time (3 days or 6 days). These results demonstrate that the cells in scaffolds possess high viability.

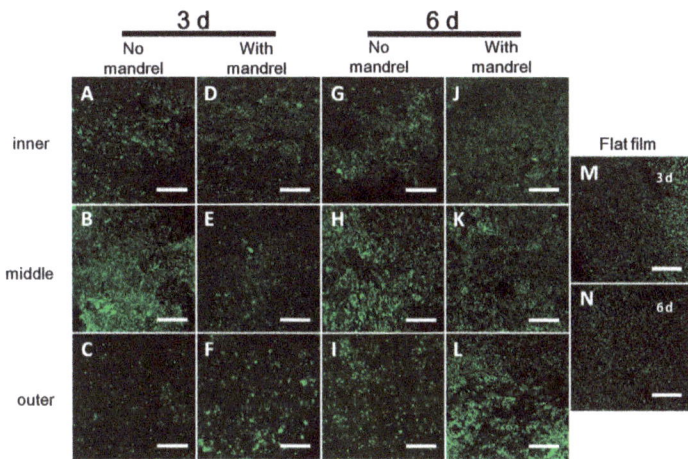

Figure 4. Cell viability at 3 (**A–F**) and 6 (**G–L**) days with (**D–F,J–L**) and without (**A–C,G–I**) mandrel. (**A,D,G,J**) inner layer. (**B,E,H,K**) middle layer. (**C,F,I,L**) outer layer. (**M**) cell viability at 3-d culture on flat film (**N**) cell viability at 3-d culture on flat film. Green: live cells. Red: dead cells. The model cell used is C2C12 cell. Each image is chosen randomly as the representative image in corresponding layers. Scale bars: 200 μm.

3.3.2. Cell Proliferation in Each Layer

We further tested the cell proliferation in each layer. The standard curve of cell number (c) and fluorescent intensity (f) shows a well linear relationship as shown in Supplementary Materials Figure S12 ($f = 0.7885c - 2994$, $R^2 = 0.9983$). Compared with cells cultured in the scaffolds for 3 days, cells cultured for 6 days in each layer have a significant growth, regardless of the scaffolds incubated with or without mandrel (Figure 5). However, it is obvious that cells growing in almost all scaffolds have less cell proliferation than cells growing on culure dishes and flat films after culture of 3 days, while no statistical significance between the cells in scaffolds and those on flat films is observed for cells after culture of 6 days (Figure 5). These results indicate that cells in different scaffolds and layers have a similar proliferation trend, but there are differences on proliferation rate. The cell proliferation in the scaffolds is also proofed by SEM image (Supplementary Materials Figures S10 and S13). With the increase of time, the cell density on the PLGA film or unrolled scaffold film increases.

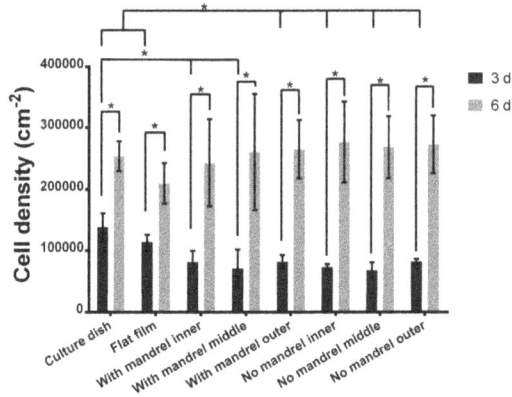

Figure 5. Cell proliferation at 3 and 6 days with or without mandrel. All tests were biological triplicates. * indicates a *p* value smaller than 0.05.

3.3.3. Cell Migration between Layers

Cell migration was assessed by the movement of pre-labeled cells between layers. If the migration occurs, some cells pre-labeled in previous layer will be found in other layers. In our case, no matter for scaffolds with or without mandrel, and no matter for scaffolds of 3 or 6 days culture, there is almost no cell migration between layers (Figure 6). The cell proliferation in the scaffolds is also proofed by SEM image (Supplementary Materials Figure S13). After culture of 3 and 6 days, very rare cells seeded in the second layer of the scaffold migrated to other layers. These results indicate that quite limited migration occurs in the scaffolds, which is largely affected by the pore size in the scaffold [18–21].

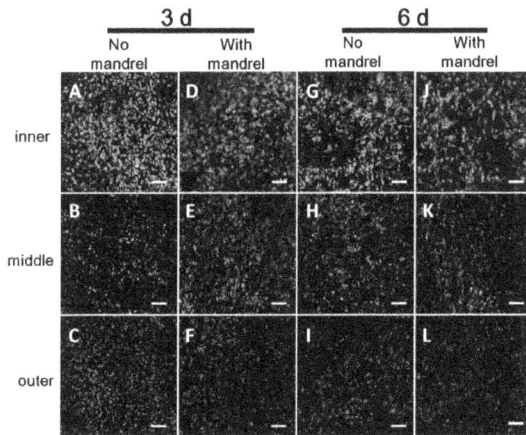

Figure 6. Cell migration at 3 (**A–F**) and 6 (**G–L**) days with (**D–F,J–L**) and without (**A–C,G–I**) mandrel. (**A,D,G,J**) inner layer. (**B,E,H,K**) middle layer. (**C,F,I,L**) outer layer. The model cell used is C2C12 cell. Each image is chosen randomly as the representative image in corresponding layer. The cells seeded on each layer were pre-labeled with different dyes. After unrolling, each image is chosen randomly as the representative image in corresponding layer. If the migration between layers occurs, the pre-labeled cells will appear on another layer. Scale bars: 50 μm.

4. Discussion

Employing the bi-component fibrin biomedical glue, we have developed a rapid and straightforward method to fabricate layered tubular scaffolds by rolling biodegradable polymer thin films around a removable smooth mandrel [12]. In order to overcome the limitations of several reported methods also based on rolling, the following efforts were made: (i) biomedical glue was used to stabilize the whole structure and reduce the layer fusion time [22]; (ii) bioabsorbable electrospinning (ES) polymer films were used to provide an easy-to-fabricate, easy-to-store, extracellular matrix (ECM)-like substrate [23]; (iii) cell suspensions were used to lessen production time for cell sheets [24]. We also inherit or imitate the advantages of the existing methods [10,22,25,26]. For example, rolling-based methods can tune the inner diameter of the tube precisely in virtue of the mandrel; the concept is simple and easy-to-learn without complex manipulations, harsh conditions, and expensive equipment; the well-designed PDMS channel is an effective tool to ensure the accurate cell distribution on the film. A number of published reports have employed rolling-based manufacture of three dimensional (3D) tubular structure from two dimensional (2D) films [24,27–29]. It is worth mentioning that, rolling 2D film into 3D structures has the intrinsic advantage that we can modify or pattern the film very conveniently with cells, molecules, and/or nanoparticles because of the booming development of various patterning techniques on 2D substrates, which may favor their future applications [30–33]. Therefore, besides the rapidness, flexibility and simplicity, simple and controlled modification of cells or other components in the specific location in 3D structures is one of the most important features of our method.

According to the scaffolds encapsulating cells or not, the prepared scaffolds can be classified into two types: cell-free or the cell-containing scaffolds. The two types of scaffolds have different advantages [6]. Generally speaking, the cell-free scaffolds possess the features of the short fabrication time and ease of storage, while cell-containing scaffolds have a more similar structure to real blood vessels and better biocompatibility. Both kinds have been intensively investigated and promising results have been acquired to further the development of the vascular tissue engineering field [34–38]. Herein, we applied the fibrin glue-assisted rolling technique to fabricate both kinds of vascular grafts.

Realizing tunable mechanical strength effectively is one of the features of our scaffolds. For cell free scaffolds, as expected, the burst pressure and suture retention increase with the increase of the layer number, while water leakage and compliance decrease with the increase of the layer number. Controlling layer number is a straightforward way to tune the mechanical strength, and the fibrin glue ensures the integrity of the stacked layers in our strategy. For cell-containing scaffolds, compared with the cell-free variant with the same layers, the water leakage significantly decreases, but there are no significant changes on other mechanical properties. Although the innermost three layers of the scaffolds contain cells (see Materials and Methods section), it is not difficult to image that the mechanical properties can also be controlled by the number of outer layers.

As a scaffold that can be implanted, the most essential requirements on mechanical strength include that: it can withstand the pressure of blood, can be easily sutured by doctors, and can confine the blood in the scaffold without severe leakage. Thus, burst pressure, suture retention, and leakage property should be the three most important mechanical properties, for either cell-free scaffolds or cell-containing scaffolds. Previous reports have demonstrated that the human saphenous vein, which is regarded as one of the most appropriate substitutes in cardiovascular clinic practice, possesses the burst pressure of larger than 1680 mmHg (0.223 MPa) [22,39–42]. In the present study, the burst pressure of 5-layered PLGA scaffold approaches this value. Another important merit of this method is the flexibility on the choice of materials. As for the material, PLGA may seem soft compared with some commonly used materials such as poly(ε-caprolactone) (PCL) and PLA. Thus, if these stronger materials are used, the layer number reaching the implantable level would be reduced. Suture retention is a basic index that ensures the scaffold can be compatible with the suture in the surgery. The reported suture retention of the human saphenous vein was ~196 gf (~1.92 N) [39,41]. However, there are examples for implantable blood vessel substitutes with much lower suture retention reported, such as the fast degradable and highly elastic poly(glycerol sebacate) (PGS) graft (suture retention: ~0.45 N) [43,44]. This value is not a challenge for our rolling-based strategy. As for the leakage property, our scaffolds have shown that the uncontrolled quick leakage can be avoided, and the value can be further reduced by increasing layer number and/or cell seeding. The cells will cover some pores on the film and thus induce the leakage decrease (Supplementary Materials Figures S8 and S10). The leakage is also greatly reduced using methyl cellulose solution as plasma mimics with a higher viscosity than water. We believe that the leakage will be further reduced when using real blood, because of its higher viscosity than plasma and the existence of the clotting mechanism. It is noted that the leakage should not be completely stopped because the leakage also reflects the free nutrient transportation in the scaffold to some extent. The literatures have documented that cells in scaffolds without pores/zero leakage have low viability [18–20].

Other mechanical properties, such as compliance and elastic modulus, are also very important. Compliance mismatch has been regarded as a key factor for vascular stenosis at post-implantation [45,46]. As a result, implantation of a scaffold with compliance and elasticity similar to those of native vessels is the most favorable [47–49]. However, whether cell-free or cell-containing scaffolds, they will undergo complex and extensive in vivo evolution, including cell infiltration, scaffold degradation, ECM deposition, and so forth. These progresses will gradually but substantially alter the structure and thus the compliance, elasticity or modulus of the scaffolds. Typically, the compliance of the scaffolds will become increasingly similar to that of the host vessels. The compliance and elasticity are mainly derived from infiltrated smooth muscle cells (SMCs) and biosynthesized elastin. Elastin is secreted by synthetic type of SMCs and fibroblasts (FBs), and thus colonization of SMCs and FBs as well as inducing the transformation of more synthetic SMCs is the pre-requisite of elastin formation. Among multiple types of ECM proteins in the blood vessel wall, elastin takes up to 50% of its dry weight. Because elastin can be stretched under load and recoil to its original configuration when the load is removed, it can determine the elasticity of the blood vessel [16,49,50]. Thus, the design for enhancing elastin formation should be considered in our future exploration. It is also noted to mention that another major ECM protein, collagen, plays a critical role in supporting the whole structure of the graft and strengthening the graft in vivo [17,51]. It possesses a structure of elongated fibrils in blood vessels. It is essential for maintaining and enhancing burst pressure and suture retention of the graft. For cell-containing scaffolds, the cell behavior in vitro can be used to predict the cell behavior in vivo [52]. Interestingly, the cells in our scaffolds live and proliferate well. These observations indicate that the nutrients can be continuously supplied to cells and the wastes can also be discharged promptly. Considering that there are no vessels to transport the nutrients and wastes, the process is free diffusion. It is known that the diffusion limit for the tissues is less than 200 μm. Beyond this value, the cells will not obtain enough nutrients and can undergo apoptosis. This phenomenon has been well documented in many literatures [53,54]. Our cell-containing scaffold contains three layers of cells in its innermost layers, and the third layer of cells has a maximal distance of less than 100 μm towards the lumen of the vessel (estimated from half of the wall thickness data of 4 layered scaffolds with cells shown in Figure 3), within the diffusion limit of 200 μm. This is the reason why the cells live and proliferate well in the scaffold. However, the cell proliferation rate in different layers is different. We speculate that it is because of the different distribution or different transportation ratio of nutrients, oxygen, and wastes as well as the different distribution of cells [18]. Cell migration is an important process for scaffold evolution. The scaffold evolution is a self-assembly process after the scaffolds are implanted into the hosts. In this stage, seeded cells and cells in hosts interact with each other, and are remodeled into an ordered structure that becomes more and more analogous to native tissue with the evolution time. The pore size plays a significant role in cell migration [18,19,55]. In our scaffold, because of the small fiber diameter-induced small pore size, the cell migration between layers has been greatly confined (Figure 6 and Supplementary Materials Figures S10 and S13). We believe that there may be some disadvantages, such as insufficient cell interactions with hosts cells and potential insufficient waste clearance. Thus, in future assays, we will use ES sheets with thicker fibers and thus larger pores to facilitate the tissue evolution.

In addition to the above discussion, there are still some issues to be addressed in our scaffold design or in our future study. First of all, an intrinsic feature of the "rolled sheet" design is a longitudinal internal ridge formed by the edge of the inner layer. The edge may foster the blood cell adhesion and complicate the endothelialization. A possible solution to smoothen this edge may be using a substrate with gradually increased thickness at its one edge (Supplementary Materials Figure S14). Second, our method faces some difficulties when fabricating very long grafts, because it is rather hard to control long films well in a certain direction when rolling manually. We believe a rolling machine can be employed to address this problem, but it will sacrifice some simplicity of our method [56]. Third, although in this study we evaluated several essential mechanical properties for the scaffold, it is far from a detailed complete evaluation. If we want to reach higher ambitions, much more complex mechanical properties should be carried out. One of the examples is that grafts need to be pliable either to lead them over the curved heart surface, or to allow movements if used as a popliteal graft. Another example is that the graft should resist kinking as this would obstruct blood flow. Unfortunately, our manufacturing cannot address these requirements in its current state. We think a graft made of some kind of highly-elastic material, such as PGS, with a wrinkled graft wall may be a potential solution for these two challenges. Fourth, in our study, we used 2 cm long grafts for most tests, and the length is too short for applications in humans. As a blood vessel scaffold in its initial stage, our short term goal is to verify its feasibility in small animal implantation, such as the rabbit. The 2 cm graft is long enough for this implantation. In the next stages of the evaluation of the blood vessel graft, such as large animal implantation and clinical trials, longer grafts should be used. Also, the applications of the graft in humans vary according to the blood vessel replaced in different body regions. These applications require the blood vessel featuring different biomechanical properties. At the current stage, we have not yet considered which specific region our graft will be used for (e.g., coronary artery bypass, renal dialysis, or popliteal graft). Our focus in this stage is characterization of the basic properties of the graft. In the future, we will concentrate on coronary artery bypass in an initial plan. Fifth, in this work, although we evaluated cell behaviors within 6 days using the well established C2C12 cells based on published reports [18,52,57], longer-term exploration should be carried out. We think in this exploration, the cells used should be blood vessel-related cells or stem cells from real patients. The graft should be incubated in a perfusion system for weeks or even longer. The formation of ECM and acquirement of contractile activity have been reported for blood vessel grafts in similar systems [58]. These provide the artificial blood vessels more analogous structure and function like the native blood vessels; although the formed thrombogenic matrix components such as collagen may be a risk factor if the grafts are exposed to blood. This issue could be addressed by seeding a confluent endothelial layer and introducing signaling factors to recruitment anti-thrombogenic cells (e.g., using CD34 antibody to home endothelial progenitor cells in circulating blood) [59]. The ECM deposition might be modulated beneficially by modifications to the polymer matrix. For example, gold nanoparticles conjugated to Type I collagen can greatly improve the resistance to collagenase of the collagen construct [60]. Sixth, it is worth investigating whether the orientation of the burst point of the scaffold in burst pressure tests depends on the orientation of the nanofibers in the ES mats. Seventh, due to inevitable blood leakage through the graft wall, it is necessary to know how the blood components interact with the polymer matrix. A clear investigation on the mechanisms, e.g., just reducing the permeability, initiating the clotting, or others, is beneficial to design an improved blood vessel graft. Finally, as we noted, the liquid flux from the graft wall probably helps to nourish the cells in the scaffold, but it will be significantly reduced by the presence of cells; therefore, it is important to know where the balance is between the cell density and the amount of liquid flux to sufficient nutriment for cells in the scaffolds.

5. Conclusions

In conclusion, we have demonstrated that the mechanical properties of our rolling-based scaffolds can be tuned by the number of layers. This feature makes our scaffolds implantable by changing the layer number according to the selected substrate materials. The scaffolds possess controlled liquid leakage with the aid of fibrin glue and cell seeding will further decrease the leakage. Cells live and proliferate well in the scaffold. These evaluations suggest that our newly designed strategy for vascular graft fabrication, featuring rolling biodegradable films around a mandrel and sealing the layers by fibrin glue, possesses satisfactory mechanical properties if using an appropriate layer number and favorable cell compatibility in vitro, which may be a promising solution to small diameter vascular grafts. In the future, we need to increase the fiber diameter to improve the cell migration between layers. The work in the next stage is an animal implantation test based on these mechanical properties and cell behavior evaluation data.

Supplementary Materials: The following are available online at www.mdpi.com/2073-4360/9/8/318/s1. Figure S1: Fabrication PDMS chambers from PMMA substrates. Figure S2: Fabrication scaffolds without cells. Figure S3: Fabrication scaffolds with cells. Figure S4: Setup for burst pressure test. Figure S5: Setup for water leakage and compliance tests. Figure S6: Fabrication scaffolds with stained cells (illustrating the cell distribution for cell migration test by a fluorensent assay). Figure S7: Fabrication scaffolds with one layer of cells seeded (illustrating the cell distribution for cell adhesion, proliferation and migration tests by SEM). Figure S8: The morphology of PLGA fibers before and after 14 days' degradation in DMEM at 37 °C with 5% CO_2. Figure S9: The 0.25% methyl cellulose solution leakage from the scaffolds. Figure S10: Cell adhesion after 12 h on PLGA ES films. Figure S11: The pressure-inner diameter curve of the scaffolds. Figure S12: The calibrated standard curve of the CyQuant Cell Proliferation Assay Kit. Figure S13: SEM images of and cell proliferation and cell migration after 3 and 6 days in scaffolds. Figure S14: A possible design for the solution to the internal edge of the rolling-based scaffolds. Table S1: The relationship between circumferential tensile strain and the Cauthy stress of the scaffolds.

Acknowledgments: We thank Yan Tian and Chuanxin Weng in National Centre for Nanoscience and Technology for the technique support on DMA tests. We thank the anonymous reviewers for the enhancement of our manuscript. We thank the Ministry of Science and Technology of China (2013YQ190467), Chinese Academy of Sciences (XDA09030305) and the National Science Foundation of China (81361140345, 51373043, 21535001) for financial support.

Author Contributions: Nuoxin Wang, Wei Zhang and Xingyu Jiang designed the experiment; Nuoxin Wang and Shiyu Cheng performed the experiment and analyzed the data; Nuoxin Wang wrote the paper; Wenfu Zheng, Wei Zhang, Shaoqin Liu and Xingyu Jiang revised the paper.

Conflicts of Interest: The authors declare no conflict of interest.

References

1. Rydén, L.; Grant, P.J.; Anker, S.D.; Berne, C.; Cosentino, F.; Danchin, N.; Deaton, C.; Escaned, J.; Hammes, H.-P.; Huikuri, H. ESC Guidelines on diabetes, pre-diabetes, and cardiovascular diseases developed in collaboration with the EASD. *Eur. Heart J.* **2013**, *34*, 3035–3087. [PubMed]

2. Kurobe, H.; Maxfield, M.W.; Breuer, C.K.; Shinoka, T. Concise review: Tissue-engineered vascular grafts for cardiac surgery: Past, present, and future. *Stem Cells Transl. Med.* **2012**, *1*, 566–571. [CrossRef] [PubMed]

3. Cleary, M.A.; Geiger, E.; Grady, C.; Best, C.; Naito, Y.; Breuer, C. Vascular tissue engineering: The next generation. *Trends Mol. Med.* **2012**, *18*, 394–404. [CrossRef] [PubMed]

4. Rocco, K.A.; Maxfield, M.W.; Best, C.A.; Dean, E.W.; Breuer, C.K. In vivo applications of electrospun tissue-engineered vascular grafts: A review. *Tissue Eng. B Rev.* **2014**, *20*, 628–640. [CrossRef] [PubMed]

5. Li, S.; Sengupta, D.; Chien, S. Vascular tissue engineering: From in vitro to in situ. *Wiley Interdisc. Rev. Syst. Biol. Med.* **2014**, *6*, 61–76. [CrossRef] [PubMed]

6. Sengupta, D.; Waldman, S.D.; Li, S. From in vitro to in situ tissue engineering. *Ann. Biomed. Eng.* **2014**, *42*, 1537–1545. [CrossRef] [PubMed]

7. Chlupac, J.; Filova, E.; Bacakova, L. Blood vessel replacement: 50 years of development and tissue engineering paradigms in vascular surgery. *Physiol. Res.* **2009**, *58*, S119. [PubMed]

8. Dong, C.; Lv, Y. Application of Collagen Scaffold in Tissue Engineering: Recent Advances and New Perspectives. *Polymers* **2016**, *8*, 42. [CrossRef]

9. Manavitehrani, I.; Fathi, A.; Badr, H.; Daly, S.; Negahi Shirazi, A.; Dehghani, F. Biomedical Applications of Biodegradable Polyesters. *Polymers* **2016**, *8*, 20. [CrossRef]

10. Yuan, B.; Jin, Y.; Sun, Y.; Wang, D.; Sun, J.; Wang, Z.; Zhang, W.; Jiang, X. A strategy for depositing different types of cells in three dimensions to mimic tubular structures in tissues. *Adv. Mater.* **2012**, *24*, 890–896. [CrossRef] [PubMed]

11. Bartalena, G.; Loosli, Y.; Zambelli, T.; Snedeker, J.G. Biomaterial surface modifications can dominate cell-substrate mechanics: The impact of PDMS plasma treatment on a quantitative assay of cell stiffness. *Soft Matter* **2012**, *8*, 673–681. [CrossRef]

12. Wang, N.; Tang, L.; Zheng, W.; Peng, Y.; Cheng, S.; Lei, Y.; Zhang, L.; Hu, B.; Liu, S.; Zhang, W.; Jiang, X.Y. A strategy for rapid and facile fabrication of controlled, layered blood vessel-like structures. *RSC Adv.* **2016**, *6*, 55054–55063. [CrossRef]

13. Hueper, W.C.; Martin, G.J.; Thompson, M.R. Methyl cellulose solution as a plasma substitute. *Am. J. Surg.* **1942**, *56*, 629–635. [CrossRef]

14. Lee, K.-W.; Stolz, D.B.; Wang, Y. Substantial expression of mature elastin in arterial constructs. *Proc. Natl. Acad. Sci. USA* **2011**, *108*, 2705–2710. [CrossRef] [PubMed]

15. Fung, Y.C.; Fronek, K.; Patitucci, P. Pseudoelasticity of arteries and the choice of its mathematical expression. *Am. J. Physiol. Heart Circ. Physiol.* **1979**, *237*, H620–H631.

16. Soletti, L.; Hong, Y.; Guan, J.; Stankus, J.J.; El-Kurdi, M.S.; Wagner, W.R.; Vorp, D.A. A bilayered elastomeric scaffold for tissue engineering of small diameter vascular grafts. *Acta Biomater.* **2010**, *6*, 110–122. [CrossRef] [PubMed]

17. Burton, A.C. Relation of structure to function of the tissues of the wall of blood vessels. *Physiol. Rev.* **1954**, *34*, 610–642.

18. Sarkar, S.; Lee, G.Y.; Wong, J.Y.; Desai, T.A. Development and characterization of a porous micro-patterned scaffold for vascular tissue engineering applications. *Biomaterials* **2006**, *27*, 4775–4782. [CrossRef] [PubMed]

19. Hou, Q.; Grijpma, D.W.; Feijen, J. Porous polymeric structures for tissue engineering prepared by a coagulation, compression moulding and salt leaching technique. *Biomaterials* **2003**, *24*, 1937–1947. [CrossRef]

20. Flemming, R.G.; Murphy, C.J.; Abrams, G.A.; Goodman, S.L.; Nealey, P.F. Effects of synthetic micro-and nano-structured surfaces on cell behavior. *Biomaterials* **1999**, *20*, 573–588. [CrossRef]

21. Diban, N.; Haimi, S.; Bolhuis-Versteeg, L.; Teixeira, S.; Miettinen, S.; Poot, A.; Grijpma, D.; Stamatialis, D. Development and characterization of poly (ε-caprolactone) hollow fiber membranes for vascular tissue engineering. *J. Membr. Sci.* **2013**, *438*, 29–37. [CrossRef]

22. L'Heureux, N.; Pâquet, S.; Labbé, R.; Germain, L.; Auger, F.A. A completely biological tissue-engineered human blood vessel. *FASEB J.* **1998**, *12*, 47–56. [PubMed]

23. Gauvin, R.; Ahsan, T.; Larouche, D.; Lévesque, P.; Dubé, J.; Auger, F.A.; Nerem, R.M.; Germain, L. A novel single-step self-assembly approach for the fabrication of tissue-engineered vascular constructs. *Tissue Eng. Part A* **2010**, *16*, 1737–1747. [CrossRef] [PubMed]

24. Kikuchi, T.; Shimizu, T.; Wada, M.; Yamato, M.; Okano, T. Automatic fabrication of 3-dimensional tissues using cell sheet manipulator technique. *Biomaterials* **2014**, *35*, 2428–2435. [CrossRef] [PubMed]

25. Gong, P.; Zheng, W.; Huang, Z.; Zhang, W.; Xiao, D.; Jiang, X. A strategy for the construction of controlled, three-dimensional, multilayered, tissue-like structures. *Adv. Funct. Mater.* **2013**, *23*, 42–46. [CrossRef]

26. Jin, Y.; Wang, N.; Yuan, B.; Sun, J.; Li, M.; Zheng, W.; Zhang, W.; Jiang, X. Stress-Induced Self-Assembly of Complex Three Dimensional Structures by Elastic Membranes. *Small* **2013**, *9*, 2410–2414. [CrossRef] [PubMed]

27. Matsuda, T.; Shirota, T.; Kawahara, D. Fabrication Factory for Tubular Vascular Tissue Mimics based on Automated Rolling Manipulation and Thermo-Responsive Polymers. *J. Tissue Sci. Eng.* **2014**, *5*, 1–10. [CrossRef]

28. Lee, Y.B.; Jun, I.; Bak, S.; Shin, Y.M.; Lim, Y.-M.; Park, H.; Shin, H. Reconstruction of Vascular Structure with Multicellular Components using Cell Transfer Printing Methods. *Adv. Healthc. Mater.* **2014**, *3*, 1465–1474. [CrossRef] [PubMed]

29. Shukla, A.; Almeida, B. Advances in cellular and tissue engineering using layer-by-layer assembly. *Wiley Interdiscip. Rev. Nanomed. Nanobiotechnol.* **2014**, *6*, 411–421. [CrossRef] [PubMed]

30. Chen, Z.L.; Li, Y.; Liu, W.W.; Zhang, D.Z.; Zhao, Y.Y.; Yuan, B.; Jiang, X.Y. Patterning mammalian cells for Mmodeling three types of naturally occurring cell-cell interactions. *Angew. Chem. Int. Ed.* **2009**, *48*, 8303–8305. [CrossRef] [PubMed]

31. Palacios-Cuesta, M.; Cortajarena, A.L.; Garcia, O.; Rodriguez-Hernandez, J. Fabrication of Functional Wrinkled Interfaces from Polymer Blends: Role of the Surface Functionality on the Bacterial Adhesion. *Polymers* **2014**, *6*, 2845–2861. [CrossRef]

32. Pimentel-Dominguez, R.; Velazquez-Benitez, A.M.; Rodrigo Velez-Cordero, J.; Hautefeuille, M.; Sanchez-Arevalo, F.; Hernandez-Cordero, J. Photothermal effects and applications of polydimethylsiloxane membranes with carbon nanoparticles. *Polymers* **2016**, *8*, 84. [CrossRef]

33. Ren, F.; Yesildag, C.; Zhang, Z.; Lensen, C.M. Surface patterning of gold nanoparticles on PEG-based hydrogels to control cell adhesion. *Polymers* **2017**, *9*, 154. [CrossRef]

34. Song, Y.; Feijen, J.; Grijpma, D.W.; Poot, A.A. Tissue engineering of small-diameter vascular grafts: A literature review. *Clin. Hemorheol. Microcirc.* **2011**, *49*, 357–374. [PubMed]

35. Sorrentino, S.; Haller, H. Tissue Engineering of Blood Vessels: How to Make a Graft. *Tissue Eng.* **2011**, 263–278.

36. Peck, M.; Gebhart, D.; Dusserre, N.; McAllister, T.N.; L'Heureux, N. The evolution of vascular tissue engineering and current state of the art. *Cells Tissues Organs* **2011**, *195*, 144–158. [CrossRef] [PubMed]

37. Krawiec, J.T.; Vorp, D.A. Adult stem cell-based tissue engineered blood vessels: A review. *Biomaterials* **2012**, *33*, 3388–3400. [CrossRef] [PubMed]

38. Menu, P.; Stoltz, J.F.; Kerdjoudj, H. Progress in vascular graft substitute. *Clin. Hemorheol. Microcirc.* **2013**, *53*, 117–129. [PubMed]

39. Konig, G.; McAllister, T.N.; Dusserre, N.; Garrido, S.A.; Iyican, C.; Marini, A.; Fiorillo, A.; Avila, H.; Wystrychowski, W.; Zagalski, K. Mechanical properties of completely autologous human tissue engineered blood vessels compared to human saphenous vein and mammary artery. *Biomaterials* **2009**, *30*, 1542–1550. [CrossRef] [PubMed]

40. L'Heureux, N.; Dusserre, N.; Marini, A.; Garrido, S.; de la Fuente, L.; McAllister, T. Technology insight: The evolution of tissue-engineered vascular grafts—From research to clinical practice. *Nat. Rev. Cardiol.* **2007**, *4*, 389–395. [CrossRef] [PubMed]

41. Bourget, J.-M.; Gauvin, R.; Larouche, D.; Lavoie, A.; Labbé, R.; Auger, F.A.; Germain, L. Human fibroblast-derived ECM as a scaffold for vascular tissue engineering. *Biomaterials* **2012**, *33*, 9205–9213. [CrossRef] [PubMed]

42. Sarkar, S.; Hillery, C.; Seifalian, A.; Hamilton, G. Critical parameter of burst pressure measurement in development of bypass grafts is highly dependent on methodology used. *J. Vasc. Surg.* **2006**, *44*, 846–852. [CrossRef] [PubMed]

43. Wu, W.; Allen, R.A.; Wang, Y. Fast-degrading elastomer enables rapid remodeling of a cell-free synthetic graft into a neoartery. *Nat. Med.* **2012**, *18*, 1148–1153. [CrossRef] [PubMed]

44. Allen, R.A.; Wu, W.; Yao, M.; Dutta, D.; Duan, X.; Bachman, T.N.; Champion, H.C.; Stolz, D.B.; Robertson, A.M.; Kim, K. Nerve regeneration and elastin formation within poly(glycerol sebacate)-based synthetic arterial grafts one-year post-implantation in a rat model. *Biomaterials* **2014**, *35*, 165–173. [CrossRef] [PubMed]

45. Abbott, W.M.; Megerman, J.; Hasson, J.E.; L'Italien, G.; Warnock, D.F. Effect of compliance mismatch on vascular graft patency. *J. Vasc. Surg.* **1987**, *5*, 376–382. [CrossRef]

46. Mehigan, D.G.; Fitzpatrick, B.; Browne, H.I.; Bouchier-Hayes, D.J. Is compliance mismatch the major cause of anastomotic arterial aneurysms? Analysis of 42 cases. *J. Vasc. Surg.* **1984**, *26*, 147–150.

47. Okuhn, S.P.; Connelly, D.P.; Calakos, N.; Ferrell, L.; Man-Xiang, P.; Goldstone, J. Does compliance mismatch alone cause neointimal hyperplasia? *J. Vasc. Surg.* **1989**, *9*, 35–45. [CrossRef]

48. Sarkar, S.; Salacinski, H.J.; Hamilton, G.; Seifalian, A.M. The mechanical properties of infrainguinal vascular bypass grafts: Their role in influencing patency. *Eur. J. Vasc. Endovasc. Surg.* **2006**, *31*, 627–636. [CrossRef] [PubMed]

49. Patel, A.; Fine, B.; Sandig, M.; Mequanint, K. Elastin biosynthesis: The missing link in tissue-engineered blood vessels. *Cardiovasc. Res.* **2006**, *71*, 40–49. [CrossRef] [PubMed]

50. Wise, S.G.; Byrom, M.J.; Waterhouse, A.; Bannon, P.G.; Ng, M.K.C.; Weiss, A.S. A multilayered synthetic human elastin/polycaprolactone hybrid vascular graft with tailored mechanical properties. *Acta Biomater.* **2011**, *7*, 295–303. [CrossRef] [PubMed]

51. Gelse, K.; Pöschl, E.; Aigner, T. Collagens-structure, function, and biosynthesis. *Adv. Drug Deliv. Rev.* **2003**, *55*, 1531–1546. [CrossRef] [PubMed]

52. Papenburg, B.J.; Liu, J.; Higuera, G.A.; Barradas, A.M.C.; de Boer, J.; van Blitterswijk, C.A.; Wessling, M.; Stamatialis, D. Development and analysis of multi-layer scaffolds for tissue engineering. *Biomaterials* **2009**, *30*, 6228–6239. [CrossRef] [PubMed]

53. Ko, H.C.; Milthorpe, B.K.; McFarland, C.D. Engineering thick tissues—The vascularisation problem. *Eur. Cell Mater.* **2007**, *14*, 1–19. [CrossRef] [PubMed]

54. Malda, J.; Klein, T.J.; Upton, Z. The roles of hypoxia in the in vitro engineering of tissues. *Tissue Eng.* **2007**, *13*, 2153–2162. [CrossRef] [PubMed]

55. Murphy, C.M.; Haugh, M.G.; O'Brien, F.J. The effect of mean pore size on cell attachment, proliferation and migration in collagen-glycosaminoglycan scaffolds for bone tissue engineering. *Biomaterials* **2010**, *31*, 461–466. [CrossRef] [PubMed]

56. Hu, J.J.; Chao, W.C.; Lee, P.Y.; Huang, C.H. Construction and characterization of an electrospun tubular scaffold for small-diameter tissue-engineered vascular grafts: A scaffold membrane approach. *J. Mech. Behav. Biomed. Mater.* **2012**, *13*, 140–155. [CrossRef] [PubMed]

57. Bettahalli, N.M.S.; Groen, N.; Steg, H.; Unadkat, H.; Boer, J.; Blitterswijk, C.A.; Wessling, M.; Stamatialis, D. Development of multilayer constructs for tissue engineering. *J. Tissue Eng. Regener. Med.* **2014**, *8*, 106–119. [CrossRef] [PubMed]

58. Martin, Y.; Vermette, P. Bioreactors for tissue mass culture: Design, characterization, and recent advances. *Biomaterials* **2005**, *26*, 7481–7503. [CrossRef] [PubMed]

59. Fernandez, C.E.; Achneck, H.E.; Reichert, W.M.; Truskey, G.A. Biological and engineering design considerations for vascular tissue engineered blood vessels (TEBVs). *Curr. Opin. Chem. Eng.* **2014**, *3*, 83–90. [CrossRef] [PubMed]

60. Grant, S.A.; Spradling, C.S.; Grant, D.N.; Fox, D.B.; Jimenez, L.; Grant, D.A.; Rone, R.J. Assessment of the biocompatibility and stability of a gold nanoparticle collagen bioscaffold. *J. Biomed. Mater. Res. Part A* **2014**, *102*, 332–339. [CrossRef] [PubMed]

MDPI

Article

Manipulating Living Cells to Construct a 3D Single-Cell Assembly without an Artificial Scaffold

Aoi Yoshida [1,†], Shoto Tsuji [1,†], Hiroaki Taniguchi [2,†], Takahiro Kenmotsu [1], Koichiro Sadakane [1,*] and Kenichi Yoshikawa [1,*]

[1]   Faculty of Life and Medical Sciences, Doshisha University, Kyoto 610-0394, Japan; aoi.yoshida342@gmail.com (A.Y.); t.shoto.bmm1109@gmail.com (S.T.); tkenmots@mail.doshisha.ac.jp (T.K.)
[2]   The Institute of Genetics and Animal Breeding, Polish Academy of Sciences, Postepu 36A, Jastrzebiec, 05-552 Magdalenka, Poland; h.taniguchi@ighz.pl
*   Correspondence: ksadakan@mail.doshisha.ac.jp (K.S.); keyoshik@mail.doshisha.ac.jp (K.Y.); Tel.: +81-774-65-6127 (K.S.); +81-774-65-6243 (K.Y.)
†   These authors contributed equally to this work.

Received: 30 June 2017; Accepted: 25 July 2017; Published: 30 July 2017

Abstract: Artificial scaffolds such as synthetic gels or chemically-modified glass surfaces that have often been used to achieve cell adhesion are xenobiotic and may harm cells. To enhance the value of cell studies in the fields of regenerative medicine and tissue engineering, it is becoming increasingly important to create a cell-friendly technique to promote cell–cell contact. In the present study, we developed a novel method for constructing stable cellular assemblies by using optical tweezers in a solution of a natural hydrophilic polymer, dextran. In this method, a target cell is transferred to another target cell to make cell–cell contact by optical tweezers in a culture medium containing dextran. When originally non-cohesive cells are held in contact with each other for a few minutes under laser trapping, stable cell–cell adhesion is accomplished. This method for creating cellular assemblies in the presence of a natural hydrophilic polymer may serve as a novel next-generation 3D single-cell assembly system with future applications in the growing field of regenerative medicine.

Keywords: 3D cellular assembly; optical tweezers; crowding effect; depletion effect; remote control

1. Introduction

Stem-cell-based tissue engineering has emerged as a promising approach for the treatment of intractable diseases [1,2]. To enhance the value of cell studies in the fields of regenerative medicine and tissue engineering, it is becoming increasingly important to create a cell-friendly technique for three-dimensional (3D) cellular assembly [3]. The development of an efficient, minimally-invasive cellular system capable of 3D assembly within an in vitro or in vivo micro environment has been challenging. We recently reported a novel system that could generate a single-cell-based 3D assembly of cells using polyethylene glycol (PEG) by establishing stable cell–cell contact even after the cell assembly was transferred to a PEG-free solution [4]. While synthetic polymers and hydrogels, including PEG, have defined chemical structures and stable physicochemical properties [5–7], they are foreign material to cells [8]. It has been reported that PEG promotes cell–cell fusion, and that it has non-negligible effects on the structure and function of living cells in a culture medium [9,10]. Recent trials aimed at producing cellular scaffolds using natural polymers including gelatin, chitosan, dextran, alginate and collagen have been gaining significant momentum [11–16]. However, while these natural polymers have also been used to promote cellular adhesion within these cell structures in a non-selective manner, further improvements in hydrogel biocompatibility and adaptability for handling 3D cellular assembly systems are needed.

Dextrans are polysaccharides that contain a linear backbone of α-linked D-glucopyranosyl repeating units [17]. Several studies have implicated dextrans in 3D cellular assembly [18,19]. In this regard, cross-linking glycidyl methacrylate derivatized dextran and dithiothreitol under physiological conditions allows 3D encapsulation of rat bone marrow mesenchymal stem cells (BM-MSCs) and NIH/3T3 fibroblasts; while maintaining high viability [20]. Although it has been proposed that dextran can be used for 3D cellular assembly in a less toxic environment, nearly all of the studies of dextran-induced 3D cellular assembly reported thus far have used a 3D mass cell culture system [21,22]. We and others have recently developed an efficient procedure for constructing cellular assemblies by arranging desired cells at desired positions using optical tweezers in a PEG solution [4,23]. The aim of the present study was to modify the aforementioned PEG protocol to establish a 3D single-cell-based manipulation system that uses a less invasive method of cellular assembly involving optical tweezers and dextran as a natural polymer. This dextran-based method may serve as a novel next-generation 3D single-cell assembling system with future applications in the growing field of regenerative medicine.

2. Materials and Methods

2.1. Cell Culture

NAMRU mouse mammary gland epithelial cells (NMuMG cells) were cultured in Dulbecco's modified Eagle's medium (DMEM) (Wako Pure Chem. Inc., Osaka, Japan) supplemented with 10% fetal bovine serum (FBS) (Cell Culture Biosci., Nichirei Biosci. Inc., Tokyo, Japan), 40 μg/mL streptomycin, and 40 units/mL penicillin (Life Tech. Corp., Carlsbad, CA, USA). Neuro2A cells were cultured in Dulbecco's modified Eagle's medium supplemented with 10% FBS (Atlas Biological., Fort Collins, CO, USA), 1% non-essential amino acids, 100 U/mL penicillin and 10 μg/mL streptomycin (Wako Pure Chem. Inc., Osaka, Japan). The cells were incubated at 37 °C in a humidified atmosphere of 5% CO_2. Sub-confluent cells were harvested with trypsin (0.25% Trypsin–EDTA (1X)) (Life Tech. Corp., Carlsbad, CA, USA) and cryopreserved with CELLBANKER1 (Nippon Zenyaku Kogyo, Koriyama, Japan). For preparation of the polymer solution, we used dextran (DEX) (200,000; molecular biology-grade, Wako Pure Chem. Inc., Osaka, Japan). We prepared DMEM solution containing 10–50 mg/mL of DEX.

2.2. Single-Beam Optical Tweezers

Optical trapping with single laser beam was carried out using an inverted microscope (TE-300, Nikon) equipped with a Charge Coupled Device (CCD) camera (WAT-120N, Watec Co., Ltd., Tsuruoka, Japan). A 1064 nm Continuous Wave (CW) laser beam (Spectra Physics, Santa Clara, CA, USA) was introduced into the microscope, and focused into the sample through an oil-immersed objective lens (100×, N.A. = 1.3). The laser power at the focal point was set between 42 and 84 mW. All experiments were carried out at room temperature, i.e., 25 °C.

2.3. Double-Beam Optical Tweezers

Optical trapping with a double laser beam was carried out using a commercial optical tweezer instrument (NanoTracker 2, JPK Instruments, Berlin, Germany), which is constructed on an inverted microscope (IX71, Olympus, Tokyo, Japan) equipped with a CCD camera (DFK 31AF03, The Imaging Source, Taipei, Taiwan). A 1064 nm CW laser beam was split into two beams by a polarization beam splitter, and both beams were focused into the sample through a water-immersed objective lens (60×, N.A. = 1.2) for independent trapping. One of these focal points can be moved by using a piezo-mirror. The laser power at each focal point was set between 40 and 130 mW. All experiments were carried out at room temperature, i.e., 25 °C.

2.4. Viscosity Measurement

The kinetic viscosities of the dextran and polyethylene glycol were measured using a vibrational viscometer (SV-10, A&D Company, Tokyo, Japan) at room temperature (between 19 and 23 °C). The measurement time for each mixture was 60 s.

2.5. Cell Viability Assay

The NMuMG cells were cultured in a 6-well dish (Thermo Fisher Scientific, Waltham, MA, USA) and treated with 10–40 mg/mL of DEX. After 24 h of culture, cell viability was verified by trypan blue staining.

3. Results

Figure 1 shows the relationship between the concentration of DEX C_{dex} and kinetic viscosity ν in a water-based dextran solution. C_{dex} denotes the concentration of dextran in solution and ν represents kinetic viscosity. ν increases linearly as C_{dex} increases. Since the slope changes at 50 mg/mL C_{dex}, the overlap concentration C^* was deduced to be ca. 50 mg/mL. This value was 2.5 times greater than that of a water-based PEG (50 K) mixture, 20 mg/mL [4].

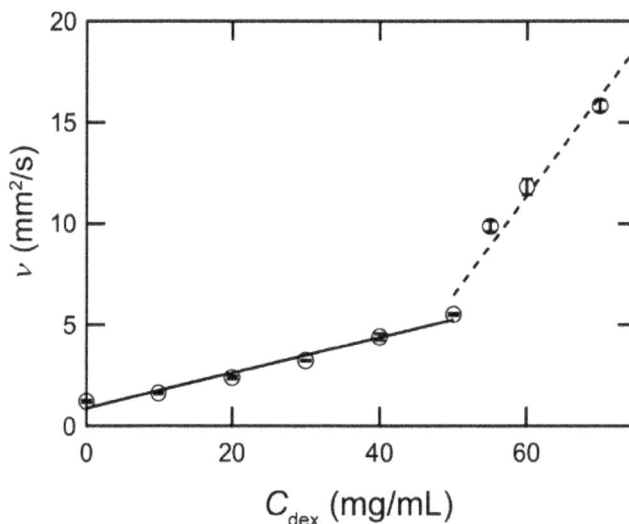

Figure 1. Relationship between C_{dex} and ν_{in} a water-based dextran solution. The solid and dashed lines represent the linear regression between a C_{dex} ranging from 0–50 and 50–70 mg/mL, respectively. Error bars represent the standard error of the mean calculated from three independent measurements.

To determine the cytotoxic effect of dextran, we adapted a trypan blue exclusion method to identify the proportion of viable cells. As a result, 40 mg/mL of dextran-containing medium was associated with a NMuMG cell viability similar to that of cells cultured in a medium without dextran (Figure 2). This suggests that, even after treatment with laser tweezers, cellular activity in the presence of dextran is maintained, without cytotoxicity.

Figure 3 examines the optical-tweezer-dependent mechanism of cell–cell contact. In this experiment, we applied single-beam optical tweezers. Figure 3A,B show cell adherence induced with optical tweezers for 5 and 300 s, respectively. As shown, 5 s of forced contact was insufficient to induce cell adherence. On the other hand, as reported previously, 300 s of forced contact produced stable cell–cell contact. This suggests that 300 s are required to produce cell adherence and in turn establish 3D cellular assembly. Figure 3C demonstrates the stability of cell–cell contact represented by the percentage of surviving cellular pairs following the indicated actions (approach or contact).

Figure 4 demonstrates the formation of various cell morphologies. Here, we applied double-beam optical tweezers.

Figure 5 confirms the formation of various cell morphologies from Neuro2A mouse brain neuroblastoma cells. In this experiment, we applied single-beam optical tweezers. Neuro2A cells were manipulated with optical tweezers to form various cellular structures. Since undifferentiated Neuro2A cells are used in models of neuronal differentiation [24], 3D assembly of these cells may be useful for modelling neuronal differentiation in 3D cell structures. This system may serve to demonstrate the relationship between 3D cell positioning of undifferentiated neuronal stem cells and neurogenesis.

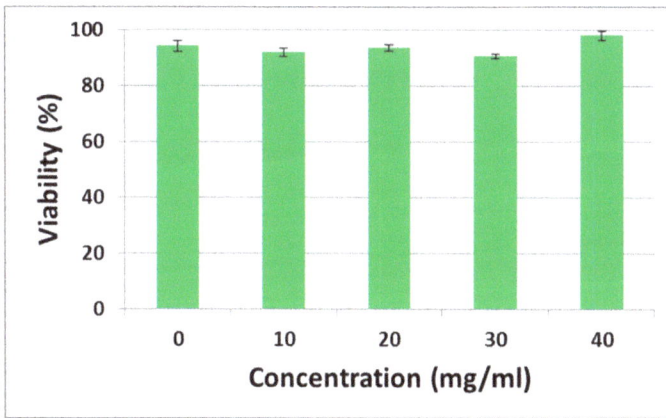

Figure 2. Viability of NMuMG cells. These cells were treated with various concentrations of dextran ranging from 0 to 40 mg/mL. Error bars represent the standard error of the mean calculated from three independent measurements.

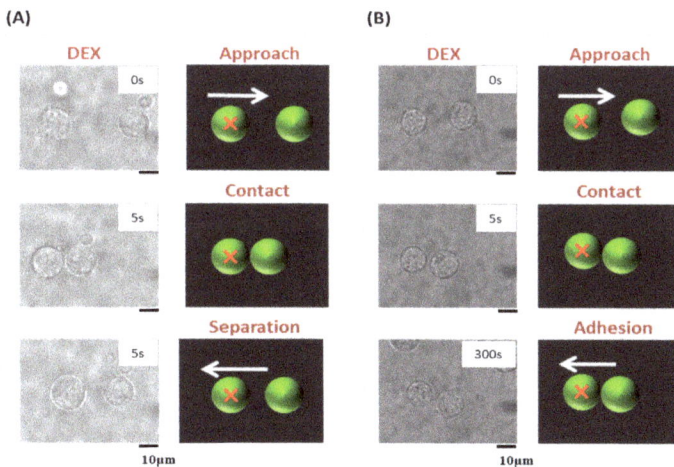

Figure 3. *Cont.*

(C)

Figure 3. Laser manipulation of a pair of epithelial cells (NMuMG); (**A**) 5 s or (**B**) 300 s in the presence of dextran (40 mg/mL): Spatio-temporal diagram illustrating the process of manipulation (Approach, Contact or Separation (**A**) or Adhesion (**B**)). (**C**) The probability that stable cell–cell contact is maintained through optical transportation for the distance of ca. 5 mm, i.e., the percentage of experimental runs to obtain the result as exemplified in (**B**), where the result as in (**A**) was counted as a failure. Error bars represent the standard error of the mean calculated from three independent measurements.

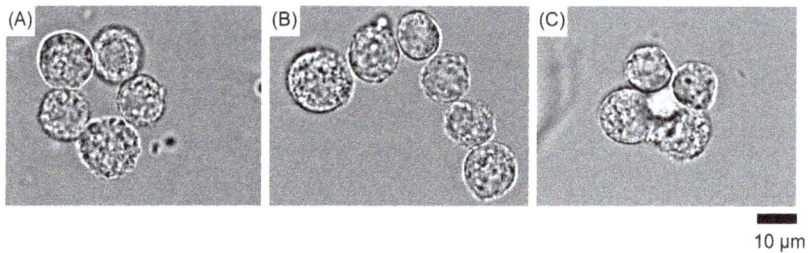

Figure 4. Assemblies of epithelial cells (NMuMG) of various shapes in a medium with DEX (40 mg/mL): the shape of a donut (**A**), letter 'L' (**B**), and tetragonal pyramid as an example of 3D cluster (**C**).

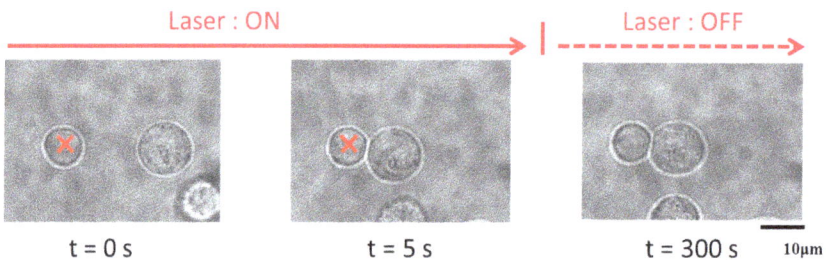

Figure 5. Stable cellular assembly of Neuro2A cells in a dextran (50 mg/mL) medium using optical laser tweezers. The focal point of the laser is marked by the red 'x'.

4. Discussion

The present study successfully extends our recent results on the use of solvable polymers for 3D single-cell-based manipulation. This system adapts an invasive cell assembly method involving optical tweezers and dextran. This dextran-based method is expected to serve as a next-generation 3D single-cell assembly system with future applications in cell biology and regenerative medicine. The results of the present study suggest further advancement of the present method for the formation of stable 3D cellular assemblies through the use of various kinds of solvable polymers, including proteins, polysaccharides, polynucleic acids, etc.

It is widely considered that cell adhesion may be engendered using the congestion or crowding effect [4], which has been attributable to the effect of higher osmotic pressure. Here, it is noted that the effect exerted by a polymer "crowding" solution on living cells is much different from the simple osmotic pressure caused by the addition of small solute molecules. With an increase in osmotic pressure due to the presence of small solutes, cells tend to shrink. However, such osmotic pressure does not directly switch cell–cell interaction from repulsive to attractive. Cell aggregation is a secondary effect caused by an increase in osmotic pressure. In contrast, a solution crowded with macromolecules or large polymer chains causes a so-called depletion effect [4,25,26]. The attractive interaction as a result of the depletion effect by polymer crowding occurs at a length scale on the order of the size of the polymer chain, for example on a scale represented by R_g for the radius of gyration or R_H for the hydrodynamic radius. In contrast, as for the osmotic effect of polymers, the increase in osmotic pressure is roughly proportional to the contour length or full-stretch length of the polymer chain, which is proportional to the volume as the product of the contour length and the cross-sectional area of the polymer, being much smaller than the volume occupied by the polymer chain. Thus, in general, the depletion effect is much more significant for polymer chains with a large contour length of more than several tens of nm. The attractive interactions between facing cells due to the entropic depletion force causes flat contact between a pair of cells by creating a gap with the size of the polymer, accompanied by slight deformation of the curvature of the facing side of the membrane on the order of the energy of thermal fluctuation. The occurrence of such a flat gap contributes to the formation of stable cellular contact, by allowing the 2D random walk of individual membrane components, such as membrane proteins and sugar lipids, so as to form attractive contact pairs upon encountering membrane surfaces. The stabilization energy between the facing membranes per unit area due to depletion interaction, ε_α, is expressed with approximation as [4],

$$\begin{aligned} \varepsilon_\alpha &= -\pi_P(2d - x) \quad \text{when} \quad x \leq 2d \\ \varepsilon_a &= 0 \qquad\qquad\quad \text{when} \quad x > 2d, \end{aligned} \tag{1}$$

where π_P is the depletion pressure due to the coexisting polymer such as DEX polymer; d is the diameter of the polymer in a random coil confirmation; and x is the distance between the facing membranes. Thus, the attractive force per surface area due to the depletion effect can be described as,

$$\begin{aligned} P_{dep} &\approx -\tfrac{\partial \varepsilon_\alpha}{\partial x} = -\pi_P \quad \text{when} \quad x \leq 2d \\ P_{dep} &\approx 0 \qquad\qquad\qquad \text{when} \quad x > 2d. \end{aligned} \tag{2}$$

Meanwhile, the repulsive interactions between a pair of cells floating in solution, such as electric interactions [27], short-range hydration repulsion force [28], and membrane duration [29,30], are inherent properties of a cell membrane [31]. The following relationship has been proposed for repulsive interaction between facing membranes [27]. The positive pressure P_{rep} can be represented as,

$$P_{rep} = P_{rep}^0 \exp\left(-\frac{x}{L}\right), \tag{3}$$

where P_{rep}^0 is a positive variable that depends on the ionic strength of the medium and the surface potential at the extreme of $x = 0$; and L is the characteristic length on the order of 1 nm for a conventional medium for cell culture.

Based on the above arguments regarding the interactions between facing membranes, the net pressure P_{net} is the sum of the depletion and repulsive interactions: $P_{net} = P_{dep} + P_{rep}$ [4]. With regard to the experimental conditions for DEX (200 k), the size of the DEX polymer chain in solution, d, is regarded as 15–20 nm [32]. We reported that the transmembrane distance is $x = 10 - 15$nm in the balance between attractive and repulsive contributions in solutions containing PEG as a crowding polymer [4]. Under similar treatment, we deduced that the width of the gap between facing cells in the present work is of a similar order of magnitude, i.e., $x = 10 - 15$nm. Thus, when cell–cell contact is achieved in the presence of a crowding polymer, there remains a void space with a width comparable to the size of the polymer between the facing membranes of the adjacent cells. As we discussed in our recent article, the existence of this void space between facing membranes provides the opportunity for the facing membranes to switch repulsive interaction to attractive interaction during cell–cell contact due to dielectric attraction under a focused laser. As a result, stable cell–cell contact could be achieved even after the removal of the coexisting polymer [4]. With regard to the generation of attractive interaction between neighboring cells, a crowding solution environment causes DNA condensation and compaction through the occurrence of the attractive interaction between negatively charged DNA segments [33]. The essential mechanism of attraction is similar to that in our observation of the formation of stable cell–cell contact in PEG solution [4]. Altogether, the overall findings suggest that repulsive single cells can be easily attracted to each other in a medium which contains water soluble polymers such as dextran and PEG, through the depletion effect. This effect facilitates us to construct 3D single cell structures and is, therefore, useful for further applications.

In this study, stable cellular adhesion of NMuMG cells was achieved at concentrations even lower than the overlap concentration C*. The fact that the polymer solution formed stable cell–cell contact below C* is of value because the viscosity of the solution is essentially the same as that of the usual cell culture medium. Under such low viscosity, transportation of individual cells by laser does not encounter any difficulty. While PEG is a synthetic polymer, we have demonstrated for the first time that a natural polymer (dextran) can precipitate cell adhesion and facilitate 3D cellular assembly. In related studies on the use of DEX solution for cell manipulation, Takayama et al. have examined cell printing on a solid substrate. With the use of a dextran solution, they successfully localized cell aggregates at desired positions [34,35]. They reported that the gene expression profile of cells that differentiated after treatment with dextran solution was almost the same as that of cells in the usual medium. In this study, we noted that 10–40 mg/mL of dextran did not have any cytotoxic effect on NMuMG cells. Moreover, since natural compounds like dextran are relatively safe and therefore easily added to the cell culture medium, our dextran-based method may be able to achieve relatively less invasive 3D single-cell assembly.

Cell–cell contact influences the ability of progenitor cells to differentiate [36,37]. Moreover, embryonic stem cells are occasionally cultured as 3D aggregates, known as embryoid bodies. This enhances the differentiation of several cell types. Similarly, neurogenesis can be triggered using sphere-cultured P19 mouse carcinoma cells [38]. Therefore, how these cells are reconstituted in a 3D structure is quite important for proper differentiation. In this regard, our dextran-based 3D cell assembly technique may enable users to control 3D-structure assembly in the hope of achieving novel advances in regenerative medicine.

Cell therapy has received great attention in the field of neuroscience. In the near future, these therapies may be used to treat neurodegenerative diseases such as Parkinson's disease. In this regard, we were able to make single cell–cell contact using Neuro2A cells. Since neuronal circuits are well organized, it is important to precisely manage the position of neuronal cells and their progenitors. Our 3D single-cell assembly method together with DEX should help to facilitate the reconstitution of damaged neuronal networks. To this end, our approach, which involves the use of optical tweezers, may contribute to the regeneration of well-functioning neuronal systems. Since neuronal systems are composed of many cell types, our next challenge will be to produce lasting cell–cell adhesion between these various cells. Moreover, although we optimized the conditions required for fine control of the position of neuronal progenitor cells in this study, a functional analysis was not performed. Therefore, it will be important to further evaluate our findings using long-term cell culture methods and various functional assays (e.g., single cell transcriptome analysis). To develop functional organoids, which has recently been used as the scientific term to indicate 3D organ-like structures composed of specific cell types or progenitor cells [39,40], it is important to establish an experimental methodology to construct 3D cellular assembly of a large number of cells (above the order of one hundred) with various morphologies. In this regard, we would like to propose the future extension of our study with the hierarchical construction of small cell assemblies (on the order of ten) into a larger assembly with specific shape where each small cell assembly exhibits suitable positioning to generate the specific function as an organoid.

Concluding the present article, we would like to stress the difference between the crowding effect or depletion effect due to macromolecules and the effect of osmotic pressure. The depletion effect induces attractive interaction between neighboring cells through the effect of the exclusion volume by macromolecules on the scale of R^3, where R is the radius of a polymer chain as represented by R_g or R_H. In contrast, the osmotic effect scales with the exclusion volume along the polymer chain, which is much smaller than the scale of R^3 for the depletion volume of macromolecules. Additionally, depletion interaction induces attractive interaction between neighboring cells as a primary, direct effect. The present study indicated that depletion interaction is promising for the formation of stable cell–cell contact. We have shown that the natural macromolecule DEX is useful for this purpose, similar to the synthetic polymer PEG. Although there is a large difference in conformational characteristics between DEX and PEG, they result in similar cell–cell attraction, suggesting that the depletion effect can generally be applicable for the construction of stable 3D cellular assemblies. Further extension of this study is expected to contribute to future developments in cellular biology and regenerative medicine. Lastly, we would like to stress that most eukaryotic cells maintain their lives surrounded by crowding body fluids.

Acknowledgments: The authors thank Hashimoto and Ohta at Doshisha University for generous assistance with the experimental setup. This work was supported by KAKENHI (15H02121, 15K05400, 25103012, 50587441) and by the MEXT-Supported Program for the Strategic Research Foundation at Private Universities.

Author Contributions: Hiroaki Taniguchi, Koichiro Sadakane, and Kenichi Yoshikawa conceived and designed the experiments; Aoi Yoshida and Shoto Tsuji performed the experiments; Aoi Yoshida, Shoto Tsuji, and Koichiro Sadakane analyzed the data; and Hiroaki Taniguchi, Koichiro Sadakane, Takahiro Kenmotsu, and Kenichi Yoshikawa wrote the paper.

Conflicts of Interest: The authors declare no conflict of interest.

References

1. Lane, S.W.; Williams, D.A.; Watt, F.M. Modulating the Stem Cell Niche for Tissue Regeneration. *Nat. Biotechnol.* **2014**, *32*, 795–803. [CrossRef] [PubMed]
2. Li, M.; Guo, K.; Ikehara, S. Intractable Diseases Treated with Intra-Bone Marrow-Bone Marrow Transplantation. *Front. Cell Dev. Biol.* **2014**, *2*, 48. [CrossRef] [PubMed]
3. Chen, Y.W.; Huang, S.X.; de Carvalho, A.L.R.T.; Ho, S.H.; Islam, M.N.; Volpi, S.; Notarangelo, L.D.; Ciancanelli, M.; Casanova, J.L.; Bhattacharya, J.; et al. A Three-Dimensional Model of Human Lung Development and Disease from Pluripotent Stem Cells. *Nat. Cell Biol.* **2017**, *19*, 542–549. [CrossRef] [PubMed]
4. Hashimoto, S.; Yoshida, A.; Ohta, T.; Taniguchi, H.; Sadakane, K.; Yoshikawa, K. Formation of Stable Cell-Cell Contact without a Solid/Gel Scaffold: Non-invasive Manipulation by Laser under Depletion Interaction with a Polymer. *Chem. Phys. Lett.* **2016**, *655*, 11–16. [CrossRef]
5. Tanaka, M.; Sackmann, E. Polymer-Supported Membranes as Models of the Cell Surface. *Nature* **2005**, *437*, 656–663. [CrossRef] [PubMed]
6. Tang, Y.; Liu, L.; Li, J.; Yu, L.; Wang, L.; Shi, J.; Chen, Y. Induction and Differentiation of Human Induced Pluripotent Stem Cells into Functional Cardiomyocytes on a Compartmented Monolayer of Gelatin Nanofibers. *Nanoscale* **2016**, *8*, 14530–14540. [CrossRef] [PubMed]

7. Masuda, S.; Yanase, Y.; Usukura, E.; Ryuzaki, S.; Wang, P.; Okamoto, K.; Kuboki, T.; Kidoaki, S.; Tamada, K. High-Resolution Imaging of a Cell-Attached Nanointerface using a Gold-Nanoparticle Two-Dimensional Sheet. *Sci. Rep.* **2017**, *7*, 3720. [CrossRef] [PubMed]

8. Li, W.A.; Lu, B.Y.; Gu, L.; Choi, Y.; Kim, J.; Mooney, D.J. The Effect of Surface Modification of Mesoporous Silica Micro-Rod Scaffold on Immune Cell Activation and Infiltration. *Biomaterials* **2016**, *83*, 249–256. [CrossRef] [PubMed]

9. Gietz, R.D.; Schiestl, R.H.; Willems, A.R.; Woods, R.A. Studies on the Transformation of Intact Yeast Cells by the LiAc/SS-DNA/PEG Procedure. *Yeast* **1995**, *11*, 355–360. [CrossRef] [PubMed]

10. Robinson, J.M.; Roos, D.S.; Davidson, R.L.; Karnovsky, M.J. Membrane Alterations and Other Morphological Features Associated with Polyethylene Glycol-Induced Cell Fusion. *J. Cell Sci.* **1979**, *40*, 63–75. [PubMed]

11. Dhandayuthapani, B.; Krishnan, U.M.; Sethuraman, S. Fabrication and Characterization of Chitosan-Gelatin Blend Nanofibers for Skin Tissue Engineering. *J. Biomed. Mater. Res. B* **2010**, *94*, 264–272. [CrossRef] [PubMed]

12. Huang, Y.; Onyeri, S.; Siewe, M.; Moshfeghian, A.; Madihally, S.V. In Vitro Characterization of Chitosan-Gelatin Scaffolds for Tissue Engineering. *Biomaterials* **2005**, *26*, 7616–7627. [CrossRef] [PubMed]

13. Liu, Y.; Chan-Park, M.B. Hydrogel Based on Interpenetrating Polymer Networks of Dextran and Gelatin for Vascular Tissue Engineering. *Biomaterials* **2009**, *30*, 196–207. [CrossRef] [PubMed]

14. Rosellini, E.; Cristallini, C.; Barbani, N.; Vozzi, G.; Giusti, P. Preparation and Characterization of alginate/gelatin Blend Films for Cardiac Tissue Engineering. *J. Biomed. Mater. Res. A* **2009**, *91*, 447–453. [CrossRef] [PubMed]

15. Sakai, S.; Hashimoto, I.; Kawakami, K. Synthesis of an Agarose-Gelatin Conjugate for use as a Tissue Engineering Scaffold. *J. Biosci. Bioeng.* **2007**, *103*, 22–26. [CrossRef] [PubMed]

16. Tan, H.; Wu, J.; Lao, L.; Gao, C. Gelatin/chitosan/hyaluronan Scaffold Integrated with PLGA Microspheres for Cartilage Tissue Engineering. *Acta Biomater.* **2009**, *5*, 328–337. [CrossRef] [PubMed]

17. Banerjee, A.; Bandopadhyay, R. Use of Dextran Nanoparticle: A Paradigm Shift in Bacterial Exopolysaccharide Based Biomedical Applications. *Int. J. Biol. Macromol.* **2016**, *87*, 295–301. [CrossRef] [PubMed]

18. Lee, V.K.; Kim, D.Y.; Ngo, H.; Lee, Y.; Seo, L.; Yoo, S.S.; Vincent, P.A.; Dai, G. Creating Perfused Functional Vascular Channels using 3D Bio-Printing Technology. *Biomaterials* **2014**, *35*, 8092–8102. [CrossRef] [PubMed]

19. Liu, Y.; Rayatpisheh, S.; Chew, S.Y.; Chan-Park, M.B. Impact of Endothelial Cells on 3D Cultured Smooth Muscle Cells in a Biomimetic Hydrogel. *ACS Appl. Mater. Interfaces* **2012**, *4*, 1378–1387. [CrossRef] [PubMed]

20. Liu, Z.Q.; Wei, Z.; Zhu, X.L.; Huang, G.Y.; Xu, F.; Yang, J.H.; Osada, Y.; Zrinyi, M.; Li, J.H.; Chen, Y.M. Dextran-Based Hydrogel Formed by Thiol-Michael Addition Reaction for 3D Cell Encapsulation. *Colloids Surf. B* **2015**, *128*, 140–148. [CrossRef] [PubMed]

21. Aloysious, N.; Nair, P.D. Enhanced Survival and Function of Islet-Like Clusters Differentiated from Adipose Stem Cells on a Three-Dimensional Natural Polymeric Scaffold: An in Vitro Study. *Tissue Eng. A* **2014**, *20*, 1508–1522. [CrossRef] [PubMed]

22. McKee, C.; Perez-Cruet, M.; Chavez, F.; Chaudhry, G.R. Simplified Three-Dimensional Culture System for Long-Term Expansion of Embryonic Stem Cells. *World J. Stem Cells* **2015**, *7*, 1064–1077. [PubMed]

23. Kirkham, G.R.; Britchford, E.; Upton, T.; Ware, J.; Gibson, G.M.; Devaud, Y.; Ehrbar, M.; Padgett, M.; Allen, S.; Buttery, L.D.; et al. Precision Assembly of Complex Cellular Microenvironments using Holographic Optical Tweezers. *Sci. Rep.* **2015**, *5*, 8577. [CrossRef] [PubMed]

24. Tremblay, R.G.; Sikorska, M.; Sandhu, J.K.; Lanthier, P.; Ribecco-Lutkiewicz, M.; Bani-Yaghoub, M. Differentiation of Mouse Neuro 2A Cells into Dopamine Neurons. *J. Neurosci. Methods* **2010**, *186*, 60–67. [CrossRef] [PubMed]

25. Asakura, S.; Oosawa, F. On Interaction between Two Bodies Immersed in a Solution of Macromolecules. *J. Chem. Phys.* **1954**, *22*, 1255–1256. [CrossRef]

26. Roth, R.; Gotzelmann, B.; Dietrich, S. Depletion Forces Near Curved Surface. *Am. Phys. Soc.* **1998**, *82*, 448–451. [CrossRef]

27. Hiemenz, P.C.; Rajagopalan, R. *Principles of Colloid and Surface Chemistry*; Marcel Dekker, Inc.: New York, NY, USA, 1997.

28. LeNeveu, D.M.; Rand, R.P.; Parsegian, V.A. Measurement of Forces between Lecithin Bilayers. *Nature* **1976**, *259*, 601–603. [CrossRef] [PubMed]

29. Chen, L.; Jia, N.; Gao, L.; Fang, W.; Golubovic, L. Effects of Antimicrobial Peptide Revealed by Simulations: Translocation, Pore Formation, Membrane Corrugation and Euler Buckling. *Int. J. Mol. Sci.* **2013**, *14*, 7932–7958. [CrossRef] [PubMed]

30. Sackmann, E.; Smith, S.A. Physics of Cell Adhesion: Some Lessons from Cell Mimetic Systems. *Soft Matter* **2014**, *10*, 1644–1659. [CrossRef] [PubMed]

31. Ohki, S. *Cell and Model Membrane Interactions*; Springer: New York, NY, USA, 1991.

32. Neul, B.; Meiselman, H.J. Depletion Interactions in Polymer Solutions Promote Red Blood Cell Adhesion to Albumin-Coated Surfaces. *Biochim. Biophys. Acta* **2006**, *1760*, 1772–1779.

33. Vasilevskaya, V.V.; Khokhlov, A.R.; Matsuzawa, Y.; Yoshikawa, K. Collapse of Single DNA Molecule in Poly(Ethylene Glycol) Solutions. *J. Chem. Phys.* **1995**, *102*, 6595–6602. [CrossRef]

34. Han, C.; Takayama, S.; Park, J. Formation and Manipulation of Cell Spheroids using a Density Adjusted PEG/DEX Aqueous Two Phase System. *Sci. Rep.* **2015**, *5*, 11891. [CrossRef] [PubMed]

35. Tavana, H.; Mosadegh, B.; Takayama, S. Polymeric Aqueous Biphasic Systems for Non-Contact Cell Printing on Cells: Engineering Heterocellular Embryonic Stem Cell Niches. *Adv. Mater.* **2010**, *22*, 2628–2631. [CrossRef] [PubMed]

36. Liu, Y.; Li, H.; Yan, S.; Wei, J.; Li, X. Hepatocyte Cocultures with Endothelial Cells and Fibroblasts on Micropatterned Fibrous Mats to Promote Liver-Specific Functions and Capillary Formation Capabilities. *Biomacromolecules* **2014**, *15*, 1044–1054. [CrossRef] [PubMed]

37. Ottone, C.; Krusche, B.; Whitby, A.; Clements, M.; Quadrato, G.; Pitulescu, M.E.; Adams, R.H.; Parrinello, S. Direct Cell-Cell Contact with the Vascular Niche Maintains Quiescent Neural Stem Cells. *Nat. Cell Biol.* **2014**, *16*, 1045–1056. [CrossRef] [PubMed]

38. Kim, B.S.; Lee, C.H.; Chang, G.E.; Cheong, E.; Shin, I. A Potent and Selective Small Molecule Inhibitor of Sirtuin 1 Promotes Differentiation of Pluripotent P19 Cells into Functional Neurons. *Sci. Rep.* **2016**, *6*, 34324. [CrossRef] [PubMed]

39. Clevers, H. Modeling Development and Disease with Organoids. *Cell* **2016**, *165*, 1586–1597. [CrossRef] [PubMed]

40. Fatehullah, A.; Tan, S.H.; Barker, N. Organoids as an in vitro Model of Human Development and Disease. *Nat. Cell Biol.* **2016**, *18*, 246–254. [CrossRef] [PubMed]

Article

Electrospinning PCL Scaffolds Manufacture for Three-Dimensional Breast Cancer Cell Culture

Marc Rabionet [1,2], Marc Yeste [3], Teresa Puig [1,*] and Joaquim Ciurana [2,*]

[1] New Therapeutic Targets Laboratory (TargetsLab)—Oncology Unit, Department of Medical Sciences, Faculty of Medicine, University of Girona, Emili Grahit 77, 17003 Girona, Spain; m.rabionet@udg.edu

[2] Product, Process and Production Engineering Research Group (GREP), Department of Mechanical Engineering and Industrial Construction, University of Girona, Maria Aurèlia Capmany 61, 17003 Girona, Spain

[3] Biotechnology of Animal and Human Reproduction (TechnoSperm), Department of Biology, Institute of Food and Agricultural Technology, University of Girona, Pic de Peguera 15, 17003 Girona, Spain; marc.yeste@udg.edu

* Correspondence: teresa.puig@udg.edu (T.P.); quim.ciurana@udg.edu (J.C.); Tel.: +34-972-419628 (T.P.); +34-972-418265 (J.C.)

Received: 27 June 2017; Accepted: 27 July 2017; Published: 1 August 2017

Abstract: In vitro cell culture is traditionally performed within two-dimensional (2D) environments, providing a quick and cheap way to study cell properties in a laboratory. However, 2D systems differ from the in vivo environment and may not mimic the physiological cell behavior realistically. For instance, 2D culture models are thought to induce cancer stem cells (CSCs) differentiation, a rare cancer cell subpopulation responsible for tumor initiation and relapse. This fact hinders the development of therapeutic strategies for tumors with a high relapse percentage, such as triple negative breast cancer (TNBC). Thus, three-dimensional (3D) scaffolds have emerged as an attractive alternative to monolayer culture, simulating the extracellular matrix structure and maintaining the differentiation state of cells. In this work, scaffolds were fabricated through electrospinning different poly(ε-caprolactone)-acetone solutions. Poly(ε-caprolactone) (PCL) meshes were seeded with triple negative breast cancer (TNBC) cells and 15% PCL scaffolds displayed significantly ($p < 0.05$) higher cell proliferation and elongation than the other culture systems. Moreover, cells cultured on PCL scaffolds exhibited higher mammosphere forming capacity and aldehyde dehydrogenase activity than 2D-cultured cells, indicating a breast CSCs enrichment. These results prove the powerful capability of electrospinning technology in terms of poly(ε-caprolactone) nanofibers fabrication. In addition, this study has demonstrated that electrospun 15% PCL scaffolds are suitable tools to culture breast cancer cells in a more physiological way and to expand the niche of breast CSCs. In conclusion, three-dimensional cell culture using PCL scaffolds could be useful to study cancer stem cell behavior and may also trigger the development of new specific targets against such malignant subpopulation.

Keywords: poly(ε-caprolactone); electrospinning; scaffolds; three-dimensional cell culture; triple negative breast cancer; breast cancer stem cells; mammospheres; aldehyde dehydrogenase

---

1. Introduction

Presently, in vitro cell culture represents a crucial tool to study cell behavior outside the organism. Most cell cultures are performed with a two-dimensional (2D) environment providing cheap and easy cell maintenance. A flat plastic surface is treated, obtaining adherent features to enable cell adhesion and proliferation. Therefore, these cells can only grow forming a monolayer, establishing interactions with surface and contiguous cells. Cells adopt a flattened morphology, which results in a modified membrane receptor polarity and cytoskeleton architecture. Different studies demonstrated that cell shape variations interfere with gene expression and protein synthesis regulation [1,2]. Consequently, the 2D cell culture model described differs from the physiological environment of living organisms. Body cells are embedded in the extracellular matrix (ECM), a three-dimensional (3D) complex constituted by fibrous proteins and molecules. This network structure provides a physical support for cell growth as well as playing a major role in cell regulation [3,4]. Cells can establish connections with their adjacent counterparts and with ECM fibrous mesh, thereby tending to adopt a more elongated morphology. This clear 3D physiological architecture contrasts with the lack of structure of two-dimensional cell culture. Hence, conclusions from in vitro monolayer cell culture experiments could be not applicable in terms of in vivo cell behavior, empowering the requirement of 3D models for cell culture.

Over recent years, various three-dimensional cell culture systems have emerged, differing in their composition, arrangement and final application. The models based on a solid, physical support are called scaffolds and are made up of a network of filaments mostly made by synthetic materials [5]. This 3D product can be manufactured by electrospinning technology using a high electric field. The polymer is dissolved and the solution is charged at high voltage. When electric force overcomes the surface tension, the polymer solution is pulled onto the target plate and the solvent is evaporated, collecting nanofibers which intersect each other [6]. The resultant architecture mimics the ECM fibrous assemblage and cultured cells are able to adopt a more in vivo shape. Obviously, all cell types possess distinct morphological characteristics which may lead to different cell culture support requirements. In this regard, it is worth noting that scaffold manufacturing techniques allow product customization, so that different process parameters, such as scaffold porosity, fiber diameter and microstructure, may be modified on the basis of the final application [7–9]. Poly(ε-caprolactone) (PCL) is a synthetic polymer that has been widely used to fabricate scaffolds, due to its viscoelastic and malleable properties, absence of isomers and low cost [10]. PCL may be processed with many technologies since it presents a low melting point around 60 °C and it is soluble in several solvents, such as chloroform, dichloromethane, benzene, acetone and dimethylformamide [11]. Moreover, PCL shows biocompatible properties, long-term biodegradability but bioresorbable [12], all these features making it a good candidate for biomedical and cell culture applications.

As previously explained, polymeric filaments offer physical support to cells to adhere and proliferate into the 3D structure. This fact enables cells to acquire a more elongated shape, close to physiological morphology, in contrast with the flatness adopted in monolayer cultures. Hence, the cancer research field is taking advantage of the three-dimensional cell culture model's benefits. Over the last two decades, cancers have not been studied as an abnormal growth of a single cell type, as cells with distinct characteristics, such as normal cancer cells and, in minor proportion, cancer stem cells (CSCs) are found in a given tumor. While the CSCs subpopulation only represents a small percentage of tumor cells, they have been demonstrated to drive cell growth in a wide variety of cancer types including leukemia [13], brain [14,15], myeloma [16] and breast [17]. Therefore, CSCs possess tumorigenic features, among other specific characteristics, useful for their identification and isolation. This subset is capable of undergoing self-renewal and differentiating into non-stem cancer cells due to their stem properties. Moreover, CSCs are able to grow in suspension and proliferate forming spheres [14,18], and can be isolated due to an enhanced activity of the aldehyde dehydrogenase (ALDH) enzyme [19]. As expected, this subpopulation with stem characteristics plays a key role in cancer development and prognosis. A link between CSCs and tumor relapse after treatment and metastasis is proven [20] since they show relative high radio- [21] and chemoresistance [22]. This fact becomes relevant in some specific cancer types with an appreciable recurrence percentage, such as breast cancer. Concretely, triple negative breast cancer (TNBC) presents the highest proportion of tumor relapse with a value around 34% and the lowest mean time to local and distant recurrence when compared with other breast cancer types [23]. TNBC is characterized by the absence of breast cancer molecular biomarker amplification, so therapeutic targets against TNBC do not exist and patients are treated with general chemotherapy [24].

Breast cancer stem cells (BCSCs) could become a potential target for future treatments against TNBC. However, BCSCs in vitro culture encounters a number of difficulties. Cancer stem subpopulation represents a low percentage within the tumor [17,25] and two-dimensional cell culture induces its differentiation losing stem features [26]. Previous investigations demonstrated that three-dimensional cell culture maintained and expanded BCSCs subset when compared with 2D culture samples [27–30]. Therefore, the present sought to test the suitability of fabricated electrospun PCL scaffolds to provide a more suitable niche for BCSCs to grow. Scaffolds from two different polymer concentrations were tested to evaluate 3D culture suitability with triple negative breast cancer cells. Cell proliferation and morphology were evaluated on different culture days and finally, BCSCs were quantified to discern the scaffold's culture effect. In agreement with literature, PCL scaffolds could be a useful tool to culture breast cancer cells in a more physiological way and expand the BCSCs subpopulation. Customizable methodologies such as electrospinning enable the production of distinct three-dimensional meshes concerning the cell of interest requests. Moreover, the BCSCs' enrichment could facilitate their study and the development of specific treatments against this malignant subpopulation. BCSCs targeted treatments could replace aggressive procedures like chemotherapy and attack the highly recurrent tumors such as triple negative breast cancer.

2. Materials and Methods

2.1. Scaffolds Fabrication

Poly(ε-caprolactone) (PCL) and acetone were chosen as biopolymer and non-toxic solvent respectively, to manufacture the scaffolds. PCL 3 mm pellets with an average molecular weight of 80,000 g/mol (Sigma-Aldrich, St. Louis, MO, USA) were dissolved in acetone (PanReac AppliChem, Gatersleben, Germany). Two different concentrations of 7.5 and 15% *w/v* PCL were achieved under 40 °C and agitation using a magnetic stirrer. Scaffolds were fabricated with an electrospinning instrument (Spraybase, Dublin, Ireland). PCL solution was placed in a plastic syringe (BD Plastipak, Franklin Lakes, NJ, USA) connected to an 18 G needle emitter with an inner diameter of 0.8 mm. A fixed voltage of 7 kV was applied and a flow rate of 6 mL/h was established by the Syringe Pump Pro software (New Era Pump Systems, Farmingdale, NY, USA). The distance between the emitter and stationary collector was 15 cm. The electrospinning process was stopped when 10 or 5 mL of solution were ejected, for 7.5 and 15% PCL concentrations respectively. The meshes were then cut into squares with a scalpel.

2.2. Scanning Electron Microscopy Analysis

Microscopic characterization was performed through scanning electron microscopy (SEM; Zeiss, Oberkochen, Germany) after carbon coating. Scaffolds were imaged on the top and bottom to confirm fibre uniformity and Image J software (National Institutes of Health, Bethesda, MD, USA) was used for image analysis. Fibre diameter, surface porosity and pore area were calculated from the top and bottom sides to calculate the average value.

2.3. Cell Line

MDA–MB–231 triple negative breast cancer cell line was obtained from the American Type Culture Collection (ATCC; Rockville, MD, USA). Cells were routinely grown in Dulbecco's Modified Eagle's Medium (DMEM; Gibco, Waltham, MA, USA) supplemented with 10% fetal bovine serum (FBS), 1% L-glutamine, 1% sodium pyruvate, 50 U/mL penicillin/streptomycin (HyClone, Logan, UT, USA). Cells were kept at 37 °C and 5% CO_2 atmosphere and culture medium was changed every 3 days.

2.4. Three-Dimensional Cell Seeding

PCL meshes were sterilized by immersion into 70% ethanol/water solution overnight, washed three times with PBS (Gibco, Waltham, MA, USA) and finally exposed to UV light for 30 min. Sterilized scaffolds were placed in non-adherent cell culture microplates (Sartstedt, Nümbrecht, Germany) and soaked in culture medium for 30 min at 37 °C before cell seeding to facilitate cell attachment. Corresponding cell density was prepared in a small volume of medium (50–100 µL). Cell suspension was pipetted drop by drop onto the scaffold centre. Then scaffolds were incubated for three hours at 37 °C and 5% CO_2 atmosphere to allow cell attachment and after that incubation period, culture medium was added.

2.5. Cell Proliferation Assay

A suspension of 100 MDA–MB–231 cells per cm^2 were seeded on adherent microplate wells (Sartstedt), 7.5% and 15% PCL scaffolds. Cell culture was maintained for 12 days. Every two days, samples were collected and 3-(4,5-dimethylthiazolyl-2)-2,5-diphenyltetrazolium bromide (MTT) assay was performed to quantify cell viability. Briefly, adherent wells and scaffolds were washed with PBS and meshes were put into new wells. Volumes of 1 mL DMEM and 100 µL MTT (Sigma-Aldrich, St. Louis, MO, USA) were added and samples were incubated for 150 min. In this test, only viable cells retain the ability of transforming yellow MTT into purple formazan crystals. After incubation, formazan crystals were dissolved with 1 mL DMSO (Sigma-Aldrich, St. Louis, MO, USA) under shaking. Four 100 µL aliquots from each well were pipetted into a 96-well plate and placed into a microplate reader (Bio-Rad, Hercules, CA, USA). Absorbance was measured at 570 nm. Culture medium of remaining samples was changed every two days.

2.6. Three-Dimensional Cell Culture

In order to evaluate the amount of BCSCs, MDA–MB–231 cells were cultured for 3, 6 and 12 days on scaffolds without passaging, changing the culture medium every three days. Considering cell growth kinetics of MDA–MD–231 cell line, 20,000, 8000 and 400 cells/cm^2 were seeded for 3, 6 and 12 days of culture respectively, achieving a similar cell confluence at the end of each culturing period. Since cell confluence affects cell behaviour and metabolism, the final cell amount was fixed to avoid variations due to this effect. In the case of 2D samples, cells were cultured on monolayer for 3 days at a cell seeding density of 20,000 cells/cm^2 in the same way as the scaffolds. Prior to the present work, cell line was grown routinely on two-dimensional plastics, thus only 3 days culture time was performed. Preceding experiments showed no differences in reference of cell behaviour between 2D cultured cells during 3, 6 and 12 days using the aforementioned initial cell densities (20,000, 8000 and 400 cells/cm^2 respectively; data not shown).

2.7. Fluorescence Microscopy Analysis

Triple negative MDA–MB–231 cells were cultured on adherent coverslips (Sarstedt) and 7.5 and 15% PCL meshes, for 2D and 3D culture respectively. After the culture period, cells were washed with PBS and fixed with 4% paraformaldehyde (PFA; Sigma, St. Louis, MO, USA) for 20 min. To permeabilize the cells, coverslips and scaffolds were washed and 0.2% Triton X-100 (Sigma) was added for 10 min. Then samples were blocked with PBS containing 3% bovine serum albumin (BSA; Sigma) as a blocking buffer for 20 min. Cells were subsequently incubated at room temperature for 20 min with rhodamine-phalloidin (Cytoskeleton Inc., Denver, CO, USA) (1:200) to stain actin cytoskeleton and then with 4,6-diamidino-2-phenylindole (DAPI; BD Pharmingen, Franklin Lakes, NJ, USA) (1:1000) also at room temperature for 10 min to stain nuclei. Fluorescence was observed under a fluorescent microscope (Zeiss Axio Imager Microscope, Carl Zeiss, Göttingen, Germany) and a Nikon DS-Ri1 coupled camera (Nikon, Tokyo, Japan) was used to acquire all images. Camera settings (illumination intensity, quality, resolution and colour) were standardised for all photographs. Rhodamine-phalloidin (red) and DAPI (blue) fluorescence were captured and merged with Image J software (National Institutes of Health, Bethesda, MD, USA). This software was also used to calculate nuclear and cytoplasmic elongation factors. In brief, five cells of ten different images were randomly selected to measure the length and width of the nucleus and cytoplasm as shown in the following formula:

$$Nuclear/Cytoplasmic\ Elongation\ Factor = \frac{length\ nucleus/cytoplasma}{width\ nucleus/cytoplasma} \geq 1$$

2.8. Mammosphere-Forming Assay

Scaffolds were washed with PBS and put into new wells to collect only those cells attached to PCL filaments. Cells from 2D culture and scaffolds were detached with trypsin-EDTA (Cultek, Madrid, Spain) at 37 °C and 5% CO_2 atmosphere. Afterwards, trypsinization cells were resuspended with DMEM/F12 medium (HyClon) containing the following supplements: B27 (Gibco, Waltham, MA, USA), EGF and FGF (20 ng/mL; Milteny Biotec, Bergisch Gladbach, Germany), 1% L-glutamine and 1% sodium pyruvate. A suspension of 2000 cells/well was seeded onto a 6-well, non-adherent cell culture microplate (Sarstedt) and incubated for 7 days at 37 °C and 5% CO_2. After this period, spherical mammospheres bigger than 50 μm were counted. The equation described below was used to calculate the Mammosphere Forming Index (MFI) of each culture condition:

$$MFI\ (\%) = \frac{no\ mammospheres}{no\ seeded\ cells} \times 100$$

2.9. ALDEFLUOR Assay

To analyze the aldehyde dehydrogenase (ALDH) activity, an ALDEFLUOR™ kit (Stem Cell Technologies, Durham, NC, USA) was used following the manufacturer indications. Cells were detached from the culture plastic (2D samples) and PCL scaffolds (3D) as explained in Section 2.8, washed with PBS and subsequently resuspended in ALDEFLUOR™ assay buffer at a concentration of 400,000 cells/mL. ALDEFLUOR™ Reagent (BODIPY-aminoacetaldehyde; BAAA) was added to each cell suspension. In order for the background fluorescence to be considered, a negative control for every sample was set by adding ALDEFLUOR™ diethylaminobenzaldehyde (DEAB), an ALDH inhibitor, to each cell suspension prior to adding BAAA in ALDEFLUOR assay buffer. All samples were incubated for 45 min at 37 °C in the dark.

Incubated samples were analyzed with a Cell Lab Quanta flow cytometer (Beckman Coulter Inc., Miami, FL, USA) to quantify the ALDH-positive cell population. The argon ion laser (488 nm) was used as a light source set at a power of 22 mW. Green fluorescence was detected with fluorescent channel 1 (FL1) optical filter (dichroic/splitter, dichroic long-pass: 550 nm, band-pass filter: 525 nm, detection width 505 to 545 nm). Information of a minimum of 10,000 events was recorded in List-mode Data files (LMD) and analyzed using FlowJo 10.2 software (FlowJo LLC, Ashland, OR, USA). Data were not compensated.

First, side-scatter (SS) and electronic volume (EV; equivalent to forward scatter) dot plots were performed and only single cells were selected, excluding debris and cells aggregates (less than 5%). Then, SS and log FL1 dot plots from DEAB samples were created to establish background fluorescence. The ALDH-positive cells' gate was traced, delimiting the rightmost area and including only the 0.5% of total cell population. BAAA samples were equally processed and ALDH-positive cells gates of respective controls were used to discern the sample percentage of cells with high ALDH activity.

2.10. Statistical Analysis

All data are expressed as mean ± standard error of the mean (SEM). Data were analyzed using IBM SPSS (Version 21,0; SPSS Inc., Chicago, IL, USA). First, normality and homoscedasticity were evaluated using Shapiro-Wilks and Levene tests, respectively. As for cell proliferation and mammosphere forming assays, data were found to present a normal distribution and variances were homogeneous, a general linear model followed by post-hoc Sidak test was run. Factors were the treatment (i.e., 2D, 3D with 7.5% PCL, and 3D with 15% PCL and the culturing time). As far as the ALDEFLUOR assay, the same tests were run following transformation of data with arcsine square root ($\arcsin\sqrt{x}$), as this was required for correcting the heteroscedasticity. Finally, as ratios between nuclear and cytoplasmic elongation factors (Fluorescence Microscopy Analysis in Section 3.3), did not fulfil with parametric assumptions, even when transformed, they were tested with Kruskal-Wallis and Mann-Whitney tests. The level of significance was set at $p < 0.05$. All observations were confirmed by at least three independent experiments.

3. Results

3.1. Electrospun Scaffolds Characterization

Once the polymer solution was electrospun, 7.5% PCL meshes showed an average thickness of 147.22 ± 5.00 μm, whereas 15% scaffolds' depth was 196.00 ± 4.65 μm. To study microscopic scaffold architecture, both specimens were imaged by Scanning Electron Microscopy (SEM; Table 1).

Table 1. Microscopic characterization of 7.5% and 15% electrospun poly (ε-caprolactone) (PCL) scaffolds (7 kV, 6 mL/h). Top and bottom sides were visualized through scanning electron microscopy micrographs at different magnifications. Both sides were used to calculate fiber diameter, surface porosity and pore area. (Scale bars: 10 μm).

	Side	Magnification		
		1500×	5000×	
7.5% PCL	Top			Fibre diameter 295.12 ± 148.45 nm Surface porosity 28.39% ± 4.53% Pore area 0.24 ± 0.42 μm
	Bottom			
15% PCL	Top			Fibre diameter 701.13 ± 401.89 nm Surface porosity 22.48% ± 7.57% Pore area 0.84 ± 1.82 μm²
	Bottom			

Meshes were visualized on the top and bottom sides so that fiber uniformity was certified, presenting similar features. Scaffolds from 7.5% PCL exhibited, apart from the filaments, spherical structures made by non-filamentous polymer. Regarding fiber diameter, 7.5% films showed an average diameter close to 300 nm, whereas the diameter of those containing 15% PCL increased up to 700 nm. Both scaffolds presented similar surface porosity, but they differed in the average pore area. Scaffolds from 15% PCL solution exhibited larger pores compared with 7.5% meshes.

3.2. Cell Proliferation

As aforementioned, scaffolds could provide a three-dimensional environment for cancer cell culture. Architecture and porosity of filaments directly interfered with cell adhesion and growth. As seen in the previous section, 7.5% and 15% PCL scaffolds were proven to exhibit distinct microscopic structures. To evaluate the influence of scaffold microenvironment on cell growth, MDA–MB–231 cells were cultured on 2D adherent surfaces and on 7.5% and 15% meshes. Cell viability was evaluated through MTT assay on successive culture days, presented in Figure 1.

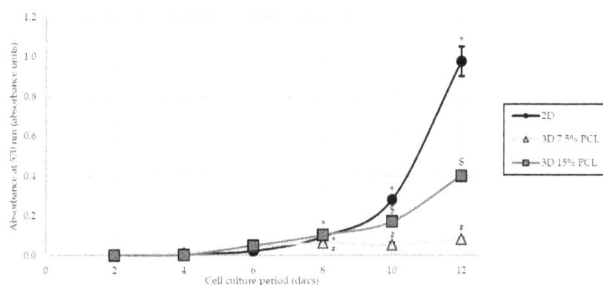

Figure 1. Cell proliferation analysis for MDA–MB–231 cells cultured on two-dimensional surfaces (2D) and on 7.5% and 15% PCL scaffolds (3D). (*), (#) and ($) symbols represent significant ($p < 0.05$) differences between groups.

On the first assay days, few differences were observed regarding cell proliferation between different culture models. Variations between cell culture supports took place when a minimum cell confluence was reached, starting from day 8. MDA–MB–231 cells cultured on 2D adherent microplates presented a higher cell proliferation ratio compared to three-dimensional scaffolds, adopting strongly exponential kinetics. Scaffolds fabricated with 15% PCL solution also showed an exponential cell growth, but with a smaller slope. In contrast, 7.5% PCL meshes exhibited the lowest cell proliferation with a fairly linear trend. Since day 8, cell proliferation of 7.5% PCL scaffolds was significantly reduced when compared with monolayer culture (p-values ranging from <0.001 to 0.014) and 15% PLC meshes (p-values ranging from 0.002 to 0.040).

3.3. Cell Morphology

MDA–MB–231 cells were cultured on adherent two-dimensional coverslips (2D) and three-dimensional PCL scaffolds (3D). Three different cell culture times (3, 6 and 12 days) were tested to evaluate whether morphology differences existed. Actin cytoskeleton and nucleus were stained to analyze possible changes in cell morphology between culture systems. MDA–MB–231 cells were routinely cultured on plastic cell culture dishes, establishing a cell monolayer where cells appeared to have a flattened structure. MDA–MD–231 cell line was also characterized to adopt a relatively lengthened cytoplasm. Fluorescent microscopy images confirmed the morphology described in 2D models (Figure 2). Some cells presented cytoplasmic prolongations, while others had a round shape and nucleus aspect was predominantly ellipsoidal. Then, MDA–MD–231 cells displayed different morphology when the two scaffold types, 7.5% and 15% PCL, were compared. Cells cultured on 7.5% PCL meshes (Figure 3a–c) exhibited similar aspects to 2D cultured ones, including nucleus and cytoplasm architecture. This trend was observed along the different days of cell culture with no noticeable differences. In contrast, a high number of MDA–MD–231 cells showed lengthened morphology when cultured on 15% PCL scaffolds (Figure 3d–f). Cytoplasm prolongations were longer than the ones from 2D and 3D 7.5% PCL cultures. When cell culture days increased, prolongations seemed to be even more extended. Moreover, some cells appeared to be unfocused on 15% PCL meshes pictures, indicating that cell culture occurred on different scaffold depth. No qualitative differences were observed concerning nuclear shape.

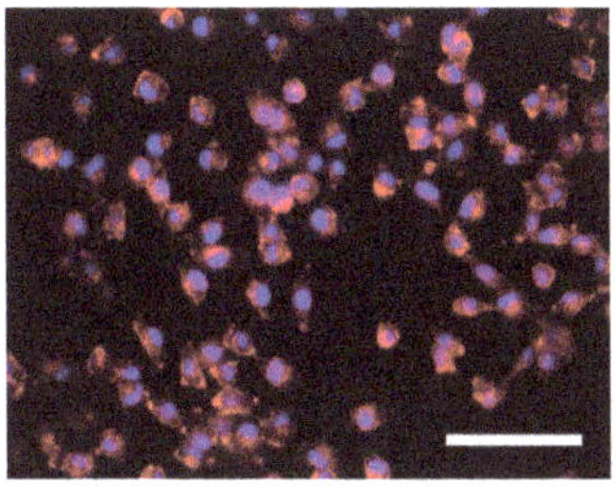

Figure 2. MDA–MB–231 cells grown in two-dimensional (2D) adherent coverslips. Actin cytoskeleton was stained with rhodamine-phalloidin (red) and nucleus was stained with 4,6-diamidino-2-phenylindole (DAPI; blue). Fluorescence microscopy images were captured at a magnification of 200× (Scale bar: 100 μm).

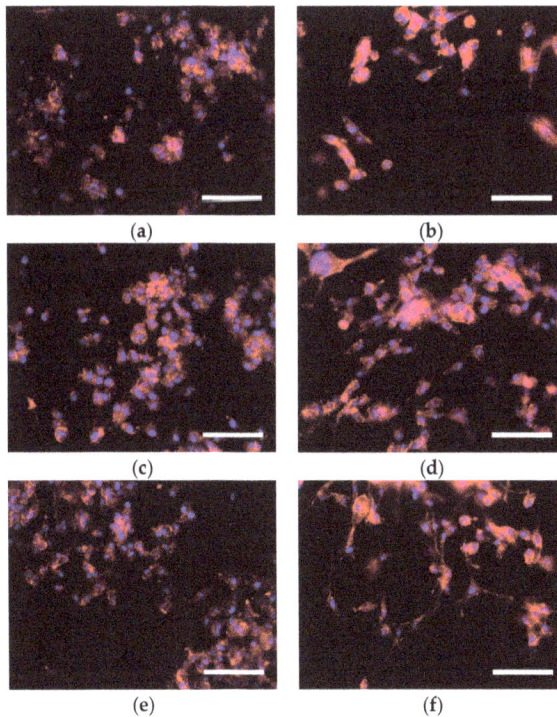

(a)

(b)

(c)

(d)

(e)

(f)

Figure 3. MDA–MB–231 cells grown in three-dimensional (3D) PCL scaffolds. Cells were seeded on 7.5% PCL (**a,c,e**) and 15% PCL meshes (**b,d,f**). Cells were cultured for 3 (**a,b**), 6 (**c,d**) and 12 days (**e,f**) without passage. Actin cytoskeleton was stained with rhodamine-phalloidin (red) and nucleus was stained with DAPI (blue). Fluorescence microscopy images were captured at a magnification of 200× (Scale bars: 100 μm).

To quantitatively evaluate cell morphology in 2D and 3D cultures, nuclear and cytoplasmic elongation were measured as described in Section 2.7. No significant changes in nucleus elongation were observed between 2D and 3D cultures (Figure 4a), which agreed with the aforementioned descriptive microscopic observation. The nuclear elongation factor of 2D cultured cells was 1.60 ± 0.12, pointing out to the ellipsoidal form of the nucleus. All 3D cultured cells factors were similar or slightly lower, between 1.47 ± 0.07 and 1.60 ± 0.08. Cytoplasm pattern was also studied (Figure 4b) and, whereas MDA–MB–231 cells cultured on 7.5% PCL scaffolds presented a similar cytoplasmic elongation factor than those cultured in 2D, with a value around 1.70, the cells on 15% PCL meshes exhibited a significantly higher cytoplasmic length.

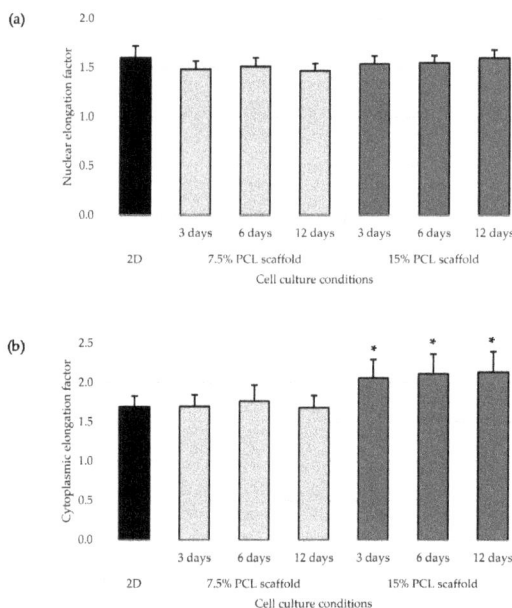

Figure 4. Cell morphology analysis for MDA–MB–231 cells cultured on two-dimensional surfaces (2D) and on 7.5% and 15% PCL scaffolds. Nuclear (**a**) and cytoplasmic elongation (**b**) were measured as explained in Section 2.7. Scaffolds values were compared with 2D culture and levels of statistically significance were indicated as * ($p < 0.05$).

As previously mentioned, SEM analysis showed that 15% PCL scaffolds were the only specimen formed exclusively by filaments. Moreover, MDA–MB–231 cells cultured on meshes of 15% PCL exhibited a high cell proliferation (Figure 1) and different cell morphology (Figure 3) compared with 7.5% PCL scaffolds. As the main aim of three-dimensional cell culture is to mimic the extracellular matrix structure and provide a comfortable support to cell growth, meshes from 15% PCL solution were chosen to conduct the further experiments of the present study. From now onwards, 3D culture samples will exclusively refer to cells cultured on 15% PCL scaffolds. As in previous experiments, the effects of culturing cells with those 15% PCL scaffolds were also tested at 3, 6 and 12 days of culture.

3.4. Mammoshperes Forming Assay

Breast cancer stem cells possess an anchorage-independent growth, proliferating into mammospheres when cultured on non-adherent surfaces. Hence, Mammospheres Forming Assay was first used to evaluate the spheres forming capacity of cells previously cultured on 2D and 3D supports. The MDA–MB–231 triple negative cell line was seeded as described in "Materials and Methods" section.

Mammospheres from all cell samples did not show qualitative differences regarding morphology and size (Figure 5a–d). However, all three-dimensional cultured cells showed a significantly higher Mammosphere Forming Index (MFI) than 2D samples, reaching the maximum value and significance on 6 days of culture (Figure 5e).

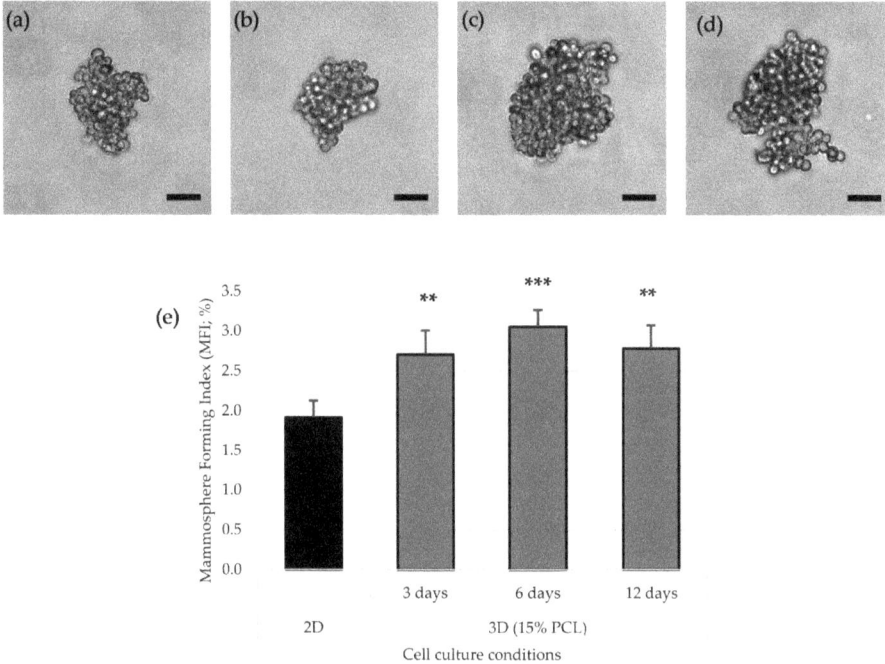

Figure 5. Mammosphere Forming Assay. Mammosphere images of MDA–MB–231 cells previously cultured on 2D surfaces (**a**) and 15% PCL scaffolds for 3 (**b**), 6 (**c**) and 12 days (**d**). (Scale bars: 50 μm). Mammosphere Forming Index (MFI) of MDA–MB–231 cell line after 2D or 3D cell culture with 15% PCL scaffolds (**e**). Significant differences of 3D with regard to 2D cultures are indicated as ** ($p < 0.01$) and *** ($p < 0.001$).

3.5. Aldehyde Dehydrogenase Activity

The aldehyde dehydrogenase (ALDH) family is composed by dehydrogenases enzymes responsible for the oxidation of retinol (vitamin A) to retinoic acid [31], the latter of which is involved in gene expression regulation [32] and embryo development [33]. Moreover, ALDH enzymes have a detoxification role, protecting organisms against damaging aldehydes [34] and cytotoxic agents [35], and have also been demonstrated to regulate hematopoietic stem cells differentiation via retinoic acids production [36]. To corroborate the relative proportion of CSCs in 2D and 3D cultured cells, ALDH activity was quantified as a measure of stem properties. Samples were assessed with ALDEFLUOR assay and the percentage of ALDH-positive cells was determined as shown in Figure 6.

The ALDH-positive subpopulation increased when cells were cultured on PCL scaffolds, compared to monolayer culture (Figure 7). ALDH activity enhancement was observed at 3 and 6 days of cell culture, the latter time point being the one with a significant major percentage. However, the proportion of ALDH + cells after 12 days of culture in 15% PCL scaffolds decreased, with figures close to those of 2D.

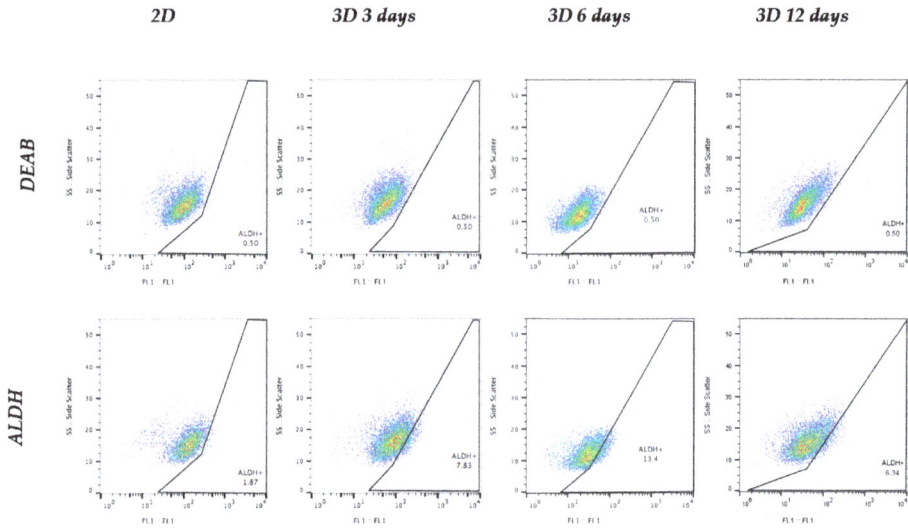

Figure 6. ALDEFLUOR assay plots of MDA–MB–231 cells cultured on monolayer (2D) and on 15% PCL scaffolds (3D). Aldehyde dehydrogenase (ALDH) inhibitor (diethylaminobenzaldehyde; DEAB) was processed to determine background fluorescence limit, gating 0.5% ALDH-positive cells for all samples.

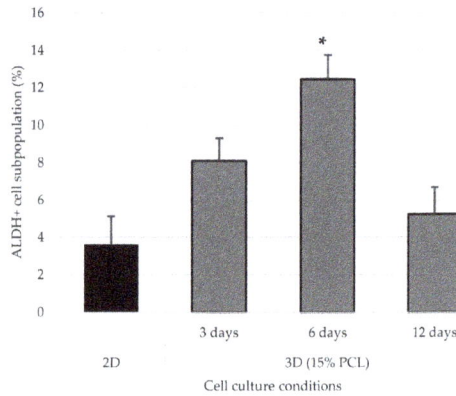

Figure 7. ALDH + cell subpopulation of MDA–MB–231 cells after monolayer culture (2D) or three-dimensional cell culture with 15% PCL scaffolds (3D). (*) Denotes significant ($p < 0.05$) differences between 3D culture at a given time point and 2D.

4. Discussion

Polymer concentration (PCL) has been found to exert a great influence on scaffold architecture. Indeed, 7.5% PCL scaffolds showed polymeric spheres connected to filaments. These structures, previously described in the literature, are called beads and result from low polymer concentration [8,37]. In contrast, no beads were observed in 15% meshes, which also showed an average fiber diameter 2.4-fold higher than that of 7.5% specimens. While performing a direct comparison with values from other studies is difficult owing to the wide range of electrospun parameters used, the significant impact of PCL concentration in acetone on fiber morphology and diameter agrees with most previous studies [7,8,38,39]. Furthermore, pore parameters were also evaluated and, while electrospun meshes presented similar surface porosity, the average pore area in 15% scaffolds was 3.5-fold larger than that of 7.5% models.

MDA–MB–231 cells were successfully expanded on electrospun mats. Both scaffolds tested (i.e., 7.5% and 15%) showed less cell proliferation compared with the homologous monolayer culture, which was in agreement with previous studies [40]. When scaffolds from different polymer concentrations were compared, microarchitecture traits seemed to influence 3D cell culture efficiency. It is worth mentioning that low filament diameters have been revealed to facilitate cell adhesion and growth [38,41,42] and collagen fibers (the main protein in the extracellular matrix) display small diameters below 100 nm [43]. However, in the current study, the scaffold with the lowest fiber diameter (i.e., 7.5%) also exhibited beads which interfered with cell proliferation. Chen et al. (2007) demonstrated that, despite having smaller fiber diameters, meshes with beads reduce fibroblast growth [38]. In our conditions, beaded scaffolds exhibited a lower surface area-to-volume ratio, providing less material for cell growth. Moreover, interconnected pores with a minimum optimal area are needed to allow cell growth and scaffold infiltration. Small pores of 7.5% scaffolds may also hinder MDA–MB–231 cells' penetration within the mesh structure. Following this hypothesis, most cells would have adhered to the surface and they would hardly have colonized different scaffold depths. With regard to cell morphology, MDA–MB–231 cells seeded on 15% PCL scaffolds presented more cytoplasmic elongations compared with round shaped cells on 7.5% PCL meshes and, specially, flat surfaces. This finding was supported by the higher cytoplasmic elongation factor of 15% PCL mat cells. Cell elongation along scaffold nanofilaments has also been observed in breast cancer cells [27,28], fibroblasts [44] and murine adult neural stem cells [45]. Accordingly, meshes from 15% PCL mimic the physiological environment better as they allow cells to set a structure that 2D monolayers are devoid of. In fact, cytoskeleton reorganization caused by 3D cell culture may regulate gene expression [2].

Taking into account cell proliferation kinetics and morphology changes, scaffold from 15% PCL solution was selected to further accommodate MDA–MB–231 cell culture and evaluate BCSCs niche expansion capacity. The MDA–MB–231 cell line presents low MFI values and can only be propagated few passages on suspension culture due to their moderate e-cadherin expression [46]. However, scaffolds culture improved mammospheres forming ability, particularly after 6 days of culture, resulting in an enlarged tumorigenic [47] and self-renewal potential [48]. Additionally, MDA–MB–231 cells cultured on electrospun mats over the same period showed a significant ALDH-positive population increase in comparison of standard culture, with a conclusive 3.4-fold increase. Greater ALDH activity indicated stem features acquisition since several studies noticed a high ALDH activity on mammary [19], hematopoietic [49] and leukemic stem cells [50]. Taking all described stem features assays into account, the present study has shown that 3D culture with electrospun scaffolds enhances triple negative MDA–MB–231 tumourigenicity and ALDH activity. These rearrangements reach the maximum significance when the culture period lasts 6 days, this time being the one that allows reaching the greatest BCSCs expansion through 3D cell culture. In agreement with our results, different breast cancer cell studies also demonstrated cancer stem cell amplification through PCL scaffolds fabricated by electrospinning [27–29] and by other methods such as additive manufacturing technologies [30].

In conclusion, the current study has revealed the vast potential of poly(ε-caprolactone) on the in vitro cell culture field. Electrospun PCL solutions resulted in nanofiber production with different architecture traits, demonstrating the high versatility of polymer and technology. Moreover, non-beaded PCL scaffolds have been proven to supply physical support for triple negative breast cancer cell proliferation and elongation. The 3D cell culture expanded the breast cancer stem cell subpopulation, which in turn expressed more malignancy markers and exhibited stem cell characteristics. Therefore, 3D culture with electrospun PCL nanofibers may be useful to maintain the in vivo structure and to culture BCSCs, making their expansion and characterization possible. Investigation of this rare subpopulation is much warranted as it could facilitate the development of new specific therapeutic approaches to prevent the high recurrence of tumors such as triple negative breast cancer.

Polymers **2017**, *9*, 328

Acknowledgments: This work was supported partially by Spanish grants from Fundación Ramón Areces, Instituto de Salud Carlos III (PI1400329) and Ministerio de Economía Y Competitividad (DPI2013-45201-P; RYC-2014-15581), and the support of the Catalonian government (2014SGR00868). The authors are grateful for the financial support from the University of Girona (MPCUdG2016/036).

Author Contributions: Joaquim Ciurana and Teresa Puig conceived and designed the experiments; Marc Rabionet and Marc Yeste performed the experiments; Marc Rabionet and Marc Yeste analyzed the data; Marc Yeste, Joaquim Ciurana and Teresa Puig contributed reagents/materials/analysis tools; Marc Rabionet, Joaquim Ciurana and Teresa Puig wrote the paper.

Conflicts of Interest: The authors declare no conflict of interest. The founding sponsors had no role in the design of the study; in the collection, analyses, or interpretation of data; in the writing of the manuscript, and in the decision to publish the results.

References

1. Thomas, C.H.; Collier, J.H.; Sfeir, C.S.; Healy, K.E. Engineering gene expression and protein synthesis by modulation of nuclear shape. *Proc. Natl. Acad. Sci. USA* **2002**, *99*, 1972–1977. [CrossRef] [PubMed]

2. Vergani, L.; Grattarola, M.; Nicolini, C. Modifications of chromatin structure and gene expression following induced alterations of cellular shape. *Int. J. Biochem. Cell Biol.* **2004**, *36*, 1447–1461. [CrossRef] [PubMed]

3. Theocharis, A.D.; Skandalis, S.S.; Gialeli, C.; Karamanos, N.K. Extracellular matrix structure. *Adv. Drug Deliv. Rev.* **2016**, *97*, 4–27. [CrossRef] [PubMed]

4. Frantz, C.; Stewart, K.M.; Weaver, V.M. The extracellular matrix at a glance. *J. Cell Sci.* **2010**, *123*, 4195–4200. [CrossRef] [PubMed]

5. Knight, E.; Przyborski, S. Advances in 3D cell culture technologies enabling tissue-like structures to be created in vitro. *J. Anat.* **2014**, *227*, 746–756. [CrossRef] [PubMed]

6. Li, W.J.; Tuan, R.S. Fabrication and application of nanofibrous scaffolds in tissue engineering. *Curr. Protoc. Cell Biol.* **2009**. [CrossRef]

7. Bosworth, L.A.; Downes, S. Acetone, a Sustainable Solvent for Electrospinning Poly(ε-Caprolactone) Fibres: Effect of Varying Parameters and Solution Concentrations on Fibre Diameter. *J. Polym. Environ.* **2012**, *20*, 879–886. [CrossRef]

8. Chen, M.; Patra, P.K.; Warner, S.B.; Bhowmick, S. Optimization of electrospinning process parameters for tissue engineering scaffolds. *Biophys. Rev. Lett.* **2006**, *01*, 153–178. [CrossRef]

9. De Ciurana, J.; Serenó, L.; Vallès, È. Selecting process parameters in RepRap additive manufacturing system for PLA scaffolds manufacture. *Procedia CIRP* **2013**, *5*, 152–157. [CrossRef]

10. Cipitria, A.; Skelton, A.; Dargaville, T.R.; Dalton, P.D.; Hutmacher, D.W. Design, fabrication and characterization of PCL electrospun scaffolds—A review. *J. Mater. Chem.* **2011**, *21*, 9419–9453. [CrossRef]

11. Coulembier, O.; Degée, P.; Hedrick, J.L.; Dubois, P. From controlled ring-opening polymerization to biodegradable aliphatic polyester: Especially poly(β-malic acid) derivatives. *Prog. Polym. Sci.* **2006**, *31*, 723–747. [CrossRef]

12. Woodruff, M.A.; Hutmacher, D.W. The return of a forgotten polymer—Polycaprolactone in the 21st century. *Prog. Polym. Sci.* **2010**, *35*, 1217–1256. [CrossRef]

13. Lapidot, T.; Sirard, C.; Vormoor, J.; Murdoch, B.; Hoang, T.; Caceres-Cortes, J.; Minden, M.; Paterson, B.; Caligiuri, M.A.; Dick, J.E. A cell initiating human acute myeloid leukaemia after transplantation into SCID mice. *Nature* **1994**, *367*, 645–648. [CrossRef] [PubMed]

14. Singh, S.K.; Clarke, I.D.; Terasaki, M.; Bonn, V.E.; Hawkins, C.; Squire, J.; Dirks, P.B. Identification of a cancer stem cell in human brain tumors. *Cancer Res.* **2003**, *63*, 5821–5828. [PubMed]

15. Kondo, T.; Setoguchi, T.; Taga, T. Persistence of a small subpopulation of cancer stem-like cells in the C6 glioma cell line. *Proc. Natl. Acad. Sci. USA* **2004**, *101*, 781–786. [CrossRef] [PubMed]

16. Matsui, W.; Huff, C.A.; Wang, Q.; Malehorn, M.T.; Barber, J.; Tanhehco, Y.; Smith, B.D.; Civin, C.I.; Jones, R.J. Characterization of clonogenic multiple myeloma cells. *Blood* **2004**, *103*, 2332–2336. [CrossRef] [PubMed]

17. Al-Hajj, M.; Wicha, M.S.; Benito-Hernandez, A.; Morrison, S.J.; Clarke, M.F. Prospective identification of tumorigenic breast cancer cells. *Proc. Natl. Acad. Sci. USA* **2003**, *100*, 3983–3988. [CrossRef] [PubMed]

18. Dontu, G.; Abdallah, W.M.; Foley, J.M.; Jackson, K.W.; Clarke, M.F.; Kawamura, M.J.; Wicha, M.S. In vitro propagation and transcriptional profiling of human mammary stem/progenitor cells. *Genes Dev.* **2003**, *17*, 1253–1270. [CrossRef] [PubMed]

19. Ginestier, C.; Hur, M.H.; Charafe-Jauffret, E.; Monville, F.; Dutcher, J.; Brown, M.; Jacquemier, J.; Viens, P.; Kleer, C.G.; Liu, S.; et al. ALDH1 is a marker of normal and malignant human mammary stem cells and a predictor of poor clinical outcome. *Cell Stem Cell* **2007**, *1*, 555–567. [CrossRef] [PubMed]

20. Abraham, B.K.; Fritz, P.; McClellan, M.; Hauptvogel, P.; Athelogou, M.; Brauch, H. Prevalence of CD44+/CD24-/low cells in breast cancer may not be associated with clinical outcome but may favor distant metastasis. *Clin. Cancer Res.* **2005**, *11*, 1154–1159. [PubMed]

21. Diehn, M.; Cho, R.W.; Lobo, N.A.; Kalisky, T.; Dorie, M.J.; Kulp, A.N.; Qian, D.; Lam, J.S.; Ailles, L.E.; Wong, M.; et al. Association of reactive oxygen species levels and radioresistance in cancer stem cells. *Nature* **2009**, *458*, 780–783. [CrossRef] [PubMed]

22. Li, X.; Lewis, M.T.; Huang, J.; Gutierrez, C.; Osborne, C.K.; Wu, M.F.; Hilsenbeck, S.G.; Pavlick, A.; Zhang, X.; Chamness, G.C.; et al. Intrinsic Resistance of Tumorigenic Breast Cancer Cells to Chemotherapy. *J. Natl. Cancer Inst.* **2008**, *100*, 672–679. [CrossRef] [PubMed]

23. Dent, R.; Trudeau, M.; Pritchard, K.I.; Hanna, W.M.; Kahn, H.K.; Sawka, C.A.; Lickley, L.A.; Rawlinson, E.; Sun, P.; Narod, S.A. Triple-Negative Breast Cancer: Clinical Features and Patterns of Recurrence. *Clin. Cancer Res.* **2007**, *13*, 4429–4434. [CrossRef] [PubMed]

24. Carey, L.A.; Dees, E.C.; Sawyer, L.; Gatti, L.; Moore, D.T.; Collichio, F.; Ollila, D.W.; Sartor, C.I.; Graham, M.L.; Perou, C.M. The triple negative paradox: Primary tumor chemosensitivity of breast cancer subtypes. *Clin. Cancer Res.* **2007**, *13*, 2329–2334. [CrossRef] [PubMed]

25. Charafe-Jauffret, E.; Ginestier, C.; Iovino, F.; Wicinski, J.; Cervera, N.; Finetti, P.; Hur, M.H.; Diebel, M.E.; Monville, F.; Dutcher, J.; et al. Breast cancer cell lines contain functional cancer stem cells with metastatic capacity and a distinct molecular signature. *Cancer Res.* **2009**, *69*, 1302–1313. [CrossRef] [PubMed]

26. Tsuyada, A.; Chow, A.; Wu, J.; Somlo, G.; Chu, P.; Loera, S.; Luu, T.; Li, A.X.; Wu, X.; Ye, W.; et al. CCL2 mediates cross-talk between cancer cells and stromal fibroblasts that regulates breast cancer stem cells. *Cancer Res.* **2012**, *72*, 2768–2779. [CrossRef] [PubMed]

27. Feng, S.; Duan, X.; Lo, P.K.; Liu, S.; Liu, X.; Chen, H.; Wang, Q. Expansion of breast cancer stem cells with fibrous scaffolds. *Integr. Biol. (Camb.)* **2013**, *5*, 768–777. [CrossRef] [PubMed]

28. Saha, S.; Duan, X.; Wu, L.; Lo, P.K.; Chen, H.; Wang, Q. Electrospun fibrous scaffolds promote breast cancer cell alignment and epithelial-mesenchymal transition. *Langmuir* **2012**, *28*, 2028–2034. [CrossRef] [PubMed]

29. Sims-Mourtada, J.; Niamat, R.A.; Samuel, S.; Eskridge, C.; Kmiec, E.B. Enrichment of breast cancer stem-like cells by growth on electrospun polycaprolactone-chitosan nanofiber scaffolds. *Int. J. Nanomed.* **2014**, *9*, 995–1003. [CrossRef] [PubMed]

30. Palomeras, S.; Rabionet, M.; Ferrer, I.; Sarrats, A.; Garcia-Romeu, M.L.; Puig, T.; Ciurana, J. Breast Cancer Stem Cell Culture and Enrichment Using Poly(ε-Caprolactone) Scaffolds. *Molecules* **2016**, *21*, 537. [CrossRef] [PubMed]

31. Duester, G. Families of retinoid dehydrogenases regulating vitamin A function: Production of visual pigment and retinoic acid. *Eur. J. Biochem.* **2000**, *267*, 4315–4324. [CrossRef] [PubMed]

32. Duester, G.; Mic, F.A.; Molotkov, A. Cytosolic retinoid dehydrogenases govern ubiquitous metabolism of retinol to retinaldehyde followed by tissue-specific metabolism to retinoic acid. *Chem. Biol. Interact.* **2003**, *143–144*, 201–210. [CrossRef]

33. Appel, B.; Eisen, J.S. Retinoids run rampant: Multiple roles during spinal cord and motor neuron development. *Neuron* **2003**, *40*, 461–464. [CrossRef]

34. Vasiliou, V.; Pappa, A. Polymorphisms of human aldehyde dehydrogenases. Consequences for drug metabolism and disease. *Pharmacology* **2000**, *61*, 192–198. [CrossRef] [PubMed]

35. Hilton, J. Role of aldehyde dehydrogenase in cyclophosphamide-resistant L1210 leukemia. *Cancer Res.* **1984**, *44*, 5156–5160. [PubMed]

36. Chute, J.P.; Muramoto, G.G.; Whitesides, J.; Colvin, M.; Safi, R.; Chao, N.J.; McDonnell, D.P. Inhibition of aldehyde dehydrogenase and retinoid signaling induces the expansion of human hematopoietic stem cells. *Proc. Natl. Acad. Sci. USA* **2006**, *103*, 11707–11712. [CrossRef] [PubMed]

37. Fong, H.; Chun, I.; Reneker, D. Beaded nanofibers formed during electrospinning. *Polymer* **1999**, *40*, 4585–4592. [CrossRef]

38. Chen, M.; Patra, P.K.; Warner, S.B.; Bhowmick, S. Role of fiber diameter in adhesion and proliferation of NIH 3T3 fibroblast on electrospun polycaprolactone scaffolds. *Tissue Eng.* **2007**, *13*, 579–587. [CrossRef] [PubMed]

39. Dias, J.; Bártolo, P. Morphological Characteristics of Electrospun PCL Meshes—The Influence of Solvent Type and Concentration. *Procedia CIRP* **2013**, *5*, 216–221. [CrossRef]
40. Bean, A.C.; Tuan, R.S. 3D cell culture and osteogenic differentiation of human bone marrow stromal cells plated onto jet-sprayed or electrospun micro-fiber scaffolds Fiber diameter and seeding density influence chondrogenic differentiation of mesenchymal stem cells seeded on electrospun poly(-caprolactone) scaffolds. *Biomed. Mater* **2015**, *10*, 1–25. [CrossRef]
41. Szot, C.S.; Buchanan, C.F.; Gatenholm, P.; Rylander, M.N.; Freeman, J.W. Investigation of cancer cell behavior on nanofibrous scaffolds. *Mater. Sci. Eng. C* **2011**, *31*, 37–42. [CrossRef]
42. Chen, M.; Michaud, H.; Bhowmick, S. Controlled Vacuum Seeding as a Means of Generating Uniform Cellular Distribution in Electrospun Polycaprolactone (PCL) Scaffolds. *J. Biomech. Eng.* **2009**, *131*, 1–8. [CrossRef] [PubMed]
43. Barton, S.P.; Marks, R. Measurement of collagen-fibre diameter in human skin. *J. Cutan. Pathol.* **1984**, *11*, 18–26. [CrossRef] [PubMed]
44. Li, D.; Wu, T.; He, N.; Wang, J.; Chen, W.; He, L.; Huang, C.; EI-Hamshary, H.A.; Al-Deyab, S.S.; Ke, Q.; Mo, X. Three-dimensional polycaprolactone scaffold via needleless electrospinning promotes cell proliferation and infiltration. *Colloids Surf. B* **2014**, *121*, 432–443. [CrossRef] [PubMed]
45. Lim, S.H.; Liu, X.Y.; Song, H.; Yarema, K.J.; Mao, H.Q. The effect of nanofiber-guided cell alignment on the preferential differentiation of neural stem cells. *Biomaterials* **2010**, *31*, 9031–9039. [CrossRef] [PubMed]
46. Manuel Iglesias, J.; Beloqui, I.; Garcia-Garcia, F.; Leis, O.; Vazquez-Martin, A.; Eguiara, A.; Cufi, S.; Pavon, A.; Menendez, J.A.; Dopazo, J. Mammosphere Formation in Breast Carcinoma Cell Lines Depends upon Expression of E-cadherin. *PLoS ONE* **2013**, *8*, 1–12. [CrossRef] [PubMed]
47. Cioce, M.; Gherardi, S.; Viglietto, G.; Strano, S.; Blandino, G.; Muti, P.; Ciliberto, G. Mammosphere-forming cells from breast cancer cell lines as a tool for the identification of CSC-like- and early progenitor-targeting drugs. *Cell Cycle* **2014**, *9*, 2950–2959. [CrossRef]
48. Shaw, F.L.; Harrison, H.; Spence, K.; Ablett, M.P.; Simoes, B.M.; Farnie, G.; Clarke, R.B. A detailed mammosphere assay protocol for the quantification of breast stem cell activity. *J. Mammary Gland Biol. Neoplasia* **2012**, *17*, 111–117. [CrossRef] [PubMed]
49. Hess, D.A.; Meyerrose, T.E.; Wirthlin, L.; Craft, T.P.; Herrbrich, P.E.; Creer, M.H.; Nolta, J.A. Functional characterization of highly purified human hematopoietic repopulating cells isolated according to aldehyde dehydrogenase activity. *Blood* **2004**, *104*, 1648–1655. [CrossRef] [PubMed]
50. Pearce, D.J.; Taussig, D.; Simpson, C.; Allen, K.; Rohatiner, A.Z.; Lister, T.A.; Bonnet, D. Characterization of cells with a high aldehyde dehydrogenase activity from cord blood and acute myeloid leukemia samples. *Stem Cells* **2005**, *23*, 752–760. [CrossRef] [PubMed]

polymers

MDPI

Article

Enhanced Cartilaginous Tissue Formation with a Cell Aggregate-Fibrin-Polymer Scaffold Complex

Soojin Lee [1,2], Kangwon Lee [2,3], Soo Hyun Kim [1,4] and Youngmee Jung [1,5,*]

[1] Biomaterials Research Center, Korea Institute of Science and Technology, 5, Hwarang-ro 14-Gil, Seoungbuk-gu, Seoul 02792, Korea; dltnwls830@kist.re.kr (S.L.); soohkim@kist.re.kr (S.H.K.)
[2] Program in Nanoscience and Technology, Graduate School of Convergence Science and Technology, Seoul National University, Seoul 08826, Korea; kangwonlee@snu.ac.kr
[3] Advanced Institutes of Convergence Technology, Gyeonggi-do 16229, Korea
[4] KU-KIST Graduate School of Converging Science and Technology, Korea University, Seoul 02841, Korea
[5] Division of Bio-Medical Science & Technology, KIST School, Korea University of Science and Technology, Seoul 02792, Korea
* Correspondence: winnie97@kist.re.kr; Tel.: +82-2-958-5348

Received: 28 June 2017; Accepted: 3 August 2017; Published: 8 August 2017

Abstract: Cell density is one of the factors required in the preparation of engineered cartilage from mesenchymal stem cells (MSCs). Additionally, it is well known for having a significant role in chemical and physical stimulations when stem cells undergo chondrogenic differentiation. Here, we developed an engineered cartilage with a cell aggregate-hydrogel-polymer scaffold complex capable of inducing the effective regeneration of cartilage tissue similar to natural cartilage while retaining a high mechanical strength, flexibility, and morphology. Cell aggregates were generated by the hanging drop method with rabbit bone marrow stromal cells (BMSCs), and poly (lactide-*co*-caprolactone) (PLCL) scaffolds were fabricated with 78.3 ± 5.3% porosity and a 300–500 µm pore size with a gel-pressing method. We prepared the cell aggregate-fibrin-poly (lactide-*co*-caprolactone) (PLCL) scaffold complex, in which the cell aggregates were evenly dispersed in the fibrin, and they were immobilized onto the surface of the polymer scaffold while filling up the pores. To examine the chondrogenic differentiation of seeded BMSCs and the formation of chondral extracellular matrix onto the complexes, they were cultured in vitro or subcutaneously implanted into nude mice for up to eight weeks. The results of the in vitro and in vivo studies revealed that the accumulation of the chondral extracellular matrices was increased on the cell aggregate-fibrin-PLCL scaffold complexes (CAPs) compared to the single cell-fibrin-PLCL scaffold complexes (SCPs). Additionally, we examined whether the mature and well-developed cartilaginous tissues and lacunae structures typical of mature cartilage were evenly distributed in the CAPs. Consequently, the cell aggregates in the hybrid scaffolds of fibrin gels and elastic PLCL scaffolds can induce themselves to differentiate into chondrocytes, maintain their phenotypes, enhance glycosaminoglycan (GAG) production, and improve the quality of cartilaginous tissue formed in vitro and in vivo.

Keywords: cell aggregate; hydrogel; poly (lactide-*co*-caprolactone); cartilage regeneration; hanging drop method

---

1. Introduction

The principal function of articular cartilage, the dense connective tissue of the diarthrodial joints, is to provide a lubricated surface as well as to bear high stress and friction loads in the body [1,2]. Articular cartilage composed of the extracellular matrix (ECM) with chondrocytes has a less intrinsic capacity for self-healing, because it does not have blood vessels, nerves, or lymphatics for promoting wound healing. Thus, when articular cartilage is damaged, it cannot repair itself spontaneously or rapidly unlike other tissues in the human body. This gives rise to severe consequences such as pain or reduced mobility caused by further degeneration [3–6].

Many studies have been performed in an effort to overcome the above-mentioned limitations of previously known therapies for cartilage regeneration. Thus, a treatment method for deep cartilage defects in the knee with autologous chondrocyte transplantation has been reported [7–9]. Because the above method has been proved successful in obtaining regenerative cartilage tissue, which is relatively close to natural cartilage, by culturing autologous chondrocytes, clinical trials using autologous chondrocyte transplantation (ACT) have been steadily rising in the United States and Northern Europe. However, because ACT injects cultured chondrocytes in a suspension directly into a cartilage defect area, ACT has several limitations, including a difficulty in obtaining a sufficient amount of chondrocytes [10], delamination, and underfilling of the defect. The injected cells are easily washed out after transplantation, and as a result, it is difficult to maintain a high cell density in the defect area [11].

Although many studies have been conducted on cartilage regeneration using techniques that include tissue engineering and ACT, some limitations still remain, such as the formation of fibrocartilage generated from the transplanted chondrocytes, and a reduced mechanical strength of the newly formed tissues compared to native cartilage, which is problematic in terms of the mid-/long-term durability of the cartilage regeneration [12–17]. The regeneration process must induce a considerable glycosaminoglycan (GAG) production to generate functional engineered cartilage tissue, in other words, to construct a structure which can bear loads in the human body [18] and avoid fibrocartilage formation [19].

To mimic better physiological tissues, three-dimensional (3D) cell cultures have been studied because cellular functions and responses in tissues are often lost in two-dimensional cell cultures [20]. In many studies, they have used the pellet culture or micro mass culture technique to induce the chondrogenic differentiation of cells or to form cartilage-like tissues. By using these culture systems, they supply a three-dimensional morphology of cartilage to cells, which has an important role in promoting cell–matrix interactions and cell–cell interactions during chondrogenesis. Furthermore, the phenotypic features of cartilage are closely related to its three-dimensional matrix of collagen and proteoglycans, which are lost in monolayer cultures [21,22].

High-density cell cultures, such as the above-mentioned methods, are widely used to induce cells to differentiate and to maintain the chondrocyte phenotype by providing the proper environment to the cells [23]. In this study, we developed cell aggregates with bone marrow stromal cells (BMSCs) that have a high density using a hanging drop culture. The cell aggregates spontaneously formed a three-dimensional structure during the culture in vitro. After that, we seeded the cell aggregates onto hybrid scaffolds of fibrin gels and elastic poly (lactide-*co*-caprolacton) (PLCL) scaffolds, which we have previously fabricated [24]. These scaffolds, called cell aggregate-fibrin-PLCL scaffold complexes (CAPs), were then used to evaluate the feasibility of chondrogenic differentiation. We also compared the CAPs to conventional single cell-fibrin-PLCL scaffold complexes (SCPs) to evaluate their capacity for cartilage regeneration in vitro and in vivo.

2. Materials and Methods

2.1. Preparation of the PLCL Scaffolds

PLCL was synthesized and subjected to a gel-pressing method to fabricate sheet-form scaffolds, as described elsewhere [25]. Briefly, PLCL (50:50) was synthesized at 150 °C for 24 h in the presence of Stannous Octoate (1 mmol, Sigma, St. Louis, MO, USA) as a catalyst. Dissolved PLCL (5% *w/v*, in Chloroform) was mixed homogeneously with approximately 300 μm sized salts to form the PLCL gels. The residual solvents were removed by placing the samples under vacuum for an additional 24 h. The salts were then leached out by immersing the samples in distilled, deionized water and subjecting them to constant shaking for 3 days. The resulting scaffolds (disk type, diameter = 9 mm and thickness = 3 mm) were freeze-dried for 24 h, and then sterilized with ethylene oxide gas. The porosity of the scaffolds was measured by a mercury intrusion porosimeter (Auropore IV; MicroMeritics, Norcross, GA, USA).

2.2. Bone Marrow Stromal Cell Isolation and Culture

Cell isolation was conducted using a modification of the method described by Dong et al. and Cho et al. [25,26]. Briefly, bone marrow cells (BMCs) obtained from the femora and tibiae of New Zealand White rabbits (250–400 g) were treated with Ficoll–Paque density gradient reagent (Sigma; St. Louis, MO, USA) for separating the bone marrow mononuclear cells (BMMNCs), and the isolated cells were cultured with culture medium (Dulbecco's Modified Eagle Medium/Nutrient Mixture Ham-12 (DMEM/F12, Sigma) containing 10% Fetal Bovine Serum (FBS, US biotechnologies, Logan, UT, USA) and 1% penicillin-streptomycin (P-S, Gibco Life Technologies, Carlsbad, CA, USA)). The medium was replaced every two or three days to remove any non-adherent cells. Prior to preparing the cell aggregates, the BMSCs were labeled with the Vybrant® CFDA SE Cell Tracer Kit (Molecular Probes, Eugene, OR, USA) at a concentration of 1 µM CFDA/mL phosphate-buffered saline (PBS) (Sigma).

2.3. Preparation of the Cell Aggregates with the Hanging Drop Method

The cultured BMSCs (passage number equal to 2) were collected by trypsin (Gibco BRL, Carlsbad, CA, USA) treatment and then resuspended in chondrogenic medium (DMEM, 1 mM Sodium Pyruvate (Sigma), 100 nM Dexamethasone (Sigma), 20 µg/mL Proline (Sigma), 37.5 µg/mL ascorbic acid 2-phosphate (Sigma), 1% P-S, 1% FBS, 1× insulin-transferrin-selenium (ITS+) (BD Bioscience, Allschwil, Switzerland), and 10 ng/mL TGF-β_1 (R&D Systems, Minneapolis, MN, USA); 3–5 × 10^5 cells/mL). Drops of 20 µL containing 6000–10,000 BMSCs suspended in culture medium were placed on the inner side of the lid of 100 mm tissue culture dishes (Corning, Corning, NY, USA). The dishes were then filled with PBS to avoid a loss of nutrients through evaporation, and incubated for 7 days in a hanging drop at 37 °C in a CO_2 incubator [27].

2.4. In Vitro and In Vivo Studies of the Cells-Scaffold Complexes with Fibrin Gels and PLCL Scaffolds

Figure 1 shows an overall schematic illustration of the preparation of a cell aggregate by the hanging drop method and the structure of a CAP. To prepare the CAPs, a suspension of cell aggregates (about 2 × 10^6 cells/scaffold) was dispersed in 35 µL of fibrinogen solution (90 mg of fibrinogen, 60 U of factor XIII, and 1000 KIU aprotinin/mL) provided in a fibrin glue kit (Greenplast®; Yongin, Korea), and then mixed homogeneously. After the addition of 35 µL of thrombin solution (20 IU/mL of 0.6% (w/v) calcium chloride solution), the mixture was rapidly incubated in a fibrin-PLCL scaffold (9 mm diameter; disk type). In addition, a single cell suspension in chondrogenic medium was inoculated in the same conditions as the cell aggregates for the formation of the SCPs. The complexes were then allowed to stabilize in the incubator for one hour, after which chondrogenic medium was added. The complexes were then cultured for 25 days in a humidified incubator at 37 °C under 5% CO_2 in vitro.

For the in vivo studies, CAPs or SCPs were implanted into mice to investigate the chondrogenic differentiation of the cell aggregates and the cartilaginous tissue formation in the complexes. The complexes were prepared with the same method that was used for the in vitro studies, and then implanted into the subcutaneous dorsum of seven-week-old male athymic mice (SLC, Hamamatsu, Japan). The implants were harvested for analysis after 4 or 8 weeks.

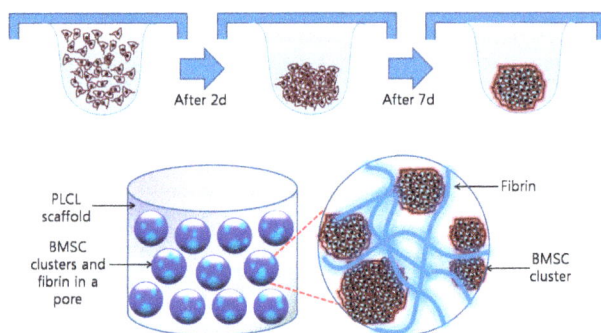

Figure 1. An overall schematic illustration of the preparation of a cell aggregate by the hanging drop method and the structure of a cell aggregate-fibrin-polymer scaffold complex. PLCL, poly (lactide-*co*-caprolacton); BMSC, bone marrow stromal cell.

2.5. Evaluation of the Cells-Scaffold Complexes

2.5.1. Scanning Electron Microscopy (SEM) Micrographs

The morphologies of the cell aggregates and the complexes were examined by scanning electron microscopy (SEM; Hitachi, Tokyo, Japan) at 15 kV. The samples were fixed in 4% (*v/v*) formaldehyde for 1 day, dehydrated with a graded ethanol series, and freeze-dried. The samples were coated with gold using a sputter-coater (Eiko IB3, Tokyo, Japan).

2.5.2. Water Soluble Tetrazolium Salts (WST) Assay

The numbers of cells in the cells-scaffold complexes were determined using a water soluble tetrazolium salt (WST) assay (Cell counting kit-8, Dojindo, Rockville, MD, USA), and cell proliferation and viability were assayed colorimetrically [28]. Briefly, 450 µL of culture medium and 50 µL of the kit solution were added to each sample containing the same amount of cells in 48-well plates (Corning), and incubated at 37 °C in a 5% CO_2 humidified incubator for 4 h. The resulting solution (100 µL per well) was transferred to 96-well µL plates (Corning), and analyzed using a microplate reader (VERSA max, Molecular Devices; San Diego, CA, USA) at 450 nm.

2.5.3. Immunofluorescent and Histological Analysis

To appraise the cellular organization and differentiation of the cell-aggregates in the fibrin-PLCL scaffolds three-dimensionally, immunofluorescence staining was performed with the cells-scaffold complexes. The complexes cultured in vitro for 1, 7, or 21 days were washed with PBS, fixed with paraformaldehyde for 1 day, permeabilized using Target Retrieval Solution (DaKoCytomation, Santa Clara, CA, USA), washed with PBS/bovine serum albumin (BSA, Sigma), and incubated with the primary antibody mouse-anti-chicken Collagen type II (Molecular Probes; Eugene, OR, USA) in 1% PBS/BSA overnight at 4 °C. The complexes were washed and labeled with the corresponding antibody Alexa Fluor®-594 conjugated anti-mouse (Molecular Probes) diluted 1:1000 in 1% BSA/PBS for 1 h at room temperature. Nuclei were counterstained with 4′,6-diamidino-2-phenylindole (DAPI, Molecular Probes), which binds to the double-strained DNA by forming a stable, blue-fluorescent complex. Finally, the complexes were mounted with Gelmount (M01, Biomeda, Foster City, CA, USA) and examined with a NiKon EZ-C1 confocal laser scanning microscope (NiKon, Tokyo, Japan) [29].

For histological analysis, the cells-scaffold complexes cultured in vitro or in vivo were fixed and embedded in paraffin. The 5 µm thick sectioned species were stained with hematoxylin and eosin (H&E). The retrieved constructs were stained with Masson's trichrome (M-T) for checking the collagen composition, and Alcian blue and Safranin O for confirming sulfated GAGs [30]. Additionally, with the sample slides, Collagen type II secreted by the seeded BMSCs was detected by immunofluorescence staining. Nuclei were counterstained with DAPI to confirm the presence of chondrocytes in the lacunae surrounded by the labeled type II collagen. The labeled tissues were examined with fluorescence microscopy (Eclipse TE2000U, Nikon, Tokyo, Japan).

2.5.4. Real-Time Polymerase Chain Reaction (Real-Time PCR)

To examine the chondrogenic differentiation of the seeded BMSCs or BMSC aggregates in the complexes, real-time polymerase chain reactions were performed [31]. The complexes were homogenized in TRIzol reagent (Gibco BRL, Carlsbad, CA, USA), after which the total RNA was extracted according to the manufacturer's instructions. Two micrograms of RNA were subsequently reversed transcribed into cDNA in a 20 μL reaction with the Omniscript® System (Qiagen, Hilden, Germany). The reverse transcription reaction was done as follows: the samples were held at 37 °C for 60 min, followed by 95 °C for 5 min in a thermal cycler. The oligonucleotide primers used in the real-time PCR are listed in Table 1 [32,33]. The real-time PCR was performed with the Applied BioSystems 7500 Real Time PCR system (Applied Biosystems, Foster City, CA, USA) and the *Power* SYBR® Green PCR Master Mix (Applied Biosystems, Foster City, CA, USA). The real-time PCR reactions were done with a final volume of 25 μL using 2 to 2.5 μL of cDNA as the template. The cDNA was amplified as follows: 45 cycles of 95 °C for 15 s and 55 °C for 60 s. A melting curve analysis was then conducted at the end of the cycles to make sure that the reaction produced only a single PCR product. The results were evaluated with the 7500 System SDS v1.4 software (Applied Biosystems). Glyceraldehyde-3-phosphate dehydrogenase (GAPDH) primers were used for the normalization of the samples. In the RNA extraction, RNase-free water (Qiagen Inc., Valencia, CA, USA) was included to monitor crossover contamination of the PCR, and it was used as a negative control. To ensure the quality of the data, a negative control was used in each run [32].

Table 1. List of primers used in the real time polymerase chain reaction (PCR) analysis of the in vitro and in vivo samples.

Primer Name	Forward Sequence	Reverse Sequence	Product Size (bp)
Aggrecan	TCGAGGACAGCGAGGCC	TCGAGGGTGTAGGCGTGTAGAGA	94
Proteogly-can	CACCTACCAGGACAAGGT	GCGCAGGCTCTGGATCTC	78
Type II collagen	CCTGTGCGACGACATAATCTGT	GCAGTGGCGAGGTCAGTAG	98
Type I collagen	GGGTTTAGACCGTCGTGAGA	TTGCCAGGAGAACCAGCAAGA	170
GAPDH	GCACCGTCAAGGCTGAGAAC	ATGGTGGTGAAGACGCCAGT	142

2.5.5. Measurement of the Glycosaminoglycan (GAG) Levels

The glycosaminoglycan (GAG) levels in the complexes were determined using the dimethyl methylene blue (DMB) assay. The complexes were homogenized and digested with 1 mL of papain solution containing 125 μg/mL papain (Sigma), 100 mM phosphate buffer (Sigma), 10 mM cysteine (Sigma), and 10 mM EDTA (Sigma) at 60 °C for 18 h. The GAG concentrations were then estimated by treating the samples with DMB dye (Sigma) [34] and comparing them against a chondroitin sulfate standard curve. All of the results are presented as the means and standard deviations (SD) ($n = 3$).

2.6. Statistical Analysis

All quantitative results were obtained from triplicate samples and all data were expressed as the means ± standard deviations. MINITAB™ (Minitab Inc., Centre County, PA, USA) was used for the statistical analysis. The samples were analyzed by one-way analysis of variance (ANOVA) followed by nonparametric LSD (Least Significant Differences) tests. A $p < 0.05$ was considered to indicate statistical significance.

3. Results and Discussion

3.1. PLCL Scaffold Characteriztion

Synthetic polymers, such as poly (ester-ether) polydioxanone (PDS), poly (ethylene glycol) (PEG), and poly (lactic-*co*-glycolic) acid (PLGA) are used as a scaffold to bear the load and to provide the three-dimensional (3D) matrix, while mesenchymal stem cells (MSCs) differentiate into chondrogenic lineages. However, PDS has poor solubility in organic solvent. PEG is a non-load-bearing scaffold, and provides no biological signals to the cells [35]. In cartilage regeneration, delivering mechanical signals to adherent cells in the body is important, and in order to transfer the improved mechanical signals more effectively, it is important that the elasticity and mechanical property of scaffold is close to those of natural cartilage [36]. Lastly, PCL is a common material to fabricate a biodegradable porous scaffold, because PCL is biocompatible while it has enough strength to bear the load [37,38].

PLCL has a flexible and elastic property which is close to natural cartilage, and it is more efficient for delivering mechanical signals to adherent cells than PCL scaffold [31]. Biodegradable porous PLCL scaffolds were fabricated by the gel-pressing method. The scaffolds had 78.3 ± 5.3%, which was measured by the mercury intrusion porosimeter, and pore size in the range of 300–500 μm. These scaffolds also had a homogeneously interconnected and open pore structure without a skin layer, and possess high elastic and flexible properties [36,39]. Moreover, the scaffolds are capable of offering a suitable environment for cell adhesion, and, in growth, maintaining mechanical integrity and delivering mechanical signals efficiently to adherent cells [40].

3.2. Charaterization of the Cell Aggregates

To facilitate and enhance the chondrogenic differentiation of BMSCs, we made the cell aggregates with the hanging drop method. It was observed that the BMSCs in the media drops aggregated over time (indicated by red arrow, Figure 2A). To investigate the morphology of the cell aggregates, the BMSCs were labeled with CFDA cell tracer, and the labeled cell aggregates were shown in green by a florescent microscope (Figure 2B). The cell aggregates were also analyzed by SEM, and it revealed that they formed spherical, dense cell clusters with a size of 182 ± 50 μm in diameter (Figure 2C–E).

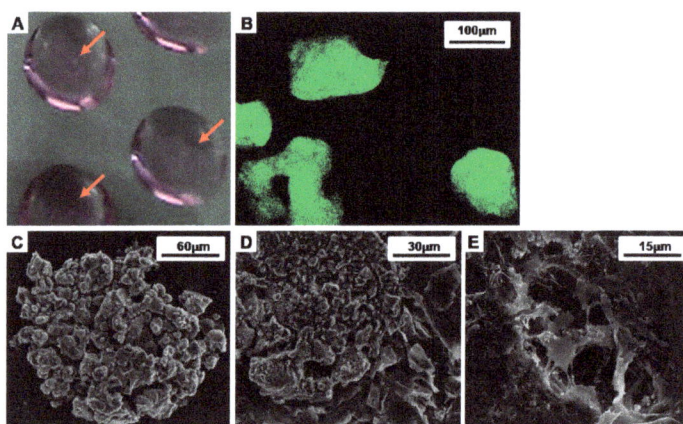

Figure 2. Images of the cell aggregates formed by the hanging drop method after 7 days. (**A**) An optical inverted microscope photograph of the cell aggregates; (**B**) A fluorescent microscope photograph of the cell aggregates in which the cells were labeled with CFDA cell tracer; (**C–E**) SEM images of a cell aggregate.

To confirm the morphology of the cell aggregates in the CAPs, we observed the surface with SEM and a florescent microscope. The images show that the cell aggregates were tightly adhered onto the fibrin-PLCL scaffolds, maintaining their morphologies (Figure 3).

Figure 3. Images of a cell aggregate-fibrin-polymer scaffold complex (**A,B**) SEM images of a cell aggregate-fibrin-polymer scaffold complex; (**C**) a confocal microscope image of a cell aggregate-fibrin-polymer scaffold complex at 2 days after seeding (Green: CFDA labeled cells; Blue: DAPI labeled nucleic acid).

3.3. Evaluation of Cells-Scaffold Complexes: In vitro and In vivo Experiments

In previous research, we developed a fibrin-PLCL complex construct seeded with BMSCs [41]. Dense collagen fibrils oriented with a relatively high density of ellipsoid-shaped flattened chondrocytes, which means that a high density of BMSCs is more effective than a low density of cells [35]. To evaluate the effect of cell aggregation on the chondrogenic differentiation of bone marrow stromal cells in vitro, the cell aggregates or single cells of BMSCs were seeded onto each fibrin-PLCL scaffold and cultured for 25 days. A cell adhesion, viability, and proliferation assay was done with the WST assay. Figure 4A shows the cell growth on each fibrin-PLCL scaffold over a 25-day period. The growth data show a significant difference in the cell proliferation rate between the CAPs and SCPs. DMB assays were also performed to determine the GAG content, which is one of the markers for chondrocyte differentiation and cartilaginous tissue formation. Figure 4B shows a plot of the GAG content, and the data exhibited a similar pattern to that of the cell proliferation. Both the cell growth rate and the GAG content of the CAPs were higher than those of the SCPs, and tended to increase over time.

Furthermore, the cells-scaffold complexes were cultured for 21 days in vitro and stained with hematoxylin and eosin (H&E) and alcian blue for histological analysis. The images of the analysis' result are presented in Figure 5. After 21 days of culturing, there were little or no lacunae in the SCPs (Figure 5A), while the cell aggregates in the CAPs formed mature and well-developed cartilage tissue, as evident by the chondrocytes within the lacunae (Figure 5C). An amount of sulfated GAG deposition was observed in the CAPs with a strong blue color for the alcian blue staining, and mature chondrocytes within the lacunae were also observed (Figure 5D). In contrast, less sGAG content from the SCPs was stained (Figure 5B). These results show that the CAPs are more effective in forming mature and well-developed cartilaginous tissues in vitro.

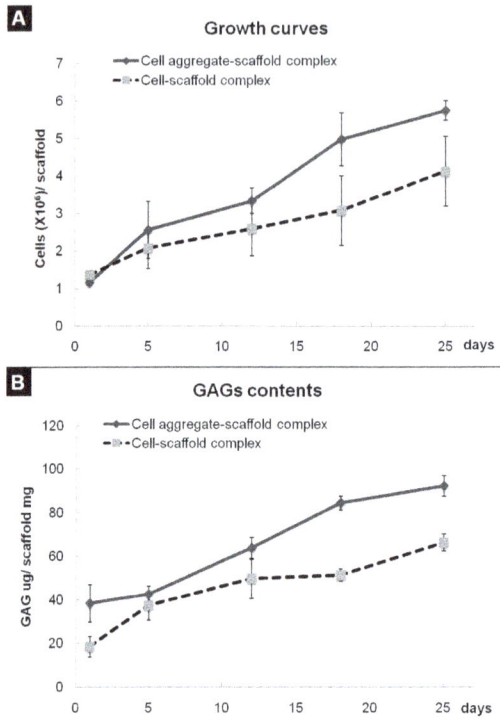

Figure 4. Quantitative analysis of cell growth (**A**) and glycosaminoglycan (GAG) levels (**B**) associated with cell aggregate-fibrin-polymer scaffold complexes and cell-fibrin-PLCL scaffold complexes cultured in vitro. The error bars show the standard deviations. (Cell-scaffold complex, complex seeded with single cells of BMSCs; Cell aggregate-scaffold complex, complex seeded with cell aggregates of BMSCs).

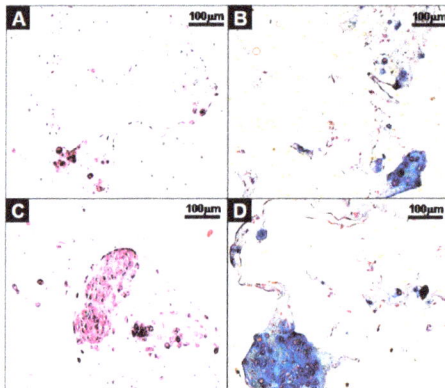

Figure 5. Histological studies of cells-scaffold constructs cultured for 21 days in vitro. The sections were stained with hematoxylin and eosin (H&E) (**A,C**), or Alcian Blue (**B,D**). The images show the cell-fibrin-PLCL scaffold complexes (**A,B**) and cell aggregate-fibrin-PLCL scaffold complexes (**C,D**).

BMSCs within the whole cells-scaffold constructs were labeled with CFDA cell tracer (Green) and cultured for 7 days or 21 days in vitro, and then the cellular organization and differentiation were examined. Immunofluorescence staining with rabbit-collagen type II (Red) and DAPI (blue) shows a clear distinction between the CAPs and SCPs. Chondrogenically differentiated BMSCs in the lacunae that are surrounded by labeled type II collagen are seen in Figure 6. The differentiation of BMSCs in the cell aggregate was already ongoing on Day 7; thus, red fluorescence for collagen type II secreted by the differentiated chondrocytes was observed (Figure 6C), whereas the fluorescence was hardly observed in the SCPs (Figure 6A). Over time, as the BMSCs differentiated, the deposition of the labeled collagen type II increased and was more clearly observed in the newly formed tissues of the CAPs (Figure 6D) than in the SCPs (Figure 6B) on Day 21. Although DAPI staining showed homogeneously distributed nuclei of the cells in the SCPs and CAPs, some of the cell aggregates in the CAPs maintained their spherical morphologies after 21 days (Figure 6D).

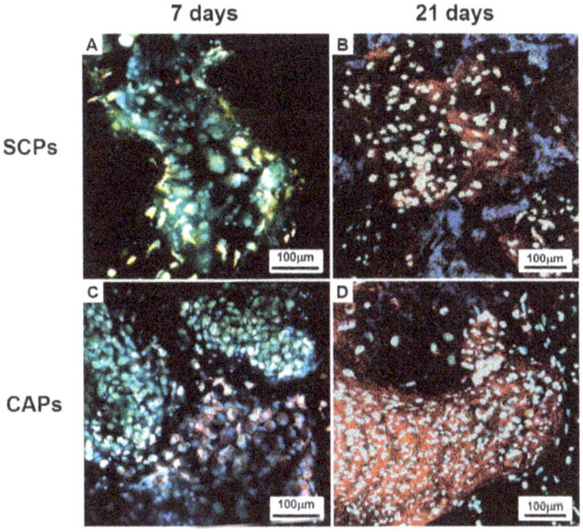

Figure 6. Immunofluorescence studies with a confocal microscope of the cells-scaffold constructs cultured for 7 days (**A,C**) or 21 days (**B,D**) in vitro. The complexes were stained for rabbit-collagen type II (Red). The cells were previously labeled with CFDA cell tracer (Green), and all samples were stained with DAPI to identify the DNA strands of the nuclei. CAP, cell aggregate-fibrin-PLCL scaffold complex; SCP, single cell-fibrin-PLCL scaffold complex.

To investigate the differentiation of the seeded BMSCs, the expression of mRNA for proteoglycan, aggrecan, type II collagen, and type I collagen was assessed by real-time PCR (Figure 7). It revealed that the expression of various mRNAs tended to increase over time in both of the cells-scaffold complex groups; however, the CAPs had a much higher expression of proteoglycan, aggrecan, and type II collagen mRNAs that that of the other group. Moreover, the ratio of type II collagen to type I collagen expression, which is an important indicator of chondrogenic differentiation, was found to be significantly greater in the complexes that were seeded with cell aggregates than in those seeded with single cells. These results showed that seeding the cell aggregates into the fibrin-PLCL scaffold promoted the upregulation of chondrogenesis marker genes and chondrogenic differentiation in vitro.

Subsequently, the CAPs and SCPs were implanted subcutaneously into nude mice for 4 or 8 weeks in vivo for preliminary observation to determine which construct has a greater capacity for chondrogenic differentiation and cartilaginous tissue formation [42]. The results of the assay for 4 and 8 weeks are presented in Figure 8. Both the SCPs and CAPs showed increased GAG contents over time; however, the GAG contents of the CAPs increased from 120.90 ± 12.84 µg/mg scaffold at 4 weeks to 146.86 ± 31.12 µg/mg scaffold at 8 weeks, while those of the SCPs only increased from

75.30 ± 15.84 µg/mg scaffold at 4 weeks to 87.80 ± 9.20 µg/mg scaffold at 8 weeks. These results show that the GAG contents of the CAPs were much higher than those of the other complexes that were seeded with single cells. These trends were consistent with those obtained from the histological studies in terms of the accumulation of the cartilaginous extracellular matrix.

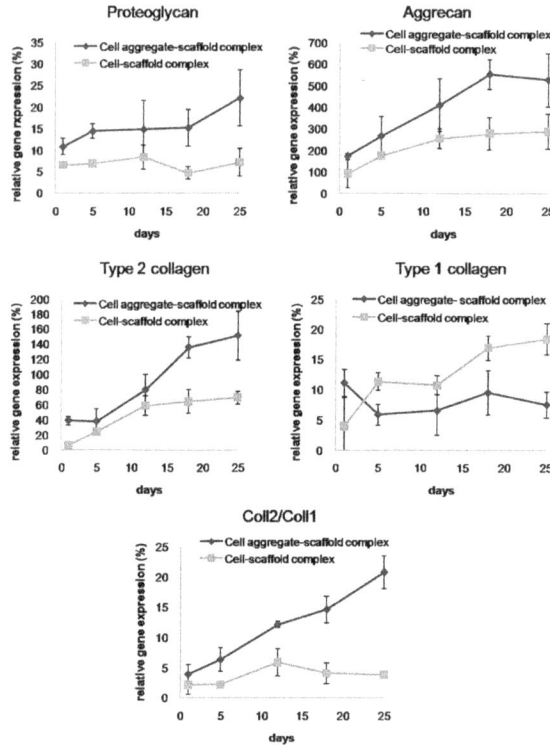

Figure 7. Relative mRNA expression of proteoglycan, aggrecan, type II collagen, type I collagen, and type II collagen to type I collagen (Coll2/Coll1) of the cell-fibrin-PLCL scaffold complexes and cell aggregate-fibrin-PLCL scaffold complexes cultured in vitro for up to 25 days. Error bars denote the standard deviation.

Figure 8. GAG levels within the cells-scaffold complexes explanted from nude mice at 4 and 8 weeks. (* = $p < 0.05$). Error bars denote the standard deviation.

A histological evaluation of the implants was carried out at the end of 4 and 8 weeks for the in vivo culture, and the images of each specimen stained with H&E, M-T, Safranin-O and Alcian blue are shown in Figure 9. H&E staining showed that cell aggregates in the CAPs formed mature and well-developed cartilaginous tissues close to native cartilage, evident by the chondrocytes within the lacunae. With Masson's trichrome, Safranin-O, and Alcian blue stain, a substantial amount of collagen and homogeneously distributed sulfated GAGs, which were present in the extracellular matrices produced by the differentiated BMSCs in the newly-formed tissues, were observed in the CAPs. Additionally, although no significant difference in the deposition of collagen was present between those two groups at 4 weeks, lacunae were partially observed in the CAPs. Additionally, the presence of more lacunae and more accumulation of sGAGs were observed in the CAPs, with strong positive staining for M-T, Safranin-O, and Alcian blue at 8 weeks. The staining and lacunae were distributed homogeneously throughout the entire tissues, and the cells within the tissues also exhibited a cartilage-like morphology. In contrast, in the case of the SCPs, there was no significant change in the sGAG contents over time, and lacunae were never or rarely detected at 4 and 8 weeks. The results of the tissue histology show that the cell aggregate promoted the BMSCs to differentiate into chondrocytes, secrete the cartilaginous extracellular matrix, and form mature cartilage tissue.

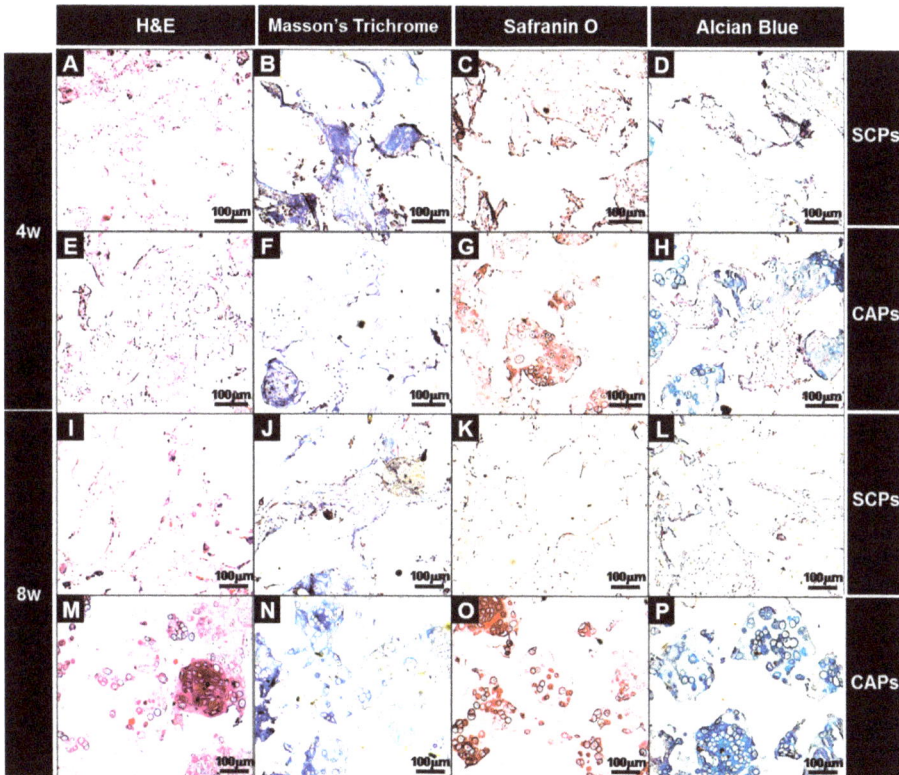

Figure 9. Histological studies of the implants at 4 (**A–H**) and 8 (**I–P**) weeks. The sections were stained with H&E (**A,E,I,M**), Masson's Trichrome (**B,F,J,N**), Safranin O (**C,G,K,O**), or Alcian Blue (**D,H,L,P**). The images show the cell-fibrin-PLCL scaffold complexes (**A–D,I–L**) and the cell aggregate-fibrin-PLCL scaffold complexes (**E–H,M–P**).

To further investigate the properties of the CAPs cultured for 8 weeks in vivo, immunofluorescence staining was conducted with Rabbit collagen type II (Red) and DAPI (blue) as a counterstain (Figure 10). Immunohistochemistry showed that the chondrogenically differentiated BMSCs within the lacunae were surrounded by labeled type II collagen (Red), and the collagen type II, an extracellular matrix material produced by differentiated chondrocytes, was clearly observed in the newly-formed tissues of the CAPs. Moreover, the magnified images show that the cell aggregates seeded onto the fibrin-PLCL scaffold were maintaining an aggregate shape (Figure 10A,B), and even single cells, which had migrated from the cell aggregate during differentiation, were also maintaining the chondrogenic phenotype (Figure 10C).

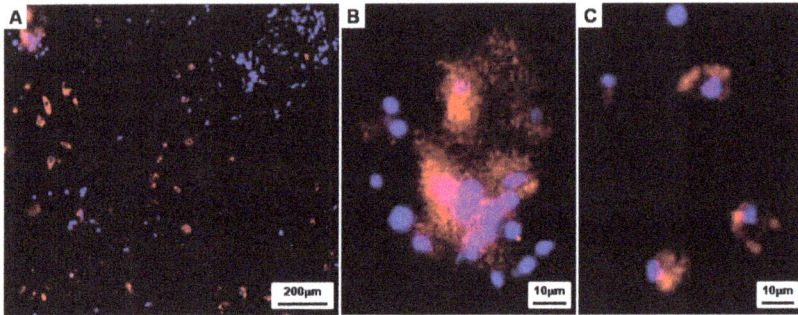

Figure 10. Immunofluorescence studies of the implants of cell aggregate-fibrin-PLCL scaffold complexes at 8 weeks. The sections were stained for rabbit-collagen type II (Red) and stained with DAPI to identify the DNA strands of the nuclei. Then, the images were merged to examine the chondrogenically differentiated BMSCs in the lacunae surrounded by the labeled type II collagen.

To analyze the mRNA expression of proteoglycan, aggrecan, type II collagen, and type I collagen in the specimens retrieved after 4 and 8 weeks, real-time PCR was performed (Figure 11). The mRNA expression of proteoglycan, aggrecan, and type II collagen within the cells-scaffold complexes increased over time. However, a significant expression of cartilage-specific markers for the CAPs was observed over time compared to that of the SCPs.

Conversely, the expression of type I collagen gene in the CAPs decreased over time. In contrast, the expression of the type I collagen gene was increased in the SCPs. Besides, the ratio of type II collagen to type I collagen expression, which is an important indicator of chondrogenic differentiation, was found to be significantly greater in the complexes that were seeded with cell aggregates than in those seeded with single cells. These results show that cartilage-specific genes are expressed much more highly in the CAPs when compared to the SCPs. In other words, seeding cell aggregates onto the fibrin-PLCL scaffold facilitated the upregulation of chondrogenesis marker genes, the secretion of chondral extracellular matrix, and chondrogenic differentiation in vivo. The results of the in vivo experiments confirmed that the cell aggregates promoted themselves to secrete cartilaginous extracellular matrix, and to form mature cartilage tissue compared with that of the single cells. Moreover, the results show the effect of cell aggregates, which affect phenotype maintenance, tissue development, and chondrogenic differentiation.

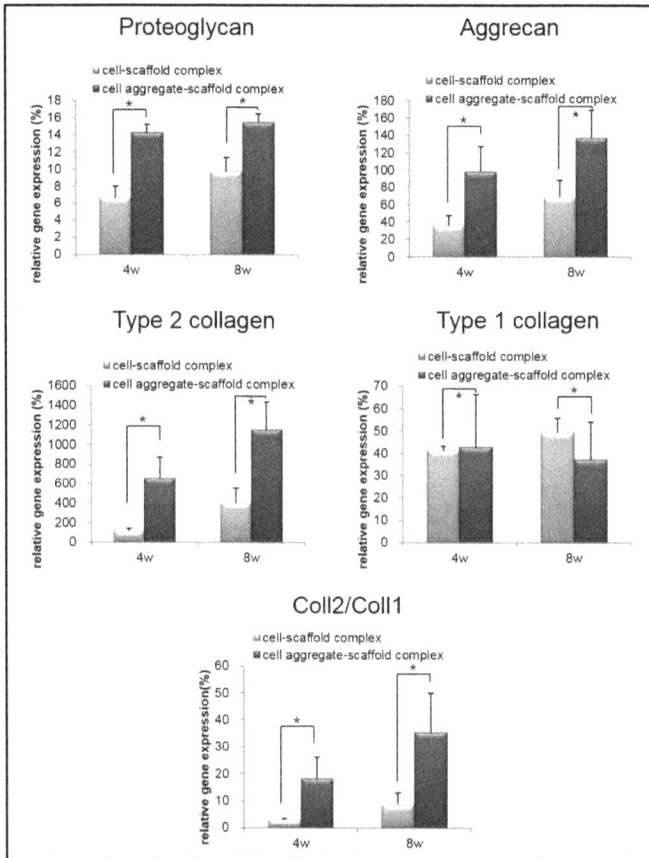

Figure 11. Relative mRNA expression of proteoglycan, aggrecan, type II collagen, type I collagen, and type II collagen to type I collagen for the cell-fibrin-PLCL scaffold complexes and cell aggregate-fibrin-PLCL scaffold complexes explanted at 4 and 8 weeks (* = *p* <0.05). Error bars denote the standard deviation.

4. Conclusions

In cartilage regeneration using stem cells, it is important to induce and maintain the differentiation of stem cells to cartilage lineages. Cell aggregates or speroids which are pre-directed for chondrogenic differentiation enhance cartilaginous tissue formation. In this study, we prepared cell aggregates using the hanging drop method and evaluated the effect of the cell aggregates in hybrid scaffolds of fibrin gels and PLCL scaffolds for the chondrogenic differentiation of BMSCs, the production of GAGs, and the formation of cartilaginous tissue. From the in vitro and in vivo studies, significantly increased amounts of GAG contents and chondral extracellular matrix, and increased gene expression, were observed in the CAP, in which mature and well-developed cartilaginous tissues were formed. Overall, these results show that cell aggregates in hybrid scaffolds can promote themselves to differentiate into chondrocytes, can maintain their phenotypes, can enhance GAG production, and can form quality cartilaginous tissue, thereby inducing the effective regeneration of cartilage tissue close to natural cartilage in vitro and in vivo.

Acknowledgments: This work was supported by a grant from Korea Health technology R&D Project through the Korea Health Industry Development Institute (KHIDI) funded by the Ministry of Health & Welfare (HI15C3060-010115), and by a grant from the Basic Science Research Program through the National Research Foundation of Korea (NRF) funded by the Ministry of Science, ICT, and future Planning (2016R1A2B2009550), Republic of Korea.

Author Contributions: Soojin Lee, Soo Hyun Kim, Kangwon Lee, and Youngmee Jung conceived and designed the experiments; Soojin Lee and Youngmee Jung performed the experiments; Soojin Lee and Youngmee Jung analyzed the data; Soojin Lee and Youngmee Jung wrote the manuscript.

Conflicts of Interest: The authors declare no conflict of interest.

References

1. Buschmann, M.D.; Gluzband, Y.A.; Grodzinsky, A.J.; Hunziker, E.B. Mechanical compression modulates matrix biosynthesis in chondrocyte/agarose culture. *J. Cell Sci.* **1995**, *108*, 1497–1508. [PubMed]

2. Sophia Fox, A.J.; Bedi, A.; Rodeo, S.A. The basic science of articular cartilage: Structure, composition, and function. *Sports Health* **2009**, *1*, 461–468. [CrossRef] [PubMed]

3. Almarza, A.J.; Athanasiou, K.A. Design characteristics for the tissue engineering of cartilaginous tissues. *Ann. Biomed. Eng.* **2004**, *32*, 2–17. [CrossRef] [PubMed]

4. Hunziker, E.B.; Quinn, T.M.; Hauselmann, H.J. Quantitative structural organization of normal adult human articular cartilage. *Osteoarthr. Cartil.* **2002**, *10*, 564–572. [CrossRef] [PubMed]

5. Solchaga, L.A.; Goldberg, V.M.; Caplan, A.I. Cartilage regeneration using principles of tissue engineering. *Clin. Orthop. Relat. Res.* **2001**, *391*, S161–S170. [CrossRef]

6. Temenoff, J.S.; Mikos, A.G. Review: Tissue engineering for regeneration of articular cartilage. *Biomaterials* **2000**, *21*, 431–440. [CrossRef]

7. Brittberg, M.; Lindahl, A.; Nilsson, A.; Ohlsson, C.; Isaksson, O.; Peterson, L. Treatment of deep cartilage defects in the knee with autologous chondrocyte transplantation. *N. Engl. J. Med.* **1994**, *331*, 889–895. [CrossRef] [PubMed]

8. Bentley, G.; Biant, L.C.; Carrington, R.W.; Akmal, M.; Goldberg, A.; Williams, A.M.; Skinner, J.A.; Pringle, J. A prospective, randomised comparison of autologous chondrocyte implantation versus mosaicplasty for osteochondral defects in the knee. *J. Bone Jt. Surg.* **2003**, *85*, 223–230. [CrossRef]

9. Fu, F.H.; Zurakowski, D.; Browne, J.E.; Mandelbaum, B.; Erggelet, C.; Moseley, J.B., Jr.; Anderson, A.F.; Micheli, L.J. Autologous chondrocyte implantation versus debridement for treatment of full-thickness chondral defects of the knee: An observational cohort study with 3-year follow-up. *Am. J. Sports Med.* **2005**, *33*, 1658–1666. [CrossRef] [PubMed]

10. Nejadnik, H.; Hui, J.H.; Feng Choong, E.P.; Tai, B.C.; Lee, E.H. Autologous bone marrow-derived mesenchymal stem cells versus autologous chondrocyte implantation: An observational cohort study. *Am. J. Sports Med.* **2010**, *38*, 1110–1116. [CrossRef] [PubMed]

11. Kim, S.H.; Park, D.Y.; Min, B.H. A new era of cartilage repair using cell therapy and tissue engineering: Turning current clinical limitations into new ideas. *Tissue Eng. Regen. Med.* **2012**, *9*, 240–248. [CrossRef]

12. Hung, C.T.; Lima, E.G.; Mauck, R.L.; Takai, E.; LeRoux, M.A.; Lu, H.H.; Stark, R.G.; Guo, X.E.; Ateshian, G.A. Anatomically shaped osteochondral constructs for articular cartilage repair. *J. Biomech.* **2003**, *36*, 1853–1864. [CrossRef]

13. Hunziker, E.B.; Rosenberg, L.C. Repair of partial-thickness defects in articular cartilage: Cell recruitment from the synovial membrane. *J. Bone Jt. Surg.* **1996**, *78*, 721–733. [CrossRef]

14. Lynn, A.K.; Brooks, R.A.; Bonfield, W.; Rushton, N. Repair of defects in articular joints. Prospects for material-based solutions in tissue engineering. *J. Bone Jt. Surg.* **2004**, *86*, 1093–1099. [CrossRef]

15. Risbud, M.V.; Sittinger, M. Tissue engineering: Advances in in vitro cartilage generation. *Trends Biotechnol.* **2002**, *20*, 351–356. [CrossRef]

16. Swieszkowski, W.; Tuan, B.H.; Kurzydlowski, K.J.; Hutmacher, D.W. Repair and regeneration of osteochondral defects in the articular joints. *Biomol. Eng.* **2007**, *24*, 489–495. [CrossRef] [PubMed]

17. Tuan, R.S. A second-generation autologous chondrocyte implantation approach to the treatment of focal articular cartilage defects. *Arthritis Res. Ther.* **2007**, *9*, 109. [CrossRef] [PubMed]

18. Pfeiffer, E.; Vickers, S.M.; Frank, E.; Grodzinsky, A.J.; Spector, M. The effects of glycosaminoglycan content on the compressive modulus of cartilage engineered in type II collagen scaffolds. *Osteoarthr. Cartil.* **2008**, *16*, 1237–1244. [CrossRef] [PubMed]

19. Peterson, L.; Minas, T.; Brittberg, M.; Nilsson, A.; Sjogren-Jansson, E.; Lindahl, A. Two- to 9-year outcome after autologous chondrocyte transplantation of the knee. *Clin. Orthop. Relat. Res.* **2000**, *374*, 212–234. [CrossRef]

20. Tung, Y.C.; Hsiao, A.Y.; Allen, S.G.; Torisawa, Y.S.; Ho, M.; Takayama, S. High-throughput 3D spheroid culture and drug testing using a 384 hanging drop array. *Analyst* **2011**, *136*, 473–478. [CrossRef] [PubMed]

21. Johnstone, B.; Hering, T.M.; Caplan, A.I.; Goldberg, V.M.; Yoo, J.U. In vitro chondrogenesis of bone marrow-derived mesenchymal progenitor cells. *Exp. Cell Res.* **1998**, *238*, 265–272. [CrossRef] [PubMed]

22. Zhang, L.; Su, P.; Xu, C.; Yang, J.; Yu, W.; Huang, D. Chondrogenic differentiation of human mesenchymal stem cells: A comparison between micromass and pellet culture systems. *Biotechnol. Lett.* **2010**, *32*, 1339–1346. [CrossRef] [PubMed]

23. Ruedel, A.; Hofmeister, S.; Bosserhoff, A.K. Development of a model system to analyze chondrogenic differentiation of mesenchymal stem cells. *Int. J. Clin. Exp. Pathol.* **2013**, *6*, 3042–3048. [PubMed]

24. Jung, Y.; Kim, S.H.; Kim, Y.H.; Kim, S.H. The effects of dynamic and three-dimensional environments on chondrogenic differentiation of bone marrow stromal cells. *Biomed. Mater.* **2009**, *4*, 055009. [CrossRef] [PubMed]

25. Dong, J.; Uemura, T.; Shirasaki, Y.; Tateishi, T. Promotion of bone formation using highly pure porous β-TCP combined with bone marrow-derived osteoprogenitor cells. *Biomaterials* **2002**, *23*, 4493–4502. [CrossRef]

26. Cho, S.W.; Park, H.J.; Ryu, J.H.; Kim, S.H.; Kim, Y.H.; Choi, C.Y.; Lee, M.J.; Kim, J.S.; Jang, I.S.; Kim, D.I.; et al. Vascular patches tissue-engineered with autologous bone marrow-derived cells and decellularized tissue matrices. *Biomaterials* **2005**, *26*, 1915–1924. [CrossRef] [PubMed]

27. Banerjee, M.; Bhonde, R.R. Application of hanging drop technique for stem cell differentiation and cytotoxicity studies. *Cytotechnology* **2006**, *51*, 1–5. [CrossRef] [PubMed]

28. Kuhn, D.M.; Balkis, M.; Chandra, J.; Mukherjee, P.K.; Ghannoum, M.A. Uses and limitations of the XTT assay in studies of Candida growth and metabolism. *J. Clin. Microbiol.* **2003**, *41*, 506–508. [CrossRef] [PubMed]

29. Hofmann, A.; Ritz, U.; Verrier, S.; Eglin, D.; Alini, M.; Fuchs, S.; Kirkpatrick, C.J.; Rommens, P.M. The effect of human osteoblasts on proliferation and neo-vessel formation of human umbilical vein endothelial cells in a long-term 3D co-culture on polyurethane scaffolds. *Biomaterials* **2008**, *29*, 4217–4226. [CrossRef] [PubMed]

30. Jung, Y.; Kim, S.H.; Kim, S.H.; Kim, Y.H.; Xie, J.; Matsuda, T.; Min, B.G. Cartilaginous tissue formation using a mechano-active scaffold and dynamic compressive stimulation. *J. Biomater. Sci. Polym. Ed.* **2008**, *19*, 61–74. [CrossRef] [PubMed]

31. Jung, Y.; Chung, Y.I.; Kim, S.H.; Tae, G.; Kim, Y.H.; Rhie, J.W.; Kim, S.H.; Kim, S.H. In situ chondrogenic differentiation of human adipose tissue-derived stem cells in a TGF-beta1 loaded fibrin-poly(lactide-caprolactone) nanoparticulate complex. *Biomaterials* **2009**, *30*, 4657–4664. [CrossRef] [PubMed]

32. Kaneshiro, N.; Sato, M.; Ishihara, M.; Mitani, G.; Sakai, H.; Mochida, J. Bioengineered chondrocyte sheets may be potentially useful for the treatment of partial thickness defects of articular cartilage. *Biochem. Biophys. Res. Commun.* **2006**, *349*, 723–731. [CrossRef] [PubMed]

33. Tonomura, H.; Takahashi, K.A.; Mazda, O.; Arai, Y.; Shin-Ya, M.; Inoue, A.; Honjo, K.; Hojo, T.; Imanishi, J.; Kubo, T. Effects of heat stimulation via microwave applicator on cartilage matrix gene and HSP70 expression in the rabbit knee joint. *J. Orthop. Res.* **2008**, *26*, 34–41. [CrossRef] [PubMed]

34. Farndale, R.W.; Buttle, D.J.; Barrett, A.J. Improved quantitation and discrimination of sulphated glycosaminoglycans by use of dimethylmethylene blue. *Biochim. Biophys. Acta* **1986**, *883*, 173–177. [CrossRef]

35. Jeuken, R.M.; Roth, A.K.; Peters, R.J.R.W.; van Donkelaar, C.C.; Thies, J.C.; van Rhijn, L.W.; Emans, P.J. Polymers in Cartilage Defect Repair of the Knee: Current Status and Future Prospects. *Polymers* **2016**, *8*, 219. [CrossRef]

36. Jung, Y.; Park, M.S.; Lee, J.W.; Kim, Y.H.; Kim, S.H.; Kim, S.H. Cartilage regeneration with highly-elastic three-dimensional scaffolds prepared from biodegradable poly(L-lactide-*co*-epsilon-caprolactone). *Biomaterials* **2008**, *29*, 4630–4636. [CrossRef] [PubMed]

37. De Santis, R.; D'Amora, U.; Russo, T.; Ronca, A.; Gloria, A.; Ambrosio, L. 3D fibre deposition and stereolithography techniques for the design of multifunctional nanocomposite magnetic scaffolds. *J. Mater. Sci. Mater. Med.* **2015**, *26*, 250. [CrossRef] [PubMed]

38. Hsieh, Y.H.; Hsieh, M.F.; Fang, C.H.; Jiang, C.P.; Lin, B.J.; Lee, H.M. Osteochondral Regeneration Induced by TGF-β Loaded Photo Cross-Linked Hyaluronic Acid Hydrogel Infiltrated in Fused Deposition-Manufactured Composite Scaffold of Hydroxyapatite and Poly (Ethylene Glycol)-Block-Poly(ε-Caprolactone). *Polymers* **2017**, *9*, 182. [CrossRef]

39. Jung, Y.; Kim, S.H.; You, H.J.; Kim, S.H.; Kim, Y.H.; Min, B.G. Application of an elastic biodegradable poly(L-lactide-*co*-ε-caprolactone) scaffold for cartilage tissue regeneration. *J. Biomater. Sci. Polym. Ed.* **2008**, *19*, 1073–1085. [CrossRef] [PubMed]

40. Chou, C.H.; Cheng, W.T.; Kuo, T.F.; Sun, J.S.; Lin, F.H.; Tsai, J.C. Fibrin glue mixed with gelatin/hyaluronic acid/chondroitin-6-sulfate tri-copolymer for articular cartilage tissue engineering: The results of real-time polymerase chain reaction. *J. Biomed. Mater. Res. A* **2007**, *82*, 757–767. [CrossRef] [PubMed]

41. Jung, Y.; Kim, S.H.; Kim, Y.H.; Kim, S.H. The effect of hybridization of hydrogels and poly(L-lactide-*co*-epsilon-caprolactone) scaffolds on cartilage tissue engineering. *J. Biomater. Sci. Polym. Ed.* **2010**, *21*, 581–592. [CrossRef] [PubMed]

42. Kundu, J.; Shim, J.H.; Jang, J.; Kim, S.W.; Cho, D.W. An additive manufacturing-based PCL-alginate-chondrocyte bioprinted scaffold for cartilage tissue engineering. *J. Tissue Eng. Regen. Med.* **2015**, *9*, 1286–1297. [CrossRef] [PubMed]

polymers

MDPI

Article

Effect of Catechol Content in Catechol-Conjugated Dextrans on Antiplatelet Performance

Yeonwoo Jeong [†], Kwang-A Kim [†] and Sung Min Kang *

Department of Chemistry, Chungbuk National University, Chungbuk 28644, Korea;
ywjeong9104@gmail.com (Y.J.); kwangakim03@gmail.com (K.-A.K.)
* Correspondence: smk16@chungbuk.ac.kr; Tel.: +82-43-261-2289
† These authors equally contributed to this work.

Received: 26 June 2017; Accepted: 17 August 2017; Published: 19 August 2017

Abstract: The surface coating of solid substrates using dextrans has gained a great deal of attention, because dextran-coated surfaces show excellent anti-fouling property as well as biocompatibility behavior. Much effort has been made to develop efficient methods for grafting dextrans on solid surfaces. This led to the development of catechol-conjugated dextrans (Dex-C) which can adhere to a number of solid surfaces, inspired by the underwater adhesion behavior of marine mussels. The present study is a systematic investigation of the characteristics of surface coatings developed with Dex-C. Various Dex-C with different catechol contents were synthesized and used as a surface coating material. The effect of catechol content on surface coating and antiplatelet performance was investigated.

Keywords: dextran; catechol conjugation; surface coating; antiplatelet

---

1. Introduction

Dextran (Dex) is a polysaccharide consisting of glucose molecules, and has been of importance due to its use in medical applications as an antiplatelet and blood volume expander [1]. Particularly, the antiplatelet properties of Dex have been used in combination with several surface coating techniques with the aim of preparing blood-compatible medical devices [2–11]. Previous studies have revealed that Dex-coated surfaces could effectively limit cell adhesion as well as protein adsorption [4,5,7,8]. For example, Massia et al. reported the covalent immobilization of Dex on a glass surface [5]. In the study, amino groups were firstly introduced on the glass surface by using 3-aminopropryltriethoxysilane, an organosilane coupling agent. Activated Dex, which is oxidized by sodium periodate, was then immobilized on the glass surface via reductive amination between amino groups of the glass surface and aldehyde groups of the activated Dex. Fibroblast cell adhesion and spreading was effectively inhibited on the Dex-immobilized glass surfaces, as opposed to untreated surfaces.

Although the covalent methods of Dex grafting on solid surfaces have been successfully employed, those methods have drawbacks, in that complicated steps and harsh reaction conditions are required. Therefore, the development of advanced grafting methods is needed. Recently, a mussel-inspired approach in which Dex can be easily grafted on solid substrates was investigated [9–11]. Given that the catechol is known to play an important role in the strong underwater adhesion of marine mussels, catechol-conjugated polymers were developed for use as functional coatings of diverse substrates [12–14]. In the case of Dex, Park et al. synthesized catechol-conjugated Dex (Dex-C) and applied it onto titanium (Ti) surfaces; Dex-C with a catechol grafting density of ~7% was prepared by a carbamate-bond forming reaction, and employed as in surface coating of Ti in anti-fouling applications [9,10]. Dex-C-coated Ti surfaces exhibited excellent resistance against protein and cell adhesions. Despite the aforementioned surface coating ability and anti-fouling properties of Dex-C, there are areas of fundamental research that have not yet been fully investigated, such as the effect of catechol content in Dex-C on surface coating and anti-fouling properties. In this respect, herein, Dex-C with various catechol contents were synthesized for coating Ti surfaces. The surface coating efficiency as a function of catechol content was investigated by spectroscopic ellipsometry, contact angle goniometry, and X-ray photoelectron spectroscopy (XPS). The anti-fouling properties of Dex-C-coated surfaces were evaluated by measuring the platelet density attached to surfaces.

2. Experimental Section

2.1. Materials

Dextran (Dex, 6 k MW, from *Leuconostoc* spp., Sigma-Aldrich, St. Louis, MO, USA), dopamine hydrochloride (98%, Sigma-Aldrich), Trizma base (99%, Sigma), Trizma·HCl (99%, Sigma), 1,1′-carbonyldiimidazole (CDI, Sigma-Aldrich), hydrochloric acid (HCl, 35~37%, Duksan, Ansan, Korea), absolute ethanol (Merck, Kenilworth, NJ, USA), dimethyl sulfoxide (DMSO, 99%, TCI, Tokyo, Japan), and acetone (99%, Daejung Chemicals & Metal, Shiheung, Korea) were used as received. A concentrated platelet solution (1.04×10^6 cells/mL) was obtained from Red Cross blood center (Chungbuk, Korea), and the platelets were used as received. The use of human platelets for this study was approved by the institutional review board (IRB) of Chungbuk National University.

2.2. Synthesis of Catechol-Conjugated Dextrans (Dex-C)

Dex-C was synthesized according to the previous report [9]. Dex (162.1 mg) was dissolved in 5 mL of DMSO at room temperature. CDI (324.3 mg) in 2 mL of DMSO was added to the Dex solution, and the resulting solution was stirred at room temperature for 30 min. Dopamine hydrochloride (189.6 mg) which was dissolved in 1 mL of DMSO was then added to the CDI-activated Dex solution. The conjugation reaction between CDI-activated Dex and dopamine hydrochloride was carried out at room temperature. After overnight reaction, 30 mL of deionized (DI) water was added to the solution. Subsequently, precipitates formed in the solution were removed, and the water-soluble part was transferred to a dialysis membrane (MWCO = 3500) and dialyzed for 24 h to remove unreacted coupling reagents. Acidified water obtained by adding 1 mL of 5 M HCl to 1 L of DI water was used for a dialysis procedure. This process was repeated with different molar ratios of reactants (monomer unit of Dex:dopamine:CDI = 9:1:1.5, 9:1:6, 1:1:1, and 1:1:2). The final products were freeze-dried and stored in a refrigerator before use.

2.3. Dex-C Coating on Solid Substrates

Ti/TiO$_2$ substrates were prepared by thermal evaporation of 100 nm of Ti onto silicon wafers. Ti/TiO$_2$ substrates (1 cm × 1 cm) were cleaned with acetone or ethanol by sonication prior to use. Dex-C coating was carried out by immersing substrates in a buffered solution (5 mg of Dex-C per 1 mL of 50 mM Tris, pH 8.5) at room temperature for 24 h. The coated substrates were rinsed with DI water and blow-dried under a steam of nitrogen gas.

2.4. Platelet Adhesion

Uncoated and Dex-C-coated Ti/TiO$_2$ substrates were incubated in 0.5 mL of platelet media (1.04×10^6 cells/mL) for 24 h. After incubation at room temperature [15], substrates were rinsed by phosphate-buffered saline (PBS, Sigma, St. Louis, MO, USA) solution (pH 7.4) and immersed into a glutaraldehyde solution (2.5%) for 24 h. After that, substrates were dehydrated by immersing into a series of ethanol solutions (25%, 50%, 75%, 95%, and 100%). The attached platelets were characterized by field emission SEM (FE-SEM).

2.5. Characterizations

XPS was carried out using a PHI Quantera II (ULVAC-PHI, Inc., Chigasaki, Japan) with an Al Kα X-ray source and ultrahigh vacuum (~10^{-10} mbar). The thickness of the organic layers on solid substrates was measured using a spectroscopic ellipsometer (Elli-SE, Ellipso Technology, Suwon, Korea). Static water contact angle measurements were carried out using a Phoenix-300 TOUCH goniometer (Surface Electro Optics Co., Ltd., Suwon, Korea). UV–Vis spectra of products (0.25 mg/mL) were obtained using a UV–Vis spectrophotometer (Libra S70, Biochrom, UK). Fourier transform infrared (FT-IR) spectra were acquired using an ALPHA FT-IR spectrometer (Bruker, Germany). FE-SEM imaging was performed by an Ultra Plus microscope (Zeiss, Germany) with an accelerating voltage of 3 kV, after sputter-coating with platinum.

3. Results and Discussion

Dex-C was synthesized via the carbamate bond forming reaction (Figure 1a) [9]. The hydroxyl groups of Dex were activated by using carbonyldiimidazole (CDI) and reacted with dopamine, which is a catecholamine. In order to synthesize Dex-C with various catechol contents, the molar ratio of reactants was varied accordingly (see the Experimental section). The conjugation of Dex with dopamine was analyzed by UV–Vis and FT-IR spectroscopy. Specifically, UV–Vis spectra of products revealed that dopamine was successfully conjugated to Dex, as evidenced by the presence of a peak at 280 nm, corresponding to the catechol of dopamine (Figure 1b) [9]. The absorbance at 280 nm was also used to quantify the extent of catechol conjugation with the hydroxyl groups of Dex. The calibration curve was generated using known concentrations of five dopamine solutions (Figure S1, Supplementary Material). The catechol content in resulting polymers was calculated by comparing with a calibration curve, and the grafting densities of catechol to glucose (the repeating unit of Dex) were found to be 1.6, 3.4, 9.0, and 16.8 mol %. The final polymers were denoted as Dex-C$_{1.6}$, Dex-C$_{3.4}$, Dex-C$_{9.0}$, and Dex-C$_{16.8}$. The synthesis of Dex-C was further analyzed by FT-IR spectroscopy. Unlike the Dex, Dex-C showed new peaks including C=O stretching (1806 and 1740 cm^{-1}) and N–H bending (1510 cm^{-1}) (Figure 1c). With an increase in catechol content in Dex-C, an increase in intensities of these characteristic peaks was also observed. Overall, these results indicated that Dex-C with various catechol contents were successfully prepared by a simple chemical reaction.

Synthesized Dex-C were subsequently used for the surface coating of solid substrates. Ti/TiO$_2$ was chosen as a model substrate, because Ti-based substrates are widely used in biomedical devices where the control of unnecessary biofouling is crucial [16]. Moreover, it is practically advantageous to use Ti/TiO$_2$, since it is compatible with conventional surface characterization techniques. After a 24-h immersion of Ti/TiO$_2$ substrates into four types of Dex-C solutions, surfaces were characterized by XPS, spectroscopic ellipsometry, and contact angle goniometry. According to the XPS analysis, all Ti/TiO$_2$ surfaces coated by Dex-C (regardless of grafting density) showed C 1s, N 1s, O 1s, and Ti 2p peaks, of which C 1s, N 1s, and O 1s peaks originated from Dex-C (Figure 2). Quantitative analysis of the surface chemical composition of surfaces was also performed in order to assess the effect of catechol content on surface coating efficiency. It was anticipated that highly more efficient surface coatings would be attained when using the Dex-C with higher catechol content. This is associated with the positive effect of catechols on the performance of surface coatings [17].

Polymers **2017**, *9*, 376

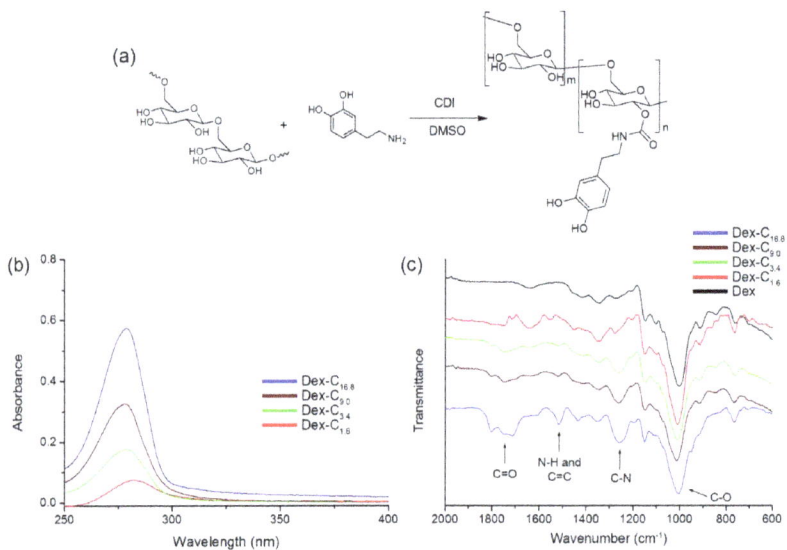

Figure 1. (**a**) Synthesis of catechol-conjugated dextran (Dex-C); (**b**) UV–Vis; and (**c**) Fourier transform infrared (FT-IR) spectra of Dex, Dex-C$_{1.6}$, Dex-C$_{3.4}$, Dex-C$_{9.0}$, and Dex-C$_{16.8}$.

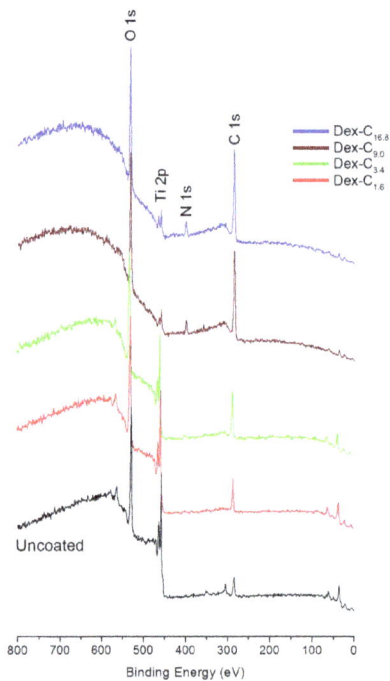

Figure 2. X-ray photoelectron spectra of uncoated, Dex-C$_{1.6}$, Dex-C$_{3.4}$, Dex-C$_{9.0}$, and Dex-C$_{16.8}$-coated Ti/TiO$_2$ surfaces.

As shown in Table 1, increased intensity of the C 1s and N 1s peaks was observed in the case of Dex-C. Decreased intensity of Ti 2p peaks was observed with increasing catechol content in the Dex-C being used in surface coating. Especially given that the intensity of peaks of the underlying substrate is sensitive to the amount of upper layers (i.e. Dex-C layer), the sequential decrease of Ti 2p peaks was direct evidence of enhanced coating efficiency when the catechol content in Dex-C was increased. The analysis of the areal ratio between O 1s and Ti 2p also gave us useful information to investigate the coating efficiency by Dex-C. The ratio (O 1s/Ti 2p) of uncoated Ti/TiO$_2$ surfaces was 2.34, and increased to 3.43, 4.67, 5.43, and 19.2 after Dex-C$_{1.6}$, Dex-C$_{3.4}$, Dex-C$_{9.0}$, and Dex-C$_{16.8}$ coatings, respectively. This implies that the contribution of Ti/TiO$_2$ surface for O 1s peak decreases by Dex-C coating, in which the coating efficiency is enhanced as the catechol content in Dex-C increases. Spectroscopic ellipsometry results also suggested that the surface coating efficiency is increased by increasing the catechol content; 0.9, 1.6, 3.8, and 5.3 nm-thick Dex-C layers were deposited on Ti/TiO$_2$ surfaces by Dex-C$_{1.6}$, Dex-C$_{3.4}$, Dex-C$_{9.0}$, and Dex-C$_{16.8}$ coatings, respectively (Table 1, Figure S2, Supplementary Material). The stability of the Dex-C coating on solid substrates is critical for practical applications. The mechanical stability of the Dex-C coating was examined by measuring the thickness change upon strong ultrasonication (40 kHz, 75 W). Dex-C$_{16.8}$-coated surfaces were used in this study, and the coating layer remained stable even after ultrasonication for 30 min, indicating the robustness of the Dex-C coating (Figure S3, Supplementary Material).

Table 1. Atomic composition (%) of uncoated, Dex-C$_{1.6}$, Dex-C$_{3.4}$, Dex-C$_{9.0}$, and Dex-C$_{16.8}$-coated Ti/TiO$_2$ surfaces. Thicknesses of Dex-C layers on Ti/TiO$_2$ surfaces after Dex-C$_{1.6}$, Dex-C$_{3.4}$, Dex-C$_{9.0}$, and Dex-C$_{16.8}$ coatings.

	C 1s	N 1s	O 1s	Ti 2p	O 1s/Ti 2p	Thickness (nm)
Uncoated	28.6	1.2	49.2	21.0	2.34	-
Dex-C$_{1.6}$	31.4	2.6	51.1	14.9	3.43	0.9
Dex-C$_{3.4}$	40.0	2.1	47.7	10.2	4.67	1.6
Dex-C$_{9.0}$	41.8	2.3	47.2	8.7	5.43	3.8
Dex-C$_{16.8}$	57.5	4.0	36.6	1.9	19.2	5.3

Prior to the use of Dex-C-coated Ti/TiO$_2$ surfaces in potential anti-fouling applications, changes in surface wettability were assessed, as this is highly related to the overall anti-fouling performance [18]. Changes in water contact angle of surfaces were comparable with results obtained by XPS and ellipsometry. The Dex-C-coated Ti/TiO$_2$ surfaces became more hydrophilic than uncoated Ti/TiO$_2$; Dex-C$_{16.8}$-coated surfaces exhibited the highest hydrophilicity (Figure 3). This suggests that the increase in coating thickness by Dex-C$_{16.8}$ provided complete coverage of Ti/TiO$_2$ surfaces, resulting in enhanced wettability. After confirming that surface wettability of Ti/TiO$_2$ substrates can be tailored by using different Dex-C in surface coatings, anti-fouling assays were conducted.

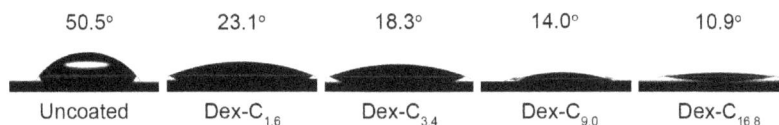

50.5°   23.1°   18.3°   14.0°   10.9°

Uncoated   Dex-C$_{1.6}$ Dex-C$_{3.4}$ Dex-C$_{9.0}$ Dex-C$_{16.8}$

Figure 3. Water contact angle images of uncoated, Dex-C$_{1.6}$, Dex-C$_{3.4}$, Dex-C$_{9.0}$, and Dex-C$_{16.8}$-coated Ti/TiO$_2$ surfaces.

Platelets were used as a model foulant, given their adverse effects on the use of blood-contacting medical implants. Given that dextran can inhibit cell adhesion on surfaces, and the fact that the antifouling performance is strongly related with surface hydrophilicity, the Dex-C$_{16.8}$ coating was expected to show superior antiplatelet properties. Platelets were seeded onto each sample and incubated for 24 h. Fouling behavior of platelets on surfaces was investigated by scanning electron microscopy (SEM). As shown in Figure 4, fouling behavior of platelets on Dex-C-coated surfaces significantly varied depending on the catechol content in Dex-C. The attached platelet density was calculated to be 1446, 997, 650, and 45 cells/image on the Dex-$C_{1.6}$-, Dex-$C_{3.4}$-, Dex-$C_{9.0}$-, and Dex-$C_{16.8}$-coated Ti/TiO$_2$ surfaces, respectively, whereas 1399 cells were attached to the uncoated Ti/TiO$_2$ surfaces. Morphological changes (spreading and pseudopodia emission) that indicate the activation of the platelets were also observed in the SEM analysis (Figure S4, Supplementary Material). Overall, Dex-C coatings conferred anti-fouling properties to the surfaces, with the only exception being Dex-$C_{1.6}$. The platelet adhesion was reduced by 28.7%, 53.5%, and 96.8% after surface coating with Dex-$C_{3.4}$, Dex-$C_{9.0}$, and Dex-$C_{16.8}$, respectively. The results suggest that both coating thickness and anti-fouling performance can be enhanced by increasing the catechol content in Dex-C. Analyzing with the results from other groups, the antiplatelet performance of the Dex-$C_{16.8}$ coating was comparable to that achieved with PEG coatings, but was weaker than that achieved with zwitterionic polymer coatings; it was reported that sulfobetaine and carboxybetaine polymer coatings reduce platelet adhesion by ~98%, whereas PEG coatings reduce the adhesion by ~95% [19,20]. Although the interaction between Dex-C and platelet is not fully understood, the anti-fouling property can be attributed to the formation of hydration layers interrupting direct contact of platelets with the surfaces [18]. Given that catechols play an important role in immobilizing polymers on surfaces as well as in intermolecular crosslinking of polymers [21,22], increase of catechol content in Dex-C can enable thicker Dex-C coating and the application of denser hydration layers on surfaces. Therefore, the Dex-$C_{16.8}$ coating exhibited the highest anti-fouling performance.

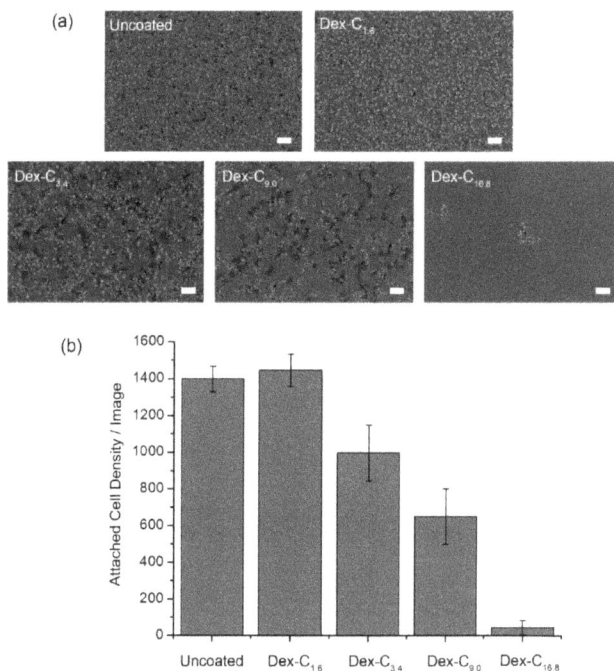

Figure 4. (**a**) SEM images and (**b**) quantification of platelets attached to uncoated, Dex-$C_{1.6}$, Dex-$C_{3.4}$, Dex-$C_{9.0}$, and Dex-$C_{16.8}$-coated Ti/TiO$_2$ surfaces. All scale bars are 10 μm. Each point indicates the mean from 15 counts from three replicate samples, and the error bars display 95% confidence limits.

Polymers **2017**, *9*, 376

4. Conclusions

In summary, catechol-conjugated dextrans (Dex-C) with different catechol content were synthesized in order to investigate the effect of catechol content on surface coating efficiency. Surface coating was carried out by immersing solid substrates into solutions of Dex-C with different grafting densities (1.6, 3.4, 9.0, and 16.8 mol %). After careful assessment of Dex-C-coated surfaces, the platelet adhesion behavior on surfaces was investigated. Increasing the catechol content in Dex-C resulted in enhanced surface coating and antiplatelet properties. In this study, the optimum grafting density of catechol to glucose was 16.8 mol %. It is hence concluded that optimization of the chemical composition of Dex-C is an area of research that merits attention, given that precise control of surface coatings can be achieved.

Supplementary Materials: Supplementary Materials are available online at www.mdpi.com/2073-4360/9/8/376/s1, Figure S1: The calibration curve of dopamine, Figure S2: Spectroscopic ellipsometry data of organic layers on Ti/TiO$_2$ surfaces after Dex-C$_{1.6}$, Dex-C$_{3.4}$, Dex-C$_{9.0}$, and Dex-C$_{16.8}$ coatings (solid line: model, dotted line: experimental data), Figure S3: Thicknesses of remaining Dex-C$_{16.8}$ on Ti/TiO$_2$ surfaces after sonication for 5, 10, and 30 min. The thickness of Dex-C$_{16.8}$ on Ti/TiO$_2$ surfaces before sonication was taken as 1. Error bars represent the standard deviation, Figure S4: Magnified SEM images of platelets attached to uncoated, Dex-C$_{1.6}$, Dex-C$_{3.4}$, Dex-C$_{9.0}$, and Dex-C$_{16.8}$-coated Ti/TiO$_2$ surfaces. All scale bars are 2 μm.

Acknowledgments: This research was supported by the Basic Science Research Program through the National Research Foundation of Korea (NRF) funded by the Ministry of Science and ICT (NRF- 2016R1C1B2008034).

Author Contributions: Yeonwoo Jeong, Kwang-A Kim, and Sung Min Kang contributed to the experiments and discussed the data. All authors wrote the manuscript and approved the final version of the manuscript.

Conflicts of Interest: The authors declare no conflict of interest.

References

1. De Belder, A.N. Medical applications of dextran and its derivatives. In *Polysaccharides in Medicinal Applications*; Dumitriu, S., Ed.; Marcel Dekker: New York, NY, USA, 1996; pp. 505–524.
2. Österberg, E.; Bergström, K.; Holmberg, K.; Riggs, J.A.; Van Alstine, J.M.; Schuman, T.P.; Burns, N.L.; Harris, J.M. Comparison of polysaccharide and poly (ethylene glycol) coatings for reduction of protein adsorption on polystyrene surfaces. *Colloids Surf. A* **1993**, *77*, 159–169. [CrossRef]
3. Marchant, R.E.; Yuan, S.; Szakalas-Gratzl, G. Interactions of plasma proteins with a novel polysaccharide surfactant physisorbed to polyethylene. *J. Biomater. Sci. Polym. Ed.* **1995**, *6*, 549–564. [CrossRef]
4. Österberg, E.; Bergström, K.; Holmberg, K.; Schuman, T.P.; Riggs, J.A.; Burns, N.L.; Van Alstine, J.M.; Harris, J.M. Protein-rejecting ability of surface-bound dextran in end-on and side-on configurations: Comparison to PEG. *J. Biomed. Mater. Res.* **1995**, *29*, 741–747. [CrossRef] [PubMed]
5. Massia, S.P.; Stark, J.; Letbetter, D.S. Surface-immobilized dextran limits cell adhesion and spreading. *Biomaterials* **2000**, *21*, 2253–2261. [CrossRef]
6. De Sousa Delgado, A.; Léonard, M.; Dellacherie, E. Surface properties of polystyrene nanoparticles coated with dextrans and dextran-PEO copolymers. Effect of polymer architecture on protein adsorption. *Langmuir* **2001**, *17*, 4386–4391. [CrossRef]
7. Martwiset, S.; Koh, A.E.; Chen, W. Nonfouling characteristics of dextran-containing surfaces. *Langmuir* **2006**, *22*, 8192–8196. [CrossRef] [PubMed]
8. Kozak, D.; Chen, A.; Bax, J.; Trau, M. Protein resistance of dextran and dextran-poly (ethylene glycol) copolymer films. *Biofouling* **2011**, *27*, 497–503. [CrossRef] [PubMed]
9. Park, J.Y.; Yeom, J.; Kim, J.S.; Lee, M.; Lee, H.; Nam, Y.S. Cell-repellant dextran coatings of porous titania using bio-inspired chemistry. *Macromol. Biosci.* **2013**, *13*, 1511–1519. [CrossRef] [PubMed]
10. Park, J.Y.; Kim, J.S.; Nam, Y.S. Mussel-inspired modification of dextran for protein-resistant coatings of titanium oxide. *Carbohydr. Polym.* **2013**, *97*, 753–757. [CrossRef] [PubMed]
11. Liu, Y.; Chang, C.P.; Sun, T. Dopamine-assisted deposition of dextran for nonfouling applications. *Langmuir* **2014**, *30*, 3118–3126. [CrossRef] [PubMed]
12. Lee, B.P.; Messersmith, P.B.; Israelachvili, J.N.; Waite, J.H. Mussel-inspired adhesives and coatings. *Annu. Rev. Mater. Res.* **2011**, *41*, 99–132. [CrossRef] [PubMed]

13. Ye, Q.; Zhou, F.; Liu, W. Bioinspired catecholic chemistry for surface modification. *Chem. Soc. Rev.* **2011**, *40*, 4244–4258. [CrossRef] [PubMed]

14. Liu, Y.; Ai, K.; Lu, L. Polydopamine and its derivative materials: Synthesis and promising applications in energy, environmental, and biomedical fields. *Chem. Rev.* **2014**, *114*, 5057–5115. [CrossRef] [PubMed]

15. Braune, S.; Fröhlich, G.M.; Lendlein, A.; Jung, F. Effect of temperature on platelet adherence. *Clin. Hemorheol. Microcirc.* **2015**, *61*, 681–688. [CrossRef] [PubMed]

16. Kang, S.M.; Kong, B.; Oh, E.; Choi, J.S.; Choi, I.S. Osteoconductive conjugation of bone morphogenetic protein-2 onto titanium/titanium oxide surfaces coated with non-biofouling poly (poly (ethylene glycol) methacrylate). *Colloids Surf. B* **2010**, *75*, 385–389. [CrossRef] [PubMed]

17. Lee, H.; Scherer, N.F.; Messersmith, P.B. Single molecule mechanics of mussel adhesion. *Proc. Natl. Acad. Sci. USA* **2006**, *103*, 12999–13003. [CrossRef] [PubMed]

18. Cho, W.K.; Kang, S.M.; Lee, J.K. Non-biofouling polymeric thin films on solid substrates. *J. Nanosci. Nanotechnol.* **2014**, *14*, 1231–1252. [CrossRef] [PubMed]

19. Zhang, Z.; Zhang, M.; Chen, S.; Horbett, T.A.; Ratner, B.D.; Jiang, S. Blood compatibility of surfaces with superlow protein adsorption. *Biomaterials* **2008**, *29*, 4285–4291. [CrossRef] [PubMed]

20. Amoako, K.A.; Sundaram, H.S.; Suhaib, A.; Jiang, S.; Cook, K.E. Multimodal, biomaterial-focused anticoagulation via superlow fouling zwitterionic functional groups coupled with anti-platelet nitric oxide release. *Adv. Mater. Interfaces* **2016**, *3*, 1500646. [CrossRef]

21. Lee, H.; Dellatore, S.M.; Miller, W.M.; Messersmith, P.B. Mussel-inspired surface chemistry for multifunctional coatings. *Science* **2007**, *318*, 426–430. [CrossRef] [PubMed]

22. Kang, S.M.; Hwang, N.S.; Yeom, J.; Park, S.Y.; Messersmith, P.B.; Choi, I.S.; Langer, R.; Anderson, D.G.; Lee, H. One-step multipurpose surface functionalization by adhesive catecholamine. *Adv. Funct. Mater.* **2012**, *22*, 2949–2955. [CrossRef] [PubMed]

polymers

MDPI

Article

Mechanical Properties of Composite Hydrogels of Alginate and Cellulose Nanofibrils

Olav Aarstad [1], Ellinor Bævre Heggset [2], Ina Sander Pedersen [1], Sindre Hove Bjørnøy [3], Kristin Syverud [2,4] and Berit Løkensgard Strand [1,*]

[1]   NOBIPOL, Department of Biotechnology and Food Sciences, NTNU Norwegian University of Science and Technology, NO-7491 Trondheim, Norway; olav.a.aarstad@ntnu.no (O.A.); inasande@stud.ntnu.no (I.S.P.)

[2]   RISE PFI, Nanocellulose and carbohydrate polymers, Høgskoleringen 6b, 7491 Trondheim, Norway; ellinor.heggset@rise-pfi.no (E.B.H.); kristin.syverud@rise-pfi.no (K.S.)

[3]   Department of Physics, NTNU Norwegian University of Science and Technology, NO-7491 Trondheim, Norway; sindre.bjornoy@ntnu.no

[4]   Department of Chemical Engineering, NTNU Norwegian University of Science and Technology, NO-7491 Trondheim, Norway

*   Correspondence: berit.l.strand@ntnu.no; Tel.: +47-7341-2243

Received: 27 June 2017; Accepted: 17 August 2017; Published: 19 August 2017

Abstract: Alginate and cellulose nanofibrils (CNF) are attractive materials for tissue engineering and regenerative medicine. CNF gels are generally weaker and more brittle than alginate gels, while alginate gels are elastic and have high rupture strength. Alginate properties depend on their guluronan and mannuronan content and their sequence pattern and molecular weight. Likewise, CNF exists in various qualities with properties depending on, e.g., morphology and charge density. In this study combinations of three types of alginate with different composition and two types of CNF with different charge and degree of fibrillation have been studied. Assessments of the composite gels revealed that attractive properties like high rupture strength, high compressibility, high gel rigidity at small deformations (Young's modulus), and low syneresis was obtained compared to the pure gels. The effects varied with relative amounts of CNF and alginate, alginate type, and CNF quality. The largest effects were obtained by combining oxidized CNF with the alginates. Hence, by combining the two biopolymers in composite gels, it is possible to tune the rupture strength, Young's modulus, syneresis, as well as stability in physiological saline solution, which are all important properties for the use as scaffolds in tissue engineering.

Keywords: alginate; TEMPO; cellulose nanofibrils; nanocellulose; composite; hydrogels; mechanical properties

---

1. Introduction

Recently, composite materials of alginate and cellulose nanofibrils (CNF) have shown promising results for bioprinting and tissue engineering applications [1–3]. In particular, the shear thinning properties of CNF combined with the viscous alginate that form hydrogels with divalent cations at physiological conditions, are attractive for bioprinting [1].

Alginates are linear copolymers of $1 \rightarrow 4$ linked β-D-mannuronic acid (M) and α-L-guluronic acid (G). The monomers are arranged in a block-wise pattern along the chain with homopolymeric regions of M and G termed M- and G-blocks, respectively, interspaced with regions of alternating structure (MG-blocks) [4]. Alginate forms hydrogels by crosslinking with divalent cations where, particularly, the G-blocks, but also the MG-blocks, are important for the mechanical properties of the resulting gel [5,6]. Alginate hydrogels can be produced under physiological conditions [7,8]. This, together with a low immunogenic profile [9], makes them popular materials for biomedical applications and, in particular, in tissue engineering [10]. Cellulose provides structural support in plant cell walls and can be processed as fibres or nanoscaled fibrils of cellulose, known as cellulose nanofibrils (CNF). Several procedures for preparation of CNF from cellulose pulp exist, among them 2,2,6,6-tetramethylpiperidine-1-oxyl radical (TEMPO)-mediated oxidation using sodium hypochlorite as oxidant [11]. By this, aldehyde and carboxyl groups are introduced on the surfaces of the cellulose fibrils and increase the negative charge of the fibrils. Aqueous dispersions of CNF have gel like character at low concentrations (approx. 0.5%) held together by fibril entanglement and hydrogen bonds. CNF hydrogels can be produced with cations, both monovalent and with higher valency [12], where gels with higher valency ions form stronger gels [12,13].

Due to their availability, renewability, biocompatibility, and low toxicity, both alginate and cellulose are attractive materials for a range of applications such as films, gels, and as viscosifiers. Although being suggested for tissue engineering applications, not much is known about the mechanical properties of composite gels of alginate and CNF. Composite gels of alginate and either oxidized CNC (cellulose nanocrystals) or CNF have shown increased compression strength, suggested to be the result of Ca^{2+} mediated crosslinking of the two components [14]. TEMPO-oxidized BNC (bacterial nanocellulose) have been used to improve the mechanical and chemical stability of an alginate hydrogel, and the composite gel was used for encapsulation of fibroblasts. The nanofibrous structure of the BNC was suggested to mimic the fibre structures of collagen and fibronectin found in the extracellular matrix [15–17], and fibroblast viability and proliferation was found to be higher in comparison to pure alginate gels [18]. For chondrocytes, the combination of sulfated alginate with nanocellulose were promising regarding cell viability and phenotype [2]. It has previously been shown that Young's modulus of the matrix is important for the development of stem cells into differentiated cell types/tissues [19]. Hence, the structure, chemistry, as well as the mechanical properties are of importance in tissue engineering applications. Although composite gels of CNF and alginate have been demonstrated in the literature, a systematic study of mechanical properties on combinations of different types of alginate and CNF has not yet been reported.

We hypothesize that by combining alginate and CNF, the advantageous properties of the individual constituents, i.e., the stiffness of CNF and the compressibility of alginate, could be preserved in the composite gel making it possible to tailor the mechanical properties.

2. Materials and Methods

2.1. Materials

Alginates: Alginates extracted from *Durvillea potatorum* and *Laminaria hyperborea* stipe were obtained from FMC Health and Nutrition (Sandvika, Norway), and *Macrocystis pyrifera* alginate were purchased from Sigma-Aldrich (St. Louis, MO, USA). Molecular weight and NMR parameters are given Table 1. For a detailed study on alginate fine structure see [20].

Table 1. M_w and sequence parameters in alginates used in this study [20] [1].

Alginate	M_w (kDa)	PI	F_G	F_M	F_{GG}	F_{GM}	F_{MM}	F_{MGM}	F_{GGG}	$N_{G>1}$
D. potatorum	163	1.76	0.32	0.68	0.20	0.12	0.56	0.07	0.16	6
M. pyrifera	177	1.94	0.41	0.59	0.21	0.20	0.40	0.18	0.17	5
L. hyperborea	200	2.23	0.67	0.33	0.56	0.11	0.23	0.08	0.52	13

[1] Molecular weight (M_w) and polydispersity index (PI = M_w/M_n) were determined from SEC-MALLS. Sequence parameters were calculated from ^{1}H-NMR spectra. F_G and F_M denotes the fraction of guluronic and mannuronic acid, respectively. Fractions of dimers and trimers of varying composition are denoted by two and three letters, respectively.

Nanocellulose: Never-dried bleached kraft softwood pulp fibres were used as the source material for production of two types of nanocellulose. The first type was produced using a mechanical pretreatment (beating in a Claflin mill; 1000 kWh/ton for 1 h) followed by homogenization (denoted as mechanically-fibrillated CNF). Production of the second type (referred to as oxidized CNF) was performed using TEMPO-mediated oxidation as pretreatment, as described by Saito and colleagues [11]. NaClO (2.3 mmol per gram of cellulose) was used in the oxidation. The fibrillation was done by using a Rannie 15 type 12.56x homogenizer (APV, SPX Flow Technology, Silkeborg, Denmark) with a pressure drop of 1000 bar in each pass. Mechanically-fibrillated CNF was homogenized using five passes, the oxidized CNF using 1 pass. The concentration of the pretreated samples was 1% (*w*/*v*) before fibrillation.

The composition of carbohydrates was determined according to the standard method NREL/TP-510-42618, using sulphuric acid hydrolysis. The composition of carbohydrate monomers produced during the hydrolysis was analysed using high-performance anion-exchange chromatography with pulsed amperometric detection (HPAEC-PAD; Dionex ICS-5000, (Thermo Fisher Scientific, Waltham, MA, USA) with a CarboPac PA1 column (Thermo Fisher Scientific, Waltham, MA, USA) and gradient elution using sodium hydroxide for elution. The carbohydrate composition of the two CNF samples is shown in Appendix A.

The intrinsic viscosity was determined as described in ISO-standard 5351:2010, using cupriethylenediamine (CED) as a solvent. The oxidized CNF was subjected to a second oxidation before the analysis. This is due to a significant amount of aldehyde groups in the oxidized samples which induce depolymerization through a β-elimination reaction [21]. In order to avoid this, a selective oxidation of the aldehyde groups was performed using sodium chlorite under acidic conditions.

The carboxylate content was determined using conductometric titration as previously described [22,23]. The equipment used was a 902 Titrando, an 856 conductivity module and Tiamo software (Metrohm). Residual fibres were quantified with a FibreTester device, as previously described by Chinga-Carrasco and colleagues [24]. The fibrillated structure and analyses of the fibre structure from atomic force microscopy (AFM) imaging with resulting characteristics of the cellulose nanofibrils have previously been described [25]. A summary of nanocellulose characteristics is shown in Appendix A.

Control polymers: Hyaluronic acid (M_w = 8.7 × 10^5 Da) was a gift from Novamatrix (Norway), xanthan (M_w = 9.6 × 10^5 Da) was a gift from CP Kelco (USA), and dextran (M_w = 2.0 × 10^6 Da) was purchased from Pharmacia AB (Stockholm, Sweden). The molecular weight of the samples were determined using a HPLC system consisting of serially connected TSK G6000 PWxl and G5000PWxl columns (Tosoh Bioscience LLC, Tokyo, Japan) followed by a Dawn Helios II MALLS photometer and an Optilab T-rEX differential refractometer (Wyatt Technology, Santa Barbara, CA, USA). The elution buffer (0.15 M NaNO$_3$ with 1 mM EDTA, pH 6.0) was applied at a flow rate of 0.5 mL/min. Data were collected and processed using ASTRA 6.1 software. Refractive index increment values (dn/dc) used in the calculations were set to 0.148 for dextran and 0.150 for alginate, hyaluronic acid and xanthan.

2.2. Preparation of Gel Cylinders

Alginate gels were prepared by internal gelling with CaCO$_3$ and D-glucono δ-lactone (GDL) as previously described [26]. Briefly, CaCO$_3$ (56.25 mg, particle size 4 μm) suspended in 5 mL MQ water was added to 25 mL, 1.5% (*w*/*v*) of alginate solution. The mixture was degassed under vacuum suction before addition of 7.5 mL of a freshly made solution of GDL (200.36 mg). The mixture was immediately poured into 24-well tissue culture plates (16/18 mm, Costar, Cambridge, MA, USA) and left at room temperature for minimum 20 h. The gel cylinders were removed from the tissue culture plates and weight was measured on the calcium unsaturated gels before being immersed in a solution of 50 mM CaCl$_2$ and 200 mM NaCl (100 mL per gel) and left for 24 h at 4 °C for calcium saturation before syneresis and compression measurement.

Two-component gels of alginate and either xanthan, dextran or hyaluronic acid were made by a similar procedure as for pure alginate gels. The two components were dissolved together in 25 mL MQ water, yielding a final concentration in the gelling solution of 1% (*w*/*v*) *M. pyrifera* alginate and 0.30% or 0.75% (*w*/*v*) of the other component.

Alginate—CNF composite gels were made by mixing 450 mg alginate powder with MQ water and CNF dispersion corresponding to final concentrations of 1% (w/v) alginate and 0.15–0.75% (w/v) CNF and a total volume of 30 mL. The blends were dissolved for at least 18 h under constant stirring and homogenized for 5 min with a Vdi 12 Ultra Turrax at 11,500 rpm. Twenty-five grams of the blend was weighed out and added $CaCO_3$ and GDL as described above.

2.3. Syneresis and Gel Strength Measurements

Syneresis was determined after calcium saturation as $(W_0 - W)/W_0 \times 100$, where W_0 and W are the initial weights of the samples before calcium saturation and final weights of the calcium saturated gels, respectively. Force/deformation curves was recorded at 22 °C on a Stable Micro Systems TA-XT2 texture analyser equipped with a P/35 probe and at a compression rate of 0.1 mm/s [26]. The gels were subjected to uniaxial compression surpassing the point of rupture, where force and deformation was recorded. Young's modulus was calculated as $G \cdot (h/A)$ where G is the initial slope (N/m) of the stress deformation curve and h and A is the height and area of the gel cylinder, respectively. In order to compare Young's modulus of gels with different degree of syneresis, the relationship $E = k \cdot c^2$ was applied [27] where k is characteristic for each type of alginate, and c is the weight concentration.

2.4. Volume Stability and Calcium Binding

The weight and calcium content of pure and composite gels of *M. pyr.* alginate and oxidized CNF were measured before and after saline treatments. Gel cylinders immersed in a solution of 50 mM $CaCl_2$ and 200 mM NaCl for 24 h were cut in three approximately equal pieces, weighed and immersed separately in 40 mL of 150 mM NaCl solution at room temperature for 24 h. The gel pieces were gently removed from the solution, weighed and subjected to a second saline treatment. After weighing, the gels were added 0.5 mL of 100 mM Na-EDTA, pH = 7.0. All samples were vortexed and centrifuged (2400 rpm, 5 min) before they were diluted $100 \times$ in 5% HNO_3 and calcium content analysed with ICP-MS (type 8800 Triple Quadrupole ICP-MS with ASX-520 Autosampler from Agilent Technologies, Santa Clara, CA, USA). As internal standards [115]In and [89]Y were used.

2.5. Light Microscopy

Homogeneity of the gels were assessed on a Nikon Eclipse TS100 microscope (Nikon Instruments, Tokyo, Japan) by placing a drop of gelling solution between two microscopy slides and observing the gellation in the microscope using $40 \times$ magnification.

2.6. Preparation of Aerogels for Scanning Electron Microscopy (SEM)

After the gels had been cast, the cylinders where cut, in the hydrated state, into 200 μm thick sections using a vibrating blade microtome (VT1000S, Leica Biosystems, Nussloch GmBH, Wetzlar, Germany). The sections where dehydrated in increasing concentrations of ethanol, before substitution with acetone and finally critical-point dried using liquid CO_2 (Emitech K850 critical point dryer, Quorum Technologies, Lewes, UK). The sections were mounted on aluminum stubs using carbon tape and sputter coated (208 HR, Cressington, Watford, UK) with a 5 nm layer of Pt/Pd (80/20). SEM analysis was performed with an acceleration voltage of 5 kV (S-5500 S(T)EM, Hitachi, Tokyo, Japan).

2.7. Fourier Transform Infrared Spectroscopy (FTIR) Characterization

The infrared spectra of vacuum dried samples were collected on a Bio-Rad Excalibur series FTS 3000 spectrophotometer (Bio-Rad, Hercules, CA, USA). The spectra were acquired in transmission mode on the films at a spectral range of 4000−500 cm^{-1}.

2.8. Statistical Analysis

All values are expressed as means ± standard deviations. Comparisons between groups were made using a two-sided Student's t-test and Microsoft Excel worksheet (2011). Statistical outcomes (*p*-values) are presented in Appendix B.

3. Results and Discussion

3.1. Composite Hydrogels of Alginate and CNF

Composite gels of *D. pot.*, *M. pyr.*, and *L. hyp.* alginates and CNF (mechanically-fibrillated or oxidized) saturated with calcium were measured with respect to gel rigidity (Young's modulus) and volume reduction upon gel formation (syneresis, Figure 1). Secondly, resistance to breakage and deformation at breakage was determined (Figure 2).

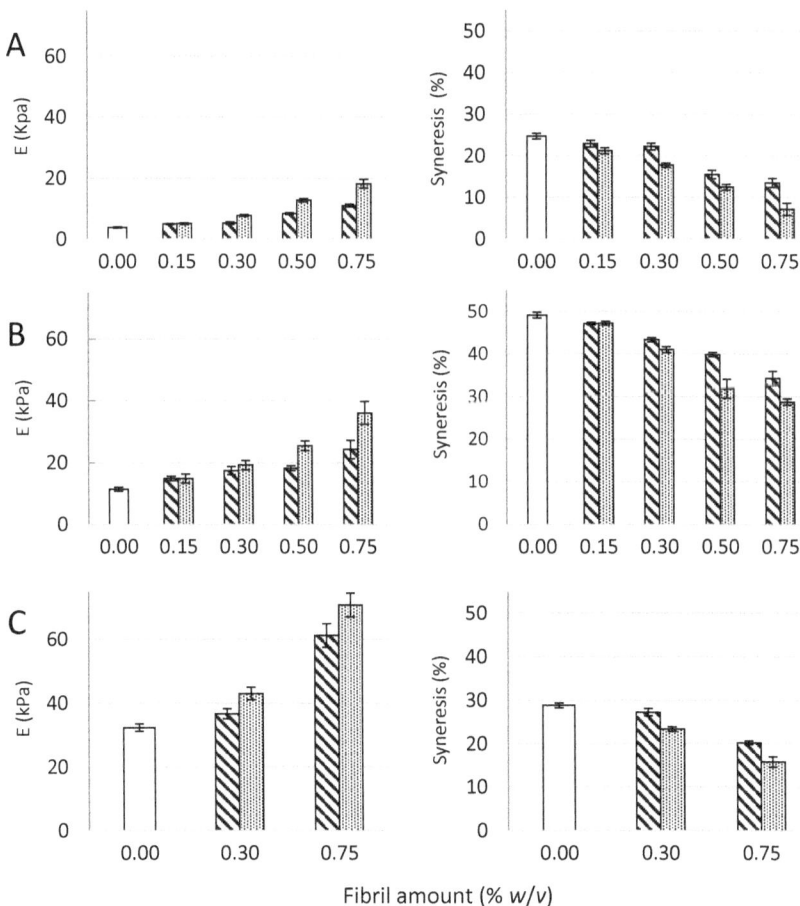

Figure 1. Young's modulus (E, left panel) and syneresis (right panel) of alginate-CNF composite Ca-gels made from 1% (*w/v*) *D. potatorum* (**A**), *M. pyrifera* (**B**), and *L. hyperborea* (**C**) alginates and 0–0.75% (*w/v*) mechanically-fibrillated (◣) and oxidized CNF (▦). Bars are means ± standard deviations, shown for *n* = 5–8 gels. *P*-values are presented in Tables A3 and A4 in Appendix B.

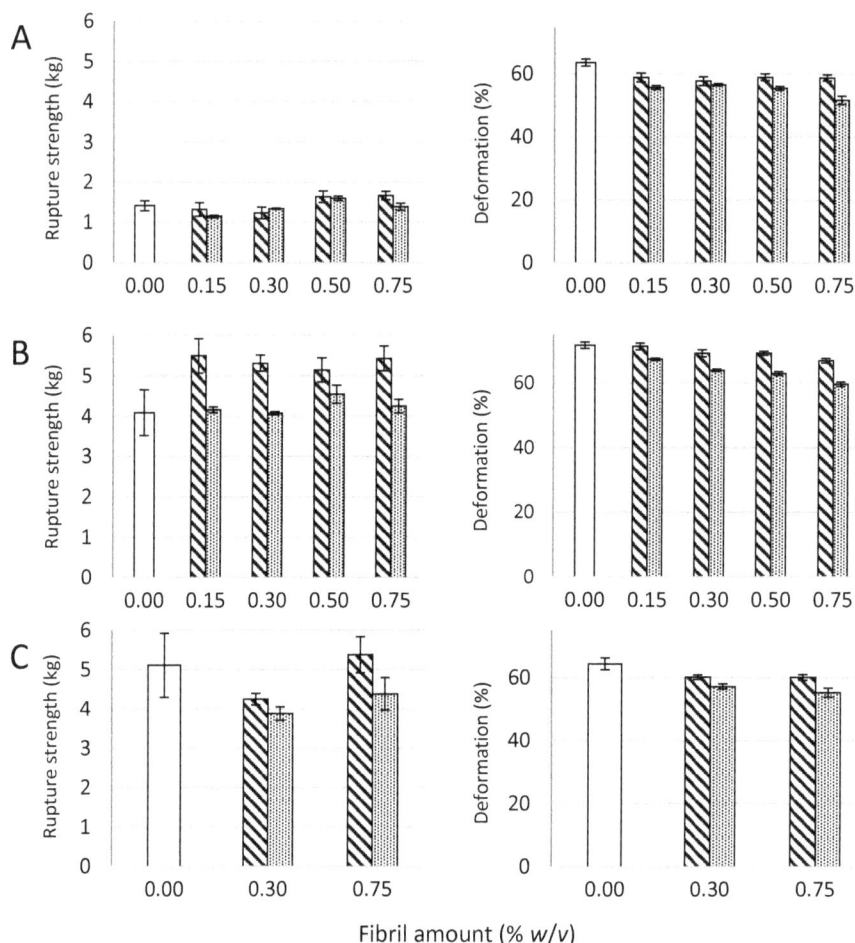

Figure 2. Rupture strength measured as the force at rupture (left panel) and compression at rupture (right panel) for *D. potatorum* (**A**), *M. pyrifera* (**B**), and *L. hyperborea* (**C**) Ca-alginate gels and mechanically-fibrillated (**◤**) and oxidized (▦) CNF. Bars are means +/− standard deviations, shown for *n* = 5–8 gels. *P*-values are presented in Tables A3 and A4 in Appendix B.

The composite gels of alginate and CNF showed increased Young's modulus and reduced syneresis compared to pure alginate gels. An increase in Young's modulus was observed for composite gels of both mechanically-fibrillated CNF and oxidized CNF relative to the alginate gels alone. The largest effect was seen for composite gels of oxidized CNF and alginates with intermediate to low G-content (*M. pyr.* and *D. pot.*) with a 3–5 fold increase in Young's modulus for the gels containing 0.75% (w/v) oxidized CNF relative to no addition of CNF. Additionally, for the alginate with a high G-content (*L. hyp.* stipe) that is known to form stiff gels with calcium [27], the effect of addition of CNF on Young's modulus was profound with about two times increase for the highest concentration. Syneresis was reduced for the composite gels relative to the alginate gels, meaning that the gels shrunk less after saturation with calcium when CNF was added to the hydrogels. As seen from Figure 1, the alginate gels are highly syneretic when saturated with calcium. The syneresis is known to be linked to the content of G-blocks, but indeed also to MG-blocks in the alginate [5,26], hence, the high degree of syneresis for the alginate from *M. pyr.* is as expected. Regardless of the composition of the alginate, CNF reduced the syneresis with increasing effect at increasing concentrations. Although the reduction in syneresis was highest for the oxidized CNF—*M. pyr.* alginate composite gel (49% to 29%), the largest relative decrease was observed for *D. pot.* alginate where the syneresis was reduced from 25% to 7% with the addition of 0.75% (*w/v*) oxidized CNF.

No significant effect of the presence of fibrils was seen on the rupture strength of the gels, except for the increase in rupture strength for the gels of alginate from *M. pyr.* upon the addition of mechanically-fibrillated CNF (Figure 2). An overall trend was a slight reduction in the rupture strength when oxidized CNF was added to the alginate relative to the mechanically-fibrillated CNF. However, the alginate concentration is higher in the pure alginate gels than in the composite gels, due to the large differences in syneresis. The rupture strength per concentration unit alginate is, therefore, increased in the composite gels. Calcium alginate gels with a high content of G-blocks, such as the alginate from *L. hyp* stipe, are known to form strong gels due to the high numbers of crosslinks, as also reflected in the Young's modulus (Figure 1). Alginate from *M. pyr.* has a high content of MG-blocks and is known to withstand large deformation. This has been explained in terms of energy dissipating effects of collapsing MG block junction zones during deformation [28]. Surprisingly, the force at rupture increased by about 30% when mechanically-fibrillated CNF was added to this alginate gel, independently of the concentration of fibrils.

3.2. Composite Hydrogels of Oxidized CNF and Alginate

In order to better evaluate separate and combined effects of the two biopolymer components, calcium crosslinked gels with either oxidized CNF or alginate solely (0.75% (*w/v*)) were prepared and measured with respect to gel rigidity, syneresis, deformation, and force at rupture (Figure 3). Hydrogels of TEMPO-oxidized CNF have previously been produced with both monovalent and divalent ions [12]. Although the Young's modulus of alginate and oxidized CNF gels were similar in our study, large differences were seen for the syneresis and the rupture strength underlining the contribution of the two components to the composite system. Oxidized CNF gels displayed almost no syneresis in contrast to the highly syneretic gels from *M. pyr.* alginate. For the rupture strength, the alginate contributed with a high degree of compressibility as shown by the compression to 70% before rupture at about 4 kg force. The oxidized CNF, on the other hand, had a rupture strength of less than 0.1 kg. A dose-dependent increase in syneresis and rupture strength upon the addition of alginate to oxidized CNF was observed. The potential effect on the properties by increasing the dry weight content in the hydrogels were studied by a control containing half concentration of the two components and thus same dry matter content as in the systems of alginate and CNF alone. Additionally, here, a combined effect of the two components could be seen on the syneresis and rupture strength. A combination of 0.75% alginate with 0.75% oxidized CNF resulted in a more than four-fold increase in Young's modulus, exceeding a solely additive effect of the two separate systems. Representative force-deformation curves of composite gels of alginate and oxidized CNF is shown in Appendix D.

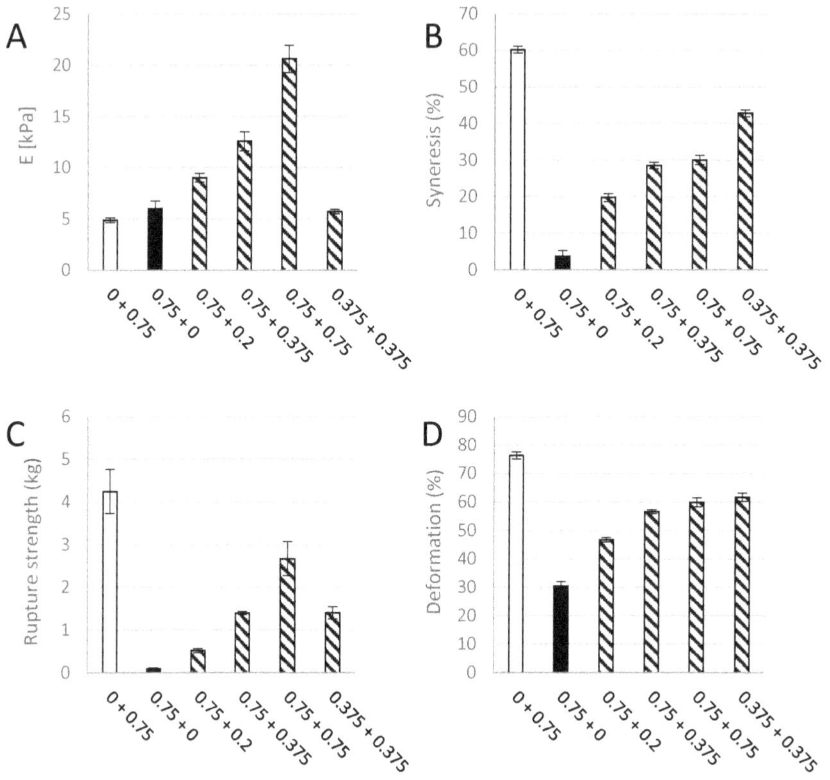

Figure 3. Young's modulus (E, (**A**)), syneresis (**B**), rupture strength (**C**) and deformation at rupture (**D**) of a pure 0.75% (*w/v*) *M. pyrifera* alginate gel (▢), a pure (0.75% *w/v*) oxidized CNF gel (■), and composite gels of 0.75% (*w/v*) of each of the two materials (◣). Mean +/− standard deviation shown for *n* = 5–8 gels.

3.3. Composite Hydrogels of Alginate and Other Polysaccharides

To study more specifically the effect of addition of the long cellulose nanofibrils to alginate, control experiments using other polysaccharides in composite alginate hydrogels were conducted. Polysaccharides with approximately the same molecular weight, but different charge density and flexibility were added to calcium-alginate hydrogels of *M. pyr.* alginate to study the effect on mechanical properties. Dextran, having an α-(1,6)-linked glucose backbone with varying degree of α-(Glc-1,3) branching, is a highly-flexible polysaccharide without charges. Sodium hyaluronate is a stiffer molecule, with alternating 4-linked β-D-glucuronic acid and 3-linked *N*-acetyl-β-D-glucosamine, with negative charges due to the guluronic acid, but with no specific calcium affinity. Xanthan consists of a (1,4)-β-D-Glc main chain where every second unit is (3,1) linked to a D-Man-(1,4)-β-D-GlcA-(1,2)-α-D-Man side chain with varying degree of acetyl and pyruvate groups on the internal and terminal mannose. Xanthan has the ability to form very stiff double-stranded structures. Finally, increasing concentrations of alginate were investigated to compare the effects relative to the effects of addition of CNF on Young's modulus, syneresis, and resistance to rupture (Figure 4).

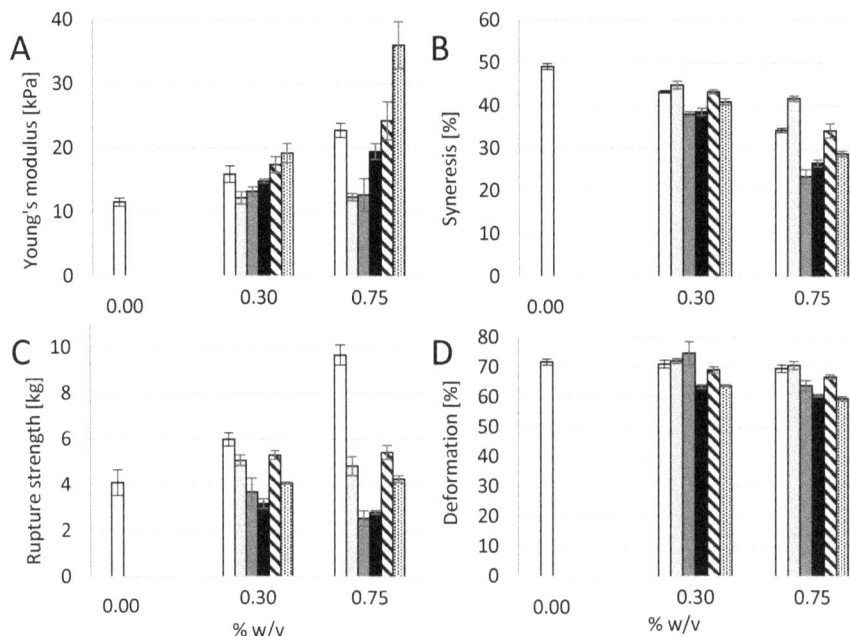

Figure 4. Young's modulus (E, (**A**)), syneresis (**B**), rupture strength (**C**), and deformation at rupture (**D**) of *M. pyrifera* alginate gels (1% *w/v*) with 0.30 and 0.75% (*w/v*) addition of *M. pyrifera* alginate (▢), dextran (▢), hyaluronic acid (▨), xanthan (▨), mechanically-fibrillated CNF (▨) and oxidized CNF (▨). Bars are means +/− standard deviations, shown for *n* = 5–9 gels. *P*-values are presented in Tables A5 and A6 in Appendix B.

Addition of dextran and sodium hyaluronate to the alginate did not affect the Young's modulus for any concentration showing that neither the increase in dry matter nor net charge by itself result in an increase in gel stiffness. A concentration-dependent increase in stiffness was seen for alginate from *M. pyr.* Although having a relatively high content of MG-blocks, the alginate from *M. pyr.* also contains a fraction of very long G-blocks [20]. Both G-blocks and MG-blocks are known to contribute to the crosslinking with calcium [5], hence, an increase in stiffness upon increase in concentration of alginate for the calcium saturated gels is as expected with increasing numbers of crosslinks per unit volume. For 0.75% (*w/v*) added polysaccharides, xanthan, and mechanically-fibrillated CNF gave about a doubling in the Young's modulus, compared to the pure alginate gel (1% (*w/v*)). Xanthan, which can be viewed as rigid rods at the length scale of a G-block junction zone, shows similarities with the addition of long G-blocks to alginate gels that results in stiffer gels with reduced syneresis [29]. The very long G-blocks have been suggested to act as reinforcement bars in the alginate hydrogel network. However, the largest increase was obtained by the addition of oxidized CNF with a 3.5 times increase in Young's modulus compared to the 1% alginate gel and also a 50% increase in Young's modulus compared to the 1.75% alginate gel for the composite gel of 1.0% alginate and 0.75% oxidized CNF. This emphasizes the combined effect of both stiffness and length of the fibres combined with the charge being important for the final stiffness of the composite gels.

In general, all the added biopolymers caused a concentration-dependent reduction in syneresis (Figure 4). Least effect was seen for the dextran, which caused slightly more syneresis than increasing the concentration of alginate. The syneresis was markedly reduced by adding xanthan and hyaluronic acid, compared to alginate alone (Figure 4). This suggests that osmotic pressure, rather than interference with G-block junction zones, affects the degree of syneresis for these composite gels. The addition of mechanically fibrillated CNF resulted in same syneresis as for the increased alginate concentrations, however, again the oxidized CNF caused less syneresis.

The rupture strength of the alginate gels was highly influenced by the alginate concentration (Figure 4). By increasing the concentration of alginate from 1.0% to 1.75% (w/v), the rupture strength increased from 4 kg to 9 kg. None of the other additives were near this increase of rupture strength. Again, there was a slight positive effect on the rupture strength by the addition of mechanically-fibrillated CNF, and no effect was seen on the addition of oxidized CNF although a slight reduction in deformation at rupture could be observed. Hence, ionic interactions between the alginate and cellulose may be weak, although significant, for the initial resistance against compression, however, they are not comparable to the alginate crosslinking when it comes to resisting the force at high deformations. Dextran had a slight positive effect on the rupture strength, however, both hyaluronan and xanthan caused a reduction in the rupture strength.

3.4. Composite Gel Morphology and Homogeneity

Investigation of the composite gels with microscopy revealed that the cellulose was homogeneously distributed in the alginate gel that appear as transparent in the light microscope (Figure 5). No phase separation was seen either by optical microscopy or scanning electron microscopy, which further indicates a good integration between the alginate and the cellulose fibrils. The oxidized CNF has a higher (negative) charge than the fibrils in the mechanically-fibrillated quality (Table A2 in Appendix A). Additionally, the cellulose fibrils differed in size as seen in Figure 5, Appendix A Table A2, and from [25]. Although the oxidized CNF sample contains a large fraction of nanosized fibrils (optically inactive in Figure 5), it contains a major fraction of residual fibres (Figure 5, Appendix A, Table A2). It is, thus, a potential for further improvements of the mechanical properties of the composite hydrogels by increasing the yield of the fibrillation. However, the oxidized fibrils are much thinner and more homogeneous than the mechanically-fibrillated fibrils. Previous studies on composite films based on Ca^{2+} crosslinked alginate and cellulose showed increasing mechanical properties by increasing the amount of the fibrils in the biocomposite and by decreasing the fibril size. However, the introduction of negative charges on the surface of the cellulose fibrils increased the mechanical crosslinking even further [30]. Hence, this suggests that the charges, more than the size of the fibrils, influence the mechanical properties of the composite material.

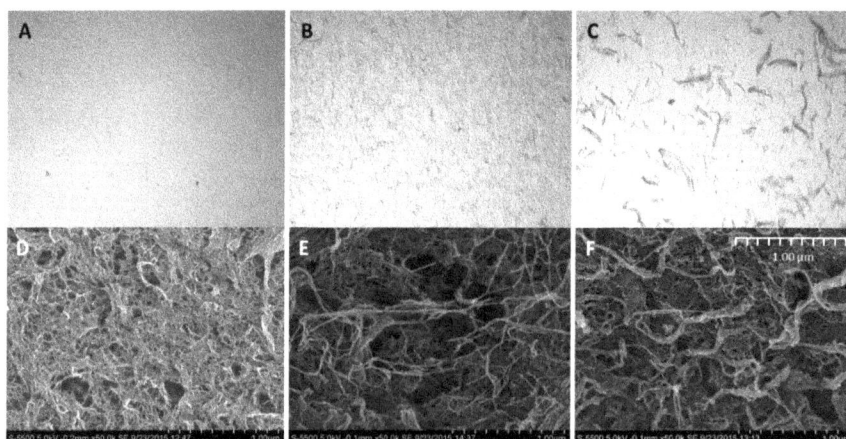

Figure 5. Microscopy of hydrogels of *M. pyrifera* alginate (1% (w/v)) (**A**), and composite gels with 0.30% mechanically fibrillated (**B**), and 0.30% oxidized CNF (**C**) at 40 × magnification. SEM images of supercritically-dried gels of the same composition (**D–F**).

In SEM, all the samples, including the alginate sample, show tread-like structures caused by the drying step which is necessary before SEM imaging (Figure 5). This is commonly seen for alginate hydrogels also when using supercritical drying, intended to preserve the original structure in the samples [31]. No formation of honeycomb structure, which is typically formed when freeze-drying CNF-samples, is observed [32]. A structural difference is observed upon the addition of CNF where some larger fibre structures was observed, both for the mechanically-produced and oxidized CNF. Additionally, here, for both types of cellulose fibrils, the fibrils seem to be well-integrated with the alginate with no indication of phase separation.

3.5. Volume Stability and Calcium Binding

The alginate and oxidized CNF gels were finally exposed to physiological saline solution (150 mM NaCl) to assess the volume stability and the binding of calcium in the gels (Figure 6). Surprisingly, the weak Ca-CNF gels (Figure 4) and all gels of alginate and oxidized CNF were stable during the exposure to NaCl solution. This is in contrast to the well-known swelling of Ca-alginate gels [33] recognized by the exchange of calcium ions with sodium ions and ingress of water into the gel. The stability in physiological solutions is important both for handling and in vitro studies of the biomaterial construct as well as for in vivo use. When the content of calcium was measured in the gels, an accumulation of calcium was found in the gels containing alginate for the Ca-saturated systems, with more calcium in the gels containing alginate solely than in the mixed system of same concentrations of alginate and oxidized CNF. Upon treatment with saline solution, increasing concentrations of calcium were found for increasing alginate concentrations. Similar calcium content was found in the gels containing alginate alone and alginate with oxidized CNF, pointing towards a more complex explanation of the increase in Young's modulus (Figure 1, Figure 3, and Figure 4) than calcium binding solely.

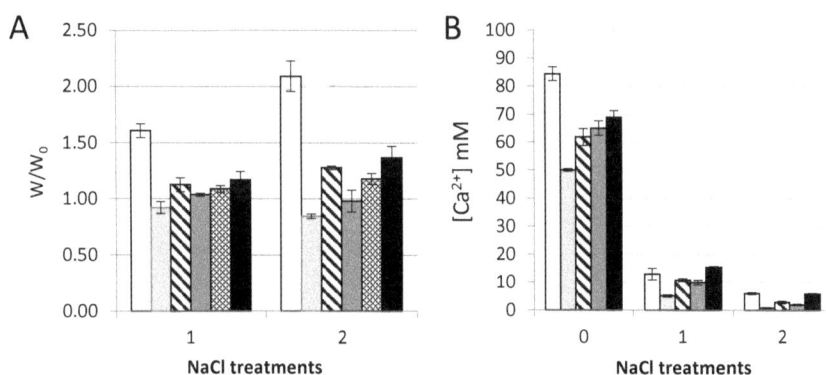

Figure 6. Calcium gels of *M. pyr.* alginate and oxidized CNF subjected to treatments with 0.9% (*w/v*) NaCl. Stability of gels (w/w$_0$, (**A**)) and calcium concentration (**B**) before and after dialysis against 150 mM NaCl solution. 0.75% (*w/v*) *M. pyr.* alginate (▢), 0.75% (*w/v*) oxidized CNF (▢), 0.375% (*w/v*) oxidized CNF + 0.375% (*w/v*) *M. pyr.* (◩), 0.75% (*w/v*) oxidized CNF + 0.2% (*w/v*) *M. pyr.* (▨), and 0.75% (*w/v*) oxidized CNF + 0.75% (*w/v*) *M. pyr.* (■). Mean ± standard deviation are shown for three gels.

3.6. Intermolecular Interaction Studied by FT-IR

FT-IR spectra of selected samples were obtained in an attempt to detect intermolecular interactions (Figure 7 and Appendix D). The spectra were normalized with respect to the characteristic $\nu(COO)_{asym}$ vibration band around 1600 cm^{-1}. This was done for better comparison of the wavenumber region where calcium-polymer interaction is most likely to be observed. Both calcium-saturated and

-unsaturated composite gels were studied to see if the calcium content influenced the carboxylic groups, as well as to compare our results to previous observations. Calcium-unsaturated gels contained only the calcium released from $CaCO_3(s)$ as a result of the hydrolysis of GDL during gel formation, whereas the calcium saturated gels were obtained after incubation in 50 mM $CaCl_2$ + 200 mM NaCl for 24 h. As the spectra of the hydrogel composite itself are disturbed by the high water content, FTIR spectra of dried samples were obtained. Spectra of alginate powder and oxidized CNF powder were also compared with the dried CNF-alginate composite hydrogels.

It has previously been reported that the (C=O) stretching band from alginate is shifted from 1602 to 1612 cm^{-1} in oxidized CNF-alginate composite films with excess calcium removed. This has been attributed to the carboxylic groups on oxidized CNF linking alginate molecules to form a crosslinked network [34]. In the calcium-unsaturated samples we do not observe the same shift in signals. The calcium unsaturated samples gives $\nu(COO)_{asym}$ signals from 1602 to 1604 cm^{-1}, for the pure alginate gel and for composite gel of alginate and oxidized CNF, respectively. The $\nu(COO)_{asym}$ signal from *M. pyrifera* alginate powder devoid of calcium is at 1583 cm^{-1} and oxidized CNF devoid of calcium is 1606 cm^{-1}. The samples are, however, not directly comparable with [34], due to differences in sample preparation, charge density on oxidized CNF, fibril characteristics, alginate composition, and molecular weight. From our measurements, wedo not have strong evidence to support intermolecular interactions between oxidized CNF and *M. pyrifera* alginate mediated by calcium in the unsaturated calcium gels.

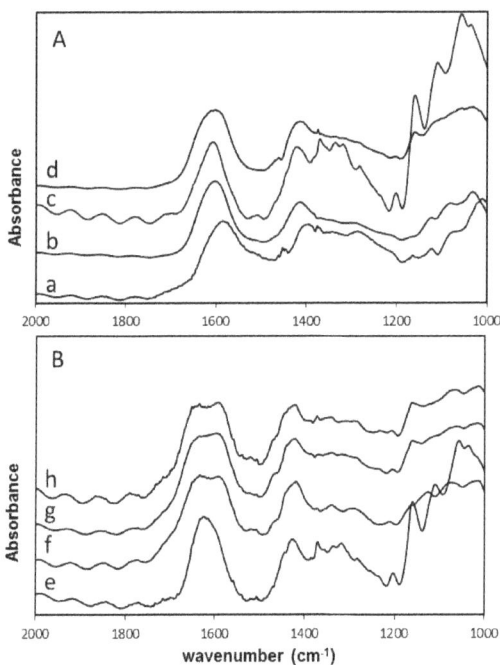

Figure 7. FT-IR spectra of freeze dried gel cylinders of 1.0% *M. pyr.* alginate and 0.75% CNF. (**A**) Calcium unsaturated samples and starting materials; (**B**) calcium saturated samples. *M. pyr.* powder, no calcium. (**a**); *M. pyr*, Ca-unsaturated (**b**); ox. CNF powder, no calcium (**c**); *M. pyr.* + ox CNF, Ca-unsaturated (**d**); Ox. CNF, Ca-saturated (**e**); *M. pyr.*, Ca-saturated (**f**); *M. Pyr.* + ox. CNF, Ca-saturated (**g**); and *M. Pyr.* + mechanically-fibrillated CNF, Ca-saturated (**h**).

The calcium saturated composite samples were strikingly different, displaying a split in the carboxyl groups (C=O) stretching band around 1600 cm^{-1}. This is not seen for the spectra of pure, calcium saturated oxidized CNF, but is observed for the pure alginate gel saturated with calcium. Our hypothesis is that the splitting of the bands for the alginate-containing samples is due to differences in Ca^{2+} binding of carboxylic acid groups from mannuronic acid and guluronic acid. The G-block junction zones will tend to associate laterally at high calcium concentrations [35], and this may shift the signal from guluronic acid towards higher wavenumbers while the carboxylic group in mannuronic acid does not bind calcium in a specific manner and, therefore, does not shift.

As a last comment one should be careful to extrapolate results obtained on dried composite films to composite hydrogels as there will be large differences in both polymer concentration and ionic strengths. It would, therefore, be desirable to be able to study molecular interactions with methods where water interference is not an issue.

4. Conclusions

This study shows that tightly integrated composite hydrogels with tailored mechanical properties can be made from alginate and nanofibrillated cellulose in calcium saturated gels. The alginate contributes with elastic properties and increased mechanical resistance at large deformations. The cellulose nanofibrils reduces the syneresis of the alginate gels and contributes to increased resistance against compression at small deformations as seen by an increase in Young's modulus. The effect was increased using oxidized nanofibrils where composite gels also were volume stable upon saline treatments. No net increase in calcium concentration was found in the composite gels relative to the alginate gel alone. Additionally, no change in FT-IR spectra of the (C = O) stretching band was found for dried composite gels relative to dried alginate gel alone. Although no specific calcium binding event could be identified, we cannot rule out possible calcium mediated CNF-alginate or CNF-CNF interactions, as the oxidized cellulose fibrils by themselves form gels with calcium ions and as more than an additive effect on Young's modulus was seen for the combined system. The increase in Young's modulus was not seen when other relevant biopolymers were added to the alginate gel. Hence, mechanical properties of the composite hydrogels can be tailored by selecting alginates with varying block compositions and charge/concentration of cellulose fibrils. This is relevant for the use of this composite system in films and membranes in addition to the direct use of these composite hydrogels in, e.g., bioprinting and tissue engineering applications.

Acknowledgments: The authors thank Ingebjørg Leirset and Anne Marie Reitan (at RISE PFI) for skillful laboratory work. We would also like to thank Finn L. Aachman at NTNU-NT-Biotechnology for initiating the collaboration between PFI and NTNU. The work has been funded by the Research Council of Norway through the NANO2021 program, grant no. 228147-NORCEL: The Norwegian Nanocellulose Technology Platform and grant No. 221576-MARPOL: Enzymatic Modification and Upgrading of Marine Polysaccharides.

Author Contributions: Olav Aarstad, Berit Løkensgard Strand, Kristin Syverud and Ellinor Bævre Heggset conceived and designed the experiments; Ina Sander Pedersen, Olav Aarstad, and Sindre Hove Bjørnøy performed the experiments; Ina Sander Pedersen, Olav Aarstad, Sindre Hove Bjørnøy, Berit Løkensgard Strand, Ellinor Bævre Heggset, and Kristin Syverud analysed the data; Berit Løkensgard Strand, Olav Aarstad, Kristin Syverud, and Ellinor Bævre Heggset contributed with materials; and Berit Løkensgard Strand, Olav Aarstad, Ellinor Bævre Heggset and Kristin Syverud wrote the paper.

Conflicts of Interest: The authors declare no conflict of interest.

Appendix A

Table A1. Carbohydrate composition in the different CNF-samples.

Sample	Carbohydrate Composition (Weight %)				
	Glucan	Xylan	Mannan	Arabinan	Galactan
Softwood pulp	85.32	8.48	6.63	0.62	0.59
Mechanically fibrillated CNF	84.46	7.98	5.96	0.61	0.67
Oxidized CNF	66.25	6.63	2.67	0.03	0.00

The discrepancy in total monosaccharides between mechanically fibrillated CNF and oxidized CNF is due to conversion to uronic acids upon axidation in the oxidized sample, which were not quantified.

Table A2. Nanocellulose characteristics. All data, except residual fibres, are previously published in [25].

Sample	Charge Density (mmol/g)	Intrinsic Viscosity (mL/g)	Fibril Diameter (nm)	Fibril Length (μm)	Residual Fibres (%)
Mechanically fibrillated CNF	0.1	620	<100	>1	1.3
Oxidized CNF	0.9 [1]	450	<20	>1	28.5

[1] Aldehyde content was determined for the oxidized quality; 0.2 mmol/g CNF.

Appendix B

Table A3. Statistical outcomes, Figure 1: E = Young's modulus, standard deviation (SD), number of samples (*n*), and *p*-value (*p*) of samples compared to the respective pure alginate sample.

Sample	*n*	E (kPa), Mean	SD	*p* (%)	Syneresis (%), Mean	SD	*p* (%)
D. pot.	9	3.794	0.256	-	24.74	0.72	-
0.15% CNF	7	4.930	0.152	<0.01	22.95	0.79	0.08
0.30% CNF	6	5.134	0.399	<0.01	22.27	0.81	0.01
0.50% CNF	7	8.318	0.356	<0.01	15.60	1.01	<0.01
0.75% CNF	5	11.017	0.546	<0.01	13.53	1.11	<0.01
0.15% ox. CNF	7	5.036	0.272	<0.01	21.25	0.72	<0.01
0.30% ox. CNF	5	7.704	0.294	<0.01	17.78	0.50	<0.01
0.50% ox. CNF	6	12.786	0.549	<0.01	12.46	0.76	<0.01
0.75% ox. CNF	5	18.243	1.380	<0.01	7.17	1.45	<0.01
M. pyr.	7	11.492	0.633	-	49.09	0.65	-
0.15% CNF	8	14.895	0.743	<0.01	47.16	0.31	<0.01
0.30% CNF	7	17.437	1.259	<0.01	43.27	0.49	<0.01
0.50% CNF	5	18.273	0.779	<0.01	39.81	0.52	<0.01
0.75% CNF	5	24.297	2.950	<0.01	34.26	1.66	<0.01
0.15% ox. CNF	6	14.884	1.465	0.02	47.24	0.44	0.01
0.30% ox. CNF	5	19.210	1.504	<0.01	40.95	0.69	<0.01
0.50% ox. CNF	5	25.444	1.533	<0.01	31.83	2.20	<0.01
0.75% ox. CNF	5	36.060	3.666	<0.01	28.74	0.69	<0.01
L. hyp.	7	32.369	1.181	-	28.81	0.57	-
0.30% CNF	8	36.668	1.551	<0.01	27.26	0.87	<0.01
0.75% CNF	6	61.167	3.694	<0.01	20.20	0.42	<0.01
0.30% ox. CNF	8	42.991	1.970	<0.01	23.39	0.54	<0.01
0.75% ox. CNF	6	70.956	3.735	<0.01	15.82	1.18	<0.01

Table A4. Statistical outcomes, Figure 2: Rupture strength (mean), deformation (mean) standard deviation (SD), number of samples (n), and p-value (p) of samples compared to the respective pure alginate sample.

Sample	n	Rupture Strength (kg), Mean	SD	p (%)	Deformation (%), Mean	SD	p (%)
D. pot.	9	1.42	0.12	-	63.55	1.13	-
0.15% CNF	7	1.32	0.17	22.7	58.89	1.45	<0.01
0.30% CNF	6	1.23	0.14	2.45	57.70	1.39	<0.01
0.50% CNF	7	1.63	0.14	0.79	58.92	1.05	<0.01
0.75% CNF	5	1.66	0.11	0.35	58.67	0.96	<0.01
0.15% ox. CNF	7	1.15	0.03	0.02	55.70	0.60	<0.01
0.30% ox. CNF	5	1.35	0.02	19.1	56.56	0.35	<0.01
0.50% ox. CNF	6	1.59	0.05	0.79	55.44	0.61	<0.01
0.75% ox. CNF	5	1.39	0.09	66.3	51.63	1.35	<0.01
M. pyr.	7	4.09	0.56	-	71.76	1.09	-
0.15% CNF	8	5.50	0.42	0.03	71.40	1.02	50.3
0.30% CNF	7	5.31	0.20	0.03	69.27	1.08	0.11
0.50% CNF	5	5.15	0.30	0.35	69.25	0.62	0.06
0.75% CNF	5	5.43	0.31	0.07	66.99	0.68	<0.01
0.15% ox. CNF	6	4.16	0.07	75.4	67.41	0.38	<0.01
0.30% ox. CNF	5	4.07	0.04	95.2	63.99	0.34	<0.01
0.50% ox. CNF	5	4.54	0.22	12.1	62.96	0.59	<0.01
0.75% ox. CNF	5	4.25	0.17	55.0	59.71	0.67	<0.01
L. hyp.	7	5.11	0.81	-	64.38	1.89	-
0.30% CNF	8	4.25	0.15	<0.01	60.16	0.70	<0.01
0.75% CNF	6	5.38	0.45	<0.01	60.07	0.93	<0.01
0.30% ox. CNF	8	3.88	0.17	<0.01	57.18	0.81	<0.01
0.75% ox. CNF	6	4.38	0.41	<0.01	55.32	1.43	<0.01

Table A5. Statistical outcomes, Figure 4A,B: E = Young's modulus, standard deviation (SD), number of samples (n), and p-value (p) of samples compared to the 1.0% alginate sample for control polymers. P-values for CNF (*) are comparisons to the respective dry weight concentrations of alginate (1.30% and 1.75%).

Sample	n	E (kPa)	SD	p (%)	Syneresis (%)	SD	p (%)
1.0% M.pyr.	7	11.493	633	-	49.09	0.65	-
0.30% M.pyr.	8	15.916	1.285	<0.01	43.22	0.29	<0.01
0.30% dextran	9	12.168	0.967	14.2	44.81	0.90	<0.01
0.30% hyaluronan	8	13.150	0.710	0.08	38.01	0.60	<0.01
0.30% xanthan	8	14.829	0.330	<0.01	38.59	0.80	<0.01
0.75% M.pyr.	8	22.773	1.092	<0.01	34.31	0.51	<0.01
0.75% dextran	9	12.314	0.560	2.05	41.77	0.61	<0.01
0.75% hyaluronan	9	12.674	2.555	26.3	23.42	1.65	<0.01
0.75% xanthan	9	19.483	1.237	<0.01	26.57	0.79	<0.01
0.30% CNF*	7	17.437	1.259	4.37	43.27	0.49	81.9
0.30% ox. CNF*	5	19.210	1.504	0.18	40.95	0.69	<0.01
0.75% CNF*	5	24.297	2.950	20.7	34.26	1.66	94.0
0.75% ox. CNF*	5	36.060	3.666	<0.01	28.74	0.69	<0.01

Table A6. Statistical outcomes, Figure 4C,D: Statistical outcomes, Figure 4: E = Young's modulus, standard deviation (SD), number of samples (*n*), and *p*-value (*p*) of samples compared to the 1.0% alginate sample for other polymers CNF. *P*-values for CNF (*) are comparisons to the respective dry weight concentrations of alginate (1.30% and 1.75%).

Sample	*n*	Rupture Strength (kg), Mean	SD	*p* (%)	Deformation (%), Mean	SD	*p* (%)
1.0% M.pyr.	7	4.09	0.56	-	71.76	1.00	-
0.30% M.pyr.	8	5.99	0.29	<0.01	71.17	1.25	34.0
0.30% dextran	9	5.07	0.23	0.08	72.17	0.82	38.3
0.30% hyaluronan	8	3.69	0.60	21.0	74.86	3.87	6.76
0.30% xanthan	8	3.18	0.21	0.16	63.70	0.61	<0.01
0.75% M.pyr.	8	9.67	0.44	<0.01	69.74	1.24	0.64
0.75% dextran	9	4.81	0.41	1.37	70.79	1.31	13.6
0.75% hyaluronan	9	2.54	0.33	<0.01	64.02	1.78	<0.01
0.75% xanthan	9	2.81	0.08	<0.01	60.76	0.40	<0.01
0.30% CNF*	7	5.31	0.20	0.04	69.27	1.08	1.06
0.30% ox. CNF*	5	4.07	0.04	<0.01	63.99	0.34	<0.01
0.75% CNF*	5	5.43	0.31	<0.01	66.99	0.68	0.015
0.75% ox. CNF*	5	4.25	0.17	<0.01	59.71	0.67	<0.01

Appendix C

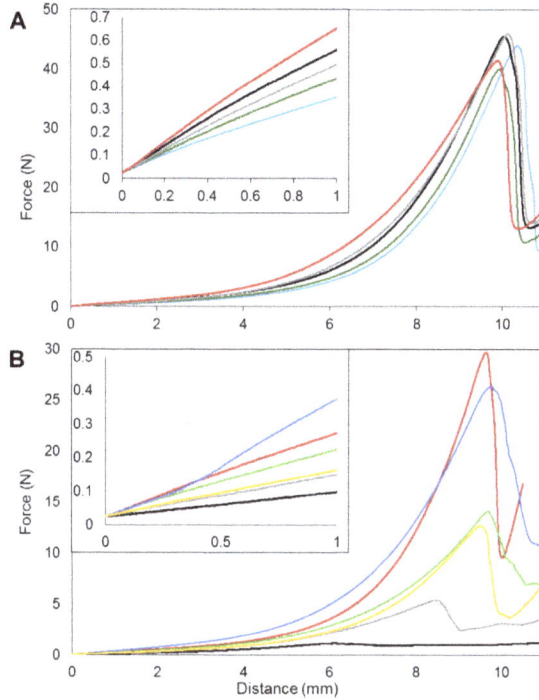

Figure A1. Stress-deformation curves of gel cylinders presented in Figures 1B and 2B (**A**) and Figure 3 (**B**). The inserted picture shows the initial compression where the gels display elastic behaviour. Young's modulus is calculated from 0.1 to 0.3 mm. (**A**) 1% (*w/v*) *M. pyr.* (——), 1% (*w/v*) *M. pyr.* + 0.15% (*w/v*) ox. CNF (——), 1% (*w/v*) *M. pyr.* + 0.30% (*w/v*) ox. CNF (——), 1% (*w/v*) *M. pyr* + 0.50% (*w/v*) ox. CNF (——) and 1% (*w/v*) *M. pyr* + 0.75% (*w/v*) ox. CNF (——); (**B**) 0.75% (*w/v*) *M. pyr* (——), 0.75% (*w/v*) ox. CNF (——), 0.375% (*w/v*) ox. CNF + 0.375% (*w/v*) *M. pyr* (——), 0.75% (*w/v*) ox. CNF + 0.2% (*w/v*) *M. pyr* (——), 0.75% (*w/v*) ox. CNF + 0.375% (*w/v*) *M. pyr* (——) and 0.75% (*w/v*) ox. CNF + 0.75% (*w/v*) *M. pyr* (——).

Appendix D

Table A7. FTIR peaks for $\nu(COO)_{asym}$ (cm^{-1}) for samples shown in Figure 7.

	Sample	Calcium Saturated Gel	$\nu(COO)_{asym}$ (cm^{-1})
a	*M. pyr.* powder, no calcium	pristine	1583
b	1.0% (*w/v*) *M. pyr*	no	1602
c	ox. CNF, no calcium	pristine	1606
d	1% (*w/v*) *M. pyr* + 0.75% (*w/v*) ox. CNF	no	1604
e	0.75% (*w/v*) ox. CNF	yes	1624
f	1.0% (*w/v*) *M. pyr.*	yes	1589/1633 *
g	1.0% (*w/v*) *M. pyr.* + 0.75% (*w/v*) ox. CNF	yes	1593/1635 *
h	1.0% (*w/v*) *M. Pyr.* + 0.75% (*w/v*) mechanically fibrillated CNF	yes	1591/1639 *

* Two unresolved signals centered around 1620 cm^{-1}.

References

1. Markstedt, K.; Mantas, A.; Tournier, I.; Martinez Avila, H.; Haegg, D.; Gatenholm, P. 3D bioprinting human chondrocytes with nanocellulose-alginate bioink for cartilage tissue engineering applications. *Biomacromolecules* **2015**, *16*, 1489–1496. [CrossRef] [PubMed]

2. Muller, M.; öztürk, E.; Zenobi-Wong, M.; Arlov, O.; Gatenholm, P. Alginate sulfate-nanocellulose bioinks for cartilage bioprinting applications. *Ann. Biomed. Eng.* **2017**, *45*, 210–223. [CrossRef] [PubMed]

3. Nguyen, D.; Enejder, A.; Nguyen, D.; Forsman, A.; Ekholm, J.; Nimkingratana, P.; Brantsing, C.; Lindahl, A.; Simonsson, S.; Hagg, D.A.; et al. Cartilage tissue engineering by the 3d bioprinting of ips cells in a nanocellulose/alginate bioink. *Sci. Rep.* **2017**, *7*, 658. [CrossRef] [PubMed]

4. Haug, A.; Larsen, B.; Smidsröd, O. Studies on the sequence of uronic acid residues in alginic acid. *Acta Chem. Scand.* **1967**, *21*, 691–704. [CrossRef]

5. Donati, I.; Holtan, S.; Morch, Y.A.; Borgogna, M.; Dentini, M.; Skjak-Braek, G. New hypothesis on the role of alternating sequences in calcium-alginate gels. *Biomacromolecules.* **2005**, *6*, 1031–1040. [CrossRef] [PubMed]

6. Grant, G.T.; Morris, E.R.; Rees, D.A.; Smith, P.J.C.; Thom, D. Biological interactions between polysaccharides and divalent cations—Egg-box model. *FEBS Lett.* **1973**, *32*, 195–198. [CrossRef]

7. Draget, K.I.; Ostgaard, K.; Smidsrod, O. Homogeneous alginate gels—A technical approach. *Carbohydr. Polym.* **1990**, *14*, 159–178. [CrossRef]

8. Martinsen, A.; Skjak-Braek, G.; Smidrsod, O. Alginate as immobilization material: I. Correlation between chemical and physical properties of alginate gel beads. *Biotechnol. Bioeng.* **1989**, *33*, 79–89. [CrossRef] [PubMed]

9. Rokstad, A.M.; Brekke, O.L.; Steinkjer, B.; Ryan, L.; Kollarikova, G.; Strand, B.L.; Skjak-Braek, G.; Lacik, I.; Espevik, T.; Mollnes, T.E. Alginate microbeads are complement compatible, in contrast to polycation containing microcapsules, as revealed in a human whole blood model. *Acta Biomater.* **2011**, *7*, 2566–2578. [CrossRef] [PubMed]

10. Lee, K.Y.; Mooney, D.J. Alginate: Properties and biomedical applications. *Prog. Polym. Sci.* **2012**, *37*, 227–258. [CrossRef] [PubMed]

11. Saito, T.; Nishiyama, Y.; Putaux, J.L.; Vignon, M.; Isogai, A. Homogeneous suspensions of individualized microfibrils from tempo-catalyzed oxidation of native cellulose. *Biomacromolecules* **2006**, *7*, 1687–1691. [CrossRef] [PubMed]

12. Dong, H.; Snyder, J.F.; Williams, K.S.; Andzelm, J.W. Cation-induced hydrogels of cellulose nanofibrils with tunable moduli. *Biomacromolecules* **2013**, *14*, 3338–3345. [CrossRef] [PubMed]

13. Zander, N.E.; Dong, H.; Steele, J.; Grant, J.T. Metal cation cross-linked nanocellulose hydrogels as tissue engineering substrates. *ACS Appl. Mater. Interfaces* **2014**, *6*, 18502–18510. [CrossRef] [PubMed]

14. Lin, N.; Huang, J.; Dufresne, A. Preparation, properties and applications of polysaccharide nanocrystals in advanced functional nanomaterials: A review. *Nanoscale* **2012**, *4*, 3274–3294. [CrossRef] [PubMed]

15. Depan, D.; Misra, R.D.K. The interplay between nanostructured carbon-grafted chitosan scaffolds and protein adsorption on the cellular response of osteoblasts: Structure-function property relationship. *Acta Biomater.* **2013**, *9*, 6084–6094. [CrossRef] [PubMed]

16. Yim, E.K.; Leong, K.W. Significance of synthetic nanostructures in dictating cellular response. *Nanomed. Nanotechnol. Biol. Med.* **2005**, *1*, 10–21. [CrossRef] [PubMed]

17. Zhang, L.J.; Webster, T.J. Nanotechnology and nanomaterials: Promises for improved tissue regeneration. *Nano Today* **2009**, *4*, 66–80. [CrossRef]

18. Park, M.; Lee, D.; Hyun, J. Nanocellulose-alginate hydrogel for cell encapsulation. *Carbohydr. Polym.* **2015**, *116*, 223–228. [CrossRef] [PubMed]

19. Engler, A.J.; Sen, S.; Sweeney, H.L.; Discher, D.E. Matrix elasticity directs stem cell lineage specification. *Cell* **2006**, *126*, 677–689. [CrossRef] [PubMed]

20. Aarstad, O.; Tøndervik, A.; Sletta, H.; Skjåk-Bræk, G. Alginate sequencing: An analysis of block distribution in alginates using specific alginate degrading enzymes. *Biomacromolecules* **2012**, *13*, 106–116. [CrossRef] [PubMed]

21. Shinoda, R.; Saito, T.; Okita, Y.; Isogai, A. Relationship between length and degree of polymerization of tempo-oxidized cellulose nanofibrils. *Biomacromolecules* **2012**, *13*, 842–849. [CrossRef] [PubMed]

22. Araki, J.; Wada, M.; Kuga, S. Steric stabilization of a cellulose microcrystal suspension by poly(ethylene glycol) grafting. *Langmuir* **2001**, *17*, 21–27. [CrossRef]

23. Saito, T.; Isogai, A. Tempo-mediated oxidation of native cellulose. The effect of oxidation conditions on chemical and crystal structures of the water-insoluble fractions. *Biomacromolecules* **2004**, *5*, 1983–1989. [CrossRef] [PubMed]

24. Chinga-Carrasco, G.; Averianova, N.; Kondalenko, O.; Garaeva, M.; Petrov, V.; Leinsvang, B.; Karlsen, T. The effect of residual fibres on the micro-topography of cellulose nanopaper. *Micron* **2014**, *56*, 80–84. [CrossRef] [PubMed]

25. Heggset, E.B.; Chinga-Carrasco, G.; Syverud, K. Temperature stability of nanocellulose dispersions. *Carbohydr. Polym.* **2017**, *157*, 114–121. [CrossRef] [PubMed]

26. Mørch, Y.A.; Holtan, S.; Donati, I.; Strand, B.L.; Skjåk-Bræk, G. Mechanical properties of c-5 epimerized alginates. *Biomacromolecules* **2008**, *9*, 2360–2368. [CrossRef] [PubMed]

27. Smidsrød, O.; Haug, A. Properties of poly(1,4-hexuronates) in the gel state. Ii. Comparison of gels of different chemical composition. *Acta Chem. Scand.* **1972**, *26*, 79–88.

28. Donati, I.; Mørch, Y.A.; Strand, B.L.; Skjåk-Bræk, G.; Paoletti, S. Effect of elongation of alternating sequences on swelling behavior and large deformation properties of natural alginate gels. *J. Phys. Chem.* **2009**, *113*, 12916–12922. [CrossRef] [PubMed]

29. Aarstad, O.; Strand, B.L.; Klepp-Andersen, L.M.; Skjak-Braek, G. Analysis of G-block distributions and their impact on gel properties of in vitro epimerized mannuronan. *Biomacromolecules* **2013**, *14*, 3409–3416. [CrossRef] [PubMed]

30. Sirvio, J.A.; Kolehmainen, A.; Liimatainen, H.; Niinimaki, J.; Hormi, O.E.O. Biocomposite cellulose-alginate films: Promising packaging materials. *Food Chem.* **2014**, *151*, 343–351. [CrossRef] [PubMed]

31. Xie, M.; Olderøy, M.Ø.; Zhang, Z.; Andreassen, J.-P.; Strand, B.L.; Sikorski, P. Biocomposites prepared by alkaline phosphatase mediated mineralization of alginate microbeads. *RSC Adv.* **2012**, *2*, 1457–1465. [CrossRef]

32. Syverud, K.; Pettersen, S.R.; Draget, K.; Chinga-Carrasco, G. Controlling the elastic modulus of cellulose nanofibril hydrogels-scaffolds with potential in tissue engineering. *Cellulose* **2015**, *22*, 473–481. [CrossRef]

33. Mørch, Y.A.; Donati, I.; Strand, B.L.; Skjåk-Bræk, G. Effect of Ca^{2+}, Ba^{2+}, and Sr^{2+} on alginate microbeads. *Biomacromolecules* **2006**, *7*, 1471–1480. [CrossRef] [PubMed]

34. Lin, N.; Bruzzese, C.; Dufresne, A. Tempo-oxidized nanocellulose participating as crosslinking aid for alginate-based sponges. *Appl. Mater. Interfaces* **2012**, *4*, 4948–4959. [CrossRef] [PubMed]

35. Stokke, B.T.; Draget, K.I.; Smidsrød, O.; Yuguchi, Y.; Urakawa, H.; Kajiwara, K. Small-angle X-ray scattering and rheological characterization of algiate gels. 1. Ca-alginate gels. *Macromolecules* **2000**, *33*, 1853–1863. [CrossRef]

polymers

MDPI

Article

Human Mesenchymal Stem Cells Differentiation Regulated by Hydroxyapatite Content within Chitosan-Based Scaffolds under Perfusion Conditions

Anamarija Rogina [1,*], Maja Antunović [2], Lidija Pribolšan [2], Katarina Caput Mihalić [2], Andreja Vukasović [3], Alan Ivković [3,4], Inga Marijanović [2], Gloria Gallego Ferrer [5,6], Marica Ivanković [1] and Hrvoje Ivanković [1]

[1] Faculty of Chemical Engineering and Technology, University of Zagreb, Marulićev trg 19, p.p.177, 10001 Zagreb, Croatia; mivank@fkit.hr (M.I.); hivan@fkit.hr (H.I.)
[2] Faculty of Science, University of Zagreb, Horvatovac102a, 10001 Zagreb, Croatia; maja.antunovic@biol.pmf.hr (M.A.); lidija.pribolsan@gmail.com (L.P.); katarina.caput.mihalic@biol.pmf.hr (K.C.M.); ingam@biol.pmf.hr (I.M.)
[3] Department of Histology and Embryology, School of Medicine, University of Zagreb, Šalata 3, 10001 Zagreb, Croatia; andreja.vukasovic@mef.hr (A.V.); alan.ivkovic@gmail.com (A.I.)
[4] Department of Orthopaedic Surgery, University Hospital, Sveti Duh, 10001 Zagreb, Croatia
[5] Centro de Biomateriales e Ingeniería Tisular, Universitat Politècnica de València, Camino de Vera s/n, 46022 Valencia, Spain; ggallego@ter.upv.es
[6] Biomedical Research Networking centre in Bioengineering, Biomaterials and Nanomedicine (CIBER-BBN), Mariano Esquillor s/n, 50018 Zaragoza, Spain
* Correspondence: arogina@fkit.hr; Tel.: +3851-45-97-253

Received: 21 July 2017; Accepted: 21 August 2017; Published: 23 August 2017

Abstract: The extensive need for hard tissue substituent greatly motivates development of suitable allogeneic grafts for therapeutic recreation. Different calcium phosphate phases have been accepted as scaffold's components with positive influence on osteoinduction and differentiation of human mesenchymal stem cells, in terms of their higher fraction within the graft. Nevertheless, the creation of unlimited nutrients diffusion through newly formed grafts is of great importance. The media flow accomplished by perfusion forces can provide physicochemical, and also, biomechanical stimuli for three-dimensional bone-construct growth. In the present study, the influence of a different scaffold's composition on the human mesenchymal stem cells (hMSCs) differentiation performed in a U-CUP bioreactor under perfusion conditioning was investigated. The histological and immunohistochemical analysis of cultured bony tissues, and the evaluation of osteogenic genes' expression indicate that the lower fraction of in situ formed hydroxyapatite in the range of 10–30% within chitosan scaffold could be preferable for bone-construct development.

Keywords: chitosan; hydroxyapatite; hMSC's; perfusion-bioreactor

---

1. Introduction

The development of functional allogeneic tissue constructs requires the coordination of cell adhesion, growth, differentiation, and organization into specific tissue architecture. Accordingly, the engineering of a suitable stem cell microenvironment becomes an important approach in regenerative medicine [1]. Mesenchymal stem cells (MSCs) isolated from adult bone marrow are recognized as self-renewing cells capable of differentiating into multiple different cell phenotypes, providing many advantages in tissue regeneration strategies [2,3]. Even with precise control over materials characteristics at the nanoscale, there is still a challenge to control the protein adsorption and cellular response due to the complexity of biological systems [4].

Biomaterial scaffolds and dynamic bioreactors, such as perfusion bioreactor, represent two important components in the formation of suitable microenvironments providing controlled patterns of biochemical and biomechanical signaling [5]. Finding the proper synergy between scaffolds features and perfusion conditioning can direct MSCs behavior during the graft development.

Predominant factors in MSCs signaling are the scaffold's composition, its surface properties, and specific cell-scaffold interaction that occurred during the implant-native tissue bonding [6]. Hydroxyapatite (HA) has already confirmed its osteoinductive properties in vitro and in vivo [7–11]. Chemical similarity to natural bone component makes it suitable for bone tissue integration and regeneration. Chitosan-based composites exhibit excellent properties: biodegradability, ability to support the bone tissue formation, possibility for modifications for specific cellular responses, inhibition of microorganisms, and oral pathogens [12–15]. Specific cationic nature of chitosan, depending on the surface modification, can positively influence protein and cell adsorption, allowing for good cell adhesion and migration, while its hydrogel properties ensure high water absorption which is important for fluid retention and nutrients transport in the biological environment. Osteogenic properties of chitosan/hydroxyapatite-based scaffolds have been previously reported by the long-term culture of human MSCs [16]. Promoted initial hMSCs adhesion on chitosan-gelatin/hydroxyapatite scaffold has maintained a higher progenicity and multi-lineage differentiation potentials with enhanced osteogenesis [6]. Osteogenic differentiation of hMSCs has been promoted by incorporation of hydroxyapatite nanoparticles into chitosan-silk fibroin matrix, as reported by Lai and Shalumon [17]. The influence of the hydroxyapatite amount was detected by extended differentiation for composites with a higher HA fraction.

The cell diffusion, cellular behavior, and tissue integration depend on scaffolds porosity, pore size, and overall pore interconnectivity. High porosity plays an important part in cell seeding, allowing for good cell distribution and tissue penetration [18]. Considerable influence of pore structure on effective diffusivity and bone ingrowth was reported by Jones et al. [19], indicating that the scaffold's porosity should also be considered and optimized in the graft engineering.

Despite scaffolds' requirements, instructive developmental stimuli can be provided by dynamic cell culture using perfusion bioreactors [20,21]. Study performed by Kim and Ma [20] has suggested that transversal media flow in the bioreactor culture of hMSCs accelerates progression from uncommitted mesenchymal stem cells, to osteoblasts as a combined result of shear stress stimulation and the depletion of mitogenic growth factors and ECM. The direct perfusion ensures exposure of the cells to interstitial fluid within the scaffold, resulting in higher seeding efficiency and even distribution of cells and newly formed tissue.

Apart from the extensive research of hydroxyapatite impact on MSCs differentiation, most of those investigations were performed in static cell cultures. According to our previous study [22], present materials have demonstrated very good osteogenic potential in the static cell culture of MC3T3-E1 preosteoblasts. In the current study, we wanted to confirm this behavior under dynamic hMSCs culture by eliminating possible hindrance in cell seeding of static conditions caused by differences in surface properties of composite scaffolds. We believe that biomechanical forces accomplished by media perfusion enhance seeding and distribution of the cells, as well as homogeneous distribution of macromolecules ensuring necessary microenvironment for hMSCs osteogenesis through a scaffolds' interior. This preliminary study indicates that the lower fraction of hydroxyapatite within chitosan-based scaffold is favorable for allogeneic graft development.

2. Materials and Methods

2.1. Materials Preparation

Chitosan (M_w = 100–300 kg/mol, DD = 0.95–0.98, Acros Organics, Geel, Belgium), calcium carbonate ($CaCO_3$, calcite; TTT, Sv. Nedjelja, Croatia), urea phosphate ((NH_2)$_2$CO–H_3PO_4; Aldrich Chemistry, St. Louis, MO, USA), acetic acid (HAc; POCH, Gliwice, Poland), sodium hydroxide (NaOH,

Carlo Erba, Val de Reuil, France), and ethanol (EtOH, 96 wt %, Kefo, Sisak, Croatia) were all of analytical grade.

The synthesis of chitosan-hydroxyapatite (Cht-HA) composites with different amount of hydroxyapatite precursors, as well as porous structures, was done according to our previous work [23]: to obtain required HA percentage, appropriate amounts of calcite were suspended in chitosan solution (1.2 wt %) prepared in dilute acetic acid solution (0.36 wt %) at an ambient temperature. Following additions of a specific amount of urea phosphate, with respect to the Ca/P ratio of 1.67, characteristic for hydroxyapatite. Temperature was set at 50 °C and the precipitation reaction was continued for 4 days. The precursor's amount was adjusted to obtain 0, 10, 30, and 50 wt % of in situ precipitated HA in final Cht-HA scaffold.

The production of porous chitosan-hydroxyapatite structures was initialized by cooling down the suspensions to room temperature, and freezing them over night at -22 °C. Frozen samples were immersed into different mediums, starting with a neutralization medium consisting of equal volume portions of 1 mol/L NaOH and EtOH, and then pure ethanol (96 wt %) at -22 °C. Finally, samples were dehydrated with ethanol at an ambient temperature and left out to dry for 24 h.

2.2. Scaffolds Characterization

Microstructure Imaging

The composite Cht-HA scaffolds were imaged by the scanning electron microscope TESCAN Vega3SEM Easyprobe at electron beam energy 10 keV. Previously to imaging, samples were sputter coated with gold and palladium for 120 s.

2.3. Dynamic hMSCs Cell Culture in Perfusion Bioreactor

2.3.1. Human Mesenchymal Stem Cells Isolation and Expansion

The bone marrow sample was collected during the surgery at the University Hospital of Traumatology in Zagreb, Croatia, with the patient's consent and approval of the Ethics Committee. The hMSCs were isolated using previously described method [24]. Briefly, 20 mL of bone marrow aspirates were added to 200 mL Dulbecco's modified Eagle medium with 1000 mg/L glucose (DMEM-low glucose) (Lonza, Basel, Switzerland) containing 10% fetal bovine serum (FBS) (Gibco, Gaithersburg, MD, USA), 100 U/mL penicillin and 100 µg/mL streptomycin (Lonza). The suspension was centrifuged at $300 \times g$ for 10 min, and pelleted cells were washed twice in phosphate-buffered saline (PBS) (Gibco). Resuspended cells were strained through a cell strainer (100 µm) (BD Biosciences, Mississauga, ON, Canada) to remove bone chips and then centrifuged at $300 \times g$ for 10 min. Cells were plated in 100 mm Petri dishes (Sarstedt, Germany) at a density of 1×10^8 in proliferation medium DMEM-low glucose, supplemented with 10% FBS, 100 U/mL penicillin and 100 µg/mL streptomycin and 10 ng/mL human fibroblast growth factor 2 (FGF2) (Gibco), and kept in a humidified incubator at 37 °C with a 5% CO_2 supply. After 24 h the non-adherent cells were removed by total media replacement, and the attached cells were grown. When hMSCs became 80% confluent, they were detached with 0.25% trypsin/EDTA (Sigma-Aldrich) and then subcultured for expansion. Proliferation medium was changed every 2–3 days. After 24 h, the number of attached hMSCs has been determined by counting the number of cells in medium (unattached cells).

2.3.2. Osteogenic Induction in Perfusion Bioreactor

Perfusion bioreactor (U-CUP Cellec Biotek, Basel, Switzerland) was used for three-dimensional bone tissue growth. Two samples of each scaffold, with diameter of 8 mm and height of 2 mm, were inserted into the bioreactor for cell seeding with a cell suspension of 1.6×10^6 cells/bioreactor. The bioreactor chamber was closed and a bioreactor with a perfusion speed of 1.7 mL/min was placed in an incubator at 37 °C and 5% CO_2 for 24 h.

The osteogenic induction medium (Minimum Essential Medium-Alpha Eagle (α-MEM) (Lonza), 10% FBS, 1% penicillin/streptomycin, 50 µg/mL ascorbic acid (Sigma-Aldrich), 4 mmol/L β-glycerophosphate (Sigma-Aldrich) and 1×10^{-7} mol/L dexamethasone (Sigma-Aldrich) was used. The total volume of medium in bioreactor was 10 mL, with a perfusion speed of 0.6 mL/min. Medium was exchanged every 2–3 days according to the manufacturer's instructions (Cellec Biotech). After 14 and 21 days of cultivation, scaffolds were removed from the bioreactor and analyzed.

2.3.3. Hoechst 33342 Staining

After removal from the bioreactor, constructs were rinsed in PBS, fixed in 4% paraformaldehyde, embedded in paraffin, cross-sectioned (5 µm thick), and deparaffinised using xylene and rehydrated through a series of ethanol washes. To visualize the cell proliferation and distribution through the cross section of the scaffolds, slides were stained with cell permeable nuclear stain Hoechst 33342 (Sigma-Aldrich). Prior to staining, all slides were permeabilized in 0.25% Triton x-100 (Sigma-Aldrich) for 10 min, followed by washing in PBS. Slides were incubated with 1 µg/mL Hoechst for 1 min in the dark and rinsed in PBS, followed by mounting in glycerol/PBS (1:1). Fluorescent staining of cell nuclei was observed under fluorescence microscope (Olympus BX51, Tokyo, Japan).

2.3.4. Histological Analysis

After removal from the bioreactor, constructs were rinsed in PBS, fixed in 4% paraformaldehyde, embedded in paraffin, cross-sectioned (5 µm thick), and stained with H&E (hematoxylin–eosin). All sections were deparaffinised using xylene, and rehydrated through a series of ethanol washes prior to staining. Detection of calcium in tissue sections was done using alizarin red S and von Kossa staining. Slides were stained with alizarin red solution (2%, pH 4.4) for 2 min. For von Kossa staining, slides were incubated with 5% silver nitrate solution in a clear glass coplin jar and placed under UVB light emitting 312 nm (UVItec Ltd., Cambridge, UK) for 2.5 min. Unreacted silver was removed with 5% sodium thiosulfate for 5 min. Mayer's hematoxylin (Biognost, Zagreb, Croatia) was used as counterstain for 10 min. All slides were dehydrated and mounted with resinous medium (Biognost). Slides were observed under microscope (Olympus BX51).

2.3.5. Isolation of Total Cellular RNA and Real-Time Quantitative Polymerase Chain Reaction (RT-qPCR) Analysis

Total cellular RNA was isolated using TRIzol reagent (Invitrogen, Carlsbad, CA, USA), according to the manufacturer's instructions. Briefly, tissues were removed from the bioreactor as well as undifferented hMSCs from 2D culture, were washed with cold PBS and homogenized in 1 mL of TRIzol. Tissues from the bioreactor were homogenized with a mixer mill (Retsch, Haan, Germany), while hMSCs were scrapped. After that, 200 µL of chloroform was added followed by centrifugation. Isopropanol was used for RNA precipitation. The RNA concentration was determined by the spectrophotometric method using a NanoVue (Thermo Fisher Scientific) at 260 nm. Total RNA was treated with DNase I (Invitrogen, Carlsbad, CA, USA) to remove genomic DNA and 1 µg of total RNA was reverse transcribed to cDNA using GeneAmp RNA PCR kit (Applied Biosystems, Waltham, MA, USA) according to the manufacturer's instructions. Reverse transcription was performed in the thermomixer (Eppendorf, Hamburg, Germany) at the following conditions: 10 min at 20 °C, 1 h at 42 °C, 5 min at 99 °C, and 5 min at 5 °C. Relative expression of collagen (ColI), bone sialoprotein (BSP), dentin matrix protein (DMP1), and osteocalcin (OC) were determined by RT-qPCR on the machine 7500 Fast Real-Time PCR System (Applied Biosystems, Foster City, CA, USA) using commercially available primers (Sigma-Aldrich; Table 1) and Power SYBR Green Mastermix (Applied Biosystems). Each reaction consisted of a duplicate in 96-well plates (ABI PRISM Optical 96-Well Plate; Applied Biosystems). The PCR reaction was conducted under the following conditions: 10 min at 95 °C for 1 cycle, 15 s at 95 °C, and 1 min at 60 °C for 40 cycles. The expression levels of osteogenic genes were normalized to β-actin as housekeeping gene, and calculated using the $^{\Delta\Delta}$Ct method.

Table 1. Human primer sequences used for determination of gene expression levels by reverse transcription-polymerase chain reaction analysis.

Primers	5'-	Sequences	Annealing Temperature (°C)
COL1A1	forward	GCTATGATGAGAAATCAACCG	61.1
	reverse	TCATCTCCATTTCCAGG	61.6
IBSP	forward	GGAGACTTCAAATGAAGGAG	57.9
	reverse	CAGAAAGTGTGGTATTCTCAG	56.4
DMP1	forward	CAACTATGAAGATCAGCATCC	58.8
	reverse	CTTCCATTCTTCAGAATCCTC	59.3
BGLAP	forward	TTCTTTCCTCTTCCCCTTG	60.8
	reverse	CCTCTTCTGGAGTTTATTTGG	59.3

2.3.6. Immunohistochemical Detection of Type I Collagen and Osteocalcin

Sections were deparaffinised, rehydrated, and then subjected to proteinase K (Proteinase K Ready to use, Dako, Togo) antigen retrieval for 12 min at room temperature. Endogenous peroxidase activity was blocked with 3% H_2O_2 in PBS for 3–6 min. Nonspecific binding sites were blocked with 10% goat serum (Dako) in PBS for 60 min at room temperature. Sections were then incubated with primary antibody (anti-collagen I, Abcam, Cambridge, UK), diluted 1:400 or anti-osteocalcin (FL-100, Santa Cruz Biotechnology, Dallas, TX, USA) diluted 1:50, respectively, with 1% goat serum in PBS, overnight at 4 °C. After washing, the signal was detected with EnVision Detection Systems Peroxidase/DAB, Rabbit/Mouse (Dako), according to the manufacturer instructions. Hematoxylin was used as a counterstain. Negative controls were processed in the same way with the omittance of primary antibody. Human bone was used as a positive control. Slides were observed under phase-contrast microscope (Olympus BX51).

2.4. Statistical Analysis

The experiments were performed in duplicate or more, therefore quantitative results were represented as a mean value with standard deviation. Significant difference between two groups was determined by statistical analysis using one-way ANOVA test, with a $p < 0.05$ value as statistically significant.

3. Results

3.1. Scaffolds' Microstructure

Microscopic analysis of a cross section of different scaffolds indicates a highly porous structure with a visually assessed pore size of up to 200 μm (Figure 1). Comparing the microstructure of pure chitosan scaffold (0% HA), the addition of HA causes alteration of ordered honeycomb-like pore structure. It is not surprising to see such changes in microstructure, taking into account the differences in solution viscosity and crystal size of hydroxyapatite of the different starting solutions. Likewise, the precursors' amount for corresponding hydroxyapatite fraction affects the final pH value of the composite solution, i.e., the solubility of chitosan that dictates the surface energy between the phases. Shape and orientation of in situ HA crystals are changing by the increase of the HA amount, leading to formation of so-called cauliflower morphology observed at 50% HA scaffold.

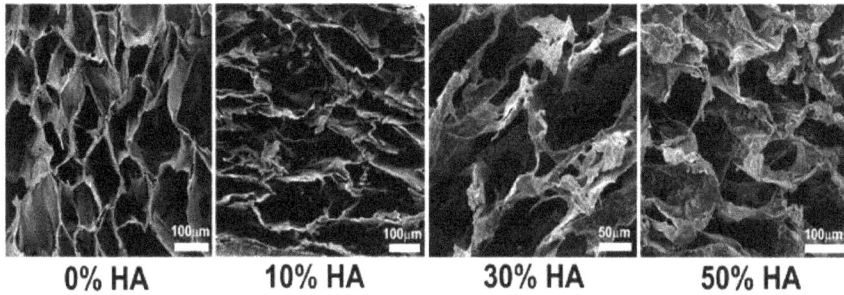

Figure 1. SEM micrographs of cross section of different chitosan-hydroxyapatite (Cht-HA) scaffolds. Pure chitosan scaffold shows smooth surface, while small crystals at the beginning of their growth were found in scaffold with 10% of HA. Petal-like hydroxyapatite crystals are formed in 30% HA scaffold, while cauliflower-like HA particles are homogeneously distributed in scaffold with 50% of HA.

3.2. Biological Evaluation of Construct (Grafts) Cultured in Perfusion Bioreactor

The regenerative medicine based on stem cell-scaffold culture has already been recognized as a promising alternative for autologous grafts in orthopedic surgery. The capability of differentiation into several cell lineages (osteoblasts, chondrocytes, adipocytes, tenocytes, and myoblasts), immunosuppressive properties, and carrying a lower risk of malignant transformation, makes mesenchymal stem cells more clinically applicable [25]. The presence of calcium phosphate, mainly hydroxyapatite, in chitosan scaffold has shown to increase cell adhesion, proliferation, alkaline phosphatase activity, protein' adsorption, type I collagen production, and expression of other osteogenic differentiation markers.

To evaluate how different HA fractions affected the cell behavior, such as the cell attachment, we determined the number of attached hMSCs by counting the number of cells in the medium (unattached cells) 24 h after seeding. High percentages of attached cells to scaffolds were confirmed: 93% ($\pm$1.80%) for 0% HA, 96.33% ($\pm$2.93%) for 10% HA, 91.83% ($\pm$1.04%) for 30% HA, and 96.67% ($\pm$1.15%) for 50% HA, respectively.

To visualize the cell distribution through the cross section of the scaffolds, slides were stained with cell permeable nuclear stain Hoechst. The cell number/mm^2 within each graft was determined after 21 days of dynamic culture by staining of cell nuclei (Figure 2A). The highest cell number/mm^2 came from grafts seeded on composite scaffolds with 30% HA, while the lowest came from grafts seeded on composite scaffolds with 50% HA (Figure 2B). The results indicate that a high percentage of HA could have unfavorable effects on cell survival during 21 days of dynamic culture.

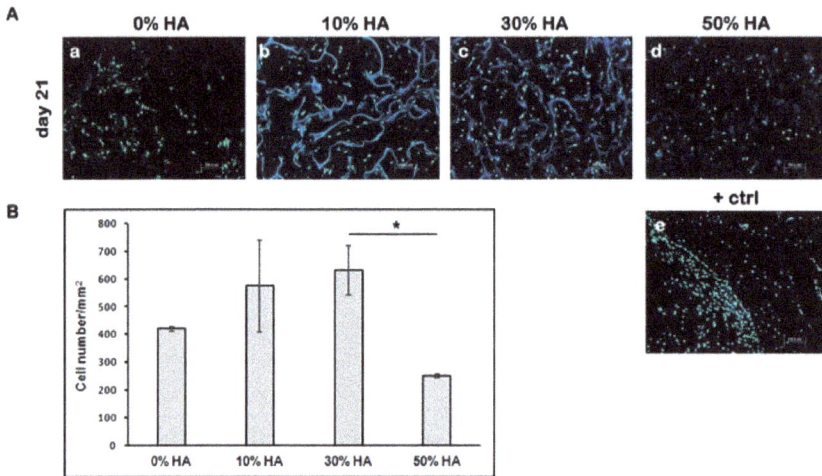

Figure 2. Cell distribution through the cross section of different Cht-HA scaffolds after 21 days of dynamic culture. (**A**) Fluorescence microscopic images of Hoechst stained cells on scaffolds composed of 0% HA (**a**), 10% HA (**b**), 30% HA (**c**) and 50% HA (**d**), respectively. Human bone sample was used as positive control (+ctrl). Representative images are shown at $100\times$ magnification ($n = 3$). Scale bar represents 100 μm. (**e**). (**B**) The cell number/mm^2 within each Cht-HA scaffold. Significant difference between two groups ($p < 0.05$) is designated as (*).

3.2.1. Histological Analysis of the Grafts

Histological analysis of grafts was done after 14 and 21 days of dynamic culture by hematoxylin-eosin staining. Figure 3 points out the significant difference in the amount of newly formed tissue on 30% HA scaffold with uniform cell distribution throughout the entire volume of the graft. Taking a look on a large area of newly formed tissue on aforementioned scaffold (asterisk designation on Figure 3), we can assume that this scaffold's composition is a suitable environment for hMSCs migration and growth after 14 days of culture.

Continued perfusion in culture has provided further extracellular matrix (ECM) synthesis, especially for constructs seeded on 10% HA scaffold, while the pure chitosan constructs has remained almost the same even after 21 days. It seems that the absence of hydroxyapatite not only influences the osteoinductive properties of chitosan scaffold, yet also the surface properties for cell adhesion and growth. Similar effects were detected on scaffolds with 50% of HA, as indicated by the weaker and non-homogeneous staining. This result is in agreement with lower number of cells counted using Hoechst staining.

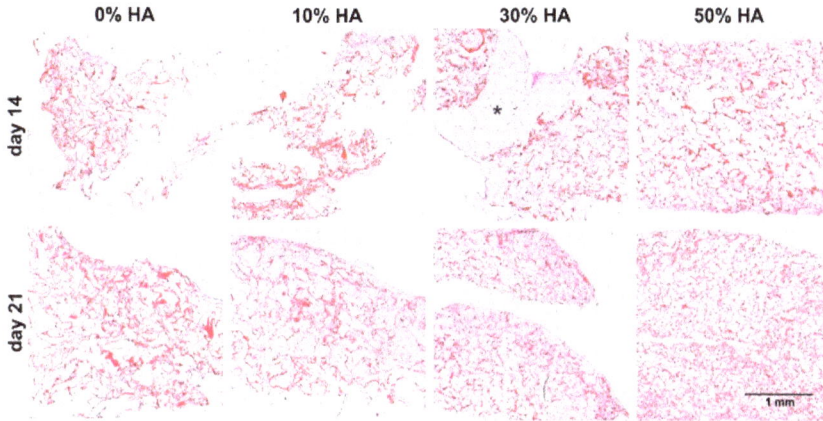

Figure 3. Histology of hMSCs cultured after 14 and 21 days. Hematoxylin-eosin staining of hMSCs-scaffold constructs after 14 and 21 days of culture. Positive influence of 30% of hydroxyapatite can be observed by large amount of newly-formed tissue with respect to the other scaffolds after 14 days of culture, designated by asterisk (*).

3.2.2. Quantitative Evaluation of Osteoinduction

The differentiation program of mesenchymal cells can be divided into three distinct stages regulated by physiological signals and based on different markers expression: proliferation regulated by vitamin D and glucocorticoids; extracellular matrix maturation characterized by alkaline phosphatase activity and extracellular matrix mineralization followed by expression of osteocalcin.

Figure 4 represents relative expression of specific osteogenic genes on different chitosan-hydroxyapatite scaffolds normalized by the gene expression of hMSCs cultured onto 2D surface, after 21 days of culture. Apart from the dexamethasone-based upregulation of some osteogenic genes, the major impact of hydroxyapatite fraction could be visible at later stage of hMSCs differentiation. Very high deviation of gene expression was found for scaffold with 10% of hydroxyapatite which could be a result of non-homogeneous distribution of smaller in situ formed HA crystals at such low fraction. On contrary, the higher expression of collagen type I, which prepare the extracellular matrix for the onset of mineralization process, was found on 30% HA scaffold regarding pure chitosan and 50% HA scaffold.

Osteocalcin is expressed immediately before the mineralization which categorized it as the late stage marker for osteoblastic differentiation, along with BSP which is highly restricted to bone and mineralized tissues [26]. Accordingly, poor mineralization follows the same trend as collagen I expression, indicated by the lower expression of osteocalcin and BSP detected on scaffolds with lowest and highest HA fraction (0% and 50%). DMP1 represents small integrin-binding ligand linked glycoprotein which is critical for regular bone mineralization [27,28]. With exception of non-detectable expression for 0% HA, all scaffolds did not exhibit significant difference in DMP1 expression after 21 days of culture.

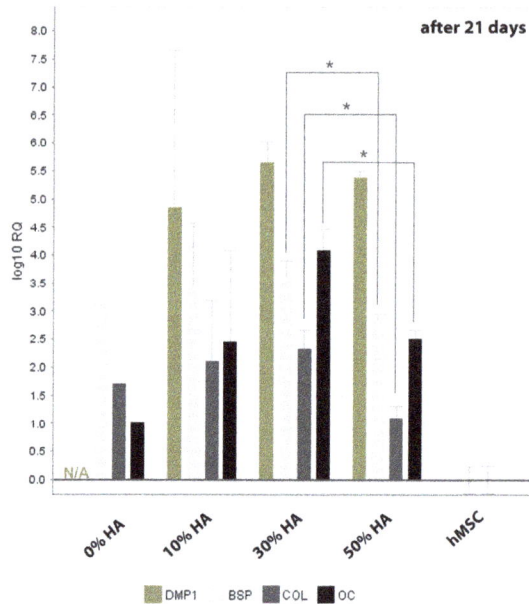

Figure 4. Relative expression (RQ) of osteogenic markers after 21 days of dynamic human mesenchymal stem cells (hMSCs) culture. The relative gene expression was analyzed by comparative cycle threshold method ($^{\Delta\Delta}$Ct) and the values were normalized to β-actin expression. Significant difference between two groups ($p < 0.05$) is designated as (*).

3.2.3. Immunohistochemical Staining of the Grafts

The production of type I collagen is one of the earlier events in MSC osteogenic differentiation [29], and it is capable of stimulating MSC osteogenic differentiation through the activation of focal adhesion kinase and extracellular signal-regulated kinase, even without additional ostoinductive factors [30]. Immunohistochemical analysis revealed weak positive staining of collagen I on all constructs after 14 days of culture, indicating early stages of osteoinduction (Figure 5), with the exception of intense localized staining for scaffold with 10% HA. However, remarkably intense and homogeneous staining of type I collagen is observed only on grafts with 30% of hydroxyapatite after 21 days, indicating a greater amount of newly deposited ECM. Enhanced production and expression of collagen and other early stage proteins, including alkaline phosphatase, set the extracellular matrix for initial mineralization. At this point, osteoblasts begin to produce calcium that creates mineralized matrix. Osteocalcin partially plays a role in the regulation of bone mineral deposition, thus its expression can be used as an indication of mineralized matrix formation [31–34].

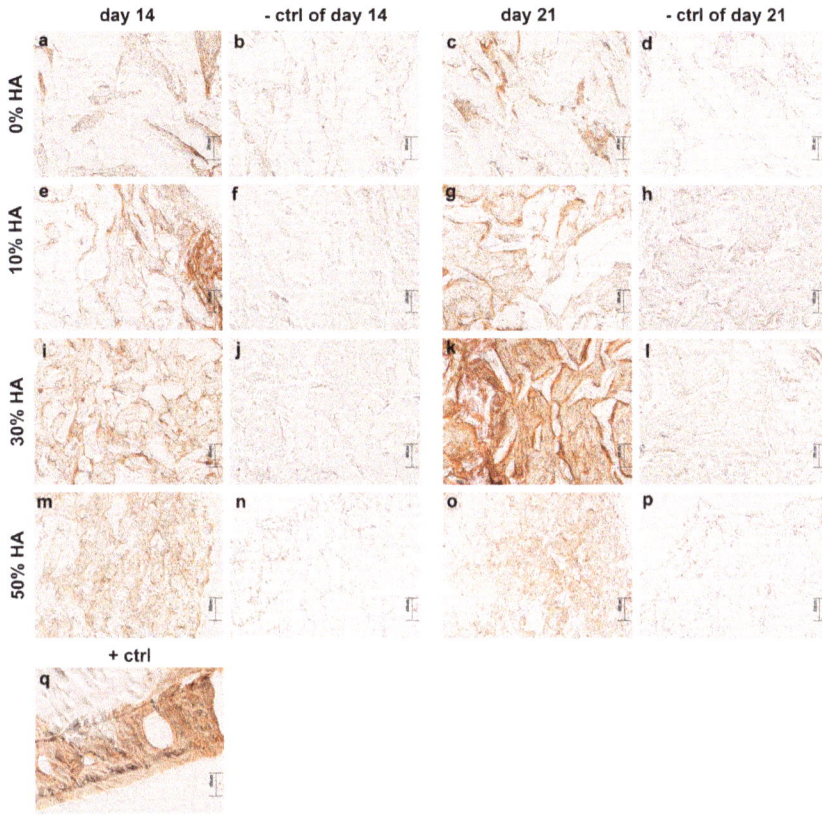

Figure 5. Immunohistochemical examination of collagen I on different scaffolds after 14 and 21 days of osteogenic induction. Brown color indicates positive staining. Weak staining is observed in samples with 0% HA (**a,c**). Very strong staining was observed in scaffold composed of 30% HA after 21 days of osteoinduction (**k**). Moderate staining for collagen I was shown for scaffolds composed of 10 and 50% HA (**e,g,m,o**), as well as for scaffold composed of 30% HA after 14 days of osteoinduction (**i**). As negative controls (−ctrl of day 14 and −ctrl of day 21), sections were processed for each sample in the absence of the suitable primary antibody (**b,d,f,h,j,l,n,p**). Human bone was used as a sample for positive control (+ctrl) (**q**). Representative images are shown at 100× magnification (*n* = 3). Scale bar represents 200 μm.

Immunohistochemistry staining of osteocalcin expression (Figure 6) is in agreement with gene expression determined by RT-qPCR. Scaffolds with 0 and 50% showed insignificant staining after 21 days of culture as compared to the dark brown staining on 30% HA scaffold. Eliminating high interference, weak brown color is detected on 10% HA scaffold, indicating a higher expression of later stage marker with respect to scaffold with 0 and 50% of HA.

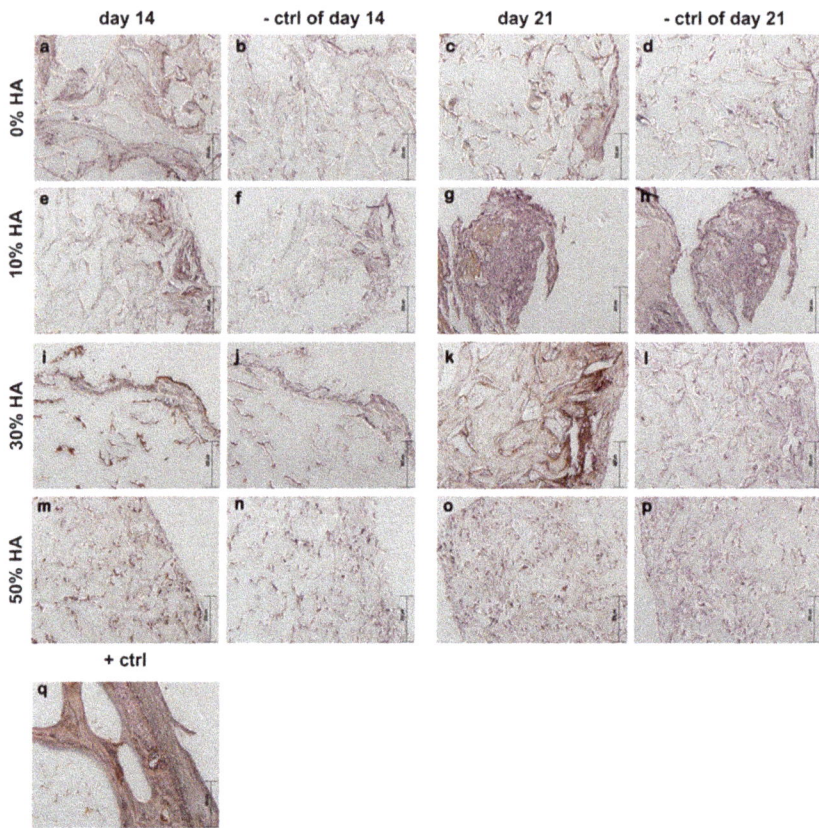

Figure 6. Immunohistochemical staining with anti-ostecalcin. Brown color indicates positive staining. Weak staining is observed in samples with 0% HA (**a,c**) and 50% HA (**m,o**). Very strong staining was observed in scaffold composed of 30% HA after 21 days of osteoinduction (**k**). Moderate staining for osteocalcin was shown for scaffold with 10% HA (**e,g**) as well as for scaffold 30% HA scaffold after 14 days of dynamic culture (**i**). As negative controls (−ctrl of day 14 and −ctrl of day 21), sections were processed for each sample in the absence of the suitable primary antibody (**b,d,f,h,j,l,n,p**). Human bone was used as a sample for positive control (+ctrl) (**q**). Representative images are shown at 100× magnification (*n* = 3). Scale bar represents 200 µm.

Von Kossa and alizarin red S staining were used to characterize the bone nodule formation of hMSCs after 14 and 21 days of dynamic culture. Although Von Kossa staining is nonspecific for calcium ion, it is positive for its carbonate or phosphate precipitates [35], while in the presence of calcium, the staining product of alizarin red S appears red.

Figure 7A,B show induced osteodifferentiation in Cht-HA composite scaffolds, which is significant in the later stage of cell culture. Localized bright staining of mineralized tissue was detected on 10% HA scaffold after 14 days, followed by more intense staining after 21 days of culture. However, mineralized bone nodules stained dark brown were clearly observed in 30% HA scaffold after 21 days, accompanied by strong alizarin red S staining, indicating a greater amount of mineralized matrix. On contrary, 50% HA scaffold exhibited significantly weaker Von Kossa staining, indicating a poor extracellular mineralization as prompted by the highest HA fraction.

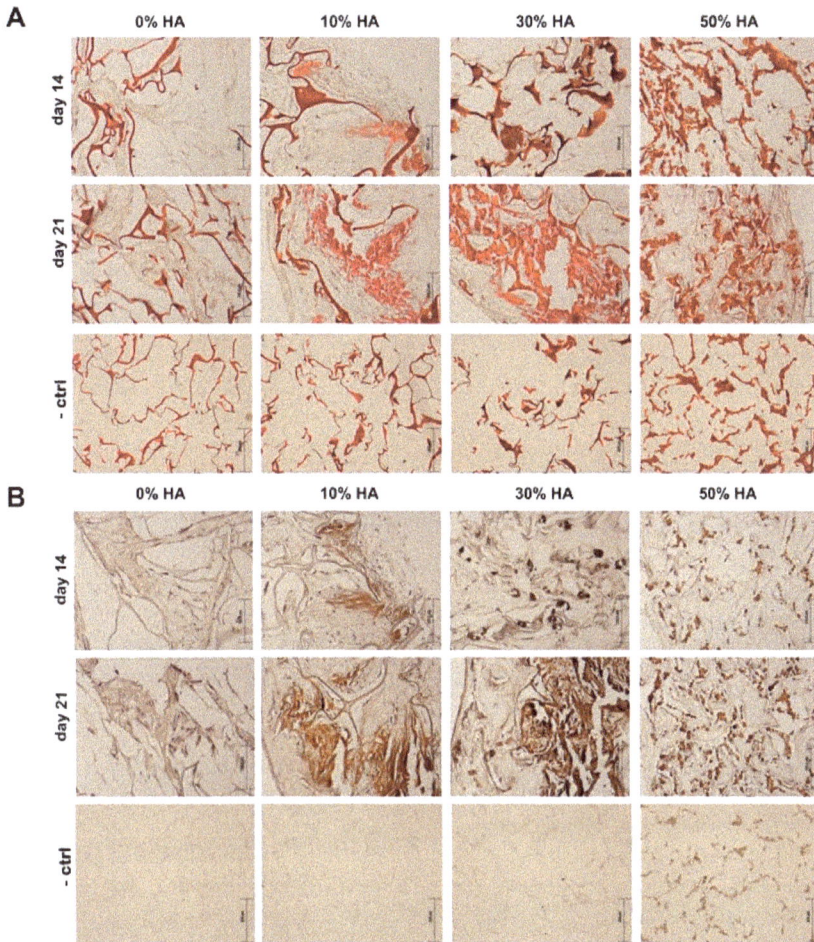

Figure 7. Assessing the mineralization of hMSCs grown in an osteogenic medium within different scaffolds after 14 and 21 days. (**A**) Mineralization of the extracellular matrix with the presence of calcium precipitates was visualized by staining with alizarin red S at days 14 and 21. Positive staining for calcium is signified by red color; (**B**) detection of mineralization using von Kossa staining (including hematoxylin counterstain). Positive staining for calcium phosphate is signified by brown color. Representative images are shown at 200× magnification (*n* = 3) and the scale bars represent 100 μm, respectively. Scaffolds without hMSCs were used as negative controls (−ctrl). Representative images of negative controls are shown at 100× magnification (*n* = 3) and the scale bars represent 200 μm.

4. Discussion

Numerous studies focusing on the importance of the preparation technique of potential bone grafts have already been emerged [36]. The in situ synthesis of hydroxyapatite as an osteoinductive inorganic component has been pointed out as a favorable method to reduce nonhomogeneity and potential localized defects in the mechanical performance of the bone graft. The higher homogeneity and better distribution preventing the agglomeration of hydroxyapatite within the sample is crucial in bioresorbability, influencing the local pH gradient of surrounding tissue and finally potential immunological reactions.

Porosity, pore size, and the interconnectivity of scaffolds play a predominant role in cell seeding and proliferation, ingrowth, and vascularization, and biomechanical properties of the scaffold during tissue formation. In fact, those features impact the direction of tissue regeneration. High porosity and pore interconnectivity dictate molecule diffusivity and fluid conductance, especially during tissue healing and integration [37]. Among the first minimum recommended pore size for a scaffold of approximately 100 μm [38], the following studies have shown better osteogenesis on scaffolds characterized with pore sizes larger than 300 μm [39]. It has been shown that relatively larger pores favor direct osteogenesis due to the easier vascularisation and high oxygen permeability, while smaller pores result in osteochondral ossification. However, the type of bone ingrowth depends on the biomaterial features and the geometry of the pores. Even though large pores provide better neotissue penetration, a certain upper limit of porosity and pore size should be considered from a mechanical and biodegradation point of view. Microscopic scanning of prepared Cht-HA scaffolds revealed large irregular pores of up to 200 μm, with good interconnectivity which has been maintained with different HA fractions. Even though pore walls have been disrupted by the increased HA amount, the size of non-defined shape pores and high interconnectivity were maintained. Former studies reported that dynamic cultures performed under stirring and perfusion flow enhance mass transfer, stimulate the cells hydrodynamically or mechanically, and positively influenced cell proliferation [40]. With high porosity, scaffolds can provide unhindered media flow during perfusion conditioning and good cell exposure to the interstitial medium within the whole scaffold, resulting in higher seeding efficiency and an even distribution of cells and newly-formed tissue.

Apart from the existing biomaterials, chitosan as a cationic polymer is extensively utilized in combination with other inorganic ceramics, especially hydroxyapatite, for the further enhancement of tissue regenerative efficacy and osteoconductivity, while chitosan-hydroxyapatite composite materials have exerted favorable bioactivity and bone-bonding ability [41]. Higher bioactivity of chitosan is detected in the presence of hydroxyapatite by the increment of cell number on composite scaffolds. This effect is shown to be limited by the HA fraction, according to the drastic decease of cells on the 50% HA scaffold. One of the important factors involved during cell adhesion and proliferation are the surface properties of the scaffold, including hydrophilicity, surface composition, charge, roughness, topography, etc. Despite bioactivity and biocompatibility, hydroxyapatite is bioresorbable under physiological conditions. More exposed HA crystals observed on 50% HA scaffold under microscopic scanning can impact the rate of HA resorption, i.e., the equilibrium rate between dissolution and precipitation of apatitic crystals during cell culture. This equilibrium could impact local changes in the pH of the medium affecting the cell adhesion and further survival [42]. Although cell seeding was very high for all of the composites, measured by the quantity of detached cells after 24 h of seeding, there is a possibility for cell detachment during culture provoked by the saturation of HA particles. Consequently, lower amounts of remained cells would generate a lower expression of osteogenic genes and poor osteogenesis.

Stem cells are regulated by physical and chemical factors localized in their extracellular surroundings [43]. The MSCs differentiation to bone cells can be induced by exposing the cells to a three-component cocktail of ascorbic acid, β-glycerophosphate, and synthetic steroid, namely dexamethasone. The influence of hydroxyapatite as an osteoinductive component of artificial bone graft has been investigated by numerous studies of hMSCs cell culture in osteogenic and basal environments [44–47]. Our Cht-HA scaffold with 30% of HA has emerged with an abundant accumulation of fibrous collagen I, and osteocalcin detected by a very strong staining after 21 days of dynamic culture, indicating its advanced osteoinductive potential by activating hMSCs differentiation into osteoblasts.

During osteogenic differentiation, the cells begin to secrete calcium phosphate minerals, referred as the mineralization phase. In this phase, the formation of bone nodule takes place, which is one of the osteodifferentation markers. The extracellular mineralization is considered to be the strongest indicator of osteoblast differentiation and osteogenesis [48]. Intense von Kossa staining in the 30% HA scaffold after 21 days suggests that this microenvironment activates and promotes the production of a mineralized matrix from the cells [49]. Such a high degree of mineralization could indicate an environment conducive to osteogenesis [48].

Biomaterial's influence on the hMSCs osteodifferentiation is greatly important, especially during in vivo bone restoration. By choosing pertinent biomaterial's composition, one could achieve simultaneous bone formation and biomaterial's degradation. We have demonstrated the existence of a critical amount of osteoinductive components, namely hydroxyapatite, for a suitable allogeneic graft development. Poor stem cell differentiation on 50% HA scaffold can be caused by lower number of cells during culture. However, this study shows paramount influence of hydroxyapatite, not only on osteoinductive properties, but also on cell-adhesive surface properties of artificial scaffold as one of primary requirement. Preliminary results presented demonstrate the important impact of hydroxyapatite fraction on hMSCs differentiation that can dictate future development of potential bone substituent, with limitations at a higher HA amount. However, concerning the variability of biological and technical sources, obtained data needs to be compared by additional dynamic cultures of cells derived from different patients.

5. Conclusions

The chitosan-hydroxyapatite scaffolds represent a favorable microenvironment for tissue penetration and ingrowth, as indicated by histological analysis of cultured grafts. Extensive influence of hydroxyapatite on hMSC proliferation and osteoinduction can be observed by an emphasized immunostaining of collagen type I and strong osteocalcin and calcium deposit staining on composite scaffolds with 30% of HA, while a higher apatite fraction indicates poor mineralization of hMSCs extracellular matrix. Despite a high percentage of attached cells (90–98%) under perfusion conditioning, poor tissue ingrowth and weak osteogenesis was observed on pure chitosan and 50% HA scaffold, pointing out that chemical composition of scaffolds is one of the dominant factors for good cell seeding and growth. The evaluation of osteonduction markers highlights the impact of hydroxyapatite fractions on osteogenesis, in terms of better ECM mineralization, as promoted by lower amounts (10–30%) of the osteoinductive phase within chitosan-based scaffolds.

Acknowledgments: BICRO PoC5_1_82, the Croatian Science Foundation under the project IP-2014-09-3752 and L'Oreal-UNESCO Foundation 'For Women in Science' are gratefully acknowledged.

Author Contributions: Alan Ivković, Inga Marijanović and Hrvoje Ivanković conceived and designed the experiments; Anamarija Rogina performed syntheses and production of investigated materials; Lidija Pribolšan performed dynamic cell culture on materials; Maja Antunović and Katarina Caput Mihalić performed experiments for biological evaluations on materials; Andreja Vukasović analyzed the data; Inga Marijanović, Gloria Gallego Ferrer, Marica Ivanković and Hrvoje Ivanković contributed reagents, materials and analysis tools; Anamarija Rogina wrote the paper.

Conflicts of Interest: The authors declare no conflict of interest.

References

1. Burdick, J.; Vunjak-Novakovic, G. Engineered microenvironments for controlled stem cell differentiation. *Tissue Eng. A* **2009**, *15*, 205–219. [CrossRef] [PubMed]
2. Ivkovic, A.; Marijanovic, I.; Hudetz, D.; Porter, R.M.; Pecina, M.; Evans, C.H. Regenerative medicine and tissue engineering in orthopaedic surgery. *Front. Biosci.* **2011**, *E3*, 923–944. [CrossRef]
3. Majumdar, M.K.; Keane-Moore, M.; Buyaner, D.; Hardy, W.B.; Moorman, M.A.; McIntosh, K.R.; Mosca, J.D. Characterization and functionality of cell surface molecules on human mesenchymal stem cells. *J. Biomed. Sci.* **2003**, *10*, 228–241. [CrossRef] [PubMed]
4. Koegler, P.; Clayton, A.; Thissen, H.; Santos, G.N.C.; Kingshott, P. The influence of nanostructured materials on biointerfacial interactions. *Adv. Drug Deliv. Rev.* **2012**, *64*, 1820–1839. [CrossRef] [PubMed]

5. Sellgren, K.L.; Ma, T. Perfusion conditioning of hydroxyapatite–chitosan–gelatin scaffolds for bone tissue regeneration from human mesenchymal stem cells. *J. Tissue Eng. Regen. Med.* **2012**, *6*, 49–59. [CrossRef] [PubMed]

6. Zhao, F.; Grayson, W.L.; Ma, T.; Brunnell, B.; Lu, W.W. Effects of hydroxyapatite in 3-D chitosan–gelatin polymer network on human mesenchymal stem cell construct development. *Biomaterials* **2006**, *27*, 1859–1867. [CrossRef] [PubMed]

7. Smith, I.O.; McCabe, L.R.; Baumann, M.J. MC3T3-E1 osteoblast attachment and proliferation on porous hydroxyapatite scaffolds fabricated with nanophase powder. *Int. J. Nanomed.* **2006**, *1*, 189–194. [CrossRef]

8. Wang, J.; De Boer, J.; De Groot, K. Proliferation and differentiation of osteoblast-like MC3T3-E1 cells on biomimetically and electrolytically deposited calcium phosphate coatings. *J. Biomed. Mater. Res. A* **2009**, *90*, 664–670. [CrossRef] [PubMed]

9. Imaizumi, H.; Sakurai, M.; Kashimoto, O.; Kikawa, T.; Suzuki, O. Comparative study on osteoconductivity by synthetic octacalcium phosphate and sintered hydroxyapatite in rabbit bone marrow. *Calcif. Tissue Int.* **2006**, *78*, 45–54. [CrossRef] [PubMed]

10. Pistone, A.; Iannazzo, D.; Panseri, S.; Montesi, M.; Tampieri, A.; Galvagno, S. Hydroxyapatite-magnetite-MWCNT nanocomposite as a biocompatible multifunctional drug delivery system for bone tissue engineering. *Nanotechnology* **2014**, *25*. [CrossRef] [PubMed]

11. Pistone, A.; Iannazzo, D.; Espro, C.; Galvagno, S.; Tampieri, A.; Montesi, M.; Panseri, S.; Sandri, M. Tethering of Gly-Arg-Gly-Asp-Ser-Pro-Lys Peptides on Mg-Doped Hydroxyapatite. *Engineering* **2017**, *3*, 55–59. [CrossRef]

12. Muzzarelli, R.A.A. Chitins and chitosans for the repair of wounded skin, nerve, cartilage and bone. *Carbohydr. Polym.* **2009**, *76*, 167–182. [CrossRef]

13. Kim, T.; Jiang, H.; Jere, D.; Park, P.; Cho, M.; Nah, J.; Choi, Y.; Akaike, T.; Cho, C. Chemical modification of chitosan as a gene carrier in vitro and in vivo. *Prog. Polym. Sci.* **2007**, *32*, 726–753. [CrossRef]

14. Muzzarelli, R.A.A. Chitosan composites with inorganics, morphogenetic proteins and stem cells, for bone regeneration. *Carbohydr. Polym.* **2011**, *83*, 1433–1445. [CrossRef]

15. Hunter, K.T.; Ma, T. In vitro evaluation of hydroxyapatite–chitosan–gelatin composite membrane in guided tissue regeneration. *J. Biomed. Mater. Res. A* **2013**, *101*, 1016–1025. [CrossRef] [PubMed]

16. Palazzo, B.; Gall, A.; Casillo, A.; Nitti, P.; Ambrosio, L.; Piconi, C. Fabrication, characterization and cell cultures on novel composite chitosan-nano-hydroxyapatite scaffold. *Int. J. Immunopathol. Pharmacol.* **2011**, *24*, 73–78. [CrossRef] [PubMed]

17. Lai, G.; Shalumon, K.T.; Chen, J. Response of human mesenchymal stem cells to intrafibrillar nanohydroxyapatite content and extrafibrillar nanohydroxyapatite in biomimetic chitosan/silk fibroin/nanohydroxyapatite nanofibrous membrane scaffolds. *Int. J. Nanomed.* **2015**, *10*, 567–584. [CrossRef]

18. Zhao, F.; Ma, T. Perfusion bioreactor system for human mesenchymal stem cell tissue engineering: Dynamic cell seeding and construct development. *Biotechnol. Bioeng.* **2005**, *91*, 482–493. [CrossRef] [PubMed]

19. Jones, A.C.; Arns, C.H.; Hutmacher, D.W.; Milthorpe, B.K.; Sheppard, A.P.; Knackstedt, M.A. The correlation of pore morphology, interconnectivity and physical properties of 3D ceramic scaffolds with bone ingrowth. *Biomaterials* **2009**, *30*, 1440–1451. [CrossRef] [PubMed]

20. Kim, J.; Ma, T. Perfusion regulation of hMSC microenvironment and osteogenic differentiation in 3D scaffold. *Biotechnol. Bioeng.* **2012**, *109*, 252–261. [CrossRef] [PubMed]

21. Zhao, F.; Chella, R.; Ma, T. Effects of shear stress on 3-D human mesenchymal stem cell construct development in a perfusion bioreactor system: Experiments and hydrodynamic modeling. *Biotechnol. Bioeng.* **2007**, *96*, 584–595. [CrossRef] [PubMed]

22. Rogina, A.; Rico, P.; Ferrer, G.G.; Ivanković, M.; Ivanković, H. In situ hydroxyapatite content affects the cell differentiation on porous chitosan/hydroxyapatite scaffolds. *Ann. Biomed. Eng.* **2016**, *44*, 1107–1119. [CrossRef] [PubMed]

23. Rogina, A.; Rico, P.; Ferrer, G.G.; Ivanković, M.; Ivanković, H. Effect of in situ formed hydroxyapatite on microstructure of freeze-gelled chitosan-based biocomposite scaffolds. *Eur. Polym. J.* **2016**, *68*, 278–287. [CrossRef]

24. Matic, I.; Antunović, M.; Brkic, S.; Josipovic, P.; Mihalic, K.C.; Karlak, I.; Ivkovic, A.; Marijanovic, I. Expression of OCT-4 and SOX-2 in bone marrow-derived human mesenchymal stem cells during osteogenic differentiation. *Open Access Maced. J. Med. Sci.* **2016**, *4*, 9–16. [CrossRef] [PubMed]

25. Valenti, M.T.; Dalle Carbonare, L.; Donatelli, L.; Bertoldo, F.; Zanatta, M.; Lo Cascio, V. Gene expression analysis in osteoblastic differentiation from peripheral blood mesenchymal stem cells. *Bone* **2008**, *43*, 1084–1092. [CrossRef] [PubMed]

26. Mizuno, M.; Fujisawa, R.; Kuboki, Y. Type I collagen-induced osteoblastic differentiation of bone-marrow cells mediated by collagen-α2β1 integrin interaction. *J. Cell. Physiol.* **2000**, *184*, 207–213. [CrossRef]

27. Qin, C.; D'Souza, R.; Feng, J.Q. Dentin matrix protein 1 (DMP1): New and important roles for biomineralization and phosphate homeostasis. *J. Dent. Res.* **2007**, *86*, 1134–1141. [CrossRef] [PubMed]

28. Feng, J.Q.; Huang, H.; Lu, Y.; Ye, L.; Xie, Y.; Tsutsui, T.W.; Kunieda, T.; Castranio, T.; Scott, G.; Bonewald, L.B.; et al. The dentin matrix protein 1 (DMP1) is specifically expressed in mineralized but not soft tissues during development. *J. Dent. Res.* **2003**, *82*, 776–780. [CrossRef] [PubMed]

29. Takeuchi, Y.; Suzawa, M.; Kikuchi, T.; Nishida, E.; Fujita, T.; Matsumoto, T. Differentiation and transforming growth factor-beta receptor down regulation by collagen-alpha 2 beta 1 integrin interaction is mediated by focal adhesion kinase and its downstreamsignals in murine osteoblastic cells. *J. Biol. Chem.* **1997**, *272*, 29309–29316. [CrossRef] [PubMed]

30. Kundu, A.K.; Putnam, A.J. Vitronectin and collagen I differentially regulate osteogenesis in mesenchymal stem cells. *Biochem. Biophy. Res. Commun.* **2006**, *347*, 347–357. [CrossRef] [PubMed]

31. Stein, G.S.; Lian, J.B.; Stein, J.L.; Van Wijnen, A.J.; Montecino, M. Transcriptional control of osteoblast growth and differentiation. *Physiol. Rev.* **1996**, *76*, 593–629. [PubMed]

32. Gundberg, C.M. Biochemical markers of bone formation. *Clin. Lab. Med.* **2000**, *20*, 489–501. [PubMed]

33. Dworetzky, S.I.; Fey, E.G.; Penman, S.; Lian, J.B.; Stein, J.L.; Stein, G.S. Progressive changes in the protein composition of the nuclear matrix during rat osteoblast differentiation. *Proc. Natl. Acad. Sci. USA* **1990**, *87*, 4605–4609. [CrossRef] [PubMed]

34. Stein, G.S.; Lian, J.B.; Van Wijnen, A.J.; Stein, J.L. The osteocalcin gene: A model for multiple parameters of skeletal-specific transcriptional control. *Mol. Biol. Rep.* **1997**, *24*, 185–196. [CrossRef] [PubMed]

35. Lee, J.H.; Shin, Y.C.; Lee, S.M.; Jin, O.S.; Kang, S.H.; Hong, S.W.; Jeong, C.M.; Huh, J.B.; Han, D.W. Enhanced osteogenesis by reduced graphene oxide/hydroxyapatite nanocomposites. *Sci. Rep.* **2015**, *5*, 1–13. [CrossRef] [PubMed]

36. Sadat-Shojai, M.; Khorasani, M.; Dinpanah-Khoshdargi, E.; Jamshidi, A. Synthesis methods for nanosized hydroxyapatite with diverse structures. *Acta Biomater.* **2013**, *9*, 7591–7621. [CrossRef] [PubMed]

37. Sundelacruz, S.; Kaplan, D. Stem cell and scaffold-based tissue engineering approaches to osteochondral regenerative medicine. *Semin. Cell Dev. Biol.* **2009**, *20*, 646–655. [CrossRef] [PubMed]

38. Hulbert, S.F.; Young, F.A.; Mathews, R.S.; Klawitter, J.J.; Talbert, C.D.; Stelling, F.G. Potential of ceramic materials as permanently implantable skeletal prostheses. *J. Biomed. Mater. Res.* **1970**, *4*, 433–456. [CrossRef] [PubMed]

39. Karageorgiou, V.; Kaplan, D. Porosity of 3D biomaterial scaffolds and osteogenesis. *Biomaterials* **2005**, *26*, 5474–5491. [CrossRef] [PubMed]

40. Takahashi, Y.; Yamamoto, M.; Tabata, Y. Osteogenic differentiation of mesenchymal stem cells in biodegradable sponges composed of gelatin and b-tricalcium phosphate. *Biomaterials* **2005**, *26*, 3587–3596. [CrossRef] [PubMed]

41. Ma, X.Y.; Feng, Y.F.; Ma, Z.S.; Li, X.; Wang, J.; Wang, L.; Lei, W. The promotion of osteointegration under diabetic conditions using chitosan/hydroxyapatite composite coating on porous titanium surfaces. *Biomaterials* **2014**, *35*, 7259–7270. [CrossRef] [PubMed]

42. Barrère, F.; Van Blitterswijk, C.A.; De Groot, K. Bone regeneration: Molecular and cellular interactions with calcium phosphate ceramics. *Int. J. Nanomed.* **2006**, *1*, 317–332.

43. Tang, S.; Tian, B.; Guo, Y.J.; Zhu, Z.A.; Guo, Y.P. Chitosan/carbonated hydroxyapatite composite coatings: Fabrication, structure and biocompatibility. *Surf. Coat. Technol.* **2014**, *251*, 210–216. [CrossRef]

44. Marinucci, L.; Balloni, S.; Becchetti, E.; Bistoni, G.; Calvi, E.M.; Lumare, E.; Ederli, F.; Locci, P. Effects of hydroxyapatite and Biostite®on osteogenic induction of hMSC. *Ann. Biomed. Eng.* **2010**, *38*, 640–648. [CrossRef] [PubMed]

45. Spadaccio, C.; Rainer, A.; Trombetta, M.; Vadalá, G.; Chello, M.; Covino, E.; Denaro, V.; Toyoda, Y.; Genovese, J.A. Poly-L-lactic acid/hydroxyapatite electrospun nanocomposites induce chondrogenic differentiation of human MSC. *Ann. Biomed. Eng.* **2009**, *37*, 1376–1389. [CrossRef] [PubMed]

46. Dimitrievska, S.; Bureau, M.N.; Antoniou, J.; Mwale, F.; Petit, A.; Lima, R.S.; Marple, B.R. Titania-hydroxyapatite nanocomposite coatings support human mesenchymal stem cells osteogenic differentiation. *J. Biomed. Mater. Res. A* **2011**, *98*, 576–588. [CrossRef] [PubMed]

47. Polini, A.; Pisignano, D.; Parodi, M.; Quarto, R.; Scaglione, S. Osteoinduction of human mesenchymal stem cells by bioactive composite scaffolds without supplemental osteogenic growth factors. *PLoS ONE* **2011**, *6*, e26211. [CrossRef] [PubMed]

48. Depan, D.; Pesacreta, T.C.; Misra, R.D.K. The synergistic effect of a hybrid grapheme oxide–chitosan system and biomimetic mineralization on osteoblast functions. *Biomater. Sci.* **2014**, *2*, 264–274. [CrossRef]

49. Lee, S.H.; Chung, H.Y.; Shin, H.I.; Park, D.J.; Choi, J.H. Osteogenic activity of chitosan-based hybrid scaffold prepared by polyelectrolyte complex formation with alginate. *Tissue Eng. Regen. Med.* **2014**, *11*, 1–7. [CrossRef]

Article

In Vitro and in Vivo Study of Poly(Lactic–*co*–Glycolic) (PLGA) Membranes Treated with Oxygen Plasma and Coated with Nanostructured Hydroxyapatite Ultrathin Films for Guided Bone Regeneration Processes

Daniel Torres-Lagares [1,*], Lizett Castellanos-Cosano [1], María Ángeles Serrera-Figallo [1,*], Francisco J. García-García [2], Carmen López-Santos [2], Angel Barranco [2], Agustín Rodríguez-Gonzalez Elipe [2], Cristóbal Rivera-Jiménez [1] and José-Luis Gutiérrez-Pérez [1]

[1]   Faculty of Dentistry, University of Seville. Avicena Street, 41009 Seville, Spain; lizettcastellanos@yahoo.es (L.C.-C.); cmriveraj@gmail.com (C.R.-J.); jlgp@us.es (J-L.G.-P.)
[2]   Institute of Materials Science of Seville (CSIC-University of Seville), Américo Vespucio Street n 49, 41092 Seville, Spain; franciscoj.garcia@icmse.csic.es (F.J.G.-G.); mclopez@icmse.csic.es (C.L.-S.); angel.barranco@csic.es (A.B.); arge@icmse.csic.es (A.R.-G.E.)
*   Correspondence: danieltl@us.es (D.T.-L.); maserrera@us.es (M.Á.S.-F.)

Received: 23 June 2017; Accepted: 30 August 2017; Published: 2 September 2017

Abstract: The novelty of this study is the addition of an ultrathin layer of nanostructured hydroxyapatite (HA) on oxygen plasma modified poly(lactic–*co*–glycolic) (PLGA) membranes (PO$_2$) in order to evaluate the efficiency of this novel material in bone regeneration. Methods: Two groups of regenerative membranes were prepared: PLGA (control) and PLGA/PO$_2$/HA (experimental). These membranes were subjected to cell cultures and then used to cover bone defects prepared on the skulls of eight experimental rabbits. Results: Cell morphology and adhesion of the osteoblasts to the membranes showed that the osteoblasts bound to PLGA were smaller and with a lower number of adhered cells than the osteoblasts bound to the PLGA/PO$_2$/HA membrane ($p < 0.05$). The PLGA/PO$_2$/HA membrane had a higher percentage of viable cells bound than the control membrane ($p < 0.05$). Both micro-CT and histological evaluation confirmed that PLGA/PO$_2$/HA membranes enhance bone regeneration. A statistically significant difference in the percentage of osteoid area in relation to the total area between both groups was found. Conclusions: The incorporation of nanometric layers of nanostructured HA into PLGA membranes modified with PO$_2$ might be considered for the regeneration of bone defects. PLGA/PO$_2$/HA membranes promote higher osteosynthetic activity, new bone formation, and mineralisation than the PLGA control group.

Keywords: guided bone regeneration; hydroxyapatite; PLGA; oxygen plasma treatment; magnetron sputtering

1. Introduction

The use of physical barriers to prevent the invasion of gingival and connective tissue cells into bone cavities during the healing process is called guided bone regeneration (GBR) [1]. Tissue engineering enables the fabrication of membranes that mimic the structural properties of the original tissues, providing stable support to the extracellular matrix [2]. The application of polymeric biodegradable materials has become a common practice, however there is limited knowledge of the behaviour of the osteoblastic cells in them [3,4]. Poly(lactic–*co*–glycolic) (PLGA) copolymers, of all of the existing membranes, show good bone adhesion, vascularity, biodegradability and the ability of osteoblastic cells to grow on the surface of in vitro cultivations [5]. PLGA is known to be degraded by hydrolysis and eliminated in the Krebs cycle in the form of carbon dioxide and water. Therefore, it is successfully used as a biodegradable membrane in GBR. Its hydrophobicity and relative low cellular affinity, however, limit its choice [6].

As is well known, internal cellular organisation and orientation are controlled by focal adhesions. These focal adhesions mediate the regulatory effects of adhesion of the extracellular matrix and the distribution of actin-myosin fibres depending on surface properties [7]. Also, a substantial increase in osteoblastic mitochondrial bioenergy is oriented to focal adhesions, which are formed during cell adhesion and are constantly inserted and uninserted with the movement of the cell. Therefore, these focal adhesions based on integrins serve as mechanosensors, converting the mechanical signals of the medium into biological signals [8,9]. This scientific and biological evidence has allowed us to consider that the induction of osteoblast cellular activity can be achieved by modifying the surface of a biomaterial.

One of the most effective processes for the modification of PLGA polymer surfaces is the oxygen plasma surface modification [10]. The oxygen plasma treatment increases the polymer roughness and introduces surface chemical functionalizations, stimulating the adhesion of osteogenic mediators and cells and speeding up the membrane degradation [11]. Therefore, we may imply that the osteoinductive capacity of the barrier could be optimised by adding fine layers of nanocomposite particles, which might promote the osteoblastic adhesion and induce a functional differentiation that enables the formation of new bone [12,13].

A nanocomposite particle such as hydroxyapatite (HA) (of synthetic origin) may act as a membrane surface modifier. HA has a composition and mechanical resistance similar to natural bone, as well as osteoconductivity, osteoinductivity and biodegradability. Furthermore, medical products such as screws, plates and cylinders made of this material have been shown to form a natural bond with the bone in vivo [14]. Therefore, including HA within a biomaterial may allow us to regulate its pH, preventing inflammatory reactions and inducing bone formation [15].

The novelty of this study is the addition of a thin nanostructured layer of HA by magnetron sputtering on oxygen plasma modified PLGA foils (PO$_2$) to evaluate the efficiency of the resulting membranes in bone regeneration. The initial hypothesis was that modified PLGA membranes were able to alter the potential for bone regeneration when compared to PLGA.

2. Materials and Methods

2.1. Preparation of the Membranes

Thirty 50 μm thick resorbable inert PLGA scaffolds based on poly(lactic–*co*–glycolic) acid were fabricated using polycondensation (Institute of Materials Science, Seville, Spain). The membranes were used to cover bone defects prepared on the skulls of eight experimental rabbits. After subduing the membranes to cell culture to identify their viability, two groups of regenerative membranes (*n* = 4 each) were prepared and tested for GBR processes: (1) PLGA (control) and (2) PLGA/PO$_2$/HA (experimental).

For the preparation of the control membrane, 10 mL of a solution of PLGA (PLGA pellets with a copolymer ratio of 75:25 (lactic/glycolic acid) from Sigma-Aldrich Inc., St. Louis, MO, USA) in 1.5% dichloromethane was prepared by evaporation of the solvent on a Teflon plate for 48 h in air at room temperature obtaining a film of suitable consistency and a thickness of about 50 μm [16]. Figure 1 shows a membrane preparation schematic, as well as the possible surface treatment with oxygen plasma and the surface coating with hydroxyapatite.

The control group of PLGA membranes was characterised by X-ray photoemission spectroscopy (XPS, ESCALAB 250Xi, ThermoFisher Scientific, Waltham, MA, USA) so that the following chemical composition was registered on the membranes' surfaces—54% carbon and 46% oxygen. The molecular weight of the PLGA copolymer was 12 kDa. The PLGA was synthesised by means of the ring-opening copolymerisation of two different monomers, the cyclic dimers (1,4-dioxane-2,5-diones) of glycolic acid and lactic acid. During polymerisation, successive monomeric units (of glycolic or lactic acid) were attached to PLGA using ester linkages, thereby yielding linear, aliphatic polyester.

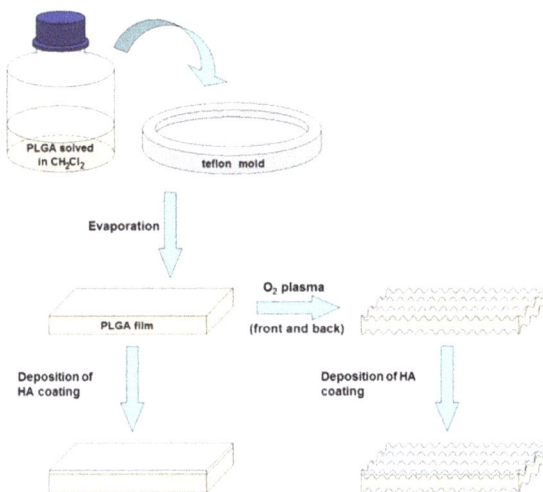

Figure 1. Manufacturing process of Poly(lactic–*co*–glycolic) PLGA membranes and subsequent surface treatment and deposition of thin layers of hydroxyapatite.

Subsequently, the membranes of Group 2 were coated with bioactive layers of HA by excited magnetron sputtering using a radiofrequency power of 50 W under an argon atmosphere at a pressure of 3×10^{-3} mbar [17]. Before deposition, the chamber was maintained at a base pressure of 1×10^{-6} mbar. The sputtering deposition was carried out using a copper coating that was 3 mm thick and 46 mm diameter white calcium phosphate. (HA) target (hydroxyapatite) was 4 mm thickness, 46 mm in diameter and 99.9% purity (Kurt J. Lesker Company, Jefferson Hills, PA, USA). The target sample distance was 10 cm and the deposition time was 40 min, obtaining a thickness of the coating of HA on PLGA of around 15 nm.

2.2. Materials Characterizations

The surface chemical composition of the samples was analysed by X-ray photoelectron spectroscopy (XPS) using an ESCALAB 210 spectrometer (Thermo Fisher Scientific, Waltham, MA, USA), operating at constant pass energy of 20 eV. Nonmonochromatized Mg Kα radiation was used as the excitation source. The atomic surface concentrations were quantitatively determined from the area of C1s, O1s, P2p and Ca2p peaks. A Shirley-type background was subtracted, and the peak areas were corrected by the electron escape depth, the spectrometer transmission and the photoelectron cross-sections.

Fourier transform infrared (FTIR) spectra were collected in a JASCO FT/IR-6200 IRT-5000 (Oklahoma, OK, USA) under vacuum conditions and specular reflectance mode.

The surface topography of the films was characterised by noncontact atomic force microscopy (AFM) with a Cervantes AFM system from NANOTEC (Feldkirchen, Germany) using commercial noncontact AFM tips from MikroMasch (Wetzlar, Germany). The surface of the membranes was also studied by scanning electron microscopy (SEM) in a Hitachi S4800 field emission microscope (Tokyo, Japan) of the coating of HA on PLGA of around 15 nm.

2.3. In Vitro Cell Cultures

The assays were performed using the human osteoblast line MG-63 acquired from the Center for Scientific Instrumentation (CIC) at the University of Granada (Granada, Spain). The MG-63 line shows faster growth than the primary bone-forming lines but retains quite a few characteristics of these, which makes it a good model in vitro.

Firstly, a control of mycoplasma contamination was performed by PCR to verify that the cells were free of contamination. A method of detection by PCR (polymerase chain reaction) was used. The amplification of a band of approximately 500 bp was performed according to the species—specific to eight species of mycoplasma (*M. hyorhinis*, *M. arginini*, *M. pneumoniae*, *M. fermentans*, *M. orale*, *M. pirum*, *Acholeplasma laidlawii* and *Spiroplasma mirum*)—using a single pair of oligonucleotides corresponding to the 16S RNA. PCR was performed by taking an aliquot of the conditioned medium from the cells in culture after at least 48 h of culture [18,19]. MG-63 cells were cultured on control and experimental membranes.

For the determination of cell adhesion and osteoblast viability, osteoblasts were cultured on the PLGA/PO$_2$/HA and the PLGA control membrane, in triplicate, planting 120,000 cells. At 24 h, the cultures were analysed by microphotography of osteoblasts [20]. Previously, the cells had been fixed with 70% ethanol for 5 min, as it was not possible to observe them without prior fixation. Bright-field microphotographs were taken of each condition at 5×, 10× and 20× with the Axio Observer A1 inverted microscope (Carl Zeiss). These images were also used to determine the average size of viable cells.

For the determination of the mitochondrial energy balance, the MitoProbe™ JC-1 Assay Kit was employed. JC-1 is a membrane permeable dye widely used for determining mitochondrial membrane potential in flow cytometry and fluorescent microscopy [21]. When mitochondria show good functioning, the probe accumulates in the mitochondria and forms aggregates, which emit in red (~590 nm). When the mitochondrial membrane potential decreases during cellular damage phenomena, the emission of the fluorescence turns green (~529 nm), decreasing the red/green ratio, because of the passage to the monomeric form of the probe. Osteoblast cultures were performed on the PLGA/PO$_2$/HA and PLGA control membrane in triplicate, planting 120,000 cells. The cultures were analysed at 24 h. The observed red and green fluorescence was due to the JC-1 probe, after a 30 min incubation of the cells in culture. Images were taken with the 40× and 20× lenses of the Axio Observer A1 (Carl Zeiss, Oberkochen, Germany) fluorescence microscope.

For the determination of osteogenesis and the morphology of adherent osteoblasts, one staining was employed. This staining consisted of the use of phalloidin-TRITC—a fluorescent phallotoxin that can be used to identify filamentous actin (F-actin) [22]—along with the use of DAPI (4′,6-diamidino-2-phenylindole)—a nuclear and chromosome counterstain emitting blue fluorescence upon binding to AT regions of DNA cells attached to the different membranes for 24 h [23]. Osteoblast cultures were performed on the PLGA/PO$_2$/HA and PLGA control membrane in triplicate, planting 120,000 cells. The cultures were analysed at 24 h. Cells were fixed with 70% alcohol. Both techniques were visualised at 20× with the fluorescence microscope Axio Observer A1 (Carl Zeiss).

2.4. Animal Experimentation Specimens

Four white, New Zealand-breed experimentation rabbits with identical characteristics (age: 6 months; weight: 3.5–4 kg) were selected for the study and fed daily with rabbit-maintenance Harlan-Teckland Lab Animal Diets (2030).

The surgical interventions were carried out at the Minimally Invasive Surgery Centre Jesús Usón (CCMI, Cáceres, Spain). The experiment was developed in accordance with the guidelines of the US National Institute of Health (NIH) and European Directive 86/609/EEC regarding the care and use of animals for experimentation. The study also complied with the European Directive 2010/63/EU about the protection of animals used for scientific purposes and with all local laws and regulations. The researchers obtained the approval of the Ethics Committee of the Minimally Invasive Surgery Centre Jesús Usón (CCMI, Cáceres, Spain). As required by the legislative framework, the minimum number of animals was used for ethical reasons [24]. Comparable models have been published concerning the histological and animal experimentation methods [6].

2.5. Surgical Procedure

The surgical procedure followed the same methodology as previously described [1,6]. The animals were immobilised, and their vital signs were checked. The anaesthesia used for induction was intravenous midazolam (0.25 mg/kg) and propofol (5 mg/kg). For maintenance, the animals inhaled 2.8% inspired sevoflurane gas. Analgesia was provided with ketorolac (1.5 mg/kg) and tramadol (3 mg/kg). After the rabbits were sedated and prepared, incisions between the bases of their ears and between their eyes were made with a No. 15 scalpel blade. After the two incisions were connected with an incision that coincided with the skull midline, a triangular field was discovered. The epithelial, connective, and muscular tissues were displaced using a Prichard periosteotome. The skull surface was washed with a sterile saline solution, maintaining aspiration. Two bone defects (diameter: 10 mm; depth: 3 mm) were created on the parietal bone, on each side of the skull midline, 3 mm apart, using a trephine (Helmut-Zepf Medical Gmbh, Seitingen, Germany) mounted on an implant micromotor operating at 2000 rpm under saline irrigation. The trephine had an internal diameter of 10 mm, a length of 30 mm, and teeth of 2.35 mm. Because the defects were created in a symmetric fashion in relation to the longitudinal axis of each rabbit skull, possible between-group variations in section thickness were minimised. For each defect, the outer table and the medullary bone were completely removed with piezosurgery, and the inner table was preserved to avoid damage to the brain tissue. The depth was controlled with a periodontal probe. A randomly assigned membrane was used to cover each bone defect. The randomisation sequence was generated using specific software (Research Randomizer, V. 4.0, Urbaniak GC & Plous S, 2013) [2]. The PLGA fibres of the barriers were also randomly oriented. The membranes were fixed with the fibrin tissue adhesive Tissucol (Baxter, Hyland S.A. Immuno, Rochester, MI, USA), which was placed on the bone rims adjacent to the defects. Proper adhesion and limited mobility of the membranes were confirmed when the flaps were moved back to their initial positions. Sutures were made on the following planes using resorbable material: periosteal (4/0), sub-epidermal (4/0) and skin (2/0). Simple stitches were used as close as possible to the edge. The wound was carefully cleaned with a sterile saline solution. Anti-inflammatory analgesia (buprenorphine 0.05 mg/kg and carprofen 1 mL/12.5 kg) was administered. The animals were sacrificed two months after surgery using an intravenous overdose of potassium chloride solution. Samples were obtained from the skull of each specimen, cutting them in an anatomical sagittal plane. After the brain mass was separated and the skull was washed with a sterile saline solution, the tissue samples were cut and marked individually. The complete surgical sequence is shown in Figure 2.

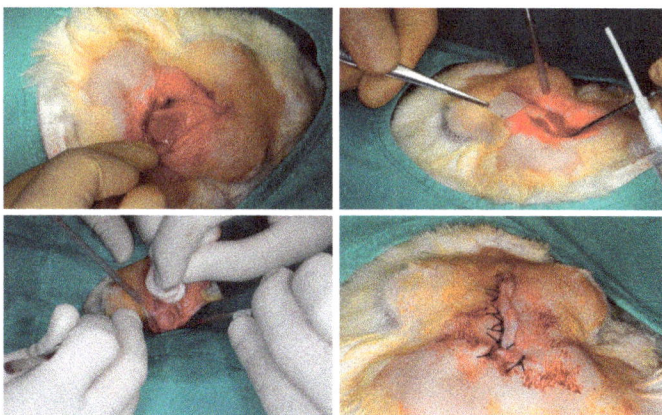

Figure 2. Surgical procedures in model animal.

2.6. Comparison of Bone Density (BoneJ)

The mean bone density of the surgical bone defect created and then covered with a random membrane was assessed by micro computed tomography (micro-CT) [25].

To determine variations in bone density measurements due to manual positioning of sampling cylinders, three cylinders of identical shape were positioned independently according to a central localisation criterion within each lesion and to maximise the overlap with bony structures. Spheres of 2 mm in diameter were placed in a rosette arrangement with a manually adjusted arrangement to match the largest possible bone structure (Figure 3).

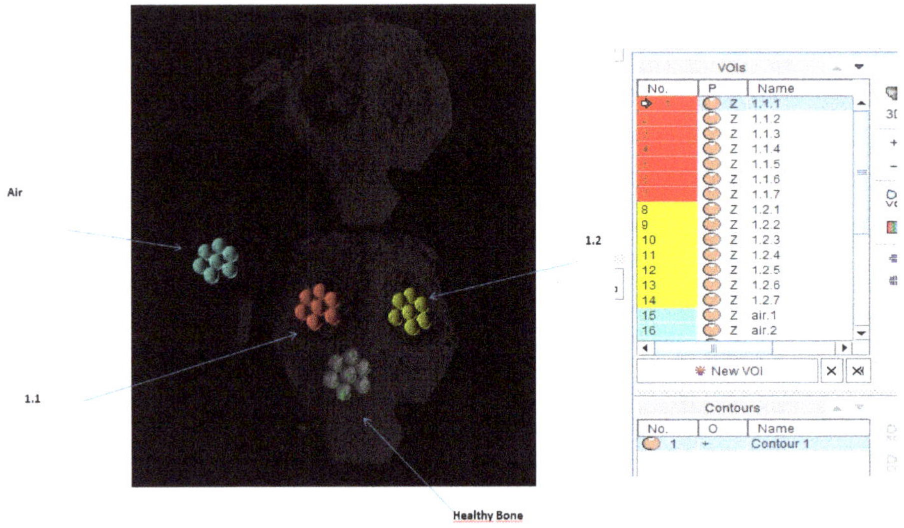

Figure 3. Microtomography study on animal skull.

The application of a binary threshold (intensity of pixels) enables us to define structures of interest. For the initial test, a threshold value of 6500 was chosen for all samples. This value is probably quite high and may lead to the detection of more individual trabecular structures while ignoring structures with less density. Changes in this threshold can have significant effects on the results obtained. Bone image analysis in ImageJ can therefore be used to evaluate different properties of the bone through a series of BoneJ plugins. The BoneJ plugin provides free, open-source tools for trabecular geometry and whole bone shape analysis [26].

2.7. Histological Processing of Samples

The processing of samples followed the protocol previously described [27]. Cranial blocks from containing fixtures were retrieved and stored in a 5% formaldehyde solution (pH 7). The blocks were retrieved from the regenerated bone defect using an oscillating autopsy saw (Exakt, Kulzer, Wehrheim, Germany). The dissected specimens were immediately immersed in a solution of 4% formaldehyde and 1% calcium and processed for ground sectioning following the Donath and Breuner method [28].

For histological staining and rapid contrast tissue analysis (Merck Toluidine Blue-Merck, Darmstadt, Germany), a metachromatic dye was used to assess the percentage of new bone formation. A 1% toluidine blue (TB) solution with a pH of 3.6 was chosen and adjusted with HCl 1 N. The samples were exposed to the dye for 10 minutes at RT, rinsed with distilled water, and air-dried. The von Kossa (VK) silver nitrate technique (Sigma–Aldrich Chemical Co., Poole, UK) was applied to visualise the mineralised bone [6].

2.8. Statistical Analysis

Means and standard deviations (SD) were calculated. The intraexaminer reliability was assessed using the Kappa test. Since the Kolmogorov–Smirnov test demonstrated that the data were not normally distributed, the Kruskal–Wallis test was run for post hoc comparisons. The level of significance was set in advance at $p \leq 0.05$ [6]. The Statview F 4.5 Macintosh software (Abacus Concepts, Berkeley, CA, USA) was used for the analysis [2]. All the statistical probes applied in this study adhere to the requirements for oral and dental research [29].

3. Results

The C1s spectrum of the bare PLGA consisted of three well-defined bands at 284.6, 287.5 and 288.5 eV—these bands are attributed to C–C/C–H, C–O and C=O/COOH functional groups in the polymer [30,31]. In addition, these membranes have a hydrophobic character demonstrated by a contact angle with water of 99° [32]. The morphology of PLGA membranes is quite flat, with an RMS roughness value of less than 1 nm obtained by analysis of surface topography through atomic force microscopy (AFM).

The membrane surfaces in Group 2 were first exposed to pure oxygen plasma for periods of 30 min in a parallel plate capacitive RF reactor working at a pressure of 0.1 mbar. An RF power of 10 W was applied to the top electrode and a self-induced negative bias voltage of 200 V was generated on the bottom electrode, acting as sample holder. This treatment modifies the surface tension state, rendering the PLGA membrane hydrophilic [33]. The treatment of PLGA substrates with oxygen plasma took place at close to ambient temperatures (RT: $23.0 \pm 1.0\,^{\circ}\text{C}$) and did not affect their structural integrity [33]. This process generates an engraving effect, improving the roughness of the surface and favouring the adhesion of oxide layers to PLGA substrates. Figure 4 shows the macroscopic appearance of the PLGA membranes before and after surface treatment with oxygen plasma. Additional information about the PLGA membranes after and before the oxygen plasma treatments can be found elsewhere [1,2,30–33].

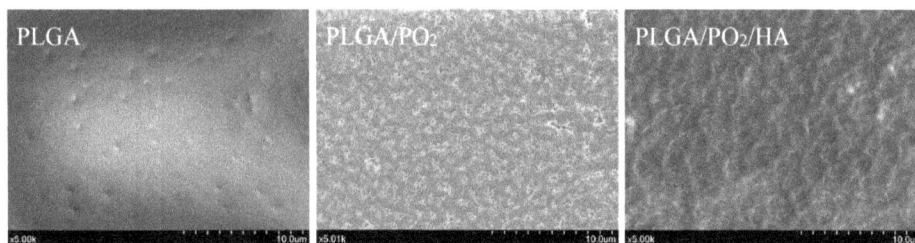

Figure 4. Scanning Electron Microscopy (SEM) images of the three steps of the membrane modification scheme.

Surface topography analysis also enables us to appreciate the considerable increase in surface roughness of the PLGA membrane after exposure to oxygen plasma, with an RMS roughness value greater than 300 nm, without affecting structural integrity. This was an intermediate step to enhance the adhesion of the HA nanostructured film [31].

A thickness of the coating of HA on PLGA around 15 nm and a RMS roughness coefficient around 2 nm was obtained, as shown in Figure 5. Additionally, the AFM surface topography image of the HA coating shows a smooth and uniform microstructure composed of grain-like features (Figure 5). Fourier transform infrared spectroscopy (FTIR) has confirmed the presence of phosphate functional groups (Figure 6). In particular, the broad absorbance band at 1031–1042 cm^{-1} corresponding to the (v_3) asymmetric vibration lines of $(PO_4)^{3-}$ group indicates an amorphous Ca–P structure [34]. Moreover, the study of the HA surface chemical composition by means of X-ray photoelectron spectroscopy (Figure 7) enables identification of the typical chemical states of the components in an ideal hydroxyapatite surface. Also, the estimated Ca/P ratio of 1.34 (see Table 1 with the atomic composition percentages) is in good agreement with other quasi-stoichiometric HA structures obtained by magnetron sputtering techniques in an inert atmosphere [35].

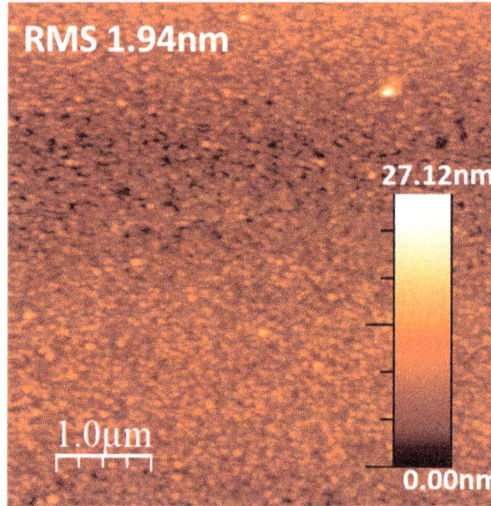

Figure 5. 1 µm × 1 µm AFM micrograph of the hydroxyapatite (HA) surface with a RMS roughness coefficient of 1.94 nm.

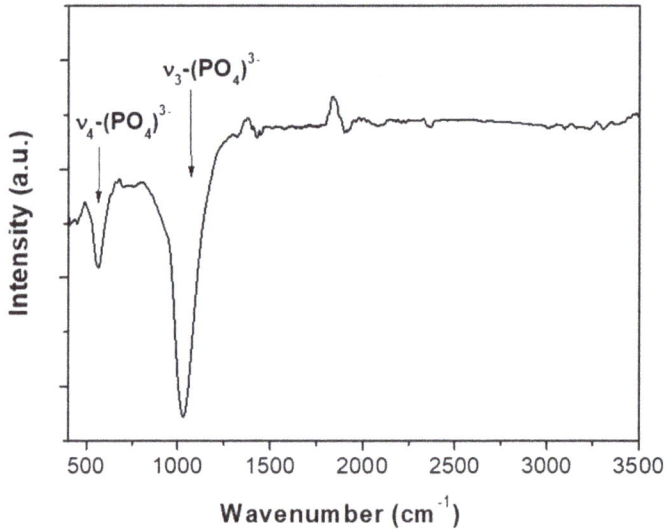

Figure 6. Fourier transform infrared (FTIR) spectrum of the HA coating.

Table 1. Atomic chemical composition percentages of the HA surface.

Atomic	C1s	O1s	Ca2p	P2p
Composition (%)	23.6	44.8	18.1	13.5

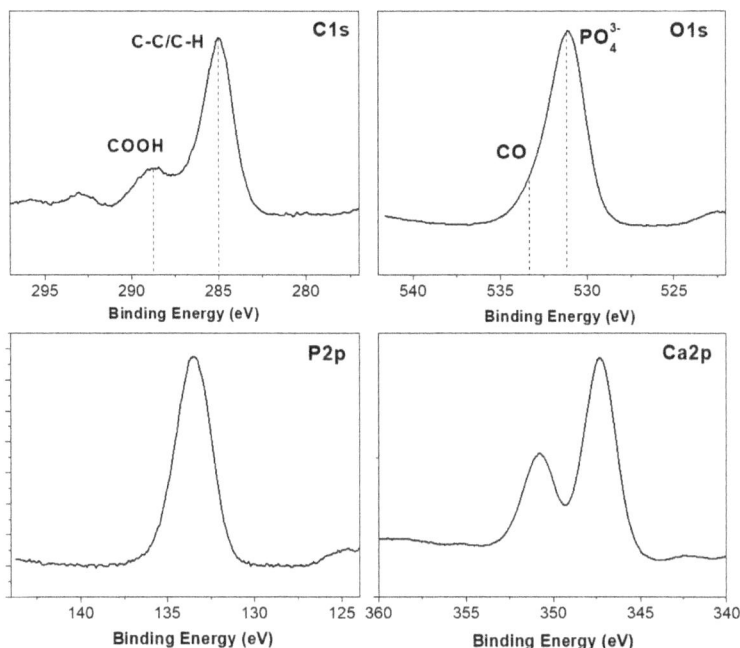

Figure 7. High resolution XPS spectra of the HA surface: carbon C1s, oxygen O1s, phosphorus P2p and calcium Ca2p regions. Guidelines mark the main functional groups.

The main study findings on cell cultures are shown in Table 2 and Figure 8. The results obtained in relation to cell morphology and adhesion of osteoblasts to the membranes show that the osteoblasts bound to PLGA were smaller in size and with a lower number of adhered cells, with fewer extensions and filopodia than the osteoblasts bound to the $PLGA/PO_2/HA$ membrane ($p < 0.05$). Also, it was observed that the $PLGA/PO_2/HA$ membrane had a higher percentage of viable cells bound than did the PLGA control membrane ($p < 0.05$).

Table 2. Studied variables in vitro cell cultures.

	PLGA (Control)	PLGA/PO₂/HA (Experimental)	Statistical Significance
Cell area (μm^2)	288 ± 124	379 ± 110	
DAPI (nuclei area) (μm^2)	2.82 ± 2.0	2.55 ± 1.9	
Probe JC-1 (red/green ratio)	1.69 ± 0.41	2.57 ± 0.09	$p = 0.06$
Cells Viability (%)	62.4 ± 6.2	78.2 ± 3.4	$p < 0.05$
Total cells (cells)	$2.1 \times 10^5 \pm 0.3$	$4.6 \times 10^5 \pm 0.4$	$p < 0.05$

Quantification by flow cytometry of the JC-1 marker was used as a measure of mitochondrial potential. The red/green ratio for the osteoblasts that adhered to the $PLGA/PO_2/HA$ membrane (2.57) was higher when compared to the PLGA control membrane (1.69), which would indicate a better energy balance of the cells that adhered to the $PLGA/PO_2/HA$ membrane, but no statistical significance was observed ($p > 0.05$) (Figure 8).

When determining the degree of osteogenesis and morphology of osteoblasts adhered by F-actin staining and bright-field images, it was observed that the $PLGA/PO_2/HA$ membrane showed cells with a larger area, although no statistically significant differences were obtained ($p > 0.05$).

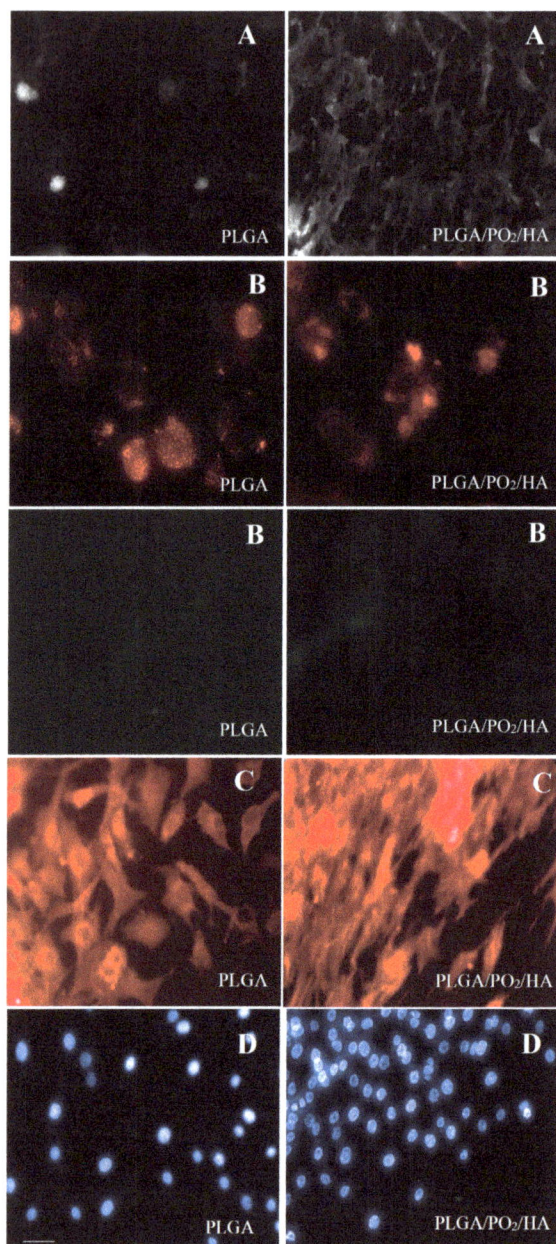

Figure 8. (**A**) Microphotography of osteoblasts on the membranes of PLGA and PLGA/PO₂/HA; (**B**) JC-1 Probe Marking Microphotography; (**C**) Staining with phalloidina-TRITC 50 μg/mL; (**D**) Staining with DAPI. Bar in D-PLGA = 10 microns (same scale for all images). Images taken with the fluorescence microscope Axio Observer A1 (Carl Zeiss).

When comparing the obtained data from the BoneJ trabecular analysis plugins, statistically significant differences were observed for numerous variables between the PLGA control membrane and the PLGA/PO$_2$/HA experimental membrane ($p < 0.05$). The data obtained from the bone density analysis are summarised in Table 3.

Table 3. Bone J analysis plugins. (* $p \leq 0.05$).

	PLGA (Control)	PLGA/PO$_2$/HA (Experimental)	Statistical Significance
Bone density (HU)	969.51 ± 145.7	1036.71 ± 241.3	
Bone density (%)	0.59 ± 0.08	0.63 ± 0.14	
Bone Surface (pixels ^{2})	5079.09 ± 1779.49	11,049.51 ± 4304.57	*
Mean trabecular thickness (pixels)	5.74 ± 1.24	6.29 ± 1.52	
Max. trabecular thickness (pixels)	9.13 ± 1.60	11.30 ± 1.75	*
Bone volume (pixels ^{2})	25,016.20 ± 9922.46	45,526.20 ± 15,275.48	*
Total volume (pixels ^{2})	8,120,601.20 ± 16,432.30	8,090,300.45 ± 17,742.30	*
Bone volume/Total volume	0.003 ± 0.001	0.005 ± 0.001	*
Euler characteristic	−34.05 ± 17.49	−74.55 ± 36.65	*
Maximum branch length (pixels)	38.89 ± 7.55	35.47 ± 12.38	
Connectivity (mm $^{-3}$)	35.25 ± 17.45	75.75 ± 36.63	*
Number of branches (branches)	152.45 ± 83.62	265.70 ± 109.02	*
Number of junctions (junctions)	77.45 ± 43.02	139.70 ± 58.92	*
Number of end-point voxels (voxels)	41.20 ± 23.63	51.70 ± 17.78	0.08
Number of junctions voxels (voxels)	180.45 ± 97.38	323.70 ± 132.25	*
Number of slab voxels (voxels)	761.95 ± 434.86	1269.95 ± 467.48	*
Average branch length (pixels)	8.87 ± 1.67	8.64 ± 1.62	
Number of triple points (points)	97.20 ± 40.15	53.450 ± 30.40	*
Number of quadruple points (points)	17.45 ± 10.96	31.95 ± 14.74	*

The histology analysis (Figure 9) only showed a statistically significant difference for the percentage of osteoid area in relation to the total area between the PLGA control membrane and the PLGA/PO$_2$/HA membrane (Table 4).

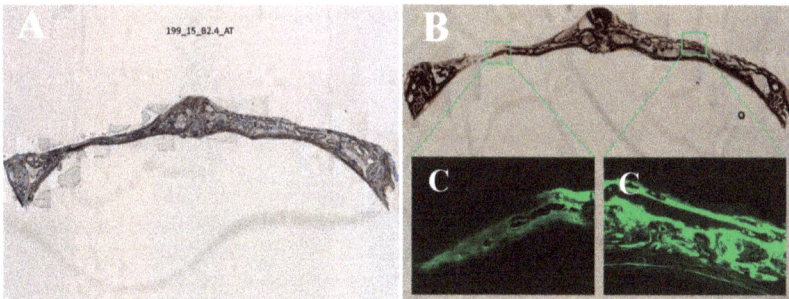

Figure 9. Histology images. (**A**) Toluidine blue; (**B**) Von Kossa and (**C**) Fluorescence. Distance from one side of the bone fragment to the other ≈35 mm. Left side is control (PLGA).

Table 4. Histomorphometric study (* $p \leq 0.05$).

	PLGA (Control)	PLGA/PO$_2$/HA (Experimental)	Statistical Significance
Bone height (μm)	1718 ± 775	1729 ± 700	
Trabecular area (μm^2)	129,558.45 ± 619,632.90	160,339.46 ± 654,020.23	
Trabecular perimeter (μm)	1282.57 ± 2525.41	1491.46 ± 2747.67	
Number of trabeculae (trabeculae)	27.75 ± 18.81	33.50 ± 33.40	
Fluorescence (μm^2)	12.43 ± 6.61	14.11 ± 4.82	
Median trabecular area (μm^2)	129,139.75 ± 618,508.04	160,339.47 ± 654,020.24	
Total trabecular area (μm^2)	235,428.77 ± 638,030.19	186,611.42 ± 398,159.18	
Osteoid area (μm^2)	4369.21 ± 8129.87	5974.17 ± 10,159.57	
% Osteoid area/ Total area (%)	0.0501 ± 0.0675	0.0854 ± 0.1172	*
% Bone area von Kossa / Total area (%)	18.72 ± 12.27	17.63 ± 10.60	
Mean trabecular width (μm)	98.19 ± 172.98	106.70 ± 191.84	
% Bone area von Kossa / Total area (%)	18.72 ± 12.27	17.63 ± 10.60	
Mean trabecular width (μm)	98.19 ± 172.98	106.70 ± 191.84	

4. Discussion

Cells that depend on their adhesions and the interactions they have with the underlying substrate structures and extracellular matrix maintain their functionality [36]. The underlying microenvironment provides a means by which the cells move orient and differentiate to form different types of cells and therefore tissues [37]. The structural and topographic characteristics of PLGA provide improved adhesive potential and ability to migrate and bond at the cell-substrate interphase on osteoblasts [1].

Previous studies have shown that modifying PLGA membranes with oxygen plasma may improve the degradation degree of the PLGA scaffold [2,6,10,38].

Other studies have found that HA increases the biodegradation rate of the membranes due to higher hydrophilicity and neutralisation against acidic degradation products of PLGA, and it has also shown excellent capabilities of calcium collection and bone-like apatite formation with great osteoconductivity [14,39,40]. Therefore, both modifier mechanisms might improve the degradation of the barriers enhancing bone healing. In our study, the functionalised membranes with PO_2 and HA particles show a significantly superior efficacy for bone regeneration to the untreated barriers, confirming the initial hypothesis.

Over the past few decades, calcium phosphate ($Ca_3(PO_4)_2$) ceramics have been widely used as bone graft substitutes, mostly due to the similarity of their chemical composition with the mineral phase of bone. Many types of calcium phosphates have been considered as biomaterials for bone reconstruction in dentistry, orthopaedics and maxillofacial surgery due to the different behaviours that each one exhibits in the living organism, including bioactivity, biodegradability and biological response [41]. The bioactivity, degradation behaviour and osteoconductivity/osteoinductivity of calcium phosphate ceramics generally depend on the calcium/phosphate ratio, crystallinity and phase composition [42]. The synthetic HA ($Ca_{10}(PO_4)_6(OH)_2$) shows good stability in the body, whereas tricalcium phosphates (α-TCP, β-TCP, $Ca_3(PO_4)_2$) are more soluble. BCP (a mixture of HA and β-TCP) has intermediate properties depending on the weight ratio of stable/degradable phases. Therefore, the dissolution rate decreases in the following order: α-TCP > β-TCP > BCP > HA [42]. Due to their nature, $Ca_3(PO_4)_2$ ceramics also exhibit high biocompatibility and ability to bind with bone tissue under certain conditions—however, given their fragility, their clinical applications have been limited to non-carrier or low-load parts of the skeleton [43]. In fact, it is thought that nanoparticles of hydroxyapatite (nHA) are one of the most promising bone graft materials due to their ability to mimic the structure and composition of natural bone [44]. The HA used in this study is a synthetic material of the same composition as the HA present in the human organism $Ca_5(PO_4)_3$ (OH). Synthetic HA as a material for GBR has a long history of use, and its results are excellent [45]. Almost all materials or scaffolds used in GBR base their composition on HA, the main component of mineralised connective tissue [46].

In vitro cell cultures in our study showed osteoblasts with a larger size, a higher number of adhered cells with more extensions and filopodia, and a higher percentage of viable cells bound to the PLGA/PO_2/HA membrane ($p < 0.05$). These results are consistent with those of previous published studies, obtaining an improved in vitro mineralisation and in vivo osteogenesis capacity of composite scaffolds with an increase in both the viability and proliferation rate of cells [5,14].

In this study, all surgical procedures of bone extraction and membrane placement were performed by two surgeons. During the procedure, the membranes' stabilities were guaranteed and did not show any displacement after implantation. The implanted biomaterial was well tolerated by the surrounding soft tissues and no evidence of necrosis, allergic reactions, immune reactions or incompatibility was observed after two months.

Both micro-CT and histological evaluations confirmed that the PLGA/PO_2/HA membranes enhance bone regeneration. Our results agree with those of other previously published studies, where they even observed the formation of new cortex and recanalisation of the marrow cavity via inspection [15].

Polymers **2017**, 9, 410

5. Conclusions

Within the limitations of this study, the following conclusions may be drawn:

(1) We have verified the incorporation of nanometric layers of nanostructured HA films into PLGA membranes modified with PO_2 are effective for the regeneration of bone defects when applied to skull defects in an animal model. We have verified the incorporation of nanometric layers of nanostructured HA films into PLGA membranes modified with PO_2. These membranes showed good potential for the regeneration of bone defects when applied to skull defects in an animal model.

(2) Compared to the untreated PLGA barriers, $PLGA/PO_2/HA$ membranes promote higher osteosynthetic activity, new bone formation and mineralisation levels that are comparable to those of the original bone tissue.

(3) Further investigations of the new membranes in humans are required to develop new techniques that might improve the aesthetic and functional features of future restorations.

Acknowledgments: This study has been subsidised by the Consejería de Salud of the Junta de Andalucia through project PI-0047-2013: "Innovation in Nanomedicine: Guided Bone Regeneration with Nanofunctionalised Resorbable Membranes (Nanorog)". The authors also thank the AEI (EU FEDER program Project MAT2016-79866-R).

Author Contributions: Daniel Torres-Lagares, María Ángeles Serrera-Figallo, José-Luis Gutiérrez-Pérez, Angel Barranco and Agustín Rodríguez-Gonzalez Elipe conceived and designed the experiments; Daniel Torres-Lagares, María Ángeles Serrera-Figallo, Francisco J. García-García, Lizett Castellanos-Cosano, Cristóbal Rivera-Jiménez and Carmen López-Santos performed the experiments; Daniel Torres-Lagares, María Ángeles Serrera-Figallo and Angel Barranco analyzed the data; Daniel Torres-Lagares, María Ángeles Serrera-Figallo, Lizett Castellanos-Cosano, Cristóbal Rivera-Jiménez and Angel Barranco wrote the paper.

Conflicts of Interest: The authors declare no conflicts of interest.

References

1. Castillo-Dalí, G.; Castillo-Oyagüe, R.; Batista-Cruzado, A.; López-Santos, C.; Rodríguez-González-Elipe, A.; Saffar, J.L.; Lynch, C.D.; Gutiérrez-Pérez, J.L.; Torres-Lagares, D. Reliability of new poly (lactic–*co*–glycolic acid) membranes treated with oxygen plasma plus silicon dioxide layers for pre-prosthetic guided bone regeneration processes. *Med. Oral Patol Oral Cir. Bucal* **2017**, 22, e242–e250. [CrossRef] [PubMed]

2. Castillo-Dalí, G.; Castillo-Oyagüe, R.; Terriza, A.; Saffar, J.L.; Batista, A.; Barranco, A.; Cabezas-Talavero, J.; Lynch, C.D.; Barouk, B.; Llorens, A.; et al. In vivo comparative model of oxygen plasma and nanocomposite particles on PLGA membranes for guided bone regeneration processes to be applied in pre-prosthetic surgery: A pilot study. *J. Dent.* **2014**, 42, 1446–1457. [CrossRef] [PubMed]

3. Liu, X.; Ma, P.X. Polymeric scaffolds for bone tissue engineering. *Ann. Biomed. Eng.* **2004**, 32, 477–486. [CrossRef] [PubMed]

4. Townsend-Nicholson, A.; Jayasinghe, S.N. Cell Electrospinning: A Unique Biotechnique for Encapsulating Living Organisms for Generating Active Biological Microthreads/Scaffolds. *Biomacromolecules* **2006**, 7, 3364–3369. [CrossRef] [PubMed]

5. Wang, D.X.; He, Y.; Bi, L.; Qu, Z.H.; Zou, J.W.; Pan, Z.; Fan, J.J.; Chen, L.; Dong, X.; Liu, X.N.; et al. Enhancing the bioactivity of poly(lactic–*co*–glycolic acid) scaffold with a nano-hydroxyapatite coating for the treatment of segmental bone defect in a rabbit model. *Int. J. Nanomed.* **2013**, 8, 1855–1865. [CrossRef] [PubMed]

6. Castillo-Dalí, G.; Castillo-Oyagüe, R.; Terriza, A.; Saffar, J.L.; Batista-Cruzado, A.; Lynch, C.D.; Sloan, A.J.; Gutiérrez-Pérez, J.L.; Torres-Lagares, D. Pre-prosthetic use of poly(lactic–*co*–glycolic acid) membranes treated with oxygen plasma and TiO_2 nanocomposite particles for guided bone regeneration processes. *J. Dent.* **2016**, 47, 71–79. [CrossRef] [PubMed]

7. Cavalcanti-Adam, E.A.; Volberg, T.; Micoulet, A.; Kessler, H.; Geiger, B.; Spatz, J.P. Cell spreading and focal adhesion dynamics are regulated by spacing of integrin ligands. *Biophys. J.* **2007**, 92, 2964–2974. [CrossRef] [PubMed]

Polymers **2017**, *9*, 410

8. Lim, J.Y.; Dreiss, A.D.; Zhou, Z.; Hansen, J.C.; Siedlecki, C.A.; Hengstebeck, R.W.; Cheng, J.; Winograd, N.; Donahue, H.J. The regulation of integrin-mediated osteoblast focal adhesion and focal adhesion kinase expression by nanoscale topography. *Biomaterials* **2007**, *28*, 1787–1797. [CrossRef] [PubMed]

9. Riveline, D.; Zamir, E.; Balaban, N.Q.; Schwarz, U.S.; Ishizaki, T.; Narumiya, S.; Kam, Z.; Geiger, B.; Bershadsky, A.D. Focal contacts as mechanosensors: Externally applied local mechanical force induces growth of focal contacts by an mdia1-dependent and rock-independent mechanism. *J. Cell Biol.* **2001**, *153*, 1175–1186. [CrossRef] [PubMed]

10. López-Santos, C.; Terriza, A.; Puértolas, J.; Yubero, F.; González-Elipe, A.R. Physiological Degradation Mechanisms of PLGA Membrane Films under Oxygen Plasma Treatment. *J. Phys. Chem. C* **2015**, *119*, 20446–20452. [CrossRef]

11. Shen, H.; Hu, X.; Yang, F.; Bei, J.; Wang, S. Combining oxygen plasma treatment with anchorage of cationized gelatin for enhancing cell affinity of poly(lactide–*co*–glycolide). *Biomaterials* **2007**, *28*, 4219–4230. [CrossRef] [PubMed]

12. Chen, F.M.; Liu, X. Advancing biomaterials of human origin for tissue engineering. *Prog. Polym. Sci.* **2016**, *53*, 86–168. [CrossRef] [PubMed]

13. Ngiam, M.; Liao, S.; Patil, A.J.; Cheng, Z.; Chan, CK.; Ramakrishna, S. The fabrication of nano-hydroxyapatite on PLGA and PLGA/collagen nanofibrous composite scaffolds and their effects in osteoblastic behaviour for bone tissue engineering. *Bone* **2009**, *45*, 4–16. [CrossRef] [PubMed]

14. Fu, L.; Wang, Z.; Dong, S.; Cai, Y.; Ni, Y.; Zhang, T.; Wang, L.; Zhou, Y. Designed Bilayer Poly(Lactic–*co*–Glycolic Acid)/nano-hydroxyapatite Membrane with Double Benefits of Barrier Function and Osteogenesis Promotion. *Materials* **2017**, *10*, 257. [CrossRef] [PubMed]

15. Wang, Z.; Xu, Y.; Wang, Y.; Ito, Y.; Zhang, P.; Chen, X. Enhanced in Vitro Mineralization and in Vivo Osteogenesis of Composite Scaffolds through Controlled Surface Grafting of *L*–Lactic Acid Oligomer on Nanohydroxyapatite. *Biomacromolecules* **2016**, *17*, 818–829. [CrossRef] [PubMed]

16. Witt, C.; Kissel, T. Morphological characterization of microspheres, films and implants prepared from poly(lactide–*co*–glycolide) and ABA triblock copolymers: Is the erosion controlled by degradation, swelling or diffusion? *Eur. J. Pharm. Biopharm.* **2001**, *51*, 171–181. [CrossRef]

17. Nieh, T.G.; Jankowski, A.F.; Koike, J. Processing and characterization of hydroxyapatite coatings on titanium produced by magnetron sputtering. *J. Mater. Res.* **2001**, *16*, 3238–3245. [CrossRef]

18. Wong-Lee, J.G.; Lovett, M. Rapid and sensitive PCR method for identification of Mycoplasma species in tissue culture. In *Diagnostic Molecular Microbiology: Principles and Applications*; Persing, D.H., Smith, T.F., Tenover, F.C., White, T.J., Eds.; American Society for Microbiology: Washington, DC, USA, 1993; pp. 257–260.

19. Paz-Pumpido, F. Biocompatibilidad de los adhesivos dentinarios. *Avances Odontoestomatol.* **2005**, *21*, 339–345. [CrossRef]

20. Di Toro, R.; Betti, V.; Spampinato, S. Biocompatibility and integrin-mediated adhesion of human osteoblasts to poly(dl–lactide–*co*–glycolide) copolymers. *Eur. J. Pharm. Sci.* **2004**, *21*, 161–169. [CrossRef] [PubMed]

21. Huang, L.; Zhang, Z.; Lv, W.; Zhang, M.; Yang, S.; Yin, L.; Hong, J.; Han, D.; Chen, C.; Swarts, S.; et al. Interleukin 11 protects bone marrow mitochondria from radiation damage. *Adv. Exp. Med. Biol.* **2013**, *789*, 257–264. [CrossRef] [PubMed]

22. Waggoner, A.; DeBiasio, R.; Conrad, P.; Bright, G.R.; Ernst, L.; Ryan, K.; Nederlof, M.; Taylor, D. Multiple spectral parameter imaging. *Methods Cell Biol.* **1989**, *30*, 449–478. [PubMed]

23. Kubista, M.; Aakerman, B.; Norden, B. Characterization of interaction between DNA and 4′,6-diamidino-2-phenylindole by optical spectroscopy. *Biochemistry* **1987**, *26*, 4545–4553. [CrossRef] [PubMed]

24. Bornstein, M.M.; Reichart, P.A.; Buser, D.; Bosshardt, D.D. Tissue response and wound healing after placement of two types of bioengineered grafts containing vital cells in submucosal maxillary pouches: An experimental pilot study in rabbits. *Int. J. Oral Maxillofac. Implants* **2011**, *26*, 768–775. [PubMed]

25. Landis, E.N.; Keane, D.T. X-ray microtomography. *Mater. Charact.* **2010**, *61*, 1305–1316. [CrossRef]

26. Doube, M.; Kłosowski, M.M.; Arganda-Carreras, I.; Cordeliéres, F.; Dougherty, R.P.; Jackson, J.; Schmid, B.; Hutchinson, J.R.; Shefelbine, S.J. BoneJ: Free and extensible bone image analysis in ImageJ. *Bone* **2010**, *47*, 1076–1079. [CrossRef] [PubMed]

27. Lopez-Píriz, R.; Fernández, A.; Goyos-Ball, L.; Rivera, S.; Díaz, L.A.; Fernández-Domínguez, M.; Prado, C.; Moya, J.S.; Torrecillas, R. Performance of a New Al$_2$O$_3$/Ce–TZP Ceramic Nanocomposite Dental Implant: A Pilot Study in Dogs. *Materials* **2017**, *10*, 614. [CrossRef] [PubMed]

28. Donath, K.; Breuner, G. A method for the study of undecalcified bones and teeth with attached soft tissues. The säge-schliff (sawing and grinding) technique. *J. Oral Pathol.* **1982**, *11*, 318–326. [CrossRef] [PubMed]

29. Hannigan, A.; Lynch, C.D. Statistical methodology in oral and dental research: Pitfalls and recommendations. *J. Dent.* **2013**, *41*, 385–392. [CrossRef] [PubMed]

30. González-Padilla, D.; García-Perla, A.; Gutiérrez-Pérez, J.L.; Torres-Lagares, D.; Castillo-Dalí, G.; Salido-Peracaula, M.; Vilches-Troya, J.; Vilches-Perez, J.I.; Terriza-Fernandez, A.; Barranco-Quero, A.; et al. Membrana Reabsorbible Para Regeneración Ósea 2012. Spanish Patent P-0201232018, 24 December 2012.

31. Terriza, A.; Vilches-Pérez, J.; González-Caballero, J.L.; de la Orden, E.; Yubero, F.; Barranco, A.; Gonzalez-Elipe, A.R.; Vilches, J.; Salido, M. Osteoblasts interaction with PLGA membranes functionalized with titanium film nanolayer by PECVD. In vitro assessment of surface influence on cell adhesion during initial cell to material interaction. *Materials* **2014**, *7*, 1687–1708. [CrossRef] [PubMed]

32. Terriza, A.; Vilches-Pérez, J.I.; de la Orden, E.; Yubero, F.; Gonzalez-Caballero, J.L.; González-Elipe, A.R.; Vilches, J.; Salido, M. Osteoconductive potential of barrier nanoSiO$_2$ PLGA membranes functionalized by plasma enhanced chemical vapour deposition. *Biomed. Res. Int.* **2014**, *2014*, 253590. [CrossRef] [PubMed]

33. Wan, Y.; Qu, X.; Lu, J.; Zhu, C.; Wan, L.; Yang, J.; Bei, J.; Wang, S. Characterization of surface property of poly(lactide–*co*–glycolide) after oxygen plasma treatment. *Biomaterials* **2004**, *25*, 4777–4783. [CrossRef] [PubMed]

34. Socol, G.; Macovei, A.M.; Miroiu, F.; Stefan, N.; Duta, L.; Dorcioman, G.; Mihailescu, I.N.; Petrescu, S.M.; Stan, G.E.; Marcov, D.A.; et al. Hydroxyapatite thin films synthesized by pulsed laser deposition and magnetron sputtering on PMMA substrates for medical applications. *Mater. Sci. Eng. B* **2010**, *169*, 159–168. [CrossRef]

35. Yamaguchi, T.; Tanaka, Y.; Ide-Ektessabi, A. Fabrication of hydroxyapatite thin films for biomedical applications using RF magnetron sputtering. *Nucl. Instrum. Methods Phys. Res. Sect. B* **2006**, *249*, 723–725. [CrossRef]

36. Gumbiner, B.M. Cell adhesion: The molecular basis of tissue architecture and morphogenesis. *Cell* **1996**, *84*, 345–357. [CrossRef]

37. Barthes, J.; Özçelik, H.; Hindié, M.; Ndreu-Halili, A.; Hasan, A.; Vrana, N.E. Cell microenvironment engineering and monitoring for tissue engineering and regenerative medicine: The recent advances. *Biomed. Res. Int.* **2014**, *2014*, 921905. [CrossRef] [PubMed]

38. Gentile, P.; Chiono, V.; Carmagnola, I.; Hatton, P.V. An overview of poly(lactic–*co*–glycolic) acid (PLGA)-based biomaterials for bone tissue engineering. *Int. J. Mol. Sci.* **2014**, *15*, 3640–3659. [CrossRef] [PubMed]

39. Kasuga, T.; Hosoi, Y.; Nogami, M.; Niinomi, M. Apatite formation on calcium phosphate invert glasses in simulated body fluid. *J. Am. Ceram. Soc.* **2001**, *84*, 450–452. [CrossRef]

40. Wong, K.L.; Wong, C.T.; Liu, W.C.; Pan, H.B.; Fong, M.K.; Lam, W.M.; Cheung, W.L.; Tang, W.M.; Chiu, K.Y.; Luk, K.D.; et al. Mechanical properties and in vitro response of strontium-containing hydroxyapatite/polyetheretherketone composites. *Biomaterials* **2009**, *30*, 3810–3817. [CrossRef] [PubMed]

41. Canillas, M.; Pena, P.; de Aza, A.H.; Rodríguez, M.A. Calcium phosphates for biomedical applications. *Boletín SECV* **2017**, *56*, 91–112. [CrossRef]

42. Bose, S.; Tarafder, S. Calcium phosphate ceramic systems in growth factor and drug delivery for bone tissue engineering: A review. *Acta Biomater.* **2012**, *8*, 1401–1421. [CrossRef] [PubMed]

43. Habraken, W.; Habibovic, P.; Epple, M.; Bohner, M. Calcium phosphates in biomedical applications: Materials for the future? *Mater. Today* **2016**, *19*, 69–87. [CrossRef]

44. Baino, F.; Novajra, G.; Vitale-Brovarone, C. Bioceramics and scaffolds: A winning combination for tissue engineering. *Front. Bioeng. Biotechnol.* **2015**, *3*, 202. [CrossRef] [PubMed]

45. Liu, J.; Kerns, D.G. Mechanisms of guided bone regeneration: A review. *Open Dent. J.* **2014**, *8*, 56–65. [CrossRef] [PubMed]

46. Palmer, L.C.; Newcomb, C.J.; Kaltz, S.R.; Spoerke, E.D.; Stupp, S.I. Biomimetic systems for hydroxyapatite mineralization inspired by bone and enamel. *Chem. Rev.* **2008**, *108*, 4754–4783. [CrossRef] [PubMed]

polymers

MDPI

Communication

Multilayered Films Produced by Layer-by-Layer Assembly of Chitosan and Alginate as a Potential Platform for the Formation of Human Adipose-Derived Stem Cell aggregates

Javad Hatami [1,2,†], Sandra G. Silva [1,2,‡], Mariana B. Oliveira [1,2,†], Rui R. Costa [1,2], Rui L. Reis [1,2] and João F. Mano [1,2,†,*]

[1] 3B's Research Group, Biomaterials, Biodegradables and Biomimetics, University of Minho, Headquarters of the European Institute of Excellence of Tissue Engineering and Regenerative Medicine, Avepark—Parque de Ciência e Tecnologia, Zona Industrial da Gandra, 4805-017 Barco GMR, Portugal; jhatami@ua.pt (J.H.); sg82silva@gmail.com (S.G.S.); mboliveira@ua.pt (M.B.O.); rui.costa@dep.uminho.pt (R.R.C.); rgreis@dep.uminho.pt (R.L.R.)

[2] ICVS/3B's, PT Government Associated Laboratory, Braga/Guimarães, Portugal

* Correspondence: jmano@ua.pt; Tel.: +351-234-370-733; Fax: +351-234-401-470

† Current address: Department of Chemistry, CICECO—Aveiro Institute of Materials, University of Aveiro, 3810-193 Aveiro, Portugal

‡ Current address: LAQV/REQUIMTE, Department of Chemistry and Biochemistry, Faculty of Science, University of Porto, Rua do campo Alegre s/n, 4169-007 Porto, Portugal

Received: 10 August 2017; Accepted: 6 September 2017; Published: 13 September 2017

Abstract: The construction of multilayered films with tunable properties could offer new routes to produce biomaterials as a platform for 3D cell cultivation. In this study, multilayered films produced with five bilayers of chitosan and alginate (CHT/ALG) were built using water-soluble modified mesyl and tosyl–CHT via layer-by-layer (LbL) self-assembly. NMR results demonstrated the presences of mesyl (2.83 ppm) and tosyl groups (2.39, 7.37 and 7.70 ppm) in the chemical structure of modified chitosans. The buildup of multilayered films was monitored by quartz-crystal-microbalance (QCM-D) and film thickness was estimated using the Voigt-based viscoelastic model. QCM-D results demonstrated that CHT/ALG films constructed using mesyl or tosyl modifications (mCHT/ALG) were significantly thinner in comparison to the CHT/ALG films constructed with unmodified chitosan ($p < 0.05$). Adhesion analysis demonstrated that human adipose stem cells (hASCs) did not adhere to the mCHT/ALG multilayered films and formed aggregates with sizes between ca. 100–200 µm. In vitro studies on cell metabolic activity and live/dead staining suggested that mCHT/ALG multilayered films are nontoxic toward hACSs. Multilayered films produced via LbL assembly of ALG and off-the-shelf, water-soluble modified chitosans could be used as a scaffold for the 3D aggregates formation of hASCs in vitro.

Keywords: layer by layer assembly; chitosan; alginate; cytotoxicity; multilayered film; adhesion; 3D culture; spheroid

1. Introduction

Mesenchymal stem cells (MSCs), which are capable of self-renewal and multilineage differentiation, are becoming increasingly important for the development of cell therapeutics in regenerative medicine. Adipose-derived stem cell (ASC) represents a very attractive cell type of MSCs that are easily accessible, abundant and rich source of adult stem cells [1]. ASCs are able to self-replicate and can differentiate into osteogenic, adipogenic, and chondrogenic lineages under specific conditions [2]. These potentials, along with their easy accessibility, made ACS a good candidate for many cell-based therapies. The stem cell niche and its microenvironment have important effects on the function and biology of stem cells [3]. For example, the cellular phenotype and biological response of cells are different in a monolayer and 3D culture [4]. Approaches providing a 3D culture environment are becoming popular for cell cultures because they mimic in vivo condition. It was demonstrated that aggregates of bone-marrow-derived MSCs could have therapeutic potential [5,6]. ASCs could represent an ideal candidate for a 3D culture of MSCs because of their abundant autologous cells and the ease of their access. Different techniques, such as non-adherent culture condition, nutrient deprivation, air–liquid surface, spinner flask and hanging drop, have been used to form 3D spheroids [4,6,7].

The layer-by-layer (LbL) deposition of materials by electrostatic assembly has often been used to modify the surface of several materials [8–10]. This technique has seen application in the construction of nanostructured and easily tailorable two-dimensional and three-dimensional self-standing structures [11]. Due to its compatibility with the construction of highly organized materials and its high versatility, the LbL deposition method was used for the building of new biomaterials and has seen promising applications in biological field [11,12].

Developing a natural-based multilayer film with tunable size and properties has many applications in the healthcare field [11]. A wide plethora of synthetic and natural origin polyelectrolytes were used as building blocks for LbL construction [13–15]. The use of synthetic polymers allows precise control of the polymers' physicochemical properties, such as molecular weight, while working with low batch-to-batch variation. Moreover, the properties of the polymers may be controlled to withstand a large range of processing parameters, including ionic strength and pH values of assembly, and their chemical modification is usually easily achievable. The main reason for the great potential shown by natural-origin polymers in the biomedical field is their chemical similarity with the native animal extracellular matrix (ECM) [16,17]. Moreover, a great part of such polymers shows a biodegradable behavior and, importantly for electrostatic-driven LbL assembly, most of them are ionizable (i.e., they are polyelectrolytes) [18].

Polysaccharides are an interesting type of polymers found in nature. They often allow high degrees of hydration, are biocompatible, and are often biodegradable [15,16,19,20]. Polysaccharides extracted from marine sources are a particular case of these polymers [16,21], which have seen application in the construction of LbL films. Alginate (ALG), chitosan (CHT), chondroitin sulfate and carrageenans are examples of polysaccharides that have found an application in the construction of nanostructured coatings in two-dimensional materials, self-standing micrometric membranes and three-dimensional structures as hollow tubes, capsules for drug release or cell encapsulation, and scaffolds [22–26]. This array of works supported the wide applicability of such polyelectrolytes, and enhanced their added-value, by incorporating these materials in high-end technological approaches.

ALG and CHT are two marine-origin polysaccharides with high applicability in the development of biomaterials. ALG is extracted from brown algae and regularly used for cell encapsulation as it is well-known for its lack of cell toxicity [27,28]. CHT is a cationic polyelectrolyte which may be extracted from different sources including shells of crustaceous. Its popularity in the biomedical field is related with its immunological, antibacterial and wound healing properties [29]. Moreover, CHT shares structural similarities with many Glycosaminoglycans (GAGs) present in native ECM [30]. The chitosan/alginate (CHT/ALG) polyelectrolyte pair has been used to construct LbL-based coatings on several materials [31] and a wide plethora of self-standing structures [11,32,33]. The nanostructured multilayered films obtained from the LbL assembly of CHT/ALG polymers are well-characterized regarding thickness variations, permeability to gases and glucose, mechanical properties and cellular response [23]. Importantly, the properties of this system can be modulated by varying LbL deposition conditions including pH, during deposition, and ionic strength. The films may also be post-processed by the chemical or ionic crosslinking of both components, which affects some system's properties, including their mechanical performance and ability to sustain cell adhesion [23,24].

The application of multilayered films using LbL deposition of CHT/ALG in biological filed is limited because it requires the use of acidic pH values due to the insolubility of CHT in water, at the neutral pH. Martins et al. [34] suggested the use of a water-soluble, commercially available chloride-salt chitosan to prepare CHT/ALG films via LbL assembly in water-soluble conditions, at a neutral pH in order to extend the application of CHT/ALG film in biological fields. Although that work opened up the possibility of processing CHT/ALG films in water-soluble conditions, the modification of CHT with an inorganic salt limits the use of characterization methods, such as NMR, for the full characterization of the modified chitosan. As an example, the presence of the pick of inorganic salt of CHT in NMR spectra interferes with the detection of the degree of substitution of CHT.

In the current work, we suggest the use of multilayered films produced by LbL assembly of mCHT/ALG for spheroid formation of hACSs. Therefore, two chitosan derivatives, prepared by chemoselective functionalizing as an amino group of parent chitosans, were used as cationic polyelectrolyte for LbL construction of multilayered films. Two different synthetized modified chitosans (mCHT) with mesyl and tosyl organic salts were obtained. These modifications conferred chitosan with water and organic solvent (e.g., dimethyl sulfoxide) solubility [35]. Moreover, the use of organic medium resulted in a well-controlled and regioselective modification leading to homogenous products that the degree of modification of CHT could be monitored by NMR. We hypothesized that the modification of chitosan with organic salts, due to their chemistry or size, could tailor the properties of the LbL films. This strategy would allow working with off-the-shelf water-soluble chitosans that could be selected prior to the film deposition according to the desired properties of the film, such as its thickness and viscoelasticity. Though chemoselective functionalizing of chitosan was reported previously for different purposes [36,37], the current work demonstrated the possibility to construct LbL multilayer films and to tailor its thickness and viscoelastic properties using water-soluble modified chitosan with organic salt and alginate polymers for the first time, to the knowledge of the authors, as a supporting scaffold for the 3D culture and aggregate production of hASCs. This approach could be interesting for the immediate adaptation of the properties of the films by maintaining their processing conditions, while simply adjusting the type of chitosan used for their construction, in a user-friendly approach.

2. Materials and Methods

2.1. Chitosan Purification, Modification and Analysis by ^{1}H NMR

CHT of medium molecular weight (MMW–CHT, M_W = 190–310 kDa, 75–85% degree of deacetylation, viscosity 200–800 cP) was purchased from Sigma-Aldrich (St. Louis, MO, USA). CHT was purified by a sequential filtration and precipitation steps in distilled water, sodium hydroxide and ethanol, as described previously [38], followed by freeze-drying. It was then grounded and stored as a powder until further usage.

Chitosan derivatives were obtained by chemoselective functionalizing as an amino group of parent chitosans. In this work, two organic acids were used to modify the chitosan according to a previously reported method [37]. The result of chitosan modification with these organic acids, methanosulfonic and *p*-toluenosulfonic acids, were mesyl–CHT and tosyl–CHT, respectively (Scheme 1). Briefly, methanosulfonic acid or *p*-tolunesulfonic acid monohydrate was added (1:1), drop-by-drop at 10 °C, to a suspension of pre-purified MMW–CHT until clear homogeneous solutions were obtained. The solutions were then stirred for 2–4 h (10 °C) then washed several times with acetone, to precipitate the salts, and diethyl ether and then, dried in a vacuum oven at 25 °C for 8 h. Before use, the materials were dissolved in distilled water and freeze-dried. The corresponding salts (mesylates and tosylates), which derived from organic acid, was detected by ^{1}H NMR (Bruker Avance III operating at 400 MHz, Billerica, MA, USA) through the peaks of the organic anion. To perform NMR analysis, the samples were prepared in D_2O in the concentration range of 10–15 mg/mL. All the reagents and solvents (p.a. quality) used in the preparation of mCHTs were purchased from Sigma.

A **B**

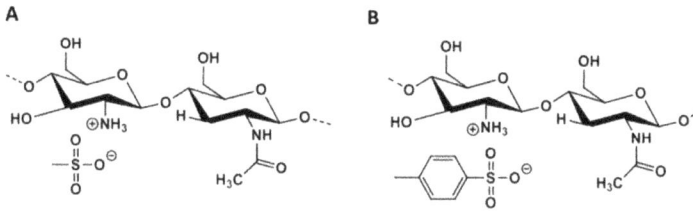

Scheme 1. Chemical structure of mesyl–chitosan (CHT) (**A**) and tosyl–CHT (**B**).

2.2. Construction and Characterization of mCHT/ALG Multilayered Films by Quartz-Crystal Microbalance

Alginic acid sodium salt from brown algae (Sigma-Aldrich, St. Louis, MO, USA ref. 71238) was dissolved in a solution of NaCl (0.15 M, Sigma) to obtain a final 0.1% (*w/v*) alginate solution. Solutions of mesyl–CHT and tosyl–CHT (all at 0.1% *w/v*) were obtained by dissolving the mesyl–CHT and tosyl–CHT in a solution of NaCl (0.15 M). A solution of unmodified chitosan (here, normal–CHT) at a concentration of 0.1% (*w/v*) was obtained by dissolving normal–CHT in 1% (*v/v*) acidic acid (Sigma) followed by the addition of NaCl to a final concentration of 0.15 M. The pH of all solutions was then adjusted to 5.5 using 2 M NaOH and 1% (*v/v*) acetic acid. Laser doppler electrophoresis is often used to measure the magnitude of the charge of colloidal suspensions, but is also appropriate to determine the polycationic and polyanionic behavior of solubilized polymeric materials. Therefore, the zeta (ζ)-potential values of solutions were measured using Nano-ZS (Malvern, Worcestershire, UK), at 25 °C.

Deposition of CHT/ALG multilayers on the gold-coated crystals was monitored by a Q-Sense E4 quartz-crystal microbalance (Q-Sense AB) with dissipation monitoring system [39]. Briefly, AT-cut quartz crystal was excited at 5, 25, 35 and 45 MHz (fundamental frequency, 5th, 7th, and 9th overtones, respectively). The initial crystal cleaning was performed in an ultrasound bath at 30 °C followed by immersion in acetone, ethanol and isopropanol. Starting with CHT solution, deposition of polyelectrolytes took place at a constant flow rate of 50 mL/min, 25 °C and pH 5.5 for 22 min. Following the deposition of each polyelectrolyte, a rinsing step with NaCl (0.15 M) at pH 5.5 for 12 min was performed. Changes in frequency (ΔF) and dissipation (ΔD) were monitored in real-time.

2.3. Estimation of the Film Thickness and Properties

The film thickness was estimated using the Voigt-based viscoelastic model [40], integrated in QTools (version 3.1.25.604) provided by Q-Sense (Biolin Scientific, Gothenburg, Sweden). The model is represented by Equations (1) and (2).

$$\Delta F \approx -\frac{1}{2\pi\rho_0 h_0}\left\{\frac{\eta_3}{\delta_3} + \sum_{j=k}\left[h_j\rho_j\omega - 2h_j\left(\frac{\eta_3}{\delta_3}\right)^2\frac{\eta_j\omega^2}{\mu_j^2 + \omega^2\eta_j^2}\right]\right\} \tag{1}$$

$$\Delta D \approx \frac{1}{2\pi f\rho_0 h_0}\left\{\frac{\eta_3}{\delta_3} + \sum_{j=k}\left[2h_j\left(\frac{\eta_3}{\delta_3}\right)^2\frac{\mu_j\omega}{\mu_j^2 + \omega^2\eta_j^2}\right]\right\} \tag{2}$$

In Equations (1) and (2), ρ_0 and h_0 are the density and thickness of the quartz crystal, k represents total number of thin viscoelastic layers, η_3 is the viscosity of the bulk liquid, δ_3 is the viscous penetration depth of the shear wave in the bulk liquid, ρ_3 is the density of liquid, μ is the elastic shear modulus of an over layer, and ω is the angular frequency of the oscillation. Estimations were made considering a fixed solvent viscosity of 0.001 Pa (the same as for water) and film density of 1200 kg/m^3. The density of solvent was changed, by trial and error, between 1000 and 1015 kg/m^3 until the total error, χ^2, was minimized. Three overtones (5th, 7th and 9th) were used for the calculations.

2.4. Human Adipose Stem Cell (hASC) Culture

To evaluate the biological performance of CHT/ALG multilayered films, cell culture studies were performed using hASCs. Human abdominal subcutaneous adipose tissue samples were obtained after informed consent from patients undergoing lipoaspiration procedure. The retrieval and transportation of the samples has been performed under a valid cooperation protocol established on 12 February 2007 by the 3B's research group and the Hospital da Prelada (Porto, Portugal), which is approved by the ethical committees of both institutions, and does not have an associated number. This protocol contains all guidelines to retrieve, transport and discard biological samples in accordance with National and European recommendations, which include the signature of an Informed consent form and a standard anonymization procedure for all samples. Isolation and process of samples were carried out within 24 h after surgical procedure according to a protocol previously established [41]. Isolated cells were cultured under basal condition, using minimum essential alpha medium (α-MEM, Sigma), supplemented with 10% (v/v) fetal bovine serum (FBS, ThermoFisher Scientific, Waltham, MA, USA) and 1% (v/v) penicillin-streptomycin, until maximum passages of three. Cell culture medium was changed 48 h after initial plating and every 3 days thereafter.

2.5. Preparation of Multilayered Films for Cell Culture Studies

Glass surfaces (1.0 cm^2) were prepared from coverslips (Fisher Scientific) as substrates to build normal, mesyl and tosyl–CHT/ALG multilayered films. Glass surfaces were kept inside six-well plates (Falcon, Corning, NY, USA) and covered by solutions of CHTs and ALG (all 0.1% in aqueous solution with NaCl 0.15 M). LbL assembly started with CHTs solution deposition and followed by ALG solution deposition (each layer with 22 min deposition time), in a sequential manner with intermediate washing step (10 min) with NaCl (0.15 M) to prepare five bilayers of films.

To study the cytotoxicity and effect of the direct contact of mCHT layer with hASCs, an additional layer of CHTs was deposited on top of five bilayers of CHT/ALG films via LbL technique. Samples were sterilized for subsequent cell culture using exposure to UV (1 h).

2.6. Cytotoxicity and Viability Analysis

Glass substrates coated with mesyl, tosyl and normal–CHT/ALG multilayered films, prepared according to the previous section, were used for viability and cytotoxicity testing according to ISO 10993-5 guidelines similar to the method described elsewhere [42]. Samples were incubated in triplicate in 2 ml α-MEM with 10% (v/v) FBS and 1% (v/v) penicillin-streptomycin in six-well plates at 37 °C, 5% CO$_2$, and fully humidified air for 3 days. The resulting mediums, enriched in lixiviates potentially released from the multilayer films, was used for cytotoxicty and viability studies. Therefore, hASCs with an initial density of 8×10^4 cells/cm^2 was cultured in resulting medium, in 24-well plates, for 3 days. Metabolic activity of cultured cells was determined using a MTS Cell Proliferation Assay Kit (Abcam, Cambridge, UK) at three time points (days 1, 2 and 3). The results were normalized to the negative control (fresh α-MEM with 10% (v/v) FBS and 1% (v/v) penicillin–streptomycin) for cytotoxicity analysis and compared to the positive control medium pre-incubated with latex.

Cell viability was assessed by incubating hACSs, pre-cultured in 24-well plates with resulting mediums for 3 days, with a live/dead assay at three time points (day 1, 2 and 3). In brief, the hASCs were incubated for 15 min with 150 μL of calcein AM (Invitrogen, Carlsbad, CA, USA) solution (2 μL calcein/mL DMEM without phenol red) and 150 μL of propidium iodide (PI, Invitrogen, Carlsbad, CA, USA) working solution (2 μL PI stock solution, 1 mg/mL in distilled water) and finally rinsed with PBS. The cells were then immediately visualized in the dark by fluorescence microscopy (Axio Imager Z1m, Zeiss, Oberkochen, Germany).

2.7. Cell Adhesion Studies on mCHT/ALG Multilayered Film

Cell morphology and adhesion were assessed according to the protocol previously reported [43]. Briefly, a cell suspension of hASCs (10^6 cells/mL) was prepared using Tryple Express (ThermoScientific). Cell suspensions (passage 3–5) with the density of 2.0×10^4 cell/cm^2 were cultured on top of glass surfaces pre-coated with five bilayers of CHT/ALG films using α-MEM supplemented with 10% (v/v) FBS and 1% (v/v) penicillin-streptomycin in humidified atmosphere (37 °C, 5% CO$_2$). After 24 h, culture medium was removed, samples were washed with PBS and incubated with 10% of formalin at room temperature (RT) for 1 h. Samples were washed with PBS and incubated with 0.1% Triton X for 5 min at RT to permeabilize cells. Samples were washed again with PBS and solutions of 4,6-Diaminidino-2-phenylindole-dilactate (DAPI, 20 mg/mL, Sigma-Aldrich) and phalloidin tetramethylrhodamine B isothiocyanate dyes (phalloidin, 10 mg/mL, Sigma-Aldrich) were added to the samples to stain the nuclei and cytoskeleton F-actin of the hASCs, respectively. After 1 h at RT and protected from light, cells were washed three times with PBS and immediately visualized in the dark condition by reflected light fluorescence microscopy (Axio Imager Z1m, Zeisss, Oberkochen, Germany) to assess cellular adhesion, and morphology.

2.8. Statistical Analysis

Statistical analysis was performed by use of IBM SPSS statistics (v. 20, Chicago, IL, USA) with the use of Mann–Whitney U analysis. Differences in results were considered statistically significant at a value of $p < 0.05$. Results are presented as a mean $\pm$ standard deviation (SD).

3. Results and Discussion

3.1. Synthesis and Characterization of the Mesyl–CHT and Tosyl–CHT

The NMR characterization of mesyl–CHT polymer demonstrated that a peak corresponding to the mesyl group appears at 2.83 ppm as a singlet (Figure 1A) which is in line with results previously described [37]. In the spectrum of the tosylated deriavative, the peaks of the aromatic ring are present at 2.39, 7.37 and 7.70 ppm (Figure 1B). These results confirm chemoselective modification of chitosan with mesyl and tosyl groups. The results also demonstrated that water solubility can be obtained by modification of chitosan with mesyl and tosyl groups.

Since CHT and ALG polyelectrolytes are only partially charged at moderate pH near their pK_as, working pH and ionic strength are expected to influence film growth and properties. In this work, the pH of CHT was adjusted to 5.5, below its pK_a value of 6.5 [44], to ensure its performance as a polycation. ζ-potential values show the cationic nature of normal and modified CHTs, as well as the anionic nature of ALG (Table 1) in aqueous solution with 0.15 M NaCl (pH = 5.5). This ensures the existence of electrostatic interactions between CHTs and ALG in order to build LbL films. The charge sign is the same for all chitosans, thus the modification does not change the polycationic character of this polysaccharide. Nonetheless, the results obtained herein provide a qualitative indication that the magnitude of the electrostatic density among normal and modified chitosans was not affected pronouncedly (Table 1).

Table 1. ζ-Potential values of CHT (0.1%) and ALG (0.1%) in aqueous solution with 0.15 M NaCl (pH = 5.5).

Polyelectrolyteolyte	ζ-Potentialtial
Normal–CHT	+20 $\pm$ 1.3
Mesyl–CHT	+22.3 $\pm$ 1.6
Tosyl–CHT	+19.8 $\pm$ 1.1
Alginate	−28.6 $\pm$ 1.7

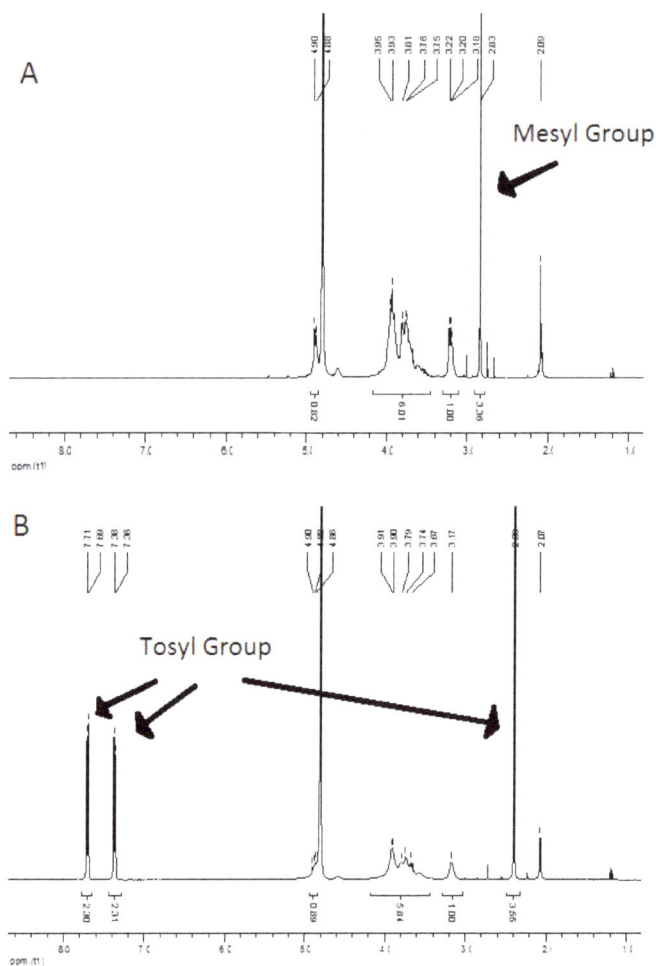

Figure 1. NMR characterization of mesyl–CHT (**A**) and tosyl–CHT (**B**). Peaks of mesyl (2.83 ppm) and tosyl groups (2.39, 7.37 and 7.70 ppm) are highlighted by arrows.

3.2. LbL Build-Up of mCHT/ALG Multilayered Films at pH = 5.5

mCHT and ALG were deposited in a sequential way to build multilayered films. Cationic solutions of mCHTs were deposited on the surface of a gold-coated quartz crystal (negatively charged) and subsequently anionic solution of ALG was deposited in a similar way, with an intermediate washing step using 0.15 M NaCl buffer. Electrostatic interactions between cationic–anionic polyelectrolytes ensures the LbL buildup of multilayered films.

The LbL assembly of the CHT and ALG was monitored by a quartz microbalance with dissipation monitoring (QCM-D). QCM-D analysis proved the effective interaction between the two polyelectrolytes, chitosans and alginate, at pH 5.5. Therefore, it can be said that the assembly of mCHT, and normal-CHT with ALG resulted in the construction of multilayered polymeric films. Alginate, as a polyanion, interacts with the positively charged chitosan. It is assumed that the carboxylate moieties on alginate ionically interact with the protonated amine groups on chitosan to form a matrix.

The frequency of QCM-D crystal decreases by the deposition of a thin film. The decrease in frequency (ΔF) is proportional to the mass of the film when the film is thin and rigid. However, this relation is not valid once a soft film is constructed as a result of the deposition of a polymeric material. In this situation, the energy stored in each vibrational cycle is lost and changes in dissipation (ΔD) represent a typical viscoelastic behavior. Frequency (ΔF) and dissipation (ΔD) variations normalized to the 5th overtone during the construction of 5 bilayers are shown in Figure 2A. The decrease in frequency observed after the adsorption of each successive layers of polyelectrolytes suggests that there was a gradual growth of the polymeric films, using normal and modified chitosans. In general, dissipation values increased with time, revealing that the films are not rigid and can dissipate energy, evidencing their viscoelastic behavior (Figure 2A). ΔD provides evidence about film's viscoelastic properties in which deposition of a soft component often leads to an increase of the dissipation values due to energy loss from the crystal's oscillation, whereas smaller dissipation values are obtained for rigid components [45,46]. Using the data acquired from the QCM-D experiments, the thickness of the multilayered films was estimated using the Voigt-based viscoelastic model (Figure 2B). The assembly of CHT and ALG followed a linear regime growth (Figure 2C), resulting in films with thickness of 87.4 ± 5.3, 52.3 ± 2.1 and 61.2 ± 1.4 nm for normal, mesyl and tosyl–CHT/ALG multilayers, respectively, after the construction of 5 bilayers. Statistical analysis demonstrated a significance difference between the thicknesses of normal–CHT/ALG and mCHT/ALG films ($p < 0.05$). Thinner mCHT/ALG films obtained by LbL process might be related to the presence of organic salts, mesylate and tosylate, in the chemical structure of mCHTs. The potential to make ionic interactions and form a polyelectrolyte complex (PEC) are among the important reasons that enabled the development of tailored biomaterials using ALG and CHT. The incorporation of modified chitosan in the LbL structure, with meticulous and selectivity practice, would give us the capability to control the thickness and mechanical properties of multilayered films. The possibility to change the physiochemical properties of CHT–ALG PECs by controlling the degree of association between functional groups provides an opportunity to design complex and tailored biopolymer scaffolds.

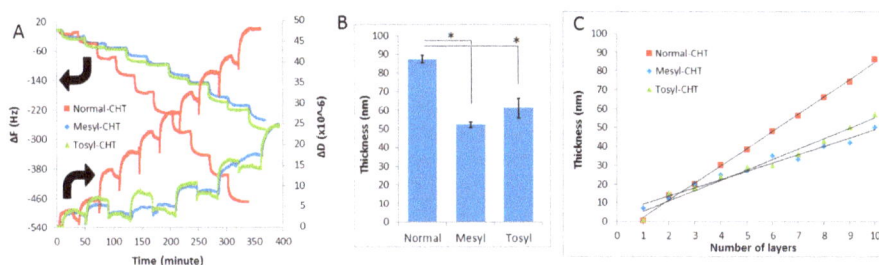

Figure 2. Representative quartz microbalance with dissipation monitoring (QCM-D) acquisition graphics depicting 5th overtone variations of frequency (descending curves) and dissipation (ascending curves) during construction of the multilayered films using normal–CHT/alginate (ALG) (normal), mesyl–CHT/ALG (mesyl) and tosyl–CHT/ALG (tosyl) (**A**); Thickness of films after buildup of 5 bilayers was estimated using Voigt-based viscoelastic model (Average $\pm$ SD; * $p < 0.05$) (**B**); Representative cumulative thickness evolution of the multilayer films as a function of the number of deposition layers. Lines represent linear trend lines with $R^2 = 0.99$ (normal-CHT), $R^2 = 0.97$ (tosyl–CHT) and $R^2 = 0.97$ (mesyl–CHT) (**C**).

3.3. Cytotoxicity of mCHT/ALG Multilayered Films

In vitro studies were performed to investigate the cytotoxicity of mCHT/ALG multilayered films using hASCs. Image analysis of live/dead staining by ImageJ (V. 1.51p, Bethesda, MD, USA) displayed a uniform distribution of viable cells (as shown by cells stained with Calcein AM, in green) with viability of $96\% \pm 4.2\%$ cultured on top of 24-well plate using the resulting culture medium with extract of mCHT/ALG films (Figure 3A). The metabolic activity of hASCs, as a result of incubating with extracts of mesyl–CHT/ALG and tosyl–CHT/ALG multilayered films, was similar to thenormal–CHT/ALG film and in the range of negative control (Figure 3B). However, the pre-incubation of the culture medium with latex resulted in lower metabolic activity and 58% $\pm$ 2.3% viability of cells (Figure 3B, positive control). These results further confirmed the cytocompatibility of mCHT/ALG multilayered films for cell culture studies.

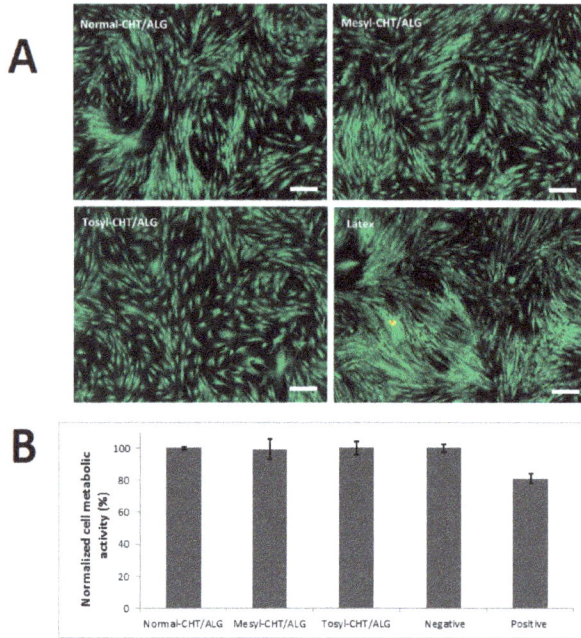

Figure 3. Representative fluorescence images of live (green) and dead (red) hASCs cultured on 24-well plates with the extract of mesyl–CHT/ALG, tosyl–CHT/ALG and normal–CHT/ALG multilayered films and latex. (**A**) (Scale bar = 200 μm); Normalized metabolic activity of hASCs was measured in contact with the extract of CHT/ALG multilayered films as well as negative (24-well plate) and positive (Latex) controls (**B**). Results are presented as average $\pm$ SD.

3.4. Aggregate Formation of hASCs on the mCHT/ALG Multilayered Films

hACSs cultured on normal and mCHT/ALG films did not show the ability to attach and spread to the multilayered films. They assembled in the form of spheroids with sizes between ca. 100 to 200 μm (Figure 4A–C). Low adhesion of the cells in such non-cytotoxic substrate can be used to stimulate cell aggregation. For example, chitosan films were used to prepare spheroids of melanocytes [47] and keratocytes [48]. In contrast, once cultured on glass substrates, hASCs exhibited common spindle shape morphology (Figure 4D). Cells need anchor points to be adhered to the substrate. Some substrates, such as tissue culture plates, are able to absorb the adhesion-related proteins, such as fibronectin, and this allows the attachment of cells to that specific substrate [49,50]. It was previously reported that the amount of adhesive-related proteins was far less in chitosan compared to tissue culture substrates [51]. This result may be attributed to the monopolar nature of chitosan, which is not able to interact with the bipolar extracellular matrix proteins presented in the culture medium. Our result also corroborates previous findings that non-crosslinked free standing CHT/ALG multilayered membranes could not support the adhesion of L929 cells [23]. One could also envisage using the developed multilayers as reservoirs of bioactive agents that could also extend the applicability of these substrates as supports to control cell behavior [52].

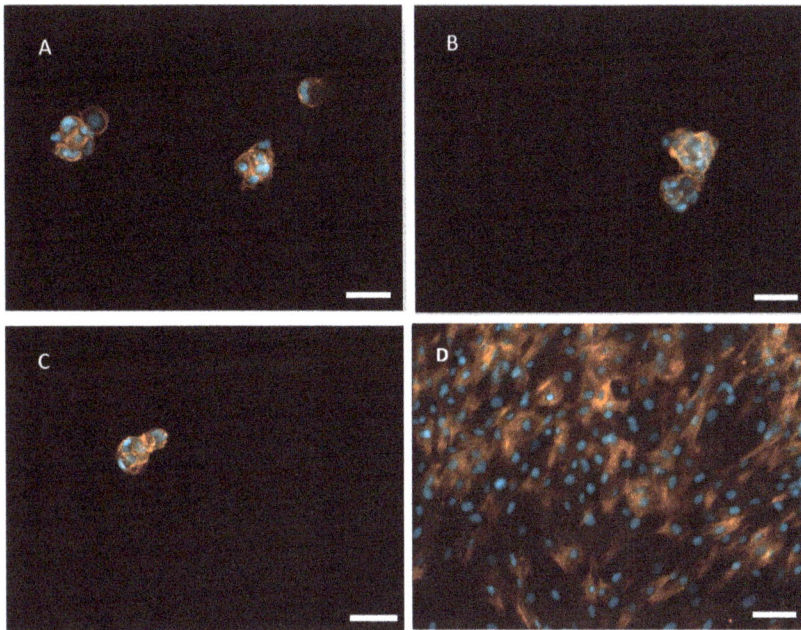

Figure 4. Representative fluorescence images of the hASCs cultured on top of normal–CHT/ALG (**A**); mesyl–CHT/ALG (**B**); tosyl–CHT/ALG (**C**); multilayered films and glass substrate (**D**). Multilayered films included subsequent layers of CHT and ALG with top layer (11th layer) was CHT. Scale bar: 200 μm. Cell nuclei appeared bright blue and F-actin filaments appeared bright red due to DAPI and phalloidin dyes, respectively.

4. Conclusions

Chitosan and alginate are two natural-based polymers with proved application in developing biomaterials for tissue engineering and regenerative medicine. In this study, we suggest the use of multilayered film produced by LbL assembly of modified chitosan/alginate as a scaffold to produce 3D aggregates of hASCs. We investigated how the modification of chitosan could affect the mCHT/ALG film properties and the corresponding biological performance. Therefore, two water-soluble cationic derivatives of chitosan, namely mesyl and tosyl chitosans, were synthesized and nanostructured multilayered LbL films were created using ALG as an anionic polyelectrolyte. mCHT/ALG films showed lower thickness compared to the normal–CHT/ALG films. Therefore, modification of chitosan with organic salts provided the ability to tailor the physiochemical properties of the multilayered films. The result demonstrated mCHT/ALG multilayered films could be used as a platform to produce 3D aggregates of hACSs. This approach may offer a new route to prepare off-the-shelf chitosan products which have the advantage of solubility in water at a neutral pH, and may be used for customizing the properties of films without the adjustment of processing parameters. Moreover, one could envisage incorporating biological signals in the multilayer films for a variety of applications, such as controlling the behavior of cells. It would be important to further investigate the effect of multilayer properties and spheroid formation on the differentiation potential and stemness of hACSs. This information is useful to develop tailored biomaterials to regulate the function and behavior of hACSs for varieties of applications in tissue engineering and regenerative medicine.

Polymers **2017**, *9*, 440

Acknowledgments: This paper was partially financed by the ELASTISLET H2020-NMP-2014-646075 project, Fundo Social Europeu (FSE) and Programa Operacional de Potencial Humano (POPH). Javad Hatami, Rui R. Costa and Mariana B. Oliveira acknowledge the Fundação para a Ciência e Tecnologia (FCT) for grants SFRH/BPD/117202/2016, SFRH/BPD/95446/2013 and SFRH/BPD/111354/2015, respectively. Sandra G. Silva acknowledges the grant from ComplexiTE project.

Author Contributions: Javad Hatami, Sandra G. Silva, Mariana B. Oliveira, Rui R. Costa and João F. Mano conceived and designed the experiments; Javad Hatami and Sandra G. Silva performed the experiments; Javad Hatami, Sandra G. Silva, Mariana B. Oliveira, Rui R. Costa, Rui L. Reis and João F. Mano analyzed the data, wrote and revised the paper.

Conflicts of Interest: The authors declare no conflict of interest. The founding sponsors had no role in the design of the study; in the collection, analyses, or interpretation of data; in the writing of the manuscript, and in the decision to publish the results.

References

1. Gimble, M.J.; Guilak, F. Differentiation potential of adipose derived adult stem (ADAS) cells. *Curr. Top. Dev. Biol.* **2003**, *58*, 137–160. [PubMed]
2. Guilak, F.; Lott, K.E.; Awad, H.A.; Cao, Q.; Hicok, K.C.; Fermor, B.; Gimble, J.M. Clonal analysis of the differentiation potential of human adipose-derived adult stem cells. *J. Cell Physiol.* **2006**, *206*, 229–237. [CrossRef] [PubMed]
3. Jones, L.D.; Wagers, A.J. No place like home: Anatomy and function of the stem cell niche. *Nat. Rev. Mol. Cell Biol.* **2008**, *9*, 11–21. [CrossRef] [PubMed]
4. Frith, J.E.; Thomson, B.; Genever, P.G. Dynamic three-dimensional culture methods enhance mesenchymal stem cell properties and increase therapeutic potential. *Tissue Eng. C* **2010**, *16*, 735–749. [CrossRef] [PubMed]
5. Bartosh, T.J.; Ylöstalo, J.H.; Mohammadipoor, A.; Bazhanov, N.; Coble, K.; Claypool, K.; Lee, R.H.; Choi, H.; Prockop, D.J. Aggregation of human mesenchymal stromal cells (MSCs) into 3D spheroids enhances their antiinflammatory properties. *Proc. Natl. Acad. Sci. USA* **2010**, *107*, 13724–13729. [CrossRef] [PubMed]
6. Wang, W.; Itaka, K.; Ohba, S.; Nishiyama, N.; Chung, U.I.; Yamasaki, Y.; Kataoka, K. 3D spheroid culture system on micropatterned substrates for improved differentiation efficiency of multipotent mesenchymal stem cells. *Biomaterials* **2009**, *30*, 2705–2715. [CrossRef] [PubMed]
7. Oliveira, M.B.; Neto, A.I.; Correia, C.R.; Rial-Hermida, M.I.; Alvarez-Lorenzo, C.; Mano, J.F. Superhydrophobic chips for cell spheroids high-throughput generation and drug screening. *ACS Appl. Mater. Interfaces* **2014**, *6*, 9488–9495. [CrossRef] [PubMed]
8. Borges, J.; Mano, J.F. Molecular Interactions Driving the Layer-by-Layer Assembly of Multilayers. *Chem. Rev.* **2014**, *114*, 8883–8942. [CrossRef] [PubMed]
9. Richardson, J.J.; Björnmalm, M.; Caruso, F. Technology-driven layer-by-layer assembly of nanofilms. *Science* **2015**, *348*. [CrossRef] [PubMed]
10. Xiao, F.X.; Pagliaro, M.; Xu, Y.J.; Liu, B. Layer-by-layer assembly of versatile nanoarchitectures with diverse dimensionality: A new perspective for rational construction of multilayer assemblies. *Chem. Soc. Rev.* **2016**, *45*, 3088–3121. [CrossRef] [PubMed]
11. Costa, R.R.; Mano, J.F. Polyelectrolyte multilayered assemblies in biomedical technologies. *Chem. Soc. Rev.* **2014**, *43*, 3453–3479. [CrossRef] [PubMed]
12. Oliveira, M.B.; Hatami, J.; Mano, J.F. Coating Strategies Using Layer-by-layer Deposition for Cell Encapsulation. *Chem. Asian J.* **2016**, *11*, 1753–1764. [CrossRef] [PubMed]
13. Boudou, T.; Crouzier, T.; Ren, K.; Blin, G.; Picart, C. Multiple Functionalities of Polyelectrolyte Multilayer Films: New Biomedical Applications. *Adv. Mater.* **2010**, *22*, 441–467. [CrossRef] [PubMed]
14. Joseph, N.; Ahmadiannamini, P.; Hoogenboom, R.; Vankelecom, I.F. Layer-by-layer preparation of polyelectrolyte multilayer membranes for separation. *Polym. Chem.* **2014**, *5*, 1817–1831. [CrossRef]
15. Banik, B.L.; Brown, J.L. Chapter 23—Polymeric Biomaterials in Nanomedicine A2—Kumbar, Sangamesh G. In *Natural and Synthetic Biomedical Polymers*; Laurencin, C.T., Deng, M., Eds.; Elsevier: Oxford, UK, 2014; pp. 387–395.

16. Mano, J.F.; Silva, G.A.; Azevedo, H.S.; Malafaya, P.B.; Sousa, R.A.; Silva, S.S.; Boesel, L.F.; Oliveira, J.M.; Santos, T.C.; Marques, A.P.; et al. Natural origin biodegradable systems in tissue engineering and regenerative medicine: Present status and some moving trends. *J. R. Soc. Interface* **2007**, *4*, 999–1030. [CrossRef] [PubMed]

17. Hynes, R.O. The Extracellular Matrix: Not Just Pretty Fibrils. *Science* **2009**, *326*, 1216–1219. [CrossRef] [PubMed]

18. Finkenstadt, V.L. Natural polysaccharides as electroactive polymers. *Appl. Microbiol. Biotechnol.* **2005**, *67*, 735–745. [CrossRef] [PubMed]

19. Cardoso, M.J.; Costa, R.R.; Mano, J.F. Marine Origin Polysaccharides in Drug Delivery Systems. *Mar. Drugs* **2016**, *14*. [CrossRef] [PubMed]

20. Basu, A.; Kunduru, K.R.; Abtew, E.; Domb, A.J. Polysaccharide-Based Conjugates for Biomedical Applications. *Bioconjug. Chem.* **2015**, *26*, 1396–1412. [CrossRef] [PubMed]

21. Senni, K.; Pereira, J.; Gueniche, F.; Delbarre-Ladrat, C.; Sinquin, C.; Ratiskol, J.; Godeau, G.; Fischer, A.M.; Helley, D.; Colliec-Jouault, S. Marine Polysaccharides: A Source of Bioactive Molecules for Cell Therapy and Tissue Engineering. *Mar. Drugs* **2011**, *9*, 1664–1681. [CrossRef] [PubMed]

22. Oliveira, S.M.; Santo, V.E.; Gomes, M.E.; Reis, R.L.; Mano, J.F. Layer-by-layer assembled cell instructive nanocoatings containing platelet lysate. *Biomaterials* **2015**, *48*, 56–65. [CrossRef] [PubMed]

23. Silva, J.M.; Duarte, A.R.C.; Caridade, S.G.; Picart, C.; Reis, R.L.; Mano, J.F. Tailored Freestanding Multilayered Membranes Based on Chitosan and Alginate. *Biomacromolecules* **2014**, *15*, 3817–3826. [CrossRef] [PubMed]

24. Caridade, S.G.; Monge, C.; Gilde, F.; Boudou, T.; Mano, J.F.; Picart, C. Free-Standing Polyelectrolyte Membranes Made of Chitosan and Alginate. *Biomacromolecules* **2013**, *14*, 1653–1660. [CrossRef] [PubMed]

25. Cuomo, F.; Lopez, F.; Miguel, M.G. Vesicle-templated layer-by-layer assembly for the production of nanocapsules. *Langmuir* **2010**, *26*, 10555–10560. [CrossRef] [PubMed]

26. Correia, C.R.; Pirraco, R.P.; Cerqueira, M.T.; Marques, A.P.; Reis, R.L.; Mano, J.F. Semipermeable Capsules Wrapping a Multifunctional and Self-regulated Co-culture Microenvironment for Osteogenic Differentiation. *Sci. Rep.* **2016**, *6*, 21883. [CrossRef] [PubMed]

27. Lee, K.Y.; Mooney, D.J. Alginate: Properties and biomedical applications. *Prog. Polym. Sci.* **2012**, *37*, 106–126. [CrossRef] [PubMed]

28. Vegas, A.J.; Veiseh, O.; Gürtler, M.; Millman, J.R.; Pagliuca, F.W.; Bader, A.R.; Doloff, J.C.; Li, J.; Chen, M.; Olejnik, K.; et al. Long-term glycemic control using polymer-encapsulated human stem cell-derived beta cells in immune-competent mice. *Nat. Med.* **2016**, *22*, 306–311. [CrossRef] [PubMed]

29. Rinaudo, M. Chitin and chitosan: Properties and applications. *Prog. Polym. Sci.* **2006**, *31*, 603–632. [CrossRef]

30. Francis Suh, J.K.; Matthew, H.W.T. Application of chitosan-based polysaccharide biomaterials in cartilage tissue engineering: A review. *Biomaterials* **2000**, *21*, 2589–2598. [CrossRef]

31. Wang, Z.; Zhang, X.; Gu, J.; Yang, H.; Nie, J.; Ma, G. Electrodeposition of alginate/chitosan layer-by-layer composite coatings on titanium substrates. *Carbohydr. Polym.* **2014**, *103*, 38–45. [CrossRef] [PubMed]

32. Silva, J.M.; Duarte, A.R.C.; Custódio, C.A.; Sher, P.; Neto, A.I.; Pinho, A.; Fonseca, J.; Reis, R.L.; Mano, J.F. Nanostructured Hollow Tubes Based on Chitosan and Alginate Multilayers. *Adv. Healthc. Mater.* **2014**, *3*, 433–440. [CrossRef] [PubMed]

33. Zhao, Q.; Han, B.; Wang, Z.; Gao, C.; Peng, C.; Shen, J. Hollow chitosan-alginate multilayer microcapsules as drug delivery vehicle: Doxorubicin loading and in vitro and in vivo studies. *Nanomed. Nanotechnol. Biol. Med.* **2007**, *3*, 63–74. [CrossRef] [PubMed]

34. Martins, G.V.; Mano, J.F.; Alves, N.M. Nanostructured self-assembled films containing chitosan fabricated at neutral pH. *Carbohydr. Polym.* **2010**, *80*, 570–573. [CrossRef]

35. Rúnarsson, Ö.V.; Malainer, C.; Holappa, J.; Sigurdsson, S.T.; Másson, M. tert-Butyldimethylsilyl O-protected chitosan and chitooligosaccharides: Useful precursors for N-modifications in common organic solvents. *Carbohydr. Res.* **2008**, *343*, 2576–2582. [CrossRef] [PubMed]

36. Song, W.; Gaware, V.S.; Rúnarsson, Ö.V.; Másson, M.; Mano, J.F. Functionalized superhydrophobic biomimetic chitosan-based films. *Carbohydr. Polym.* **2010**, *81*, 140–144. [CrossRef]

37. Sahariah, P.; Gaware, V.S.; Lieder, R.; Jónsdóttir, S.; Hjálmarsdóttir, M.Á.; Sigurjonsson, O.E.; Másson, M. The Effect of Substituent, Degree of Acetylation and Positioning of the Cationic Charge on the Antibacterial Activity of Quaternary Chitosan Derivatives. *Mar. Drugs* **2014**, *12*, 4635–4658. [CrossRef] [PubMed]

38. Signini, R.; Campana, S.P. On the preparation and characterization of chitosan hydrochloride. *Polym. Bull.* **1999**, *42*, 159–166. [CrossRef]

39. Costa, R.R.; Custódio, C.A.; Arias, F.J.; Rodríguez-Cabello, J.C.; Mano, J.F. Layer-by-Layer Assembly of Chitosan and Recombinant Biopolymers into Biomimetic Coatings with Multiple Stimuli-Responsive Properties. *Small* **2011**, *7*, 2640–2649. [CrossRef] [PubMed]

40. Voinova, M.V.; Rodahl, M.; Jonson, M.; Kasemo, B. Viscoelastic acoustic response of layered polymer films at fluid-solid interfaces: Continuum mechanics approach. *Phys. Scr.* **1999**, *59*, 391–399. [CrossRef]

41. Cerqueira, M.T.; Pirraco, R.P.; Santos, T.C.; Rodrigues, D.B.; Frias, A.M.; Martins, A.R.; Reis, R.L.; Marques, A.P. Human adipose stem cells cell sheet constructs impact epidermal morphogenesis in full-thickness excisional wounds. *Biomacromolecules* **2013**, *14*, 3997–4008. [CrossRef] [PubMed]

42. Poursamar, S.A.; Hatami, J.; Lehner, A.N.; da Silva, C.L.; Ferreira, F.C.; Antunes, A.P.M. Gelatin porous scaffolds fabricated using a modified gas foaming technique: Characterisation and cytotoxicity assessment. *Mater. Sci. Eng. C* **2015**, *48*, 63–70. [CrossRef] [PubMed]

43. Oliveira, M.B.; Custódio, C.A.; Gasperini, L.; Reis, R.L.; Mano, J.F. Autonomous osteogenic differentiation of hASCs encapsulated in methacrylated gellan-gum hydrogels. *Acta Biomater.* **2016**, *41*, 119–132. [CrossRef] [PubMed]

44. Younes, I.; Rinaudo, M. Chitin and chitosan preparation from marine sources. Structure, properties and applications. *Mar. Drugs* **2015**, *13*, 1133–1174. [CrossRef] [PubMed]

45. Cho, N.J.; Kanazawa, K.K.; Glenn, J.S.; Frank, C.W. Employing two different quartz crystal microbalance models to study changes in viscoelastic behavior upon transformation of lipid vesicles to a bilayer on a gold surface. *Anal. Chem.* **2007**, *79*, 7027–7035. [CrossRef] [PubMed]

46. Höök, F.; Kasemo, B.; Nylander, T.; Fant, C.; Sott, K.; Elwing, H. Variations in coupled water, viscoelastic properties, and film thickness of a Mefp-1 protein film during adsorption and cross-linking: A quartz crystal microbalance with dissipation monitoring, ellipsometry, and surface plasmon resonance study. *Anal. Chem.* **2001**, *73*, 5796–5804. [CrossRef] [PubMed]

47. Lin, S.J.; Jee, S.H.; Hsaio, W.C.; Lee, S.J.; Young, T.H. Formation of melanocyte spheroids on the chitosan-coated surface. *Biomaterials* **2005**, *26*, 1413–1422. [CrossRef] [PubMed]

48. Chen, Y.H.; Wang, I.J.; Young, T.H. Formation of keratocyte spheroids on chitosan-coated surface can maintain keratocyte phenotypes. *Tissue Eng. A* **2009**, *15*, 2001–2013. [CrossRef] [PubMed]

49. Neto, A.I.; Vasconcelos, N.L.; Oliveira, S.M.; Ruiz-Molina, D.; Mano, J.F. High-Throughput Topographic, Mechanical, and Biological Screening of Multilayer Films Containing Mussel-Inspired Biopolymers. *Adv. Funct. Mater.* **2016**, *26*, 2745–2755. [CrossRef]

50. Caridade, S.G.; Monge, C.; Almodóvar, J.; Guillot, R.; Lavaud, J.; Josserand, V.; Coll, J.L.; Mano, J.F.; Picart, C. Myoconductive and osteoinductive free-standing polysaccharide membranes. *Acta Biomater.* **2015**, *15*, 139–149. [CrossRef] [PubMed]

51. Cuy, J.L.; Beckstead, B.L.; Brown, C.D.; Hoffman, A.S.; Giachelli, C.M. Adhesive protein interactions with chitosan: Consequences for valve endothelial cell growth on tissue-engineering materials. *J. Biomed. Mater. Res. A* **2003**, *67*, 538–547. [CrossRef] [PubMed]

52. Costa, R.R.; Alatorre-Meda, M.; Mano, J.F. Drug nano-reservoirs synthesized using layer-by-layer technologies. *Biotechnol. Adv.* **2015**, *33*, 1310–1326. [CrossRef] [PubMed]

Article

Osteogenesis of Adipose-Derived and Bone Marrow Stem Cells with Polycaprolactone/Tricalcium Phosphate and Three-Dimensional Printing Technology in a Dog Model of Maxillary Bone Defects

Jeong Woo Lee [1], Seung Gyun Chu [1], Hak Tae Kim [1], Kang Young Choi [1], Eun Jung Oh [1,2], Jin-Hyung Shim [3], Won-Soo Yun [3], Jung Bo Huh [4], Sung Hwan Moon [5], Seong Soo Kang [6] and Ho Yun Chung [1,2,*]

1   Department of Plastic and Reconstructive Surgery, School of Medicine, Kyungpook National University, Daegu 41944, Korea; jeongwoo@hanmail.net (J.W.L.); choo870818@naver.com (S.G.C.); patraqushe@naver.com (H.T.K.); prschoi@gmail.com (K.Y.C.); fullrest74@hanmail.net (E.J.O.)
2   Cell & Matrix Research Institute, School of Medicine, Kyungpook National University, Daegu 41944, Korea
3   Department of Mechanical Engineering, Korea Polytechnic University, Siheung-Si 15073, Korea; happyshim@kpu.ac.kr (J.-H.S.); wsyun@kpu.ac.kr (W.-S.Y.)
4   Department of Prosthodontics, Dental Research Institute, Institute of Translational Dental Science, School of Dentistry, Pusan National University, Yangsan-Si 50612, Korea; neoplasia96@hanmail.net
5   Department of Medicine, School of Medicine, Konkuk University, Seoul 05029, Korea; sunghwanmoon@kku.ac.kr
6   College of Veterinary Medicine, Chonnam National University, Gwangju 61186, Korea; vetkang@chonnam.ac.kr
*   Correspondence: hy-chung@knu.ac.kr or chunghoyun@gmail.com; Tel.: +82-53-420-5692; Fax: +82-53-425-3879

Received: 25 July 2017; Accepted: 12 September 2017; Published: 15 September 2017

Abstract: Bone graft material should possess sufficient porosity and permeability to allow integration with native tissue and vascular invasion, and must satisfy oxygen and nutrient transport demands. In this study, we have examined the use of three-dimensional (3D)-printed polycaprolactone/tricalcium phosphate (PCL/TCP) composite material in bone grafting, to estimate the scope of its potential application in bone surgery. Adipose-derived stem cells (ADSCs) and bone marrow stem cells (BMSCs) are known to enhance osteointegration. We hypothesized that a patient-specific 3D-printed solid scaffold could help preserve seeded ADSCs and BMSCs and enhance osteointegration. Diffuse osteogenic tissue formation was observed by micro-computed tomography with both stem cell types, and the ADSC group displayed similar osteogenesis compared to the BMSC group. In histological assessment, the scaffold pores showed abundant ossification in both groups. Reverse transcription polymerase chain reaction (RT-PCR) showed that the BMSC group had higher expression of genes associated with ossification, and this was confirmed by Western blot analysis. The ADSC- and BMSC-seeded 3D-printed PCL/TCP scaffolds displayed promising enhancement of osteogenesis in a dog model of maxillary bone defects.

Keywords: adipose-derived stem cells; bone marrow stem cells; osteogenesis; 3D-printed scaffold; polycaprolactone; tricalcium phosphate; dog; maxillary bone

1. Introduction

Charles W. Hull, who founded 3D Systems in the United States in 1986, invented the world's first three-dimensional (3D) printer. Since then, 3D printing has revolutionized a variety of manufacturing technologies, as well as tissue engineering and regenerative medicine [1]. Due to its innovative promise,

3D printing technology has attracted increasing interest in recent years, including its use in human organ and tissue development [2].

Tissue engineering is an important medical technology that combines biological and engineering techniques for the restoration of damaged or missing tissues or organs. Materials used to construct scaffolds for tissue engineering should have biocompatibility and biodegradability, and mechanical properties that can be supported in the body. In addition, an artificial scaffold prepared according to the characteristics of the transplantation site should allow differentiation and proliferation of cells attached to the pore [3–6]. The production of scaffolds using various materials has been an area of active study in tissue engineering.

Based on computed tomography (CT) or magnetic resonance imaging (MRI) data, it is possible to make patient-customized scaffolds using 3D printing technology. An advantage of this approach is that the mechanical properties, pore size, and porosity of the scaffold can be controlled. In addition, the excellent internal connective structure formed in the 3D shape improves the penetration and differentiation of the cells into the scaffolding, resulting in excellent biocompatibility. This technology is widely used for the reconstruction of ears and craniofacial defects in plastic surgery [7–11].

We have previously published literature introducing blended polycaprolactone (PCL) and tricalciumphosphate (TCP) as a 3D printable material for guided bone regeneration [12]. The blended TCP played a role in improving PCL's deficient bone regeneration ability.

Stem cells are cells that can infinitely regenerate over a lifetime, and can differentiate into various cell types in response to specific biological signals and external stimuli. Techniques for differentiating mesenchymal stem cells (MSCs) isolated from bone marrow (bone marrow stem cells, BMSCs) and adipose tissue (adipose-derived stem cells, ADSCs), into chondrocytes, osteoblasts, and muscle cells are well established, and these cells can be used in tissue engineering to regenerate various tissues such as cartilage, bone, and muscle [13–17].

MSCs are the main constituents of bone marrow stromal cells, and are the precursors of bone, cartilage, muscle, and connective tissue [18]. Immunologically, they are rarely rejected, and proliferate readily without significant apoptosis [19]. Because of this nature, bone marrow-derived stem cells are of interest in biotechnology. ADSCs have also been actively studied in many fields because of their ease of acquisition, storage and handling of cells; however, ethical issues remain to be overcome [20–22].

We hypothesized that ossification between normal bone and a 3D-printed scaffold is an important factor in the reconstruction of maxillary bone defects, and that substances secreted from ADSCs and BMSCs could enhance osteointegration and thus increase both the degree of contact between the normal bone and the 3D-printed scaffold and bone formation inside the scaffold. This study was conducted to investigate the use of polycaprolactone/tricalcium phosphate (PCL/TCP) scaffolds, which are 3D-printed with a mixture of PCL, a polymeric material, and TCP, a bioceramic material, with ADSCs and BMSCs in bone regeneration in a dog model of maxillary bone defects.

2. Methods

2.1. Experimental Materials

2.1.1. Fabrication and Characterization of 3D-Printed PCL/TCP Scaffolds

Biodegradable PCL (Evonik Industries AG, Essen, Germany) and TCP (Sigma-Aldrich, St. Louis, MO, USA) were used as materials for the scaffolds. PCL was dried at 105 °C for 1 day before use. TCP had a particle size of 100 nm or less. The 3D defect model was created by commercial 3D medical image editing software (Mimics, Materialize NV, Leuven, Belgium), and customized to maxilla defect of animals used in this study. A steel syringe containing the PCL/TCP mixture was equipped to a 3D printer and maintained at 120 °C. The molten PCL/TCP mixture was precisely dispensed through stainless steel nozzle with diameter of 300 µm. The inner microstructure and surface composition were characterized by field-emission scanning electron microscope (FE-SEM, S-4700, Hitachi Co.,

Ltd., Tokyo, Japan) and energy dispersive spectrometer (EDS) analyses (INCA Energy 350, Oxford Instruments, Oxford, UK).

2.1.2. Harvest of ADSCs from Beagles

Adipose tissues obtained from the abdominal cavity of the beagles were washed with an equal volume of phosphate buffered saline (PBS). Then, enzymatic degradation was carried out at 37 °C for 30 min using 0.075% collagenase type I (Worthington Biochemical, Lakewood, NJ, USA). To remove debris from the connective tissues, the degraded tissue was filtered and the supernatant of the adipocyte layer was separated. The cell suspension was centrifuged for 10 min at a force of 200× *g*. Contaminating erythrocytes were removed by the addition of erythrocyte lysis buffer (Sigma-Aldrich, St. Louis, MO, USA) at pH 7.3. The stromal cells were rinsed twice with PBS, and the ADSCs were collected.

2.1.3. Harvest of BMSCs from Beagles

Under general anesthesia, the humerus of each beagle was punctured with an 18-gauge injection needle and 1 mL of bone marrow tissue was collected with a 1 mL syringe. The collected bone marrow aspirates were combined with Histopaque (Sigma-Aldrich, St. Louis, MO, USA) at a ratio of 1:1 in a 15-mL tube and centrifuged at 2000 rpm for 30 min. After centrifugation, the white layer was collected in a 15-mL tube, diluted 10-fold in PBS, and then centrifuged at 2000 rpm for 10 min. After aspiration of the supernatant, the cells were again diluted 10-fold in PBS and centrifuged as above to collect BMSCs.

2.1.4. Experimental Animals and Treatment Groups

Six healthy beagles, each 12 months old and weighing 9 kg, were included in the study. Prior to the 3D printing model experiment, spiramycin and metronidazole were injected under general anesthesia, the teeth were scaled, and the left molar and premolar were removed. Prior to implanting 3D-printed PCL/TCP scaffolds, the dogs were divided into two groups (*n* = 3 in each group) for application of ADSCs and BMSCs.

ADSCs and BMSCs were prepared at 1 μg/mL and injected into the 3D-printed PCL/TCP scaffolds in 0.01-mL increments in 10 areas of uniform distribution. Then, the 3D-printed PCL/TCP scaffolds were fixed in the maxillary bone defects of the beagles.

2.2. Experimental Methods

Thiopental sodium (Pentothal; Choong Wae Pharmacy, Seoul, Korea) was injected intravenously and general anesthesia was performed with halothane (Ilsung-halothane; Ilsung Pharmaceuticals, Seoul, Korea). During the operation, Lactated Ringer's Solution and 1 g of cephalosporin were administered. After exposure to anesthesia, the fur on the left side of the beagle's face was removed and the skin was disinfected with Betadine and alcohol. The maxillary bone was exposed through a skin incision from the zygomatic arch to the infraorbital foramen, and the maxillary bone defect was created from the infraorbital rim to the zygomatic arch in accordance with the 3D-printed scaffold prepared before surgery. The infraorbital nerve was preserved.

First, the bottom pores of the 3D-printed PCL/TCP scaffolds were blocked with fibrin glue. ADSCs and BMSCs were injected into the pores according to the experimental group, and then fibrin glue was applied to the exposed outer pores to prevent cell leakage. The prepared 3D-printed scaffold was inserted into the defect, fixed with a plate and screw, and the skin was sutured (Figure 1). Both groups were dressed once a day after surgery, and stitches were removed 7 days after surgery. All experiments conducted in this study were approved by the Institutional Animal Care and Use Committee of Chonnam National University (Approval No. CNU IACUC-YB-2016-43, The approval date is 28 September 2016) and followed their recommendations.

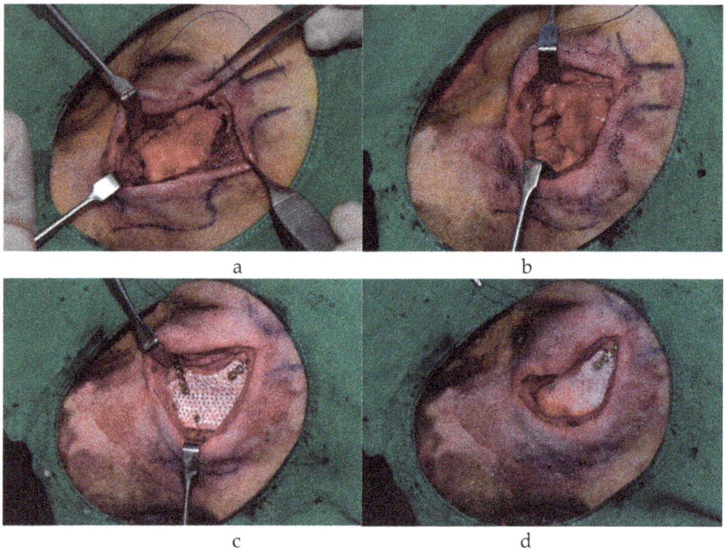

Figure 1. Experimental design: (**a**) exposed maxillary bone in a beagle model; (**b**) the defect was made; (**c**) the polycaprolactone/tricalcium phosphate (PCL/TCP) scaffold was fixed by plate and screw; and (**d**) fibrin glue seeding after adipose-derived stem cells (ADSCs) or bone marrow stem cells (BMSCs) were applied.

2.3. Experimental Evaluation

2.3.1. Evaluation using 3D CT

To evaluate the ossification of the implanted 3D-printed scaffold, CT images of coronal, axial, and sagittal views were taken immediately after surgery, and 4, 8, and 12 weeks postoperatively. The ossification activity was determined by calcification of the marginal area, increased bone density, and decreased internal pore size. The level of tissue density depending on the time was analyzed with the Hounsfield unit (HU), which is a quantity commonly used in CT imaging. The HU value of the tissue in the pore was randomly measured at 5 sites for each individual, and the average values were obtained (Mimics Innovation Suite 17.0 64-bit version; Materialise NV, Leuven, Belgium). The measured HU values were normalized and calculated. To measure the extent of ossification, 3D-CT reconstruction was used.

2.3.2. Histological Evaluation

For histological evaluation, all beagles were euthanized 12 weeks after 3D-printed scaffold implantation and the grafts were harvested, including 1 cm of marginal bone tissue. Tissues were taken from three normal bone/scaffold junctions and the center of one scaffold, and each was divided into three pieces. One third was fixed in 10% neutral formalin, and paraffin embedded tissue sections were stained with Hematoxylin and Eosin (H&E) and Goldner's trichrome (GT). New ossification of the entire graft, the thickness and extent of periosteum produced, the degree of bone ingrowth into the pores, the infiltration of inflammatory cells into the graft, and the formation of collagen fibers were observed with an optical microscope (SkyScan 1173; Bruker microCT, Kontich, Belgium). Each slide was measured for optical density using an image analysis program (iSolution Lite; Image & Microscope Technology Inc., Vancouver, BC, Canada). In each group, the optical density of normal bone, tissue in the pore, and soft tissue around the scaffold including the periosteum were measured randomly at 5 sites for each tissue sample, and average values were obtained.

2.3.3. Reverse Transcription Polymerase Chain Reaction (RT-PCR)

One third of each collected tissue was frozen in liquid nitrogen, and after grinding, total RNA was isolated using TRIzol (Thermo Fisher Scientific, Waltham, MA, USA). RNA was reverse transcribed to cDNA using a RevertAid First Strand cDNA Synthesis Kit (Thermo Fisher Scientific, MA, USA) and polymerase chain reaction (PCR) was performed using an AccuPower RT-PCR kit (Bioneer, Daejeon, Korea). Type 1 collagen (COL1), osteocalcin (OCN), runt-related transcription factor 2 (RUNX2) and glyceraldehyde-3-phosphate dehydrogenase (GAPDH) expression were measured by RT-PCR (Table 1).

Table 1. Sequences of primers used for reverse transcription polymerase chain reaction (RT-PCR).

Gene name	Sequence (5'-3')
COL1-dog-F	CTCGTCACAGTTGGGGTTGA
COL1-dog-R	GGTGCAAGTATGAAGCGGGA
OCN-dog-F	AATTGCGCTCGAGCATCTCT
OCN-dog-R	ATTGCCACGGTTGCTACTGA
RUNX2-dog-F	GGCGGCTATAACTCTTCCCA
RUNX2-dog-R	ACGCAGCGGCTTTTTATTTCA

COL1, type 1 collagen; OCN, osteocalcin; RUNX2, runt-related transcription factor 2.

2.3.4. Western Blot Analysis

Proteins were extracted from obtained skin tissue by PRO-PREP kit (iNtRON Biotechnology, Seoul, Korea) and protein concentration was determined with BCA method. Proteins and 5X Sodium dodecyl sulfate polyacrylamide gel electrophoresis (SDS-PAGE) loading buffer were mixed and boiled at 95 °C for 5 min before incubation on ice. This protein was separated with 10% SDS-polyaclamide gel at 80~100 V for 2 h and transferred to nitrocellulose membrane at 100 V for 70 min on ice. Membranes were blocked with 5% skim milk at room temperature for 1 h and probed with primary antibody such as OCN (ab13420, 1 µg/mL; Abcam, Cambridge, MA, USA), COL1 (ab6308, 1 µg/mL; Abcam, Cambridge, MA, USA), RUNX2 (ab23981, 1 µg/mL; Abcam, Cambridge, MA, USA), β-actin (ab8226, 1:10000; Abcam, Cambridge, MA, USA). β-actin was used as a housekeeping protein. After incubating the membranes with antibody solution overnight at 4 °C, the membranes were washed with Tris buffered saline with Tween-20 (pH 7.4) and probed with secondary antibody for 1 hour at room temperature. Membrane was then exposed to ECL. The blots were followed immediately by exposure to X-ray film and the bands were quantified using ImageJ software (version 1.45s; U.S. National Institute of Health, Bethesda, MD, USA).

2.4. Statistical Analysis

Statistical analysis was performed with SPSS Ver. 22.0. (IBM Corp., Armonk, NY, USA). A subgroup analysis was performed using the paired *t*-test. Statistical significance was set at $p < 0.05$.

3. Results

3.1. Characterization of the 3D-Printed Scaffolds

For fabrication of the customized scaffold, 3D defect was reconstructed using Mimics software (Figure 2a). The 3D defect model was converted into printing path data by automated printing path generation algorithm (Figure 2b). Using the printing path data, the customized scaffolds were fabricated via 3D printer (Figure 2c). The line width and pore size of the printed scaffolds were 300 and 400 µm, respectively (Figure 2d). The EDS analysis was used to confirm exposure TCP onto the printed strut. As shown in Figure 2e, there were calcium (Ca) and phosphate (P) elements detected on PCL/TCP scaffold surface, indicating TCP powder was fully incorporated within the scaffold.

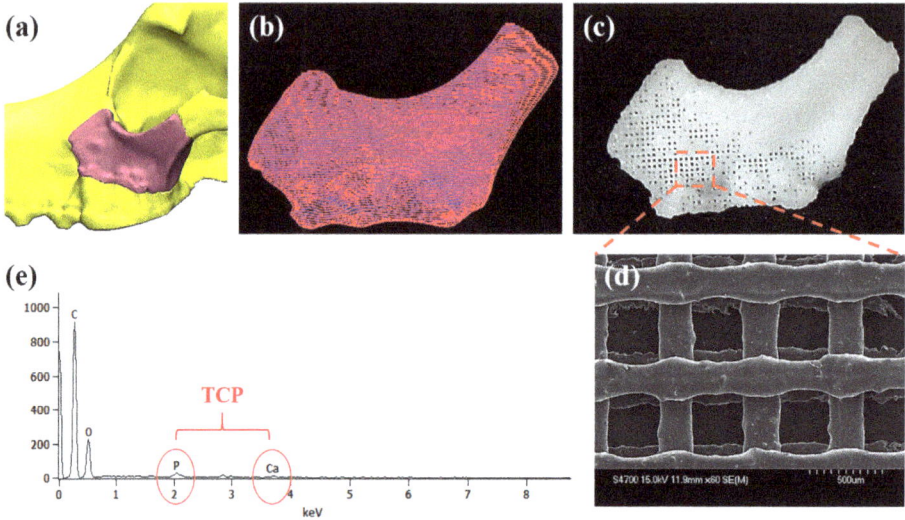

Figure 2. 3D printing procedure and characterization: (**a**) 3D reconstruction of the defect; (**b**) printing path generation; (**c**) the photograph of the printed scaffold; (**d**) SEM image; and (**e**) EDS results.

3.2. Evaluation of Results Using 3D CT

3.2.1. Results from Coronal, Axial, and Sagittal CT Views

In both groups, ossification was not observed until four weeks after surgery. After eight weeks, the bone density along the scaffold margins was similar to that of normal bone. This can be regarded as progression of ossification along the margin, and showed a marked increase in density after 12 weeks. This suggests that ossification of the margin continuously increased for up to 12 weeks after surgery.

In both groups, the size and number of pores inside the scaffold also decreased after eight weeks. However, the bone density of the scaffold was only partially increased, and did not reach the density of the surrounding normal bone. This suggests that most of the pores are formed by the increase in connective tissue and the inflammation reaction with the surrounding tissues and the ossification is partially occurring. After 12 weeks, pore size decreased further, and bone density was increased compared to eight weeks. However, at both 8 and 12 weeks, the BMSC group showed more pronounced marginal ossification and greater pore size reduction than the ADSC group (Figures 3 and 4).

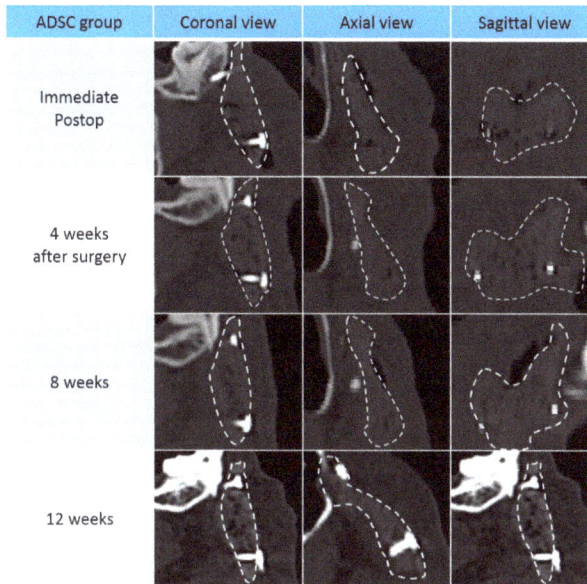

Figure 3. Computed tomography (CT) images of coronal, axial, and sagittal views from the adipose-derived stem cell (ADSC) group. Images were acquired directly after surgery and 4, 8, and 12 weeks later.

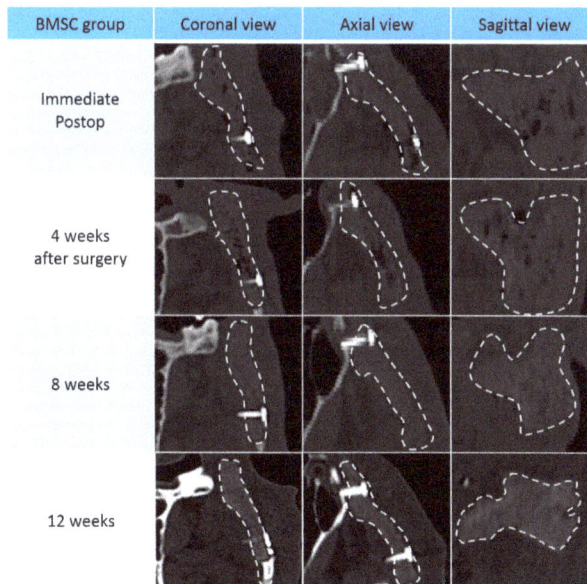

Figure 4. Computed tomography (CT) images of coronal, axial, and sagittal views from the bone marrow stem cell (BMSC) group. Images were acquired directly after surgery and 4, 8, and 12 weeks later.

When the density of the normal bone was set to 100, the normalized density value of the new tissue (including the new bone) in the pores was 36.5 and 35.7 in the ADSC and BMSC groups, respectively, which was higher than the relative density of the periosteum around the scaffold (27.3 and 24.1 for the

ADSC and BMSC groups, respectively; Figure 5). This suggests that ossification progressed sufficiently in the transplanted scaffolds in both groups. The ADSC group displayed a slightly higher density than the BMSC group. However, there was no significant difference between ADSC and BMSC groups. The normalized density of scaffold only (in vitro) was 15.

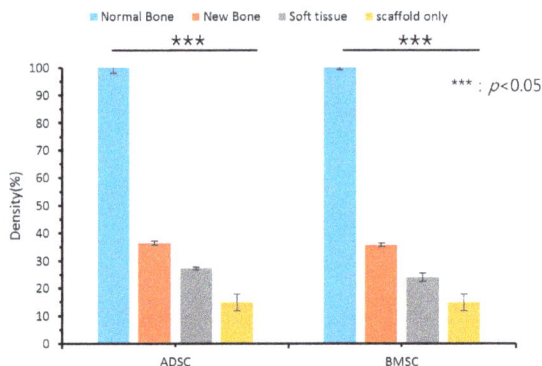

Figure 5. Gray values of normal bone, new bone, and soft tissue from the adipose-derived stem cell (ADSC) and bone marrow stem cell (BMSC) groups.

3.2.2. Results from 3D CT

In both groups, 3D CT did not display characteristic features of ossification until four weeks after surgery. After eight weeks, diffuse ossification could be observed inside the pores of the scaffold. A more prevalent ossification process was observed at 12 weeks in both groups. At 8 and 12 weeks, more pronounced ossification was observed in the ADSC group than in the BMSC group (Figure 6).

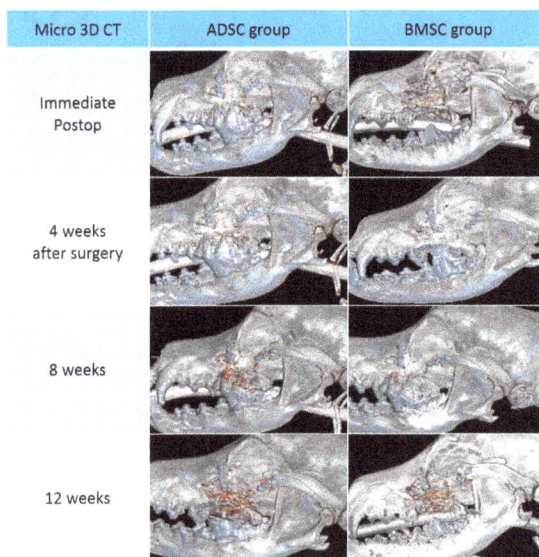

Figure 6. Three-dimensional (3D) computed tomography (CT) images of the adipose-derived stem cell (ADSC) and bone marrow stem cell (BMSC) groups. Images were acquired directly after surgery and 4, 8, and 12 weeks later.

3.3. Histological Findings

There was no gross rejection of the scaffold in either group. Upon histopathologic examination, connective tissue, such as the periosteum, was enclosed in the margin of the scaffold, and dense connective tissue was observed in the pore. These results demonstrate the excellent biocompatibility of the scaffolds. Some inflammatory findings were observed in the pore, which are thought to be due to inflammation before scaffold engraftment. Ossification was more pronounced in the center of the scaffold than in the margins. Ossification in the ADSC group was increased compared to the BMSC group (Figures 7 and 8).

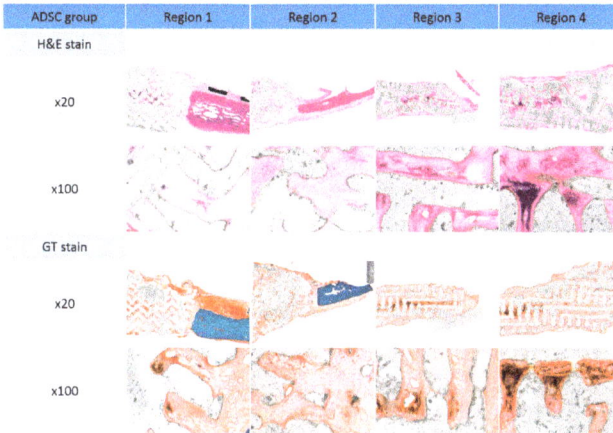

Figure 7. Histopathological analysis of the adipose-derived stem cell (ADSC) group 12 weeks after surgery by Hematoxylin and Eosin (H&E) and Goldner's trichrome (GT) staining. Original magnification: ×20 and ×100.

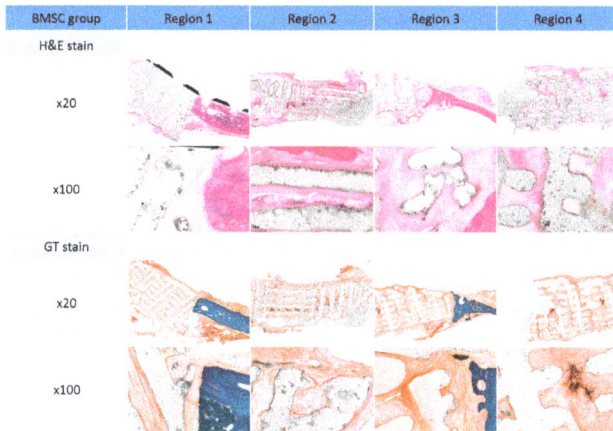

Figure 8. Histopathological analysis of the bone marrow stem cell (BMSC) group 12 weeks after surgery by Hematoxylin and Eosin (H&E) and Goldner's trichrome (GT) staining. Original magnification: ×20 and ×100.

The mean optical density values of normal bone, pore tissue, and soft tissues (including the periosteum) around the scaffold were obtained for the two groups. In GT and H&E staining, the optical density inside the pores was higher than that of the periosteum and other soft tissues around the scaffold, and was higher with GT than with H&E, however, there was no significant difference (Figure 9).

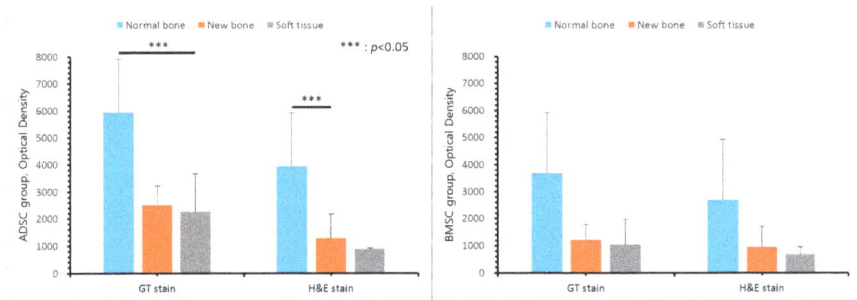

Figure 9. Optical density of Hematoxylin and Eosin stain (H&E) and Goldner's trichrome (GT) staining.

3.4. RT-PCR

Histological evaluation 12 weeks after surgery displayed ossification in the pore of the scaffold and an increase of fibrinogen in the periphery. The increase of fibrinogen was confirmed by an increase of COL1 gene expression, and ossification was confirmed by increases in OCN and RUNX2 gene expression. By RT-PCR, the band intensity of COL1, OCN, and RUNX2 was normalized to that of GAPDH. The degree of ossification-related gene expression was slightly increased in the BMSC group compared to the ADSC group, however, there was no significant difference (Figure 10).

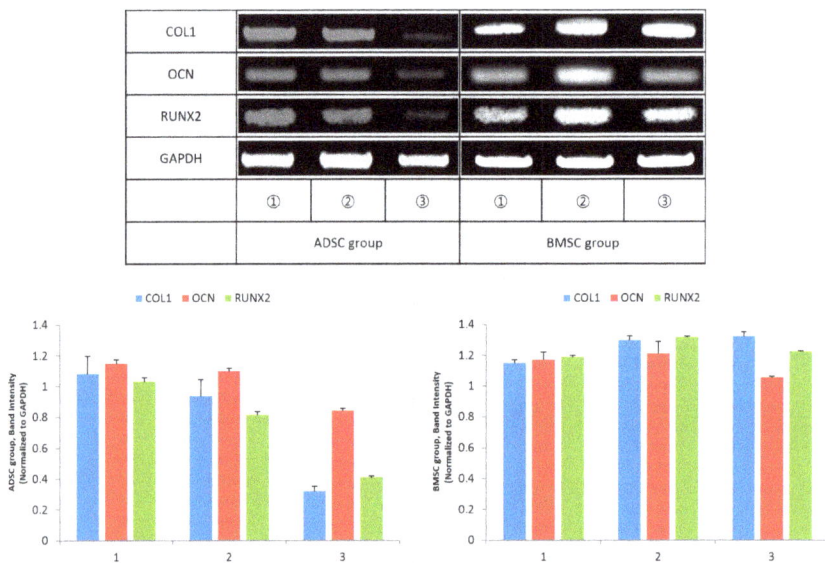

Figure 10. Expression of type 1 collagen (COL1), osteocalcin (OCN), runt-related transcription factor 2 (RUNX2) and glyceraldehyde-3-phosphate dehydrogenase (GAPDH) 12 weeks after surgery, measured by reverse transcription polymerase chain reaction (RT-PCR). ADSC, adipose-derived stem cell; BMSC, bone marrow stem cell.

3.5. Western Blot Analysis

Twelve weeks after surgery, expression of the ossification-related proteins COL1, OCN, and RUNX2 was increased in both groups. The expression of these proteins was higher in the BMSC group than in the ADSC group (Figure 11). In particular, the expression of COL1 was significantly higher in the BMSC than in the ADSC. The band intensity of COL1, OCN, and RUNX2 was normalized to that of β-actin.

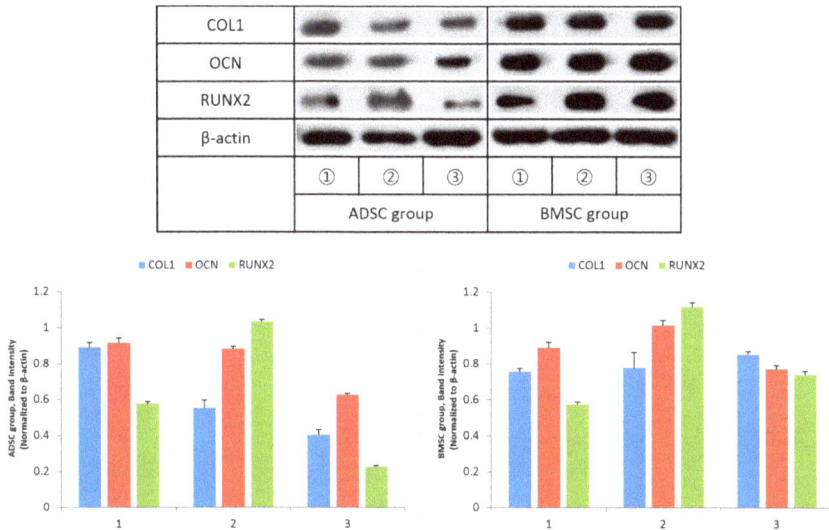

Figure 11. Expression of type 1 collagen (COL1), osteocalcin (OCN), runt-related transcription factor 2 (RUNX2) and β-actin 12 weeks after surgery, measured by Western blot analysis. ADSC, adipose-derived stem cell; BMSC, bone marrow stem cell.

4. Discussion

Due to its ease of use and adaptability, 3D printing technology is being rapidly developed and applied in various medical fields. For example, 3D printing technology can be used to visualize the complex deformities of congenital cardiovascular anomalies, to map intricate blood vessels during renal cancer surgery, and allow the development of approaches to intracranial tumor neurosurgery. These applications help surgeons analyze and evaluate the individual condition of the patient before surgery, aiding in the development of a surgical plan [23,24]. However, despite being an innovative technology that can be continuously developed in the medical field, there remains a lack of scientific data on its application.

With tissue engineering, a new era in bioengineering has begun. In general, tissue engineering involves three components: cells, scaffolds, and growth factors. Recently, 3D technologies have been used to develop cell and tissue printing technologies, allowing tissue engineering to go beyond conventional cell culture technology. Scaffold development and 3D cell culture both have remaining limitations, but it is expected that in the near future, 3D cell-printing technology will enable the simultaneous formation of 3D biological structures and scaffolds [25–30].

Currently, 3D printing technology is most often used in craniomaxillofacial surgery, and several challenges must be overcome for its widespread application. First, more specialized computer software is required. Pre-operative 3D printing is not easy to plan because it takes too much time to perform a computer simulation of a differentiated process. This requires more automated functionality and simulation via a variety of software.

Second, there must be a sufficient relationship between the pre-operative simulation and the actual operating environment, because the pre-operative simulation serves as a guide for the operation. For example, to apply 3D-printed titanium implants in surgery, bone must be resected precisely according to the preoperative plan, as 3D-printed implants are generally quite hard, and cannot easily be cut or bent.

Third, accuracy during measurement is very important. Due to technological advances, it is possible to obtain very thin CT images, but the combination of these images to form 3D images introduces error, particularly for very thin bones such as the orbital bones.

Finally, there is a possibility of artifacts due to metal materials. The use of CT scans in 3D printing for prosthetic dentistry, which mainly uses metal models, may be less accurate due to these artificial shadows. Tooth occlusion requires very fine 3D rendering, and CT may not provide images of adequate resolution because of these artifacts [31,32].

Despite these problems, 3D printing technology is attracting attention as a new medical technology. Using 3D printing, the medical industry could visualize various characteristics of individual patients and apply them to personalized medical care.

Scaffolds used in tissue engineering are generally made of natural and synthetic polymeric materials. Polymers derived from natural materials, animals, and human bodies have superior biocompatibility and are nontoxic. These include gelatin, collagen, fibrin, elastin, and alginate. Synthetic polymeric materials are relatively inexpensive, have excellent mechanical properties, and are either hydrolyzed in vivo or degraded by enzymes, making them ideal polymers for scaffolds. PCL, polylactide-co-glycolide, and polylactic acid (PLA) are typical synthetic polymer materials. In particular, various scaffolds for bone tissue regeneration are made using PLA and TCP, a bioceramic material. Since TCP is made of powder, it confers decreased mechanical strength to the scaffold in comparison with a solid polymer material that can be produced using heat, and is more difficult to form into 3D scaffold with excellent internal pore structure. To overcome this problem, many studies have fabricated scaffolds by mixing ceramic and polymeric materials [33–37].

The 3D printing system enables precise control and rapid production of desired 3D shape structures at the micron scale. It also has the advantage of determining the line width and line height of laminated materials according to pneumatic pressure, temperature, and moving speed. Generally, when a scaffold is produced by 3D printing with a mix of polymer and a bioceramic material, the ceramic material content is limited. Because of the difficulty of nozzle discharge with increased amounts of bioceramic powder, controlling the mixing method of the bioceramic powder is a very important aspect of scaffold development. In this study, a 3D printing system was used to fabricate 3D scaffolds from PCL mixed with β-TCP, and evaluated their effects on bone formation.

MSCs grow adherently in the form of fibroblasts in vitro, form cell clusters from single cells, and differentiate into osteocytes, adipocytes, and chondrocytes. Friedenstein first reported the presence of MSCs in the bone marrow in 1970 [38]. Subsequently, cells with similar characteristics have been reported in various tissues such as adipose, brain, spleen, liver, kidney, lung, bone marrow, muscle, thymus, and pancreas [39].

In order to improve osteointegration in bone grafting and the regeneration of bone defects, ADSCs and BMSCs have recently been used in tissue engineering. However, when harvesting BMSCs, complications such as pain and infection are likely to occur, and there is a limit to the amount of BMSCs taken. In contrast, ADSCs are readily available in large quantities under local anesthesia [13–16]. ADSCs are simpler to collect than BMSCs, with less donor morbidity, and adipose tissue provides more stem cells than bone marrow. However, previous studies have shown that BMSCs induce bone formation after differentiation into osteoblasts in vitro and in vivo, and thus provide favorable conditions for regeneration of bone tissue [40–43].

The use of scaffolds for the application of ADSCs and BMSCs to bone defects has many advantages. The scaffold serves as a bridge, allowing adequate time for stem cell survival and subsequent bone formation. Generally, the scaffold should be strong enough to withstand external impact, and the pores should disassemble after the bone has adequately regenerated. In this study, an artificial scaffold was prepared using biodegradable PCL/β-TCP, which is currently in clinical use as an osteoinductive material. β-TCP is highly compatible with bone morphogenetic protein 2 (BMP-2), a bone growth-promoting factor, and thus increases osteoinduction [44–47]. In addition, β-TCP degrades with time. These two properties are ideal for scaffolds used in bone regeneration.

To evaluate the bone regeneration ability of PCL/β-TCP scaffolds injected with ADSCs and BMSCs, 3D CT and histological analysis were performed after implantation of the scaffolds into maxillary bone defects in beagles. Diffuse osteogenic tissue formation was observed by micro-computed tomography with both stem cell types, and the ADSC group displayed similar osteogenesis compared to the BMSC group. RT-PCR and Western blot analysis confirmed increased expression of proteins associated with osteogenesis in both groups, with higher expression of COL1 of the BMSC group observed in Western blot. As a result, new bone formation inside the scaffold was effective in the reconstruction of maxillary bone defects in a dog model using PCL/TCP biomaterials manufactured using a 3D printing system, with ADSCs and BMSCs. However, unlike other previous studies, gross and histological results suggested that ADSC treatment was superior to BMSC treatment.

These results suggest that ADSCs induce osteogenesis similar to BMSCs when 3D-printed PCL/TCP scaffolds are implanted in bone defects. Clinically, ADSCs have many advantages over BMSCs because ADSCs can be harvested easily, quickly, safely, and in relative abundance by lipoaspiration in the office-based setting. On the other hand, BMSCs harvest is an invasive procedure with low yield, which may impart donor site morbidity, such as pain, bleeding, hematoma, and deep infection. Therefore, if used with a 3D-printed PCL/TCP scaffold, ADSC can be a good alternative to BMSC to improve osteointegration in bone grafting and the regeneration of bone defects.

Acknowledgments: This research was supported by a grant of the Korea Health Technology R&D Project through the Korea Health Industry Development Institute (KHIDI), funded by the Ministry of Health & Welfare, Republic of Korea (grant number: HI14C3309).

Author Contributions: Eun Jung Oh, Jin-Hyung Shim, Won-Soo Yun, Jung Bo Huh, Sung Hwan Moon, Seong Soo Kang, and Ho Yun Chung conceived and designed the experiments; Seung Gyun Chu and Hak Tae Kim performed the experiment; Kang Young Choi, and Ho Yun Chung analyzed the data; and Jeong Woo Lee and Ho Yun Chung wrote the paper.

Conflicts of Interest: The authors declare no conflict of interest.

References

1. Park, S.; Lee, J.H.; Kim, W.D. Development of Biomimetic Scaffold for Tissue Engineering. *Elastomers Compos.* **2009**, *44*, 106–111.
2. Sachlos, E.; Czernuszka, J.T. Making tissue engineering scaffolds work. Review: The application of solid freeform fabrication technology to the production of tissue engineering scaffolds. *Eur. Cells Mater.* **2003**, *5*, 29–39. [CrossRef]
3. Fuchs, M.; Köster, G.; Krause, T.; Merten, H.A.; Schmid, A. Degradation of and intraosseous reactions to biodegradable poly-L-lactide screws: A study in minipigs. *Arch. Orthop. Trauma Surg.* **1998**, *118*, 140–144. [CrossRef] [PubMed]
4. Päivärinta, U.; Böstman, O.; Majola, A.; Toivonen, T.; Törmälä, P.; Rokkanen, P. Intraosseous cellular response to biodegradable fracture fixation screws made of polyglycolide or polylactide. *Arch. Orthop. Trauma Surg.* **1993**, *112*, 71–74. [CrossRef] [PubMed]
5. Peltoniemi, H.H.; Hallikainen, D.; Toivonen, T.; Helevirta, P.; Waris, T. SR-PLLA and SR-PGA miniscrews: Biodegradation and tissue reactions in the calvarium and dura mater. *J. Craniomaxillofac. Surg.* **1999**, *27*, 42–50. [CrossRef]

6. Bergsma, E.J.; Rozema, F.R.; Bos, R.R.; de Bruijn, W.C. Foreign body reactions to resorbable poly(L-lactide) bone plates and screws used for the fixation of unstable zygomatic fractures. *J. Oral Maxillofac. Surg.* **1993**, *51*, 666–670. [CrossRef]

7. D'Urso, P.S.; Earwaker, W.J.; Barker, T.M.; Redmond, M.J.; Thompson, R.G.; Effeney, D.J.; Tomlinson, F.H. Custom cranioplasty using stereolithography and acrylic. *Br. J. Plast. Surg.* **2000**, *53*, 200–204. [CrossRef] [PubMed]

8. Faber, J.; Berto, P.M.; Quaresma, M. Rapid prototyping as a tool for diagnosis and treatment planning for maxillary canine impaction. *Am. J. Orthod. Dentofac. Orthop.* **2006**, *129*, 583–589. [CrossRef] [PubMed]

9. Müller, A.; Krishnan, K.G.; Uhl, E.; Mast, G. The application of rapid prototyping techniques in cranial reconstruction and preoperative planning in neurosurgery. *J. Craniofac. Surg.* **2003**, *14*, 899–914. [CrossRef] [PubMed]

10. Choi, J.W.; Kim, N. Clinical application of three-dimensional printing technology in craniofacial plastic surgery. *Arch. Plast. Surg.* **2015**, *42*, 267–277. [CrossRef] [PubMed]

11. Mannoor, M.S.; Jiang, Z.; James, T.; Kong, Y.L.; Malatesta, K.A.; Soboyejo, W.O.; Verma, N.; Gracias, D.H.; McAlpine, M.C. 3D printed bionic ears. *Nano Lett.* **2013**, *13*, 2634–2639. [CrossRef] [PubMed]

12. Shim, J.H.; Won, J.Y.; Park, J.H.; Bae, J.H.; Ahn, G.; Kim, C.H.; Lim, D.H.; Cho, D.W.; Yun, W.S.; Bae, E.B.; et al. Effects of 3D-Printed Polycaprolactone/β-Tricalcium Phosphate Membranes on Guided Bone Regeneration. *Int. J. Mol. Sci.* **2017**, *18*, 899. [CrossRef] [PubMed]

13. Zuk, P.A.; Zhu, M.; Mizuno, H.; Huang, J.; Futrell, J.W.; Katz, A.J.; Benhaim, P.; Lorenz, H.P.; Hedrick, M.H. Multilineage cells from human adipose tissue: Implications for cell-based therapies. *Tissue Eng.* **2001**, *7*, 211–228. [CrossRef] [PubMed]

14. Zuk, P.A.; Zhu, M.; Ashjian, P.; De Ugarte, D.A.; Huang, J.I.; Mizuno, H.; Alfonso, Z.C.; Fraser, J.K.; Benhaim, P.; Hedrick, M.H. Human adipose tissue is a source of multipotent stem cells. *Mol. Biol. Cell* **2002**, *13*, 4279–4295. [CrossRef] [PubMed]

15. Dragoo, J.L.; Choi, J.Y.; Lieberman, J.R.; Huang, J.; Zuk, P.A.; Zhang, J.; Hedrick, M.H.; Benhaim, P. Bone induction by BMP-2 transduced stem cells derived from human fat. *J. Orthop. Res.* **2003**, *21*, 622–629. [CrossRef]

16. Huang, J.I.; Beanes, S.R.; Zhu, M.; Lorenz, H.P.; Hedrick, M.H.; Benhaim, P. Rat extramedullary adipose tissue as a source of osteochondrogenic progenitor cells. *Plast. Reconstr. Surg.* **2002**, *109*, 1033–1041. [CrossRef] [PubMed]

17. Ogawa, R.; Mizuno, H.; Hyakusoku, H.; Watanabe, A.; Migita, M.; Shimada, T. Chondrogenic and osteogenic differentiation of adipose-derived stem cells isolated from GFP transgenic mice. *J. Nippon Med. Sch.* **2004**, *71*, 240–241. [CrossRef] [PubMed]

18. Prockop, D.J. Marrow stromal cells as stem cells for nonhematopoietic tissues. *Science* **1997**, *276*, 71–74. [CrossRef] [PubMed]

19. Quirici, N.; Soligo, D.; Bossolasco, P.; Servida, F.; Lumini, C.; Deliliers, G.L. Isolation of bone marrow mesenchymal stem cells by anti-nerve growth factor receptor antibodies. *Exp. Hematol.* **2002**, *30*, 783–791. [CrossRef]

20. Gu, J.H.; Han, S.K. Brief Review of Adipose Derived Cells. *J. Korean Soc. Aesthet. Plast. Surg.* **2009**, *15*, 192–198.

21. Kon, E.; Muraglia, A.; Corsi, A.; Bianco, P.; Marcacci, M.; Martin, I.; Boyde, A.; Ruspantini, I.; Chistolini, P.; Rocca, M.; et al. Autologous bone marrow stromal cells loaded onto porous hydroxyapatite ceramic accelerate bone repair in critical-size defects of sheep long bones. *J. Biomed. Mater. Res.* **2000**, *49*, 328–337. [CrossRef]

22. Pelegrine, A.A.; Aloise, A.C.; Zimmermann, A.; de Mello, E.; Oliveira, R.; Ferreira, L.M. Repair of critical-size bone defects using bone marrow stromal cells: A histomorphometric study in rabbit calvaria. Part I: Use of fresh bone marrow or bone marrow mononuclear fraction. *Clin. Oral Implants Res.* **2014**, *25*, 567–572. [CrossRef] [PubMed]

23. Winder, J.; Bibb, R. Medical rapid prototyping technologies: State of the art and current limitations for application in oral and maxillofacial surgery. *J. Oral Maxillofac. Surg.* **2005**, *63*, 1006–1015. [CrossRef] [PubMed]

24. Ventola, C.L. Medical Applications for 3D Printing: Current and Projected Uses. *Pharm. Ther.* **2014**, *39*, 704–711.

25. Fullerton, J.N.; Frodsham, G.C.; Day, R.M. 3D printing for the many, not the few. *Nat. Biotechnol.* **2014**, *32*, 1086–1087. [CrossRef] [PubMed]

26. Dankowski, R.; Baszko, A.; Sutherland, M.; Firek, L.; Kałmucki, P.; Wróblewska, K.; Szyszka, A.; Groothuis, A.; Siminiak, T. 3D heart model printing for preparation of percutaneous structural interventions: Description of the technology and case report. *Kardiol. Pol.* **2014**, *72*, 546–551. [CrossRef] [PubMed]

27. Rengier, F.; Mehndiratta, A.; von Tengg-Kobligk, H.; Zechmann, C.M.; Unterhinninghofen, R.; Kauczor, H.U.; Giesel, F.L. 3D printing based on imaging data: Review of medical applications. *Int. J. Comput. Assist. Radiol. Surg.* **2010**, *5*, 335–341. [CrossRef] [PubMed]

28. McGowan, J. 3D printing technology speeds development. *Health Estate* **2013**, *67*, 100–102. [PubMed]

29. Lee, H.; Fang, N.X. Micro 3D printing using a digital projector and its application in the study of soft materials mechanics. *J. Vis. Exp.* **2012**, *69*, e4457. [CrossRef] [PubMed]

30. Mironov, V.; Boland, T.; Trusk, T.; Forgacs, G.; Markwald, R.R. Organ printing: Computer-aided jet-based 3D tissue engineering. *Trends Biotechnol.* **2003**, *21*, 157–161. [CrossRef]

31. Xu, X.; Ping, F.Y.; Chen, J.; Yan, F.G.; Mao, H.Q.; Shi, Y.H.; Zhao, Z.Y. Application of CAD/CAM techniques in mandible large-scale defect and reconstruction with vascularized fibular bone graft. *Zhejiang Da Xue Xue Bao Yi Xue Ban* **2007**, *36*, 498–502. [PubMed]

32. Zhang, T.; Zhang, Y.; Li, Y.S.; Gui, L.; Mao, C.; Chen, Y.N.; Zhao, J.Z. Application of CTA and CAD/CAM techniques in mandible reconstruction with free fibula flap. *Zhonghua Zheng Xing Wai Ke Za Zhi* **2006**, *22*, 325–327. [PubMed]

33. Buddy, D.R.; Allan, S.H.; Frederick, J.S.; Jack, E.L. (Eds.) *Biomaterials Science: An Introduction to Materials in Medicine*, 3rd ed.; Academic Press: New York, NY, USA, 2012.

34. Todo, M.; Park, S.D.; Arakawa, K.; Takenoshita, Y. Relationship between microstructure and fracture behavior of bioabsorbable HA/PLLA composites. *Compos. Part A Appl. Sci. Manuf.* **2006**, *37*, 2221–2225. [CrossRef]

35. Agrawal, C.M.; Athanasiou, K.A. Technique to control pH in vicinity of biodegrading PLA-PGA implants. *J. Biomed. Mater. Res.* **1997**, *38*, 105–114. [CrossRef]

36. Habibovic, P.; de Groot, K. Osteoinductive biomaterials-properties and relevance in bone repair. *J. Tissue Eng. Regen. Med.* **2007**, *1*, 25–32. [CrossRef] [PubMed]

37. Habibovic, P.; Gbureck, U.; Doillon, C.J.; Bassett, D.C.; van Blitterswijk, C.A.; Barralet, J.E. Osteoconduction and osteoinduction of low-temperature 3D printed bioceramic implants. *Biomaterials* **2008**, *29*, 944–953. [CrossRef] [PubMed]

38. Friedenstein, A.J.; Chailakhjan, R.K.; Lalykina, K.S. The development of fibroblast colonies in monolayer cultures of guinea-pig bone marrow and spleen cells. *Cell Tissue Kinet.* **1970**, *3*, 393–403. [CrossRef] [PubMed]

39. Otley, C.C.; Gayner, S.M.; Ahmed, I.; Moore, E.J.; Roenigk, R.K.; Sherris, D.A. Preoperative and postoperative topical tretinoin on high-tension excisional wounds and full-thickness skin grafts in a porcine model: A pilot study. *Dermatol. Surg.* **1999**, *25*, 716–721. [CrossRef] [PubMed]

40. Bruder, S.P.; Jaiswal, N.; Haynesworth, S.E. Growth kinetics, self-renewal, and the osteogenic potential of purified human mesenchymal stem cells during extensive subcultivation and following cryopreservation. *J. Cell. Biochem.* **1997**, *64*, 278–294. [CrossRef]

41. Cancedda, R.; Mastrogiacomo, M.; Bianchi, G.; Derubeis, A.; Muraglia, A.; Quarto, R. Bone marrow stromal cells and their use in regenerating bone. *Novartis Found. Symp.* **2003**, *249*, 133–143. [PubMed]

42. Dieudonné, S.C.; Kerr, J.M.; Xu, T.; Sommer, B.; DeRubeis, A.R.; Kuznetsov, S.A.; Kim, I.S.; Gehron Robey, P.; Young, M.F. Differential display of human marrow stromal cells reveals unique mRNA expression patterns in response to dexamethasone. *J. Cell. Biochem.* **1999**, *76*, 231–243. [CrossRef]

43. Satomura, K.; Krebsbach, P.; Bianco, P.; Gehron Robey, P. Osteogenic imprinting upstream of marrow stromal cell differentiation. *J. Cell. Biochem.* **2000**, *78*, 391–403. [CrossRef]

44. Jingushi, S.; Urabe, K.; Okazaki, K.; Hirata, G.; Sakai, A.; Ikenoue, T.; Iwamoto, Y. Intramuscular bone induction by human recombinant bone morphogenetic protein-2 with beta-tricalcium phosphate as a carrier: In vivo bone banking for muscle-pedicle autograft. *J. Orthop. Sci.* **2002**, *7*, 490–494. [CrossRef] [PubMed]

45. Alam, I.; Asahina, I.; Ohmamiuda, K.; Enomoto, S. Comparative study of biphasic calcium phosphate ceramics impregnated with rhBMP-2 as bone substitutes. *J. Biomed. Mater. Res.* **2001**, *54*, 129–138. [CrossRef]

46. Laffargue, P.; Hildebrand, H.F.; Rtaimate, M.; Frayssinet, P.; Amoureux, J.P.; Marchandise, X. Evaluation of human recombinant bone morphogenetic protein-2-loaded tricalcium phosphate implants in rabbits' bone defects. *Bone* **1999**, *25*, 55S–58S. [CrossRef]

47. Urist, M.R.; Nilsson, O.; Rasmussen, J.; Hirota, W.; Lovell, T.; Schmalzreid, T.; Finerman, G.A. Bone regeneration under the influence of a bone morphogenetic protein (BMP) beta tricalcium phosphate (TCP) composite in skull trephine defects in dogs. *Clin. Orthop. Relat. Res.* **1987**, *214*, 295–304. [CrossRef]

![polymers logo] *polymers*

MDPI

Article

Incorporation of Calcium Containing Mesoporous (MCM-41-Type) Particles in Electrospun PCL Fibers by Using Benign Solvents

Liliana Liverani [1], Elena Boccardi [1], Ana Maria Beltrán [2,3] and Aldo R. Boccaccini [1,*]

[1]   Institute of Biomaterials, Department of Materials Science and Engineering, University of Erlangen-Nuremberg, 91058 Erlangen, Germany; liliana.liverani@fau.de (L.L.); elena.boccardi@fau.de (E.B.)

[2]   Institute de Ciencia de Materiales de Sevilla (ICMS), CSIC-Universidad de Sevilla, Seville 41092, Spain; abeltran3@us.es

[3]   Department of Engineering and Materials Science, University of Seville, Sevilla 41092, Spain

*   Correspondence: aldo.boccaccini@fau.de; Tel.: +49-(0)913-185-28-601

Received: 10 July 2017; Accepted: 26 September 2017; Published: 4 October 2017

Abstract: The electrospinning technique is a versatile method for the production of fibrous scaffolds able to resemble the morphology of the native extra cellular matrix. In the present paper, electrospinning is used to fabricate novel SiO_2 particles (type MCM-41) containing poly(epsilon-caprolactone) (PCL) fibers. The main aims of the present work are both the optimization of the particle synthesis and the fabrication of composite fibers, obtained using benign solvents, suitable as drug delivery systems and scaffolds for soft tissue engineering applications. The optimized synthesis and characterization of calcium-containing MCM-41 particles are reported. Homogeneous bead-free composite electrospun mats were obtained by using acetic acid and formic acid as solvents; neat PCL electrospun mats were used as control. Initially, an optimization of the electrospinning environmental parameters, like relative humidity, was performed. The obtained composite nanofibers were characterized from the morphological, chemical and mechanical points of view, the acellular bioactivity of the composite nanofibers was also investigated. Positive results were obtained in terms of mesoporous particle incorporation in the fibers and no significant differences in terms of average fiber diameter were detected between the neat and composite electrospun fibers. Even if the Ca-containing MCM-41 particles are bioactive, this property is not preserved in the composite fibers. In fact, during the bioactivity assessment, the particles were released confirming the potential application of the composite fibers as a drug delivery system. Preliminary in vitro tests with bone marrow stromal cells were performed to investigate cell adhesion on the fabricated composite mats, the positive obtained results confirmed the suitability of the composite fibers as scaffolds for soft tissue engineering.

Keywords: electrospinning; benign solvents; mesoporous silica calcium containing MCM-41; composites; nanofibers; poly(epsilon-caprolactone)

---

1. Introduction

The fabrication of nanocomposite fibers by using electrospinning techniques is a challenging and interesting topic for several applications, such as the development of scaffolds for tissue engineering and drug delivery [1–5]. The electrospinning process allows the fabrication of scaffolds with a fibrillar structure, by the application of high voltage between the conductive tip of a polymeric solution and a grounded fiber collector. The regulation of process parameters, polymeric solution parameters and environmental conditions affects the fiber morphology and properties [6].

It is possible to process many polymers and copolymers, and the selection of a suitable solvent system is fundamental to improving the electrospinning of a solution. In the last years, the spread of the "green electrospinning" concept, the reduction of the use of toxic solvents and the related increase of the use of benign solvents have been reported in the literature, for example in [7–9]. Limiting the use of harsh solvents is highly beneficial in terms of processing proteins, such as collagen and other sensitive biomolecules, preventing denaturation. Using benign solvents is also important to avoid the presence of toxic solvent residuals inside the mats which could limit their applications in the biomedical field, also bringing advantages in terms of lab worker safety and environmental impact [7–9].

Composite electrospun fibers can be obtained by the incorporation of nanoparticles inside the polymeric mats. This approach has been already widely investigated in the literature, in particular for the development of composite electrospun scaffolds for bone tissue engineering [10–16].

Ordered mesoporous silica particulate materials have been increasingly investigated since the first series of ordered mesoporous silica materials were developed in 1982 (known as MCM-41) [17,18]. These silica particles are characterized by ordered porosity at the nanoscale and disordered arrangement at the atomic level. After the development of MCM-41 materials, Grün et al. [19] presented the first synthesis of spherical MCM-41 nanoparticles. The most interesting features of such MCM-41 particles are the ordered mesoporous structure, hexagonally shaped pores, narrow pore size distribution (2–10 nm), and a large surface area (900–1500 $m^2 \cdot g^{-1}$) [17,18]. The synthesis procedure was a modification of the Stöber's reaction [20] for the preparation of monodispersed silica spheres. Grün et al. [19] modified this procedure by adding a cationic surfactant (cetyl trimethylammonium bromide, CTAB) to the synthesis solution, providing a source of micelles. These ordered mesoporous materials have been extensively used as molecular sieves, catalyst, adsorbents and molds [17]. In 2001 [21], mesoporous silica particles were proposed for the first time as a drug delivery system. These silica-based mesoporous materials are able to incorporate a relatively high amount of drug into the mesopores. Moreover, their silanol groups can be functionalized and the pore diameter can be modulated in order to increase the drug-uptake and to better control the drug-release kinetics [22]. Due to their pure silica network, MCM-41 particles are not bioactive [23], i.e., they are not able to form hydroxyapatite layers once in contact with biological fluids (even after two months) due to the small pore size and the lower concentration of silanol groups compared to other silica particles.

In the present work, the production of calcium-containing MCM-41 (Ca_MCM-41) particles in order to couple the drug up-take capability of the ordered mesoporous particles with the bioactivity of Ca–Si systems is reported. The obtained particles are characterized by an ordered mesoporous structure and hydroxy-carbonate apatite (HCA) formation capability after one day of immersion in simulated body fluid (SBF) solution.

A proof of concept of the incorporation of the synthesized Ca-containing MCM-41 particles in electrospun poly(epsilon-caprolactone) (PCL) fibers was performed. PCL was selected because of its biocompatibility, biodegradability, ability to be processed by electrospinning—in particular with benign solvents—and considering its FDA approval for clinical use [24–26]. The obtained composite nanofibers were characterized from the morphological, chemical and mechanical point of view. Even if the proposed approach of incorporating silica mesoporous nanoparticles into electrospun polymeric mats for the development of drug delivery systems has been already reported in the literature [27,28], the novelty of the present research work is represented by the synthesis optimization of the Ca_MCM-41 particles and the development of electrospinning using benign solvents aiming at the fabrication of optimized scaffolds for tissue engineering and drug delivery applications which have not been reported previously.

2. Materials and Methods

2.1. Ca_MCM-41 Particle Synthesis

The Ca_MCM-41 particles were produced by optimizing a method available in the literature in order to obtain spherical particles with ordered mesostructured and reduced agglomeration [29].

Briefly, 0.6 g of cetyltrimethylammonium bromide (CTAB, Sigma Aldrich, Munich, Germany) were dissolved in a solution of 300 mL of deionized water and 0.2 g of sodium hydroxide (NaOH, Sigma Aldrich, Munich, Germany) under continuous stirring for up to 30 min. 3 mL tetraethyl orthosilicate (TEOS, Sigma Aldrich, Munich, Germany) was added (rate 0.25 mL·min^{-1}) and the solution was stirred for 30 min. After 2.34 g of calcium nitrate (Ca(NO$_3$)$_2$·4H$_2$O, Merck KGaA, Darmstadt, Germany) was added and the solution was stirred for 1 h 30 min. All synthesis steps were carried out at 80 °C to avoid the agglomeration of the final product. The resulting dispersion was centrifuged and washed once with deionized water and twice with ethanol. The white precipitate was then dried for 12 h at 60 °C and heat-treated at 600 °C (1 °C/min) for 6 h.

2.2. Degradation in Simulated Body Fluid (SBF)

Simulated body fluid (SBF) was prepared by dissolving reagent grade 8.035 g·L^{-1} NaCl, 0.355 g·L^{-1} NaHCO$_3$, 0.225 g·L^{-1} KCl, 0.231 g·L^{-1} K$_2$HPO$_4$·(3H$_2$O), 0.311 g·L^{-1} MgCl$_2$ (6H$_2$O), 0.292 g·L^{-1} CaCl$_2$, and 0.072 g·L^{-1} Na$_2$SO$_4$ in deionized water and buffering at pH 7.4 at 36.5 °C with 6.118 g·L^{-1} tris(hydroxymethyl) aminomethane ((CH$_2$OH)$_3$CNH$_2$) and 1M HCl, as previously reported by Kokubo and Takadama [30]. Ca_MCM-41 particles were immersed in SBF at a 1.5 g·L^{-1} ratio [31,32]. The specimens were kept in a polypropylene container at 37 °C in an incubator on an oscillating tray for up to seven days. The solution was not renewed and a falcon tube containing SBF as a control was also used for the entire period of the experiment, in order to control over time the stability of the testing solution. At each time point, the particles were centrifuged and washed with deionized water and dried at 60 °C overnight. The microstructural changes were investigated by means of scanning electron microscopy (SEM) equipped with an Energy Dispersive X-ray Spectrometry (EDS) detector (Auriga 0750 from ZEISS, Jena, Germany). The ordered mesoporous structure was checked by high resolution transmission electron microscopy (HR-TEM), in a Tecnai G2F30 S-Twin microscope (FEI, Eindhoven, The Netherlands) with 0.2 nm point resolution, operated at 300 kV and equipped with a HAADF Fischione detector (0.16 nm point resolution, Fischione Instrument, Pittsburgh, PA, USA). For the TEM observation, particles were homogeneously dispersed in ethanol by ultrasound and dropped on a carbon film. The pore diameter analysis of the ordered mesoporous particles was conducted on HRTEM images with ImageJ2 [33] analysis software.

2.3. Electrospinning Solution and Process Parameters

Poly(epsilon-caprolactone) (PCL) (80 kDa, Sigma Aldrich, Munich, Germany) was dissolved at 15 *w/v*% in a mixture of formic acid (VWR International, Darmstadt, Germany) and acetic acid (VWR International, Darmstadt, Germany) in a ratio 1:1, according to the protocol reported elsewhere [10]. For the fabrication of the composites fibers, Ca containing MCM-41 particles were added to the PCL solution (same concentration of the neat sample) in a ratio of 30 wt % respect to the polymer amount. This ratio was selected according to the results of our previous investigations [10,11]. Electrospinning was performed by using a commercially available EC-CLI device (IME Technologies, Geldrop, The Netherlands). The device is equipped with a climate control which allows the setting and control of the temperature and relative humidity (RH) inside the electrospinning chamber. During the optimization of the neat PCL process, the temperature was set at 23 °C and different values of relative humidity were tested, namely 25%, 30%, 40%, 50% and 60%. The optimized value (40%), according to trial-and-error approach, was used for the fabrication of the composite mats. The device was also equipped with a gas shield module able to optimize the Taylor cone and the nitrogen flux was set at 8 mL/min. The applied voltage was 20 kV, in particular +18 kV were applied to the metallic nozzle with a diameter of 23 G and −2 kV were applied to a drum collector, rotating at 1000 rpm. The distance between the nozzle and the collector was 15 cm and both the solution and the suspension were dispensed at 1.3 mL/h. The duration of the electrospinning process was fixed for all samples at 30 min.

2.4. Electrospun Mats Characterization

The optimization of the RH values for the neat PCL electrospun mats was carried out by investigating the variation in fiber morphology related to the variation of the RH values. The morphology of all obtained fibers was also investigated after immersion in SBF solution using SEM after sputtering with gold (Sputter coater Q150T, Quorum Technologies, Darmstadt, Germany). Fiber average diameters and pore size were calculated by using the software ImageJ and the plugin DiameterJ [33,34] on SEM micrographs and measuring 50 fibers. ATR-FTIR spectroscopy was used for the confirmation of the incorporation of the particles inside the polymeric mats and for the bioactivity assessment. The ATR-FTIR spectra were collected in the range 4000–550 cm^{-1}, with 32 scans and a resolution of 4 cm^{-1} (Nicolet 6700, Thermo Scientific, Schwerte, Germany). The mechanical properties of the electrospun mats were assessed by performing a uniaxial tensile test at room temperature, with a load cell of 50 N and a crosshead speed of 10 mm/min (K. Frank GmbH, Mannheim, Germany). The electrospun samples were cut and fixed in a paper frame, the sample measured area was 5 mm $\times$ 20 mm, as reported in a previous work [10]. The uniaxial tensile test was also performed on the composite fibers after immersion in SBF solution for one day. The acellular bioactivity test was performed by immersing the electrospun mats in SBF solution (prepared as described above) up to seven days. The ratio between the sample surface and the SBF volume was calculated according to reference [30]. Before the immersion in SBF, the electrospun mats were fixed on round scaffold supports (Scaffdex, Sigma Aldrich, Munich, Germany). Contact angle measurements were performed to assess the wettability of the electrospun mats and they were carried out by the release of 3 μL of distilled water (Krüss DSA30, Hamburg, Germany).

Preliminary cell viability and cell morphology studies on the composite mats were performed by using bone murine stromal cells ST-2 (Leibniz-Institut DSMZ—German Collection of Microorganisms and Cell Cultures GmbH, Braunschweig, Germany). Electrospun samples were cut, fixed on scaffold supports (Scaffdex, Sigma Aldrich, Munich, Germany), placed in a 24-well plate and disinfected after UV exposure for 30 min. No preconditioning was used for the electrospun mats. Before the scaffold seeding, ST-2 cells were cultured for 24 h in RPMI 1640 medium (Gibco^{TM}, Thermo Fisher Scientific, Schwerte, Germany), supplemented with 10% fetal bovine serum (Lonza, Cologne, Germany) and 1% penicillin/streptomycin (Lonza, Cologne, Germany) and incubated at 37 °C with 5% CO$_2$. ST-2 cells were seeded onto the electrospun mats with a density of 2×10^4 cells/cm^2. After 24 h, to assess cell viability, Cell Counting Kit-8 (CCK-8) assay based on WST-8 (2-(2-methoxy-4-nitrophenyl)-3-(4-nitrophenyl)-5-(2,4-disulfophenyl)-2H-tetrazolium, monosodium salt), (Sigma Aldrich, Munich, Germany) was performed in triplicate on neat PCL nanofibers and composite electrospun mats; cells seeded on the neat PCL microfibers with micrometric diameters, obtained according to [10], were used as control. Briefly, fresh RPMI 1640 medium containing 10% WST-8 was added to all the electrospun seeded samples. After sample incubation at 37 °C and 5% CO$_2$ for 3 h, a microplate Elisa reader (PHOmo Elisa reader, Autobio Diagnostics Co. Ltd., Zhengzhou, China) was used to measure the absorbance at 450 nm. Cell morphology and adhesion on the composite electrospun mats was investigated by SEM analysis. Before performing the SEM analysis, the seeded samples were fixed using a fixation solution containing glutaraldehyde, paraformaldehyde, sucrose, and sodium cacodylate trihydrate (Sigma Aldrich, Munich, Germany). Subsequently, sample dehydration was obtained using a series of ethanol concentrations in deionized water. The samples were then dried in air and sputtered with gold (Sputter coater Q150T, Quorum Technologies, Darmstadt, Germany).

3. Results

3.1. Ca_MCM-41 Particles Characterization

The obtained material from the MCM-41 synthesis is a white fine powder characterized by low agglomeration. The morphology and the microstructure of the obtained particles were assessed by

SEM and HR-TEM analysis. The Ca_MCM-41 particles were characterized by spherical shape and homogeneous size distribution, as shown in Figure 1.

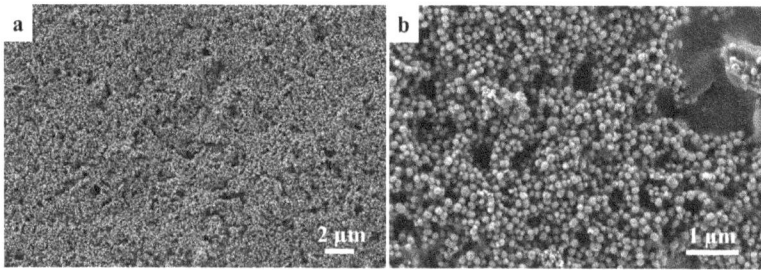

Figure 1. SEM micrographs of calcium-containing MCM-41 (Ca_MCM-41) particles at low magnification. (**a**) higher magnification; (**b**) show the homogeneity of particles size and shape.

The HR-TEM images, reported in Figure 2, showed the existence of a highly ordered hexagonal array and streaks, structural features confirming the formation of an ordered structure at the nanoscale.

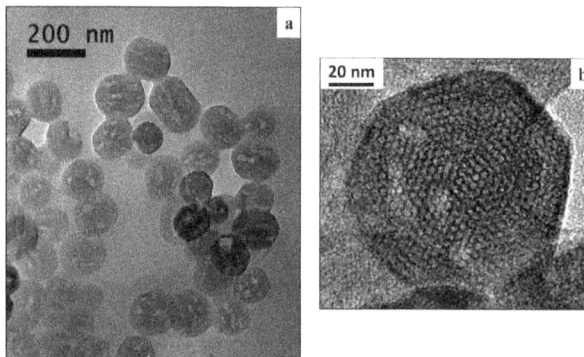

Figure 2. HR-TEM images of Ca_MCM-41 particles at different magnifications to evaluate the particle shape, dimension and the ordered structure of the obtained material: (**a**) overview of Ca_MCM-41 particles; (**b**) detailed view of a single particle.

The capability of these Ca_MCM-41 particles to develop HCA was assessed in SBF solution for up to seven days of immersion. Already after one day of immersion in SBF, the formation of HCA crystals was detected by mean of SEM and reported in Figure 3. HCA crystals do not grow homogeneously on the surface of the particles, but a single HCA layer seems to develop over time, enclosing the Ca_MCM-41 particles. This formation of the HCA layer is particularly evident when comparing the SEM micrographs of the Ca_MCM-41 particles before the immersion in SBF solution (Figure 1) and the particles after immersion (Figure 3). The presence of the HCA deposition was also investigated using FTIR spectroscopy. In particular, as showed in Figure 3d, it is possible to notice the bands related to P–O asymmetric bending around 560 and 600 cm^{-1} and the bands ascribable to H_2O molecular at 1630 and 3450 cm^{-1}, correlated to the deposition of the HCA layer [35,36].

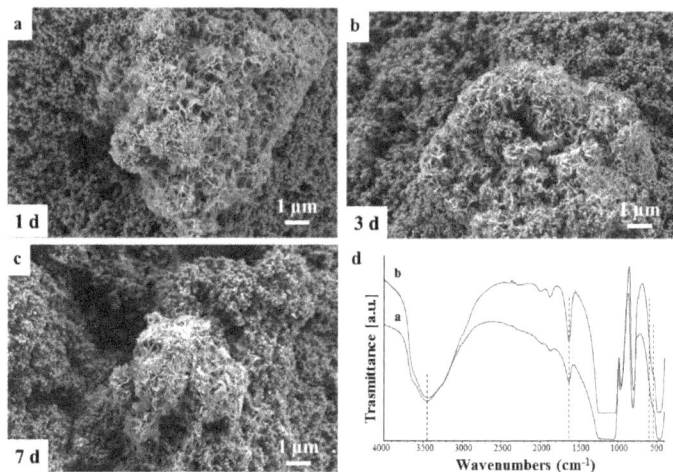

Figure 3. SEM micrographs and FTIR spectra of Ca_MCM-41 particles after immersion in simulated body fluid (SBF) at different time points. SEM image after 1 day (**a**), 3 days (**b**), 7 days of immersion (**c**); FTIR spectra (**d**) after 3 days (**a**) and 7 days (**b**) of immersion.

The pH variation of the SBF solution containing the Ca_MCM-41 particles was monitored overtime and the results are shown in Figure 4. The pH of the solution increased over time, confirming indirectly that the particles released Ca^{2+} ions once in contact with SBF solution. The phenomenon of pH-increase due to the presence of Ca^{2+} ions released from silica network-based particles and its mechanisms have been already widely investigated in the literature [31,37] and the evidence of the pH increase, confirms the release of Ca^{2+} ions in the solution. Briefly, after the immersion in SBF solution, Ca_MCM-41 particles exchange Ca^{2+} with H^+ ions from SBF solution. This exchange induces an increase in the pH of the solution and is due to the presence of mesopores, these particles showed high porosity, surface area and reactivity which accelerate the ionic exchange [37].

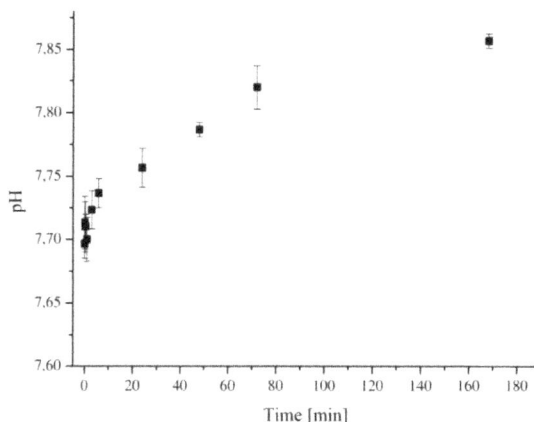

Figure 4. Graph of pH variation with time of the SBF solution after the immersion of Ca_MCM-41 particles.

3.2. Optimization of Electrospinning Parameters

The optimization of the electrospinning parameters was performed on neat PCL. Different RH values were evaluated, namely 25%, 30%, 40%, 50% and 60%. SEM analysis of the obtained electrospun mats is reported in Figure 5. It was determined that in a quite wide range of RH values, between 30% and 50%, it was possible to obtain stability in terms of fiber distribution and process yield. The lowest and highest RH values, namely 25% and 60%, introduced fiber branching and non-homogeneous fiber diameter distribution, as reported in Figure 5a,e. Based on such observations, the RH value of 40% was selected as optimal and used for the fabrication of both the composite fibers and the neat PCL fibers used as a control.

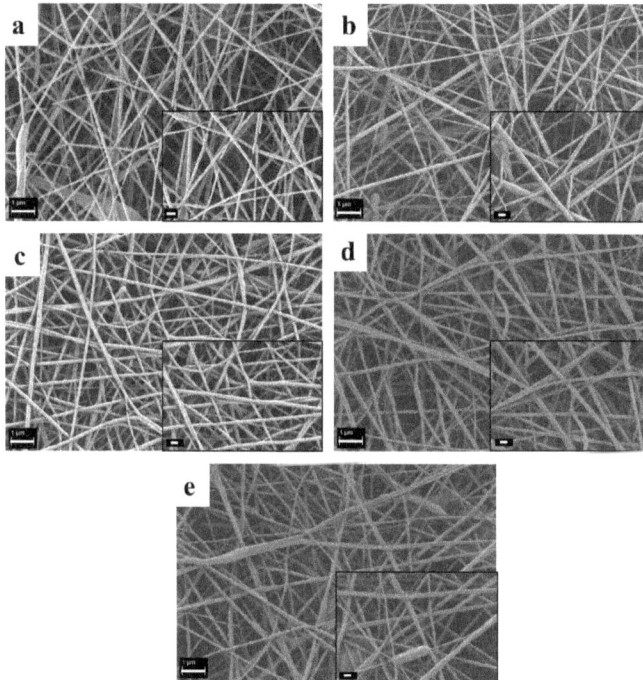

Figure 5. SEM analysis of neat PCL electrospun fibers obtained by setting different values of relative humidity (RH) : 25% (**a**), 30% (**b**), 40% (**c**), 50% (**d**) and 60% (**e**). Common scale bar 1 μm, higher magnification micrographs in the insets with common scale bar of 200 nm.

Composite electrospun fibers were characterized with SEM/EDS to identify particle distribution inside the polymeric matrix. The Si and Ca amounts were clearly detected in the polymeric PCL matrix, as reported by the map EDS analysis and the spectrum on the particle cluster, reported in Figure 6. The average fiber diameter of the composite mats was 257 ± 52 nm, comparable with the average fiber diameter of the neat PCL fibers that was 200 ± 40 nm, as reported in our previous study [10]. The calculated average pore size of the composite electrospun mats was around 0.6 μm^2. This result demonstrates that the presence of the particles did not affect fiber morphology in terms of average fiber diameter and fiber diameter distribution within the indicated ranges. The contact angle measurements showed hydrophobic behavior for both neat PCL and the composite electrospun mats, in particular, the value for the neat PCL was $141 \pm 3°$, while the value for the composite mats was $144 \pm 5°$.

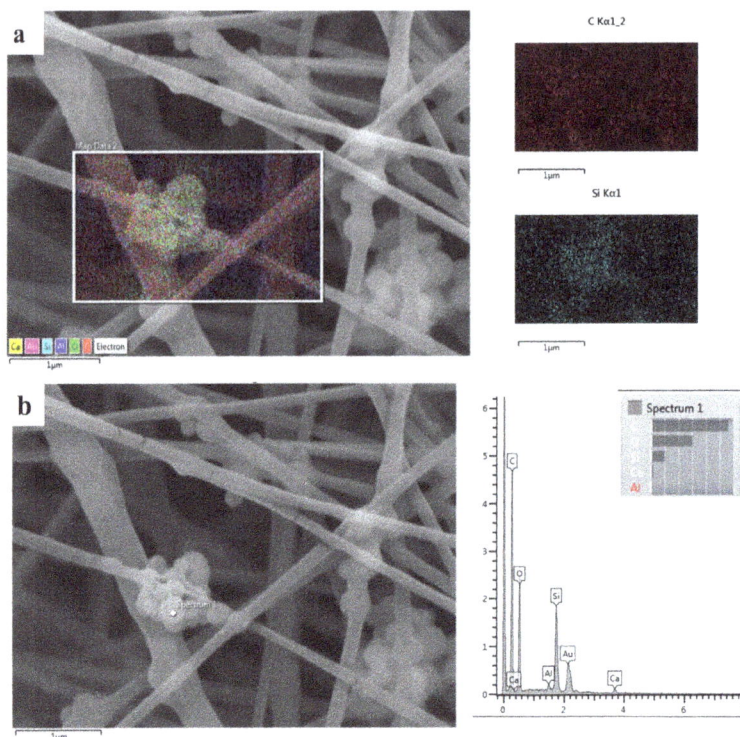

Figure 6. SEM/EDS analysis of the composite electrospun mats. SEM with maps and details for C (polymer component) and Si (from particles) (**a**) showing also the EDS spectrum on the particle cluster (**b**).

3.3. Mechanical Properties

As reported in the literature for both sample types, namely composite and neat polymeric fibers, two linear trends in the stress–strain curve could be observed. The first one is due to the load application, and the second could be ascribed to the fiber alignment before the sample fracture. The mechanical properties, such as Young's modulus, ultimate tensile strength (UTS) and tensile strain are reported in Table 1. Comparable values were obtained for the Young's modulus, even if the presence of surface roughness increased the distribution of the measured values, as indicated by the higher value of the standard deviation. After immersion in SBF for one day, a uniaxial tensile test was performed on the composite fibers, obtaining modulus and UTS values comparable to those of the sample before immersion.

Table 1. Mechanical properties of the neat PCL and PCL composite electrospun fibers.

Sample	Young's Modulus (MPa)	UTS (MPa)	Tensile Strain (%)
Neat PCL[1]	11.0 ± 0.8	6.2 ± 0.9	115 ± 2
PCL_Ca_MCM41	13 ± 4	3.3 ± 0.9	49 ± 10
PCL_Ca_MCM-41 after SBF	10 ± 2	2.4 ± 0.4	55 ± 3

[1] Values from reference [10].

A reduction in UTS and tensile strain could be related to weak adhesion at the interface between Ca_MCM-41 particles and the polymeric matrix. To confirm this relation between particle adhesion and mechanical properties, SEM analysis was performed on the composite fibers after the mechanical test, as reported in Figure 7. As shown, the lack of strong adhesion between particles and the PCL matrix can be noticed, and particles have detached from the cracks in the polymeric fibers created during the tensile test.

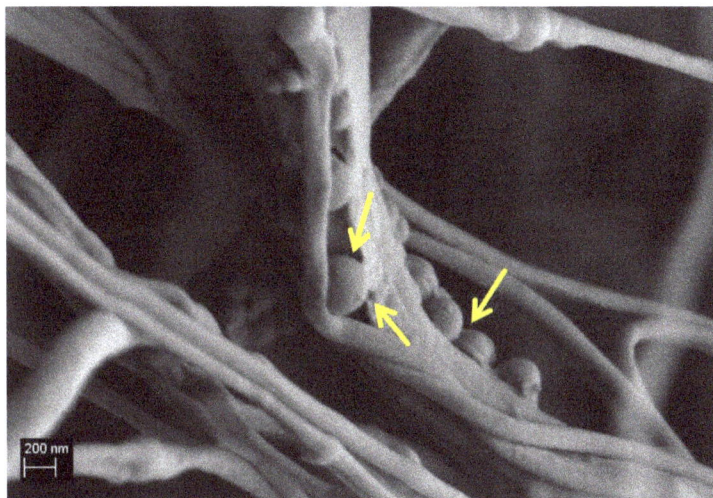

Figure 7. SEM micrograph of composite PCL_Ca_MCM-41 electrospun fibers after the tensile test, indicating lack of strong adhesion at the particle-PCL interface (arrows).

3.4. Chemical Characterization

The incorporation of Ca_MCM-41 particles inside the PCL electrospun mats was investigated using ATR-FTIR analysis. In Figure 8 the FTIR spectra of the neat PCL electrospun fibers (Figure 8a), PCL_Ca_MCM-41 electrospun fibers (Figure 8b), the subtraction spectrum (neat PCL subtracted from composite spectrum, Figure 8c) and the Ca_MCM-41 spectrum (Figure 8d) are reported.

Both the spectra of the neat PCL fibers and composite fibers are dominated by the main PCL bands, like CH_2 asymmetric and symmetric stretching vibrations around 2950 and 2865 cm^{-1}, carbonyl stretching at 1720 cm^{-1}, asymmetric and symmetric C–O–C stretching around 1240 cm^{-1} and 1160 cm^{-1} [11]. New bands appear in the spectrum of composite fibers (Figure 8b) and it is possible to compare them with the Ca_MCM-41 particle spectrum (Figure 8d). The presence of Ca_MCM-41 particles could be identified by observing the new bands in the composite fiber spectrum (Figure 8b). These bands are related to the presence of Si–O bonds, namely Si–O–Si stretching around 1100 cm^{-1}, Si–OH stretching around 1000 cm^{-1} and Si–O stretching around 800 cm^{-1}, and they are more evident in the subtraction spectrum reported in Figure 8c and indicated with dotted lines.

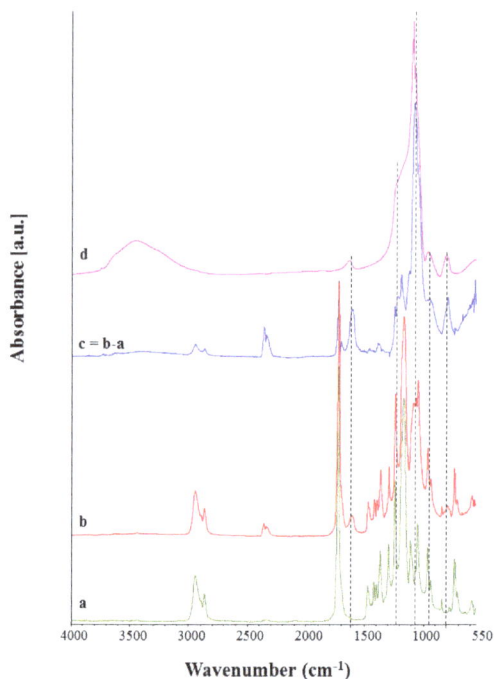

Figure 8. ATR-FTIR spectra of neat PCL electrospun fibers (**a**), PCL_Ca_MCM-41 composite electrospun fibers (**b**), the subtraction spectrum (neat PCL fibers subtracted from the composite fiber spectrum, (**c**) and the Ca_MCM-41 particle spectrum (**d**). Peaks of relevance are discussed in the text.

3.5. Bioactivity Assessment

Electrospun composite mats were immersed in SBF solution to assess the mats' bioactivity. As reported in Figure 9, it is not possible to detect any biomineralization on the fibers, because the particles were released from the fibers already after one day of immersion in SBF solution. In fact, in Figure 9a the sample morphology after one day of immersion in SBF solution is presented, showing the presence of holes in the fibers in place of the particles. The same morphology was shown after seven days of immersion in SBF solution, as reported in Figure 9b, showing that no further fiber surface modifications occurred between one and seven days of immersion.

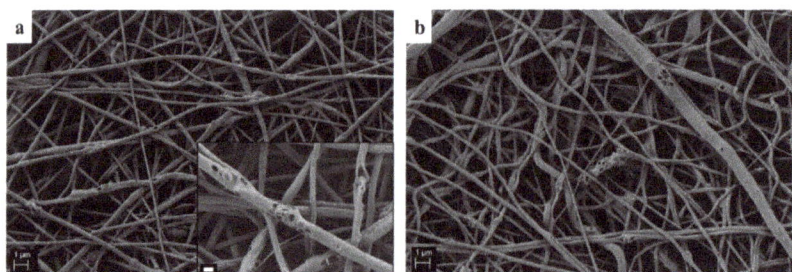

Figure 9. SEM micrograph of PCL_Ca_MCM-41 electrospun mats after 1 day (**a**) and 7 days (**b**) of immersion in SBF solution. Scale bar 1 μm and for the inset scale bar 200 nm.

FTIR analysis also confirmed that the presence of MCM-41 cannot be detected in the composite electrospun samples after the immersion in SBF solution. As reported in Figure 10, the bands in the range 1000–1200 cm^{-1}, ascribable to Si-O bonds, that are visible in PCL_Ca-MCM-41 composite fibers before immersion (Figure 10b), are not detected in the same sample after one day of immersion in SBF solution (Figure 10c). No differences were noticed after one day or seven days of immersion in SBF solution (Figure 10c,d), because already after one day of immersion all particles were released from the composite fibers.

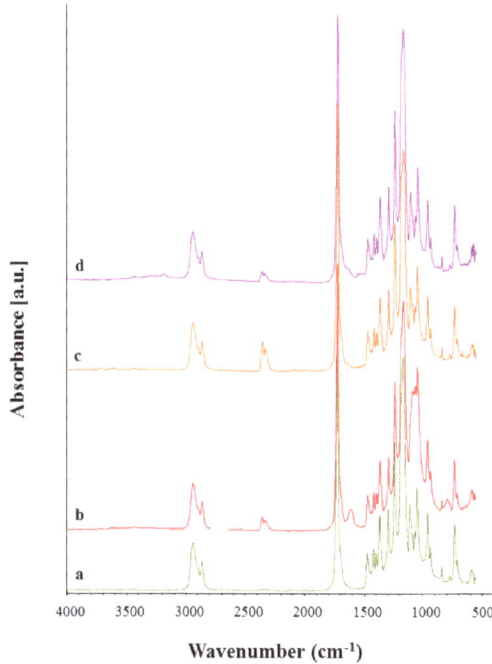

Figure 10. ATR-FTIR spectra of neat PCL fibers (**a**) and composite PCL_Ca_MCM-41 fibers (**b**) before immersion in SBF solution and PCL_Ca_MCM-41 fibers after immersion in SBF solution for 1 day (**c**) and 7 days (**d**).

3.6. Cell-Composite Scaffold Interactions: Preliminary Tests

Cell viability on the composite scaffolds was assessed by using WST-8 assay. The results showed increased cell viability for the neat PCL nanofibers and composite PCL_Ca_MCM-41 fibers with respect to the neat PCL microfibers used as a control, as reported in Figure 11. No significant differences were detected between the neat PCL nanofibers and composite PCL_Ca_MCM-41fibers, indicating that the presence of MCM-41 did not affect the cell viability on the polymeric mats.

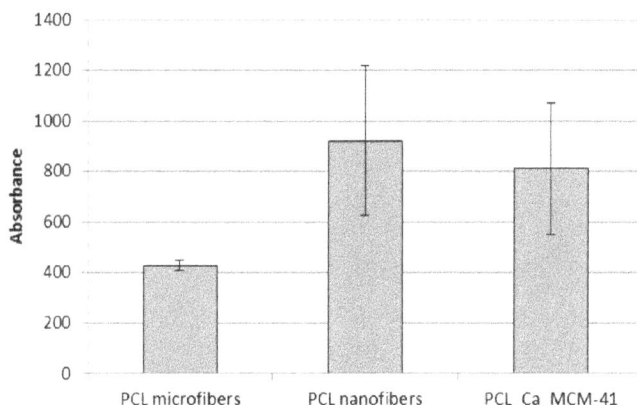

Figure 11. Cell viability measured by using WST-8 assay on seeded electrospun samples of neat PCL microfibers, neat PCL nanofibers and composite PCL_Ca_MCM-41 fibers.

SEM analysis performed on the seeded composite fibers to evaluate the expressed cell morphology, reported in Figure 12, showed a good adhesion of the cells on the electrospun composite mats with the typical phenotype of bone stromal cells; it is possible to observe this in the micrographs at higher magnification (Figure 12b and its inset). It is also possible to notice that the cell's body shows a very spread morphology, following the electrospun fiber structure and orientation, and no negative influence of the fibers on cells was detected.

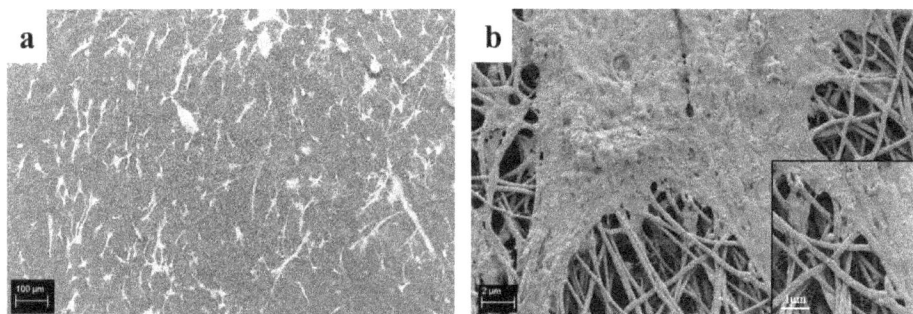

Figure 12. SEM micrographs of ST-2 cells seeded onto PCL_Ca_MCM-41 composite electrospun mats at different magnifications: 200× (**a**) and 10,000× (**b**).

4. Discussion

The development of composite organic–inorganic fibers is attractive for several applications related to the fabrication of scaffolds for tissue engineering.

The incorporation of mesoporous silica particles in electrospun polymeric mats has already been reported in the literature [27,28]. In particular, Song et al. [27] prepared rhodamine B (RHB)-loaded mesoporous silica particles and fluorescein (FLU)-loaded mesoporous silica particles. Both particle types were incorporated in a solution of poly(lactic-*co*-glycolic acid) (PLGA) dissolved in *N,N*-dimethylformamide (DMF) and tetrahydrofuran (THF), fabricating a dual-drug-loaded system. They reported a successful co-delivery system of both hydrophilic and hydrophobic drugs loaded on mesoporous silica nanoparticles. It is interesting to notice that for both drugs, the release is not related to the dissolution or detachment of the nanoparticles, since they were still visible in the

SEM micrographs after the release of the drugs. Mehrasa et al. [28] also reported the successful incorporation of mesoporous silica nanoparticles in PLGA and PLGA/gelatin electrospun nanofibers. The nanoparticles were added to the polymer and blend solution and hexafluoro-2-propanol (HFP) was used as the solvent for electrospinning. They also reported an increase in the value of the Young's modulus and the UTS related to the addition of mesoporous nanoparticles, while a reduction in the elongation was observed for the composite samples. These data are not supported by morphological analysis of fibers after tensile tests and it is not possible to compare or correlate these results with the adhesion between the particles and the polymeric matrix.

Most literature on the use of electrospinning for the incorporation of nano and/or microsized particles inside polymeric fibrous mats is related to applications to bone tissue engineering [38–40], but other applications for soft tissue regeneration or wound healing have also been reported [12,27,41].

In the present work, the optimization of the synthesis of calcium-containing MCM-41 particles was presented. The bioactivity of the obtained particles was confirmed. As an interesting application, these particles were dispersed in a PCL solution and processed with the electrospinning technique for potential applications as drug delivery systems and scaffolds for tissue engineering. The use of benign solvents is highlighted as this will enable the use of particles loaded with bioactive molecules for drug delivery.

The optimization of the electrospinning process parameters plays a pivotal role in the obtainment of bead-free fibrous mats, in particular when benign solvents are used. Besides the process parameters, environmental parameters also affect fiber morphology [6]. In the present work, stability in terms of fiber morphology was obtained at a wide range of RHs, between 30 and 50%.

After the optimization of the neat PCL fibers, Ca_MCM-41 particles were successfully incorporated, as shown in the SEM/EDS and FTIR analyses. The average fiber diameter was not affected by the addition of the particles, and only an increase in the standard deviation value was reported, likely due to the different rheological properties of the suspension compared to the neat PCL solution, as previously reported in the literature [10,14]. Usually, the differences in terms of solution or suspension properties require separated electrospinning optimization processes, obtaining different optimal parameters, but in the present work all the process parameters were kept constant and a comparable average fiber diameter was obtained. In literature, an increase in average fiber diameter related to the introduction of particles in the polymeric solution is usually reported, for example by Gönen et al. [14], but this increase should also relate to the particles size. In fact, in the present study the viscosity of the suspension was comparable to the neat PCL solution.

The mechanical properties of the obtained fibers were not enhanced by the addition of the particles. This trend has been already reported in the literature and it is highly likely related to the lack of strong adhesion of the particles to the polymer fibrous matrix [10,42], also documented by the SEM analysis performed on the composite fibers after the tensile tests (Figure 7). Another confirmation of the weak interaction between the inclusions and the PCL matrix was provided by the results of the uniaxial tensile tests performed after the immersion for one day in SBF, in fact the obtained comparable values showed that the mechanical properties were not modified after the release of the particles. This weak interaction between the particles and PCL matrix may be beneficial when the particles are loaded with drugs or biomolecules to facilitate the release of the molecules, without affecting the macroscopic mats fibrillary structure, preserving the mechanical function of the scaffold. In fact, after the release of the Ca_MCM-41, it is possible to observed cracks and holes on the fiber surface, but the fibers structure is preserved. Another advantage of this system is represented by the fact that the released particles are embedded in the polymer fibers. For this reason, it is possible to target the particle release in situ where the scaffold is implanted. It could be also considered that the release can be tailored by both the concentration of particles in the fibers and the extent of degradation of the fibers.

The acellular bioactivity of the composite electrospun fibers was assessed by immersion in SBF solution. Conversely from the results obtained on the Ca_MCM-41 particles, the composite PCL_Ca_MCM-41 fibers were not bioactive, in fact no deposits of HCA were observed on the fiber surface. All the particles incorporated inside the PCL fibers were already released after one day of immersion in SBF solution, leaving pores on the fiber surfaces. The release of the particles was confirmed by SEM and ATR-FTIR analyses.

Encouraging preliminary results about cell viability and cell adhesion on the composite electrospun mats were reported. Further studies are ongoing to investigate cell migration inside the electrospun scaffolds, and cell proliferation and differentiation induced by the electrospun composite fibers.

The obtained results are relevant and broaden the field of applications of composite electrospun mats that could be used for the fabrication of scaffolds for soft tissue engineering, but also for the fabrication of multilayered scaffolds providing layers which do not require biomineralization (no bioactivity) and as complex drug delivery systems.

5. Conclusions

Calcium-containing MCM-41 particles were successfully synthesized and a proof of concept of their use for the fabrication of composite electrospun fibers was reported. Composite electrospun fibers were obtained after the optimization of the electrospinning parameters for both the composite sample and the neat PCL solution, used as control. Benign solvents for the electrospinning were used for the fabrication of PCL and PCL_Ca_MCM-41 composite fibers, obtaining homogeneous particle distribution inside the electrospun mats. The particle incorporation in the polymeric electrospun mats was confirmed by SEM/EDS and FTIR analysis. The bioactivity of the Ca_MCM-41 particles was not preserved, because the particles were already released from the electrospun mats after one day of immersion in SBF solution. The obtained composite fibers are promising for applications in tissue engineering involving drug delivery.

Acknowledgments: Liliana Liverani acknowledges funding from the European Union's Horizon 2020 research and innovation program under the Marie Skłodowska-Curie grant agreement No 657264. The authors acknowledge DirkW. Schubert (FAU Erlangen-Nuremberg, Erlangen, Germany) for the use of his facilities for the samples' mechanical characterization. The authors acknowledge Rainer Detsch and Alina Grünewald for their support during the cell biology studies. Ana Maria Beltrán thanks the Talent-Hub Program funded by the Junta de Andalucía and the European Commission under the Co-funding of the 7th Framework Program in the People Program (Marie Curie Special Action). She also thanks the Laboratory for Nanoscopies and Spectroscopies (LANE) at the ICMS for TEM facilities.

Author Contributions: Liliana Liverani and Elena Boccardi designed, performed the experiments and wrote the first draft of the manuscript. Liliana Liverani performed the experiments and sample characterization of the electrospun mats and Elena Boccardi performed the synthesis and characterization of the particles. Ana Maria Beltrán performed the particle characterization by HR-TEM and contributed to writing the manuscript. Aldo R. Boccaccini contributed to the conception of the project, advised during the project progress and collaborated in writing the manuscript. All the authors read and approved the manuscript.

Conflicts of Interest: The authors declare no conflict of interest.

References

1. Kai, D.; Liow, S.S.; Loh, X.J. Biodegradable polymers for electrospinning: Towards biomedical applications. *Mater. Sci. Eng. C. Mater. Biol. Appl.* **2014**, *45*, 659–670. [CrossRef] [PubMed]

2. Teo, W.-E.; Inai, R.; Ramakrishna, S. Technological advances in electrospinning of nanofibers. *Sci. Technol. Adv. Mater.* **2011**, *12*, 013002. [CrossRef] [PubMed]

3. Liu, M.; Duan, X.-P.; Li, Y.-M.; Yang, D.-P.; Long, Y.-Z. Electrospun nanofibers for wound healing. *Mater. Sci. Eng. C* **2017**, *76*, 1413–1423. [CrossRef] [PubMed]

4. Prabhakaran, M.P.; Venugopal, J.R.; Chyan, T.T.; Hai, L.B.; Chan, C.K.; Lim, A.Y.; Ramakrishna, S. Electrospun biocomposite nanofibrous scaffolds for neural tissue engineering. *Tissue Eng. Part A* **2008**, *14*, 1787–1797. [CrossRef] [PubMed]

5. Shin, S.-H.; Purevdorj, O.; Castano, O.; Planell, J.A.; Kim, H.-W. A short review: Recent advances in electrospinning for bone tissue regeneration. *J. Tissue Eng.* **2012**, *3*, 2041731412443530. [CrossRef] [PubMed]

6.   Putti, M.; Simonet, M.; Solberg, R.; Peters, G.W. M. Electrospinning poly(ε-caprolactone) under controlled environmental conditions: Influence on fiber morphology and orientation. *Polymer* **2015**, *63*, 189–195. [CrossRef]

7.   Agarwal, S.; Greiner, A. On the way to clean and safe electrospinning-green electrospinning: Emulsion and suspension electrospinning. *Polym. Adv. Technol.* **2011**, *22*, 372–378. [CrossRef]

8.   Dong, B.; Arnoult, O.; Smith, M.E.; Wnek, G.E. Electrospinning of collagen nanofiber scaffolds from benign solvents. *Macromol. Rapid Commun.* **2009**, *30*, 539–542. [CrossRef] [PubMed]

9.   Van der Schueren, L.; De Schoenmaker, B.; Kalaoglu, Ö.I.; De Clerck, K. An alternative solvent system for the steady state electrospinning of polycaprolactone. *Eur. Polym. J.* **2011**, *47*, 1256–1263. [CrossRef]

10. Liverani, L.; Boccaccini, A.R. Versatile production of poly(Epsilon-caprolactone) fibers by electrospinning using benign solvents. *Nanomaterials* **2016**, *6*, 75. [CrossRef] [PubMed]

11. Liverani, L.; Lacina, J.; Roether, J.A.; Boccardi, E.; Killian, M.S.; Schmuki, P.; Schubert, D.W.; Boccaccini, A.R. Incorporation of bioactive glass nanoparticles in electrospun PCL/chitosan fibers by using benign solvents. *Bioact. Mater.* **2017**. [CrossRef]

12. Moura, D.; Souza, M.T.; Liverani, L.; Rella, G.; Luz, G.M.; Mano, J.F.; Boccaccini, A.R. Development of a bioactive glass-polymer composite for wound healing applications. *Mater. Sci. Eng. C* **2017**, *76*, 224–232. [CrossRef] [PubMed]

13. Lepry, W.C.; Smith, S.; Liverani, L.; Boccaccini, A.R.; Nazhat, S.N. Acellular Bioactivity of Sol-Gel Derived Borate Glass-Polycaprolactone Electrospun Scaffolds Biomedical Glasses. *Biomed. Glas.* **2016**, *2*, 88–98.

14. Gönen, S.Ö.; Taygun, M.E.; Küçükbayrak, S. Fabrication of Bioactive Glass Containing Nanocomposite Fiber Mats For Bone Tissue Engineering Applications. *Compos. Struct.* **2016**, *138*, 96–106. [CrossRef]

15. Liverani, L.; Abbruzzese, F.; Mozetic, P.; Basoli, F.; Rainer, A.; Trombetta, M. Electrospinning of hydroxyapatite-chitosan nanofibers for tissue engineering applications. *Asia-Pac. J. Chem. Eng.* **2014**, *9*, 407–414. [CrossRef]

16. Ding, Y.; Yao, Q.; Li, W.; Schubert, D.W.; Boccaccini, A.R.; Roether, J.A. The evaluation of physical properties and in vitro cell behavior of PHB/PCL/sol–gel derived silica hybrid scaffolds and PHB/PCL/fumed silica composite scaffolds. *Colloids Surf. B Biointerfaces* **2015**, *136*, 93–98. [CrossRef] [PubMed]

17. Zhao, D.; Wan, Y.; Zhou, W. *Ordered Mesoporous Materials*; Wiley-VCH Verlag GmbH & Co. KGaA: Weinheim, Germany, 2013.

18. Vallet-Regí, M.; García, M.M.; Colilla, M. *Biomedical Applications of Mesoporous Ceramics: Drug Delivery, Smart Materials and Bone Tissue Engineering*; CRC Press: Boca Raton, FL, USA, 2012.

19. Grün, M.; Unger, K.K.; Matsumoto, A.; Tsutsumib, K. Novel pathways for the preparation of mesoporous MCM-41 materials: Control of porosity and morphology. *Microporous Mesoporous Mater.* **1999**, *27*, 207–216. [CrossRef]

20. Stöber, W.; Fink, A.; Bohn, E. Controlled growth of monodisperse silica spheres in the micron size range. *J. Colloid. Interface Sci.* **1968**, *26*, 62–69. [CrossRef]

21. Vallet-Regi, M.; Rámila, A.; del Real, R.P.; Pérez-Pariente, J. A New Property of MCM-41: Drug Delivery System. *Chem. Mater.* **2000**, *13*, 308–311. [CrossRef]

22. Vallet-Regí, M. Ordered Mesoporous Materials in the Context of Drug Delivery Systems and Bone Tissue Engineering. *Chem. A Eur. J.* **2006**, *12*, 5934–5943. [CrossRef] [PubMed]

23. Izquierdo-Barba, I.; Colilla, M.; Manzano, M.; Vallet-Regí, M. In vitro stability of SBA-15 under physiological conditions. *Microporous Mesoporous Mater.* **2010**, *132*, 442–452. [CrossRef]

24. Qin, X.; Wu, D. Effect of different solvents on poly(caprolactone) (PCL) electrospun nonwoven membranes. *J. Therm. Anal. Calorim.* **2012**, *107*, 1007–1013. [CrossRef]

25. Soliman, S.; Pagliari, S.; Rinaldi, A.; Forte, G.; Fiaccavento, R.; Pagliari, F.; Franzese, O.; Minieri, M.; Di Nardo, P.; Licoccia, S.; et al. Multiscale three-dimensional scaffolds for soft tissue engineering via multimodal electrospinning. *Acta Biomater.* **2010**, *6*, 1227–1237. [CrossRef] [PubMed]

26. Ferreira, J.L.; Gomes, S.; Henriques, C.; Borges, J.P.; Silva, J.C. Electrospinning polycaprolactone dissolved in glacial acetic acid: Fiber production, nonwoven characterization, and In Vitro evaluation. *J. Appl. Polym. Sci.* **2014**, *131*, 37–39. [CrossRef]

27. Song, B.; Wu, C.; Chang, J. Controllable delivery of hydrophilic and hydrophobic drugs from electrospun poly(lactic-*co*-glycolic acid)/mesoporous silica nanoparticles composite mats. *J. Biomed. Mater. Res. Part B Appl. Biomater.* **2012**, *100B*, 2178–2186. [CrossRef] [PubMed]

28. Mehrasa, M.; Asadollahi, M.A.; Ghaedi, K.; Salehi, H.; Arpanaei, A. Electrospun aligned PLGA and PLGA/gelatin nanofibers embedded with silica nanoparticles for tissue engineering. *Int. J. Biol. Macromol.* **2015**, *79*, 687–695. [CrossRef] [PubMed]

29. Li, X.; Zhang, L.; Dong, X.; Liang, J.; Shi, J. Preparation of mesoporous calcium doped silica spheres with narrow size dispersion and their drug loading and degradation behavior. *Microporous Mesoporous Mater.* **2007**, *102*, 151–158. [CrossRef]

30. Kokubo, T.; Takadama, H. How useful is SBF in predicting in vivo bone bioactivity? *Biomaterials* **2006**, *27*, 2907–2915. [CrossRef] [PubMed]

31. Cerruti, M.; Greenspan, D.; Powers, K. Effect of pH and ionic strength on the reactivity of Bioglass 45S5. *Biomaterials* **2005**, *26*, 1665–1674. [CrossRef] [PubMed]

32. Maçon, A.L.B.; Kim, T.B.; Valliant, E.M.; Goetschius, K.; Brow, R.K.; Day, D.E.; Hoppe, A.; Boccaccini, A.R.; Kim, I.Y.; Ohtsuki, C.; et al. A unified in vitro evaluation for apatite-forming ability of bioactive glasses and their variants. *J. Mater. Sci. Mater. Med.* **2015**, *26*, 115. [CrossRef] [PubMed]

33. Schneider, C.A.; Rasband, W.S.; Eliceiri, K.W. NIH Image to ImageJ: 25 years of image analysis. *Nat. Methods* **2012**, *9*, 671–675. [CrossRef] [PubMed]

34. Hotaling, N.A.; Bharti, K.; Kriel, H.; Simon, C.G. DiameterJ: A validated open source nanofiber diameter measurement tool. *Biomaterials* **2015**, *61*, 327–338. [CrossRef] [PubMed]

35. Aguiar, H.; Serra, J.; González, P.; León, B. Structural study of sol–gel silicate glasses by IR and Raman spectroscopies. *J. Non. Cryst. Solids* **2009**, *355*, 475–480. [CrossRef]

36. Zheng, K.; Solodovnyk, A.; Li, W.; Goudouri, O.-M.; Stähli, C.; Nazhat, S.N.; Boccaccini, A.R. Aging Time and Temperature Effects on the Structure and Bioactivity of Gel-Derived 45S5 Glass-Ceramics. *J. Am. Ceram. Soc.* **2015**, *98*, 30–38. [CrossRef]

37. Izquierdo-Barba, I.; Arcos, D.; Sakamoto, Y.; Terasaki, O.; López-Noriega, A.; Vallet-Regí, M. High-Performance Mesoporous Bioceramics Mimicking Bone Mineralization. *Chem. Mater.* **2008**, *20*, 3191–3198. [CrossRef]

38. Liverani, L.; Roether, J.A.; Boccaccini, A.R. Nanofiber composites in bone tissue engineering. In *Nanofiber Composite Materials for Biomedical Applications*; Ramalingam, M., Ramakrishna, S., Eds.; Elsevier: Amsterdam, The Netherlands, 2016.

39. Zhou, P.; Cheng, X.; Xia, Y.; Wang, P.; Zou, K.; Xu, S.; Du, J. Organic/Inorganic Composite Membranes Based on Poly(L-lactic-co-glycolic acid) and Mesoporous Silica for Effective Bone Tissue Engineering. *ACS Appl. Mater. Interfaces* **2014**, *6*, 20895–20903. [CrossRef] [PubMed]

40. Paşcu, E.I.; Stokes, J.; McGuinness, G.B. Electrospun composites of PHBV, silk fibroin and nano-hydroxyapatite for bone tissue engineering. *Mater. Sci. Eng. C* **2013**, *33*, 4905–4916. [CrossRef] [PubMed]

41. Xu, H.; Lv, F.; Zhang, Y.; Yi, Z.; Ke, Q.; Wu, C.; Liu, M.; Chang, J. Hierarchically micro-patterned nanofibrous scaffolds with a nanosized bio-glass surface for accelerating wound healing. *Nanoscale* **2015**, *7*, 18446–18452. [CrossRef] [PubMed]

42. Jo, J.H.; Lee, E.J.; Shin, D.S.; Kim, H.E.; Kim, H.W.; Koh, Y.H.; Jang, J.H. In vitro/in vivo biocompatibility and mechanical properties of bioactive glass nanofiber and poly(ε-caprolactone) composite materials. *J. Biomed. Mater. Res. Part B Appl. Biomater.* **2009**, *91*, 213–220. [CrossRef] [PubMed]

![polymers logo] *polymers*

MDPI

Article

Surface Hydrophilicity of Poly(L-Lactide) Acid Polymer Film Changes the Human Adult Adipose Stem Cell Architecture

Chiara Argentati [1,†], Francesco Morena [1,†], Pia Montanucci [2], Marco Rallini [3], Giuseppe Basta [2], Nicolino Calabrese [4], Riccardo Calafiore [2], Marino Cordellini [5], Carla Emiliani [1], Ilaria Armentano [6] and Sabata Martino [1,*]

[1]   Department of Chemistry, Biology and Biotechnologies, Biochemistry and Molecular Biology Unit, University of Perugia, Via del Giochetto, 06126 Perugia, Italy; chiara.argentati89@gmail.com (C.A.); effemorena@gmail.com (F.M.); carla.emiliani@unipg.it (C.E.)
[2]   Section of Cardiovascular, Endocrine and Metabolic Clinical Physiology, Laboratory for Endocrine Cell Transplants and Biohybrid Organs, Department of Medicine, University of Perugia, 06126 Perugia, Italy; piamontanucci@hotmail.com (P.M.); gius.basta@gmail.com (G.B.); riccardo.calafiore@unipg.it (R.C.)
[3]   Civil and Environmental Engineering Department, UdR INSTM, University of Perugia, 05100 Terni, Italy; marcorallini@gmail.com
[4]   Private Dental Practice, 06134 Perugia, Italy; nicperiodoc@hotmail.com
[5]   Plastic and Reconstructive Surgery Unit, 06024 ASL 1 Umbria, Italy; marinocordellini@yahoo.it
[6]   Department of Ecological and Biological Sciences, Tuscia University, 01100 Viterbo, Italy; ilaria.armentano@unitus.it
*   Correspondence: sabata.martino@unipg.it; Tel.: +39-0755857442; Fax: +39-0755857443
†   These authors contributed equally to this work.

Received: 24 November 2017; Accepted: 31 January 2018; Published: 1 February 2018

Abstract: Current knowledge indicates that the molecular cross-talk between stem cells and biomaterials guides the stem cells' fate within a tissue engineering system. In this work, we have explored the effects of the interaction between the poly(L-lactide) acid (PLLA) polymer film and human adult adipose stem cells (hASCs), focusing on the events correlating the materials' surface characteristics and the cells' plasma membrane. hASCs were seeded on films of pristine PLLA polymer and on a PLLA surface modified by the radiofrequency plasma method under oxygen flow (PLLA+O_2). Comparative experiments were performed using human bone-marrow mesenchymal stem cells (hBM-MSCs) and human umbilical matrix stem cells (hUCMSCs). After treatment with oxygen-plasma, the surface of PLLA films became hydrophilic, whereas the bulk properties were not affected. hASCs cultured on pristine PLLA polymer films acquired a spheroid conformation. On the contrary, hASCs seeded on PLLA+O_2 film surface maintained the fibroblast-like morphology typically observed on tissue culture polystyrene. This suggests that the surface hydrophilicity is involved in the acquisition of the spheroid conformation. Noteworthy, the oxygen treatment had no effects on hBM-MSC and hUCMSC cultures and both stem cells maintained the same shape observed on PLLA films. This different behavior suggests that the biomaterial-interaction is stem cell specific.

Keywords: cytoskeleton architecture; stem cell fate; regenerative medicine

---

1. Introduction

A growing number of studies indicate that stem cells seeded on a biomaterial respond to the surrounding environment by activating nanoscale interactions with extracellular milieu following a scheme that recapitulates the tissue/organ interplays [1–4]. The first critical event occurs at the stem cell-material interface. This involves the materials' surface characteristics and the living cells' plasma membrane. In fact, several classes of proteins may be able to establish a bonding with the materials' surface molecules, generating a biochemical signal cascade (mechanotransduction axis), that is transmitted to the cell nucleus, modulating the chromatin conformation and inducing a peculiar gene expression profile. As an outcome, stem cells could steer the phenotype toward a different specification lineage [5–14].

Thus, understanding the molecular events taking place at the stem cell-biomaterial interface is instrumental for generating a functional biohybrid system for applications in regenerative medicine.

Within this aim, we are studying the stem cell-biomaterial interaction in ex vivo models, consisting of poly(L-lactide) acid (PLLA) polymer-based films and stem cells.

Polylactic polymer is a bio-based polymer produced from renewable resources and currently is one of the most extensively used biodegradable aliphatic polyesters [15–17]. Many research groups have been studying PLLA polymers in various formulations: both PLLA and poly(D-lactide) (PLDA)enantiomers of PLA stereo-complex structures; films or fibrous forms; and in combination with other molecules to generate nanocomposite materials [5,16–18].

In this issue, we have produced fibrous and films of PLLA polymers and nanocomposites and have investigated their effects on several types of stem cells [5,7,15,19,20]. We have demonstrated that human bone-marrow mesenchymal stem cells (hBM-MSCs), murine-induced pluripotent stem cells (miPSCs), and murine embryonic stem cells (mESCs) responded to fibrous electrospun PLLA polymer in a similar way, maintaining shape, proliferation rate, adhesion, and stemness phenotype, as on conventional tissue culture polystyrene (TCP) cultures [7]. This contrasts with the acquisition of an osteogenic-differentiation lineage observed when the above stem cell types were seeded on nanocomposite PLLA/hydroxyapatite [7]. We have also uncovered that human umbilical cord matrix stem cells (hUCMSCs) cultured on films of PLLA lost the canonical fibroblast-like morphology typically observed on TCP, acquired three-dimensional spheroid conformations, and were steered toward an Epiblast-like phenotype [5]. Conversely, murine and human BM-MSCs cultured on the same films of PLLA polymer maintained the fibroblast-like morphology, the adhesion, the proliferation rate, and the multipotential properties as typically observed when cultured on TCP [19,20].

Together these findings suggested that the stem cells respond to the PLLA polymer differently, depending on the pristine polymer, the nanocomposite, and the chemical-physical properties of the material [5,7,15,19,20].

Starting from these results, we further investigated the potentials of PLLA polymer films on more types of adult stem cells.

In this work, we present data on the effects of PLLA polymer films on adult human adipose stem cells (hASCs). These cells were selected based on the mesenchymal origin, as hBM-MSCs and hUCMSCs previously investigated, and on their potentials in regenerative medicine application [5,21–23].

We cultured hASCs on PLLA polymer films and evaluated their effect on stem cells in terms of morphology, adhesion, and stem cell markers. Moreover, to establish the relevance of the stem cells-PLLA interface interaction, comparative experiments were performed with hASCs, hBM-MSCs, and hUCMSCs cultured on the PLLA polymer films modified with oxygen-plasma treatment (PLLA+O_2).

We demonstrated that hASCs grew on PLLA as spheroids. We also showed that the oxygen-plasma treatment changed the surface hydrophilicity of PLLA and that this surface modification influenced the hASCs shape and cell adhesion. hASCs lost the capability to form spheroids and maintained their original fibroblast-like morphology, typically observed on TCP. Notably, the treatment with oxygen of PLLA had no effect on hUCMSCs and hBM-MSCs that maintained the same behavior as on PLLA.

2. Materials and Methods

2.1. Preparation and Caracterization of PLLA Polymer Films

2.1.1. Processing of PLLA Films

Poly(L-lactide)acid (PLLA) biodegradable polymer, with a molecular weight (M_n) of 120,000 g/mol and a polydispersity index (M_w/M_n) of 1.27, was supplied by Purac Biochem (Amsterdam, The Netherlands). PLLA films were prepared as previously reported by the solvent casting method, in chloroform ($CHCl_3$), by using 10% *w/v* as polymer/solvent ratio [5,20]. PLLA was completely dissolved in $CHCl_3$ by magnetic stirring for 5 h and, after the mixture was casted onto a Teflon substrate and air dried at room temperature (RT) for 24 h, stirred for a further 48 h in vacuum. Films of 60 mm in diameter and 0.2 mm in thickness were obtained.

2.1.2. PLLA Oxygen-Plasma Treatment Films

The surface of PLLA films were treated by means of the radio frequency (RF) plasma method under oxygen (O_2) flow by using a Sistec apparatus (Sistec, Binasco, Italy), with a Huttinger power supply at 13.56 MHz. The films were placed into a stainless-steel chamber, evacuated for 1 h until the pressure (P) was 9×10^{-3} Torr. The O_2 flow was maintained at 60 standard cm^3/min (sccm). The deposition conditions were: power supply: 20 W; bias voltage: 220 V; pressure: 1×10^{-1} Torr. Treatment time was 10 min. Process parameters were selected to obtain modulated surface features, specifically, morphology and wettability, without modifying the bulk PLLA chemical properties, according to our previous works [24].

2.1.3. PLLA and PLLA+O_2 Film Characterization

The surface microstructures were analyzed by field emission scanning electron microscope (FESEM Supra 25, Zeiss, Baden-Württemberg, Germany). A piece of PLLA film (1 cm × 1 cm) was gold coated with an Agar automatic sputter-coater and then analyzed.

Water static contact angle (WCA) measurements were used to measure the wettability of PLLA and plasma-treated PLLA films. The contact angles were assessed using the sessile drop method in air using a FTA1000 analyzer. Drops of 20 µL (high-performance liquid chromatography grade water) were placed on films and measurements were recorded 10 s after the liquid made contact with the surface.

PLLA bulk properties. Mechanical properties were performed by the tensile test method in a digital Lloyd testing machine, on rectangular samples. Infrared spectroscopy was carried out in ATR mode, by using a JASCO FT-IR 615 spectrometer (Cremella, Italy). Thermal properties were analyzed by differential scanning calorimeter (DSC, Mettler Toledo 822/e, Milano, Italia) and were conducted from −25 to 210 °C, at 10 °C min^{-1}, with two heating and one cooling scans.

2.1.4. Protein Adsorption

Protein adsorption assessments were performed by transferring on PLLA and on PLLA+O_2 film surfaces 200 µL of: bovine serum albumin (BSA 2 mg/mL, Sigma Aldrich, St. Louis, MI, USA), fetal bovine serum 2% (FBS, Euroclone, Pero, Italy), FBS 10% (Euroclone), and plasma from normal donors at a dilution of 1:10 (5 mg/mL). Proteins were incubated for either 30 min or 24 h at 37 °C, according to our previous work [6]. After three washing steps in H_2O, total protein content was measured by the Bradford method using BSA as reference curve standard. Absorbance (595 nm) was measured using a microtiter plate reader (ELISA reader, GDV-DV990BV6, Roma, Italy) [25]. Every sample was analyzed in three independent experiments. Data reported are the mean value ± the standard error of the mean of each group.

2.2. Isolation and Culture of Human Adult Stem Cells

2.2.1. Adipose Stem Cells

Adipose stem cells were isolated from lipoaspirate adipose tissue according to our procedure [26,27]. Lipoaspirate was obtained from healthy donor patients undergoing plastic intervention, after collecting written consent, according to Ethical Committee'. Briefly, after extensively washing in phosphate-buffered saline (PBS) containing 5% penicillin/streptomycin (EuroClone, Pero, Italy), lipoaspirate fragments were incubated 40 min at 37 °C, 5% carbon dioxide (CO_2), with 0.075% collagenase type I prepared in PBS containing 2% penicillin/streptomycin for tissue digestion, and then neutralized by adding 5 mL of DMEM (Dulbecco's Modified Eagle Medium, EuroClone, Pero, Italy) containing 20% heat inactivated fetal bovine serum (FBS, EuroClone, Pero, Italy). The digested fragments were centrifuged at 300× *g*, the pellet washed with PBS 2% penicillin/streptomycin, and centrifuged at 300× *g* for 5 min. Finally, the cell pellet was re-suspended in growth medium (DMEM) supplemented with FBS 10%, 1% L-glutamine (EuroClone, Pero, Italy), 1% penicillin/streptomycin) plated in tissue culture flasks (TCP) and incubated at 37 °C, 5% CO_2. hASCs started to grow as adherent fibroblast-like cells. The medium was changed every three days.

FACScan flow cytometry. To assess the cell phenotype, isolated stem cells underwent flow cytometry analysis. Stem cells were fixed on 3.7 paraformaldehyde and incubated with the following conjugated antibodies in PBS, pH 7.2, 0.5% BSA, and 2 mM EDTA: Integrin beta-1/CD29 antibody (MEM-101A), FITC conjugate; CD90/Thy-1 antibody (eBio5E10), FITC conjugate; CD105/Endoglin Antibody (SN6), RPE conjugate; CD44/H-CAM antibody (IM7), PerCP-Cy5.5 conjugate; CD45/PTPRC antibody (HI30), Pacific Orange conjugate, all from Molecular Probes. The cells were analyzed with the Attune® Acoustic Focusing Cytometer (Thermo Fisher Scientific, Waltham, MA, USA) cytofluorometer. The same procedure was used to analyze the phenotype of hASCs after culture on PLLA and on PLLA+O_2 films. Cells that acquired spheroid conformation were previously disaggregated with 0.05% Collagenase P (Roche, Basel, Switzerland), 2′ RT, and mechanical pipette action. Cells were electronically gated according to light-scattering properties to discriminate cell debris. Isotype-matched nonspecific antibodies were used as a negative control and 5000 events were recorded per each condition.

2.2.2. Bone Marrow-Mesenchymal Stem Cells

hBM-MSCs were isolated from bone marrow obtained during washouts of the medullary cavities of the femurs of patients undergoing primary total hip replacement and cultured as previously described [6,7,28]. Informed consent was obtained from all donors and the institutional ethical committee approved the procedures. Mononuclear cells were isolated according to density gradient on lympholyte (Cedarlane Laboratories Limited, Hornby, ON, Canada) and were seeded in culture flasks at a density of 2.5×10^6 cells in control medium consisting of RPMI-1640 (Euroclone, Milano, Italy) medium containing FBS 10%, 2 mM l-glutamine, and 1% penicillin–streptomycin (Euroclone) in a humidified atmosphere and 5% CO_2 at 37 °C. After 5 to 7 days, the non-adherent cells were removed, and fresh medium was added to the flasks. After 15 days, a fibroblast-like colony started to grow. The medium was changed every three days.

2.2.3. Umbilical Blood Matrix Stem Cells

hUCMSCs were isolated from matrix of cord blood and expanded in culture according to our protocol [5,29]. At the end of gestation, human umbilical cords, collected after caesarean deliveries, under official consent of the Hospital Board at University of Perugia (Perugia, Italy) and patient's own informed consent was sent us from the Department of Obstetrics and Gynecology. The cord was cut into pieces (about 10 cm) and injected by a syringe with the digestion solution: 77 mM NaCl, 0.1 mg/mL bovine serum albumin (BSA, Biochrom, Biopsa, Milan, Italy), 1.5 mg/mL hyaluronidase (Sigma Aldrich), 0.5 mg/mL liberase purified enzyme blend (Roche, Milan, Italy), in 0.02 M phosphate buffer at pH 7, containing 77 mM NaCl and 0.01% BSA, pre-warmed at 37 °C. The tissue digest was first spun at 1500 rpm for 5 min at 4 °C, and thereby resuspended in DMEM with antibiotics. Residual RBC's were removed by Lymphoprep^{TM} gradients at 1850 rpm for 20 min at 4 °C. hUCMSCs were seeded at a concentration of 250,000 per flask pre-treated with hyaluronic acid, in FBS 2% culture medium, according to Weiss [30]. After 24 h, EGF (1 ng/mL) (Peprotech, LiStarFish, Milano, Italy) and PDGF-BB (10 ng/mL) (Peprotech) were added to the culture flasks. The cells were maintained at 37 °C in a humidified atmosphere with 5% CO_2. Cell expansion throughout 80% confluence was achieved by treatment with 0.05% trypsin/EDTA (Gibco, Invitrogen, Milan, Italy) for 3 min at 37 °C. The medium was changed every three days.

2.3. Culture of Human Adipose Stem Cells on PLLA and PLLA+O$_2$ Films

PLLA and PLLA+O$_2$ films were cut into 1 cm^2 snippets squares, sterilized through immersion in pure ethanol for 30 min, followed by five rinses in PBS, and then deposited in a 24-well plate. Stem cell suspensions of 3×10^3 stem cells were seeded dropwise, on sterilized films, and 500 µL of culture medium was gradually added to each snippet. Stem cell-PLLA and -PLLA+O$_2$ cultures were incubated at 37 °C in a humidified atmosphere with 5% CO_2. The medium was changed every three days. As an internal control, experiments were performed seeding stem cells on tissue culture polystyrene (TCP). Cultures were conducted at different time points (from day 0 to day 21) and evaluated for proliferation, viability, morphology, and expression of stem cell markers.

2.4. Culture of Human BM-MSCs and Human UCMSCs on PLLA and PLLA+O$_2$ Films

Stem cell suspensions of 3×10^3 stem cells (both hBM-MSCs and hUCMSCs) were seeded dropwise on 1 cm^2 snippet squares of PLLA and PLLA+O$_2$ films, as above described. Stem cell-PLLA and -PLLA+O$_2$ cultures were incubated at 37 °C in a humidified atmosphere with 5% CO_2. The medium was changed every three days. As an internal control, experiments were performed by seeding stem cells on TCP.

2.5. Cells Viability Assay

Stem cell viability was evaluated by incubating 3×10^3 cells/mL cultured on PLLA and PLLA+O$_2$ at different time points (1, 2, 3, 5, 7, 14, and 21 days) with XTT (2,3-bis[2-Methoxy-4-nitro-5-sulfophenyl]-2H-tetrazolium-5-carboxyanilide inner salt) (Sigma Aldrich) according to the manufacturer's recommendation and our previous works [5,26]. As an internal control, parallel experiments were conducted by seeding stem cells on TCP. Interference effects of PLLA and PLLA+O$_2$ snippet squares without cells on XTT assay were also considered.

The absorbance of the samples was measured using a microtiter plate reader (ELISA reader, GDV, Roma, Italy) at 450 nm with a reference wavelength at 650 nm.

The presence of dead cells was monitored by counting stem cells in a haemocytometer by using the Trypan Blue reagent.

2.6. Immunofluorescences

Immunostaining was performed as previously described [5,7,28]. Briefly, cells on PLLA and on PLLA+O$_2$ snippet squares were rinsed twice with PBS, fixed in 4% paraformaldehyde for 20 min and, after PBS washing, permeabilized and incubated in blocking solution (PBS + FBS 10%, 0.1% Triton X-100) for 1 h at room temperature (RT). Samples were incubated with phalloidin (Alexa-fluor-488 phalloidin, Invitrogen, Grand Island, NY, USA) for 20 min, or overnight at 4 °C with several primary antibodies: anti-Vinculin (Abcam, Cambridge, UK), anti-Myosin-IIA (Santa Cruz Biotechnology, Inc., Dallas, TX, USA), anti-Filamin-A (Santa Cruz Biotechnology, Inc., Dallas, TX, USA), and anti-Lamin-B (Santa Cruz Biotechnology, Inc., Dallas, TX, USA). In the latter cases, after being washed with PBS, they were stained with Alexa-Fluor 488-nm conjugated secondary antibodies (Invitrogen) for 1 h at room temperature.

Finally, after being washed with PBS, samples were mounted and nuclei were counterstained with Vectashield® with DAPI (4,6-diamidino-2-phenylindole; Vector Laboratories Inc., Burlingame, CA, USA).

Immunofluoresces were evaluated by fluorescence microscope Eclipse-TE2000-S (Nikon, Tokyo, Japan) equipped with the F-ViewII FireWire camera (Soft Imaging System, Olympus, Germany).

Interference of PLLA and PLLA+O_2 snippet squares without cells was evaluated to the fluorescence microscope.

2.7. Image Analysis

The nuclear shape index [5] was evaluated in cells cultured on each substrate. Five different areas were photographed (20× magnification, 20× Plan Fluo NA0.5) and an average of 300 nuclei were analyzed. To quantify the variation of nuclear shape index (NSI), the area and perimeter of nuclei were measured by Fiji (Fiji Life-Line version, 22 December 2015) on fluorescent-stained DAPI images and used to calculate the NSI from the relationship:

$$NSI = \frac{(4\pi \times \text{area})}{\text{perimeter}^2} \tag{1}$$

NSI values range from 0 (elongated, elliptic morphology) to 1 (circular shape).

The edge detection method was used to detect the location of the leading edge with a custom-written automated method in MATLAB software that automatically generate the Prewitt image.

Briefly, the image was imported in '*RGB.bmp*' format *(imread)* and doubled *(im2double)*. The Prewitt operator was applied to the doubled image by specifying a 5 × 5 mask, with the two convolution kernels Gx = [2 2 4 2 2; 1 1 2 1 1; 0 0 0 0 0; −1 −1 −2 −1 −1; −2 −2 −4 −2 −2] and Gy = [2 1 0 −1 −2; 2 1 0 −1 −2;4 2 0 −2 −4; 2 1 0 −1 −2; 2 1 0 −1 −2], which are convolved with the original image to calculate approximations of the derivatives (one for horizontal and one for vertical changes). The final image G=$\sqrt{Gx^2 + Gy^2}$ for each RGB channel was saved.

2.8. Field Emission Scanning Electron Microscopy (FESEM)

Stem cell–film interaction was evaluated by FESEM at each time point of culture. Samples were rinsed twice with PBS and fixed in 2.5% glutaraldehyde for 30 min at RT, then dehydrated by adding progressively more concentrated ethanol (5–100% *v/v*) every 5 min, and finally dried by the critical point machine (CPD, Emitech K850). Once dried, the samples were gold sputter-coated before examination by FESEM (Supra 25 Zeiss), at an accelerating voltage of 5 kV.

2.9. Statistical Analysis

Data analyses were reported as the mean ± SEM (GraphPad 4.03 Software, San Diego, CA, USA, 2011). $p \leq 0.05$ was considered statistically significant.

3. Results

3.1. PLLA and PLLA+O_2 Film Characterization

In this work we developed films of pristine poly(L-lactide)acid (PLLA) polymer and films of PLLA polymer treated with oxygen-plasma (PLLA+O_2) (Figure 1). Films of PLLA were made according to our previous works by using the solvent casting procedure [5,20]. PLLA polymer films were optically transparent and have a thickness of 250 μm. Mechanical property measurements showed a Young's modulus of 680 MPa and an elongation at break of 144%. No variations were observed in oxygen treated samples.

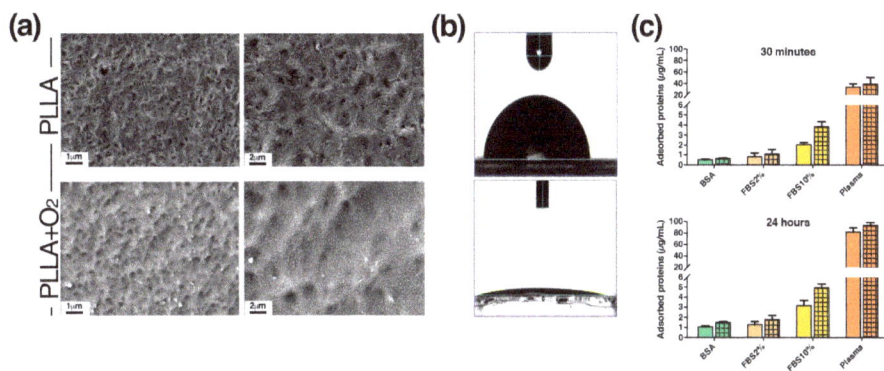

Figure 1. PLLA and PLLA+O$_2$ polymer films. (**a**) Representative FESEM images of PLLA and PLLA+O$_2$ surface morphology; (**b**) WCA images of PLLA and PLLA+O$_2$ surfaces (see method for details); and (**c**) protein adsorption is reported as mean $\pm$ SEM. Clear fill bar: PLLA; patterned fill bar: PLLA+O$_2$.

DSC analysis revealed that the plasma surface treatment did not affect the thermal properties of the PLLA polymer film. The melting temperature was about 172 °C for neat PLLA and for PLLA oxygen treated samples, with a percentage of crystallinity around 50% for all systems. Thus, the oxygen plasma process did not affect the characteristic thermal transitions of the polymer matrix.

The overall data confirmed our previous findings [24,31] and indicated that the treatment with oxygen-plasma modify the surface of the pristine PLLA without changing PLLA's bulk properties.

3.1.1. Surface Morphology Characterization of PLLA and PLLA+O$_2$

FESEM images show the surface morphology of PLLA and PLLA+O$_2$ films. Pristine PLLA presented a detectable roughness with a specific topology (e.g., parallel lines) that was induced by the Teflon topography, used for the casting samples' preparation (Figure 1a). Images also revealed the etching effect of the oxygen plasma treatment of PLLA (Figure 1a). It is well known that the plasma treatment of surfaces gave a mass reduction due to ion etching [31]. Increasing the magnification, the images show the presence of nanostructured surfaces, induced by the oxygen treatment.

3.1.2. Wettability Measurements

Water static contact angle (WCA) measurements were performed to study the effects of plasma treatment on surface wettability. Due to PLLA's hydrophobic nature, PLLA films have hydrophobic behaviors. Ten minutes of radio frequency oxygen-plasma treatment with RF fixed at 30 W of power decreased the water contact angle of the pristine PLLA from 90° to less than 10°, changing the surface's properties from hydrophobic to hydrophilic (Figure 1b).

Modulating the parameters of the oxygen treatment we may control the wettability properties. Low power supply values (i.e., 10 W, [24]) permit increasing the wettability, obtaining materials with 50° WCA. Hence, the power level and treatment time were selected in order to obtain lower values of WCA and, at the same time, to maintain the polymer's chemical stability. According to our study [24], spectra analysis confirmed that this treatment did not induce chemical changes in the carbonyl (1756 cm^{-1}) and in the C–O–C (1080 cm^{-1}) stretching peaks (data not shown).

3.1.3. Protein Adsorption

Figure 1c shows the adsorption of BSA (2 mg/mL), FBS 2%, FBS 10%, and plasma (5 mg/mL) on PLLA and on PLLA+O$_2$ films at two different time points, 30 min and 24 h, of protein incubation.

We observed an increase of adsorption level of FBS 2%, FBS 10%, BSA, and plasma on PLLA+O_2 compared to PLLA. The increase was higher at 24 h compared to 30 min of incubation on both polymer films and was more evident for FBS 10% and plasma (Figure 1c). The highest increase in protein adsorption on both films of plasma and FBS 10% (>FBS 2%) is a consequence of the highest protein concentration in these samples.

3.2. Culture of hASCs on PLLA and PLLA+O_2

hASCs were seeded on films snipped into squares (Figure 2a). hASCs grew generating spheroids attached to the PLLA surface. On the contrary, hASCs cultured on PLLA+O_2 films maintained the canonical monolayer fibroblast-like morphology, typically observed on TCP (Figure 2b).

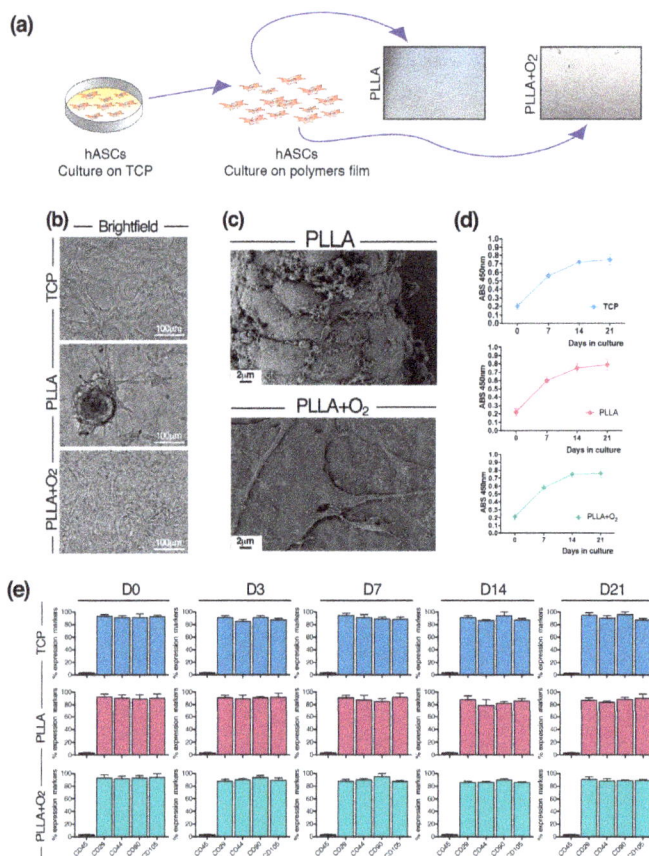

Figure 2. hASCs cultured on PLLA and PLLA+O_2 polymer films. (**a**) Schematic of hASCs on PLLA and PLLA+O_2 culture platforms; (**b**) brightfield representative images of hASCs on TCP, PLLA, and PLLA+O_2; (**c**) FESEM representative images highlight the hASCs spheroids organization on PLLA film and the monolayer fibroblast-like shape on PLLA+O_2 film; (**d**) hASCs viability (XTT assay) on TCP, on PLLA, and on PLLA+O_2; and (**e**) expression of mesenchymal stem cell markers in hASCs at different time points of culture (0, 3, 7, 14, 21) on TCP, on PLLA, and on PLLA+O_2. Average values for mesenchymal phenotype markers shown by hASC (*n* = 3) are reported as the mean ± SEM.

The different cellular organization of hASCs on PLLA and on PLLA+O_2 films is also revealed by FESEM images. While on films of PLLA hASCs were strictly packed together in a spheroid conformation, on PLLA+O_2 films the stem cells grew adherent to the polymer surface in a monolayer organization (Figure 2c).

hASCs viability was not affected by the characteristics of both PLLA films. We observed a similar dehydrogenase activity in spheroids hASCs on PLLA, in fibroblast-like hASCs on PLLA+O_2, and on TCP (Figure 2d). No dead cells were observed during in vitro cultures' maintenance, as confirmed by Trypan Blue assay.

We investigated the expression of stem cell markers in hASCs cultured on PLLA and on PLLA+O_2. We found a similar expression level of mesenchymal stem markers in hASCs cultured on TCP, on PLLA, and on PLLA+O_2 (Figure 2e). The levels of CD29, CD44, CD90, and CD105 markers were comparable in stem cells at the time point of seeding on each substrate (day 0) and remained stable over the time in culture (days 3, 7, 14, and 21) (Figure 2e). No expression of the CD45 marker was detected in all samples at each time point (Figure 2e), confirming the non-hematopoietic origin of hASCs [21,22]. These results indicate that the oxygen treatment did not influence the stemness property of hASCs.

3.3. Morphology of hASCs on PLLA and PLLA+O_2

Stem cell shape was revealed by cytoskeleton F-actin staining. The analysis was performed at days 3, 7, 14, and 21. Reported images are representative of these time points.

In hASCs on TCP and on PLLA+O_2 films F-actin stress fibers crossed the cells traversing the cytoplasm and were almost oriented parallel to the main cellular longitudinal axis (Figure 3). Conversely, in hASCs on PLLA polymer films, F-actin fibers showed a circular organization, following the perimeter of the cells that had changed shape from elliptical to rounded (Figure 3).

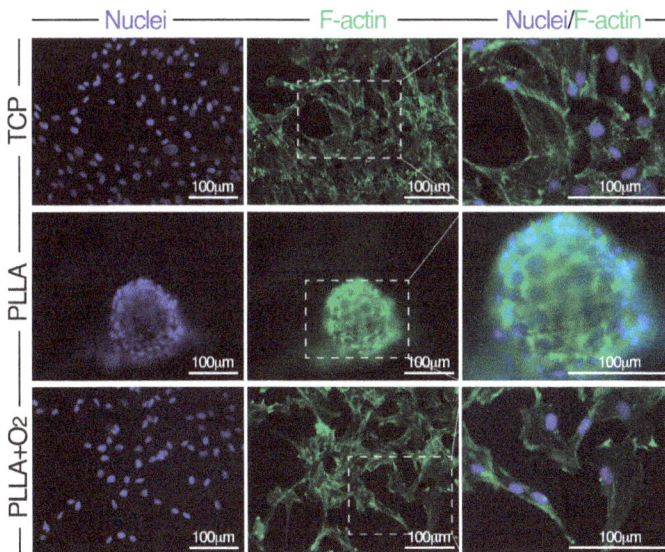

Figure 3. Morphology of hASCs on PLLA and on PLLA+O_2 cultures. Representative fluorescence images of nuclei (DAPI, blue) and F-actin (Phalloidin-FITC, green) in hASCs on TCP, on PLLA, and on PLLA+O_2. High magnification of selected dashed squares (merged DAPI-Phalloidin) is reported in the right panel.

We also monitored the expression of the cytoskeleton actin-linking proteins Filamin-A and Myosin-IIA (MHIIA). These proteins specifically interact with F-actin fibers to generate a three-dimensional structure determining cells' shape [32,33]. Moreover, together with the F-actin, Filamin-A and Myosin-IIA are the targets of cues generated by the adhesion complexes [32,33] and, therefore, are informative on the interaction between polymers and cells.

We found different expression of Filamin-A and Myosin-IIA proteins between spheroids and fibroblast-like monolayer hASCs (Figure 4a). In hASCs organized in spheroids on PLLA, both proteins depicted a circular organization of the cytoskeleton; on the contrary, they organized as elliptic structures in fibroblast-like hASCs on PLLA+O_2 (Figure 4a).

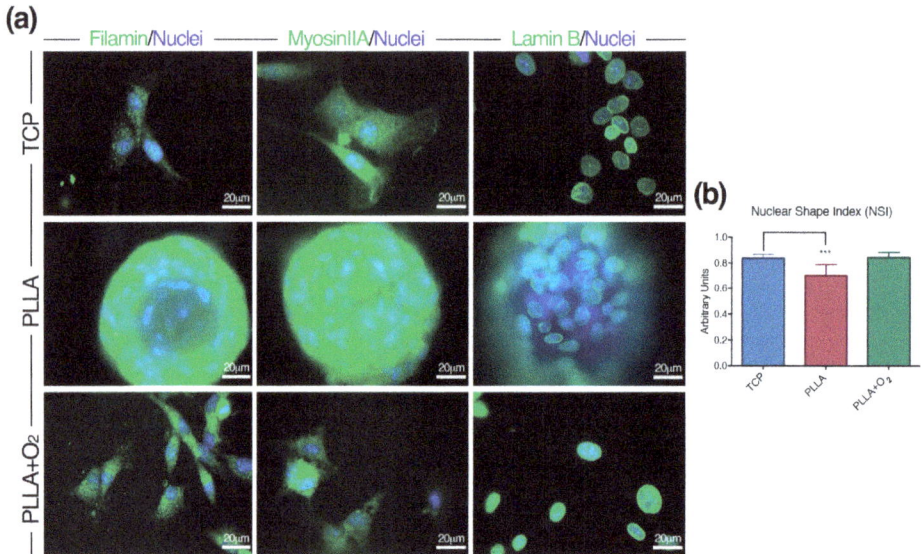

Figure 4. hASCs architecture on PLLA and on PLLA+O_2. (**a**) Representative immunofluorescence images of Filamin-A, Myosin-IIA and Lamin-B proteins in hASC spheroids on PLLA, and fibroblast-like hASCs on PLLA+O_2 and on TCP; and (**b**) NSI in hASCs on PLLA, PLLA+O_2, and TCP (see methods for details). *** $p \leq 0.001$.

Finally, we monitored the expression of Lamin-B, one of the nucleoskeleton proteins connecting the cytoskeleton proteins and the nucleus [34,35]. Lamin-B expression followed the change of the nuclear shape as a consequence of the spheroid organization on pristine PLLA. On the contrary, Lamin-B designed a canonical nuclear shape in ASCs on PLLA+O_2 and on TCP (Figure 4a).

We also measured the NSI in stem cells cultured on pristine PLLA and on oxygen-plasma treated PLLA polymer films (Figure 4b). We observed a reduction of NSI in hASCs organized in spheroids on PLLA. This finding indicated a nuclear stretching, possibly influencing chromatin conformation. No variations of NSI in hASCs growing on PLLA+O_2 or on TCP were detected (Figure 4b).

3.4. Adhesion of hASCs on PLLA and on PLLA+O_2 Surfaces

We studied the adhesion of hASCs on PLLA and on PLLA+O_2 surfaces analyzing the expression of the focal adhesion protein Vinculin, classically involved on the cellular-matrix interaction [5,36,37]. The analysis was performed at days 3, 7, 14, and 21. Reported images are representative of these time points (Figure 5).

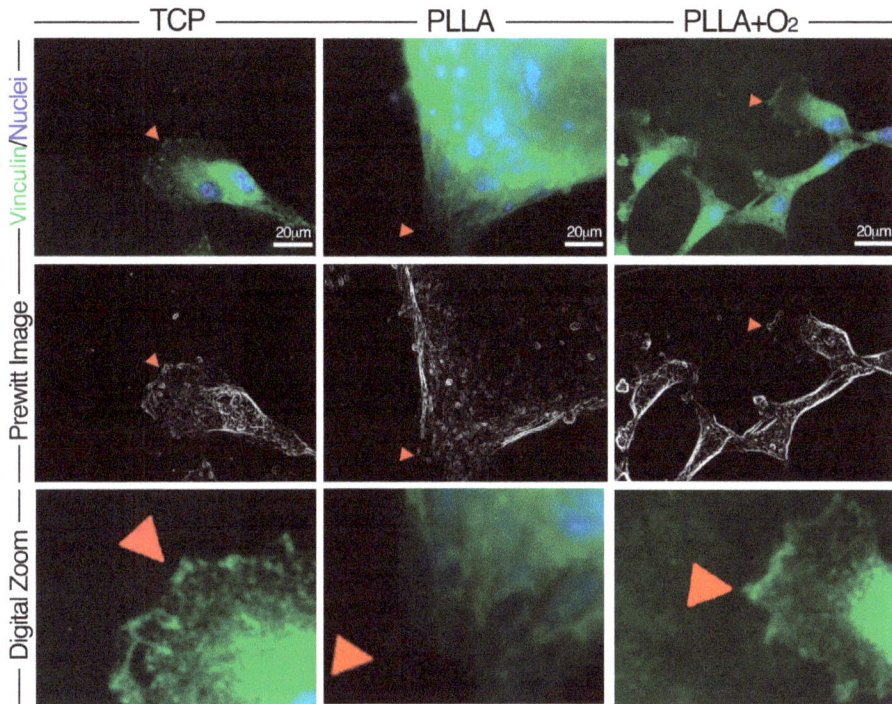

Figure 5. Distribution of Vinculin Focal Adhesion Spots in hASCs on PLLA and on PLLA+O$_2$. Immunofluorescence representative images show the expression of Vinculin (green) and VFASs (red arrow) and the related Prewitt images elaboration by MATLAB software. Image magnification: 60×. Digital zoom 3× from the original 60× images.

Within spheroids' conformation, Vinculin focal adhesion spots (VFASs) converged in small areas allowing the robust adhesion of hASCs to the surface of PLLA polymer films. Oppositely, monolayer fibroblast-like hASCs engaged VFASs in large areas on the surface of PLLA+O$_2$ polymer films as on TCP (Figure 5). The VFASs organization at the hASC-polymer film interface is highlighted by the corresponding Prewitt images obtained by MATLAB software. Images revealed a fan-shaped contact area distribution of VFASs in stem cells on PLLA+O$_2$ films and on TCP while they showed an organization in a small convergent contact area in hASCs on PLLA polymer films (Figure 5).

3.5. hASCs Spheroids Formation on PLLA

In Figure 6 are reported the time-to-time formation of hASC spheroids on PLLA polymer films. hASCs acquired spheroids conformation rapidly, indicating that stem cells responded to the surface characteristic of PLLA immediately after cell seeding. Spheroids appeared after 24 h and their number and dimension increased during culture maintenance, suggesting that spheroids are consequence of cell aggregation (Figure 6). This contrasted with the culture of hASCs on PLLA+O$_2$ where stem cells retained the native morphology as on TCP (Figure 6).

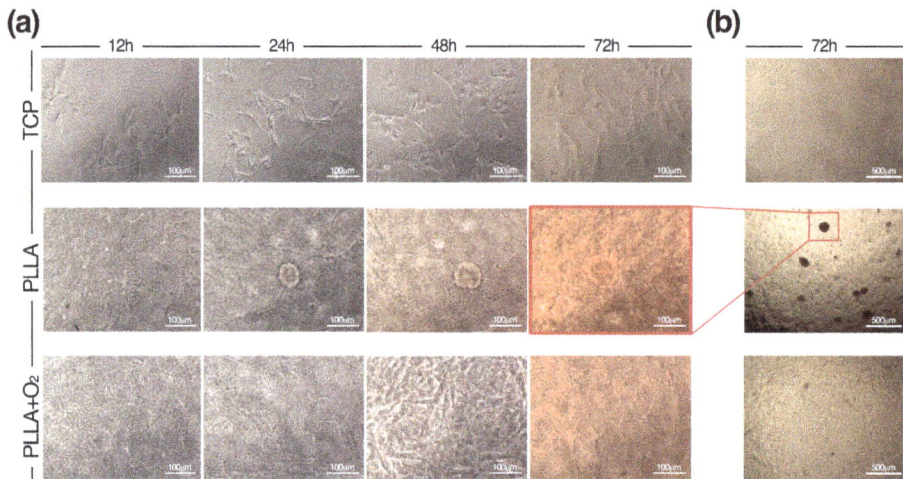

Figure 6. Time-to-time formation of hASC spheroids on PLLA. (**a**) Representative brightfield images by an Eclipse-TS100 microscope equipped with a Nikon camera (Nikon, Tokyo, Japan), show the spheroids' organization of hASCs during the culture period on PLLA films and their canonical fibroblasts-like morphology on PLLA+O_2 film and on TCP; and (**b**) representative overview of hASCs on TCP, on PLLA, and on PLLA+O_2 films.

3.6. *Effect of PLLA Oxygen-Plasma Treatment on hBM-MSCs and hUCMSCs Cultures*

In order to establish whether the oxygen-plasma treatment to PLLA influnces the adhesion of other adult human stem cell types we perfomed comparative experiments using hBM-MSCs and hUCMSCs. These stem cells were selected based on our previuos studies exploring their interaction with the surface of PLLA polymer films [5,20], as well as on their mesenchymal origin [21–23].

The oxygen treatment of PLLA did not change hBM-MSCs performances. hBM-MSCs maintained the same fibroblast-like shape and viability behaviours on PLLA and on PLLA+O_2 as on TCP (Figure 7). Maintainace of the canonical mesenchymal shape by hBM-MSCs on both PLLA films as typically observed on TCP was confirmed at molecular level by F-actin immunostaining (Figure 7).

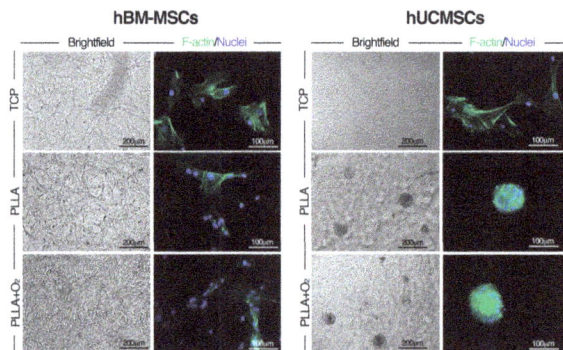

Figure 7. Effect of oxygen-plasma treatment to PLLA on hBM-MSCs and hUCMSCs cultures. Representative brightfield and immunofluorescence images of hBM-MSCs and hUCMSCs cultured on TCP, PLLA, and PLLA+O_2, respectively. F-actin (Phalloidin-FITC, green) and nuclei (DAPI, blue).

No changes in shape were observed when hUCMSCs were cultured on PLLA+O_2 polymer films. In fact, immufluorescences staining and brightfield inspection revealed that the spheroid conformation, typically generated by hUCMSCs on PLLA polymer films [5], was also acquired by hUCMSCs cultured on PLLA+O_2 (Figure 7). hUCMSCs cultured on TCP as reference, mainatined their fibroblast-like morphology (Figure 7).

These results suggest that the wettability properties of PLLA surfacs are not involved in the formation of hUCMSC spheroids nor in the hBM-MSCs adhesion.

4. Discussion

In this work, we demonstrated that PLLA polymer films' surface properties guide hASCs toward a multicellular three-dimensional spheroid conformation instead of a monolayer fibroblast-like morphology as typically observed on TCP. Noteworthy, we demonstrated that the PLLA film surface hydrophilicity is involved in the acquisition of a spheroid conformation by hASCs, since the treatment of the PLLA film surface with plasma-oxygen maintained the fibroblast-like morphology of hASCs. Interestingly, the oxygen treatment had no effect on the performance of hBM-MSCs and hUCMSCs cultured on PLLA+O_2 as they maintained the same cell shape observed on PLLA polymer film.

Modification of the PLLA surface is commonly used to modulate the affinity of this polymer with the cells [38–42]. Many different functional groups, including diethylenetriamine, 2-(2-aminoethoxy)ethanol or GRGDS (cell adhesion peptide) [40], aldehyde and amine groups [41], and poly(sulfobetaine methacrylate)-catechol conjugates [42] have been used to improve stem cells adhesion and proliferation on PLLA polymer (both fibrous and film forms).

The plasma oxygen treatment is a useful method to spread hydroxyl groups on the surface of several polymers [43], including PLLA [44–47], in order to improve cellular adhesion to their surface.

In our study, after oxygen-plasma treatment, the surface of PLLA films (PLLA+O_2) became hydrophilic and its surface nanoroughness increased, whereas the hydrolytic degradation properties of PLLA were not affected [31]. In agreement with other studies [24,31,44–46] we demonstrated the increase of BSA, FBS 2%, FBS 10%, and plasma adsorption on the hydrophilic PLLA+O_2 surface compared to protein adsorption on the hydrophobic surface of PLLA polymer films. Moreover, according to our previous work, RF oxygen-plasma surface modification did not affect PLLA bulk properties [24,31]. These data confirmed our previous evidence on the adsorption of FITC-BSA, showing how plasma treatment modulates the PLLA surface properties and induces the formation of the surface nanostructure of different shapes affecting the hydrophilic/hydrophobic behaviors without changing the PLLA bulk properties [24,31]. Furthermore, in vitro degradation studies performed in physiological conditions have showed that plasma treatment does not affect the PLLA bulk and the hydrolytic degradation properties [31].

Thus, oxygen treatment allows the production of two distinct PLLA polymer films, differing for hydrophilicity and WCA (PLLA > PLLA+O_2), representing a suitable model for studying the effect of the polymer film on stem cells.

Compared to the in vitro culture on TCP, where hASCs adhere and display a monolayer fibroblast-like shape, adipose stem cells cultured on PLLA grew generating three-dimensional multicellular organizations (spheroids), strongly attached to the film's surface. We suggest that this new cellular organization is the consequence of the adipose stem cell-material interaction. Several studies have suggested that stem cells may change their shape depending on signals they receive from external milieu through the focal adhesion (FA) proteins, directly involved in tethering with the extracellular matrix [10–13,48,49]. In this contest, Vinculin plays a central role in organizing the FA complexes (VFASs) [36] and contributes to the organization of the cytoskeleton architecture [37]. In our system, the different VFASs' distribution of hASCs on PLLA (fan-shaped) and on PLLA+O_2 (convergent area) significantly influenced the cytoskeleton F-actin organization, along with Filamin-A and Myosin-IIA (two F-actin-linking proteins involved on the control of the cell stiffening [32,33] and, therefore, in stem cells morphology acquisition). It is likely that the difference in hydrophilicity between the two surfaces interferes with hASCs' adhesion to the polymer films surface and drives the stem cells to acquire different conformation: spheroids on PLLA and monolayer fibroblasts-like on PLLA+O$_2$.

According with other reports [43,44,47], our results confirmed that the oxygen-plasma treatment improves the adhesion and spreading of hASCs on the polymer films. Nevertheless, surface modification neither changed the stem cells' proliferation rate, nor stemness markers' expression compared to hASCs at the time of seeding (day 0). Levels of mesenchymal stem cell markers were comparable in hASCs seeded on PLLA and on PLLA+O$_2$ during the culture period (from day 0 to day 21) and were similar to those measured in hASCs cultured on TCP [26,27]. These results are in agreement with our previous finding, showing that the PLLA polymers maintain the stem cells' properties [5,7,19,20].

Of note, our work adds new finding on the effects of the hydrophilicity on the final cellular organization and points out the role of the stem cell's type with respect to this phenomenon.

The results that we obtained by culturing adult human UCMSCs and BM-MSCs on PLLA and PLLA+O$_2$ polymer films may be explained in this context. In fact, after seeding on PLLA+O$_2$, hUCMSCs maintained the spheroid conformation they had on PLLA polymer film, however, different from the fibroblast-like monolayer morphology typically observed on TCP [5]. No variations in cellular organization were found in hBM-MSCs cultured on PLLA+O$_2$. These cells maintained the canonical monolayer fibroblast-like shape as on PLLA film and on TCP [20]. Together, these findings indicate that PLLA+O$_2$ surface hydrophilicity is not directly involved in driving the architecture of hBM-MSCs and the hUCMSCs on the polymer films. We may explain the formation of spheroids by hASCs and hUCMSCs on PLLA polymer films and, foremost, the absence of hASC spheroids rather than the maintenance of hUCMSC spheroids on PLLA+O$_2$ polymer films, as the consequence of specific molecular differences at the stem cell interface with the film. More experiments are needed to identify the origin of this phenomenon. However, according to other studies [5,8–12], these findings support the hypothesis that, at the cell-material interface, stem cells and the material assist each other. This causes an active cross-talk linking the molecular cellular mechanisms with the intrinsic materials' properties to generate a selected cell response (e.g., stem cells shape, adhesion, or functions).

Collectively, our results highlight the role of surface modifications as suitable tools to study stem cells-material interfaces. Moreover, as the interaction might be stem cells specific, they point out the need to perform such experiments considering more types of stem cells.

Author Contributions: Sabata Martino conceived and designed the experiments; Chiara Argentati and Francesco Morena isolated hASCs and hBM-MSCs and performed the experiments; Ilaria Armentano designed, developed, and characterized the material. Pia Montanucci isolated UCMSCs, and performed flow cytometry analyses; Marco Rallini performed FESEM analyses; Francesco Morena has developed custom scripts in MATLAB for image elaboration and analysis; Giuseppe Basta, Riccardo Calafiore, Nicolino Calabrese, Carla Emiliani, Ilaria Armentano, Pia Montanucci, Francesco Morena, and Sabata Martino analyzed the data; Marino Cordellini contributed reagents; and Sabata Martino wrote the paper.

Conflicts of Interest: The authors declare no conflict of interest.

References

1. Mager, M.D.; La Pointe, V.; Stevens, M.M. Exploring and exploiting chemistry at the cell surface. *Nat. Chem.* **2011**, *3*, 582–589. [CrossRef] [PubMed]
2. Martino, S.; D'Angelo, F.; Armentano, I.; Kenny, J.M.; Orlacchio, A. Stem cell-biomaterial interactions for regenerative medicine. *Biotechnol. Adv.* **2012**, *30*, 338–351. [CrossRef] [PubMed]
3. Wheeldon, I.; Farhadi, A.; Bick, A.G.; Jabbari, E.; Khademhosseini, A. Nanoscale tissue engineering: Spatial control over cell-materials interactions. *Nanotechnology* **2011**, *22*, 212001. [CrossRef] [PubMed]
4. Martino, S.; Morena, F.; Barola, C.; Bicchi, I.; Emiliani, C. Proteomics and Epigenetic Mechanisms in Stem Cells. *Curr. Proteom.* **2014**, *11*, 193–209. [CrossRef]

5. Morena, F.; Armentano, I.; Montanucci, P.; Argentati, C.; Fortunati, E.; Montesano, S.; Bicchi, I.; Pescara, T.; Pennoni, I.; Mattioli, S.; et al. Design of a nanocomposite substrate inducing adult stem cell assembly and progression toward an Epiblast-like or Primitive Endoderm-like phenotype via mechanotransduction. *Biomaterials* **2017**, *144*, 211–229. [CrossRef] [PubMed]

6. D'Angelo, F.; Armentano, I.; Mattioli, S.; Crispoltoni, L.; Tiribuzi, R.; Cerulli, G.G.; Palmerini, C.A.; Kenny, J.M.; Martino, S.; Orlacchio, A. Micropatterned hydrogenated amorphous carbon guides mesenchymal stem cells towards neuronal differentiation. *Eur. Cells Mater.* **2010**, *20*, 231–244. [CrossRef]

7. D'Angelo, F.; Armentano, I.; Cacciotti, I.; Tiribuzi, R.; Quattrocelli, M.; Del Gaudio, C.; Fortunati, E.; Saino, E.; Caraffa, A.; Cerulli, G.G.; et al. Tuning multi/pluri-potent stem cell fate by electrospun poly(L-lactic acid)-calcium-deficient hydroxyapatite nanocomposite mats. *Biomacromolecules* **2012**, *13*, 1350–1360. [CrossRef] [PubMed]

8. William, W.L.; McDevitt, T.C.; Engler, A.J. Materials as stem cell regulators. *Nat. Mater.* **2014**, *13*, 547–557. [CrossRef]

9. Kai, Y.; Xuan, W.; Luping, C.; Shiyu, L.; Zhenhua, L.; Lin, Y.; Jiandong, D. Matrix Stiffness and Nanoscale Spatial Organization of Cell-Adhesive Ligands Direct Stem Cell Fate. *Nano Lett.* **2015**, *15*, 4720–4729. [CrossRef]

10. D'Angelo, F.; Tiribuzi, R.; Armentano, I.; Kenny, J.M.; Martino, S.; Orlacchio, A. Mechanotransduction: tuning stem cells fate. *J. Funct. Biomater.* **2011**, *2*, 67–87. [CrossRef] [PubMed]

11. Downing, T.L.; Soto, J.; Morez, C.; Houssin, T.; Fritz, A.; Yuan, F.; Chu, J.; Patel, S.; Schaffer, D.V.; Li, S. Biophysical regulation of epigenetic state and cell reprogramming. *Nat. Mater.* **2013**, *12*, 1154–1162. [CrossRef] [PubMed]

12. Trappmann, B.; Gautrot, J.E.; Connelly, J.T.; Strange, D.G.; Li, Y.; Oyen, M.L.; Cohen Stuart, M.A.; Boehm, H.; Li, B.; Vogel, V.; et al. Extracellular-matrix tethering regulates stem-cell fate. *Nat. Mater.* **2012**, *9*, 642–649. [CrossRef] [PubMed]

13. Li, W.; Wang, J.; Ren, J.; Qu, X. Endogenous signalling control of cell adhesion by using aptamer functionalized biocompatible hydrogel. *Chem. Sci.* **2015**, *6*, 6762–6768. [CrossRef] [PubMed]

14. Rood, M.T.; Spa, S.J.; Welling, M.M.; Ten Hove, J.B.; van Willigen, D.M.; Buckle, T.; Velders, A.H.; van Leeuwen, F.W. Obtaining control of cell surface functionalizations via Pre-targeting and Supramolecular host guest interactions. *Sci. Rep.* **2017**, *7*, 39908. [CrossRef] [PubMed]

15. Armentano, I.; Bitinis, N.; Fortunati, E.; Mattioli, S.; Rescignano, N.; Verdejo, R.; Lopez-Manchadob, M.A.; Kenny, J.M. Multifunctional nanostructured PLA materials for packaging and tissue engineering. *Prog. Polym. Sci.* **2013**, *38*, 1720–1747. [CrossRef]

16. Garlotta, D. A literature review of poly (lactic acid). *J. Polym. Environ.* **2001**, *9*, 63–84. [CrossRef]

17. Bayer, I.S. Thermomechanical properties of polylactic acid-graphene composites: A state-of-the-art review for biomedical applications. *Materials* **2017**, *10*, 748. [CrossRef] [PubMed]

18. Tsuji, H. Poly(lactic acid) stereocomplexes: A decade of progress. *Adv. Drug Deliv. Rev.* **2016**, *107*, 97–135. [CrossRef] [PubMed]

19. Bianco, A.; Bozzo, B.M.; Del Gaudio, C.; Cacciotti, I.; Armentano, I.; Dottori, M.; D'Angelo, F.; Martino, S.; Orlacchio, A.; Kenny, J.M. Poly (L-lactic acid)/calcium-deficient nanohydroxyapatite electrospun mats for bone marrow stem cell cultures. *J. Bioact. Compat. Polym.* **2011**, *26*, 225–241. [CrossRef]

20. Lizundia, E.; Sarasua, J.R.; D'Angelo, F.; Orlacchio, A.; Martino, S.; Kenny, J.M.; Armentano, I. Biocompatible poly(L-lactide)/MWCNT nanocomposites: Morphological characterization, electrical properties, and stem cell interaction. *Macromol. Biosci.* **2012**, *12*, 870–881. [CrossRef] [PubMed]

21. Bertozzi, N.; Simonacci, F.; Grieco, M.P.; Grignaffini, E.; Raposio, E. The biological and clinical basis for the use of adipose-derived stem cells in the field of wound healing. *Ann. Med. Surg.* **2017**, *20*, 41–48. [CrossRef] [PubMed]

22. Choi, J.R.; Yong, K.W.; Wan Safwani, W.K.Z. Effect of hypoxia on human adipose-derived mesenchymal stem cells and its potential clinical applications. *Cell. Mol. Life Sci.* **2017**, *74*, 2587–2600. [CrossRef] [PubMed]

23. Montanucci, P.; Alunno, A.; Basta, G.; Bistoni, O.; Pescara, T.; Caterbi, S.; Pennoni, I.; Bini, V.; Gerli, R.; Calafiore, R. Restoration of T cell substes of patients with type 1 diabetes mellitus by microencapsulated human umbilical cord Wharton jelly-derived mesenchymal stem cells: An in vitro study. *Clin. Immunol.* **2016**, *163*, 34–41. [CrossRef] [PubMed]

24. Armentano, I.; Ciapetti, G.; Pennacchi, M.; Dottori, M.S.; Devescovi, V.; Granchi, D.; Baldini, N.; Olalde, B.; Jurado, M.J.; Alava, J.I.; et al. Role of PLLA plasma surface modification in the interaction with human marrow stromal cells. *J. Appl. Polym. Sci.* **2009**, *114*, 3602–3611. [CrossRef]

25. Morena, F.; Argentati, C.; Trotta, R.; Crispoltoni, L.; Stabile, A.; Pistilli, A.; di Baldassarre, A.; Calafiore, R.; Montanucci, P.; Basta, G.; et al. A comparison of lysosomal enzymes expression levels in peripheral blood of mild- and severe-Alzheimer's disease and MCI patients: Implications for regenerative medicine approaches. *Int. J. Mol. Sci.* **2017**, *18*, 1806. [CrossRef] [PubMed]

26. Morena, F.; Argentati, C.; Calzoni, E.; Cordellini, M.; Emiliani, C.; D'Angelo, F.; Martino, S. Ex-vivo tissues engineering modeling for reconstructive surgery using human adult adipose stem cells and polymeric nanostructured matrix. *Nanomaterials* **2016**, *6*, 57. [CrossRef] [PubMed]

27. Tarpani, L.; Morena, F.; Gambucci, M.; Zampini, G.; Massaro, G.; Argentati, C.; Emiliani, C.; Martino, S.; Latterini, L. The influence of modified silica nanomaterials on adult stem cell culture. *Nanomaterials* **2016**, *6*, 104. [CrossRef] [PubMed]

28. Martino, S.; D'Angelo, F.; Armentano, I.; Tiribuzi, R.; Pennacchi, M.; Dottori, M.; Mattioli, S.; Caraffa, A.; Cerulli, G.G.; Kenny, J.M.; et al. Hydrogenated amorphous carbon nanopatterned film designs drive human bone marrow mesenchymal stem cell cytoskeleton architecture. *Tissue Eng. Part A* **2009**, *15*, 3139–3149. [CrossRef] [PubMed]

29. Montanucci, P.; Basta, G.; Pescara, T.; Pennoni, I.; Di Giovanni, F.; Calafiore, R. New simple and rapid method for purification of mesenchymal stem cells from the human umbilical cord Wharton jelly. *Tissue Eng. Part A* **2011**, *17*, 2651–2661. [CrossRef] [PubMed]

30. Weiss, M.L.; Troyer, D.L. Stem cells in the umbilical cord. *Stem Cell Rev.* **2006**, *2*, 155–162. [CrossRef] [PubMed]

31. Mattioli, S.; Kenny, J.M.; Armentano, I. Plasma surface modification of porous PLLA films: Analysis of surface properties and in vitro hydrolytic degradation. *J. Appl. Polym. Sci.* **2012**, *125*, E239–E247. [CrossRef]

32. Kasza, K.E.; Nakamura, F.; Hu, S.; Kollmannsberger, P.; Bonakdar, N.; Fabry, B.; Stossel, T.P.; Wang, N.; Weitz, D.A. Filamin A is essential for active cell stiffening but not passive stiffening under external force. *Biophys. J.* **2009**, *96*, 4326–4335. [CrossRef] [PubMed]

33. Reymann, A.C.; Boujemaa-Paterski, R.; Martiel, J.L.; Guérin, C.; Cao, W.; Chin, H.F.; De La Cruz, E.M.; Théry, M.; Blanchoin, L. Actin network architecture can determine myosin motor activity. *Science* **2012**, *336*, 1310–1314. [CrossRef] [PubMed]

34. Stewart, R.M.; Zubek, A.E.; Rosowski, K.A.; Schreiner, S.M.; Horsley, V.; King, M.C. Nuclear–cytoskeletal linkages facilitate cross talk between the nucleus and intercellular adhesions. *J. Cell Biol.* **2015**, *209*, 403–418. [CrossRef] [PubMed]

35. Isermann, P.; Lammerding, J. Nuclear mechanics and mechanotransduction in health and disease. *Curr. Biol.* **2013**, *23*, R1113–R1121. [CrossRef] [PubMed]

36. Geiger, B.; Spatz, J.P.; Bershadsky, A.D. Environmental sensing through focal adhesions. *Nat. Rev. Mol. Cell Biol.* **2009**, *10*, 21–33. [CrossRef] [PubMed]

37. Carisey, A.; Tsang, R.; Greiner, A.M.; Nijenhuis, N.; Heath, N.; Nazgiewicz, A.; Kemkemer, R.; Derby, B.; Spatz, J.; Ballestrem, C. Vinculin regulates the recruitment and release of core focal adhesion proteins in a force-dependent manner. *Curr. Biol.* **2013**, *23*, 271–281. [CrossRef] [PubMed]

38. Armentano, I.; Rescignano, N.; Fortunati, E.; Mattioli, S.; Morena, F.; Martino, S.; Torre, L.; Kenny, J.M. Multifunctional nanostructured biopolymeric materials for therapeutic applications. In *Nanostructures for Novel Therapy: Synthesis, Characterization and Applications*; Grumezescu, A.M., Ficai, D., Eds.; Elsevier: Amsterdam, The Netherlands, 2017; pp. 107–135. ISBN 978-0-323-46142-9.

39. Armentano, I.; Mattioli, S.; Morena, F.; Argentati, C.; Martino, S.; Torre, L.; Kenny, J.M. Recent Advances in Nanostructured Polymeric Surface: Challenges and Frontiers in Stem Cells. In *Advanced Surfaces for Stem Cell Research*; Tiwari, A., Bora Garipcan, B., Uzun, L., Eds.; Wiley: Hoboken, NJ, USA, 2016; pp. 141–164. ISBN 978-1-119-24250-5.

40. Schaub, N.J.; Le Beux, C.; Miao, J.; Linhardt, R.J.; Alauzun, J.G.; Laurencin, D.; Gilbert, R.J. The effect of surface modification of aligned poly-L-lactic acid electrospun fibers on fiber degradation and neurite extension. *PLoS ONE* **2015**, *10*, e0136780. [CrossRef] [PubMed]

41. Li, H.; Xia, Y.; Wu, J.; He, Q.; Zhou, X.; Lu, G.; Shang, L.; Boey, F.; Venkatraman, S.S.; Zhang, H. Surface modification of smooth poly(L-lactic acid) films for gelatin immobilization. *ACS Appl. Mater. Interfaces* **2012**, *4*, 687–693. [CrossRef] [PubMed]

42. Yang, W.; Sundaram, H.S.; Ella, J.R.; He, N.; Jiang, S. Low-fouling electrospun PLLA films modified with zwitterionic poly(sulfobetaine methacrylate)-catechol conjugates. *Acta Biomater.* **2016**, *40*, 92–99. [CrossRef] [PubMed]

43. Hanson, A.D.; Wall, M.E.; Pourdeyhimi, B.; Loboa, E.G. Effects of oxygen plasma treatment on adipose-derived human mesenchymal stem cell adherence to poly(L-lactic acid) scaffolds. *J. Biomater. Sci. Polym. Ed.* **2007**, *18*, 1387–1400. [CrossRef] [PubMed]
44. Yamaguchi, M.; Shinbo, T.; Kanamori, T.; Wang, P.C.; Niwa, M.; Kawakami, H.; Nagaoka, S.; Hirakawa, K.; Kamiya, M. Surface modification of poly(L-lactic acid) affects initial cell attachment, cell morphology, and cell growth. *J. Artif. Organs* **2004**, *7*, 187–193. [CrossRef] [PubMed]
45. Wan, Y.; Qu, X.; Lu, J.; Zhu, C.; Wan, L.; Yang, J.; Bei, J.; Wang, S. Characterization of surface property of poly(lactide-*co*-glycolide) after oxygen plasma treatment. *Biomaterials* **2004**, *25*, 4777–4783. [CrossRef] [PubMed]
46. Liu, W.; Zhan, J.; Su, Y.; Wu, T.; Wu, C.; Ramakrishna, S.; Mo, X.; Al-Deyab, S.S.; El-Newehy, M. Effects of plasma treatment to nanofibers on initial cell adhesion and cell morphology. *Colloids Surf. B Biointerfaces* **2014**, *113*, 101–106. [CrossRef] [PubMed]
47. Song, W.L.; Veiga, D.D.; Custodio, C.A.; Mano, J.F. Bioinspired degradable substrates with extreme wettability properties. *Adv. Mater.* **2009**, *21*, 1830–1834. [CrossRef]
48. Rosch, J.C.; Hollmann, E.K.; Lippmann, E.S. In vitro selection technologies to enhance biomaterial functionality. *Exp. Biol. Med.* **2016**, *241*, 962–971. [CrossRef] [PubMed]
49. Ayala, R.; Zhang, C.; Yang, D.; Hwang, Y.; Aung, A.; Shroff, S.S.; Arce, F.T.; Lal, R.; Arya, G.; Varghese, S. Engineering the cell-material interface for controlling stem cell adhesion, migration, and differentiation. *Biomaterials* **2011**, *32*, 3700–3711. [CrossRef] [PubMed]

MDPI

St. Alban-Anlage 66

4052 Basel

Switzerland

Tel. +41 61 683 77 34

Fax +41 61 302 89 18

www.mdpi.com

Polymers Editorial Office

E-mail: polymers@mdpi.com

www.mdpi.com/journal/polymers

www.ingramcontent.com/pod-product-compliance
Lightning Source LLC
Chambersburg PA
CBHW051707210326
41597CB00032B/5394